International Conference on Heavy Ion Collisions at Near-Barrier Energies (FUSION17)

EPJ Web of Conferences Volume 163 (2017)

Hobart, Australia
20-24 February 2017

Editors:

E.C. Simpson
K.J. Cook
C. Simenel
I.P. Carter

ISBN: 978-1-5108-5253-2

Printed from e-media with permission by:

Curran Associates, Inc.
57 Morehouse Lane
Red Hook, NY 12571

Some format issues inherent in the e-media version may also appear in this print version.

Printed by Curran Associates, Inc. (2018)

For additional information, please contact EDP Sciences – Web of Conferences at the address below.

EDP Sciences – Web of Conferences
17, Avenue du Hoggar
Parc d'Activité de Courtabœuf
BP 112
F-91944 Les Ulis Cedex A
France

Phone: +33 (0) 1 69 18 75 75
Fax: +33 (0) 1 69 28 84 91

contact-edps@webofconferences.org

Additional copies of this publication are available from:

Curran Associates, Inc.
57 Morehouse Lane
Red Hook, NY 12571 USA
Phone: 845-758-0400
Fax: 845-758-2633
Email: curran@proceedings.com
Web: www.proceedings.com

TABLE OF CONTENTS

Fusion and elastic scattering of ^{6}Li + ^{58}Ni at low energies

Elí F. Aguilera[1,*], *Paulina* Amador-Valenzuela[1], *Enrique* Martinez-Quiroz[1], *David* Lizcano[1], *Araceli* Garcia-Flores[1], and *James J.* Kolata[2]

[1]*Departamento de Aceleradores, Instituto Nacional de Investigaciones Nucleares, Apartado Postal 18-1027, Código Postal 11801, México, Distrito Federal, México*
[2]*Physics Department, University of Notre Dame, Notre Dame, Indiana, 46556-5670, USA*

Abstract. Sub-barrier fusion cross sections (σ_{fus}) for the ^{6}Li + ^{58}Ni system, obtained from the respective evaporation protons, are examined in the present work. With respect to expectations of a simple one-dimensional barrier penetration model, a large enhancement of the data is observed. Good consistency with equivalent data reported previously for similar systems is found. A comparison with total reaction cross sections (σ_R), deduced from elastic scattering measurements reported previously, indicates that σ_{fus} is close to σ_R within the measured energy range. To estimate the contribution of complete fusion (CF), an optical model analysis of the elastic scattering data is performed where CF is identified with the absorption in a short range volume potential. A surface polarization potential is added to the bare nuclear potential to simulate the effect of peripheral reactions. The results obtained indicate that other mechanisms different from CF may be dominant, especially in the lower energy region.

1 Introduction

Several measurements have been done lately to study different reactions induced by the stable weakly-bound nucleus ^{6}Li [1–5]. However, the different data sets are still far from being completely understood. An analysis involving the ^{6}Li + ^{58}Ni system is presented here.

Total fusion cross sections were recently measured for this system [6], which complement the elastic scattering data previously reported [1]. Evaporated protons were measured to obtain the respective proton cross sections and then statistical model calculations were used to map them into fusion cross sections [6]. The results are shown in figure 1 with red circles. The total reaction cross sections obtained earlier at similar energies for the same system [1], are shown with black squares. The barrier is indicated with the vertical arrow, so these are mostly sub-barrier data. The barrier parameters (V_B = 12.36 MeV, R_B = 9.02 fm, $\hbar\omega_0$ = 3.63 MeV) were obtained from the Sao Paulo Potential (SPP) [7] and used in the single barrier penetration model (BPM) of Wong [8] to obtain the dashed line. The solid curve is the result of adding a short range imaginary potential to the real SPP and doing optical model calculations. The absorption in such imaginary potential is then identified with complete fusion. Because of the approximations involved in Wong's model, this should be a more realistic BPM prediction. One can see that the data present a big enhancement with respect to this curve. The dotted line is the result of coupled channel calculations including inelastic excitation of the first 2+ and 3- states in ^{58}Ni. The conclusion is that these inelastic couplings cannot explain the enhancement, i.e., additional channels would be needed in order to describe the data.

Using a well-known prescription to reduce data [9–12] (see axis labels on Fig. 2), a comparison was made in Fig. 2 with data reported for the same projectile and similar-mass targets. The present data are fairly consistent with results for ^{59}Co [13], ^{64}Ni [14], and ^{64}Zn [3], and one can see from figure 2 that the three data sets show a large sub-barrier enhancement with respect to BPM predictions. Therefore, it would seem that the origin for the enhancement can be ascribed to the projectile, independent of the target, although this conclusion is not quite supported by the wide spread of the data at the lower energies. Such spread was ascribed in Ref. [6] to a possible effect of direct reactions, which could be contaminating the fusion data reported for some of the systems. For the case of the ^{6}Li + ^{64}Zn system, for instance, it is possible that the data include contributions from one-neutron and/or one-proton transfer reactions [3]. The present data do not include such contributions (protons from neutron stripping were estimated and corrected for by doing appropriate DWBA calculations, while no protons are expected from proton stripping [6]), but could in principle contain contributions from certain direct processes which cannot be distinguished with the experimental technique used. To be more specific, if a deuteron is transferred to the target and the resulting product becomes excited above the proton emission threshold, then some evaporation protons would be emitted in this process which cannot be distinguished from protons coming from fusion. To estimate

*e-mail: eli.aguilera@inin.gob.mx

possible contributions from direct or peripheral processes, a reanalysis of the previous data for elastic scattering was done.

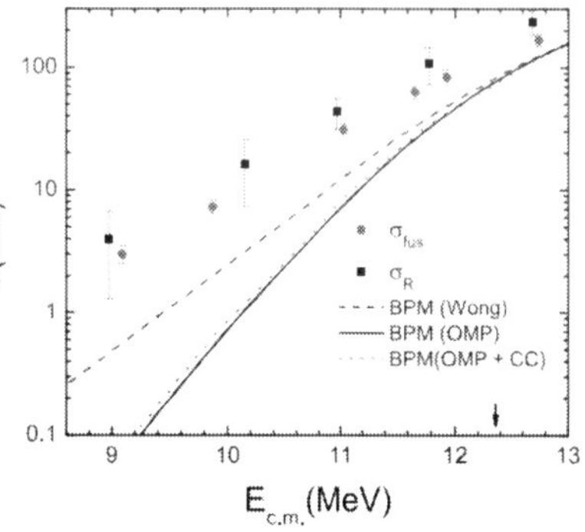

Figure 1. Fusion ([6]) and total reaction ([1]) cross section data for ^{6}Li + ^{58}Ni. The curves correspond to calculations described in the text. Figure adapted from Ref. [6]

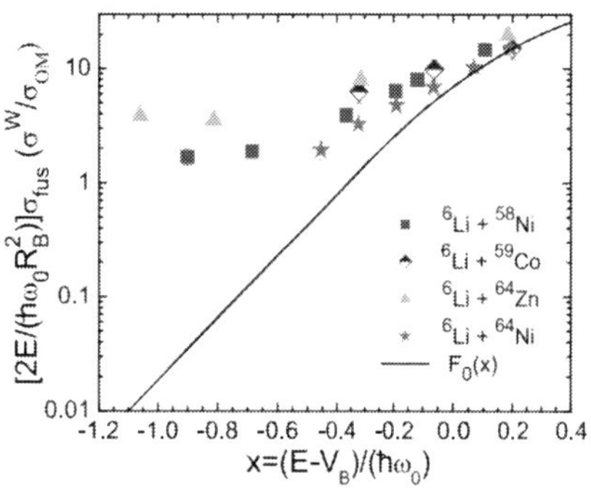

Figure 2. Comparison of reduced cross sections for four systems. The curve represents the Universal Fusion Function, which corresponds to expectations for normal systems. Figure taken from Ref. [6]

2 Analysis of elastic scattering

The five elastic scattering angular distributions for ^{6}Li + ^{58}Ni that were published in Ref. [1] were taken to make an optical model analysis. A surface polarization potential was introduced to simulate direct reactions, using in addition the SPP as the bare potential and a short-range imaginary potential (W_0 = 50 MeV, r_0 = 1.0 fm, a_0 = 0.1 fm) to describe fusion. The surface polarization potential was taken to have a derivative Woods-Saxon shape with 4 parameters: the real and imaginary depths (V_S, W_S), one common radius and one common diffuseness. From a first search, we were able to fix the radius at 1.47 fm and the diffuseness at 0.68 fm for all energies, so only the depths V_S, W_S were varied for each energy.

Figure 3 shows the results of the final fits. One can see that a very good description of the data is achieved. Also, the obtained total reaction cross sections were consistent with those reported in the original paper [1], plotted in Fig. 1. Figure 4 shows the values of depths obtained from the fit, as a function of the energy. The black squares are the imaginary depths and the red circles are the real depths. The imaginary depths were approximated by straight line segments, as shown by the black lines, applying then the dispersion relation, given by the following formula:

$$V(E) = V(E_S) + (E - E_S)\frac{1}{\pi}\wp \int_0^\infty \frac{W(r, E')}{(E' - E_S)(E' - E)} dE'. \quad (1)$$

The result is the red curve, which describes well the red points. So, the real and imaginary polarization potentials satisfy the dispersion relation, as they should.

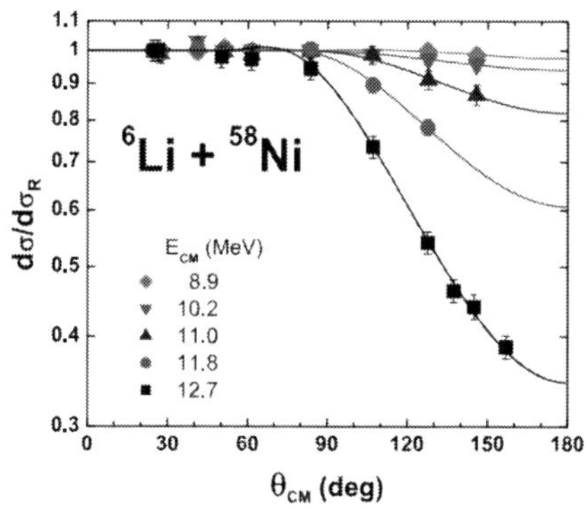

Figure 3. Elastic scattering angular distributions for ^{6}Li + ^{58}Ni (symbols), from Ref. [1], and results of the Optical Model fits obtained in the present work (curves).

In this analysis, complete fusion (CF) is identified with the absorption in W_0, and this absorption is represented in figure 5. The empty circles represent the complete fusion cross sections predicted in this analysis. It is quite surprising because the "experimental" values are much larger, represented by the filled circles. The effect of changing the radius of the surface potential was investigated. When a value of 1.61 fm instead of 1.47 fm was used for that radius, the results are shown by the blue diamonds. There is some sensitivity to the radius, but the changes are small, so there is no real qualitative change. Anyhow, these results

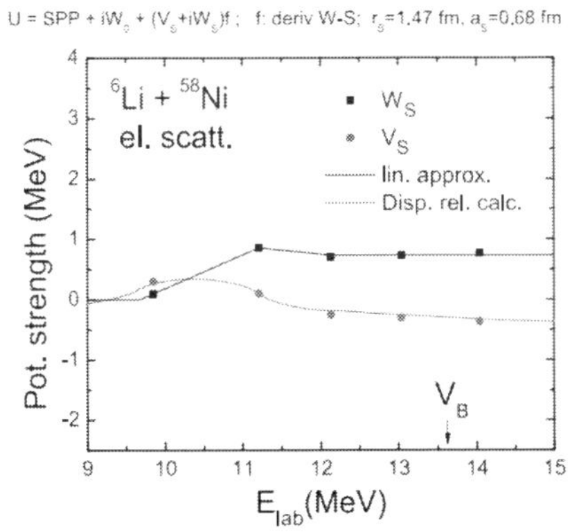

Figure 4. Real (V_S) and imaginary (W_S) potential depths obtained from the fits in Fig. 3

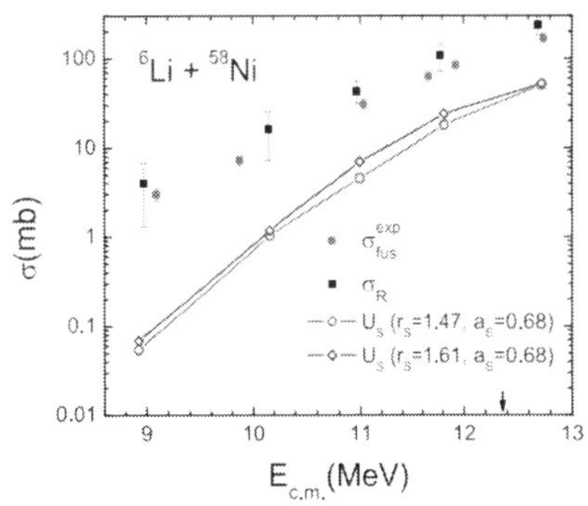

Figure 5. CF cross sections obtained from the present Optical Model analysis correspponding to two different values of r_S, the radius of the surface polarization potential.

should be considered as preliminary as there might be ambiguities in the potential that still need to be investigated. If the results are confirmed, this would mean that some processes different from CF dominate the data at these low energies. These processes could be incomplete fusion (ICF) or even some of the direct processes mentioned in Section 1. It is worth mentioning that Di Pietro and collaborators arrived at similar conclusions in the case of the ^{6}Li + ^{64}Zn data [3]. They showed that their relative populations of heavy residues could not be explained by CF calculations alone but exhibited evidence for an important contribution of ICF and/or Direct Cluster Transfer.

Finally, it may be pointed out that figure 5 is reminiscent of the situation observed for the neutron halo projectile ^{6}He. Figure 6 shows a similar plot but for the ^{6}He + ^{64}Zn system. In this case also the fusion cross sections are much lower than the total reaction cross sections. Both data sets in figs. 5 and 6 correspond to comparable energy regions in terms of the respective barrier, and there is a remarkable similarity between them. It is possible, therefore, that between the weakly-bound neutron-halo nucleus ^{6}He and the stable weakly-bound isotope ^{6}Li there are more similarities than are usually conceded.

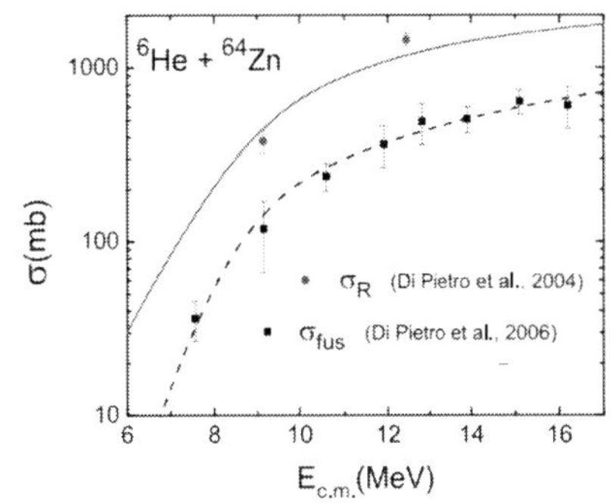

Figure 6. Fusion and total reaction cross sections for ^{6}He + ^{64}Zn Data from Refs. [15] and [16], respectively. Figure adapted from Ref. [17].

3 Conclusions

Fusion cross sections (σ_{fus}) derived for ^{6}Li + ^{58}Ni from measurements of proton evaporation yields are presented. σ_{fus} is seen to be close to σ_R, i.e., fusion appears to be dominant in the respective energy region. A good global consistency is observed with data for the same projectile but targets of ^{59}Co, ^{64}Ni, and ^{64}Zn, but considerable data spread is noted in the lower energy region. A surface polarization potential was introduced which describes well the elastic scattering data for ^{6}Li + ^{58}Ni and satisfies the dispersion relation. The respective Optical Model calculation predicts a low contribution of CF to TF, so ICF or other direct (distant) processes would be important. If the above conclusion is confirmed, ^{6}Li could behave more similar to ^{6}He than is usually thought.

4 Acknowledgments

Work partially supported by CONACYT (México) and by NSF (USA) under grant No PHY14-01343.

References

[1] E. F. Aguilera, E. Martinez-Quiroz, D. Lizcano, A. Gómez-Camacho, J. J. Kolata, L. O. Lamm, V. Guimarães, R. Lichtenthäler, O. Camargo, F. D. Becchetti, H. Jiang, P. A. DeYoung, P. J. Mears, and T. L. Belyaeva, Phys. Rev. C **79**, 021601(R) (2009)

[2] M. Zadro, *et al.*, Phys. Rev. C **87**, 054606 (2013)
[3] A. DiPietro, *et al.*, Phys. Rev. C **87**, 064614 (2013)
[4] D. R. Otomar, *et al.*, Phys. Rev. C **87**, 014615 (2013)
[5] S. Kalkal, *et al.*, Phys. Rev. C **93**, 044605 (2016)
[6] E. F. Aguilera, E. Martinez-Quiroz, P. Amador-Valenzuela, D. Lizcano, A. García-Flores, J. J. Kolata, A. Roberts, G. V. Rogachev, G. F. Peaslee, V. Guimarães, F. D. Becchetti, A. Villano, M. Ojaruega, Y. Chen, H. Jiang, M. Febbraro, P. A. DeYoung, and T. L. Belyaeva, Phys. Rev. C **96**, 024616 (2017).
[7] L. C. Chamon, B. V. Carlson, L. R. Gasques, D. Pereira, C. De Conti, M. A. G. Alvarez, M. S. Hussein, M. A. Cândido Ribeiro, E. S. Rossi, Jr., and C. P. Silva, Phys. Rev. C **66**, 014610 (2002)
[8] C. Y. Wong, Phys. Rev. Lett. **31**, 766 (1973)
[9] N. V. S. V. Prasad, A. M. Vinodkumar, A. K. Sinha, K. M. Varier, D. L. Sastry, N. Madhavan, P. Sugathan, D. O. Kataria, and J. J. Das, Nucl. Phys. **A603**, 176 (1996)
[10] L. R. Gasques, L. C. Chamon, D. Pereira, M. A. G. Alvarez, E. S. Rossi Jr., C. P. Silva, and B. V. Carlson, Phys. Rev. C **69**, 034603 (2004)
[11] L. F. Canto, P. R. S. Gomes, J. Lubian, L. C. Chamon, and E. Crema, J. Phys. G: Nucl. Part. Phys. **36**, 015109 (2009)
[12] L. F. Canto, P. R. S. Gomes, J. Lubian, L. C. Chamon, and E. Crema, Nucl. Phys. **A821**, 51 (2009)
[13] C. Beck, *et al.*, Phys. Rev. C **67**, 054602 (2003).
[14] M. M. Shaikh, *et al.*, Phys. Rev. C **90**, 024615 (2014)
[15] A. Di Pietro *et al.*, Phys. At. Nucl. **69**, 1366 (2006)
[16] A. Di Pietro *et al.*, Phys. Rev. C **69**, 044613 (2004)
[17] E. F. Aguilera and J. J. Kolata, Phys. Rev. C **85**, 014603 (2012)

Estimation of evaporation residue cross section for the synthesis of superheavy nuclei.

Yoshihiro Aritomo[1,*], Shyou Sawada[1], Nur Liyana[1], Jin Nakagawa[1], Shoya Tanaka[1], and Ken'etsu Hanabusa[2]

[1] Faculty of Science and Engineering, Kindai University, Higashi-Osaka, Japan
[2] HPC Systems Inc., Minatoku, Tokyo, Japan

Abstract. To estimate the evaporation residue cross section of superheavy nuclei, we clarify the parameter dependence of the models. At the first attempt, we focus our attention on the parameters of the statistical model, and investigate the parameter dependence of the survival probability. Then we discuss the the evaporation residue cross section for each parameter.

1 Introduction

Since 1966, according to macroscopic-microscopic calculations [1, 2], the existence of the island of stability in the nuclear chart surrounding the doubly magic superheavy nucleus containing 114 protons and 184 neutrons has been predicted. The property and structure of nuclei in the superheavy mass region have been investigated, taking into account a large multidimensional deformation for the ground state [3–5]. Recently, within the relativistic mean-field model [6] and nonrelativistic Skyrme-Hartee-Fock approach [7], some other spherical magic numbers have been found, such as Z=120 and N=172 [8].

Attempts for synthesizing heavy elements beyond the atomic number $Z \sim 100$ have become active since the 1970s by means of various developments in experimental techniques [9–12] and have recently succeeded in identifying the element till $Z = 118$.

The synthesis of these superheavy elements has been carried out using heavy-ion fusion reactions between stable nuclei, in which two different types of reaction have been employed. In the cold fusion reactions, lead and bismuth targets are used [13, 14]. The element with Z=113 reported by RIKEN used this type of reaction [12]. The superheavy nuclei (SHN) synthesized in the cold fusion reaction produce nuclei with relatively small number of neutrons.

The other type of reaction, called the hot fusion reaction, on the other hand, uses actinide nuclei as targets. With this type of reaction, production of elements till $Z = 118$ were reported by the Flerov Laboratory of Nuclear Reactions (FLNR) [15]. Recently, other laboratories than FLNR also performed experiments of hot fusion reactions and obtained results that are consistent with the data by FLNR [16–21].

For synthesis of new nuclei and elements, an accurate prediction of the production cross sections is an important issues in the superheavy elements research.

Many theoretical studies on the synthesis of superheavy elements have been published and the evaporation residue cross section corresponding to the above experiments has been estimated [22]. In the theoretical calculation, the evaporation residue cross section is obtained as the product of the fusion probability forming a compound nucleus and its survival probability in the competition with the fission process [23]. Since a substantial uncertainty is involved in each stage and model, we discuss uncertainties of parameters. Here, we focus our attention on the parameters of the statistical model.

In Sec. 2, we describe in detail the framework of the model. We discuss uncertainty of parameters in the models in Sec. 3. In Sec. 4, we investigate the parameter dependence of the survival probability. In consequence, we show the evaporation residue cross section in the reaction ^{48}Ca+^{244}Pu which is calculated with the several parameters in the statistical model. In the last section, we gave also an outlook to our next step for improvements of calculations.

2 Estimation of evaporation residue cross section

To estimate the evaporation residue cross section, the whole fusion process is divided into the three stages depend on the reaction time scale t. The first stage is the approaching process, $t < 10^{-22} - 10^{-21}$s, and the capture probability is denoted by T_l. The second stage corresponds to the competition between the fusion and quasi-fission processes, $10^{-21} \le t \le 10^{-18}$s, and the formation probability of forming a compound nucleus is denoted by P_{CN}. The decay process of the compound process is presented as the third stage $t \ge\sim 10^{-18}$s, and the survival probabil-

* e-mail: aritomo@ele.kindai.ac.jp

ity of compound nuclei during de-excitation is denoted by W_{sur}.

Using these probabilities, the evaporation residue cross section σ_{EV} is estimated as

$$\sigma_{EV} = \frac{\pi\hbar^2}{2\mu_0 E_{c.m.}} \sum_{l=0}^{\infty}(2l+1)T_l(E_{CM}, l)P_{CN}(E^*, l)W_{sur}(E^*, l), \tag{1}$$

where μ_0 denotes the reduced mass in the entrance channel. E_{CM} and E^* denote the incident energy in the center-of-mass frame and the excitation energy of the compound nucleus, respectively. E^* is given as $E^* = E_{CM} - Q$, where the latter denotes the Q-value of the reaction. $T_l(E_{cm}, l)$ is the capture probability of the l-th partial wave, which is calculated with the coupled channel model [24, 25]. To estimate $P_{CN}(E^*, l)$, we use the dynamical model and employ the Langevin equation [26, 27]. $P_{CN}(E^*, l)$ is obtained to To estimate P_{CN}, the definition of the fusion area in the deformation space is very important. It is reasonable to define the fusion area around the pocket near the ground state in the deformation space. Here, we define the fusion area (fusion box) as the inside of the fission saddle point in the system, which is $\{z < 0.8, \delta < 0.3, |\alpha| < 0.3\}$ [26]. $W_{sur}(E^*, l)$ is calculated using a statistical model [28, 29].

3 Uncertainty of parameters in the theoretical models

Many theoretical models have been developed and applied to estimate the evaporation residue cross section [22]. The results show rather good agreement with the experimental data. However, for the synthesis of unknown elements, $Z = 119, 120$ etc, the predictions from different theoretical approaches are quite different [22].

Inevitably, a substantial uncertainty is involved in each stage. In the first stage, there are few parameters in the coupled channel model. Though the parameters of the potential (potential depth, diffuseness) of the compound nuclei are defined to reproduce the capture cross section in experimental data, these parameters are not clear for unknown nuclei which have not been measured. In the second stage, the Langevin calculation includes parameters: potential energy (parameters of liquid drop model and shell correction energy), nuclear shape parametrization and the number of the dynamical variables (shape parameters), the transport coefficients (using the macroscopic or microscopic models). The definition of the fusion region is also unclear. In the third stage, the statistical model includes the uncertainty of the fission barrier height of the compound nucleus, the friction parameter, the level density parameter (a_f/a_n), etc. [28, 29]. Moreover, the reaction Q-value has also an uncertainty, because the masses of superheavy nuclei are not determined experimentally yet.

We should define the values of unknown parameters in each stage, or at least we would like to show explicitly and specify the values of parameters which are used in the calculation. As the first step, we have to know the parameter dependence of the evaporation residue cross section within the model calculation.

In the next section, we demonstrate the dependence using the present model.

4 Results

To know the parameter dependence of our model, first we focus our attention on the survival probability in the third stage.

4.1 Survival probability

We calculate the survival probability using the statistical code MASADEC, which is developed by M. Ohta [29] based on the idea in the reference [28]. In the code, we change the values of parameters and investigate the influence of the survival probability. The fission barrier height of compound nucleus B_f, the friction parameter γ, and the level density parameter a_f/a_n are changed for the compound nucleus ^{292}Fl with the angular momentum $\ell = 30$. Figure 1 shows the results for each parameter. In the model, we calculate the fission barrier height B_f [29] based on the table [30]. We introduce the adjustment parameter of B_f, and the results are shown in Fig. 1(a). We can see that the survival probability changes by one order of magnitude as the B_f changes by 1 MeV. It seems the survival probability is moved in parallel in this case.

Fig. 1(b) shows the dependence of γ parameters. When the γ is large, the fission is surprised. In this case, as changing the γ values, the slope of the survival probability changes. The dependence of the parameter a_f/a_n is the same as γ, that causes to change the slope of the survival probability, which is shown in Fig. 1(c). In the figure, a_f/a_n is represented by `fact`.

Due to the calculation results, we can see the tendency of the parameter dependence of the survival probability and the statistical model.

4.2 Evaporation residue cross section

Under the parameter dependence in the survival probability, we obtain the evaporation residue cross section. Figure 2 shows the evaporation residue cross section for the reaction ^{48}Ca + ^{244}Pu. The dependence of B_f and the γ parameter in the statistical model are shown in Fig. 2(a) and (b), respectively. The evaporation residue cross section changes significantly depending on the values of parameters, and we can clearly understand the dependence.

By the systematical calculation and comparison with the experimental data, we try to find the reasonable parameter sets. As the first attempt, to reproduce the excitation function of the evaporation residue cross section for the reactions with ^{48}Ca-projectile leading the heavy elements from $Z = 102$ to 118, we change the B_f and γ in the statistical model. In the systematical investigation, we found the reasonable values to reproduce the experimental data. To reproduce the experimental data, we introduce the additional value of B_f, which is denoted as ADDS. When we plot the correlation between the Z number of the compound nuclei and ADDS, we just find the linear correlation.

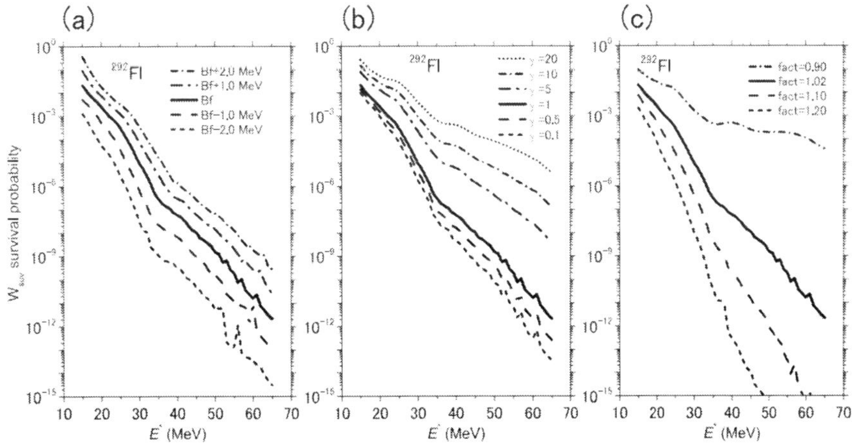

Figure 1. Parameter dependence of survival probability of ^{292}Fl calculated by the statistical code `MASADEC`. (a) Fission barrier height B_f, (b) Friction parameter γ, (c) Level density parameter a_f/a_n.

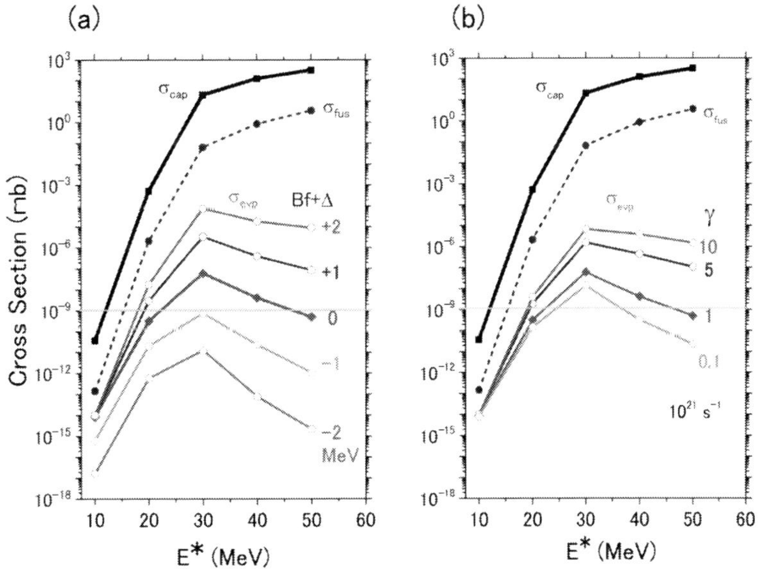

Figure 2. Evaporation residue cross section for the reaction ^{48}Ca + ^{244}Pu. The dependence of B_f and γ in the statistical model are shown in (a) and (b), respectively.

It is rough estimation and seems to be the model dependence. Also, we could show that FRIC has the exponential correlation of Z. These correlations show as following,

$$\mathrm{ADDS} = 0.218 \times Z - 23.2 \tag{2}$$

$$\mathrm{FRIC} = 8.84 \times \exp[0.169 \times Z]. \tag{3}$$

Though they are a rough estimation and we have to explain the reasons, it seems important to find the correlation among the parameters. For the future work, using the more accurate mathematical methods, for example, the covariance function, we try to predict the possibility of synthesis of new elements more accurately.

The author is grateful to Prof. K. Hagino (Tohoku University), Dr. K. Nishio (JAEA) and Mr. T. Tanaka (Kyushu University, RIKEN), Dr. Y. Kim (IBS) and J.Y. Lee (IBS) for their helpful suggestions and valuable discussion throughout the present work. This work was supported by the Sasakawa Scientific Research Grant from The Japan Science Society.

References

[1] W.D. Myers and W.J. Swiatecki, Nucl. Phys. **81**, 1 (1966).

[2] A. Sobiczewski, F.A. Gareev, and B.N. Kalinkin, Phys. Lett. **22**, 500 (1966).

[3] P. Möller and J.R. Nix, J. Phys. G. **20**, 1681 (1994).

[4] R. Smolanczuk, J. Skalski, and A. Sobiczewski, Phys. Rev. C **52**, 1871 (1995).

[5] H. Koura, M. Uno, T. Tachibana, and M. Yamada, Nucl. Phys. **A674**, 47 (2000).

[6] P.G. Reinhard, Rep. Prog. Phys. **52**, 439 (1989).

[7] P. Quentin and H. Flocard, Annu. Rev. Nucl. Part. Sci. **28**, 523 (1978).

[8] K. Rutz, M. Bender, T. Burvenich, T. Schilling, P.G. Reinhard, J.A. Maruhn, and W. Greiner, Phys. Rev. C **56**, 238 (1997).

[9] Yu.Ts. Oganessian and Y.A. Lazarev, in Treatise on Heavy-Ion Science, edited by D. A. Bromley (Plenum, New York, 1985), p. 3.

[10] G. Münzenberg, Rep. Prog. Phys. **51**, 57 (1988).

[11] P. Armbruster, Annu. Rev. Nucl. Sci. **35**, 135 (1985).

[12] .K. Morita *et al.*, J. Phys. Soc. Jpn. **73**, 2593 (2004); K. Morita *et al.*, J. Phys. Soc. Jpn. **76**, 043201 (2007).

[13] Yu.Ts. Oganessian, *Lecture Notes in Physcs (Springer Heidelberg)* **33**, 221 (1975).

[14] S. Hofmann and G. Münzenberg, Rev. Mod. Phys. **72**, 733 (2000); S. Hofmann et al., Eur. Phys. J. A **14**, 147 (2002).

[15] Yu.Ts. Oganessian et al., Nature **400**, 242 (1999); Phys. Rev. Lett. **83**, 3154 (1999); Phys. Rev. C **62**, 041604 (2000); Phys. Rev. C **63**, 011301 (2000); Phys. Rev. C **69**, 021601 (2004); Phys. Rev. C **72**, 034611 (2005); Phys. Rev. C **74**, 044602 (2006); Phys. Rev. C **76**, 011601 (2007); Phys. Rev. Lett. **104**, 142502 (2010).

[16] S. Hofmann and G. Münzenberg, Rev. Mod. Phys. **72**, 733 (2000); S. Hofmann et al., Eur. Phys. J. A **14**, 147 (2002); S. Hofmann et al. Eur. Phys. J. A **31**, 251 (2007).

[17] L. Stavsetra, K.E. Gregorich, J. Dvorak, P.A. Ellison, I. Dragojević, M.A. Garcia, and H. Nitsche, Phys. Rev. Lett. **103**, 132502 (2009).

[18] P.A. Ellison et al., Phys. Rev. Lett. **105**, 182701 (2010).

[19] Ch.E. Düllmann et al., Phys. Rev. Lett. **104**, 252710 (2010).

[20] K. Morita et al., RIKEN Accel. Prog. Rep. **47**, p.xi (2014).

[21] D. Kaji et al., RIKEN Accel. Prog. Rep. 48, (2015) in print.

[22] R.S. Naik, W. Loveland, P.H. Sprunger, and A.M. Vinodkumar, Phys. Rev. C **76**, 054604 (2007).

[23] Y. Aritomo, T. Wada, M. Ohta, and Y. Abe, Phys. Rev. C **59**, 796 (1999).

[24] K. Hagino and N. Rowley, Phys. Rev. C **69**, 054610 (2004).

[25] K. Hagino, N. Rowley, and A.T. Kruppa, Comput. Phys. Comm. **123**, 143 (1999).

[26] Y. Aritomo and M. Ohta, Nucl. Phys. A **744**, 3 (2004).

[27] Y. Aritomo,K. Hagino, K. Nishio, and S. Chiba, Phys. Rev. C **85**, 044614 (2012).

[28] R. Vandenbosch, J.R. Huizenger, Nuclear Fission, Academic Press, New York, 1973, p. 233.

[29] M. Ohta, in: Proceedings of Fusion Dynamics at the Extremes, Dubna, 25-27 May 2000, World Scientific, Singapore, 2001, p. 110.

[30] P. Möller, J.R. Nix, W.D. Myers, and W.J. Swiatecki, Atomic Data and Nuclear Data Tables **59**, 185 (1995).

EPJ Web of Conferences **163**, 00003 (2017)
FUSION17 DOI: 10.1051/epjconf/201716300003

The AGFA and AIRIS separators at ATLAS

B. B. Back[1]

[1] *Physics Division, Argonne National Laboratory, Argonne, IL 60439, USA*

Abstract. The ATLAS facility is being augmented with two new separators in order to take full advantage of recent and future anticipated beam intensity upgrades to the accelerator. A novel concept is used for the Argonne Gas-Filled Analyzer (AGFA), which employs only two magnetic elements; a quadrupole singlet for vertical focusing and a multipole magnet that provides a dipole field for the separation and a quadrupole field for horizontal focusing of the reaction products. The design allows for placing Gammasphere or GRETINA at the target position. This arrangement enables studies of prompt gamma-ray emission from weakly populated trans-fermium nuclei and those near the doubly-magic N=Z=50 shell closures measured in coincidence with the recoils registered in the AGFA focal plane. The Argonne In-flight Radioactive Ion Separator (AIRIS) is a magnetic chicane that will be installed immediately downstream of the last ATLAS cryostat. It will be used to separate radioactive ion beams generated in flight by the primary beam at an upstream high-intensity production target. These beams will be further purified by a downstream RF sweeper and transported to a number of target stations including HELIOS, the Enge spectrograph, the FMA and Gammasphere.

1 Introduction

With the recent accelerator upgrades at ATLAS, which were achieved by replacing several cryostats with the original split-ring resonators with modern quarter-wave resonators, it is now possible to accelerate high-intensity beams of stable heavy ions. In anticipation of this development, we have undertaken a program of instrumentation upgrades in order to utilize these beams in an effective manner and to expand the research program at ATLAS. First, we have developed a novel design for a gas-filled separator, the Argonne Gas-Filled Analyzer, AGFA, chiefly to separate and study the structure of transuranic nuclei that can be produced in heavy-ion fusion reactions, but only with small cross sections. This endeavor requires both intense heavy-ion beams and efficient collection of the reaction products at the focal plane of the separator in order to characterize their decay properties.

Secondly, we have for two decades carried out a successful research program with light, radioactive beams produced in flight using various light charged-particle reactions, such as (d,p), (d,n), (p,n) *etc.*, in inverse kinematics by focusing the primary, stable heavy-ion beam onto a production gas cell containing hydrogen or deuterium gas. The radioactive reaction products of interest are focused by a solenoid and subsequently separated from the primary beam by a magnetic bend downstream before being focused onto the target of the experimental setup. However, thus far this production facility is limited by being able to deliver radioactive beams only to the Enge spectrograph and HELIOS beam lines and by the fact that the gas cell can only tolerate primary beams of moderate intensity. Since an enhanced facility for producing in-flight radioactive beams will be able to support a wide range of experimental research, most in nuclear astrophysics and nuclear structure studies, we have designed a dedicated separator, the Atlas In-flight Radioactive Ion Separator, AIRIS, which will be installed immediately downstream from the last cryostat of the ATLAS linac such that all experimental stations in the facility can accept these beams and the full intensity of the primary beam can be utilized by providing effective shielding and increasing the distance between the primary beam dump and the experimental areas.

2 AGFA

The technique of collecting reaction products from fusion evaporation and separating them from the primary beam using the gas-filled separator technique [1] has proven very successful and it is now used almost exclusively in experiments aimed at strudying the structure of transactinide nuclei [2, 3] and synthesizing the heaviest elements [4–6]. We have developed a new, compact design [7] for such a separator with the dual functions of re-focusing the evaporation residues onto a limited area of the focal plane and separating them from the beam. With the AGFA design (see table 1), these functions are achieved by using only two magnetic components, 1) a vertically focusing quadrupole magnet, Q_v, and 2) a dipole magnet, D_m, that incorporates also a quadrupole component for horizontal focusing of the evaporation residues. This design concept, illustrated in Fig.1, has several attractive features, chiefly the possibility to place the Gammasphere array at the target position ~84 cm upstream of the quadrupole magnet to allow for detection of promptly emitted γ rays. In this

 DOI: 10.1051/epjconf/201716300003

Table 1. Some relevant parameters characterizing the AGFA separator. Note that the total flight-path length and the solid angle depends on the location of the target upstream of the quadrupole.

Parameter	Value	
Configuration	$Q_v D_m$	
Maximum bending power, $B\rho$	2.5 Tesla-m	
Maximum field at Q_v pole tip	1.24 Tesla	
Maximum field at D_m pole tip	1.7 Tesla	
Bend angle	38 degrees	
Target to Q_v distance	40 cm	84 cm
Solid angle, Ω	44 msr	22 msr
Target to focal plane distance	3.7 m	4.3 m

position, the solid angle of the separator is about 22 msr. Alternatively, using a target position 40 cm upstream of the front of the quadrupole for experiments where prompt γ-ray detection is not needed, increases the solid angle to 44 msr. In typical experiments studying transuranic nuclei via heavy-ion fusion reactions, such as *e.g.* ^{48}Ca+^{208}Pb $\rightarrow^{254}$No +2n, simulations show that there is a very good separation between the ^{254}No recoils and the beam. In this case, ~70% of the ^{254}No recoils fall within an area of 64×64 mm^2, the size of a typical Double Sided Strip Detector, (DSSD). This relatively narrow focus allows for an efficient detection of subsequent γ decays by using the Argonne X-array, which consists of five large HPGe Clover detectors that can be mounted around the focal plane chamber containing the 64 ×64 mm^2 DSSD implantation detector.

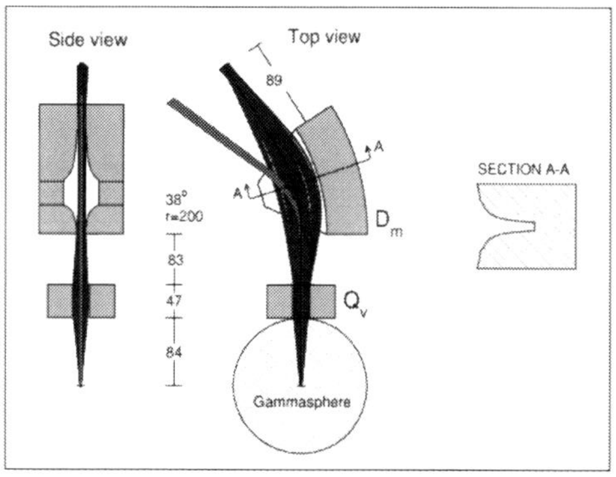

Figure 1. Schematic views of the AGFA separator with Gammasphere at the target position. The non-uniform character of the gap in the D_m magnet provides the horizontally focus. Dimensions are given in units of cm.

An important feature of the design is the fact that the pole pieces of the D_m magnet constitute part of the vacuum envelope. The vacuum chamber inside the gap contains O-ring seals that mate with the pole surfaces. This vacuum chamber was fabricated by the magnet vendor, Buckley Systems [8], to ensure the effectiveness of this seal. Fig. 2 shows the two magnets mounted on the support stand in the final location at ATLAS and with the dipole chamber outfitted with two blank-off flanges in order to verify that good vacuum could be obtained also after re-assembly of the D_m magnet after shipment to Argonne.

Figure 2. Photograph of the AGFA magnets at ATLAS taken during the vacuum test of the D_m vacuum chamber seal.

2.1 Planned research program

A very diverse research program is envisioned for the AGFA separator. It is anticipated that a main focus of this future research program will be in the study of the structure of the transuranic elements, which can be populated in heavy-ion fusion reactions involving beams of *e.g.* ^{48}Ca, ^{50}Ti, and ^{54}Cr impinging on targets around Pb and Bi. The AGFA separator operated in conjunction with Gammasphere will provide a world-leading setup for this type of studies. In addition, it is anticipated that studies of exotic, heavy proton emitters will constitute a large part of the research program. A number of heavy proton emitters have been predicted in the At-Bk range of the most neutron-deficient nuclei. These have not yet been identified experimentally because of very low cross sections and possibly short half-lives. With the efficient collection of evaporation residues with AGFA, these studies can be pushed toward lower cross sections and the implementation of a digital readout system for the focal plane detector system and the short flight path of AGFA allows for more effective searches for weakly populated, short-lived proton emitters expected in this area. A long-standing area of interest in nuclear structure is the region around the doubly-magic ^{100}Sn nucleus. Again, studies in this area have been hampered by the very small cross sections for populating the most neutron-deficient nuclei in this region, so the superior collection efficiency of AGFA coupled with the high γ-detection efficiency of Gammasphere offers hope that these studies can proceed at a more rapid pace. As always, a range of other possibilities may be realized once the capabilities of the AGFA-Gammasphere setup become better understood. Already, several new applications have been suggested.

2.2 Present status

At present, April 2017, the AGFA separator is nearing completion. Several simple experimental tests are planned in order to understand and verify the expected capabilities of the instrument as part of the commissioning process. It is expected that this will be completed in early summer 2017 such that the experimental program can start immediately thereafter.

3 AIRIS

The possibility of producing a range of radioactive beams with masses up to about A=60 removed by one to two nucleons from the stable nuclei is being realized by installing the AIRIS separator immediately downstream of the last cryostat of the ATLAS linac. Radioactive beams produced and separated from the primary beams and other contaminants can be transported to all downstream experimental stations. A schematic of the separator setup is given in Fig. 3. Some of the relevant design parameters are given in Table 2.

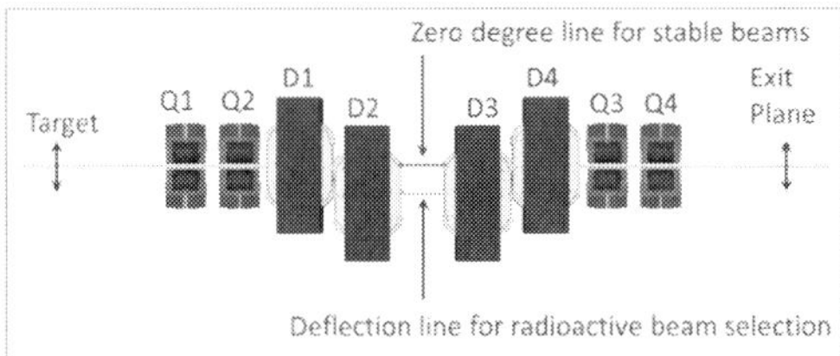

Figure 3. Schematic of the AIRIS separator consisting of four quadrupole singlets and four dipole magnets. The radioactive beam selection occurs at the mid-plane of the setup where the primary beam is stopped.

Radioactive beams produced by the interaction of the primary beams with the upstream high-intensity target are focused by a quadrupole doublet, Q1-Q2, to pass through a vertical slit opening at the mid-plane after having been bent about 25 cm down from the straight-through beam path. The primary beam will be stopped on either the upper or the lower slit depending on its magnetic rigidity relative to that of the selected radioactive beam. Just downstream of the slits, several diagnostics elements, such as a Faraday cup or a Si detector telescope, can be inserted into the radioactive beam path in order to ensure that the correct beam has been selected for further transport to the exit of the device.

Further downstream, the radioactive beam is brought back to the straight-through beam path by a second pair of dipole magnets and focused onto the exit plane by a second set of quadrupole magnets, Q3-Q4. At the exit plane, a second set of slits will further help to eliminate beam contaminants that managed to be transported up to this point. Here, again, various diagnostic elements can be used to characterize the beam particles that are allowed to proceed downstream. In cases where beam contaminants are not

Table 2. List of some relevant design parameters for the AIRIS separator.

Parameter	Value
Dipole bend angle	22.5deg
Dipole radius	1 m
Dipole full gap	10 cm
Max. Dipole field	1.75 Tesla
Quadrupole full aperture	15 cm
Max. quadrupole tip field	1.0 Tesla
Offset at mid-plane	25 cm
Total length: Target to Exit slits	7.1 m
Geometric acceptance	75 mrad
Energy dispersion at mid-plane	1.25 mm/%

fully eliminated by the methods described above, a final clean-up may be necessary by using an existing RF (radio-frequency) sweeper. This device is currently installed in the Enge Spectrograph beam line, but will be re-located further upstream such that it can serve a larger number of experimental stations.

While the AIRIS separator is intended for selecting and transporting secondary, in-flight beams, it is important that it also allows for the transport of stable beams directly from the ECR source or radioactive beams of ^{252}Cf fission fragments from the CARIBU source directly to the experimental stations as such beams are used in the majority of experiments performed at ATLAS. This is achieved by de-energizing the four dipole magnets in AIRIS and using only the two quadrupole doublets Q1-Q2 and Q3-Q4 for focusing of these beams. The upper slits at the mid-plane are, therefore, fully retractable to allow for the transport of straight-through beams.

The selectivity and efficiency of this design [9] has been characterized by a large number of transport simulations using the TRACK beam transport code [10] in order to arrive at an optimal design solution that can be realized within the spatial constraints of the ATLAS tunnel.

In the particle transport simulations, we concentrated on following: the trajectories of the in-flight beam, the primary beam including its small angle scattering in the production target (beam core), the elastically scattered beam particles (beam tail) as well as possile contaminants produced in he reaction. As an example, Fig. 4 illustrates the separation of a ^{56}Ni beam produced in a two-proton adding reaction, namely ^{54}Fe(^{12}C,^{10}Be), using a 2 mg/cm^2 production target. At the mid-plane focus there is good separation between the primary beam core and the in-flight ^{56}Ni beam in the 28$^+$ charge state, but the ^{55}Co^{27+} contaminant, produced in the one-proton adding reaction, clearly extends into the acceptance region of the y= -5 cm to 5 cm opening of the mid-plane slits. However, this contaminant follows trajectories that end up above the in-flight beam at the exit focal plane, where they are easily removed by a set of vertical slits located at this position.

In order to characterize the capabilities of AIRIS, we have performed a large number of TRACK simulations for representative in-flight beams in the mass range up to A~60. The assumptions concerning the angle-integrated

DOI: 10.1051/epjconf/201716300003

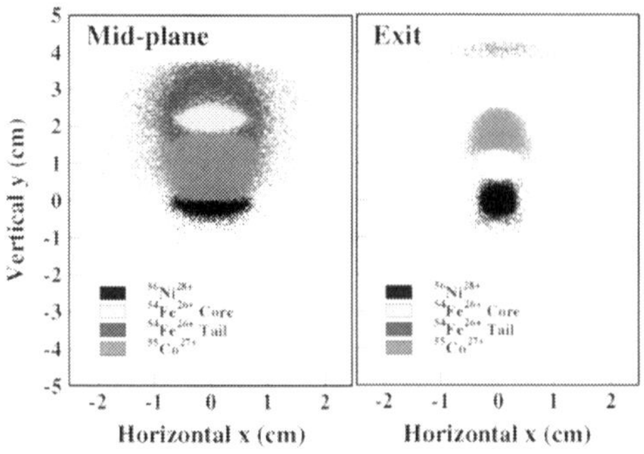

Figure 4. The left panel shows the transverse distribution of secondary $^{56}Ni^{28+}$ ions at the AIRIS mid-plane (black dots) along with those of the $^{54}Fe^{26+}$ beam core (green dots) and the $^{54}Fe^{26+}$ beam tail (blue dots). A secondary beam contaminant, $^{55}Co^{27+}$ (red dots), is also transported to the AIRIS mid-plane, but this component is not completely removed by the mid-plane slits which in this simulation select ions within the range -5 mm < y < 5 mm. The right panel shows the transverse distribution at the exit (focal plane) of particles that have passed through the mid-plane slits.

production cross section, σ_{pr}, the charge state fraction, $F(q)$, and the primary beam energy are listed in Table 3 along with the resulting secondary beam energy, the transmission, T, through AIRIS to the exit plane, and the resulting in-flight beam intensity. These simulations are all based on an effective target thickness of 2 mg/cm^2 and a primary beam intensity of 1 pμA. Further details on these simulations and the relevant input parameters are given in Ref. [11]

3.1 High-intensity production target

Several options exist for the design of a production target that can tolerate heavy-ion beam intensities in the range of 1 pμA. For solid target materials, such as C or Be, rotating target wheels are well proven to be able to accomodate beam intensities of this magnitude. However, concerning gaseous targets such as deuterium, hydrogen, and helium, it is not possible to employ the present method with gas cells because of the fragility of the entrance and exit windows. Instead, we have developed a high intensity target that generates a thin, liquid film by impinging two oil jets of ~0.5 mm diameter on each other inside a vacuum chamber. This mechanism generates a thin oil film of areal thickness ~ 1-2 mg/cm^2 needed for the primary target. The low vapor pressure of regular pump oil ensures that high vacuum can be maintained in the beam line. Several baffles and an in-line cold trap further ensure that the oil does not migrate along the beamline. The hydrogen content of the oil makes this target ideal for proton-induced reactions, *e.g.* (p,n) and (p,d). However, in order to use deuteron initiated reactions, such as (d,n) and (d,p), a deuterated pump oil is required, an option that we're presently researching. In addition, it is possible that the liquid film target can also be used in carbon induced reactions.

3.2 Planned research program

One of the research areas that will benefit from the development of the AIRIS separator is nuclear astrophysics. Over the last two decades, such studies have been carried out with great success using in-flight beams with a gas-cell production target, a superconducting collector solenoid and a single magnetic bend in the beam line leading to either the Enge Splitpole Spectrograph or HELIOS. With the implementation of AIRIS, higher beam intensities, heavier and more exotic beams can be provided for such studies. In addition, these beams are expected to enable a strong research program in structure studies of exotic nuclei by using the HELIOS spectrometer for studies of transfer reactions or inelastic scattering. Nuclear structure studies based on γ-ray detection using either Gammasphere of GRETINA will also benefit from the in-flight beams that AIRIS wil be able to provide.

3.3 Status

The implementation of the AIRIS separator is presently at the following stage. All the main components are under contract with outside companies with expected delivery to Argonne in the early summer of 2017. The installation into the ATLAS tunnel is, however, very disruptive to the ongoing research program and detailed pre-planning and scheduling of this phase is of paramount importance. This phase is, thus, scheduled to coincide with the annual month-long accelerator maintenance period that includes the last week of December 2017 and January 2018. However, we anticipate that this down time has to be extended into February or early March 2018 before the beam transport through the ATLAS tunnel can be resumed. The commissioning of AIRIS will then follow and the first radioactive beams from this facility can be used in research in Spring 2018.

4 Summary

Two new devices that will substantially enhance the research capabilities of the ATLAS facility have been presented. The AGFA gas-filled separator is presently under construction and will support a wide-ranging research program in studies of trans-uranium nuclei, exotic proton emitters and very neutron deficient nuclides near the ^{100}Sn doubly-closed shell. The combination of this device with high-efficiency γ detector arrays, such as Gammasphere and GRETINA, holds promise for extending nuclear structure studies into exotic regions of the nuclear chart that were previously inaccessible because of small cross sections for producing the nuclei of interest. It is clear, however, that the general property of very high collection efficiency of this device will also inspire research in areas that have not yet been foreseen. The AIRIS separator is being implemented in order to substantially enhance the current capabilities for in-flight production of

Table 3. List of in-flight radioactive beams, reactions used to produce them, primary and secondary beam energies, assumed angle-integrated cross sections, σ_{pr}, radioactive beam transmission to the AIRIS exit plane, T, the percentage of the charge state used, $F(q)$, and secondary beam intensities for 19 representative reactions using a 2 mg/cm^2 target and 1 pμA primary beam.

Second. Beam	Reaction	Sec.[Pri.] Energy (MeV/u)	σ_{pr} (mb)	T (%)	$F(q)$ (%)	Rate Exit (ions/s)
$^{11}Li^{3+}$	$^{14}C(^{50}Ti,^{53}Mn)$	7.6 [9]	0.01	50	100	4.0×10^3
$^{12}Be^{4+}$	$^{14}C(^{26}Mg,^{28}Si)$	15.9 [15]	0.2	50	100	8.0×10^4
$^{15}C^{6+}$	$^{14}C(d,p)$	4.1 [5]	50	60	100	2.4×10^7
$^{16}C^{6+}$	$^{14}C(^{26}Mg,^{24}Mg)$	12.0 [15]	1	80	100	6.4×10^5
$^{16}C^{6+}$	$^{18}O(^9Be,^{11}C)$	10.4 [10]	0.2	50	100	8.0×10^4
$^{14}O^{8+}$	$^{14}N(p,n)$	14.0 [15]	30	50	100	1.2×10^7
$^{19}O^{8+}$	$^{18}O(d,p)$	14.1 [15]	50	60	100	2.4×10^7
$^{18}Ne^{10+}$	$^{16}O(^{12}C,^{10}Be)$	7.1 [10]	1	40	100	3.2×10^5
$^{25}Al^{13+}$	$^{24}Mg(d,n)$	3.8 [5]	50	80	42	1.3×10^7
$^{26}Al^{13+}$	$^{26}Mg(p,n)$	9.0 [10]	50	80	86	2.8×10^7
$^{34}Si^{14+}$	$^{36}S(^9Be,^{11}C)$	9.7 [10]	0.2	50	81	6.5×10^4
$^{37}S^{16+}$	$^{36}S(d,p)$	8.9 [10]	50	40	72	1.1×10^7
$^{33}Cl^{17+}$	$^{32}S(d,n)$	9.1 [10]	50	80	67	2.1×10^7
$^{38}Ca^{20+}$	$^{36}Ar(^{12}C,^{10}Be)$	8.2 [10]	1	80	52	3.3×10^5
$^{49}Ca^{20+}$	$^{48}Ca(d,p)$	9.2 [10]	50	25	52	5.2×10^6
$^{50}Ca^{20+}$	$^{48}Ca(^{18}O,^{16}O)$	9.1 [10]	1	80	52	3.4×10^5
$^{41}Sc^{21+}$	$^{40}Ca(d,n)$	8.7 [10]	50	80	48	1.5×10^7
$^{56}Ni^{27+}$	$^{54}Fe(^{12}C,^{10}Be)$	8.6 [10]	1	80	44	2.8×10^5
$^{56}Ni^{28+}$	$^{54}Fe(^{12}C,^{10}Be)$	8.6 [10]	1	80	14	9.0×10^4

radioactive beams close to the line of stability. This development builds on our experience of producing such beams over the past two decades using a relatively simple setup, which has still proven very productive in terms of physics results. Using high-intensity beams from the upgraded ATLAS linac and the dedicated AIRIS separator higher intensity in-flight beams can be produced and used in a broad research program.

The AFGA collaboration includes: B.B. Back, W.F. Henning, R.V.F. Janssens, T.L. Khoo, J.A. Nolen , D.H. Potterveld , J. Rohrer, G. Savard, and D. Seweryniak, Argonne National Laboratory, M. Paul, Hebrew University, Jerusalem, P. Chowdhury and C.J. Lister, University of Massachusetts - Lowell, W.B. Walters, University of Maryland, P.J. Woods, University of Edinburgh, K. Gregorich, Lawrence Berkeley National Laboratory, and W. Loveland, Oregon State University.

The AIRIS collaboration consists of: M. Alcorta (now at Triumf, Canada), B. B. Back, C. Dickerson, C. R. Hoffman, B. P. Kay, A. Levand, S. Manikonda (now at AML, Florida), B. Micklich, B. Mustapha, J. Nolen, P. Ostroumov (now at Michigan State University), R. C. Pardo, K. E. Rehm, G. Savard, J. P. Schiffer and D. Seweryniak, Argonne National Laboratory.

This work was supported by the U.S. Department of Energy, Office of Nuclear Physics, under Contract No. DE-AC02-06CH11357 and uses resources from ATLAS, an Office of Science user facility.

References

[1] M. Paul *et al.*, Nucl. Inst. Meth. A **277**, 418 (1989)
[2] M. Leino *et al.*, Nucl. Inst. Meth. B **99**, 653 (1995)
[3] V. Ninov and K.E. Gregorich, in ENAM 98, *Exotic Nuclei and Atomic Masses*, edited by B. M. Sherrill, D. J. Morrissey, and C. N. Davids, AIP Conf. Proc. No. 455 AIP, Woodbury, 1998, p. 704
[4] K. Subotic, Yu. Ts. Oganessian, V. K. Utyonkov, Yu. V. Lobanov, F. Sh. Abdullin, A. N. Polyakov, Yu. S. Tsyganov, and O. V. Ivanov, Nucl. Instrum. Methods Phys. Res. Sec. A **481**, 71 (2002)
[5] A. Semchenkov *et al.* Nucl. Instr. Meth. B, **266**, 4153 (2008)
[6] K. Morita *et al.*, Nucl. Inst. Meth. B **70**, 220 (1992)
[7] The AGFA design and associated simulations are due to David Potterveld, Argonne National Laboratory.
[8] Buckley Systems, Mt Wellington, Auckland, New Zealand, http://www.buckleysystems.com/
[9] The AIRIS design and associated ion-beam simulations are due to Brahim Mustapha, Argonne National Laboratory.
[10] V.N. Aseev, P.N. Ostroumov, E.S. Lessner and B. Mustapha in Proc. 2005 Part. Accel. Conf., Knoxville, TN, TPAT029, p. 2053 (2005)
[11] http://www.phy.anl.gov/airis/index.html

The Active Target Time Projection Chamber at NSCL

D. Bazin[1,2,*], J. Bradt[1,2], Y. Ayyad[3], W. Mittig[1,2], T. Ahn[4], S. Beceiro-Novo[1,2], L. Carpenter[1,2], M. Cortesi[1], A. Fritsch[5], J. J. Kolata[4], W. Lynch[1,2], and N. Watwood[1,6]

[1]National Superconducting Cyclotron Laboratory, Michigan State University, East Lansing, MI 48824-1321, USA
[2]Department of Physics and Astronomy, Michigan State University, East Lansing, MI 48824-1321, USA
[3]Department of Applied Nuclear Physics, Lawrence Berkeley National Laboratory, Berkeley, CA 94720, USA
[4]Department of Physics, University of Notre Dame, Notre Dame, IN 46556, USA
[5]Department of Physics, Gonzaga University, Spokane, WA 99258, USA
[6]Department of Chemistry, Michigan State University, East Lansing, MI 48824-1321, USA

Abstract. Reactions in inverse kinematics close to the Coulomb barrier offer unique opportunities to study exotic nuclei, but they are plagued by the difficulty to efficiently and precisely measure the characteristics of the emerging particles. The Active Target Time Projection Chamber (AT-TPC) offers an elegant solution to this dilemma. In this device, the detector gas of the time projection chamber is at the same time the target in which nuclear reactions take place. The use of this new paradigm offers several advantages over conventional inert target methods, the most significant being the ability to increase the luminosity of experiments without loss of resolution. The AT-TPC and some results obtained on resonant α scattering to explore the clustering properties of neutron-rich nuclei are presented, as well as fusion cross section results using a ^{10}Be radioactive beam. In addition, the first re-accelerated radioactive beam experiment using the fully commissioned ReA3 linac was conducted recently at the NSCL with the AT-TPC, where proton resonant scattering of a 4.6 MeV/u ^{46}Ar beam was used to measure the neutron single-particle strength in ^{47}Ar.

1 Introduction

Reactions close to the Coulomb barrier have been used with light beam probes such as protons, deuterons or alphas since the dawn of the first accelerators. The properties of the majority of stable isotopes found in the chart of the nuclides have been studied using a wide variety of reactions such as resonant scattering or transfer reactions. However, extending these techniques to radioactive nuclei produced as beams has proven to be a significant experimental challenge. The main reason lies with inverse kinematics where the light probe nuclei are at rest in a target and have much wider kinetic energy range after the reaction than in direct kinematics. Measuring the properties of the scattered probe or its remnant becomes much more challenging, and in most cases a compromise between target thickness and resolution becomes necessary.

The concept of active target is an attempt at avoiding this compromise by turning the target into a detector medium where all the energy dissipated during the reaction can be measured and the location of each reaction vertex determined experimentally [1]. This is best done in a gas volume where the radioactive beam can react with the gas atomic nuclei as it slows down. Since the energy of the reaction products can be measured as they emerge from the reaction vertex, there is no loss of resolution regardless of the amount of material traversed by the incoming beam. The gas pressure can be adjusted to vary the effective target thickness and measure reactions in a particular range of energies. From the reaction vertex location measurement, the energy of the beam at the time of the reaction can be determined for each event and an excitation function of the cross section can be deduced. This feature is particularly useful when studying reactions such as fusion or resonant scattering for instance.

The AT-TPC and its smaller scale prototype (pAT-TPC) have been used at NSCL and other laboratories to conduct these types of experiments at energies close to the Coulomb barrier. Most of the experiments conducted so far have used low energy resonant scattering and fusion reactions. A few examples of the results obtained are presented in the following, after a brief description of the detector and its associated technology.

2 Detector

The AT-TPC consists of a 250 liter cylindrical volume closed by a cathode plate at the upstream end, and a sensor plane (labeled Micromegas) at the downstream end (see fig. 1). Upon applying a voltage at the cathode end, electrons freed by charged particles along their track drift towards the sensor plane at a constant velocity. The electric field is kept uniform along the main axis of the cylinder by a series of concentric rings located both inside and outside the field cage, and polarized via a linear resistor

* e-mail: bazin@nscl.msu.edu

 DOI: 10.1051/epjconf/201716300004

chain. Upon arriving at the sensor plane, the electrons

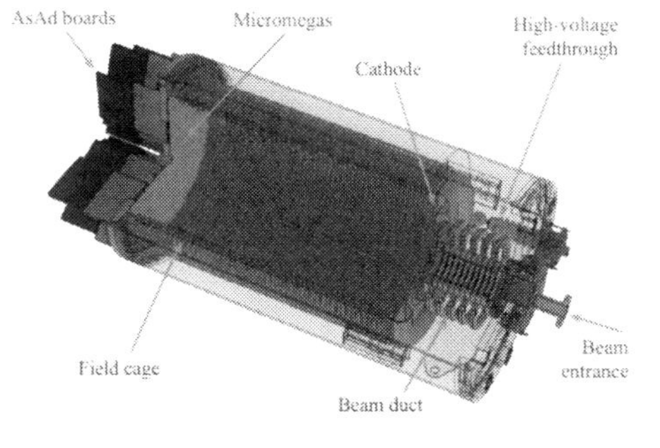

Figure 1. Schematic view of the AT-TPC. The field cage is made of a non-conductive material, fiberglass and epoxy. The outer chamber shields the field cage from the exterior and holds the beam duct that couples the high voltage of the cathode to the grounded beam pipe. The beam enters the field cage through a small 3 μm-thick window placed at the center of the cathode. The front-end cards (AsAd boards) of the digital electronics are directly connected to the back of the sensor plane via small feed-through boards equipped with o-rings for sealing the inner volume to the exterior.

are multiplied by a Micromegas device [2] consisting of a micro mesh mounted on a pad plane made of a Printed Circuit Board (PCB) where a pattern of 10,240 triangular pads have been etched. On the other side of this PCB high density connectors are used to transfer the signals to the digital electronics. A 3-dimensional image of the tracks left by the charged particles inside the field cage can be reconstructed from a time recording of the signals on each pad. The AT-TPC can be placed inside a large bore MRI solenoid to apply a magnetic field superimposed on the electric field. The resulting curved trajectories can be analyzed to measure the magnetic rigidity of the emitted particles for identification as well as momentum determination. The prototype version of the AT-TPC was built first and is very similar, except for its size, the geometry of the sensor plane and the number of channels [3].

The high density electronics used with the AT-TPC has been specially developed by a collaboration (GET for General Electronics for TPC) [4], that features for each channel a charge preamplifier and shaping amplifier followed by a 512-cell analog memory and 12-bit ADC for the digitizing section. In addition, this electronics has the unique feature of implementing discriminators on each channel and combining them into a running multiplicity that can be used to generate an internal trigger. This feature is particularly important in an active target as all charged particles entering the field cage do not escape, therefore no external detector can be used to generate a trigger from them. A complete description of the full scale AT-TPC will be available in a forthcoming publication [5].

3 Results

A few results obtained with the AT-TPC detectors are presented. They involve reactions around the Coulomb barrier which is the domain most adapted to the active target concept.

3.1 α resonant scattering

α resonant scattering at low energy is an efficient reaction to populate virtual states (also referred to as resonances) of the target+projectile composite system. In the case of α particles, these resonances may exhibit α clustering effects that are of high interest in the nuclear physics community. The most famous α cluster state is the so-called Hoyle state of ^{12}C, that is critical in the description of ^{12}C nucleo-synthesis via the triple α reaction. α cluster states are sought in other light nuclei, that may have equally important impacts on nucleo-synthesis, but are also of great interest in structure models of the nucleus.

The α clustering resonances in the nuclei ^{10}Be [6], ^{14}C [7] and more recently ^{12}Be have been investigated with the pAT-TPC using radioactive beams of ^{6}He, ^{10}Be and ^{8}He respectively. The target was composed of ^{4}He gas mixed with a 10% concentration of CO_2 as quencher. The radioactive beams were produced at the Notre Dame tandem facility using the Twinsol device [8] for [6, 7], and the TRIUMF facility for the ^{8}He beam.

The results shown below were obtained with the ^{10}Be beam, and show a clear separation between the elastic scattering events and the inelastic events where ^{10}Be was excited to its first 2^{+} state (panel b of fig. 2). The tracks from ^{4}He and ^{10}Be were clearly identified from the respective energy losses of the two particles, helped by polarizing every 5th ring of the pad plane to a higher gain for the Micromegas [7], as seen on panel a of the same figure 2. From the track reconstruction, the vertex of the reactions

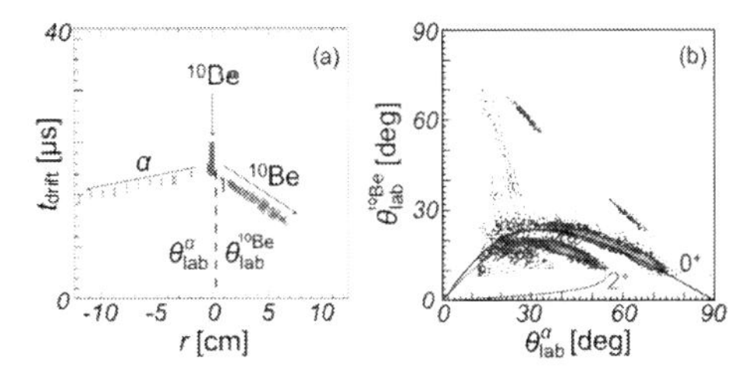

Figure 2. Panel a: typical scattering event recorded in the pAT-TPC, showing the double-gain setup of the Micromegas, by polarizing every 5th ring of the pad plane to a higher gain in order to better cover the dynamic range differences between the ^{4}He and ^{10}Be particles. Panel b: kinematics plot of the angle of ^{10}Be vs the angle of ^{4}He, where the elastic and inelastic components of the cross section are clearly separated (from [7]).

can be determined event by event, then used to calculate the reaction energy. Excitation functions can then be deduced as a function of the center-of-mass angle, as shown in panels a and c of fig. 3. The excitation functions show a

DOI: 10.1051/epjconf/201716300004

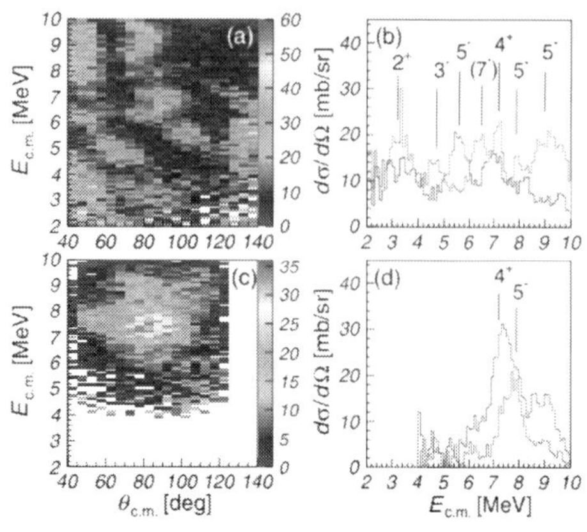

Figure 3. Panels a and c: excitation functions plotted for different center-of-mass angle of the reaction for the elastic and inelastic components of the cross section, respectively. The projections of the excitation functions between angles 70° - 90° (shaded spectrum) and 90° - 110° (clear spectrum) (from [7]).

number of resonances which can be identified via R-matrix analysis of their angular distributions. In this work 2^+ and 4^+ resonances were identified at 3.0 MeV and 7.0 MeV, respectively and tentatively assigned to a rotational band corresponding to a predicted linear chain structure in ^{14}C [9]. The predicted 0^+ band head resonance was not observed likely due to the 2 MeV experimental trigger cutoff.

3.2 Proton resonant scattering

Proton resonant scattering has been used extensively using proton beams to extract spectroscopic information on the analog states of the populated resonances in the neutron plus target system of the same isospin. The idea is to populate $T_0 + \frac{1}{2}$ compound nucleus states of a T_0 ^AZ nucleus plus a proton that are isospin analogs of bound levels in the ^{A+1}Z nucleus. An example can be found in [10] for the case of ^{41}Ar where the authors find good agreement between their deduced spectroscopic factors and those obtained from a ^{40}Ar(d,p) transfer reaction. The $T_>$ resonances observed in [10] were identified and analyzed by averaging the excitation functions with a width of about 20 keV, in order to smooth out the superimposed $T_<$ resonances of the ^{40}K compound nucleus.

Although many stable isotopes have been studied via this method using intense proton beams of very well defined energies, only a few radioactive isotopes with long enough half-lives to make target material could be measured. Carrying this methodology to short-lived radioactive nuclei implies the use of radioactive beams, therefore inverse kinematics. The motivation for measuring proton resonant scattering on ^{46}Ar stems from the observation of the weakening of the N=28 shell gap in isotones of ^{48}Ca such as ^{44}S and ^{42}Si [11, 12]. Located between the doubly magic ^{48}Ca and highly deformed ^{44}S, ^{46}Ar could exhibit reduced single-particle strength [13], and impose meaningful constraints on the effective interactions used in shell model calculations. Investigations using Coulomb excitation [14] and transfer reactions [15] indicate such a reduction might be small, and using an orthogonal method such as proton resonant scattering offers an alternative way of measuring spectroscopic factors in ^{47}Ar.

Performing this type of measurement in inverse kinematics on radioactive beams requires good quality beam properties. The energy resolution in particular is critical because the reaction energy is deduced from the position of the reaction vertex and the corresponding energy loss of the beam. The re-accelerated ^{46}Ar beam used to perform proton resonant scattering on this nucleus had an energy of 4.6 MeV/u with an energy width below 1%, hence an energy resolution around 40 keV/u. This radioactive beam was first produced via projectile fragmentation of a 140 MeV/u ^{48}Ca primary beam on a ^{9}Be target, filtered by the A1900 fragment separator, then stopped in a gas cell before being injected into an electron beam charge breeder, accelerated by the ReA3 linac and delivered to the AT-TPC. The average beam intensity was around 1,500 particles per second, with a small admixture (around 10%) of the stable ^{60}Ni and a few % of the ^{46}Ar daughter and grand-daughter ^{46}K and ^{46}Ca. A small ion chamber was placed in front of the AT-TPC to measure the energy loss of the incoming particles for identification. Fig. 5 shows the vertex position distribution observed in this experiment, where the sharp falloff corresponds to the entrance window of the field cage. The position resolution obtained from fitting the falloff edge of the distribution is 14.1 mm FWHM, which corresponds to an energy resolution of 43.3 keV/u FWHM.

Several challenges were encountered during this experiment. In particular, the large difference in energy loss between the ^{46}Ar and proton caused saturation and distortion effects in the electronics that required extensive cleaning of the data during the analysis phase. Unlike in the ^{10}Be + ^{4}He experiment presented above, individual polarization of the 10,240 pads of the AT-TPC was not available to reduce the avalanche gain of the pads on which the ^{46}Ar beam tracks were projected. This shortcoming is now being addressed via electronically controlled protection boards that allow the polarization of individual pads to be set programmatically.

The proton scattering event displayed in fig. 4 has been processed using image recognition algorithms such as the Hough transform to remove noise and spurious data. The left-hand side shows the pad plane projection of the event, while the right-hand side shows projections of the same event together with the result of the Monte-Carlo fitting used to model the event and extract its physical parameters such as vertex energy, scattering and azimuthal angles. More details on the analysis method can be found in [5]. The gas target used in this experiment was C_4H_{10} at about 20 Torr, therefore events corresponding to scattering on ^{12}C nuclei had to be rejected from the analysis. The final phase of this complex analysis is still ongoing and will be the subject of a publication in the near future.

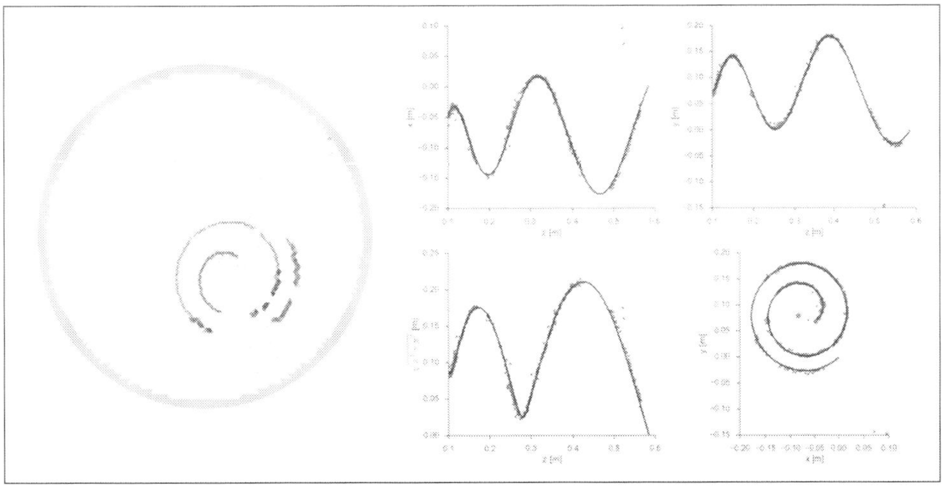

Figure 4. Typical event observed in ^{46}Ar + p scattering. The spiral track corresponds to the recoil proton orbiting in the 1.7 Tesla solenoidal field and slowing down in the gas. The hit pattern (left) and projections (right) show the data after processing designed to remove noise and cross talk, resulting in significant fragmentation of the track. The right-hand side projections have the results of the Monte-Carlo fit superimposed on the data. The red dot indicates the center of the spiral as guessed from the initial circular Hough transformation algorithm.

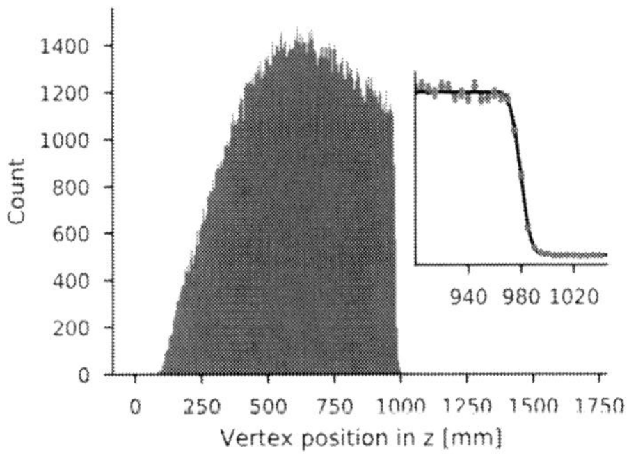

Figure 5. Vertex position distribution measured for the ^{46}Ar + p reaction. The sharp falloff corresponds to the entrance window of the field cage. The inset shows the detail of the falloff and the fit function used to estimate the resolution.

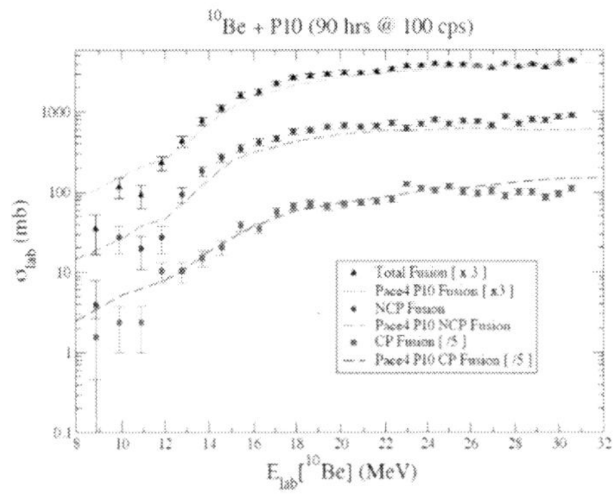

Figure 6. Excitation function of the ^{40}Ar + ^{10}Be fusion cross section. The total fusion cross section, as well as fusion cross section where a charged particle was emitted (CP) or not (NCP) are compared to Pace4 calculations (from [16]).

3.3 Fusion

For the same reasons as explained in the previous sections, the AT-TPC is particularly well adapted to measure fusion cross section excitation functions. The following results were obtained using a ^{10}Be radioactive beam produced by the Twinsol device at the University of Notre Dame [16]. Using P10 gas (90% Ar and 10% CH_4) in the pAT-TPC, the fusion cross section between ^{10}Be and ^{40}Ar was measured with very low beam intensity (100 counts per second). The selection and identification of fusion events was achieved using the range and energy loss information recorded by the detector. The observation of events where a charged particle was emitted during the reaction enabled us to separate fusion cross sections in different categories, as shown in fig. 6.

4 Conclusion

The active target concept combined with a time projection chamber detector shows promising prospects for experiments using rare radioactive beams in inverse kinematics at energies close to the Coulomb barrier. Several experiments have already shown interesting results in the study of α clustering states of neutron-rich nuclei and fusion cross section measurements. Resonant proton scattering turns out to be more challenging experimentally, in particular due to the large size difference between the beam and recoil particle, but the first experiment conducted with a re-accelerated beam of ^{46}Ar from the ReA3 linac at NSCL shows the potential of this method for exploring the single-particle structure of nuclei far from stability. Several improvements aimed at mitigating the shortcomings observed during this commissioning run are well under way, and will greatly benefit future experiments.

5 Acknowledgements

This work is supported by the US National Science Foundation under Grants No. MRI09-23087, No. PHY09-69456 and No. PHY14-01343.

References

[1] S. Beceiro-Novo et al., *Prog. in Part. and Nucl. Phys.* **84**, 124 (2015).

[2] Y. Giomataris et al., *Nucl. Instrum. Methods Phys. Res. A* **376**, 29 (1996).

[3] D. Suzuki et al., *Nucl. Instrum. and meth. in Phys. Res. A* **691**, 39 (2012).

[4] E. Pollacco et al., *Physics Procedia* **37**, 1799 (2012).

[5] J. Bradt et al., *submitted to Nucl. Instrum. Methods Phys. Res. A* (2017).

[6] D. Suzuki et al., *Phys. Rev. C* **87**, 054301 (2013).

[7] A. Fritsch et al., *Phys. Rev. C* **93**, 014321 (2016).

[8] F. Becchetti et al., *Nucl. Instrum. Methods Phys. Res. A* **505**, 377 (2003).

[9] T. Suhara et al., *Phys. Rev. C* **82**, 044301 (2010).

[10] H. L. Scott et al., *Phys. Rev.* **172**, 1139 (1968).

[11] C. Force et al., *Phys. Rev. Lett.* **105**, 102501 (2010).

[12] B. Bastin et al., *Phys. Rev. Lett.* **99**, 022503 (2007).

[13] A. Gade et al., *Phys. Rev. C* **93**, 054315 (2016).

[14] S. Calinescu et al., *Phys. Rev. C* **93**, 044333 (2016).

[15] L. Gaudefroy et al., *Phys. Rev. Lett.* **97**, 092501 (2006).

[16] J. J. Kolata et al., *Nucl. Instrum. Methods Phys. Res. A* **830**, 82 (2016).

Determination of Precision Fusion Cross Sections Using a High Efficiency Superconducting Solenoidal Separator

L. T. Bezzina[1,*], E. C. Simpson[1], I. P. Carter[1], M. Dasgupta[1], T. Ebadi[1], D. J. Hinde[1], and D. C. Rafferty[1]

[1]Department of Nuclear Physics, Research School of Physics and Engineering, The Australian National University, Acton ACT 2601, Australia

Abstract. A novel fusion product separator based on a gas-filled superconducting solenoid has been developed at the Australian National University. Though the transmission efficiency of the solenoid is very high, precision cross sections measurements require this efficiency to be estimated accurately. A Monte Carlo simulation for the efficiency can be performed, but in turn requires knowledge of the angular distribution of the evaporation residues. We have developed a method to deduce the angular distribution of the evaporation residues from the laboratory-frame velocity distribution $\sigma_{exp}(v_l)$ of the evaporation residues measured after the solenoid. The method will be discussed, focussing on the example of $^{34}S+^{89}Y$, where the angular distribution has been independently measured using a velocity filter [1]. The establishment of this method now allows the novel solenoidal separator to be used to obtain the most reliable, precision fusion cross-sections.

1 Introduction

The study of heavy-ion fusion gives us insights into nuclear structure and many-body quantum dynamics. With applications from the production of medical isotopes to the creation of superheavy elements, accurate knowledge of the cross sections of such fusion reactions is vital in order to optimise the product yield.

For light nuclei, good agreement between experiment and theory is found utilising simple barrier-passing models of fusion. However, studies of the fusion of heavier nuclei revealed an enhanced cross section at below-barrier energies [2–5]. The explanation of this enhancement was provided by the quantum coupled-channels model of fusion [6–9], which introduced the concept of multiple barriers resulting fromthe quantum superposition of target and projectile states. Measurements presented in 2007 [10], however, revealed the shortcomings of current coupled-channels models, notably the failure to provide simultanaeous prediction of above- and below-barrier cross sections.

This inconsistency of the coupled-channels model suggests a fundamental process is being overlooked theoretically. In order to accommodate this inconsistency, a number of explanations have been put forward, including effects of energy dissipation both below and above the barrier [10], and the damping of quantum vibrations as the single-particle wavefunctions of the two nuclei mix [11]. The role of the incompressibility of nuclear matter has also been explored [12], while more recently, a microscopic approach to include the effects of Pauli repulsion has also been developed [13]. As modifications to the existing theory are suggested and new models of fusion arise, verification of these theories via experimental measurements will be required.

The most strenuous experimental tests for new theoretical models will be provided by measurements of the fusion barrier distribution, requiring extremely precise cross section measurements. The novel fusion product separator based on a gas-filled superconducting 8 T solenoid, developed at the Australian National University (ANU), is well-placed to make these measurements.

This paper describes the development of a method to extract the angular distribution of fusion products, correcting for the transmission efficiency of the solenoid. Section 2 introduces the solenoidal separator, giving context to the iterative correction procedure described in section 3. Comparison of the deduced angular distribution with dedicated measurements and statistical model calculations is covered in section 4. Final conclusions and future directions of study conclude this paper in section 5.

2 Superconducting Solenoidal Separator

The 8 T superconducting SOLenoid for In-beam Transport And Identification of Recoiling Evaporation-residues, SOLITAIRE, is a new high efficiency fusion product separator at the ANU, used to ensure spatial separation of the high flux of elastically scattered beam particles from the evaporation residue (ER) fusion products. A schematic of the SOLITAIRE system is presented in figure 1.

Following the interaction of the beam with the target foil, the elastically scattered beam and ERs must first pass through the soft iron nose cone making up part of the iron shielding that encases the solenoid. The nose cone limits

*e-mail: lauren.bezzina@anu.edu.au

DOI: 10.1051/epjconf/201716300005

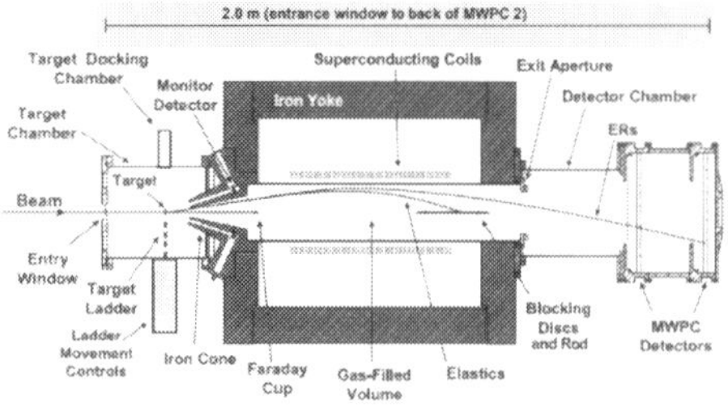

Figure 1: Schematic cross section of SOLITAIRE [14] showing radial trajectories of elastically scattered beam and ERs.

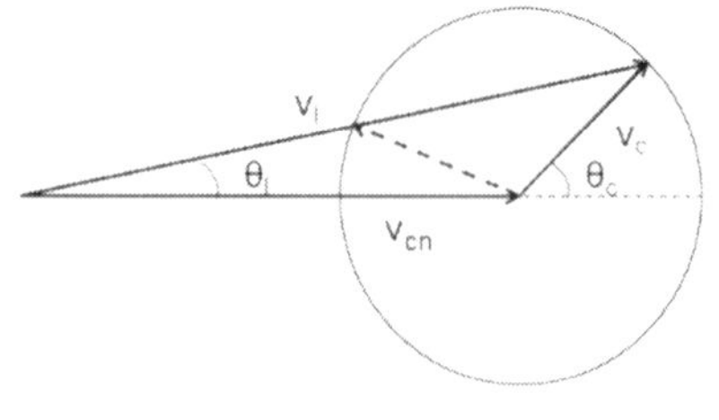

Figure 2: Velocity vector diagram illustrating the relationship between evaporation residue laboratory velocity (v_l) and angle (θ_l) and centre of mass velocity (v_c) and angle (θ_c), through the laboratory-frame compound nucleus velocity v_{cn}.

the angular acceptance of SOLITAIRE to 9.5°. Beyond this nose cone, the unscattered beam that passes through the target is collected by a Faraday Cup at the solenoid entrance which subtends an angle of less than 1°. The ERs are deflected away from the beam axis by the recoil effect of the evaporation process, ensuring the loss of ERs to the Faraday Cup is minimal.

Those elastically scattered beam particles and ERs not collected by the Faraday Cup continue to travel through the gas-filled volume of the solenoid. Prior to entering the solenoid, these particles have a distribution of charge states due to interactions with the thin target foil. In the solenoid, interactions with the dilute helium gas mean that an equilibrium mean charge state is quickly reached. It is the difference in average velocity (lower for the ERs compared to the elastically scattered beam), which allows the elastically scattered particles to be brought to a focus before the ERs, allowing them to be physically blocked using a blocking rod and series of blocking discs on the solenoid axis.

The ERs are brought to a focus further downstream, beyond the exit of the solenoid bore. Here they are detected using SOLITAIRE's two large (20 x 20 cm) position sensitive multiwire proportional gas counters (MWPCs). These detectors allow the identification of the detected particles from their energy loss and time of flight information, from which the velocity distribtion of the ERs can be determined.

3 Iterative Procedure to Deduce Angular Distribution

The extraction of ER cross sections requires a precise determination of the solenoid's efficiency. In turn, this requires an understanding of the angular distribution of the evaporation residues exiting the target. A method has been developed to deduce the angular distribution from the measured laboratory-frame velocity distribution [15] of the ERs exiting the solenoid.

This method is based on the relationship between the velocity vectors of the evaporation residue following emission in the laboratory frame, and centre-of-mass frame, depicted in figure 2, and denoted with a subcript l and c, respectively. These are related through the laboratory frame compound nucleus velocity, v_{cn}. It is clear from this diagram that both the laboratory-frame velocity distribution, $\sigma(v_l)$ and angular distribution, $\sigma(\theta_l)$ are related by the velocity distribution in the centre-of-mass frame, $\sigma(v_c)$. By assuming an isotropic form of the centre-of-mass angular distribution $\sigma(\theta_c)$, extraction of $\sigma(v_c)$ from $\sigma(v_l)$ can be performed. Once $\sigma(v_c)$ is known, it can be used to construct a variety of cross section expressions, including the 2D cross section as a function of laboratory angle and velocity, $\sigma(v_l, \theta_l)$ [15].

The measured $\sigma_{exp}(v_l)$, however, is moderated by the transmission efficiency of the solenoid. The efficiency will clearly depend on θ_l due to the physical obstruction of the Faraday Cup and the nose cone. Since changes in v_l change the corresponing magnetic rigidity, there is also a potential v_l dependence. It is thus necessary to correct for the efficiency of the solenoid as a function of both v_l and θ_l.

Testing of this method to include the effects of transmission efficiency was achieved using existing measurments of the velocity distribution of the reaction of a 124 MeV ^{34}S beam on a ^{89}Y target of thickness 100 μg/cm^2. This reaction was chosen because it not only has a well-known fusion cross section measured previously, but there are also existing, reliable direct measurements of the angular distribution. These measurements, made by Mukherjee et al. [1], were made possible by scanning in angle using a velocity filter, as detailed by Wei et al. [16]. While more direct, measuring the angular distribution in this way is far more time-consuming, taking a number of days to collect enough counts for satistfactory statistics at all angles.

In order to take the transmission efficiency of the solenoid into account, simulations of the transport of the ERs through the solenoid were performed using the simulation software Solix. Solix is a Monte Carlo based C++ code developed at the Australian National University [17], designed to replicate as closely as possible the SOLITARE setup. These simulations take as input the laboratory-frame angle and velocity of the ERs, and can be used to generate a two-dimensional distribution of the transmission efficiency as a function of these two parame-

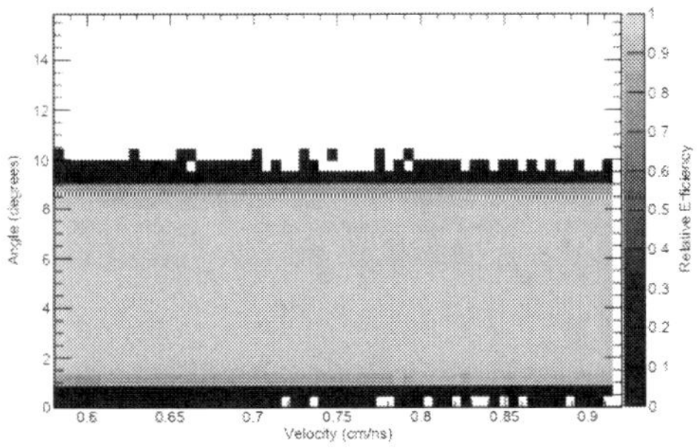

Figure 3: Transmission efficiency distribution as a function of v_l and θ_l used for the corrective iteration procedure.

ters, $\varepsilon(v_l, \theta_l)$. The efficiency distribution for the 124 MeV $^{34}S+^{89}Y$ reaction is shown in figure 3. The uniformity in v_l is a direct result of the presence of the gas filling the solenoid; since the mean charge state is assumed to depend linearly on v_l [18], the magnetic rigidity is independent of v_l, and thus the efficiency is also. This distribution can then be used in an iterative correction procedure to moderate the 2D cross section $\sigma(v_l, \theta_l)$.

The basis of this procedure is a parameterised form of the centre-of-mass velocity distribution, $\sigma(v_c)$. In the first instance, this distribution is formed from the laboratory-frame velocity distribution $\sigma_{exp}(v_l)$ using the analytical transformations of Reference [15].

The $\sigma(v_c)$ distribution is then parameterised using the sum of a number of Maxwell-Boltzmann functions. This distribution was chosen following the arguments established by Weisskopf [19] to model the deexcitation of the compound nucleus using the statistical model. Initial calculation of $\sigma(v_c)$ from the experimental $\sigma(v_l)$ (seen in figure 4(a)) suggests that, despite statistical fluctuations giving some irregularity, the distribution takes the shape of the sum of two Maxwell-Boltzmann-like distrbutions, one a true Maxwell-Boltzmann distribution, the other with a similar shape, but a non-zero offset of the distribution. This form suggests two dominant emission modes, one with no threshold to emission, and the other with a potential threshold to emission, resulting in the nonzero offset. These distributions can be related to neutron and charged particle evaporation respectively [20].

The parameterised $\sigma(v_c)$ is then used to generate the two-dimensional distribution $\sigma(v_l, \theta_l)$. This distribution is filtered by the transmission efficiency $\varepsilon(v_l, \theta_l)$ by mutiplying the two distributions in a bin-by-bin manner. Following this moderation by efficiency, $\sigma(v_l, \theta_l)$ is then projected onto the v_l axis to produce the distribtion $\sigma(v_l)$. This derived distribution can then be compared to the experimental velocity distribution $\sigma_{exp}(v_l)$ numerically using the χ^2 parameter.

This enables the design of an iterative procedure, based on the minimisation of the χ^2 parameter by varying the input parameters of the original centre-of-mass velocity distribution $\sigma(v_c)$. Once the minimum χ^2 has been reached, the v_c distribution produced can be taken as the true $\sigma(v_c)$ distribution, as it correctly reproduces the measured $\sigma_{exp}(v_l)$ distribution after being passed through the efficiency of the solenoid. Figure 5 shows the deduced $\sigma(v_l)$ that gave the minimum χ^2 value alongside the experimental $\sigma_{exp}(v_l)$, while figure 4(b) shows the parameterised $\sigma(v_c)$ which gave this best fit.

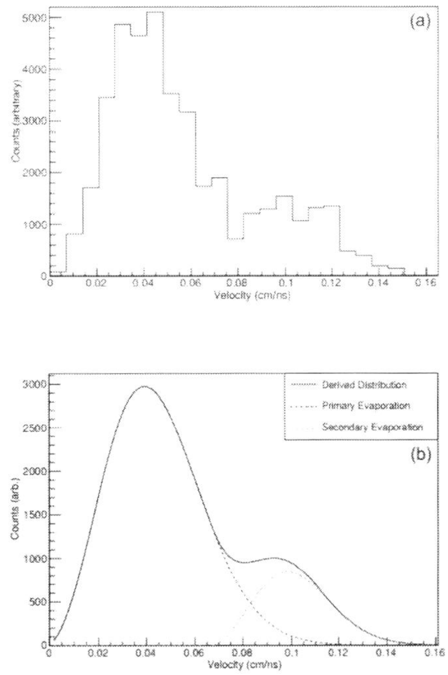

Figure 4: Comparison of initial extracted and final parameterised $\sigma(v_c)$ distribution. (a) $\sigma(v_c)$ distribution obtained by boot-strap extraction from the measured $\sigma_{exp}(v_l)$ data. (b) Final parameterised $\sigma(v_c)$ produced by minimising the χ^2 between $\sigma_{exp}(v_l)$ and calculated $\sigma(v_l)$. Note that the y-axis scales are arbitrary, and differ due to the moderation of (a) by the transmission efficiency of the solenoid.

4 Deduced Angular Distribution

Following the derivation of this true $\sigma(v_c)$, a final calculation can be performed to obtain the two-dimensional cross section $\sigma(v_l, \theta_l)$ which can then be projected onto the θ_l axis to obtain the laboratory-frame angular distribution $\sigma(\theta_l)$. This distribution can then be normalised in order to present the absolute fusion cross sections of the reaction in question. This normalisation is made possible through SOLITAIRE's silicon monitor detectors, placed at fixed, known angles before the entrance of the solenoid in order to detect the elastically scattered beam particles. Nor-

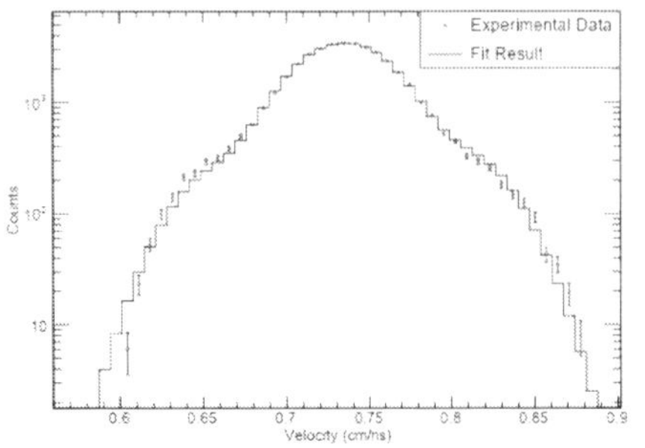

Figure 5: $\sigma(v_l)$ distribution deduced by minimisation routine, shown along with the measured $\sigma_{exp}(v_l)$ data set.

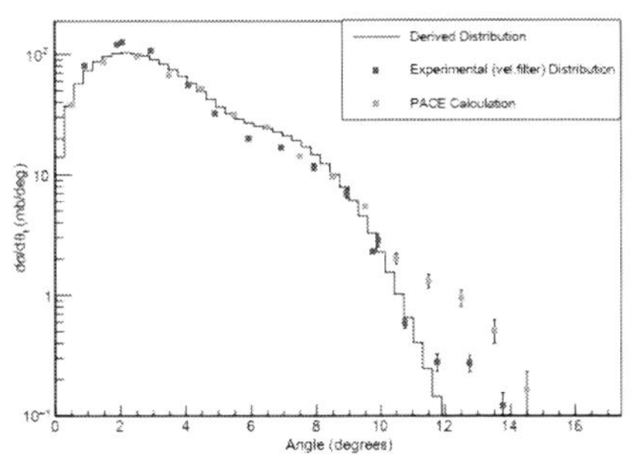

Figure 6: Final $\sigma(\theta_l)$ produced by the corrective minimisation routine, shown alongside direct angular distribution measurements (red squares) [1] and PACE4 statistical model calculations (green squares). Despite being too small to see, the measurements from reference [1] include experimental error bars, while the PACE4 data points include error bars defined by the square root of the counts, as is standard for Monte Carlo simulations.

malisation of the distribution allows the derived angular distribution to be compared with the dedicated experimental measurements made by Mukherjee et al. [1]. An additional comparison with the statistical model deexcitation code PACE4 [21] (included in the program LISE++ [22]) is included, as seen in figure 6.

It is immediately clear that there is good agreement of the primary peak position (at low angles) between the data and two sets of calculations. The magnitude of the primary peak, however, is underestimated in both the derived distribution and in the PACE4 calculations when compared to the experimental data. There is also a significant overestimation of the experimental cross section between 5° and 10°. This suggests an issue with the parameterised $\sigma(v_c)$, the efficiency calculation, or with the measured velocity distribution $\sigma_{exp}(v_l)$.

Underestimation of the peak at small angles suggests that counts from the high- and low-velocity edges of the distribution are missing, which could be put down to the initial graphical cuts made to the energy vs. time of flight data set made to select out the ERs for the creation of the $\sigma_{exp}(v_l)$ distribution. Considering also the overestimation of the cross section at the mid-range angles, however, suggests that the counts missing from the edges of the velocity distribution may actually be present at velocities near v_{cn}, and a systematic issue exists.

This systematic issue may be related to assumptions made about how much energy is lost by the ERs as they traverse the solenoid. The current assumption is that there is negligible energy loss, resulting in $\sigma_{exp}(v_l)$ being centred precisely on v_{cn}. Considering energy losses through the gas will result in a centroid smaller than v_{cn}, as well as compressing the distribution at higher velocities (higher energy ERs are subject to greater energy loss than lower energy ERs [23]). This suggests that an investigation into quantifying and correcting for the energy losses through the solenoid is a worthwhile direction for future studies.

The position of the shoulder at 7-8° is also in good agreement across all three distributions. The magnitude of this second feature, however, again differs across the distributions. There is better agreement between the PACE4 simulation and the distribution produced in this study than with the experimental data. One explanation for this could be the assumption of isotropic evaporation of the particles. While the calculated distribution assumes isotropic evaporation, there may be some anisotropy present in the real evaporation process. For anisotropy favouring the forward and backward emission of particles in the centre-of-mass frame, the number of ERs detected at lower angles will be enhanced compared to isotropic emission. If this assumption was incorporated into the calculated distribution, the peak at low angles would be enhanced compared to that at around 8°, and it would be expected that the agreement between the calculated and experimental data sets would be improved.

The final feature of note is the identification of a third Maxwell-Boltzmann component of the θ_l distribution in the PACE4 calculations. This is prominent in the region of 11-16°, and it indicates the contribution of a third evaporation mode to the overall distribution. Analysis of the PACE4 output data suggests that this third component arises from the emission of two alpha particles [24]. An argument could also be made for this presence of this third component in the experimental distribution, too, but it is not entirely clear whether this shape is a genuine feature or just a statistical fluctuation in the data. During this study, a third Maxwell-Boltzmann was added to the parameterisation in an attempt to test whether this component was indeed contributing in a significant way. It was found, however, that since this component was contributing to the distribution at larger angles, where the efficiency of the solenoid was found to be low, this third Maxwell-Boltzmann was not well-constrained enough to give a reliable contribution to the overall distribution.

5 Conclusions and Future Work

The large ER transmission efficiency of the Australian National University's fusion product separator SOLITAIRE has been characterised as a function of laboratory angle and velocity using Monte-Carlo simulations. This efficiency distribution was used alongside a parameterised centre-of-mass frame velocity distribution in the form of a double Maxwell-Boltzmann and the appropriate mathematical transformations to extract the angular distribution of evaporation residues entering the solenoid. The generated angular distribution compares quite well in magnitude and has excellent agreement in overall shape to both dedicated measurements and PACE4 calculations.

In order to improve the agreement between the derived and directly-measured angular distributions, this method will be refined by taking into account energy loss of the ERs with the fill gas as they traverse the solenoid. Additionally, the effect of anisotropy of particle emission in the centre-of-mass frame will be investigated using a parameterised anisotropy dsitribution, where the parameters will be set and varied following physical arguments, and changes to the derived angular distribution will be examined.

References

[1] A. Mukherjee, M. Dasgupta, D.J. Hinde, K. Hagino, J.R. Leigh, J.C. Mein, C.R. Morton, J.O. Newton, H. Timmers, Phys. Rev. C. **66**, 034607 (2002)

[2] R.G. Stokstad, Y. Eisen, S. Kaplanis, D. Pelte, U. Smilansky, I. Tserruya, Phys. Rev. C **21**, 2472 (1980)

[3] J.R. Leigh, M. Dasgupta, D.J. Hinde, J.C. Mein, C.R. Morton, R.C. Lemmon, J.P. Lestone, J.O. Newton, H. Timmers, J.X. Wei et al., Phys. Rev. C **52**, 3151 (1995)

[4] M. Beckerman, J. Ball, H. Enge, M. Salomaa, A. Sperduto, S. Gazes, A. DiRienzo, J.D. Molitoris, Phys. Rev. C **23**, 1581 (1981)

[5] M. Beckerman, M. Salomaa, A. Sperduto, J.D. Molitoris, A. DiRienzo, Phys. Rev. C **25**, 837 (1982)

[6] C.H. Dasso, S. Landowne, A. Winther, Nucl. Phys. A **405**, 381 (1983)

[7] C.H. Dasso, S. Landowne, A. Winther, Nucl. Phys. A **407**, 221 (1983)

[8] R.A. Broglia, C.H. Dasso, S. Landowne, A. Winther, Phys. Rev. C **27**, 2433 (1983)

[9] R.A. Broglia, C.H. Dasso, S. Landowne, G. Pollarolo, Phys. Lett. B **133**, 34 (1983)

[10] M. Dasgupta, D.J. Hinde, A. Diaz-Torres, B. Bouriquet, C.I. Low, G.J. Milburn, J.O. Newton, Phys. Rev. Lett. **99**, 192701 (2007)

[11] T. Ichikawa, K. Matsuyanagi, Phys. Rev. C **92**, 021602 (2015)

[12] S. Mişicu, H. Esbensen, Phys. Rev. Lett. **96**, 112701 (2006)

[13] C. Simenel, A.S. Umar, K. Godbey, M. Dasgupta, D.J. Hinde, Phys. Rev. C **95**, 031601 (2017)

[14] M.D. Rodríguez, M.L. Brown, M. Dasgupta, D.J. Hinde, D.C. Weisser, T. Kibedi, M.A. Lane, P.J. Cherry, A.G. Muirhead, R.B. Turkentine et al., Nucl. Instrum. Meth. A **614**, 119 (2010)

[15] I.P. Carter, M.L. Brown, M. Dasgupta, D.J. Hinde, M. Evers, D.H. Luong, A. Wakhle, E. Williams, *Determination of the angular distribution of evaporation residues following transmission through the superconducting solenoidal separator SOLITAIRE*, in *EPJ Web. Conf.* (EDP Sciences, 2012), Vol. 35, p. 05003

[16] J.X. Wei, J.R. Leigh, D.C. Weisser, J.O. Newton, S. Elfstrom, J.P. Lestone, J.X. Chen, D.G. Popescu, D.J. Hinde, Nucl. Instrum. Meth. A **306**, 557 (1991)

[17] M.L. Brown, *Simulation of a gas-filled solenoid: Optimising functionality and expanding applicability* (2011)

[18] H.D. Betz, Rev. Mod. Phys. **44**, 465 (1972)

[19] V. Weisskopf, Phys. Rev. **52**, 295 (1937)

[20] K. Schmidt, *Deexcitation mechanisms in compound nucleus reactions* (2009)

[21] A. Gavron, Phys. Rev. C **21**, 230 (1980)

[22] O.B. Tarasov, D. Bazin, Nucl. Instrum. Meth. B **204**, 174 (2003)

[23] H.A. Bethe, J. Ashkin, Experimental nuclear physics **1**, 309 (1953)

[24] M. Wang, G. Audi, A.H. Wapstra, F.G. Kondev, M. MacCormick, X. Xu, B. Pfeiffer, Chinese Phys. C **36**, 1603 (2012)

 DOI: 10.1051/epjconf/201716300006

Laser spectroscopy studies on nobelium

Michael Block[1,2,3,*]

[1]*GSI Helmholtzzentrum für Schwerionenforschung, Planckstrasse 1, 64291 Darmstadt, Germany*
[2]*Helmholtz Institut Mainz, Staudingerweg 18, 55128 Mainz, Germany*
[3]*Johannes Gutenberg-Universität Mainz, 55099 Mainz, Germany*

Abstract. Laser spectroscopy of the heaviest elements provides high-precision data on their atomic and nuclear properties. For example, atomic level energies and ionization potentials allow us to probe the influence of relativistic effects on their atomic structure and to benchmark state-of-the-art atomic structure calculations. In addition, it offers an alternative route to determine nuclear properties like spins, magnetic moments and quadrupole moments in a nuclear model-independent way. Recently, a sensitive method based on resonant laser ionization has been applied to nobelium isotopes around $N = 152$ at GSI Darmstadt. In pioneering experiments, several atomic states have been identified extending the reach of laser spectroscopy beyond fermium. In this contribution, the main achievements and future perspectives are briefly summarized.

1 Introduction

The atomic and chemical properties of the heaviest elements ($Z \gtrsim 100$) are affected by strong relativistic effects and quantum electrodynamics [1–3]. Relativistic effects increase approximately with the square of the atomic number and are responsible for the distinct color of gold and for the liquid state of mercury at room temperature. For example, these effects stabilize s and $p_{1/2}$ orbitals, whereas $p_{3/2}$ and d orbitals are destabilized in energy. This leads to changes of the atomic ground state configurations compared to the trend observed in the elements we find in nature. In lawrencium ($Z = 103$), for example, a $7p_{1/2}$ ground state is predicted instead of a 6d state [4]. At some point, a deviation from the regular pattern that governs the ordering of elements in the periodic table is expected. Such deviations were predicted for the superheavy elements copernicium ($Z = 112$) and flerovium ($Z = 114$) in the mid 1970s [5]. According to Pitzer these elements should already show properties of noble gases even though the next nobel gas in a regular periodic table would be oganesson ($Z = 118$).

Experimentally, atomic and chemical properties of the heaviest elements have been mainly studied by chemistry techniques either in gas phase or in liquid phase [6–8]. Copernicium was actually found to be a rather regular member of group twelve [9], while the properties of flerovium are still a matter of ongoing research [10, 11].

An alternative approach to study atomic properties of heavy elements is through laser spectroscopy. Well-established techniques provide high-precision data on atomic properties such as atomic level energies, lifetimes and ionization potentials that allow us to benchmark theoretical predictions and to probe the influence of relativistic effects. Also, nuclear properties are reflected in the atomic spectrum. This enables studies of the evolution of nuclear structure features in a complementary way to traditional nuclear spectroscopy. For example, hyperfine spectroscopy give access to the nuclear spin and to nuclear moments. The shift of an atomic transition in different isotopes allows determining changes in mean square charge radii of nuclei and hence, the nuclear size and deformation in a nuclear model-independent way. However, accurate calculations of the hyperfine parameters A and B or the field shift factor F are required inputs from atomic theory.

*e-mail: m.block@gsi.de

2 Resonant ionization laser spectroscopy of heavy nuclides

The method of choice for optical spectroscopy of rare isotopes is resonant laser ionization spectroscopy (RIS). This technique features high sensitivity, in particular if the laser-created ions are detected by their characteristic radioactive decay. The method can even be applied to atoms where no experimental information on atomic transitions exists, which had been illustrated in fermium [12] and, more recently, in astatine [13], for example.

For more than a decade, fermium ($Z = 100$) was the heaviest element that had been studied by laser spectroscopy [12]. A tiny quantity of 46 pg of the long-lived isotope ^{255}Fm ($T_{1/2} \approx 20$ h) was prepared in the high-flux reactor at Oak Ridge National laboratory. This enabled offline experiments in which several atomic transitions were identified. Good agreement with predictions by multi-configuration Dirac-Fock (MCDF) calculations [14] was observed. In addition, broadband spectroscopy of the complex hyperfine spectrum of ^{255}Fm with a nuclear spin of $I = 7/2$ was feasible and the hyperfine factors A and B were obtained from a fit to the data [15].

The heavier members of the actinide series beyond fermium have to be studied online. However, these elements cannot be produced at isotope separator online (ISOL) facilities, but in fusion-evaporation reactions. This is experimentally challenging due to low production rates on the order of at most few particles per second. In addition, the radionuclides have to be slowed down from tens of MeV to rest and they have to be neutralized to perform RIS. This is nowadays accomplished using gas cells filled with an inert buffer gas like argon.

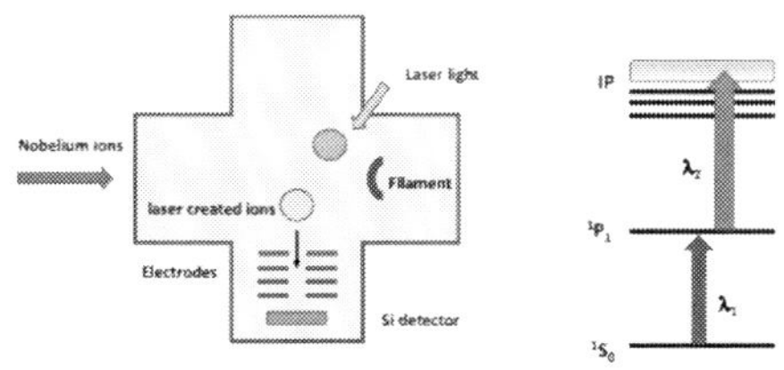

Figure 1. Left: Schematic view of the buffer gas cell for laser spectroscopy. Right: Simplified laser excitation scheme.

A dedicated RIS version, the so-called radiation detected resonance ionization spectroscopy method (RADRIS), has been developed for nobelium laser spectroscopy at GSI Darmstadt [16, 17]. The setup is schematically shown in 1. In the RADRIS method, nobelium ions are separated from the primary beam by the separator for heavy ion reaction products (SHIP [18, 19]) and slowed down in 100 mbar ultra-pure argon gas. The fraction of the stopped ions that remains in a charged state is accumulated on a tantalum filament. Neutral nobelium atoms are then evaporated by heating this filament. The atoms are ionized in a two-step laser excitation scheme with pulsed lasers. For the initial level search the first step is provided by a tunable dye laser that is scanned across the spectral region in which atomic states were predicted. The second excitation step into the continuum is provided by a fixed-frequency excimer laser (at 351 nm) featuring high power to compensate the lower cross section of a non-resonant excitation. The laser-created ions are detected by their characteristic α decay registered by a silicon detector to which the ions are transported by electric fields. Details on the method and the employed setup have been described elsewhere [20].

3 Scientific questions addressed by laser spectroscopy

Detailed studies of the nuclear structure and its evolution in regions of enhanced shell stabilization are of great importance for a better understanding of the heaviest elements. Their very existence is intimately linked to nuclear shell effects. Ultimately, such investigations will shed light on the nature of the underlying strong interaction at one of the extremes of the nuclear chart. However, the nuclides at the predicted spherical closed shells with $N = 184$ and $Z = 114, 120$, or 126 [21] are experimentally still inaccessible at present. The most neutron-rich nuclide known to date contains only 177 neutrons [22]. In addition, detailed studies of the elements with $Z \geq 110$ are hampered by limited statistics due to low cross sections on a level of 1-10 picobarn.

Lighter nuclides in the region around $Z = 100$, $N = 152$ can be produced with higher rates that allow Penning trap-mass spectrometry [23], laser spectroscopy [24] and α-γ spectroscopy [25]. Indeed, some of the single-particle orbitals that are responsible for the spherical shell gap at $Z = 114$ appear at low excitation energy in deformed nuclei around $Z \approx 100$, $N = 152$. Thus, by fixing the position of these orbitals at a given deformation allows discriminating models predicting the spherical gap. This region also forms the backbone for spin assignments in heavier nuclides in decay spectroscopy experiments based on systematics along isotones.

A complementary determination of nuclear spins by laser spectroscopy is desirable. In addition, measurements of dipole and quadrupole moments allow us to determine single-particle configurations in specific nuclei. This would, for example, provide an alternative to determine the configuration of the long-lived $K = 8^-$ isomer in ^{254}No for which different interpretations have been put forward based on decay spectroscopy [26–30]. The magnetic moment that could be measured by laser spectroscopy could answer the questions whether this isomer is based on a quasi-neutron or quasi-proton configuration that can both lead to $K = 8^-$ states at similar energy.

Nobelium atoms have a favorable atomic structure for laser spectroscopy: the 5f shell is filled by 14 electrons and the two valence electrons in the 7s orbital result in a ground-state configuration [Rn] $5f^{14}$ $7s^2$ $^1S_0$. This facilitates accurate calculations of atomic properties. Modern calculations of the electronic structure in the heaviest elements require the explicit treatment of relativistic and quantum electrodynamics effects. In addition, electron correlations have to considered. This is accomplished employing many-body approaches like multi-configuration Dirac-Fock or relativistic coupled cluster methods [1–3, 14]. They can reach high accuracy with a typical precision of meV for level energies. The accuracy of such calculations is often estimated by a comparison of calculated properties to experimental data in the lanthanide region since the available data for the heaviest elements are limited. According to theoretical predictions nobelium features a strong ground-state transition $^1S_0$-$^1P_1$ around 30,000 cm^{-1} [14, 31, 32].

4 Status and future perspectives

A practical advantage for studies of nobelium isotopes around $N = 152$ is their rather high yield: ^{254}No can be produced in the reaction ^{208}Pb(^{48}Ca,2n) with a cross section of about 2 μb corresponding to a yield of few particles per second at an accelerator facility like GSI Darmstadt.

Applying the RADRIS method to the nobelium isotope ^{254}No, the strong ground state transition $^1S_0$-$^1P_1$ has been searched for in a tedious effort [16, 17]. Such experiments depend crucially on the guidance by theoretical predictions. Different atomic transitions in the nobelium atom have been predicted prior to the GSI experiment [14, 31, 32]. The typical uncertainties and the scattering of the predictions resulted in scan range on the order of $\geq 2,000$ cm^{-1} that had to be covered. For a typical line width of in-gas cell spectroscopy of about 0.5 cm^{-1} [16] this corresponds to a few thousand frequency steps for the initial search. In the case of ^{254}No the cycle time for one such step sums up to about five minutes equivalent to a minimum search time of more than 200 hours to cover the range of the predicted atomic transitions.

Nonetheless, the $^1S_0$-$^1P_1$ transition as well as several Rydberg states in the nobelium isotope ^{254}No have been identified for the first time recently [33]. An overall efficiency about six percent was achieved using a two-step laser excitation that could be boosted for a limited time by conditioning the filament. Based on the observed Rydberg series in ^{254}No the first ionization potential could already be derived with rather high precision. However, quenching collisions led to the population of a metastable state, most likely the $^3D_1$ state that is close to the $^1P_1$ state that was excited by the first laser pulse. This behavior was explained by a rate equation model that describes the experimental data well and allows determining the energy of the $^3D_1$ state indirectly [33, 34]. However, Rydberg states could be excited by the second laser from either of the two states resulting in different series. In 2016, different Rydberg series were unambiguously identified, including a series originating form the $^1P_1$ state from which the first ionization potential of nobelium could be determined with high precision. The data analysis is close to completion and the results will be published shortly. All in all, the results for atomic properties of nobelium showed good agreement with theoretical predictions [14, 31, 32].

The measurements were extended to the nobelium isotopes ^{252,253}No. The lowest yield was available for ^{252}No with a cross section of about 400 nanobarn in the reaction ^{208}Pb(^{48}Ca,2n)^{252}No. Decay losses on the filament reduced efficiency to about three precent for this shorter-lived nobelium isotope [33]. The isotope shift of the $^1S_0$-$^1P_1$ transition between ^{252,253,254}No was measured and the hyperfine splitting in ^{253}No was observed. The data analysis is ongoing and the results will be subject of forthcoming publications. The experiments will provide the change of the nuclear charge radius between the isotopes ^{252,254}No as well as the magnetic moment and the ground-state spin assignment in ^{253}No.

Resonant laser ionization spectroscopy in a gas cell is a powerful method to identify atomic transitions in rare isotopes where little is known experimentally. However, the line width that can be achieved with in-gas cell spectroscopy is limited by pressure and Doppler broadening to typically about 5 GHz. In some cases higher resolution spectroscopy is required, for example to resolve individual hyperfine components for the determination of the magnetic moment. A higher resolution was recently demonstrated with a new technique, the so-called gas-jet laser spectroscopy [35]. The method was recently developed at KU Leuven and was applied to actinium isotopes where a gain in resolution by almost one order of magnitude compared to in-gas cell spectroscopy without sacrificing the high efficiency.

Future efforts will be dedicated to extend the reach of laser spectroscopy even further towards heavier elements. For a first foray into an unexplored area, a broadband level search with high-power lasers in a gas cell, for example with the RADRIS technique, will be favorable. Once suitable transitions have been identified, high-resolution spectroscopy can be performed by in-gas jet spectroscopy. In any case the steeply dropping cross section for the production of heavier elements are challenging. In this respect, laser spectroscopy will profit from new powerful stable-beam accelerators. Such machines that are anticipated to deliver 10–100 times higher primary beam intensities are presently planned or under construction in several facilities worldwide [36–38].

For the first exploration of the element lawrencium, the RADRIS method can be adapted in a straightforward manner. A key question in Lr that can be answered by laser spectroscopy concerns the atomic ground-state configuration as discussed above. Theoretical models predict a $7p_{1/2}$ ground state, but the 6d state is rather close in energy [4]. Recent experiments performed in Tokai, Japan used a surface ionization technique to determine the first ionization potential of Lr [39]. Their result agrees with the theoretical prediction well and confirms that Lr terminates the actinide series. However, the uncertainty of the employed method does not allow distinguishing between the two different ground state configurations unambiguously. Thus, an experimental verification by laser spectroscopy is still of interest.

Acknowledgements

The experiments on nobelium discussed in this contribution were performed by the RADRIS collaboration comprising scientists from GSI Darmstadt, Helmholtz Institute Mainz, Mainz University, Technical University Darmstadt, KU Leuven, University of Liverpool, and TRIUMF Vancouver. Their contributions are gratefully acknowledged.

References

[1] P. Schwerdtfeger, L.F. Pašteka, A. Punnett, P.O. Bowman, Nuclear Physics A **944**, 551 (2015)

[2] E. Eliav, S. Fritzsche, U. Kaldor, Nucl. Phys. **A944**, 518 (2015)

[3] V. Pershina, Nuclear Physics A **944**, 578 (2015)

[4] A. Borschevsky, E. Eliav, M. Vilkas, Y. Ishikawa, U. Kaldor, Eur. Phys. J. D **45**, 115 (2007)

[5] K.S. Pitzer, The Journal of Chemical Physics **63**, 1032 (1975)

[6] M. Schädel, Phil. Trans. R. Soc. A **373**, 20140191 (2015)

[7] A. Türler, R. Eichler, A. Yakushev, Nuclear Physics A **944**, 640 (2015)

[8] Y. Nagame, J.V. Kratz, M. Schädel, Nucl. Phys. **A944**, 614 (2015)

[9] R. Eichler, N. Aksenov, A. Belozerov, G. Bozhikov, V. Chepigin, S. Dmitriev, R. Dressler, H. Gäggeler, V. Gorshkov, F. Haenssler et al., Nature **447**, 72 (2007)

[10] R. Eichler, N. Aksenov, Y.V. Albin, A. Belozerov, G. Bozhikov, V. Chepigin, S. Dmitriev, R. Dressler, H. Gäggeler, V. Gorshkov et al., Radiochimica Acta International journal for chemical aspects of nuclear science and technology **98**, 133 (2010)

[11] A. Yakushev, J.M. Gates, A. Türler, M. Schädel, C.E. Düllmann, D. Ackermann, L.L. Andersson, M. Block, W. Brüchle, J. Dvorak et al., Inorganic chemistryü **53**, 1624 (2014)

[12] M. Sewtz, H. Backe, A. Dretzke, G. Kube, W. Lauth, P. Schwamb, K. Eberhardt, C. Grüning, P. Thörle, N. Trautmann et al., Phys. Rev. Lett **90**, 163002 (2003)

[13] S. Rothe, V.N. Fedosseev, T. Kron, B.A. Marsh, R.E. Rossel, K.D.A. Wendt, Nucl. Instrum. Meth. **B317**, 561 (2013)

[14] S. Fritzsche, C. Dong, F. Koike, A. Uvarov, Eur. Phys. J. D **45**, 107 (2007)

[15] H. Backe, A. Dretzke, R. Haire, P. Kunz, W. Lauth, M. Sewtz, N. Trautmann et al., Hyperfine interactions **162**, 3 (2005)

[16] H. Backe, P. Kunz, W. Lauth, A. Dretzke, R. Horn, T. Kolb, M. Laatiaoui, M. Sewtz, D. Ackermann, M. Block et al., Eur. Phys. J. D **45**, 99 (2007)

[17] M. Laatiaoui, H. Backe, M. Block, P. Chhetri, F. Lautenschläger, W. Lauth, T. Walther, Hyperfine Interact. **227**, 69 (2014)

[18] G. Münzenberg, W. Faust, S. Hofmann, P. Armbruster, K. Güttner, H. Ewald, Nucl. Instrum. and Meth. **161**, 65 (1979)

[19] S. Hofmann, G. Münzenberg, Reviews of Modern Physics **72**, 733 (2000)

[20] F. Lautenschläger, P. Chhetri, D. Ackermann, H. Backe, M. Block, B. Cheal, A. Clark, C. Droese, R. Ferrer, F. Giacoppo et al., Nuclear Instruments and Methods in Physics Research Section B: Beam Interactions with Materials and Atoms **383**, 115 (2016)

[21] A. Sobiczewski, K. Pomorski, Progress in Particle and Nuclear Physics **58**, 292 (2007)

[22] Yu.T. Oganessian, V.K. Utyonkov, Nucl. Phys. **A944**, 62 (2015)

[23] M. Block, Nuclear Physics A **944**, 471 (2015)

[24] H. Backe, W. Lauth, M. Block, M. Laatiaoui, Nuclear Physics A **944**, 492 (2015)

[25] M. Asai, F.P. Heßberger, A. Lopez-Martens, Nucl. Phys. **A944**, 308 (2015)

[26] R.D. Herzberg, P. Greenlees, P. Butler, G. Jones, M. Venhart, I. Darby, S. Eeckhaudt, K. Eskola, T. Grahn, C. Gray-Jones et al., Nature **442**, 896 (2006)

[27] S. Tandel, T. Khoo, D. Seweryniak, G. Mukherjee, I. Ahmad, B. Back, R. Blinstrup, M. Carpenter, J. Chapman, P. Chowdhury et al., Phys. Rev. Lett. **97**, 082502 (2006)

[28] H. Jeppesen, R. Clark, K. Gregorich, A. Afanasjev, M. Ali, J. Allmond, C. Beausang, M. Cromaz, M. Deleplanque, I. Dragojević et al., Physical Review C **80**, 034324 (2009)

[29] R. Clark, K. Gregorich, J. Berryman, M. Ali, J. Allmond, C. Beausang, M. Cromaz, M. Deleplanque, I. Dragojević, J. Dvorak et al., Physics Letters B **690**, 19 (2010)

[30] F. Heßberger, S. Antalic, B. Sulignano, D. Ackermann, S. Heinz, S. Hofmann, B. Kindler, J. Khuyagbaatar, I. Kojouharov, P. Kuusiniemi et al., Eur. Phys. J. A **43**, 55 (2010)

[31] A. Borschevsky, E. Eliav, M.J. Vilkas, Y. Ishikawa, U. Kaldor, Phys. Rev. A **75**, 042514 (2007)

[32] V.A. Dzuba, M.S. Safronova, U.I. Safronova, Phys. Rev. **A90**, 012504 (2014)

[33] M. Laatiaoui et al., Nature **538**, 495 (2016)

[34] P. Chhetri et al., Eur. Phys. J. D (2017 (under review))

[35] R. Ferrer, A. Barzakh, B. Bastin, R. Beerwerth, M. Block, P. Creemers, H. Grawe, R. de Groote, P. Delahaye, X. Fléchard et al., Nature Communications **8**, 14520 (2017)

[36] S. Dmitriev, M. Itkis, Y. Oganessian, *Status and perspectives of the Dubna superheavy element factory*, in *EPJ Web of Conferences* (EDP Sciences, 2016), Vol. 131, p. 08001

[37] F. Déchery, A. Drouart, H. Savajols, J. Nolen, M. Authier, A. Amthor, D. Boutin, O. Delferriére, B. Gall, A. Hue et al., The European Physical Journal A **51**, 66 (2015)

[38] W. Barth, K. Aulenbacher, M. Basten, F. Dziuba, V. Gettmann, M. Miski-Oglu, H. Podlech, S. Yaramyshev, *A superconducting CW-LINAC for heavy ion acceleration at GSI*, in *EPJ Web of Conferences* (EDP Sciences, 2017), Vol. 138, p. 01026

[39] T. Sato, M. Asai, A. Borschevsky, T. Stora, N. Sato, Y. Kaneya, K. Tsukada, C.E. Düllmann, K. Eberhardt, E. Eliav et al., Nature **520**, 209 (2015)

 DOI: 10.1051/epjconf/201716300007

Nuclear Fission: from more phenomenology and adjusted parameters to more fundamental theory and increased predictive power

Aurel Bulgac[1,*], *Shi* Jin[1], *Piotr* Magierski[2,1], *Kenneth* Roche[3,1], *Nicolas* Schunck[4], and *Ionel* Stetcu[5]

[1] *Department of Physics, University of Washington, Seattle, WA 98195-1560, USA*
[2] *Faculty of Physics, Warsaw University of Technology, ulica Koszykowa 75, 00-662 Warsaw, POLAND*
[3] *Pacific Northwest National Laboratory, Richland, WA 99352, USA*
[4] *Nuclear and Chemical Science Division, Lawrence Livermore National Laboratory, Livermore, CA 94551, USA*
[5] *Theoretical Division, Los Alamos National Laboratory, Los Alamos, NM 87545, USA*

Abstract. Two major recent developments in theory and computational resources created the favorable conditions for achieving a microscopic description of fission dynamics in classically allowed regions of the collective potential energy surface, almost eighty years after its discovery in 1939 by Hahn and Strassmann [1]. The first major development was in theory, the extension of the Time-Dependent Density Functional Theory (TDDFT) [2–5] to superfluid fermion systems [6]. The second development was in computing, the emergence of powerful enough supercomputers capable of solving the complex systems of equations describing the time evolution in three dimensions without any restrictions of hundreds of strongly interacting nucleons. Thus the conditions have been created to renounce phenomenological models and incomplete microscopic treatments with uncontrollable approximations and/or assumptions in the description of the complex dynamics of fission. Even though the available nuclear energy density functionals (NEDFs) are phenomenological still, their accuracy is improving steadily and the prospects of being able to perform calculations of the nuclear fission dynamics and to predict many properties of the fission fragments, otherwise not possible to extract from experiments.

Meitner and Frisch [7] and Bohr and Wheeler [8, 9] interpreted the neutron induced fission of uranium observed by Hahn and Strassmann [1] as the Coulomb-driven division of a classically charged liquid drop in competition with the surface tension of the liquid drop and they obtained a good back-of-the-envelope estimate of the total kinetic energy (TKE) of the fission fragments (FFs). Unlike the α-decay and spontaneous fission observed later [10], induced fission occurs entirely above the barrier, and therefore is to a large extent a classical process. The impinging neutron on uranium leads to the formation of an excited state, the compound nucleus, in a region of the nuclear spectrum where the level density is extremely high and the nucleus is thus relatively hot. The nucleus evolves for a very long time from its ground state shape towards the top of the outer fission barrier, where the Coulomb repulsion overcomes the surface tension and the nucleus starts its descent to scission, where the two FFs are formed. The top of the fission barrier is energetically barely below the neutron threshold. Although the nucleus is already quite elongated at this point, it has relatively little intrinsic excitation energy and is relatively cold. It was soon realized that in nuclei nucleons form shells, nucleons live on quantized orbits [11, 12], and the spin-orbit interaction plays a critical role in the formation of the nuclear shells and ultimately in the formation of the FFs [13]. Hill

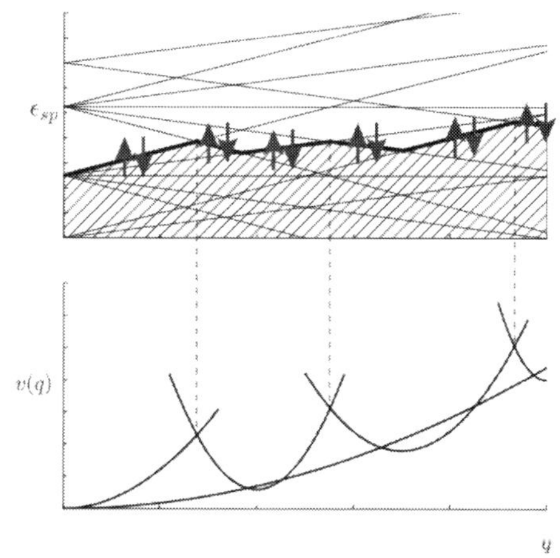

Figure 1. The qualitative evolution of the single-particle levels (upper panel) and of the total nuclear energy (lower panel) as a function of nuclear deformation [14, 15]. The Fermi level is shown with a thick line.

*e-mail: bulgac@uw.edu

and Wheeler [14] discussed how the liquid drop deformation energy emerges from a quantum mechanical approach based on considering the quantized single-particle motion of nucleons in a slowly deforming potential well. In their approach, the liquid drop potential deformation energy was, in the first approximation, an envelope of many intersecting parabolas caused by single-particle level crossings, see Figure 1. At single-particle level crossings of the last occupied level, nucleons jump from one level to another, in order to maintain the sphericity of the Fermi sphere. Without such a redistribution of nucleons at the Fermi level, the Fermi sphere would become oblate when a nucleus deforms on the way to scission into two fragments, while the spatial shape of the nucleus would become ever more prolate. This would lead to a volume excitation energy of the nucleus. This is inconsistent with the fact that nuclei are saturating systems with a large surface tension that can only deform by changing the shape of their surface, while maintaining their volume constant. Hence, only the Coulomb and the surface contributions to the total energy change. Each single-particle level is typically double degenerate, due to the Kramers degeneracy, and nucleons have to jump in pairs, otherwise the nuclear shape evolution towards scission would be hindered [15, 16]. In this case, nucleons would follow diabatic levels instead of adiabatic ones, as was indeed indirectly established in recent TDHF simulations of fission dynamics [17, 18]. Pairing correlations, while relatively weak in nuclei, are very effective at promoting simultaneously two nucleons from time-reversed orbitals into other time-reversed orbitals and thus they greatly facilitate the evolution of the nuclear shape towards scission. An approximate treatment of the pairing correlations in fission was implemented using either a frozen [19] or time-dependent [20] BCS approximation in conjunction within the TDHF approach, but only at relatively large elongations, where FFs were practically formed. It was established in the 1960's and 1970's that single-particle level bunching exists in nuclear systems not only in the case of spherical nuclei (as in the case of atoms), but also in deformed and highly deformed nuclei. At first this phenomenon was experimentally observed in the case of fission isomers at very large elongations [21–23] and subsequently in the case of superdeformed nuclei [24]. The existence of nucleonic shells at large deformations results in a potential energy deformation surface with local maxima and minima, which are otherwise absent in the case of a classical charged liquid drop. The aforementioned level crossings cause discontinuities in the potential energy surface. Some of these discontinuities can be smoothed out when pairing correlations are present, leading to avoided level crossings. Overall fission dynamics is a very complex process, which still did not reach a fully microscopic description [25] in spite of almost eight decades of effort. The formation and the properties of FFs was and still is the most debated and arguably the central topic in phenomenology of nuclear fission. In the absence of a microscopic description and because of the enormous challenges arising from solving the necessarily time-dependent coupled nonlinear Schrödinger equations for hundreds of strongly interacting nucleons, many phenomenological models, both quantum and classical in nature, have been suggested over the years. Unfortunately most, if not all, of them lack a solid microscopic underpinning and for that reason their predictive power is limited.

The bulk behavior of the deformation potential energy surface is determined by the surface tension and Coulomb energy, which are rather well-known and understood, and their competition is the main driving force leading to fission as in liquid drop models. The underlying shell structure imprints hills and valleys on the overall charged liquid drop energy landscape [22, 23]. On the way to scission configuration, nucleons have to go through a large number of redistributions at the single-particle levels crossing around the Fermi level, in order to maintain the spherical symmetry of their local momentum distribution. Overall, the deformation potential energy surface acquires a profile somewhat similar to that of an uneven mountain, with little hills and valleys and covered by trees, and the evolution of the nuclear shape is in the end similar to the erratic motion of a pinball, not straight down the hill, but rather left and right, bouncing (mostly elastically) from the many obstacles on the way to the bottom of the valley, where the pinball eventually breaks up. In the last stages of this complex evolution, the independent character of the nucleons inside the nucleus plays again a critical role, since magic closed shells control the nuclear shape evolution and the FF formation.

Since fission dynamics is a truly non-stationary phenomenon, an extension of the TDDFT [2–5] in the Local Density Approximation (LDA) implementation of Kohn and Sham [3] was needed, and the Superfluid Local Density Approximation (SLDA) [6] was developed and validated quite extensively against experiment and theoretical results obtained using the Schrödinger description. In the case of many strongly interacting particles the Schrödinger equation for a many-body system has to be replaced with the TDDFT, which is mathematically equivalent to the Schrödinger equation for one-body observables and the only theoretical framework suitable to describe the structure and dynamics of quantum many-body systems, if the corresponding energy density functional (the equivalent of the potential in the Schrödinger equation) is known in sufficient detail. TDSLDA allows the evaluation of FFs properties, which are otherwise impossible to either measure or compute, such as FFs excitation energies immediately after scission.

As our simulations within TDSLDA demonstrate [26], see Figure 2, the nucleus separates typically into two unequal FFs. In our calculations, we used in the particle-hole channel a phenomenological NEDF [27], complemented with pairing correlations described as in Refs. [28, 29]. The larger FF, which has properties very similar to the double-magic ^{132}Sn, emerges with an almost spherical shape, while the lighter fragment emerges in an elongated configuration, with a ratio of the major to minor axes close to 3/2. To much of our surprise, the properties of the FFs in these simulations [26] are very close to experimental measurements, even though no effort was put into reproducing observations. We attribute this to the fact that the NEDF reproduces bulk nuclear properties relatively well.

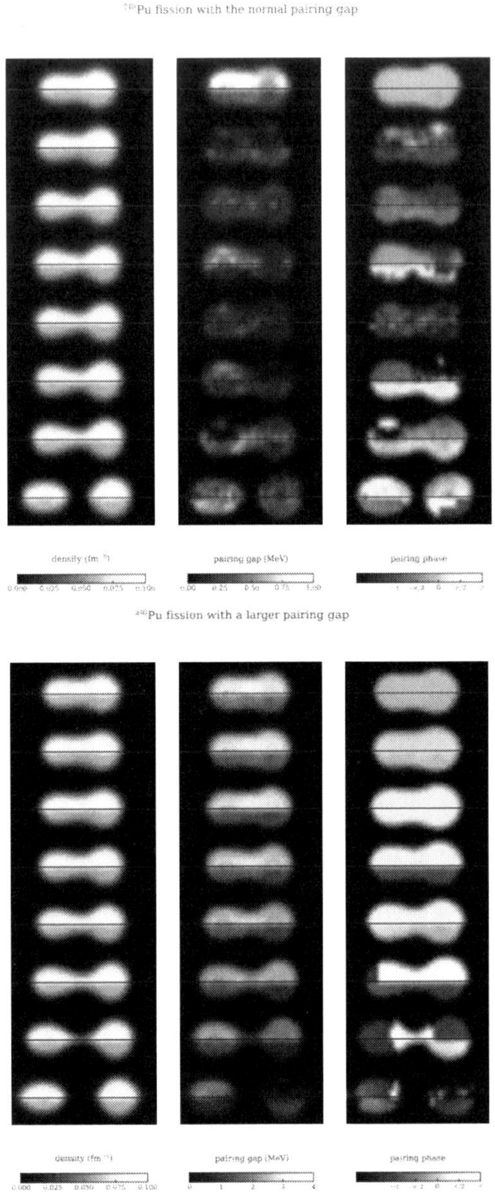

Figure 2. In the upper frames is the induced fission of ^{240}Pu with normal pairing strength last up to 14,000 fm/c ($\approx 47 \times 10^{-21}$ s) from saddle-to-scission. The columns show sequential frames of the density (1st column), the magnitude of the pairing field (2nd column), and the phase of the pairing field (3rd column). In each frame the upper/lower part of each frame shows the neutron/proton density, the magnitude of neutron/proton pairing fields, and of the phase of the pairing field respectively [26]. In the lower frames is the induced fission of ^{240}Pu with enhanced pairing strength lasts about 1,400 fm/c.

The biggest surprise, however, was the observation that the time from saddle-to-scission was about an order of magnitude larger than ever estimated previously in the literature. This was ulteriorly confirmed by independent simulations [31]. The lower frames of Figure 2 serves as an illustration of the crucial role played by pairing correlations in fission dynamics. In the upper frames one can see that the pairing field on the way from saddle-to-scission does fluctuate noticeably in magnitude and phase. Therefore, strictly speaking the pairing field during its time evolution stops being a superfluid condensate of Cooper pairs, which otherwise would show a uniform phase. By increasing artificially the strength of the pairing field, we observe that fission dynamics proceeds approximatelyt ten times faster, see Figure 2. In that case, the pairing field shows the expected characteristics of a slowly-evolving superfluid condensate, the nuclear fluid behaving almost like a perfect or ideal fluid, where dissipation is practically absent, and the two FFs emerge macroscopically entangled now [30].

At FUSION17 workshop W. Mittig [32] discussed the very intriguing possibility of the existence of long-lived nuclear systems with very large charges, $Z \approx 200$ or even higher. One can contemplate the collision of two heavy nuclei with an energy just at the Coulomb barrier, when fusion is likely to occur. As in the case of fission dynamics, immediately after fusion many collective degrees of freedom of the combined system are likely to get excited, leading to a depletion of the energy from the incoming collision channel. This combined nuclear system might thus survive for a relatively long time, but eventually the system will decay into two almost equal fragments, as the combined system would likely equilibrate, even in the case of asymmetric collisions.

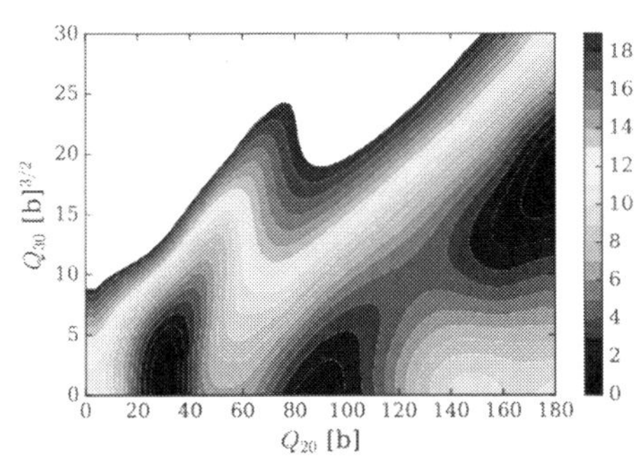

Figure 3. The potential energy surface of ^{240}Pu evaluated with the new energy density functional SeaLL1 [36].

A major limitation of the DFT is a lack of a method to evaluate two-body observables, such as the width of the FFs mass and charge yields and of the total TKE of the FFs. Within the TDHF approach one can use the Balian and Vénéroni (BV) [33] prescription or, alternatively, the more involved stochastic mean-field model introduced by Ayik [34]. Ayik's model is computationally more costly than the TDHF with the BV prescription included. The widths estimated using either the BV or the Ayik prescription are mathematically identical under certain conditions [34]. Ayik's prescription has been applied recently to the spontaneous fission of ^{258}Fm [31] in an approach characterized by the authors somewhat optimistically as "the first time a fully microscopic description of the fragment TKE distribution after fission and gives unique microscopic information on the fission process." While appealing, Ayik's model is nevertheless phenomenological in nature (as is random matrix theory), it relies on a several assumptions, and it is prone to large numerical noise. The main difference of the approach of Ref. [31] and TDHF calculations performed in the past is in choosing an *ad hoc* ensemble of stochastic initial conditions, instead of a single initial nuclear shape, at the point where the nucleus exits the outer fission barrier. In the case of spontaneous fission the nucleus emerges from under the barrier likely in its ground state corresponding to that instantaneous shape with $Q_{20} \approx 100$ barn in case of ^{258}Fm [31]. These authors [31] however chose to start their fission dynamics much further, when Q_{20} is either 125 or 160 barn, at which point the authors impose a thermal excitation of the nucleus to an energy corresponding to the ground state. This stochastic ensemble has large unphysical fluctuations of all possible observables, since most of the single-particle occupation probabilities obtained after diagonalizing the stochastic one-body density matrix $\rho(t) = \sum_{ij} \phi_i(t)\rho_{ij}\phi_j^*(t)$ acquire unphysical values, either with very large negative or very large positive values outside the interval [0, 1]. (Here ρ_{ij} are Gaussian random complex numbers.) Ref. [31] and Ref. [20] using the BV-prescription are so far the only attempts to describe quantum mechanically within TDHF the distribution of the FFs mass, charge yields, and TKE distributions. The mass widths using the BV prescription [20] appear visually narrower than those estimated in Ref. [31]. There is no theoretical argument presented in Ref. [31] to choose as a starting point of the dynamical simulations the arbitrary deformations Q_{20} = 125 or 160 barn. It would be more natural to start the simulation at the exact configuration where the nucleus emerged from under the barrier at $Q_{20} \approx 100$ barn. In that case the size of the "quantum fluctuations" would be zero and the nucleus would have no intrinsic excitation energy. But that would deprive the authors from the ability to generate a desired FFs distribution. The authors even establish that if they were to start their simulations closer to the scission configurations their results would be quite different, thus precluding this approach of its predictive power. One can safely conclude that the FFs distributions definitely depend strongly on the choice of the initial conditions within such a stochastic approach. A more natural approach would be a TDGCM approach advocated by Regnier *et al.* [35], in which one generates the distribution of the initial conditions in a TDDFT simulation of fission dynamics.

The predictive power of a TDDFT approach depends critically on the quality of the NEDFs, which are phenomenological so far and depend on a rather large number of fitting parameters (typically 14 and sometimes even more), some of them rather strongly correlated with each other. This points to the fact that there are elements of NEDF that are still poorly understood and there is a need for better constrained NEDFs, either phenomenologically or microscopically. We have developed a qualitatively new NEDF [36], which contains terms previously not considered in literature, which depends on only 7 uncorrelated parameters, with a rms for 602 masses of even-even nuclei from the AME2012 [37] of 1.74 MeV and for 345 corresponding charge radii of 0.034 fm. The potential energy surface for ^{240}Pu predicted by SeaLL1, see Figure 3, has potential energy barriers comparable to those obtained in FRLDM [38], though a relatively low energy of the fission isomer, and a nice single-particle spectrum of ^{208}Pb, see Figure 4.

This work was supported in part by U.S. Department of Energy (DOE) Grant No. DE-FG02-97ER41014 and by the U.S. Department of Energy under Contract Nos. DE-AC52-07NA27344 (Lawrence Livermore National Laboratory). This research used resources of the Oak Ridge Leadership Computing Facility, which is a DOE Office of Science User Facility supported under Contract DE-AC05-00OR22725. It also used resources of the National Energy Research Scientific Computing Center, a DOE Office of Science User Facility supported by the Office of Science of the U.S. Department of Energy under Contract No. DE-AC02-05CH11231. Some of the calculations were performed on Moonlight at LANL. Computing support for this work also came from the Lawrence Livermore National Laboratory (LLNL) Institutional Computing Grand Challenge program.

References

[1] O. Hahn and F. Strassmann, Naturwissenschaften **27**, 11 (1939).
[2] P. Hohenberg and W. Kohn, Phys. Rev. **136**, B864 (1964).
[3] W. Kohn and L.J. Sham, Phys. Rev. **140**, A1133 (1965).
[4] R.M. Dreizler and E.K.U. Gross, *Density Functional Theory: An Approach to the Quantum Many-Body Problem*, (Springer-Verlag, Berlin, 1990).
[5] *Fundamentals of Time-Dependent Density Functional Theory*, Eds. M.A.L. Marques, *et al.*, Lecture Notes in Physics, vol. 837 (Springer-Verlag, Berlin, 2012).
[6] A. Bulgac, Ann. Rev. Nucl. Part. Sci. **63**, 97 (2013).
[7] L. Meitner and O.R. Frisch, Nature (London), **143**, 239 (1939).
[8] N. Bohr, Nature (London), **143**, 330 (1939).
[9] N. Bohr and J.A. Wheeler, Phys. Rev. **56**, 426 (1939).
[10] G.N. Flerov and K.A. Petrzhak, Phys. Rev. **58**, 275 (1940).
[11] Maria Goeppert Mayer, Phys. Rev. **75**, 1969 (1949).
[12] O. Haxel, J.H.D. Jensen, and H.E. Suess, Phys. Rev. **75**, 1766 (1949).
[13] L. Meitner, Nature, **165**, 561 (1950).
[14] D.L. Hill and J.A. Wheeler, Phys. Rev. **89**, 1102 (1953).

 DOI: 10.1051/epjconf/201716300007

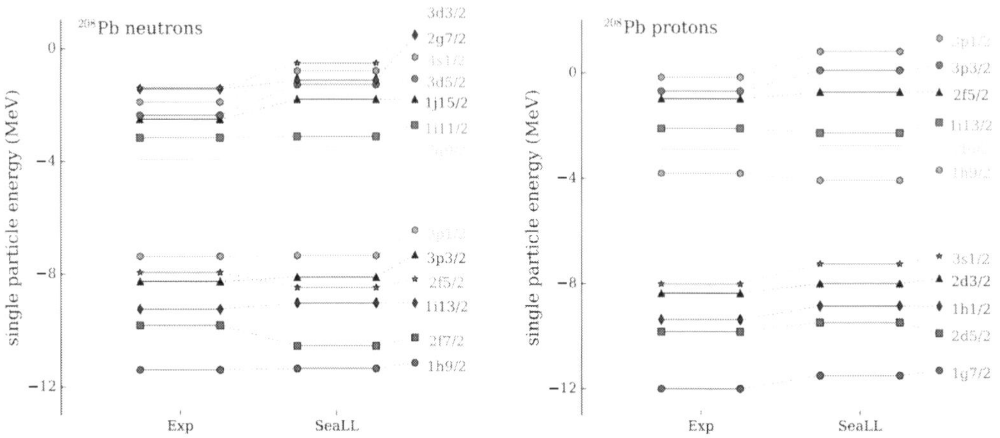

Figure 4. The single-particle energy levels ^{208}Pb evaluated with the new energy density functional SeaLL1 [36] versus experiement.

[15] G.F. Bertsch, Phys. Lets. **B 95**, 157 (1980).

[16] G.F. Bertsch and A. Bulgac, Phys. Rev. Lett. **79**, 3539 (1997).

[17] P. Goddard, P. Stevenson, and A. Rios, Phys. Rev. C **92**, 054610 (2015).

[18] P. Goddard, P. Stevenson, and A. Rios, Phys. Rev. C **93**, 014620 (2016).

[19] C. Simenel and A.S. Umar, Phys. Rev. C **89**, 031601(R) (2014).

[20] G. Scamps, C. Simenel, and D. Lacroix, Phys. Rev. **92**, 011602(R) (2015).

[21] S.M. Polikanov, *et al.*, JETP (Sov. Phys.) **42**, 1016 (1962)

[22] M. Brack, *et al.*, Rev. Mod. Phys. **44**, 32 (1972).

[23] S. Bjørnholm and J.E. Lynn, Rev. Mod. Phys. **52**, 725 (1980).

[24] P.J. Twin, *et al.*, Phys. Rev. Lett. **57**, 811 (1986).

[25] N. Schunck and L.M. Robledo, Rep. Prog. Phys. **79**, 116301 (2016).

[26] A. Bulgac, P. Magierski, K.J. Roche, and I. Stetcu, Phys. Rev. Lett. **116**, 122504 (2016).

[27] E. Chabanat, P. Bonche, P. Haensel, J. Meyer, R. Schaeffer, Nucl. Phys. **A635**, 231 (1998).

[28] A. Bulgac and Y. Yu, Phys. Rev. Lett. **88**, 042504 (2002)

[29] Y. Yu and A. Bulgac, Phys. Rev. Lett. **90**, 222501 (2003).

[30] A. Bulgac and S. Jin, an updated version of arXiv:1701.06683 to be resubmitted.

[31] Y. Tanimura, D. Lacroix, and S. Ayik, Phys. Rev. Lett. **118**, 152501 (2017).

[32] W. Mittig, talk presented at FUSION17, Hobart, Tasmania, February 20-24, 2017.

[33] R. Balian and M. Vénéroni, Phys. Lett. **B136**, 301 (1984).

[34] S. Ayik, Phys. Lett. **B658**, 174 (2008).

[35] D. Regnier, N. Dybray, N. Schunck, and M. Verrière, Phys. Rev. C **93**, 054611 (2016).

[36] A. Bulgac, M.M. Forbes, S. Jin, N. Schunck, and R. Perez Navarro, to be submitted.

[37] M. Wang, *et al.*, Chinese Phys. C. **36**, 1603 (2012).

[38] P. Möller, *et al.*, Phys. Rev. C **79**, 064304 (2009).

EPJ Web of Conferences **163**, 00008 (2017) DOI: 10.1051/epjconf/201716300008
FUSION17

Fission in Inverse Kinematics

A path to new experimental observables

Manuel Caamaño[1,*], *Fanny* Farget[2], and *Diego* Ramos[1,**]

[1] *Universidade de Santiago de Compostela, E-15706 Santiago de Compostela, Spain*
[2] *GANIL, CEA/DSM-CNRS/IN2P3, BP 55027, F-14076 Caen Cedex 5, France*

Abstract. Historically, experimental fission studies were based on reactions in direct kinematics with fixed target-like fissioning systems. Besides its advantages, this technique suffers from some drawbacks such as the difficulty of producing exotic fissioning systems and the seldom measurement of the fragment atomic number. Inverse kinematic provides an alternative to ease these issues and offers a new set of experimental observables that improves our level of information about the fission process, including an unprecedented access to the scission point. In this document, we review some of the observables obtained from the experimental campaign based on inverse kinematics, performed at VAMOS/GANIL.

1 Introduction

Nuclear fission is a complex phenomenon in which the nucleus is deformed and heated, collective energy is dissipated in the form of intrinsic degrees of freedom, while the pairing and shell structure guide the process until two excited fragments are produced and accelerated due to their Coulomb repulsion.

Since its discovery, almost 80 years ago [1, 2], our current knowledge of fission was built on the interpretation of observables almost exclusively gathered in direct kinematics experiments: a heavy-nuclide target is bombarded by lighter projectiles that produce a target-like fissioning system. The properties of the fission products are measured to form the ensemble of experimental observables. Among them, direct kinematics allows the study of the fission probability, the fragments total kinetic energy, TKE, and mass distributions, their angular distribution, or the neutron evaporation. Together with the characterization of the fissioning system in terms of isotopic identification, excitation energy, etc., they shape the current image of fission: a process that runs along a potential energy formed by macroscopic and microscopic features of the nucleus, which depends sensitively on the number of protons and/or neutrons involved, and opens isolated deformation paths leading to different configurations at the scission point, called modes or channels [3, 4].

In spite of a wide opportunity of the fission setup in the direct kinematics method, it suffers from the following drawbacks: the small velocities of the emitted fragments disturb a good measurement of their the atomic number Z and impose restrictions to the amount of material in the target, while the use of direct kinematics limits systematic measurements to fissioning systems close in proton and neutron numbers to those of the target species. These difficulties can be overcome with the use of inverse kinematics: exotic nuclei can be produced with in-flight techniques and later guided to a secondary reaction target where fission can be induced; the measurement of the fragments Z distribution is also facilitated by their higher velocity when emitted from the moving fissioning systems.

The pioneering fission experiments with inverse kinematics at GSI [5] offered the measurement of Z and TKE distributions for a large number of neutron-deficient actinides [6], along the systematic measurement of the even-odd staggering [7], which is another scarce observable. Together, these observables revealed a fresh look on the role of protons and dissipation in the fission process. This is particularly interesting considering that shells are revealed in the number of protons and neutrons, and their effect is less evident in the mass distributions.

The last upgrade in the use of inverse kinematics was to include also the mass of the fragments, A, in the set of observables. The use of magnetic spectrometers allowed such improvement within two of the current experimental campaigns: SOFIA at GSI, which measures electromagnetic-induced fission of neutron-deficient systems [8, 9] with the help of the large dipole magnet ALADIN; and the fission campaign at GANIL, where fissioning systems around ^{238}U are studied through transfer- and fusion-induced fission using the VAMOS spectrometer [10, 11]. In the following, we discuss the collection of observables measured in the VAMOS/GANIL experiments.

* e-mail: manuel.fresco@usc.es
** Present address: IPN Orsay, IN2P3/CNRS-UPS, F-91406 Orsay Cedex, France

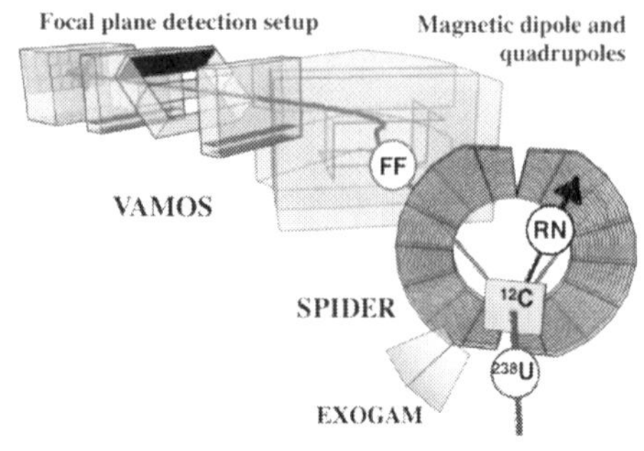

Figure 1. Schematic layout of the VAMOS setup during the first run of the fission campaign. After the transfer reaction, the target-like recoil (RN, black arrow) is measured in SPIDER, while one of the fission fragments (FF, red arrow) emitted after fission of the beam-like system is detected in the detector setup placed at the focal plane of the VAMOS spectrometer.

2 Transfer- and Fusion-Induced Fission in Inverse Kinematics at GANIL

As mentioned in the previous section, the fission campaign carried out at GANIL focused on the study of fission induced by direct reactions measured in the VAMOS variable mode spectrometer. A ^{238}U beam was shoot towards a ^{12}C target with 6.1 AMeV, an energy slightly above the Coulomb barrier, assuring the cross sections for transfer and fusion channels to reach tenths and hundreds of mb, respectively. Transfer reactions produced beam-like fissioning systems from U to Cm within a range of excitation energy, E^*_{FS}, below some 30 MeV, while fusion reactions resulted in ^{250}Cf with a fixed excitation energy of around 45 MeV.

The experimental setup can be laid out in two stages: the first one includes SPIDER, a set of two annular, double-sided, stripped silicon detectors devoted to the reconstruction of the direct reaction through the identification of the target-like recoil and the measurement of its angle and energy. A small part of the EXOGAM gamma detector is also placed to measure prompt gamma emission from the fragments. The second stage is based on the VAMOS spectrometer and its focal detectors, which detect one of the fission fragments for the isotope identification as well as the flight angle and velocity measurements. Figure 1 shows the VAMOS setup during the first run of the campaign. Further details can be found in Refs. [10–12]

The VAMOS setup offers a collection of observables carrying information from both the fissioning system and the produced fragments. For each fission event, the fissioning system is identified, and its E^*_{FS}, angle, and velocity are measured with the reconstruction of the direct reaction. Concerning the fission fragments, the use of inverse kinematics allowed, for the first time, the isotopic identification of the full fission fragment distribution [10]. Taking advantage of the features of transfer reactions, the evolution of the fission-fragment distributions on the *Z*-*A* plane can be studied as a function of the fissioning nuclides and their excitation energies [13]. In addition to *Z* and *A*, the fragments velocities and angles with respect to the beam direction are also measured, which, together with the identification and kinematics of the fissioning system, can be translated to the fissioning system reference frame. The simultaneous measurement of these characteristics allows to build new observables, offering new insights into the fission process. In the following sections we review and discuss some of these new observables.

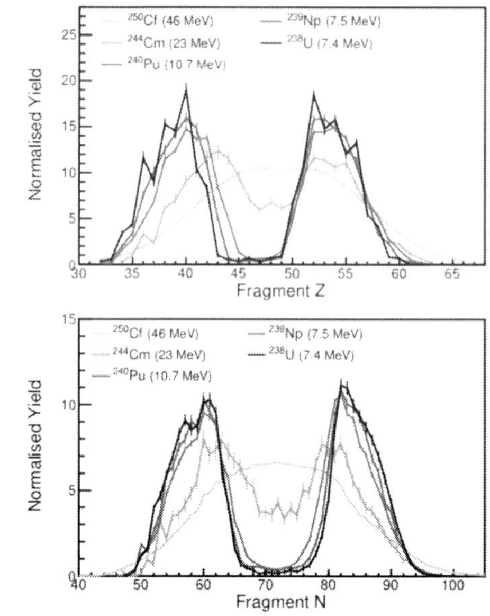

Figure 2. Fragment *Z* (top panel) and *N* (bottom panel) distributions for the fissioning systems measured in the second run of the VAMOS/GANIL campaign. The parentheses show the average excitation energy.

2.1 Neutron distribution and average neutron excess

The simultaneous measurement of the fragments *Z* and *A* distributions makes it possible to correlate both observables and build the neutron-number distribution for each event *j*, as N_j=A_j-Z_j. Such distribution corresponds to the fragments after post-scission neutron evaporation and thus it reflects the effect of neutron shells modified by the release of evaporated neutrons. Figure 2 shows the *Z* and post-evaporation *N* distributions for a set of systems with different average E^*_{FS}. The *Z* distributions of these systems show very similar features on the heavy fragment, with a clear preference for producing *Z* ~52, 54 fragments. However, the *N* distributions are more difficult to interpret. The pronounced peak around *N*=82 might be a combination of the influence of the neutron shell and the accumulation

DOI: 10.1051/epjconf/201716300008

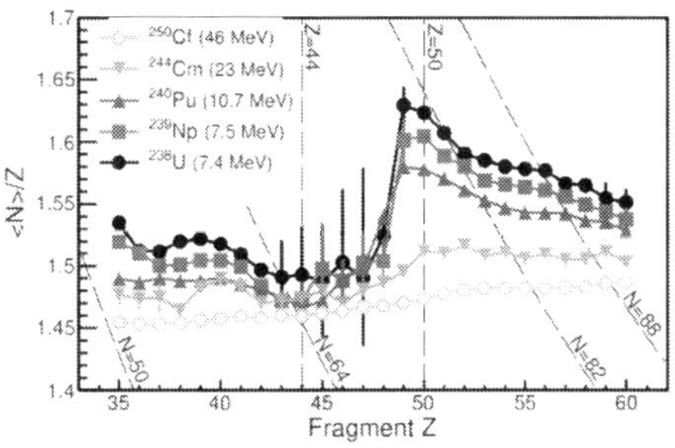

Figure 3. Fragment $\langle N\rangle/Z$ distribution for the fissioning systems measured in the second run of the VAMOS/GANIL campaign. The parentheses show the average excitation energy. The long-dashed lines correspond to proton and neutron numbers related to spherical and deformed shells.

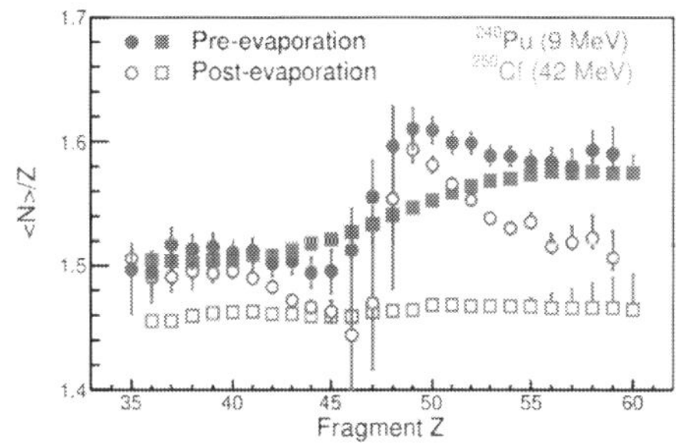

Figure 4. Pre-evaporation (open symbols) and post-evaporation (solid symbols) $\langle N\rangle/Z$ distributions for ^{240}Pu (blue dots) and ^{250}Cf (red squares) measured in the first run of the VAMOS/GANIL campaign. The parentheses show the average excitation energy.

of the end point of the neutron evaporation chains of its neighbors, which is coherent with the known minimum of evaporation for $A\sim130$ (see [14, 15], for instance).

A degree of the magicity of nuclei affecting the fragments distribution can be represented by the neutron excess, defined as:

$$\langle N\rangle/Z=\frac{\sum(A-Z)\cdot Y(A,Z)}{\sum Y(A,Z)}, \qquad (1)$$

where $Y(A,Z)$ is the normalized yield of the A, Z fragment.

Figure 3 shows the $\langle N\rangle/Z$ distribution of the systems measured in the second run of the campaign. Despite the effect of neutron evaporation, a clear polarization is observed. Its shape evolve with E^*_{FS} until disappearing for ^{250}Cf, produced with E^*_{FS}=46 MeV, suggesting that these features might be the result of the shell structure of the fragments. To explore this possibility we plot, in the same figure, the lines corresponding to proton and neutron numbers of relevant spherical and deformed shells [16].

We can see that the low-E^*_{FS} systems tend to approach the crossing of Z=50, N=82 spherical shells in the heavy fragments, and the crossing of Z=44, N=64 deformed shells in the light fragments. It is also worthy of note an accumulation around this point for all the systems, regardless of their E^*_{FS}. The region around Z=44, N=64 is known to yield a high neutron multiplicity in the post-scission evaporation [14] but, at the same time, it was also observed to remain almost unaffected by the E^*_{FS} of the fissioning system [17].

2.2 Observables at the scission point

As mentioned in Sec. 2, the velocity of the fragments in the reference frame of the fissioning system can be determined with the reconstruction of the direct reaction producing the fissioning system and the measurement of the velocities of the fragments in the laboratory frame. Assuming that post-scission evaporation does not modify the velocity of the fragments, and applying the momentum conservation, the masses of the fragments at scission can be deduced [18]. In the case of the VAMOS/GANIL campaign, one fragment per fission event is measured and thus the observables calculated at scission are expressed as average quantities as a function of the fragment Z. Consequently, and unless stated otherwise, all quantities referred in this work to fragment properties are functions of the fragment Z. A full discussion of the method applied to these data can be found in [19].

Figure 4 shows the difference between the post-evaporation and pre-evaporation $\langle N\rangle/Z$ for ^{240}Pu with E^*_{FS}=9 MeV and ^{250}Cf with E^*_{FS}=42 MeV fission, both measured in the first run of the campaign. As we discussed in the previous section, the main features of pre-evaporation $\langle N\rangle/Z$ can be observed for low-E^*_{FS} systems. High-E^*_{FS} systems present a complete different picture before and after evaporation: the almost flat $\langle N\rangle/Z$ distribution of the post-evaporation ^{250}Cf data (E^*_{FS}=42 MeV) is an outcome of the subtle counterbalance between the pre-evaporation $\langle N\rangle/Z$ value and the neutron emission, since both of them increase with Z. We can observe a clear change of slope around Z=54 that might be associated with the different evolution of deformed single-particle states in the regions between $44<Z<54$ and $55<Z$.

The access to Z, A, and velocity of the fragments at scission, combined with the identification of the fissioning system and its E^*_{FS}, also permits to perform the energy balance between the input and output of the fission reaction. The input energy corresponds to the sum of the mass of the fissioning system, M_{FS}, and its E^*_{FS}. Considering no emission from saddle to scission, this is transformed into the ground-state masses of the fragments, M_1 and M_2, their total excitation energy, TXE, defined as the sum of the intrinsic and deformation energy of the fragments, and the total kinetic energy TKE, which is the sum of the Coulomb repulsion energy and the pre-scission kinetic energy. The energy balance is then

$$M_{\mathrm{FS}}+E^*_{\mathrm{FS}}=M_1+M_2+TKE+TXE. \qquad (2)$$

It is important to note that while $M_{\mathrm{FS}}+E^*_{\mathrm{FS}}$ is obtained from the reconstruction of the direct reaction, this quantity is, in

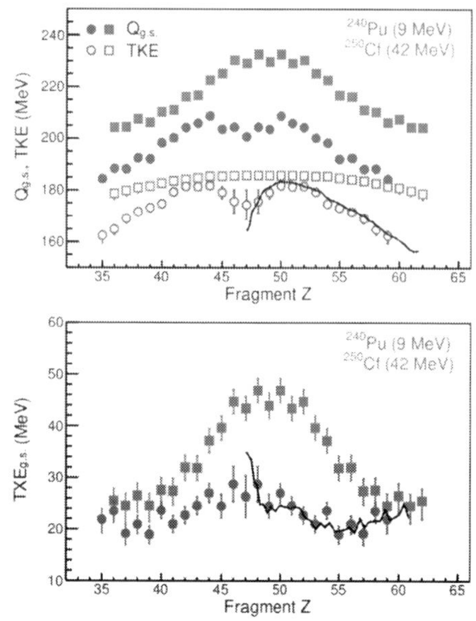

Figure 5. The figure shows the energy components for ^{240}Pu with E^*_{FS}=9 MeV (blue dots) and ^{250}Cf with E^*_{FS}=42 MeV (red squares). Top panel: TKE (open symbols) and $Q_{g.s.}$ (solid symbols); the black line is the TKE measured from ^{239}Pu(n$_{th}$,f) [15]. Bottom panel: $TXE_{g.s.}$; the black line is the $TXE_{g.s.}$ estimated from ^{239}Pu(n$_{th}$,f) [15].

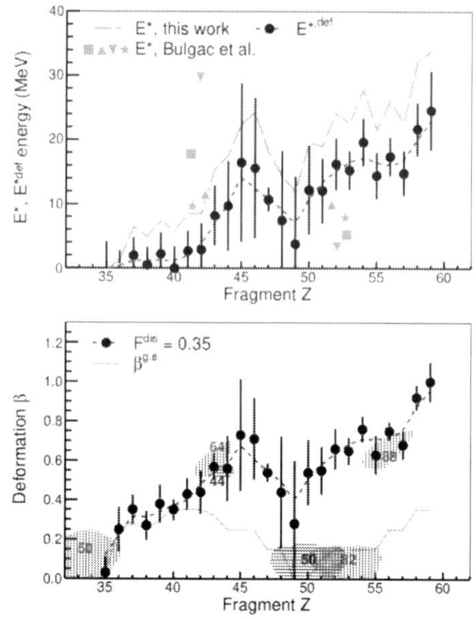

Figure 6. Top panel: E^* (green long-dashed line) and $E^{*,def}$ (black dots) of fragments produced in low-energy fission of ^{240}Pu. The blue dashed line is a moving average of the $E^{*,def}$ data. The green symbols correspond to results of the energy density functional model used in [28]. Bottom panel: Quadrupole deformation β (black dots) displayed with its moving average (blue dashed line). The hatched areas correspond to neutron (red) and proton (blue) spherical and deformed shells. The short-dashed black line is the average deformation at the ground state.

general, accurate for E^*_{FS} below the sum of the neutron separation energy and the fission barrier. For higher energies, an estimation of the pre-saddle evaporation is needed. As explained, M_1 and M_2 are obtained from the momentum conservation in the fissioning system reference frame, while TKE is calculated with the velocities in the same reference frame. From these quantities is possible to calculate the distribution of both the average $TXE_{g.s.}$ and the $Q_{g.s.}$ value at the ground state of the reaction, defined as

$$TXE_{g.s.} = M_{FS} - M_1 - M_2 - TKE = Q_{g.s.} - TKE. \quad (3)$$

Figure 5 shows the resulting TKE, $Q_{g.s.}$, and $TXE_{g.s.}$ at scission for ^{240}Pu with E^*_{FS}=9 MeV and ^{250}Cf with E^*_{FS}=42 MeV, both from the first run of the campaign. A first glance at the top panel shows a very similar behavior for both TKE and $Q_{g.s.}$ in the case of the low-energy fission of ^{240}Pu. This similarity was already hinted in previous studies of cold fission in light actinides, where the contribution of TXE is at minimum [20]. In the case of high-energy fission of ^{250}Cf, the similarity between TKE and $Q_{g.s.}$ disappears around the symmetry, where a difference builds up. Concerning $TXE_{g.s.}$, the bottom panel of Fig. 5 shows an almost flat distribution for low energy fission, with an increase of almost 10 MeV around the symmetry. However, the case of high-energy fission shows the $TXE_{g.s.}$ growing from ~ 25 MeV at asymmetric splits to ~ 45 MeV around the symmetry.

The results of TKE and $TXE_{g.s.}$ in ^{240}Pu are also compared in Fig. 5 with previous studies of ^{239}Pu(n$_{th}$,f) [15]. The overall values of TKE and $TXE_{g.s.}$ are in agreement with the present data with the exception of the symmetric split, where a larger TKE and a smaller $TXE_{g.s.}$ are obtained in the present data. This difference is likely to be a combination of the larger excitation energy of the fissioning system and the limited resolution of the present experiment for symmetric splits due to the lower statistics.

2.3 Energetics and configuration at scission

The fission observables discussed so far permit, combined with reasonable assumptions, to extract more information on the scission configuration of the fragments, such as their excitation energy, deformation, and distance. The details of such calculation can be found in [23]

The measured TXE is the sum of the excitation energy E^*_i of each fragment i.[1] These can be separated into each fragment deformation energy $E^{*,def}_i$ and intrinsic energy, which is the sum of the initial excitation energy above the

[1] As previously mentioned, with this setup we deal with average quantities per fragment Z. Since we detect one fragment per fission, the average properties of fragment Z are correlated with those of fragment Z_{FS}-Z

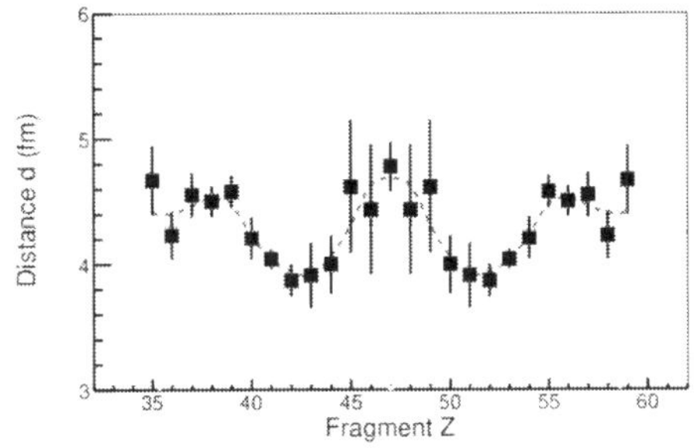

Figure 7. Tip distance between the fragments (symbols). The red dashed line is a moving average of the data.

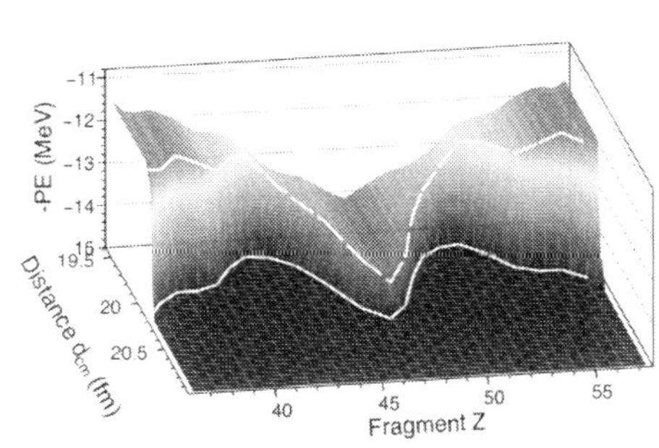

Figure 8. Potential landscape at scission. The white dashed line corresponds to a moving average of the scission point as a function of PE, d_{cm}, and Z. The white solid line shows the projection onto the d_{cm}-Z plane.

barrier $E_i^{*,Bf}$ and the dissipated energy along the process $E_i^{*,diss}$:

$$TXE = \sum_{i=1}^{2} E_i^* = \sum_{i=1}^{2} \left(E_i^{*,Bf} + E_i^{*,dis} + E_i^{*,def}\right). \quad (4)$$

Experimentally, TXE is fully released in the form of neutron and γ evaporation:

$$TXE = \sum_{i=1}^{2} \left(Q_i^n + \nu_i\varepsilon_i + E_i^\gamma\right), \quad (5)$$

where ν_i is the neutron multiplicity, ε_i is the neutron kinetic energy, E_i^γ is the energy released by γ emission, and Q_i^n is the difference between pre- and post-evaporation fragment masses. Previous data show that the average ε is approximately symmetric with respect to the fragments split [14], and that E_i^γ follows the behavior of ν_i with a total value similar to the neutron separation energy S_n [24]. These properties allow to express Eq. 5 as

$$TXE \approx \sum_{i=1}^{2} Q_i^n + \nu_i\left(\varepsilon + S_n/(\nu_1 + \nu_2)\right). \quad (6)$$

ν_i and Q_i^n are measured observables, while S_n is calculated as the average neutron separation energy for each split.

The combination of Eqs. 4 and 6 permits to deduce $E_i^{*,def}$, once the part $E_i^{*,Bf}+E_i^{*,dis}$ is shared between the fragments in a regime of statistical equilibrium, following the prescription of Refs. [25, 26]. Concerning the dissipation energy, in this work we followed the prescription of Ref. [27], where the dissipation is a fixed fraction of the available TXE.[2] These approximations and prescriptions are valid for low-energy fission of actinides and, in our case, we applied them to the ^{240}Pu data at E^*_{FS}=9 MeV. The top panel in Fig. 6 shows the distribution of the excitation energy E_i^* and the fraction dedicated to deformation, $E_i^{*,def}$. The E_i^* data is compared with a recent calculation based on a real-time, microscopic framework of energy density functional [28], also applied to low-energy fission of ^{240}Pu. We can see what the authors interpret as a quasi-spherical slightly-excited heavy fragment around Z=52 and a highly-deformed highly-excited light one around Z=42. There is a fair discrepancy with our results: at Z ~52 we find deformed fragments excited up to 20 MeV, while at Z ~42 we have similarly deformed fragments with an excitation energy of ~10 MeV.

It is also worthy to mention the relatively high multiplicity around Z=50: the literature shows multiplicities of the order of ν ~0.5 around the corresponding masses (see [15, 17] for example). In the present case, the excess of ν can be explained, at least in its most part, with the E^*_{FS} induced by the transfer reaction. Other factors, such as an increase of the deformation or a lower fission barrier due to the angular momentum, might also play a role.

The bottom panel of Fig. 6 shows the translation of $E^{*,def}$ into quadrupole deformation β by computing the increase in energy of the Weizsäcker liquid-drop mass-formula for variations in the surface and Coulomb terms due to small quadrupole deformations, following the prescription of Swiatecki [29]. We can see that β retains the well-known sawtooth shape already observed in ν and E^*, while it runs through Z=44, N=64, 88 deformed shells. It is also interesting that, although β is "pulled" towards lower values around Z=50, N=82 spherical shells, the deformation is relatively high. The competition between single particle structure and the macroscopic potential, which is particularly strong close to symmetry, seems to hinder the effect of these spherical shells.

Once the deformation of the fragments are known, the tip distance between the fragments, d, can be extracted with the measured TKE, which is the sum of the contribution from the Coulomb repulsion between the fragments, $E^{k,C}$, and the pre-scission kinetic energy, $E^{k,pre}$:

$$TKE = E^{k,C}(Z_1, Z_2, \beta_1, \beta_2, d) + E^{k,pre}(Z_1, Z_2), \quad (7)$$

In this case, we use $E^{k,pre}$ from a two-center shell-model parameterization, calculated as a function of the fragment [30]. The tip distance d is then deduced from $E^{k,C}$ following the Cohen-Swiatecki formula [31]. The resulting distance d is displayed in Fig. 7. We observe values between 4 and 5 fm, further than the "standard" 2-3 fm of low-energy fission of actinides [3, 16, 32–35] but similar

[2] A complete discussion of the convenience of these prescriptions can be found in [23].

 DOI: 10.1051/epjconf/201716300008

to the values used in recent scission-point models [36]. We can also see that configurations around Z_1=42, Z_2=50 are more compact than those at symmetric and very asymmetric splits. This minimum coincides with a maximum in TKE, suggesting a correlated behavior between both the TKE and d, and also with shell effects.

The TXE discussed so far corresponds to the total excitation energy gained by the fragments at scission with respect to their ground states. We can recover the potential energy surface by expressing the total excitation energy with respect to the deformed ground states of the fragments (see for example [21, 22]). Together with the calculated $E^{*,\mathrm{def}}$, β_i, and d, they can be combined to outline the shape of the potential energy PE landscape that the system experiences at scission, defined as

$$PE = TXE - E_1^{*,\mathrm{def}} - E_2^{*,\mathrm{def}}, \quad (8)$$

with the distance between the centers of the fragments d_{cm}:

$$d_{\mathrm{cm}} = d + r_0 A_1^{1/3}\left(1 + \sqrt{\frac{5}{4\pi}}\beta_1\right) + r_0 A_2^{1/3}\left(1 + \sqrt{\frac{5}{4\pi}}\beta_2\right), \quad (9)$$

to obtain the PE-d_{cm}-Z correlation at scission. The result is displayed in Fig. 8. The potential energies at scission appear with a prominent minimum of some 10 MeV smaller than the asymmetric splits, which coincides with an elongated shape. At the same time, the distance between fragments is still smallest around Z_1=42, Z_2=50, meaning the system breaks sooner when affected by microscopic effects. The fact that the macroscopic potential at symmetry is observed in the depth of the landscape with no apparent structure effects suggests a strong connection between these effects and the elongation that the system can reach before splitting.

3 Conclusions

The use of inverse kinematics in new experimental campaigns allows to obtain simultaneously a number of observables that were historically difficult to access and always in separate experiments, such as the fragment A, Z, and velocity distributions. The correlation of these observables permitss to construct new quantities with improved sensitivity to fission properties. The neutron excess, and the TXE and TKE at scission are now available for different systems and as a function of the fission energy. In some cases, such as low-energy fission of actinides, the new observables can be extended to reveal the energy sharing and the configuration at scission, in terms of deformation and distance of the fragments. These results help to better understand the role of structure effects in the process and contribute to the pool of experimental fission data to be compared with state-of-the-art models.

Acknowledgements

M.C. acknowledges the financial support of the Program "Ramón y Cajal" of the Ministerio de Economía y Competitividad of Spain through the grant RYC 2012-11585.

References

[1] O. Hahn and F. Strassmann, Naturwiss **27**, 11 (1939)
[2] L. Meitner and O. Frisch, Nature **143**, 239 (1939)
[3] U. Brosa *et al.*, Phys. Reports **197**, 167 (1990)
[4] M.G. Itkis *et al.*, Z. Phys. A - Atoms and Nuclei **320**, 433 (1985)
[5] K.-H. Schmidt *et al.*, Nucl. Phys. A **665**, 221 (2000)
[6] C. Böckstiegel *et al.*, Nucl. Phys. A **802**, 12 (2008)
[7] S. Steinhäuser *et al.*, Nucl. Phys. A **634**, 89 (1998)
[8] J.-F. Martin *et al.*, Eur. Phys. J. A **51**, 174 (2015)
[9] E. Pellereau *et al.*, Phys. Rev. C **95**, 054603 (2017)
[10] M. Caamaño *et al.*, Phys. Rev. C **88**, 024605 (2013)
[11] C. Rodríguez-Tajes *et al.*, Phys. Rev. C **89**, 024614 (2014)
[12] D. Ramos, "Fragment Distributions of Transfer- and Fusion-Induced Fission from ^{238}U+^{12}C Reactions Measured Through Inverse Kinematics" PhD (Univ. of Santiago de Compostela, 2016)
[13] D. Ramos *et al.*, EPJ Web Conf. **111**, 10001 (2016)
[14] K. Nishio *et al.*, Nucl. Phys. A **632**, 540 (1998)
[15] C. Tsuchiya *et al.*, J. Nucl. Sci. Technol. **37**, 941 (2000)
[16] B.D. Wilkins *et al.*, Phys. Rev. C **14**, 1832 (1976)
[17] K.-H. Schmidt and B. Jurado, Phys. Rev. Lett. **104**, 212501 (2010)
[18] H.C. Britt *et al.*, Nucl. Instrum. Methods **24**, 13 (1963) and references therein
[19] M. Caamaño *et al.*, Phys. Rev. C **92**, 034606 (2015)
[20] F. Gönnenwein, Z. Phys. A **349**, 259 (1994)
[21] P. Möller *et al.*, Phys. Rev. C **79**, 064304 (2009)
[22] J. Randrup and P. Möller, Phys. Rev. Lett. **106**, 132503 (2011)
[23] M. Caamaño and F. Farget, Phys. Lett. B **770**, 72 (2017)
[24] R. Capote *et al.*, Nucl. Data Sheets **131**, 1 (2016), and references therein.
[25] R. Capote *et al.*, Nucl. Data Sheets **110**, 3107 (2009)
[26] K.-H. Schmidt and B. Jurado, Phys. Rev. C **83**, 061601(R) (2011)
[27] F. Rejmund *et al.*, Nucl. Phys. A **678**, 215 (2000)
[28] A. Bulgac *et al.*, Phys. Rev. Lett. **116**, 122504 (2016)
[29] W.J. Swiatecki, Phys. Rev. **104**, 993 (1956)
[30] F.A. Ivanyuk *et al.*, Phys. Rev. C **90**, 054607 (2014)
[31] S. Cohen and W.J. Swiatecki, Annals of Physics **19**, 67 (1962)
[32] T.-S. Fan *et al.*, Nucl. Phys. A **591**, 161 (1995)
[33] C. Böckstiegel *et al.*, Nucl. Phys. A **802**, 12 (2008)
[34] H. Goutte *et al.*, Phys. Rev. C **71**, 024316 (2005)
[35] K.-H. Schmidt *et al.*, Nucl. Data Sheets **131**, 107 (2016)
[36] J.-F. Lemaître *et al.*, Phys. Rev. C **92**, 034617 (2015)

EPJ Web of Conferences **163**, 00009 (2017)
FUSION17

DOI: 10.1051/epjconf/201716300009

Nuclear Astrophysics in underground laboratories: the LUNA experiment

Antonio Caciolli for the LUNA Collaboration[1,*]

[1] *University and INFN of Padua, via Marzolo 8, 35131, Padua, Italy*

Abstract. One of the main ingredients of nuclear astrophysics is the knowledge of the thermonuclear reactions responsible for powering the stellar engine and for the synthesis of the chemical elements. At astrophysical energies the cross section of nuclear processes is extremely reduced by the effect of the Coulomb barrier. The low value of cross sections prevents their measurement at stellar energies on Earth surface and often extrapolations are needed. The Laboratory for Underground Nuclear Astrophysics (LUNA) is placed under the Gran Sasso mountain and thanks to the cosmic-ray background reduction provided by its position can investigate cross sections at energies close to the Gamow peak in stellar scenarios. Many crucial reactions involved in hydrogen burning has been measured directly at astrophysical energies with both the LUNA-50kV and the LUNA-400kV accelerators, and this intense work will continue with the installation of a MV machine able to explore helium and carbon burnings. Based on this progress, currently there are efforts in several countries to construct new underground accelerators. In this talk, the typical techniques adopted in underground nuclear astrophysics will be described and the most relevant results achieved by LUNA will be reviewed. The exciting science that can be probed with the new facilities will be highlighted.

Nuclear processes generate the energy that makes stars shine. Moreover they are responsible of the synthesis of the elements (and isotopes) in stars. As a matter of fact, hydrogen, helium and all isotopes until lithium and beryllium are synthesised during the Big Bang Nucleosynthesis (BBN). All other nuclei are produced during the different characteristic phases of the star evolution [1].

The understanding of these nuclear processes is the goal of nuclear astrophysics and, in particular, the knowledge of the nuclear cross-sections involved in that processes. The astrophysical relevant energies is given by the folding of two strongly energy dependent functions [1, 2]: the Maxwell-Boltzmann velocity distribution and the cross section of charged particle induced reactions. The folding results in a structure called the Gamow peak, which peaks at the energy E_G (called Gamow energy):

$$E_G \approx 0.1220(Z_1^2 Z_2^2 \mu)^{\frac{1}{3}} T_9^{\frac{2}{3}} \qquad (1)$$

where $Z_{1,2}$ are the charges of the two reaction partners, $\mu = m_1 m_2/(m_1 + m_2)$ their reduced mass, and $T_9 = T/10^9$ K is the temperature of the astrophysical scenario under study. The Gamow energy of nuclear reactions taking place in the Sun, with its core temperature of $T_9 \approx 0.016$, is typically 20 keV depending on the precise reaction, leading to cross sections in the range of pbarn and below. As matter of fact, in this energy range the cross section is highly reduced by the effect of the Coulomb repulsion and the nuclear reactions can occur only via tunnel effect. In particular the cross section can be written as:

$$\sigma(E) = \frac{S(E)}{E} exp(-2\pi\eta), \qquad (2)$$

where $S(E)$ is the astrophysical factor (which contains the nuclear physics information) and η is the Sommerfeld parameter, given by:

$$2\pi\eta = 31.29 Z_1 Z_2(\mu/E)^{1/2}, \qquad (3)$$

μ is the reduced mass (in units of amu), and E is the center of mass energy (in units of keV).

Due to these small cross section values, the rate of the reactions, characterised by a typical energy release of a few MeV, is too low, down to a few events per year, in order to stand out from the laboratory background. In many cases it is not possible even to reach energy values close to the Gamow peak and extrapolations are needed, leading to substantial uncertainties. A way to handle that background problem is to go in an underground environment [3]. As a matter of fact the natural shielding provided by an underground site will guarantee a reduction of the cosmic flux of orders of magnitude leading to the success of experimental nuclear physics. LUNA (Laboratory for Underground Nuclear Physics) [4, 5] is placed under the Gran Sasso National Laboratories of INFN. Two accelerators were used during years. First a 50 kV accelerator (hereafter LUNA1)[6] and then a 400kV accelerator (hereafter LUNA2)[7].

Under the Gran Sasso National Laboratory (LNGS) the muon flux is reduced by a factor 10^6 and the neutron flux by a factor of 1000 [8, 9]. Further background re-

* e-mail: caciolli@pd.infn.it

 DOI: 10.1051/epjconf/201716300009

duction in the region below 3 MeV in the gamma spectrum can be achieved by implementing a shielding made by copper and lead [10, 11]. A comparison between two gamma spectra acquired with the same detector in surface and at LNGS is shown in Figure 1. In the figure, the detector was also shielded by few cm of OFHC copper and 25 cm of low level background lead when placed in the underground laboratory to obtain almost 4 order of magnitude reduction in the γ-ray background below 3 MeV [10]. A review of the results achieved by the LUNA collaboration will be presented in this paper combined with a discussion on the future projects for nuclear astrophysics in underground with a MV accelerator.

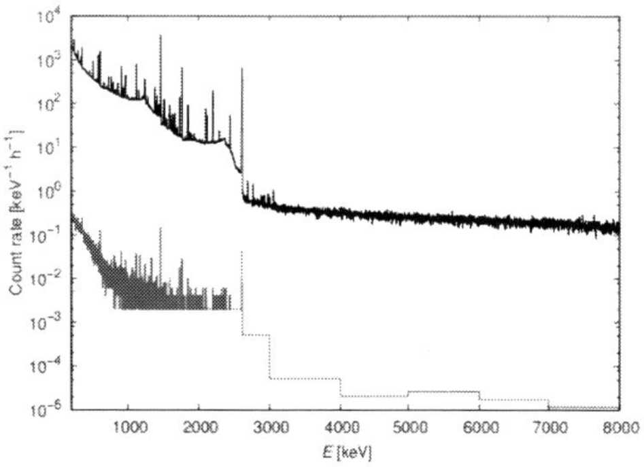

Figure 1. The spectrum acquired with an HPGe on the surface in Padua (black) is compared with the one acquried with same detector in the underground laboratory of Gran Sasso and with a full shield as reported in [10] (red).

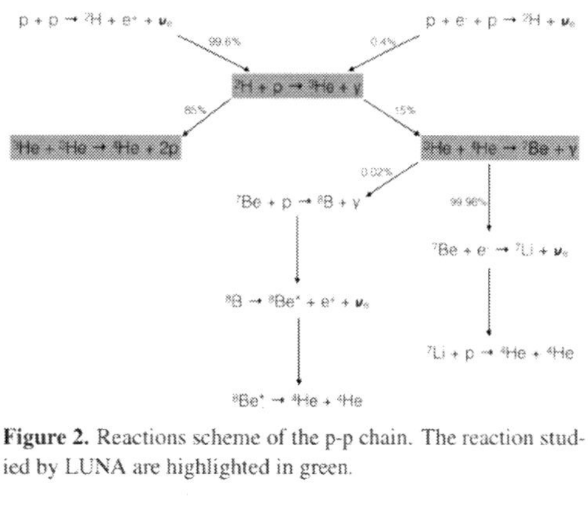

Figure 2. Reactions scheme of the p-p chain. The reaction studied by LUNA are highlighted in green.

1 Solar Hydrogen Burning

Hydrogen burning in the Sun proceeds mainly by the proton-proton chain (pp-chain, Fig. 2), with a 0.8% contribution from the carbon-nitrogen-oxygen cycle (CN cycle) [12]. The basic processes are by now well understood, leading to the so-called standard solar model [13] that explains both helioseismological data and neutrino observations.

LUNA started its work studying the ^{3}He(^{3}He,2p)^{4}He reaction. This is a key reaction of the hydrgen burning proton proton chain and at the beginning of the seventies it was believed to exist a possible resonance in the ^{3}He(^{3}He,2p)^{4}He Gamow peak that could have explained the low measured neutrino flux measured by the Homestake experiment [14]. The experiment was performed at the LUNA1 accelerator by using a windowless gas target, filled with ^{3}He, and eight 1 mm thick silicon detectors of 5x5 cm^2 area. The cross section was measured in an energy range from 16.5 to 24.4 keV, totally covering the Gamow peak for the Sun [15, 16]. The debated resonance [17, 18] was not observed, ruling out the possible nuclear explanation of the so called Solar Neutrino Problem.

The neutrino fluxes emitted by the Sun are strictly correlated with other nuclear processes involved in the hydrogen burning. In particular, the ^{3}He(α,γ)^{7}Be reaction reaction controls the ^{7}Be and ^{8}B neutrinos and it is also fundamental for the production of ^{7}Li during the Big Bang Nucleosynthesis (BBN). The ^{3}He(α,γ)^{7}Be reaction was studied at the LUNA2 accelerator by using both the gamma prompt detection [19, 20] and the activation techniques [21, 22] finding a perfect agreement within the two methods. This result was important not only to reduce the systematics, but also to solve the discrepancy previously shown in experiments based on the two different techniques.

The ^{14}N(p,γ)^{15}O reaction is the bottleneck of the CN cycle and therefore the ^{13}N and ^{15}O neutrinos are controlled by this reaction. The LUNA collaboration started the study of this reaction by performing two different experiments: one characterised by solid target setup and HPGe detector. This kind of setup allows to study separately all branchings and to perform R-Matrix analysis [23, 24]. A second setup, made of a windowless gas target and a 4π-BGO detector, was used to measure the total cross section in a wide range of energy (down to 70 keV in the center of mass) thanks to the high efficiency provided by the BGO [25, 26], totally covering the nova explosion scenario with experimental data. In those experiments the cross section was found lower by a factor of 2 with respect to the reported value in NACRE [27] with considerably effects on the neutrino fluxes from the Sun and the age of globular clusters[28].

Recently, a discrepancy between helioseismology and the new metal abundances (i.e. the amount of elements different from hydrogen and helium) that emerged from improved modelling of the photosphere has been reported [29]. This so called Solar Composition Problem [12] can be addressed by the possibility to detect the CNO neutrino with new experiment and to parametrise with extreme precision the Solar models. For this reason the ^{14}N(p,γ)^{15}O cross section is requested to be known with a precision of 5%, almost factor of two lower than what reported in the last database [30]. In order to reduce this nuclear uncertainties in the solar model a new measurement was performed at LUNA by measuring three energy point above the 259 keV resonance and leading to and extrapolated value $S_{1,14}(0)$=1.57±0.13 keV barn [31, 32]. Another recent measurements has been performed at the Notre Dame University [33] at higher energies obtaining an extrapolated value higher than the one reported by LUNA. In this context a new measurement is requested in a wide energy range and it will be done as first measurement of the future LUNA MV accelerator, which will cover an energy range from 3 MeV down to 200 keV with an unique set of data, leading to precise extrapolation.

1.1 The other CNO cycles

The $^{15}N(p,\gamma)^{16}O$ is the link from the first to the second CNO cycle and it was studied intensively at the LUNA accelerator. Data acquired with a gas target setup and natural nitrogen (which contains a 0.37% of ^{15}N) to study the $^{14}N(p,\gamma)^{15}O$ reaction were analysed to determine the $^{14}N(p,\gamma)^{15}O$ cross section in the energy range from 230 keV down to 90 keV in the center of mass[34]. After that, two new experiments were performed by using ^{15}N enriched solid targets [35, 36]. The LUNA data cover totally the Gamow peak in nova explosion where the $^{15}N(p,\gamma)^{16}O$ is mainly important and the cross section was found to be lower than a factor of 2 with respect what reported in the NACRE database [27]. This leads to a reduction of ^{16}O produced by nova of a 40%.

The $^{17}O(p,\gamma)^{18}F$ reaction was investigated from 2011 to 2013. The ratio between the reaction rates of $^{17}O(p,\alpha)^{14}N$ (Q = 1.2 MeV) and $^{17}O(p,\gamma)^{18}F$ (Q = 5.6 MeV) channels is one of the most important parameters for the galactic synthesis of ^{17}O, the stellar production of radioactive ^{18}F, and for predicted O isotopic ratios in premolar grains [37, 38].

Since the ^{18}F is a radionuclides with a half life of ≈110 min, the cross section has been determined both by detecting the prompt gamma rays and by counting the 511 keV γs emitted by the ^{18}F decay. The results are in perfect agreement reducing considerably the systematics [39]. New results on the non resonant component and on the 183.3 keV resonance have been obtained ($\omega\gamma$ = (1.67 ± 0.12)μeV [39, 40]) reducing the uncertainties of isotopes production in nova explosion models to a 10% level [40]. The very low uncertainty obtained in this experiment was possible thank to an intensive study of the target, realised and tested directly by the LUNA group with ion beam analysis and secondary induced mass spectroscopy techniques [41].

An intense study on the (p,α) channel has been followed: a detection system composed by eight silicon detectors was used in this study to detect with high efficiency the α particles emitted. This experimental system is described in [42] and was used to measure directly the low energy resonance at 64.5 keV finding an $\omega\gamma = (8.7 \pm 1.2_{stat} \pm 0.6_{syst})$neV for the bare nuclei [43]. The LUNA value is a factor of 2 higher than what previously reported in literature [44, 45] bringing to important consequences in the AGB [46] and, in particular, in finding a possible explanation for the production sites of pre-solar grains enriched in ^{17}O [47].

1.2 The Ne-Na and Mg-Al cycles at LUNA

The nucleosynthesis in the neon sodium cycle occurs via many reactions. In particular, the uncertainty on the $^{22}Ne(p,\gamma)^{23}Na$ dominates the uncertainty of the entire cycle. As a matter of fact this reaction has several resonances below 400 keV and many of them were only given as upper limits based on indirect measurements. In particular, the databases [27, 48] reported a discrepancy of a factor one thousand. A fully shielded setup has been implemented [11] to observe for the first time three resonances and to improve the upper limits for other three resonances [49, 50]. Finally, an updated reaction rate has been obtained, which is significantly higher than the one given in the most recent compilation of reaction rates [48]. As a consequence, new values for the ejected mass of ^{22}Ne and ^{23}Na in thermally pulsing AGB stars have been obtained with much reduced uncertainties [51].

The study of the CNO cycles is the natural precursor for the hydrogen burning in Mg-Al cycles. The problem of the ^{26}Al production is one of the most interesting cases [52]. LUNA measured precisely several resonances for $^{24,25}Mg(p,\gamma)^{26,27}Al$ in order to reduce the uncertainties on those reactions as requested in the astrophysical models [53, 54]. The impact of the LUNA results is discussed in details in a recent work [55].

2 Big Bang Nucleosynthesis

The $^{3}He(\alpha,\gamma)^{7}Be$ reaction has an important role in solving the problem of the Spite plateau [56]. LUNA measured this reaction in the Gamow peak for Big Bang Nucleosynthesis reducing the uncertainties to 3% overall. Another problem concerning lithium isotopes as been raised recently: the ^{6}Li has been measured to be 3 order of magnitude higher than what expected in BBN [57]. While it is generally believed that ^{6}Li is produced by cosmic-ray spallation over the lifespan of the universe, the ^{6}Li observations in halo stars seem to suggest a primordial origin instead. For ^{6}Li production in the Big Bang, the main nuclear physics unknown is the $^{2}H(\alpha,\gamma)^{6}Li$ reaction rate. The setup used to study this reaction is already described in details [58]. The scattering chamber was filled with high purity deuterium gas. The accelerated deuterium particles, after Rutherford scattering with the α-beam, interact with the gas itself producing the two reactions: $^{2}H(d,n)^{3}He$ and $^{2}H(d,p)^{3}H$. In particular, the neutrons emitted by the $^{2}H(d,n)^{3}He$ were responsible for a high beam induced background in the γ-spectrum. A long and detailed study of this background was required in order to perform the analysis of the data [58–60].

3 Science case for a future higher-energy accelerator underground

Recent advances in astronomy and astrophysics require nuclear data at energies that are higher than the high-energy limit of LUNA2.

To explore the helium and carbon burnings a new MV accelerator will be installed at LNGS during the 2018.

With the new accelerator, a long list of reactions will be studied. A first 5 years program is detailed discussed in [61]. The first reaction will be the $^{14}N(p,\gamma)^{15}O$. The solar composition problem has been already addressed above and recently a new measurement has been performed at high energy founding discrepancies with the LUNA data [33]. With the new LUNA MV it will be possibile to cover with an unique set of data a wide energy range from 2 MeV down to the LUNA2 energies, allowing to reduce the systematics down to the goal of 5%.

Another important open issue of nuclear astrophysics is the neutron source reactions. In particular the $^{13}C(\alpha,n)^{16}O$ and $^{22}Ne(\alpha,n)^{25}Mg$ reactions. They are responsible for the production on neutrons involved in the slow neutron capture process, called the astrophysical s-process. Whereas those reactions are the subject of intensive experimental study [62, e.g.], so far the reactions actually producing the neutrons have not yet been measured in the relevant energy range since they should be addressed by an underground accelerator.

The $^{12}C+^{12}C$ is the trigger of the carbon burning and is a key reaction to predict the final stage of a star. It determines which is the mass required to a star to ignite this burning stage. In addition, this reaction also affects the outcome of type Ia supernovae. Both the two channels (proton and α) have never been studied down below 2 MeV. At the LUNA MV it will be possible to measure the cross section down to 1.7 MeV very close to cover the whole Gamow peak.

A third topic is to complement some of the proton- and α-capture reactions studied at the LUNA2 accelerator at higher energy. Such a continuation is particularly important for the Big Bang reactions $^{3}He(\alpha,\gamma)^{7}Be$ and $^{2}H(\alpha,\gamma)^{6}Li$, where the present LUNA2 400 kV machine can only cover the lower part of the relevant energy region.

4 Future underground accelerator facilities

Based on the successes of the LUNA collaboration, around the world several efforts are underway to install high-current, stable-beam accelerators in underground sites. LUNA MV project was started in order to install a 3.5 MV machine in the underground laboratories of Gran Sasso. The new accelerator has already been financed by the Italian government and it should be installed in the next year at the Gran Sasso Laboratory. The synergy between the existing LUNA2 and the new LUNA MV accelerator will allow to perform reaction studies in a wide range of energies with complete understanding of the setups involved. Other facilities have been proposed for installing an accelerator in deep underground laboratory in US [63] and in China [64]. Another project is under discussion at the Canfranc underground laboratory in the Pyrenees, Spain. As part of a staged approach, even an accelerator laboratory in a shallow-underground facility such as Felsenkeller (Dresden, Germany) is under consideration.

At present, the existing 400 kV LUNA2 machine continues its scientific program outlined in [65]. The next few years will show where this highly successful approach will eventually be complemented by one or more higher-energy accelerators underground. The technique is sufficiently mature to address not only the data needs of the astrophysics community, but it has the potential to benefit also the astroparticle and other communities.

Acknowledgments

The work at LUNA is carried out in the framework of the LUNA collaboration; see e.g. Ref. [43] for an author list.

References

[1] C. Iliadis, *Nuclear Physics of Stars* (Wiley-VCH, Weinheim, 2007)
[2] C. Rolfs, W. Rodney, *Cauldrons in the Cosmos* (University of Chicago Press, Chicago, 1988)
[3] Best, A., Caciolli, A., F|"ulüp, Zs., Gy|"orky, Gy., Laubenstein, M., Napolitani, E., Rigato, V., Roca, V., Sz|"ucs, T., Eur. Phys. J. A **52**, 72 (2016)
[4] H. Costantini, A. Formicola, G. Imbriani, M. Junker, C. Rolfs, F. Strieder, Reports on Progress in Physics **72**, 086301 (2009)
[5] C. Broggini, D. Bemmerer, A. Guglielmetti, R. Menegazzo, Annu. Rev. Nucl. Part. Sci. **60**, 53 (2010)
[6] U. Greife, C. Arpesella, C. Barnes, F. Bartolucci, E. Bellotti, C. Broggini, P. Corvisiero, G. Fiorentini, A. Fubini, G. Gervino et al., Nucl. Inst. Meth. A **350**, 327 (1994)
[7] A. Formicola, G. Imbriani, M. Junker, D. Bemmerer, R. Bonetti, C. Broggini, C. Casella, P. Corvisiero, H. Costantini, G. Gervino et al., Nucl. Inst. Meth. A **507**, 609 (2003)
[8] D. Bemmerer, F. Confortola, A. Lemut, R. Bonetti, C. Broggini, P. Corvisiero, H. Costantini, J. Cruz, A. Formicola, Z. Fülöp et al., Eur. Phys. J. A **24**, 313 (2005)
[9] T. Szücs, D. Bemmerer, C. Broggini, A. Caciolli, F. Confortola, P. Corvisiero, Z. Elekes, A. Formicola, Z. Fülöp, G. Gervino et al., Eur. Phys. J. A **44**, 513 (2010)
[10] A. Caciolli, L. Agostino, D. Bemmerer, R. Bonetti, C. Broggini, F. Confortola, P. Corvisiero, H. Costantini, Z. Elekes, A. Formicola et al., Eur. Phys. J. A **39**, 179 (2009)
[11] F. Cavanna, R. Depalo, M.L. Menzel, M. Aliotta, M. Anders, D. Bemmerer, C. Broggini, C.G. Bruno, A. Caciolli, P. Corvisiero et al., Eur. Phys. J. A **50**, 179 (2014), 1411.2888
[12] A.M. Serenelli, S. Basu, J.W. Ferguson, M. Asplund, Astrophys. J. Lett. **705**, L123 (2009), 0909.2668
[13] J.N. Bahcall, M.H. Pinsonneault, Phys. Rev. Lett. **92**, 121301 (2004)
[14] R. Davis, D.S. Harmer, K.C. Hoffman, Phys. Rev. Lett. **20**, 1205 (1968)

[15] M. Junker, A. D'alessandro, S. Zavatarelli, C. Arpesella, E. Bellotti, C. Broggini, P. Corvisiero, G. Fiorentini, A. Fubini, G. Gervino et al., Phys. Rev. C **57**, 2700 (1998), nucl-ex/9707003
[16] R. Bonetti, C. Broggini, L. Campajola, P. Corvisiero, A. D'Alessandro, M. Dessalvi, A. D'Onofrio, A. Fubini, G. Gervino, L. Gialanella et al., Phys. Rev. Lett. **82**, 5205 (1999), arXiv:nucl-ex/9902004
[17] W.A. Fowler, Nature **238**, 24 (1972)
[18] V.N. Fetisov, Y.S. Kopysov, Phys. Lett. B **40**, 602 (1972)
[19] D. Bemmerer, F. Confortola, H. Costantini, A. Formicola, G. Gyürky, R. Bonetti, C. Broggini, P. Corvisiero, Z. Elekes, Z. Fülöp et al. (LUNA Collaboration), Phys. Rev. Lett. **97**, 122502 (2006)
[20] G. Gyürky, F. Confortola, H. Costantini, A. Formicola, D. Bemmerer, R. Bonetti, C. Broggini, P. Corvisiero, Z. Elekes, Z. Fülöp et al. (LUNA Collaboration), Phys. Rev. C **75**, 035805 (2007)
[21] F. Confortola, D. Bemmerer, H. Costantini, A. Formicola, G. Gyürky, P. Bezzon, R. Bonetti, C. Broggini, P. Corvisiero, Z. Elekes et al. (LUNA Collaboration), Phys. Rev. C **75**, 065803 (2007)
[22] H. Costantini, D. Bemmerer, F. Confortola, A. Formicola, G. Gyürky, P. Bezzon, R. Bonetti, C. Broggini, P. Corvisiero, Z. Elekes et al., Nucl. Phys. A **814**, 144 (2008), 0809.5269
[23] A. Formicola, G. Imbriani, H. Costantini, C. Angulo, D. Bemmerer, R. Bonetti, C. Broggini, P. Corvisiero, J. Cruz, P. Descouvemont et al., Phys. Lett. B **591**, 61 (2004)
[24] G. Imbriani, H. Costantini, A. Formicola, A. Vomiero, C. Angulo, D. Bemmerer, R. Bonetti, C. Broggini, F. Confortola, P. Corvisiero et al., Eur. Phys. J. A **25**, 455 (2005)
[25] A. Lemut, D. Bemmerer, F. Confortola, R. Bonetti, C. Broggini, P. Corvisiero, H. Costantini, J. Cruz, A. Formicola, Z. Fülöp et al., Phys. Lett. B **634**, 483 (2006), nucl-ex/0602012
[26] D. Bemmerer, F. Confortola, A. Lemut, R. Bonetti, C. Broggini, P. Corvisiero, H. Costantini, J. Cruz, A. Formicola, Z. Fülöp et al., Nucl. Phys. A **779**, 297 (2006)
[27] C. Angulo, M. Arnould, M. Rayet, P. Descouvemont, D. Baye, C. Leclercq-Willain, A. Coc, S. Barhoumi, P. Aguer, C. Rolfs et al., Nucl. Phys. A **656**, 3 (1999)
[28] Imbriani, G., Costantini, H., Formicola, A., Bemmerer, D., Bonetti, R., Broggini, C., iero, P. Corvi, Cruz, J., Fülop, Z., Gervino, G. et al., A&A **420**, 625 (2004)
[29] C. Peña-Garay, A. Serenelli, ArXiv e-prints (2008), 0811.2424
[30] E. Adelberger, A. García, R.G.H. Robertson, K.A. Snover, A.B. Balantekin, K. Heeger, M.J. Ramsey-Musolf, D. Bemmerer, A. Junghans, C.A. Bertulani et al., Rev. Mod. Phys. **83**, 195 (2011)
[31] M. Marta, A. Formicola, G. Gyürky, D. Bemmerer, C. Broggini, A. Caciolli, P. Corvisiero, H. Costantini, Z. Elekes, Z. Fülöp et al., Phys. Rev. C **78**, 022802(R) (2008)
[32] M. Marta, A. Formicola, D. Bemmerer, C. Broggini, A. Caciolli, P. Corvisiero, H. Costantini, Z. Elekes, Z. Fülöp, G. Gervino et al., Phys. Rev. C **83**, 045804 (2011), 1103.5393
[33] Q. Li, J. Görres, R.J. deBoer, G. Imbriani, A. Best, A. Kontos, P.J. LeBlanc, E. Uberseder, M. Wiescher, Phys. Rev. C **93**, 055806 (2016)
[34] D. Bemmerer, A. Caciolli, R. Bonetti, C. Broggini, F. Confortola, P. Corvisiero, H. Costantini, Z. Elekes, A. Formicola, Z. Fülöp et al., J. Phys. G **36**, 045202 (2009), 0902.0783
[35] P.J. LeBlanc, G. Imbriani, J. Görres, M. Junker, R. Azuma, M. Beard, D. Bemmerer, A. Best, C. Broggini, A. Caciolli et al., Phys. Rev. C **82**, 055804 (2010), 1011.2518
[36] A. Caciolli, C. Mazzocchi, V. Capogrosso, D. Bemmerer, C. Broggini, P. Corvisiero, H. Costantini, Z. Elekes, A. Formicola, Z. Fülöp et al., Astron. Astrophys. **533**, A66 (2011), 1107.4514
[37] C. Fox et al., Phys. Rev. C **71**, 055801 (2005)
[38] J.R. Newton, C. Iliadis, A.E. Champagne, J.M. Cesaratto, S. Daigle, R. Longland, Phys. Rev. C **81**, 045801 (2010)
[39] D.A. Scott, A. Caciolli, A. Di Leva, A. Formicola, M. Aliotta, M. Anders, D. Bemmerer, C. Broggini, M. Campeggio, P. Corvisiero et al., Phys. Rev. Lett. **109**, 202501 (2012), 1210.6529
[40] A. Di Leva, D.A. Scott, A. Caciolli, A. Formicola, F. Strieder, M. Aliotta, M. Anders, D. Bemmerer, C. Broggini, P. Corvisiero et al., Phys. Rev. C **89**, 015803 (2014)
[41] A. Caciolli, D.A. Scott, A. Di Leva, A. Formicola, M. Aliotta, M. Anders, A. Bellini, D. Bemmerer, C. Broggini, M. Campeggio et al., Eur. Phys. J. A **48**, 144 (2012), 1210.0327
[42] C.G. Bruno, D.A. Scott, A. Formicola, M. Aliotta, T. Davinson, M. Anders, A. Best, D. Bemmerer, C. Broggini, A. Caciolli et al., Eur. Phys. J. A **51**, 94 (2015)
[43] C.G. Bruno, D.A. Scott, M. Aliotta, A. Formicola, A. Best, A. Boeltzig, D. Bemmerer, C. Broggini, A. Caciolli, F. Cavanna et al., Phys. Rev. Lett. **117**, 142502 (2016), 1610.00483
[44] J. Blackmon et al., Phys. Rev. Lett. **74**, 2642 (1995)
[45] M.L. Sergi, C. Spitaleri, M. La Cognata, A. Coc, A. Mukhamedzhanov, S.V. Burjan, S. Cherubini, V. Crucillá, M. Gulino, F. Hammache et al., Phys. Rev. C **82**, 032801 (2010)
[46] Straniero, O., Bruno, C. G., Aliotta, M., Best, A., Boeltzig, A., Bemmerer, D., Broggini, C., Caciolli, A., Cavanna, F., Ciani, G. F. et al., A&A **598**, A128 (2017)
[47] M. Lugaro, A.I. Karakas, C.G. Bruno, M. Aliotta, L.R. Nittler, D. Bemmerer, A. Best, A. Boeltzig, C. Broggini, A. Caciolli et al., Nature Astronomy **1**, 0027 (2017)

[48] C. Iliadis, R. Longland, A.E. Champagne, A. Coc, R. Fitzgerald, Nucl. Phys. A **841**, 31 (2010), 1004.4517

[49] F. Cavanna, R. Depalo, M. Aliotta, M. Anders, D. Bemmerer, A. Best, A. Boeltzig, C. Broggini, C.G. Bruno, A. Caciolli et al., Phys. Rev. Lett. **115**, 252501 (2015), 1511.05329

[50] R. Depalo, F. Cavanna, M. Aliotta, M. Anders, D. Bemmerer, A. Best, A. Boeltzig, C. Broggini, C.G. Bruno, A. Caciolli et al. (LUNA Collaboration), Phys. Rev. C **94**, 055804 (2016)

[51] A. Slemer, P. Marigo, D. Piatti, M. Aliotta, D. Bemmerer, A. Best, A. Boeltzig, A. Bressan, C. Broggini, C.G. Bruno et al., Monthly Notices of the Royal Astronomical Society **465**, 4817 (2017)

[52] C. Iliadis, A. Champagne, A. Chieffi, M. Limongi, The Astrophysical Journal Supplement Series **193**, 16 (2011)

[53] B. Limata et al., Phys. Rev. C **82**, 015801 (2010)

[54] F. Strieder, B. Limata, A. Formicola, G. Imbriani, M. Junker, D. Bemmerer, A. Best, C. Broggini, A. Caciolli, P. Corvisiero et al., Phys. Lett. B **707**, 60 (2012)

[55] O. Straniero, G. Imbriani, F. Strieder, D. Bemmerer, C. Broggini, A. Caciolli, P. Corvisiero, H. Costantini, S. Cristallo, A. DiLeva et al., Astrophys. J. **763**, 100 (2013), 1211.6661

[56] M. Spite, F. Spite, Nature **297**, 483 (1982)

[57] M. Asplund, D.L. Lambert, P.E. Nissen, F. Primas, V.V. Smith, Astrophys. J. **644**, 229 (2006)

[58] M. Anders, D. Trezzi, A. Bellini, M. Aliotta, D. Bemmerer, C. Broggini, A. Caciolli, H. Costantini, P. Corvisiero, T. Davinson et al., Eur. Phys. J. A **49**, 28 (2013), 1301.7185

[59] M. Anders, D. Trezzi, R. Menegazzo, M. Aliotta, A. Bellini, D. Bemmerer, C. Broggini, A. Caciolli, P. Corvisiero, H. Costantini et al., Phys. Rev. Lett. **113**, 042501 (2014)

[60] D. Trezzi, M. Anders, M. Aliotta, A. Bellini, D. Bemmerer, A. Boeltzig, C. Broggini, C.G. Bruno, A. Caciolli, F. Cavanna et al., Astropart. Phys. **89**, 57 (2017)

[61] *LUNA MV Scientific Proposal*, https://luna.lngs.infn.it/images/LUNA-MV-5y-proposal.pdf

[62] M. Heil, R. Detwiler, R.E. Azuma, A. Couture, J. Daly, J. Görres, F. Käppeler, R. Reifarth, P. Tischhauser, C. Ugalde et al., Phys. Rev. C **78**, 025803 (2008)

[63] D. Robertson, M. Couder, U. Greife, F. Strieder, M. Wiescher, *Underground nuclear astrophysics studies with CASPAR*, in *European Physical Journal Web of Conferences* (2016), Vol. 109 of *European Physical Journal Web of Conferences*, p. 09002

[64] W. Liu, Z. Li, J. He, X. Tang, G. Lian, Z. An, J. Chang, H. Chen, Q. Chen, X. Chen et al., *Progress of Jinping Underground laboratory for Nuclear Astrophysics (JUNA)*, in *European Physical Journal Web of Conferences* (2016), Vol. 109 of *European Physical Journal Web of Conferences*, p. 09001

[65] *LUNA 400kV 2015 - 2018 Proposal*, https://luna.lngs.infn.it/images/proposal2008.pdf

 DOI: 10.1051/epjconf/201716300010

Sub-barrier fusion of Si+Si systems

G. Colucci[1,⋆], *G. Montagnoli*[1], *A. M. Stefanini*[2], *D. Bourgin*[4], *P. Čolović*[5], *L. Corradi*[2], *S. Courtin*[4], *M. Faggian*[1], *E. Fioretto*[3], *F. Galtarossa*[2,6], *A. Goasduff*[2], *F. Haas*[4], *M. Mazzocco*[1], *F. Scarlassara*[1], *C. Stefanini*[1], *E. Strano*[1], *M. Urbani*[1], *S. Szilner*[5], and *G. L. Zhang*[2]

[1]*Dipartimento di Fisica e Astronomia, Universitá di Padova, and INFN, I-35131 Padova, Italy*
[2]*INFN, Laboratori Nazionali di Legnaro, I-35020 Legnaro, Italy*
[3]*IPHC, CNRS-IN2P3, Université de Strasbourg, F-67037 Strasbourg Cedex 2, France*
[4]*Ruder Bošković Institute, HR-10002 Zagreb, Croatia*
[5]*Dipartimento di Fisica e Scienze della Terra, Univ. di Ferrara, Ferrara, Italy*

Abstract. The near- and sub-barrier fusion excitation function has been measured for the system ^{30}Si+^{30}Si at the Laboratori Nazionali di Legnaro of INFN, using the ^{30}Si beam of the XTU Tandem accelerator in the energy range 47 - 90 MeV. A set-up based on a beam electrostatic deflector was used for detecting fusion evaporation residues. The measured cross sections have been compared to previous data on ^{28}Si+^{28}Si and Coupled Channels (CC) calculations have been performed using M3Y+repulsion and Woods-Saxon potentials, where the low-lying 2^+ and 3^- excitations have been included. A weak imaginary potential was found to be necessary to reproduce the low energy ^{28}Si+^{28}Si data. This probably simulates the effect of the oblate deformation of this nucleus. On the contrary, ^{30}Si is a spherical nucleus, ^{30}Si+^{30}Si is nicely fit by CC calculations and no imaginary potential is needed. For this system, no maximum shows up for the astrophysical S-factor so that we have no evidence for hindrance, as confirmed by the comparison with CC calculations. The logarithmic derivative of the two symmetric systems highlights their different low energy trend. A difference can also be noted in the two barrier distributions, where the high-energy peak present in ^{28}Si+^{28}Si is not observed for ^{30}Si+^{30}Si, probably due to the weaker couplings in last case.

1 Introduction

It is well known that nuclear structure strongly influences fusion reaction dynamics at energies near and below the Coulomb barrier. The comparison of data for neighboring isotopes is a sensitive tool to evidence this influence.
Through the application of this method it has been possible to discover the influence of transfer channels on fusion reactions, as in Ni + Ni [1, 2] and Ca +Ca [3] systems, and the effect of changing structure from spherical to strongly deformed nuclei, as in ^{16}O + ^{148,154}Sm [4, 5].
A similar comparative study has been performed also for the Si+Si systems [6], where the interest originates from the different shape of the silicon isotopes: ^{30}Si is nearly spherical, whereas ^{28}Si is strongly deformed with an oblate shape. Previous studies [6, 7] revealed the influence of transfer on fusion of the asymmetric system ^{28}Si+^{30}Si which has been explained by considering one- and successive two-neutron transfer channels in the coupling scheme.
The excitation functions of ^{28}Si+^{28}Si and ^{30}Si+^{30}Si have different trends (see Fig. 1a).The case of ^{28}Si+^{28}Si involving deformed nuclei shows an unusual behaviour, where the cross section is hindered just below the barrier and then enhanced at lower energies, as shown in the comparison with the CC calculations of Fig. 1b. It was further surprising that the low-energy data were well reproduced only by artificially applying a weak, short-ranged imaginary potential, probably simulating the effect of the oblate deformation. This feature has to be further investigated and a complete study of ^{30}Si+^{30}Si appeared to be very important in this respect.
The spherical shape of ^{30}Si together with the absence of transfer channels with positive Q-values allows to only consider couplings to the vibrational states. The lack of experimental fusion data below 4 mb for ^{30}Si+^{30}Si prevented a meaningful comparison between the two systems. Therefore, a fusion experiment has been recently carried out at Laboratori Nazionali di Legnaro (LNL) with the purpose to extend the data of ^{30}Si+^{30}Si down to energies deeply below the Coulomb barrier.
In this contribution we present the results of this full measurement from well below to well above the Coulomb barrier, and we perform a comparison with the existing data for the ^{28}Si+^{28}Si system.

2 Experimental set-up

Fusion cross sections for the ^{30}Si+^{30}Si system have been determined by direct detection of the evaporation residues (ER) at very forward angles by exploiting their different

⋆e-mail: giulia.colucci@studenti.unipd.it

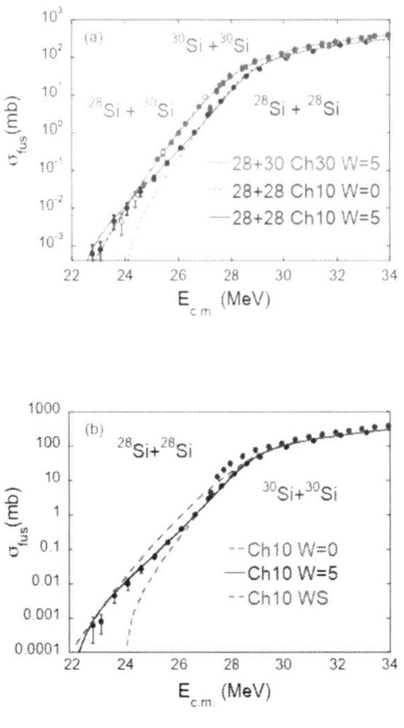

Figure 1: (a) Measured fusion excitation function of ^{28}Si+^{28}Si, ^{28}Si+^{30}Si [6] and previous data for ^{30}Si+^{30}Si [8]; the lines are the results of CC calculations. (b) Comparison of ^{28}Si+^{28}Si with CC calculations involving a Woods-Saxon potential (Ch10 WS) and a double folding M3Y+rep. potential with (Ch10 w5) and without (Ch10 w0) an imaginary short ranged potential.

electrical rigidity with respect to beam and beam-like particles, using an electrostatic deflector. This experimental set-up is shown in Fig. 2, it allows fast and reliable measurements of relative and absolute cross sections and has been employed in several sub- and near-barrier fusion experiments at LNL in recent years. A subsequent E-TOF-ΔE telescope allows the ER identification. This telescope consists of two micro-channel plate (MCP) detectors, a transverse field ionization chamber (IC) and a silicon detector placed in the same gas (CH_4) volume. The IC provides an energy loss signal ΔE, while the two MCPs yield the time of flights (TOF) together with the silicon detector. In this configuration the silicon detector measures the residual energy of the ER and gives both the trigger for data acquisition and the start signal for the TOF. Typical spectra of TOF as a function of the residual energy E are shown in Fig. 3, where a good separation of ER events from the residual beam can be appreciated at energies both above and below the Coulomb barrier. Analogous two-dimensional spectra of TOF vs. ΔE and ΔE vs. E were used in the data analysis.

Four silicon detectors were placed symmetrically at θ_{lab}=16° around the beam direction, in the sliding seal reaction chamber. These detectors were used to monitor the beam and normalize to the Mott scattering cross section. The XTU Tandem accelerator at LNL provided ^{30}Si beams in the energy range of 47-90 MeV, with intensities of 15-30 pnA. The targets consisted of 50 μg/cm^2 metallic ^{30}Si evaporated on 30 μg/cm^2 carbon backings facing the beam. The isotopic enrichment of ^{30}Si was 99.64 %, where the small amounts of ^{29}Si and ^{28}Si did not affect the fusion cross section because of their higher Coulomb barriers. The carbon backing and the silicon target itself introduced an average beam energy loss of around 750-850 keV, which was taken into account in the analysis.
Three ER angular distributions were measured at the energies of 58, 72 and 80 MeV in the range from -6° to +9°. The total fusion cross section was derived by integrating those distributions, and by simple interpolations or extrapolations for all other energies, where measurements were taken only at 2° (or 3° for low energies).
The measured energy range allowed to extend the excitation function down to $\simeq$4 μb, as shown Fig. 4a. The reported errors are statistical uncertainties, which are 1-2 % at the higher energies and increase to 10-20 % at energies below the barrier. The systematic errors on the absolute cross sections are estimated ± 7-8 %, due to the geometrical solid angle uncertainties, angular distribution integrations and mainly to the deflector transmission.
The fusion cross sections of ^{30}Si+^{30}Si and ^{28}Si+^{28}Si are reported in Fig. 4b. From this first comparison it appears that the two excitation functions behave differently at low energies with a flatter slope for ^{28}Si+^{28}Si. A CC analysis has been performed to understand the origin of this difference.

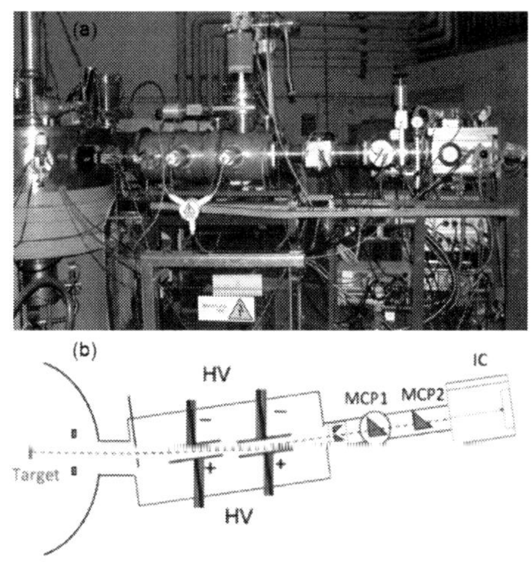

Figure 2: The experimental set-up (a) and its scheme (b). The reaction chamber, the electrostatic deflector and the telescope E-TOF-ΔE are drawn from the left.

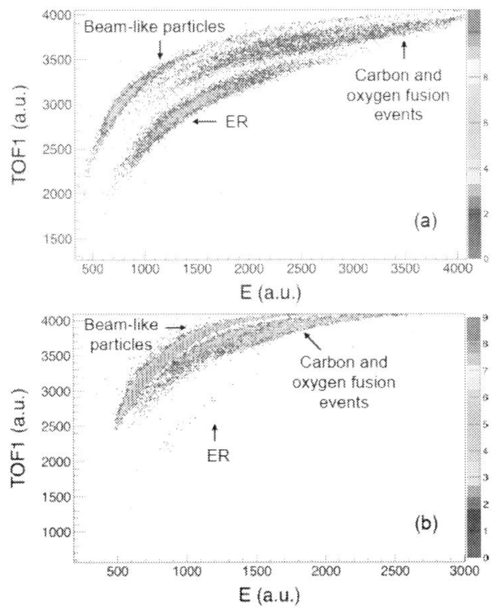

Figure 3: Time of flight TOF1 (ordinate) as a function of the residual energy E (abscissa). The spectra were obtained above the Coulomb barrier (80 MeV, σ_{fus}=527 mb, (a)) and at a sub-barrier energy of 50 MeV (σ_{fus}=115 μb, (b)). We can see, besides fusion events from ^{30}Si+^{30}Si and beam-like particles, the residues produced by the fusion of ^{30}Si with the carbon and oxygen present in the target.

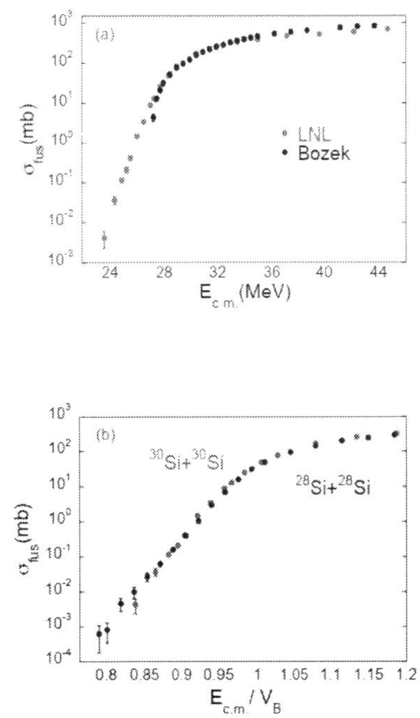

Figure 4: Fusion excitation function of ^{30}Si+^{30}Si compared to previous experimental data [8] (a) and to the ^{28}Si+^{28}Si system (b).

3 Coupled channel calculations

The CC analysis of the sub-barrier excitation function employed, in a first step, a double-folding M3Y+repulsion potential [9]. The calculation makes use of the isocentrifugal approximation and includes one- and two-phonon vibrational modes as well as mutual excitations of the low-lying 2^+ and 3^- states in both projectile and target (this calculation is called Ch10 in Fig. 5). The adopted coupling strengths are reported in Table I of Ref. [6].

The ^{30}Si nucleus is nearly spherical, indeed the measured quadrupole moment of the 2^+ state is Q_2 = −0.05(6) b, which is consistent with zero. As a consequence, the location of minimum of channels potentials is essentially the same in all channels, as discussed in detail in Ref. [6]. As anticipated in Sec. 2, the two excitation functions of ^{28}Si+^{28}Si and ^{30}Si+^{30}Si have different trends below the barrier. This difference was suggested to be due to the different shape of the two silicon isotopes. The effect of the oblate deformation of ^{28}Si was simulated in Ref. [6] by applying a weak imaginary potential at small ion-ion distances (inside the Coulomb barrier), and this allowed to obtain a good data fit. On the contrary, it appears that in the case of ^{30}Si+^{30}Si no imaginary potential is needed below the barrier, simply because that nucleus is spherical. This is shown in Fig. 5a where the results of CC calculations without and with the imaginary potential are reported. It is evident that there is essentially no difference between the two calculations.

4 Barrier distributions

Fusion barrier distributions [10] often give a significant insight into the role of the target and projectile structure. Therefore the barrier distribution of ^{30}Si+^{30}Si has been extracted and compared to the one of ^{28}Si+^{28}Si in order to obtain more information on the possible role of the deformation of ^{28}Si.

The barrier distributions were obtained using the three-point difference formula [11], with two energy steps of 1.0 MeV and 1.5 MeV, and a meaningful difference between the two systems has been observed. This is shown in Fig. 6, where ^{30}Si+^{30}Si has a single well-defined peak, while a double-peak structure appears for ^{28}Si+^{28}Si with a clearly defined high-energy peak. These dissimilar behaviours arise from the different structures of the two nuclei and from the stronger couplings present in ^{28}Si+^{28}Si due to the oblate shape of ^{28}Si.

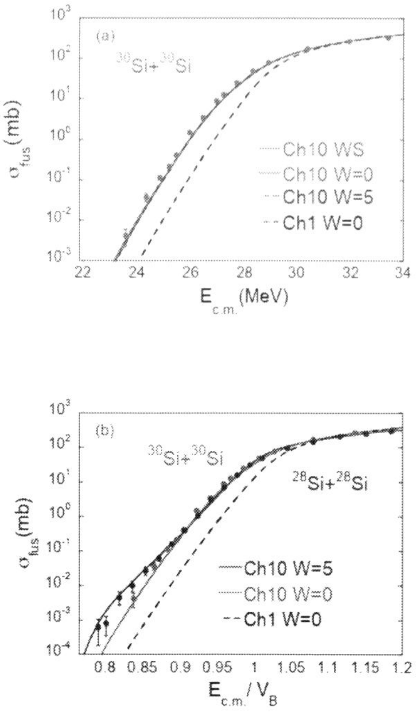

Figure 5: (a) Measured fusion cross section for ^{30}Si+^{30}Si compared to the no-coupling limit and to the CC calculations, based on three different potentials: a Woods-Saxon potential (Ch10 WS), and a M3Y+rep. potential with (Ch10 W5) and without (Ch10 W0) an imaginary potential. (b) Comparison of ^{30}Si+^{30}Si with ^{28}Si+^{28}Si. CC Calculations for the second system [6] are also reported, based on a M3Y+rep. potential with an imaginary part.

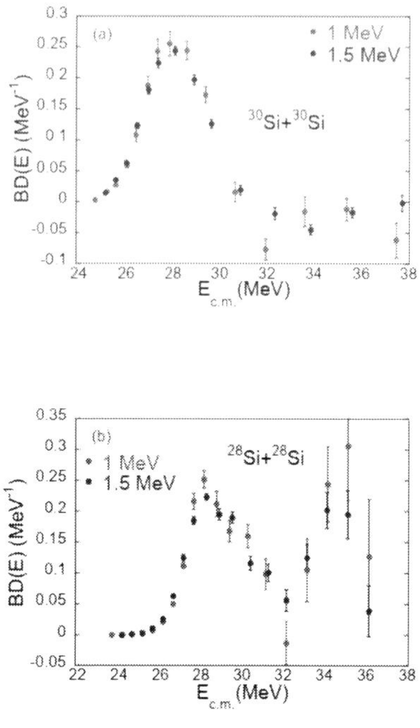

Figure 6: Barrier distributions extracted for ^{30}Si+^{30}Si (a) and ^{28}Si+^{28}Si (b), with energy steps of 1.0 MeV (red points) and 1.5 MeV (blue points).

5 Hindrance

Hindrance is a suppression of fusion cross section with respect to standard CC calculations, occurring far below the Coulomb barrier [12, 13]. The investigation of hindrance usually makes use of two model-independent representations of the excitation functions which allow to reveal the presence of the phenomenon: the astrophysical S-factor [14] and the logarithmic derivative L(E). Indeed, the S factor shows a maximum vs. energy in the presence of hindrance at the energy where the logarithmic slope L(E) of the excitation function reaches the value L_{CS} expected for a constant astrophysical S factor.
In the case of ^{30}Si+^{30}Si the S factor does not show any maximum vs. energy (see Fig. 7b), even if it appears to saturate at the lowest measured energies and the logarithmic derivative does not cross the L_{CS} curve (Fig. 7a). These features suggested the absence of hindrance in the measured energy range, but a confirmation was necessary. Indeed, it has been pointed out that in systems with a positive Q-value for fusion the S factor does not necessarily develop a maximum [13]. A comparison with the theoretical model was then performed by CC calculations with a standard Woods-Saxon potential. These calculations nicely fit the experimental excitation function, as can be seen in Fig. 5a. The analogous calculations using the M3Y+rep. potential give very similar results. It is evident that the excitation function is not suppressed at energies below the Coulomb barrier, so that there is no evidence of hindrance down to around 4 μb.

6 Summary

The fusion excitation function of ^{30}Si+^{30}Si has been measured in a wide energy range down to 4 μb. The comparison with ^{28}Si+^{28}Si shows that the two systems behave differently near the barrier where the high energy peak observed in the barrier distribution of ^{28}Si+^{28}Si is not found for ^{30}Si+^{30}Si.
Below the barrier the heavier system exhibits a regular trend in contrast with the unusual behaviour of ^{28}Si+^{28}Si. The weaker couplings in ^{30}Si+^{30}Si and its spherical shape could be responsible for this difference. This is confirmed by CC calculations using the M3Y+rep. and a

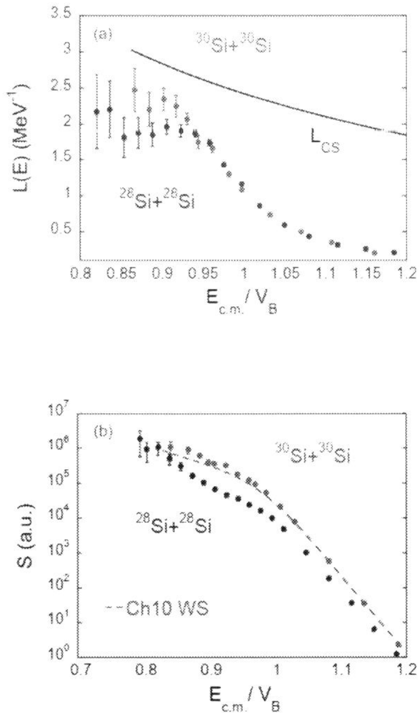

Figure 7: (a) Logarithmic derivative of the excitation function of the two systems, where L_{CS} (black line) is the slope expected for a constant astrophysical S-factor. (b) S-factor extracted from experimental data for both cases and, for ^{30}Si+^{30}Si, from CC calculations based on a Woods - Saxon potential (green line), well reproducing the data.

Woods-Saxon potential, that are in good agreement with experimental data. At variance with the case of ^{28}Si+^{28}Si, adding a weak short-ranged imaginary potential is not necessary to reproduce the sub-barrier fusion cross sections of ^{30}Si+^{30}Si. When observing the logarithmic slope of the excitation function and the S-factor, there is no evidence for hindrance in the measured energy range for this system.

7 Acknowledgments

The collaboration with H.Esbensen who performed the CC calculations with the M3Y+rep. potential is gratefully acknowledged. This work has been supported in the part by the Croatian Science Foundation under the project 7194.

References

[1] M. Beckerman, M. Salomaa, A. Sperduto, J. D. Molitoris, and A. Di Rienzo, Phys. Rev. C **25**, 837 (1982).
[2] M. Beckerman et al., Phys. Rev. Lett. **45**, 1472 (1980).
[3] G. Montagnoli et al., Phys. Rev. C **85**, 024607 (2012).
[4] R. Stokstad et al., Z. Phys. A **295**, 269 (1980).
[5] H. Esbensen, Nucl. Phys. A **352**, 147 (1982).
[6] G. Montagnoli et al., Phys. Rev. C **90**, 044608 (2014).
[7] C. L. Jiang, et al., Phys. Rev. C **78**, 017601 (2008).
[8] E. Bozek et al., Nucl. Phys. A **451**, 171 (1986) .
[9] Ş. Mişicu and H. Esbensen, Phys. Rev. C **75**, 034606 (2007).
[10] N. Rowley, G. R. Satchler, and P. H. Stelson, Phys. Lett. B **254**, 25 (1991).
[11] H. Timmers et al., Nucl. Phys. A**633**, 421 (1998).
[12] C. L. Jiang, et al., 2002, Phys. Rev. Lett. **89**, 052701(2002).
[13] C. L. Jiang, H. Esbensen, B. B. Back, R. V. F. Janssens, and K. E. Rehm, Phys. Rev. C **69**, 014604 (2004).
[14] E. M. Burbidge, G. Burbidge, W. Fowler, and F. Hoyle, Rev. Mod. Phys. **29**, 547 (1957)

How well do we understand the reaction rate of C burning?

S. Courtin[1,2,3,*], C.L. Jiang[4], G. Fruet[1,2], K. Auranen[4], M.L. Avila[4], A.D. Ayangeakaa[4], B.B. Back[4], S. Bottoni[4], M. Carpenter[4], C. Dickerson[4], B. DiGiovine[4], J.P. Greene[2], D.J. Henderson[4], C.R. Hoffman[4], R.V.F. Janssens[4], B.P. Kay[4], S.A. Kuvin[4], T. Lauritsen[4], R.C. Pardo[4], K.E. Rehm[4], D. Santiago-Gonzalez[4], J. Sethi[4], D. Seweryniak[4], R. Talwar[4], C. Ugalde[4], S. Zhu[4], C.M. Deibel[5], S.T. Marley[5], D. Bourgin[1,2], F. Haas[1,2], M. Heine[1,2], D. Montanari[1,2,3], D.G. Jenkins[6], L.G. Morris[6], A. Lefebvre-Schuhl[7], S. Almaraz-Calderon[8], X. Fang[9], X.D. Tang[10], M. Alcorta[11], B. Bucher[12], M. Albers[4], and P. Bertone[4]

[1] *IPHC, Université de Strasbourg, F-67037 Strasbourg, France*
[2] *CNRS, UMR7178, F-67037 Strasbourg, France*
[3] *USIAS, F-67083 Strasbourg, France*
[4] *Physics Division, Argonne National Laboratory, Argonne, IL 60439, USA*
[5] *Department of Physics and Astronomy, Louisiana State University, Baton Rouge, LA70803, USA*
[6] *Department of Physics, University of York, Heslington, York YO10 5DD, UK*
[7] *CSNSM IN2P3-CNRS and University of Paris Sud, F-91405 Orsay, France*
[8] *Department of Physics, Florida State University, Tallahassee, FL 32306, USA*
[9] *University of Notre Dame, Notre Dame, IN 46556, USA*
[10] *Institute of Modern Physics, Lanzhou, China*
[11] *TRIUMF, Vancouver, BC V6T 2A3, Canada*
[12] *Lawrence Livermore National laboratory, Livermore, CA 94551, USA*

Abstract. Carbon burning plays a crucial role in stellar evolution, where this reaction is an important route for the production of heavier elements. A particle-γ coincidence technique that minimizes the backgrounds to which this reaction is subject and provides reliable cross sections has been used at the Argonne National Laboratory to measure fusion cross-sections at deep sub-barrier energies in the $^{12}C+^{12}C$ system. The corresponding excitation function has been extracted down to a cross section of about 6 nb. This indicates the existence of a broad S-factor maximum for this system. Experimental results are presented and discussed.

1 Introduction

Reaction rates for C burning are essential ingredients to understand the production of chemical elements heavier than carbon as well as the evolution of massive stars. Carbon burning processes determine whether a star will join to the heavy-ion burning branches following hydrogen and helium burning and if white dwarfs will evolve into type Ia supernovae. It is thus of very high importance to know the $^{12}C+^{12}C$ fusion cross section with good accuracy from the Coulomb barrier (CB) down to the Gamow window which is centered around $E_G = 1.5 \pm 0.3$ MeV at a temperature of $T = 5x10^8$ K [1]. In a stellar environments, C burning occurs essentially via the $^{12}C+^{12}C$ fusion reaction. The exit channels for this reaction are : $^{12}C(^{12}C,\alpha)^{20}Ne$, $^{12}C(^{12}C,p)^{23}Na$ and $^{12}C(^{12}C,n)^{23}Mg$. The associated Q-values are 4.62 MeV, 2.24 MeV and -2.62 MeV MeV respectively. The ^{23}Mg channel with negative Q-value is essentially closed at deep sub-barrier energies.

One of the most striking results obtained in the early studies of heavy-ion collisions is the observation of resonant structures in the reaction cross-sections : i.e. elastic, inelastic and fusion channels of some light heavy-ion systems. These structures were found especially strong in the fusion cross section of the $^{12}C+^{12}C$ system at energies above the CB down to sub barrier energies. These resonances have often been attributed to ^{12}C-^{12}C molecular configurations of the ^{24}Mg compound nucleus, and their strength related to the number of open channels in the reaction which is minimal for this system at the CB [2]. The possible persistance of these resonances at astrophysical energies is still a debated question. For example, the resonance phenomena in $^{12}C+^{12}C$ have been explained through the impact on the cross section of the relatively large spacings and the narrow widths of ^{24}Mg compound levels in the corresponding excitation-energy region [3] years. The $^{12}C+^{12}C$ fusion reaction has been the object of a realm of experimental investigations in the past [4-10]. The lowest-energy measured resonance in this reaction is at $E_{c.m.} = 2.1$ MeV [9], only partially overlapping with the high-energy part of the Gamow window. Some of the previous experimental results are presented in Fig. 1.

Techniques used to measure these S factors where based on the identification of charged particles, i.e. p and α, or of the γ-rays emitted from the evaporation residues,

* e-mail: sandrine.courtin@iphc.cnrs.fr

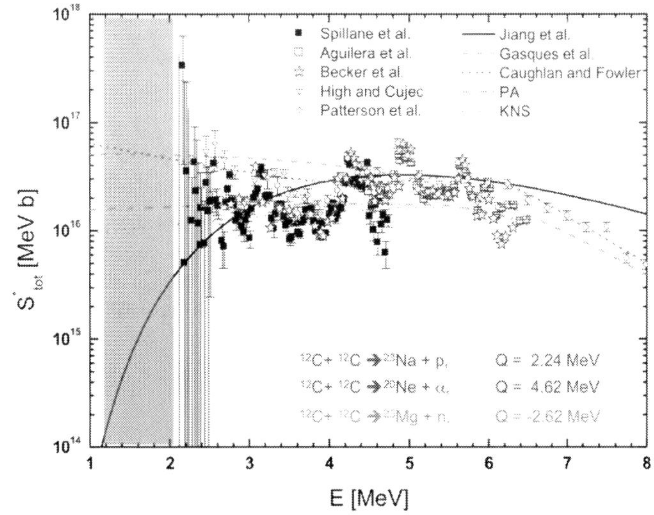

Figure 1. Experimental S factor for the $^{12}C+^{12}C$ fusion reaction from refs. [4,6,7,9,10]. The astrophysics region is indicated by the green area.

^{20}Ne and ^{23}Na. It should be noted that at the lowest investigated energies, the error bars are large and large discrepancies appear between the different measurements. Interestingly enough, the different extrapolations based on different potentials differ from more than 2 orders of magnitude in the Gamow region. The ubiquitus contamination of helium and deuterium in the target can indeed lead to severe background at low energies for both techniques. Moreover measurements based on γ-ray detection are subject to room and cosmic γ backgrounds. To suppress these backgrounds, a new technique has been developed at the Argonne National Laboratory recently, based on γ-particle coincidences. Details about the technique as well as spectra describing the drastic suppression of background are given in Ref. [11]. This method was used in the present work.

2 Experimental set-up

The experiment has been performed at the Argonne National Laboratory using a ^{12}C beam delivered by the ATLAS facility impinging on a highly enriched (99.9 %) ^{12}C target. The beam intensity was ~ 600 pnA and the target thickness was ~ 50 μg.cm^{-2}.

Ten energy points have been measured between $E_{c.m.}$ = 4.93 and 2.68 MeV. Gamma transitions from the evaporation residues were measured using the Gammasphere 100 Ge detectors array in coincidence with charged particles (p and α) from 3 annular double sided silicon detectors. Fig. 2 shows the target chamber with 2 annular detectors at backward angles and 1 at forward angles covering in total ~ 25 % of 4π. Normalization of the beam current was obtained using a Faraday cup and two surface barrier Si monitor detectors identifiying scattered ^{12}C nuclei, at 45 ° forward angles.

The $^{12}C(^{12}C,\alpha)^{20}Ne$ and $^{12}C(^{12}C,p)^{23}Na$ were identified by gating on the characteristic γ transitions 1635 keV $2^+ \rightarrow 0^+$ of ^{20}Ne and 440 keV $\frac{5}{2}^+ \rightarrow \frac{3}{2}^+$ of ^{23}Na. The associated α and p particles were identified in coincidence in the annular DSSDs.
The measured cross-sections converted into S factors are presented in the next section.

3 Results and discussion

Figure 3 shows $^{12}C+^{12}C$ S factors as a function of $E_{c.m.}$ measured in the present work in the type IA supernova Gamow energy region (indicated by the yellow region on the figure) together with the most recent results for the same system [12]. The present data, which lowest point corresponds to a cross section of 6 nb, is in fairly good agreement with this measurement but shows smaller error bars. It should be noted that at the lowest measured energies, the data seem to indicate a decreasing S factor, which would be in agreement with the Jiang extrapolation

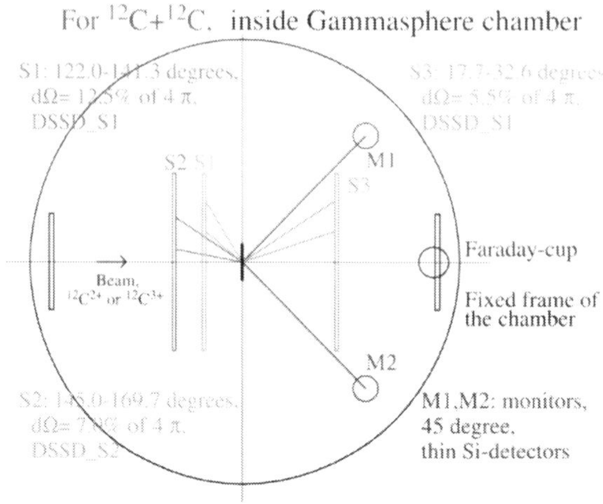

Figure 2. Schematic view of the target chamber showing the 3 annular DSSDs, the Si monitors and the Faraday cup.

[15]. Extrapolations predicting an increasing S factor or a slight decrease at low energies are based on standard potentials and use the sudden model or coupled-channel approach whereas the Jiang extrapolation takes into account the fusion hindrance phenomenon. This phenomenon was introduced by Jiang et al. fifteen years ago in fusion reactions of medium-mass systems, measured below the 0.1 mb regime [17]. Indeed, at low energies, the fusion cross sections are smaller than predicted by coupled-channel calculations using standard Woods-Saxon potentials. Several theoretical descriptions were proposed for this phenomenon : i. a consequence of the saturation properties of the nuclear matter introduced in calculations using a double-folding potentiel + repulsive core [18], ii. the effects of a 2 steps process involving the capture in a 2 body potential pocket and then the penetration of a one body potential to reach a compound nucleus state [19], iii. Effects of the Pauli repulsion during the fusion process introduced recently by C. Simenel et al. and presented at this Fusion 17 conference.

Two representations have been used to discuss signatures of fusion hindrance, the logarithmic derivative of the energy-weighted cross section and the S factor. In the latter, often used for systems of astrophysics relevance, a maximum and a decrease of S when going down in energy is taken as a signature of fusion hindrance. Interestingly enough, the present data seem to be in agreement with the presence of a broad S factor maximum. This phenomenon may have dramatic consequences on the reaction rate of C burning and the subsequent nucleosynthesis of heavier chemical elements in massive stars.

4 Acknowledgements

This work was supported by the US Department of Energy, Office of Nuclear Physics, under Contract No. DE-AC02-06CH11357 and by USIAS, Strasbourg Fellowship of S. Courtin.

References

[1] C. E. Rolfs, and W. S. Rodney, Cauldrons in the Cosmos, The University of Chicago Press, 1988.

[2] F. Haas and Y. Abe, Phys. Rev. Lett. **46** (1981).

[3] C.L. Jiang et al., Phys. Rev. Lett. **110** 072701 (2013).

[4] Patterson L J et al., The Astrophys. J. **157** 367 (1969).

[5] Mazarakis M et al. Phys. Rev. C **7** 1280 (1973).

[6] High M D and Cujec B, Nucl. Phys. A **282** 181 (1977).

[7] H.W. Becker et al., ZPA **303**, 305312 (1981).

[8] Barron-Palos L R et al., Nucl. Phys. A **779** 218 (2006).

[9] Aguilera E F et al., Phys. Rev. C **73** 064 (2006).

[10] Spillane T et al., Phys. Rev. Lett. **98** 122 (2007).

[11] C.L. Jiang et al., NIMA **682**, 1215 (2012).

[12] J. Zickefoose, PhD Thesis, U. Connecticut, (2011) .

[13] W. Fowler, G. Caughlan and B. Zimmerman, Annu. Rev. Astrophys. **13**, 69 (1975).

[14] L.R. Gasques et al., Phys. Rev.. C **72**, 025806 (2005).

[15] C.L. Jiang, K.E. Rehm, B.B. Back and R.V.F. Janssens, Phys. Rev. C **75**, 015803 (2007), and references therein.

[16] H. Esbensen, X.D. Tang, and C.L. Jiang, Phys. Rev. C **84**, 064613 (2011).

DOI: 10.1051/epjconf/201716300011

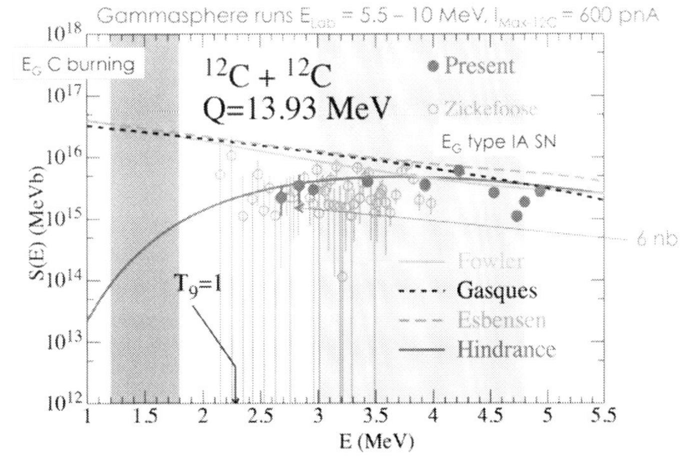

Figure 3. Red points : $^{12}C+^{12}C$ S factors measured in this work. Green circles : results from ref. [12]. The blue, black, red and green dashed lines correspond to extrapolations from Fowler [13], Gasques [14], Jiang [15] and Esbensen [16].

[17] C.L. Jiang et al., Phys. Rev. Lett. **89**, 052701 (2002).
[18] S. Misicu et al., Phys. Rev. Lett. **96**, 112701 (2006).
[19] T. Ichikawa, K.Hagino and A. Iwamoto et al., Phys. Rev. C**75**, 064612 (2007) and Phys. Rev. Lett. **103**, 202701 (2009).

 DOI: 10.1051/epjconf/201716300012

Understanding the fusion cross section among light nuclei around the Coulomb barrier

Antonio Del Zoppo[1,*] and Marco La Cognata[1,**]

[1] Istituto Nazionale di Fisica Nucleare – Laboratori Nazionali del Sud, via S. Sofia 62, 95123 Catania, Italy

Abstract. In this work we investigate fusion induced by a radioactive ^{8}Li projectile on a ^{4}He gas target, at center-of-mass energies between 0.6 and 5 MeV. The main result is the tendency of the dimensionless fusion cross section $\frac{\sigma}{\pi\lambda^2}$ to form well visible plateaus alternated to steep rises. This is likely to be the most genuine consequence of the discrete nature of the intervening angular momenta observed so far in fusion reactions right above the Coulomb barrier. A partial-wave analysis, exclusively based on a pure quantal penetration fusion model, identifies a remarkably low-height barrier. Indeed, these plateaus allow enhanced experimental sensitivity to the fusion barrier given that the most barrier-sensitive lowest partial waves are well separated. We expect that the present results for ^{8}Li+^{4}He will promote further investigations of the fusion reaction mechanism between very light ions at energies much below the interaction barrier. For the moment, we believe that understanding the plateau origin in the cross section above the barrier will almost certainly be useful to corroborate the extrapolation to the important astrophysical region below the Coulomb barrier, not only in the case of the ^{8}Li+^{4}He fusion but also for other systems, such as the ^{12}C+^{12}C.

1 Introduction

Fusion between light ions plays an important role in many astrophysical sites such as, for instance, quiescent-burning stars, including massive stars beyond H-burning, type Ia supernovae and surface explosions of neutron stars. Exotic scenarios like pycnonuclear reactions occurring in high density matter, when nuclei are frozen in lattice structures, are also dominated by ^{12}C and ^{16}O induced fusion processes [1]. A exhaustive review can be find, for instance, in Ref.[2]. Fusion between light nuclei is also of growing importance in the field of energy production by means, among others, of inertial confinement [3]. However, energies of astrophysical interest reach at most few MeV even in the case of explosive scenarios, when temperatures as large as 10^9 K are achieved [2], which are usually lower or comparable with the corresponding Coulomb barriers. Therefore, extrapolation sometimes supported by nuclear reaction theory is often necessary to establish the low-energy trend of the cross section at astrophysical energies. This can introduce large systematic errors if the underpinning theory turns out to be an inaccurate description of the physical processes, possibly causing large deviations of the theoretical models using the extrapolated cross sections from observations. Therefore, the problem of extrapolation has to be challenged to provide as much accurate as possible cross section previsions for astrophysics and applied physics. In this work we investigate the low energy trend of the ^{8}Li + α fusion reaction. The need of a realistic extrapolation of the cross section to energies below about 1 MeV and the existence of some incompatible data sets have triggered a close examination of the origin of the oscillations characterising its cross section. A detailed discussion on the ^{8}Li + α cross section measurement and on the discrepancy between the different data sets can be found in Refs.[4–10].

* e-mail: delzoppo@lns.infn.it
** e-mail: lacognata@lns.infn.it

2 The ^{8}Li + α fusion cross section

The ^{8}Li + α fusion has important astrophysical applications in explosive scenarios such as the big bang nucleosynthesis and the heavy-element production through the r-process. If neutron-rich regions formed before the onset of nucleosynthesis (inhomogeneous model, see, for instance, Refs.[11, 12]), the ^{8}Li + α capture might have acted as a bridge towards the production of elements heavier than lithium, otherwise synthesised in negligible amount in standard big band scenarios. Similarly, in neutron rich environments where r-process can take place [13, 14], seed nuclei eventually burnt to heavier elements in the fast neutron captures could be obtained through α-capture reactions, in particular on ^{8}Li.

Fig.1a) displays the ^{8}Li + α concordant data set as established in [7]. Such fusion cross section clearly shows a oscillating pattern, with a period of some MeV, recalling the ^{12}C + ^{12}C data [17], the literature case where such oscillations are most evident, given in Fig.1b). In Figs.1c) and d) exhibit the same data expressed in the form of dimensionless cross section in units of $\pi\lambda^2$. Both figures indicate the existence of nearly horizontal plateaus clearly

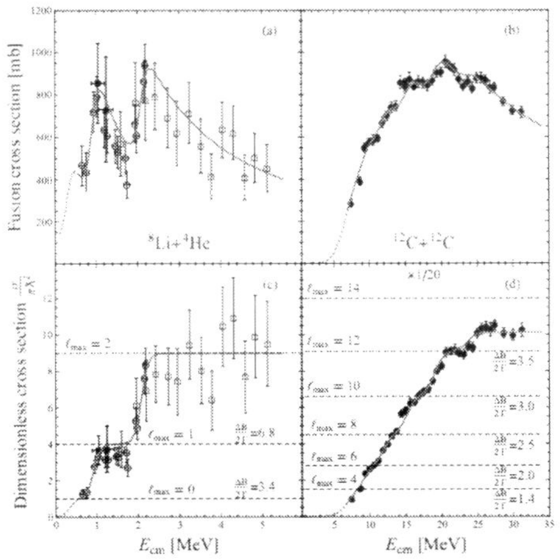

Figure 1. Panel a) ^{8}Li+α fusion cross section. Experimental data are taken from: Ref.[15] open squares, Ref.[16] open circles, Ref.[4] filled circle, Ref.[6, 8] filled square. Panel b) ^{12}C +^{12} C fusion cross section. Experimental data are taken from Ref.[17]. Panel c) and panel d) dimensionless cross sections for the ^{8}Li + α and ^{12}C +^{12} C channels, respectively. In panel d) all ordinates are divided by 20 to keep the same vertical axis. The solid curves are the results of the MINUIT data fits using Eq.1. Dashed curves are extrapolation to zero energy

alternate with steep rises that, in the case of ^{12}C +^{12} C data are apparent only above about 20 MeV.

As already done for the ^{12}C +^{12} C fusion cross section, we assume that the oscillatory structures arise from the progressive addition of higher partial waves with increasing energy, as already discussed in Refs.[18–20]. The detailed analysis is reported in Ref.[21]. To probe such interpretation for the ^{8}Li + α data, we make use of a formalism that explicitly takes into account the angular momentum of the relative motion l and introduce a sharp angular momentum cut-off that allows all of the flux crossing the interaction barrier to fuse for values of $l \leq l_{max}$. Accordingly, the dimensionless cross section takes the form:

$$\frac{\sigma_f}{\pi \lambda\!\!\!^{-2}} = \Delta l \sum_{l=0,\Delta l}^{l_{max}} (2l+1)T_l \qquad (1)$$

where Δl = 1 for ^{8}Li + α, whereas Δl = 2 for identical even-even nuclei, for which all odd partial waves vanish. In Eq.1, T_l is the energy-dependent penetration probability of the $l-th$ partial wave through the interaction barrier. In particular, when all T_l competing to a given energy tend to unity, plateaus may appear taking the following values:

$$\frac{\sigma_f}{\pi \lambda\!\!\!^{-2}} \rightarrow (l_{max}+1)^2 , \qquad (2)$$

for Δl = 1 and

$$\frac{\sigma_f}{\pi \lambda\!\!\!^{-2}} \rightarrow (l_{max}+1)(l_{max}+2) , \qquad (3)$$

Table 1. Barrier height, intrinsic width and moment of inertia parameter sets determined for both reactions using MINUIT data fitting procedure.

Reaction	B_0 (MeV)	Γ (MeV)	I (MeV^{-1})
^{8}Li + α	0.34 ± 0.05	0.08 ± 0.02	1.72 ± 0.07
^{12}C +^{12} C	5.70 ± 0.12	0.65 ± 0.14	5.88 ± 0.10

for Δl = 2. The plateau values calculated with Eq.2 and Eq.3 for the ^{8}Li + α and ^{12}C +^{12} C systems, respectively, are given in Fig.1c) and d) for relevant values of l_{max} and marked with short-dashed lines.

Very good agreement is found in the ^{8}Li+α dimensionless cross section, while only for l_{max} = 10 and 12 there is significant evidence of the occurrence of the plateaus for the ^{12}C +^{12} C system. This is a clear sign that the oscillatory structures plausibly arise from the progressive addition of higher partial waves with increasing energy, rather than from properties of the compound nuclear system in the ^{8}Li + α case under examination. Indeed, the results for both processes are absolutely compatible if the $\Delta B/2\Gamma$ parameter is considered for both cases, where ΔB is the difference of barrier heights for two subsequents allowed values of the orbital angular momenta and Γ is the barrier width assumed constant [21]. To obtain them, we fitted the dimensionless cross sections using Eq.1, taking

$$T_l(E_{cm}) = \left[1+\exp\left(\frac{B_l - E_{cm}}{\Gamma}\right)\right]^{-1} , \qquad (4)$$

where $B_l = B_0 + \frac{l(l+1)}{2I}$. The fitting parameter are given in Tab.1. The resulting $\Delta B/2\Gamma$ parameters are shown in Fig.1c) and d); the comparison of the values for the ^{8}Li + α and ^{12}C +^{12} C systems demonstrates that plateaus are already visible in the former case for l_{max} = 0 since the indicator value $\Delta B/2\Gamma$ = 3.4, a number obtained only for l_{max} = 12 for ^{12}C +^{12} C fusion.

3 Concluding remarks

In this work we obtain unprecedented experimental information on the ^{8}Li + α fusion at center-of-mass energies E_{cm} between 0.6 and 5 MeV, right above the Coulomb barrier. Experimental dimensionless cross section exhibits two well visible plateaus alternated to steep rises; a fusion model that solely includes the action of the relative motion angular momenta turns out to be a proper description of the reaction mechanism. Therefore, this appears as the most genuine consequence of the discrete nature of subsequent angular momenta observed so far in fusion reactions.

Concerning barrier shape assessment, for ^{8}Li + α neither coupled-channel effects nor collision partner deformations should significantly contribute [22], while for ^{12}C +^{12} C coupled channels calculations were not able to reproduce fusion data [18]. Consequently, a pure single barrier penetration fusion model is adopted here for both reactions. Hill-Wheeler barrier penetration formula is used and most of the barrier parameters determined here for ^{8}Li + α fusion are consistent with those of ^{12}C +^{12} C, with the possible exception of B_0. This work identifies a

DOI: 10.1051/epjconf/201716300012

notably low Coulomb barrier that can be possibly linked to the presence of the loosely bound ^{8}Li.

References

[1] D.G. Yakovlev, L.R. Gasques, A.V. Afanasjev, M. Beard, M. Wiescher, Phys. Rev. C **74**, 035803 (2006), astro-ph/0608488
[2] C. Iliadis, *Nuclear Physics of Stars* (Wiley-VCH Verlag, 2007)
[3] R. Betti, O.A. Hurricane, Nature Physics **12**, 435 (2016)
[4] S. Cherubini, P. Figuera, A. Musumarra, C. Agodi, R. Alba, L. Calabretta, L. Cosentino, A. Del Zoppo, A. Di Pietro, M. La Cognata et al., European Physical Journal A **20**, 355 (2004)
[5] A. Del Zoppo, P. Figuera, A. Musumarra, N. Colonna, R. Alba, C. Bonomo, S. Cherubini, L. Cosentino, A. Di Pietro, M. Gulino et al., Nuclear Instruments and Methods in Physics Research A **581**, 783 (2007)
[6] M. La Cognata, A. Del Zoppo, P. Figuera, A. Musumarra, R. Alba, S. Cherubini, N. Colonna, L. Cosentino, V. Crucillà, A. di Pietro et al., Physics Letters B **664**, 157 (2008)
[7] M. La Cognata, A. Del Zoppo, R. Alba, S. Cherubini, N. Colonna, A. Di Pietro, P. Figuera, M. Gulino, L. Lamia, A. Musumarra et al., Astrophys. J. Lett. **706**, L251 (2009), 0911.2585
[8] M. La Cognata, A. Del Zoppo, R. Alba, S. Cherubini, N. Colonna, A. Di Pietro, P. Figuera, M. Gulino, L. Lamia, A. Musumarra et al., Journal of Physics G Nuclear Physics **37**, 105105 (2010)
[9] M. La Cognata, A. Del Zoppo, Astrophys. J. **736**, 148 (2011)
[10] M. La Cognata, A. Del Zoppo, Physica Scripta Volume T **150**, 014019 (2012)
[11] R.A. Malaney, W.A. Fowler, Astrophys. J. **333**, 14 (1988)
[12] J.F. Lara, T. Kajino, G.J. Mathews, Phys. Rev. D **73**, 083501 (2006), astro-ph/0603817
[13] S.E. Woosley, R.D. Hoffman, Astrophys. J. **395**, 202 (1992)
[14] T. Sasaqui, K. Otsuki, T. Kajino, G.J. Mathews, Astrophys. J. **645**, 1345 (2006), astro-ph/0507393
[15] R.N. Boyd, I. Tanihata, N. Inabe, T. Kubo, T. Nakagawa, T. Suzuki, M. Yonokura, X.X. Bai, K. Kimura, S. Kubono, Physical Review Letters **68**, 1283 (1992)
[16] X. Gu, R.N. Boyd, M.M. Farrell, J.D. Kalen, C.A. Mitchell, J.J. Kolata, M. Belbot, K. Lamkin, K. Ashktorab, F.D. Becchetti et al., Physics Letters B **343**, 31 (1995)
[17] D.G. Kovar, D.F. Geesaman, T.H. Braid, Y. Eisen, W. Henning, T.R. Ophel, M. Paul, K.E. Rehm, S.J. Sanders, P. Sperr et al., Phys. Rev. C **20**, 1305 (1979)
[18] H. Esbensen, Phys. Rev. C **85**, 064611 (2012), 1206.1068
[19] C.Y. Wong, Phys. Rev. C **86**, 064603 (2012)
[20] N. Rowley, K. Hagino, Phys. Rev. C **91**, 044617 (2015)
[21] A. Del Zoppo, M. La Cognata, Physics Letters B **753**, 449 (2016)
[22] K. Hagino, N. Takigawa, Progress of Theoretical Physics **128**, 1061 (2012), 1209.6435

 DOI: 10.1051/epjconf/201716300013

Fusion of neutron-rich oxygen nuclei

Romualdo deSouza[1,2,*], *Justin* Vadas[1,2], *Varinderjit* Singh[1,2], *Blake* Wiggins[1,2], *Tracy* Steinbach[1,2], *Zidu* Lin[2,3], *Chuck* Horowitz[2,3], *Lagy* Baby[4], *Sean* Kuvin[4], *Vandana* Tripathi[4], *Ingo* Wiedenhover[4], and *Sait* Umar[5]

[1] *Department of Chemistry, Indiana University, Bloomington, IN 47405 USA*
[2] *Center for Exploration of Energy and Matter, Indiana University, Bloomington, IN 47408 USA*
[3] *Department of Physics, Indiana University, Bloomington, IN 47405 USA*
[4] *Department of Physics, Florida State University, Tallahassee, FL, 32306 USA*
[5] *Department of Physics and Astronomy, Vanderbilt University, Nashville, TN 37235 USA*

Abstract. Measurement of the fusion excitation function for ^{18}O + ^{12}C and ^{19}O + ^{12}C is described. The fusion cross-section is extracted through the direct measurement of evaporation residues resulting from the fusion process. At near barrier energies, the single additional neutron present in ^{19}O results in an enhancement in the fusion cross-section by a factor of over three as compared to ^{18}O.

1 Introduction

Measuring the fusion excitation function for an isotopic chain of projectile nuclei presents an unique opportunity to examine the character of neutron-rich matter. For a given element, with increasing neutron number the neutron density distribution usually extends further out while the proton distribution remains largely unaffected. Hence, the repulsive Coulomb potential is largely unchanged while the attractive nuclear potential changes. As fusion at near barrier energies is sensitive to the interplay between the repulsive Coulomb and attractive nuclear potentials, the comparison of the fusion excitation function for isotopically related projectiles provides a sensitive probe of the change in the attractive nuclear potential. This change in the attractive potential can be related to changes in both the structure and dynamics of the neutron density distribution as the number of neutrons increases.

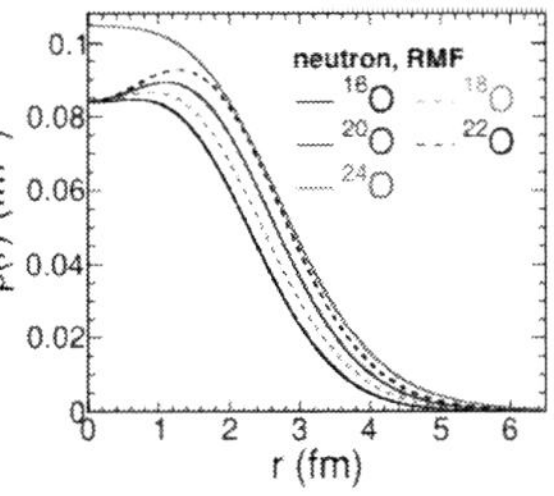

Figure 1. Neutron-density distribution of oxygen isotopes calculated within a relativistic mean field theory (RMF).

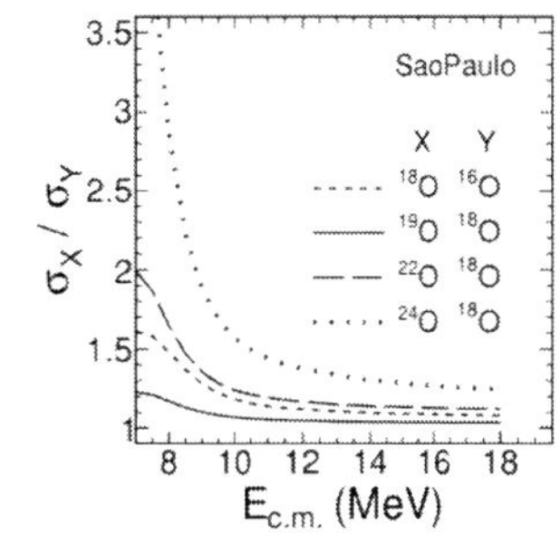

Figure 2. Relative cross-section for Oxygen isotopes using a simple barrier penetration model with the RMF density distributions.

Presented in Fig. 1 are the neutron density distributions for oxygen isotopes calculated within the context of a relativistic mean field theory [1, 2]. As expected, with increasing neutron number the tail of the density distribution extends further out. It is perhaps surprising however that the tail of the neutron density distribution for ^{22}O is so close to that of the drip-line nucleus ^{24}O. This observation is particularly noteworthy as beams of ^{20,21}O are presently available at GANIL and a beam of ^{22}O is anticipated in the near future. In this paper, we describe the first step towards the systematic measurement of the fusion excitation functions of the oxygen isotopes namely fusion in ^{18}O+^{12}C and ^{19}O+^{12}C.

An extended neutron distribution could impact fusion both through its static spatial extent as well as through its dynamics (e.g. polarization effects). The influence of the static contribution can be estimated by using a one-dimensional barrier penetration model such as the Sao-

* e-mail: desouza@indiana.edu

Paulo model [3]. For the density distributions presented in Fig. 1, we have calculated the relative fusion cross-sections for several oxygen isotopes and present the results in Fig. 2. Evident in the figure is the fact that at all energies as the neutron-richness increases the fusion cross-section increases. At energies well above the barrier ($E_{c.m.}$ ~ 8 MeV) this enhancement is relatively constant and can be considered geometric. Near and below the barrier however, the enhancement increases rapidly with decreasing incident energy. In this energy domain the enhancement reflects the increased importance of the nuclear potential due to the neutron-skin. It should be emphasized that as this is a purely static calculation, the inclusion of fusion dynamics could result in an enhancement larger than that depicted in Fig. 2.

In order to measure the fusion excitation function for ^{18}O + ^{12}C and ^{19}O + ^{12}C two separate experiments were performed at the John D. Fox accelerator laboratory at Florida State University. In the initial experiment we measured the fusion excitation function for ^{18}O + ^{12}C, extending the measured cross-section down to the sub 1 mb level, a factor of 30 lower than previously measured. Having established the technique, we subsequently measured the fusion excitation function for ^{19}O + ^{12}C. The experimental details of both measurements are summarized below.

2 The ^{18}O + ^{12}C experiment

2.1 Experimental setup

The experimental setup for this experiment consisted of two ExB microchannel plate detectors and two annular silicon detectors as depicted in Fig. 3. The beam first passes through a ExB microchannel plate detector, designated MCP_{US}, situated approximately 1.3m upstream of the target position. For each ion traversing this detector a fast timing signal is generated. The beam subsequently encounters a second ExB microchannel plate detector, designated MCP_{TGT}. The 100 μg/cm^2 thick carbon foil of MCP_{TGT} is dual function. Not only does it serve as a secondary emission for the detector, but it also serves as the target for the experiment. The coincidence of these two ExB MCP detectors with the appropriate time-of-flight provided the incident beam count. The typical intensity of the ^{18}O beam incident on the target was ~2x10^5 ions/s.

2.2 Measuring the fusion products

Fusion of a ^{18}O nucleus in the beam together with a ^{12}C nucleus in the target foil results in the production of an excited ^{30}Si nucleus. For collisions near the Coulomb barrier the excitation of the fusion product is relatively modest, E* ≈ 35 MeV. De-excitation of this fusion product by evaporation of a few neutrons, protons, and α particles results in an evaporation residue (ER). Statistical model calculations [4] indicate that for a ^{30}Si compound nucleus, the nuclei ^{29}Si, ^{28}Si, ^{28}Al, ^{27}Al, and ^{25}Mg account for the bulk of the ERs. Emission of the light particles deflects the ER from the beam direction allowing their detection and identification in two annular silicon detectors, designated T2 and T3, that are situated downstream of the MCP_{TGT}. These detectors subtend the angular range 3.5° < θ_{lab} < 25° allowing detection of the majority of the ERs produced [5]. By measuring the time-of-flight of particles between the MCP_{TGT} detector and the silicon detectors [6] together with the energy deposit in the Si detector, evaporation residues are distinguished from scattered beam, as well as emitted light particles. By utilizing the measured energy deposit and time-of-flight, the mass of the ion can be calculated providing clear separation of ERs from the incident beam [7].

2.3 Fusion excitation function for ^{18}O + ^{12}C

The fusion cross-section is extracted by summing the total number of evaporation residues observed. Comparison of this yield with the number of incident ^{18}O ions while accounting for the target thickness and the geometric efficiency of the experimental setup yields the absolute fusion cross-section [7]. The measured excitation function is displayed in Fig. 4a together with previously published results [9–11]. Vertical error bars on the new data reflect both the statistical uncertainties as well as a 2% systematic error associated with the analysis. Horizontal error bars represent the uncertainty in whether the fusion occurs at the front or back of the target foil. While prior measurements using the direct measurement of evaporation residues only measured the fusion cross-section down to the 25 mb level [9], in this work the fusion cross-section is measured down to the 820 μb level, a factor of approximately 30 lower in cross-section. In the energy region where the present data overlaps with published data, overall agreement of the cross-sections is good, close to the statistical uncertainties. This overall agreement indicates both that our approach in extracting the fusion cross-section is sound and that there are no significant uncertainties in the values of the target thickness or detector efficiency. Closer comparison of the present dataset with the data of Ref. [9] indicates that the presently measured cross-sections are approximately 3-5% lower for $E_{c.m.} \geq$ 10 MeV. This is within the statistical uncertainties of the data reported by Eyal et al. In addition, the present data are higher in their statistical quality.

We have compared the experimental fusion excitation function with the predictions of a microscopic model. Over the past several years, the density constrained TDHF (DC-TDHF) method for calculating heavy-ion potentials [12] has been employed to calculate heavy-ion fusion cross-sections with remarkable success [13, 14]. While most applications have been for systems involving heavy nuclei, recently the theory was used to study above and below barrier fusion cross-sections for lighter systems, specifically for reactions involving various isotopes of O+O and O+C [15, 16]. One general characteristic of TDHF and DC-TDHF calculations for light systems is that the fusion cross-section at energies well above the barrier are usually overestimated [17, 18], whereas an excellent agreement is found for sub-barrier cross-sections [15].

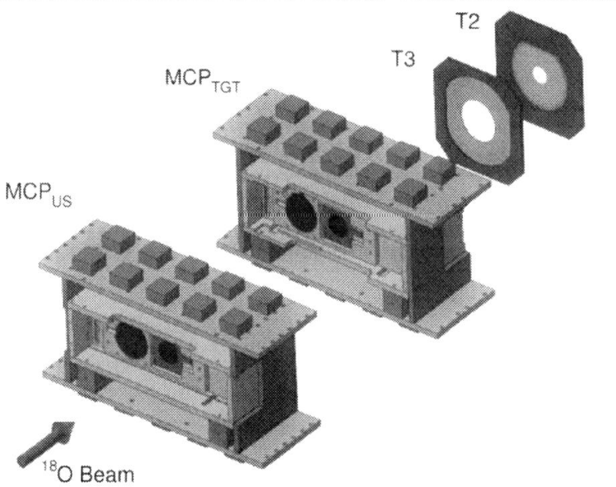

Figure 3. Experimental setup utilized to directly measure the evaporation residues and extract the fusion cross-section

Comparison of the experimental fusion excitation function with the DC-TDHF microscopic model is presented in Fig. 4a. Overall comparison of the experimental cross-sections with the DC-TDHF calculations indicate that for energies ≳7 MeV the experimental cross-sections (symbols) are lower than the theoretical predictions (solid line). In order to facilitate a quantitative comparison of the experimental excitation function with the theoretical predictions, we have fit the experimental cross-sections with a functional form [19] that describes the penetration of an inverted parabolic barrier. The values of the fit parameters are presented in Table 1.

$$\sigma = \frac{R_C^2}{2E}\hbar\omega \cdot ln\left\{1 + exp\left[\frac{2\pi}{\hbar\omega}(E - V_C)\right]\right\} \qquad (1)$$

where E is the incident energy, V_C is the barrier height, R_C is the radius of interaction, and $\hbar\omega$ is the barrier curvature.

Shown in Fig. 4b is the ratio of the fit of the experimentally measured cross-sections to the DC-TDHF calculations. For energies 7.5 MeV < $E_{c.m.}$ < 14 MeV, the ratio $\sigma_{Experiment}/\sigma_{DC-TDHF}$ is ≈ 0.75 and is relatively constant with respect to energy. As the incident energy decreases below 7.5 MeV, the ratio increases reaching a value of over 10 at the lowest energy measured, $E_{c.m.}$ ≈ 5 MeV. The fact that the ratio is only ≈ 0.75 at energies above the barrier can be understood as due to the presence of breakup reactions in this energy range. With decreasing incident energy, the role of breakup reactions diminshes hence the ability of the DC-TDHF model to describe fusion is expected to improve. We therefore focus our attention on the comparison of the model and experiment in the sub-barrier region. The key feature in the ratio is therefore its change with decreasing incident energy in the sub-barrier domain, specifically its increase from a value smaller than unity to a value larger than unity. This trend emphasizes that the experimental and theoretical excitation functions

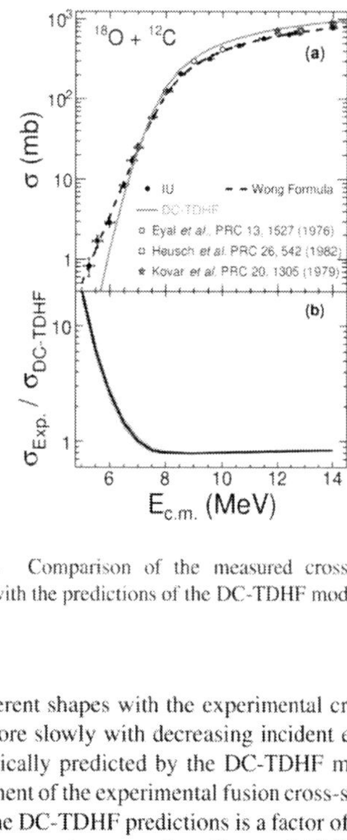

Figure 4. Comparison of the measured cross-section for $^{18}O+^{12}C$ with the predictions of the DC-TDHF model.

have different shapes with the experimental cross-section falling more slowly with decreasing incident energy than is theoretically predicted by the DC-TDHF model. This enhancement of the experimental fusion cross-sections relative to the DC-TDHF predictions is a factor of ≈10 as the incident energy decreases from $E_{c.m.}$ = 7 MeV to $E_{c.m.}$ = 5 MeV. We have assessed the impact of the experimental uncertainties on the ratio presented and display the result as a shaded band in Fig. 4b. The trends exhibited by the ratio are significantly larger than the magnitude of the uncertainties.

The fact that the experimental fusion cross-sections decrease more slowly with decreasing energy than the calculated cross-sections can be interpreted as a larger tun-

 DOI: 10.1051/epjconf/201716300013

neling probability for the experimental data as compared to the theoretical calculations. This enhanced tunneling probability can be associated with a narrower, lower barrier. The underlying reason that the barrier determined from the experimental data is weaker than in the model is presently unclear.

3 The $^{19}O + ^{12}C$ experiment

3.1 Producing and Chararacterizing a ^{19}O beam

A beam of ^{18}O ions at an energy of 80.7 MeV was used to bombard a deuterium gas cell at a pressure of 350 torr to produce the ^{19}O beam. To increase the gas density, the gas cell was cooled to a temperature of 77 K. Ions of ^{19}O produced via a (d,p) reaction were separated from the incident beam by the electromagnetic spectrometer RESOLUT [20]. Despite the rejection of most of the unreacted beam by RESOLUT, the beam exiting the spectrometer consisted of both ^{19}O and ^{18}O ions. It was therefore necessary to identify each ion incident on the target. By accomplishing this it was possible to simultaneously measure the fusion excitation function for $^{18}O + ^{12}C$ and $^{19}O + ^{12}C$ which provided an important consistency check. Comparison of the $^{18}O + ^{12}C$ with the prior high statistics measurement of $^{18}O + ^{12}C$ demonstrated that there were no systematic differences between the two measurements. This dual measurement thus provided confidence that any observed fusion enhancement was a robust signal.

3.2 Experimental setup

Although the experimental setup for the ^{19}O beam was largely the same as in the prior experiment, a few changes were made to handle the radioactive beam. In order to identify beam particles, the energy deposit (ΔE) and time-of-flight (TOF) of each particle was measured prior to the target. After exiting RESOLUT particles traversed a thin foil (0.5 μm thick aluminized mylar) which served as the electron emission foil for an MCP detector. Approximately 3.5 m downstream of this thin foil the oxygen ions passed through a compact ionization detector (CID) depositing an energy, ΔE. CID served two roles: to help identify the ion as well as to reduce its energy. While the production of ^{19}O is favored at higher energies, the measurement of the fusion cross-section at near barrier energies requires reducing the energy of the radioactive ion after its production. To perform the excitation function measurement, the energy of the incident beam was decreased by adjusting the gas pressure in CID. Upon exiting CID the ions were incident on a 105 μg/cm^2 carbon foil. This carbon foil, as in the prior experiment, served both as a secondary electron emission foil for the target microchannel plate detector (MCP_{TGT}) and as the target for the fusion experiment [8]. To measure the energy distribution of ^{19}O and ^{18}O ions incident on the target, a surface barrier, silicon detector was periodically inserted into the beam path just prior to the target.

The timing signals from both microchannel plate detectors together with the energy deposit in the ionization chamber allowed identification of ions in the beam through measurement of the ΔE-TOF. Ions of $^{19}O^{7+}$, $^{18}O^{7+}$, and $^{18}O^{6+}$ are clearly identified with ^{19}O ions corresponding to approximately 31 % of the beam intensity. The two charge states of ^{18}O contributed to approximately 50 % of the beam intensity on target. The intensity of the ^{19}O beam incident on the target was 1.5 - 4 x 10^3 ions/s.

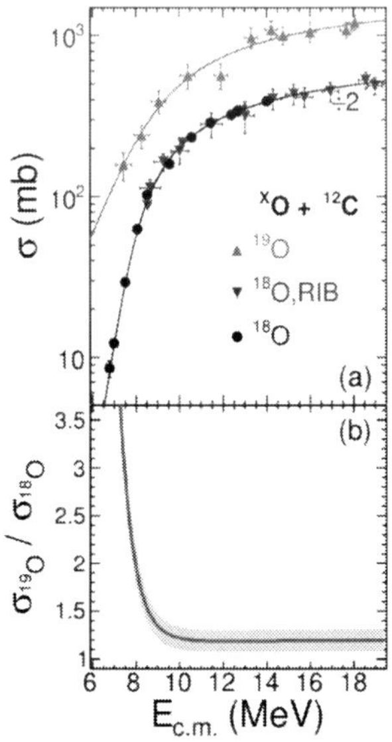

Figure 5. Comparison of the measured cross-section for $^{18}O + ^{12}C$ and $^{19}O + ^{12}C$. The cross-sections for the ^{18}O reaction have been scaled by a factor of two for clarity.

3.3 Fusion excitation function for $^{19}O + ^{12}C$

The same general trend is observed for both of the excitation functions depicted in Fig. 5a. With decreasing incident energy the cross-section decreases as expected for a barrier controlled process. At essentially all energies measured the ^{19}O data exhibits a larger fusion cross-section as compared to the ^{18}O data.

Table 1. Fit parameters for the $^{x}O + ^{12}C$ fusion excitation functions. See text for details.

	V_C (MeV)	R_C (fm)	$\hbar\omega$ (MeV)
$^{18}O + ^{12}C$	7.66 ± 0.10	7.39 ± 0.11	2.90 ± 0.18
$^{19}O + ^{12}C$	7.73 ± 0.72	8.10 ± 0.47	6.38 ± 1.00

To quantitatively examine the differences in the two excitation functions we have fit the excitation functions with the Wong formalism of penetration of an inverted parabolic barrier [19]. The fit of the high resolution ^{18}O data is indicated as the solid black line in Fig. 5a. The solid red curve in Fig. 5a depicts the fit of the ^{19}O data. A reasonable fit of the measured fusion cross-sections is achieved in both cases. The extracted parameters for the

^{18}O and ^{19}O reactions are summarized in Table 1. Since the charge density distribution is essentially unchanged, it is unsurprising that the barrier height, V_C, remains essentially the same for both of the reactions examined. Moreover, as expected with increasing neutron number an increase in R_C is observed. This increase in the radius can be viewed by calculating the quantity $R_C/A^{1/3}$ where A is the mass number of the compound nucleus. This quantity has a value of 2.38 for the ^{18}O induced reaction, while it is 2.58 for the ^{19}O induced reaction. The most significant change in the fit parameters is a substantial increase in the magnitude of $\hbar\omega$ for the ^{19}O case corresponding to a narrower barrier, reflecting an increase in the attractive nuclear potential.

Depicted in Fig. 5b as the solid (red) line is the dependence of the measured ratio of $\sigma(^{19}\mathrm{O})/\sigma(^{18}\mathrm{O})$ on $E_{c.m.}$. At energies well above the barrier $\sigma(^{19}\mathrm{O})/\sigma(^{18}\mathrm{O})$ is essentially flat at a value of $\approx$ 1.2. As one approaches the barrier it rapidly increases to a value of approximately 3.5. Hence, the addition of a single additional neutron in ^{19}O as compared to ^{18}O results in a dramatic enhancement in the fusion cross-section at sub-barrier energies.

4 Summary

We have measured the fusion excitation functions for ^{18}O + ^{12}C and ^{19}O + ^{12}C using low intensity beams. Comparison of these excitation functions indicates a significant enhancement at near barrier energies for the neutron-rich projectile. The addition of a single neutron increases the fusion cross-section by more than a factor of three at the lowest energy measured. This enhancement may reflect the increased role of neutron transfer or coupling to collective degrees of freedom. These measurements represent the first step in the measurement of the fusion excitation function for an isotopic chain of oxygen nuclei. Acquiring a systematic, high quality dataset of this type, coupled with microscopic calculations of the fusion process has considerable promise in elucidating the nature of neutron-rich nuclear matter.

5 Acknowledgements

The support of the staff at Florida State University's John D. Fox accelerator in providing the ^{18}O and ^{19}O beams made this experiment feasible. We gratefully acknowledge their efforts. Development of the Sao Paulo model code by Dr. Helber Dussan is gratefully acknowledged. This work was supported by the U.S. Department of Energy under Grant Nos. DE-FG02-88ER-40404 (Indiana University), DE-FG02-87ER40365 (Indiana University Nuclear Theory), DE-SC0008808 (NUCLEI SciDAC Collaboration), DE-FG02-02ER-41220 (Florida State University), DE-SC0013847 (Vanderbilt University) and the National Science Foundation under Grant No PHY-1491574 (Florida State University). J.V. acknowledges the support of a NSF Graduate Research Fellowship under Grant No. 1342962.

References

[1] P. Ring, Prog. Part. Nucl. Phys. **37**, 193 (1996)
[2] B. D. Serot and J. D. Walecka, Adv. Nucl. Phys. **16**, 1 (1986)
[3] L. R. Gasques et al., Phys. Rev. C **69**, 034603 (2004)
[4] N. G. Nicolis and J. R.Beene, unpublished, 1993
[5] T. K. Steinbach, PhD Thesis, Indiana University (2016)
[6] R. T. deSouza et al., Nucl. Instr. and Meth. **A632**, 133 (2011)
[7] T. K. Steinbach et al., Phys. Rev. C. **90**, 041603(R)(2014)
[8] T. K. Steinbach et al., Nucl. Instr. and Meth. **A743**, 5 (2014)
[9] Y. Eyal and M. Beckerman and R. Checkhik and Z. Fraenkel and H. Stocker, Phys. Rev. C **13** 1527 (1976)
[10] D. Kovar, D.F. Geesaman, T.H. Braid, Y. Eisen, W. Henning, T.R. Ophel, M. Paul, K.E. Rehm, S.J. Sanders, P. Sperr, J.P. Schiffer, S.L. Tabor, S. Vigdor, B. Zeidman and F.W. Prosser Jr., Phys. Rev. C **20**, 1305 (1979)
[11] B. Heusch, C. Beck, J.P. Coffin, P. Engelstein, R. M. Freeman, G. Guillaume, F. Haas, P. Wagner, Phys. Rev. C **26**, 542 (1982)
[12] A. S. Umar and V. E.Oberacker, Phys. Rev. C **74** 021601(R) (2006)
[13] R. Keser and A. S. Umar and V. E.Oberacker, Phys. Rev. C **85** 044606 (2012)
[14] Back, B. B., Esbensen, H., Jiang, C. L. and Rehm, K. E., *Recent developments in heavy-ion fusion reactions* Rev. Mod. Phys. **86**, 317 (2014)
[15] A. S. Umar, V. E.Oberacker and C. J. Horowitz, Phys. Rev. C **85** 055801 (2012)
[16] R. T. deSouza, S. Hudan and V. E. Oberacker and A. S. Umar", Phys. Rev. C **88**, 014602 (2013)
[17] C. Simenel, R. Keser, A. S. Umar and V. E. Oberacker, Phys. Rev. C, **88**, 024617 (2013)
[18] A. S. Umar, C. Simenel and V. E. Oberacker, Phys. Rev. C **89**, 034611 (2014)
[19] C. Y. Wong, Phys. Rev. Lett.**31**, 766 (1973)
[20] I. Wiedenhöver et al., *Fifth International Conference on Fission and Properties of Neutron-rich Nuclei*, Eds. J.H. Hamilton and A.V. Ramayya, World Scientific, Sanibel Island, FL 2012

DOI: 10.1051/epjconf/201716300014

A new analysis procedure to extract fusion excitation function with large beam energy dispersions: application to the ^{6}Li+^{120}Sn and ^{7}Li+^{119}Sn

Alessia Di Pietro[1,*], *Pierpaolo* Figuera[1], *Maria* Fisichella[1], *Alan C.* Shotter[2], *Juan Pablo* Fernandez-Garcia[1,3], *Jonas L.* Ferreira[4], *Marcello* Lattuada[1,3], *Paolo* Lotti[5], *Jesus* Lubian[4], *Chris* Ruiz[6], *Domenico* Torresi[1,3], and *Mile* Zadro[7]

[1] *INFN, Laboratori Nazionali del Sud, via S. Sofia 62, I-95123*
[2] *School of Physics and Astronomy, University of Edinburgh, JCMB, Mayfield Road, Edinburgh EH9 3JZ, UK*
[3] *Dipartimento di Fisica ed Astronomia, Università di Catania, via S. Sofia 64, I-95123 Catania, Italy*
[4] *Instituto de Física, Universidade Federal Fluminense, Avenida Litoranea s/n, Gragoatá, Niterói, Rio de Janeiro 24210-340, Brazil*
[5] *INFN, Sezione di Padova, Via F. Marzolo 8, 35131, Padova, Italy*
[6] *TRIUMF 4004 Wesbrook Mall Vancouver BC V6T 2A3, Canada*
[7] *Ruđer Bošković Institute, Bijenička cesta 54,HR-10000 Zagreb, Croatia*

Abstract.
In the present paper it is described an analysis procedure suited for experiments where cross-sections strongly varying with energy are measured using beams having large energy dispersion. These cross-sections are typically the sub-barrier fusion excitation function of reactions induced by radioactive beams. The large beam energy dispersion, typical of these experiments, can lead to ambiguities in the association of the effective beam energy to the reaction product yields and consequently to an error in the determination of the excitation function. As a test case, the approach is applied to the experiments ^{6}Li+^{120}Sn and ^{7}Li+^{119}Sn measured in the energy range 14 MeV $\leq E_{c.m.} \leq$28 MeV. The complete fusion cross sections are deduced from activation measurements using the stacked target technique. The results of these experiments, that employ the two weakly-bound stable Li isotopes, show that the complete fusion cross sections above the barrier are suppressed of about 70% and 85% with respect to the Universal Fusion Function, used as a standard reference, in the ^{6}Li and ^{7}Li induced reactions respectively. Moreover, the excitation functions of the two systems at energies below the barrier, do not show significant differences, despite the two systems have different *n*-transfer Q_{value}.

1 Introduction

The stacked target technique has been used in many experiments to measure excitation functions of reactions, such as for example fusion, induced by low energy radioactive beams e.g. [1–5]. In order to increase the production yields of the reaction products, thick targets have also been employed in some of these experiments. The advantage of using the stacked target technique consists in measuring the cross-section at several energies simultaneously; as a drawback, however, there is the degradation of the beam quality as it passes through the foils of the stack, due both, to the statistical nature of the energy loss process, and to any target non-uniformities. Ambiguities of associating effective beam energies to reaction product yields for the targets within the stack may arise and, as a consequence, an error in the extraction of the excitation function for the reaction under study. In [6] the possible sources of ambiguities arising when performing experiments with large beam energy spread were investigated thoroughly. Typically, the large beam energy dispersions are due either to the use the stacked target technique and/or thick-non-uniform targets, or to the beam production technique. Moreover, in [6], a best practice procedure to extract cross-sections strongly varying with energy obtained with large beam energy dispersions was proposed. In the present paper the technique proposed in [6] will be summarized and compared to the iterative process suggested in [7]. In addition, the analysis procedure proposed in [6] is applied to the measurements of ^{6}Li+^{120}Sn and ^{7}Li+^{119}Sn fusion excitation functions, measured in the energy range from 14 MeV to 28 MeV in the center of mass. The aim of these experiments is twofold: to investigate the effect of neutron transfer channels with positive Q_{value} on the sub-barrier fusion probability e.g. [8–12], and the suppression of the complete fusion cross-section above the barrier in reactions induced by weakly-bound beams, with respect to one-dimensional Barrier Penetration Model (1D-BPM), or coupled-channel (CC) calculations which do not take coupling to continuum or transfer into account (see e.g. [13–17]).

* e-mail: dipietro@lns.infn.it

 DOI: 10.1051/epjconf/201716300014

2 Experimental determination of the fusion excitation function

The mean cross-section, σ_{mean}, of a given reaction process, is experimentally deduced by using the equation:

$$\sigma_{mean} = \frac{Y}{N_{t_0} \cdot N_B} \quad (1)$$

where N_{t_0} is the number of atoms per unit area of the target, and N_B the number of beam particles passing through the target. If one has to derive an excitation function, the problem is how to relate these measured mean cross-sections to effective beam energies. Generally it is assumed $\sigma_{mean} = \sigma(\bar{E})$ where $\bar{E}$ is the energy in the centre of the target, as calculated by energy-loss programs. Alternatively, it has been assumed [5] that $\sigma_{mean} = \sigma(E_{eff})$, where E_{eff} is the weighted mean energy defined by eq. 2:

$$E_{eff} = \frac{\int_0^\infty E\sigma(E)D(E,t_0)dE}{\int_0^\infty \sigma(E)D(E,t_0)dE} \quad (2)$$

The weighting factors are the cross-section σ(E) and the beam energy probability distribution function $D(E,t_0)$ inside the target. $D(E,t_0)$ represents the probability to find a beam particle with energy E inside a given target with average thickness t_0.

In the case of large beam energy dispersion, and when the cross section is known to vary rapidly over the energy range explored by the beam in the target, the above methods of relating σ_{mean} to σ(E) can result in misinterpretation of the real excitation function.

The measured mean cross sections σ_{mean}, are energy mean values of the real excitation function σ(E) and, as discussed in [6], can be specified for a given measurement i as:

$$\sigma_{mean,i} = \frac{\int_0^\infty \sigma(E)D_i(E,t_0)dE}{\int_0^\infty D_i(E,t_0)dE}, \quad (3)$$

The problem of deducing the correct σ(E) from the measured σ_{mean} can be solved by choosing a suitable function $\hat{g}(E,\mu)$ (with μ parameters to be determined) which shows the same expected energy behavior as $\sigma(E)$. In [6] it was suggested an unfolding procedure to deduce the correct σ(E). The mean value of the function $\hat{g}(E,\mu)$ over the energy range explored by the beam in the target is given by eq. 4:

$$\hat{g}_{mean,i} = \frac{\int_0^\infty \hat{g}(E,\mu)D_i(E,t_0)dE}{\int_0^\infty D_i(E,t_0)dE}, \quad (4)$$

The function $\hat{g}(E,\mu)$ can be obtained by minimizing the following expression with respect to μ:

$$S = \sum_i \left(\frac{\sigma_{mean,i} - \hat{g}_{mean,i}}{\beta_i}\right)^2, \quad (5)$$

where β_i is the experimental error associated with $\sigma_{mean,i}$.

In order to apply the above procedure, in addition to finding the proper $\hat{g}(E,\mu)$, it is required the precise knowledge of $D_i(E,t_0)$: this can be deduced, with a good degree of approximation, using the code SRIM [18, 19].

3 Determination of the energy probability distribution for uniform and non uniform foils

The beam energy probability distribution is given by eq. 6:

$$D(E,t_0) = \int_0^\infty g(E_0,E_i) \int_0^{t_0} f(E_i,t_0,E,x')dx'dE_i \quad (6)$$

this, as mentioned above, represents the probability to have a beam particle with energy E inside the considered target of thickness t_0. $f(E_i,t_0,E,x')$ represents the probability that a projectile incident on the target of thickness t_0 at an incident energy E_i has an energy E at a depth x' inside the target, with $0 < x' < t_0$. $g(E_0,E_i)$ represents the probability that a particle, as part of the entrance beam of mean energy E_0, has an energy E_i. In many cases, $g(E_0,E_i)$ is usually considered a Gaussian distribution, however, as shown in [6], this is not always the case.

The SRIM code can be used for determining $f(E_i,t,E,x)$ and $g(E_0,E_i)$ with a procedure discussed in detail in [6].

In general, targets are non-uniform, so to explicitly account for target non-uniformity Eq. (6) should include another term, the thickness probability distribution function, $w(t)$, which represents the probability that a particle entering a non-uniform target actually experiences a thickness t. In this case the energy distribution inside the target will be defined as follows:

$$D(E,t_0) = \int_0^{t_0} w(t) \int_0^\infty g(E_0,E_i) \int_0^t f(E_i,t,E,x')dx'dE_idt \quad (7)$$

with $0 < x' < t$, and t_0 represents the maximum target thickness. Therefore, for determining $D_i(E,t_0)$ it is necessary to know $w(t)$. The probability distribution function $w(t)$ of a given target can be determined experimentally using the residual energy spectra of α-particles emitted from a ^{241}Am source passing through the target, and SRIM simulations. Using SRIM, the residual energy spectra of the ^{241}Am α-particles, passing though a ^{93}Nb foil plus a ^{120}Sn foil whose thickness was varied from 0 mg/cm^2 to 1.5 mg/cm^2, were calculated. A weight, $w(t)$, was attributed to each of the individual calculated spectra such that their weighted sum was reproducing the experimental residual energy spectrum. Fig. 1 represents the experimental residual energy spectrum of α-particles crossing a ^{120}Sn target evaporated onto a ^{93}Nb substrate, used in the ^{6}Li+^{120}Sn experiment, compared with the spectrum obtained with the SRIM simulations. In Fig. 2 the corresponding $w(t)$ is reported as function of the target thickness. In this specific case the ^{120}Sn targets used in the experiment were highly non-uniform. It seems that high non-uniformities are typical of Sn targets, unless special care is taken in the target production procedure to

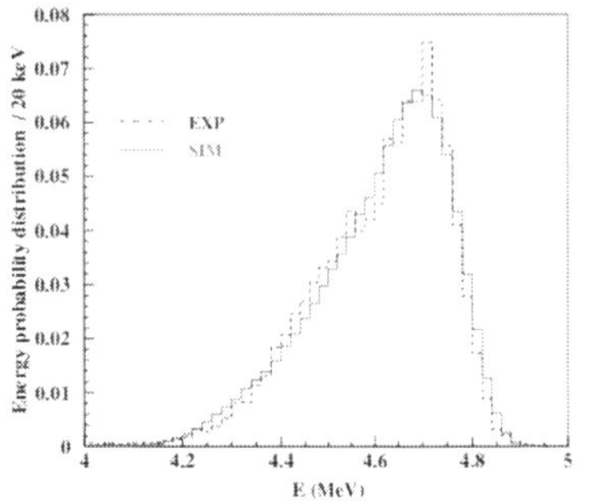

Figure 1. (Color online). Experimental (black dashed line) residual energy spectrum of α-particles passing-through a ^{120}Sn+^{93}Nb foil. The red spectrum is the result of the fit procedure used to extract the target thickness probability distribution. Figure from [6].

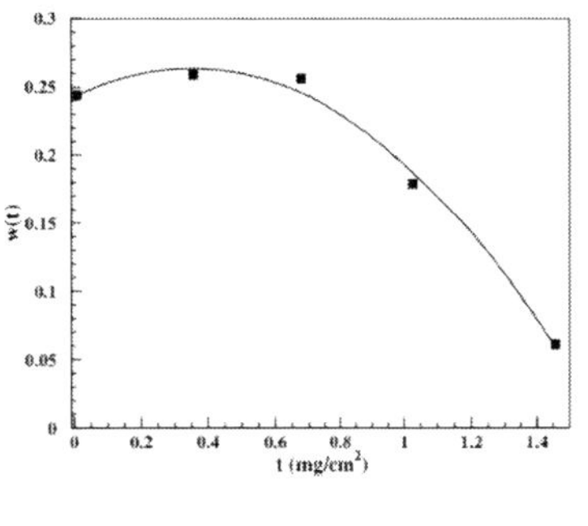

Figure 2. ^{120}Sn target thickness distribution. The continuous line is a polynomial fit of the the $w(t)$. Figure from [6].

prevent their non-uniformity.

4 An alternative procedure to deduce σ(E)

The fitting procedure reported in section 2 is one of the possible ways to solve the problem of finding the best approximation of σ(E). An alternative solution to the problem of deducing the excitation function from mean cross-sections measured over a broad energy range is suggested in [7]. In [7], an iterative method is used to obtain the fusion excitation function in the case of large beam energy-loss in the target. In the experiment reported in [7] the beam energy loss in the target was as large as 40 MeV in the laboratory system. The starting point was plotting the measured cross-section as a function of $\bar{E}$. The obtained curve was fitted with a tensioned spline. The resulting curve was used to calculate σ_{mean} for each measured point i using eq. 8.

$$\sigma_{mean,i} = \int \frac{\sigma(E)}{\frac{dE}{dx}} \frac{dE}{\rho}, \qquad (8)$$

where $\frac{dE}{dx}$ is the stopping power of the beam in the target and ρ is the target thickness. Each energy E$_i$, corresponding to the cross section, $\sigma_{mean,i}$,was obtained by interpolation using the fitting curve. These new points ($\sigma_{mean,i}$, E$_i$) were used as inputs for the next iteration of the fit. The procedure was repeated iteratively until convergence was reached. Contrary to what suggested in the present paper, the procedure of [7] does not take account of the correct energy probability distribution of the beam in the target $D_i(E,t_0)$, but only of the effect of the energy-loss. Moreover, it is opinion of the present authors that eq. 8, is formally inaccurate; a more formally correct form of eq. 8 reads as:

$$\sigma_{mean,i} = \frac{\int \sigma(E) \frac{1}{\frac{dE}{dx}(E)} dE}{\int \frac{1}{\frac{dE}{dx}(E)} dE}. \qquad (9)$$

In order to check if the iterative procedure gives the same results as minimising eq. 5, the former is applied to the simulated study of ^{9}Li+^{120}Sn fusion reaction reported in [6]. The function σ(E) is assumed to be given by a Wong-like function [20].
Simulations are performed considering a ^{9}Li beam of about 28 MeV impinging on a stack of 5 foils made of 5 mg/cm^2 ^{120}Sn, each followed by 1.5 mg/cm^2 ^{93}Nb foil. The correct beam energy probability distribution in the target $D_i(E,t_0)$ is calculated in the case of uniform foils and non-uniform foils having Gaussian type non-uniformities of FWHM=20% for ^{120}Sn and 15% for ^{93}Nb. The results are shown in fig. 3 in the case of uniform foils and in fig. 4 for non-uniform foils. Moreover, a comparison is made using the beam energy probability distribution used in [7] i.e. $\frac{1}{\frac{dE}{dx}(E)}$, also shown in figures 3 and 4.

As can be seen from the figures, the iteration process will give the correct excitation function (red symbols overlap with the full line). However, this is true only if the correct beam energy distributions inside the targets are used. In fact, the points at the lowest energies deviate from the curve in the case $\frac{1}{\frac{dE}{dx}(E)}$ rather than the correct $D_i(E,t_0)$ is used, and these deviations are larger for non-uniform foils.
A problem of the iterative approach is, however, the error estimate in the deduced excitation function. In the procedure suggested in the present paper the uncertainties are deduced from the errors in the function parameters. It is not so obvious how to deduce them in the case of the iterative procedure. No mention is made in [7].

5 The ^{6}Li+^{120}Sn and ^{7}Li+^{119}Sn fusion excitation functions

5.1 Experiment

^{6}Li+^{120}Sn and ^{7}Li+^{119}Sn fusion cross-sections were measured at INFN-Laboratori Nazionali del Sud (Catania), by using an activation technique based on the off-line detection of the atomic X-rays emitted after the electron capture (EC) decay of the evaporation residues (ER).

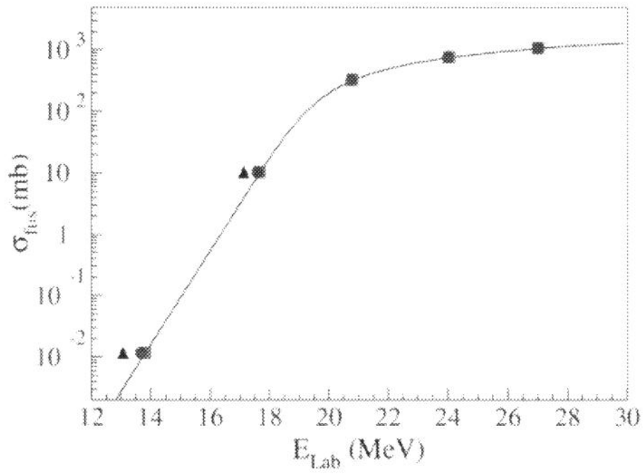

Figure 3. (Color online). Results of the ^{9}Li+^{120}Sn simulations using the iterative procedure to deduce the fusion excitation function: continuous line, Wong-like function; red squares, excitation function obtained considering the correct $D_i(E,t_0)$ in the target for the case of uniform foils; blue circles, excitation function obtained considering only the effect of energy-loss in the target as in [7]. The black triangles correspond to σ_{mean} versus $\bar{E}$.

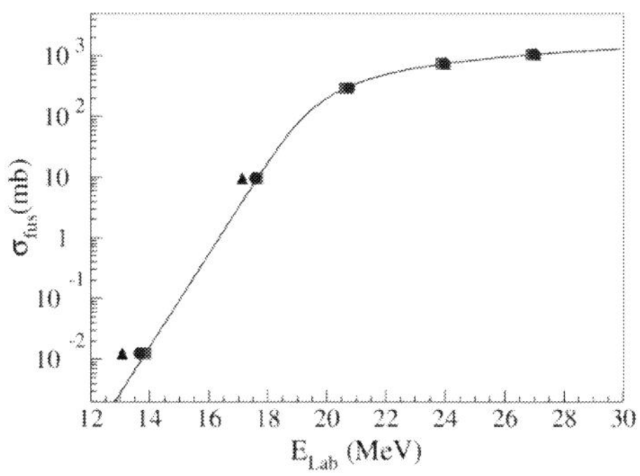

Figure 4. (Color online). Results of the ^{9}Li+^{120}Sn simulations using the iterative procedure to deduce the fusion excitation function: continuous line, Wong-like function; red squares, excitation function obtained considering the correct $D_i(E,t_0)$ in the target for the case of non-uniform foils; blue circles, excitation function obtained considering only the effect of energy-loss in the target as in [7]. The black triangles correspond to σ_{mean} versus $\bar{E}$.

This technique is particularly suitable for these systems since, according to statistical model calculations, the compound nucleus (^{126}I) decays predominantly by the 3n and 2n evaporation channels, producing ^{123}I ($t_{1/2}$ = 13 h) and ^{124}I ($t_{1/2}$ = 4 days), decaying by EC. Thus, in this case, the complete fusion process can be clearly discriminated against any incomplete fusion or transfer process. The same technique has been used successfully to measure the fusion excitation functions of the systems ^{6,7}Li,^{4,6}He+^{64}Zn [1, 2, 21, 22].

In the present case, it would have been possible to perform the experiment using single target/catcher foil at each beam energy, due to the high intensity of the stable beams involved. However, as a test for a future experimental program with unstable lithium isotopes, the stacked target technique is used instead. The unfolding procedure is therefore required to deduce the complete fusion excitation function. Details of the experiment are reported in [23].

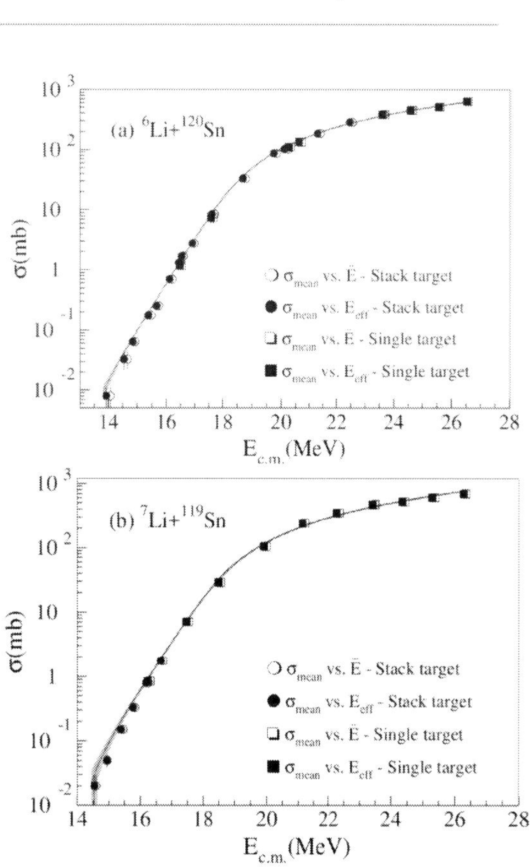

Figure 5. (Color online) Fusion excitation function for a) ^{6}Li+^{120}Sn and b) ^{7}Li+^{119}Sn systems extracted as described in detail in [23] (continuous line) and its associated uncertainty (coloured band). The symbols represent the measured cross-sections plotted in the usual way i.e. as (σ_{mean}, $\bar{E}$) and (σ_{mean}, E_{eff}). Circles are used for points measured with stacked targets and squares are used for single foil measurements. Figure from [23].

For each irradiated foil the thickness probability distribution function $w(t)$ is extracted. Known $w(t)$ it is possible to calculate, using SRIM, the beam energy probability distribution $D_i(E,t_0)$ for each measured point of the excitation function. Using the unfolding procedure described in section 2 the fusion excitation function for the two reactions is extracted, and it is shown in fig. 5. An important check to be done in order to verify that the functional form obtained using the unfolding procedure is a good representation of the fusion excitation function, is to plot the ratio between the $\sigma_{calc-mean,i}$, calculated from

DOI: 10.1051/epjconf/201716300014

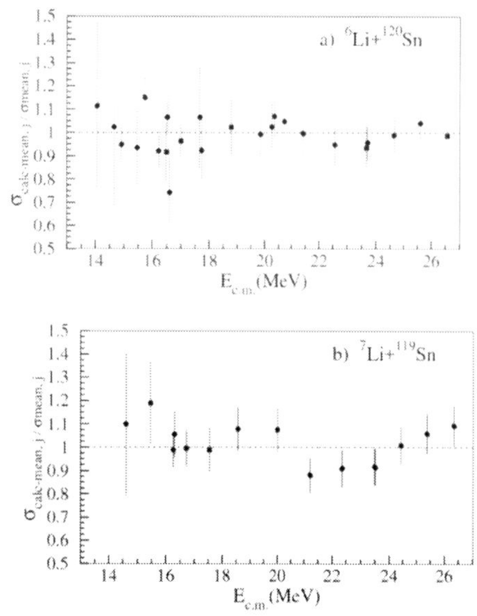

Figure 6. Ratio between $\sigma_{calc-mean,i}$ and $\sigma_{mean,i}$ for the reactions: ^{6}Li+^{120}Sn a), and ^{7}Li+^{119}Sn b). Figure from [23]

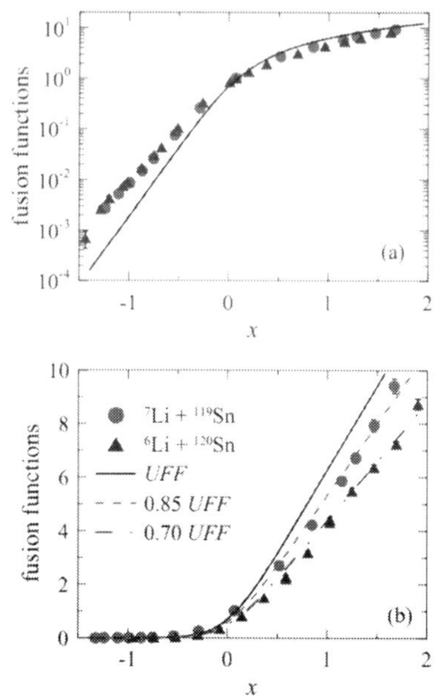

Figure 7. (Color online) Comparison of CF reduced cross section for the ^{6}Li +^{120}Sn and ^{7}Li +^{119}Sn systems with the UFF. The experimental data reported in the figure correspond to ($\sigma_{mean,i}$, $E_{dec,i}$), where $E_{dec,i}$ is the energy point of deconvolution curve which corresponds to the measured cross-section $\sigma_{mean,i}$. Figure from [23].

the functional form $\hat{g}(E,\mu)$ averaged over the beam energy distribution in the target $D_i(E,t_0)$, and the measured one $\sigma_{mean,i}$. If this ratio is ≈ 1 in the whole energy range then the procedure is self consistent, and the extracted functional form is a good representation of the fusion excitation function. This was verified to be the case of the present analysis, and the ratio for the two reactions is shown in fig. 6.

5.2 Results

At energies above the Coulomb barrier the complete fusion cross-section, in the case of reactions induced by weakly bound beams, shows a suppression with respect to the 1D-Barrier Penetration Model (BPM) and/or the Coupled Channel (CC) calculations which do not take coupling to continuum or transfer into account ([16] and reference therein). It is found that the suppression factor depends on the breakup threshold of the projectile e.g. [16].

In order to deduce the suppression factors of the complete fusion cross-section for the two systems ^{6}Li+^{120}Sn and ^{7}Li+^{119}Sn, the cross-sections were reduced according to the reduction procedure introduced by Canto et al. [14, 15]. Once the data are reduced, the complete fusion suppression factor is extracted by comparing them with a benchmark function, called Universal Fusion Function (UFF) defined in [14, 15]. In fig. 7 such a comparison is shown. In fig. 7 a, the reduced cross-sections are shown in logarithmic scale and in fig. 7 b in linear. The Coulomb barrier corresponds to $x = 0$. From fig. 7, it can be seen that by comparing the experimental reduced fusion cross-section with the UFF, for both reactions, the complete fusion is enhanced at energies below the barrier and hindered above. The deduced suppression factor above the barrier for the ^{6}Li +^{120}Sn system is 0.70±0.05, and for the ^{7}Li+^{119}Sn system is 0.85±0.05. Such values, while confirming the breakup threshold dependence of the suppression factors, appear to be slightly nearer to unity than the ones found in the systematic of Wang et al. [17] for other systems.

At energies below the barrier, as it was already observed in collisions of ^{6,7}Li with different targets e.g. [16], the UFF underestimates the data. Since coupling to continuum and transfer channels are not taken into account in the reduction of the data, they are responsible for the underestimation of the UFF.

What is interesting to note, is that below the barrier the two sets of reduced data (fig. 7 a) are very close in spite of the different n-transfer Q_{value}. This behavior seems to be consistent with the one observed in fusion reactions of heavier systems [24–26] and appears to confirm the idea, proposed by [27–29], that the mechanism that produces the enhancement of the sub-barrier fusion cross-section may be rather complex, and that the relevant effects is not limited to the difference in the n-transfer Q_{value}. CRC calculations were also performed for the ^{7}Li+^{119}Sn which has the largest positive Q_{value} for n-transfer (details can be found [23]), and they confirm that the effect of the coupling with n-transfer channel produces negligible effects.

6 Summary and conclusions

This paper highlights the problem associated with the extraction of the correct excitation function in the case where cross-sections strongly varying with energy, with beams having large energy dispersion, are measured. These experiments are typically the sub-barrier fusion excitation function measurements with radioactive beams, where it is necessary to maximise the experimental yields due to the low intensity of the beams combined with the small cross-sections involved.

It has been demonstrated that in the case of large beam energy dispersion, there are ambiguities in associating the effective beam energy to the measured mean cross-section and this can lead to an error in the determination of the excitation function. In fact, plotting σ_{mean} either versus E_{mean} or E_{eff} will not give the correct excitation function, and then a significant misinterpretations of the data may arise. Accurate information can be obtained from the experimentally measured σ_{mean}, provided that a full characterisation of the targets is done. In such a way, it is possible to calculate for every target the beam energy probability distribution $D(E,t_0)$ function, which specifies how the cross-section will be integrated over the beam energy.

This deconvolution procedure is used to deduce complete fusion excitation functions for the two reactions, ^{6}Li+^{120}Sn and ^{7}Li+^{119}Sn in the c.m. energy range from 14 to 28 MeV. The measured cross-sections, renormalised according to [14, 15], were compared to the UFF. The experimental cross sections are found to be suppressed with respect to the UFF at energies above the Coulomb barrier. The suppression factor for ^{6}Li+^{120}Sn is found 0.7±0.05, and for ^{7}Li+^{119}Sn is found 0.85±0.05. These factors are a little smaller than the ones obtained with heavier targets.

Below the Coulomb barrier, the renormalized complete fusion cross-sections are very similar even though the Q_{value} for neutron transfer are substantially different. Therefore, the mechanism that produce the sub-barrier enhancement, experimentally observed in some reactions, must be more complicated than just the difference in the Q_{value}.

References

[1] A. Di Pietro et al. Phys. Rev. C **69**, 044613 (2004)
[2] V. Scuderi et al. et al. Phys. Rev. C **84**, 064604 (2011)
[3] J. J. Kolata et al. Phys. Rev. Lett. **81**, 4580 (1998)
[4] C. Signorini et al. Nucl. Phys. A **735**, 329 (2004)
[5] R. Wolski et al. Eur. Phys. Jour. A **47**, 111 (2011)
[6] M. Fisichella et al. Phys. Rev. C **92**, 064611 (2015)
[7] J. F. Liang et al. Phys. Rev. C **75**, 054607 (2007)
[8] T. H. Stelson et al. Phys. Rev. C **41**, 1584 (1990)
[9] R. A. Broglia et al. Phys. Rev. C **27**, 2443 (1983)
[10] S. Kalkal et al. Phys. Rev. C **81**, 044610 (1990)
[11] H. Q. Zhang et al. Phys. Rev. C **82**, 054609 (2010)
[12] A. M. Stefanini et al. Phys. Rev. C **52**, R1727 (1995)
[13] M. Dasgupta et al. Phys. Rev. C **70**, 024606 (2004)
[14] L. F. Canto et al. Jour. Phys. G **36**, 015109 (2009)
[15] L. F. Canto et al. Nucl. Phys. A **821**, 51 (2009)
[16] L. F. Canto et al. Phys. Rep. **596**, 1 (2015)
[17] B. Wang et al. Phys. Rev. C **90**, 034612 (2014)
[18] BJ. F. Ziegler et al. Nucl. Inst. Meth. B **268**, 1818 (2010)
[19] J. F. Ziegler, *SRIM code* (http://www.srim.org)
[20] C. Y. Wong Phys. Rev. Lett. **31**, 766 (1973)
[21] A. Di Pietro et al. Eur. Phys. Jour. S. T. **150**, 15 (2007)
[22] A. Di Pietro et al. Phys. Rev. C **87**, 064614 (2013)
[23] M. Fisichella et al. Phys. Rev. C **95**, 034617 (2017)
[24] Z. Kohley et al. Phys. Rev. Lett. **107**, 202701 (2011)
[25] H. M. Jia et al. Phys. Rev. C **86**, 044621 (2012)
[26] A. M. Stefanini et al. Eur. Phys. Jour. A **49**, 1 (2013)
[27] V. V. Sargsyan et al. Phys. Rev. C **85**, 024616 (2012)
[28] V. V. Sargsyan et al. Phys. Rev. C **91**, 014613 (2015)
[29] V. A. Rachkov et al. Phys. Rev. C **90**, 014614 (2014)

How elements up to 118 were reached and how to go beyond

Christoph E. Düllmann[1,2,3,4,⋆]

[1]*Institute of Nuclear Chemistry, Johannes Gutenberg University, 55128 Mainz, Germany*
[2]*GSI Helmholtzzentrum für Schwerionenforschung, 64291 Darmstadt, Germany*
[3]*Helmholtz Institute Mainz, 55099 Mainz, Germany*
[4]*Australian National University, Canberra, ACT 2601, Australia*

Abstract. The new superheavy elements with Z=113, 115, 117, and 118 were recently accepted into the periodic table and have been named. Elements with Z≥112 are predominantly produced in ^{48}Ca-induced fusion reactions on actinide targets. This pathway is exhausted at Z=118 due to the lack of target materials with sufficiently high proton number to reach elements with Z≥119. Search experiments for yet heavier elements were performed at GSI Darmstadt and FLNR Dubna. The reactions ^{50}Ti + ^{249}Bk, which leads to Z=119, as well as ^{64}Ni + ^{238}U, ^{58}Fe + ^{244}Pu, ^{54}Cr + ^{248}Cm, and ^{50}Ti + ^{249}Cf, leading to Z=120, have been studied. Despite a total duration of these experiments of more than one year, neither succeeded in the identification of a new element. To obtain improved guidance for better-informed search experiments, nuclear reaction studies appear necessary and have recently started. Also technical advances will be an important pillar to this end. At GSI, work towards a new continuous-wave linear accelerator is ongoing and is briefly described.

1 Introduction

Recently, new elements with Z=113, 115, 117, and 118 were officially accepted [1, 2] and were named in November 2016 [3]. With this, the periodic table of the elements (cf. Fig. 1) is complete to the end of the 7th period. Fig. 2 shows the corresponding cut-out of the chart of nuclei (with decay properties updated according to [4]), where all known or claimed nuclei are indicated in colored boxes, overlaying the calculated shell correction energy landscape according to a macroscopic-microscopic model [5]. Also indicated (by white/black boxes for yet unknown isotopes) are the hypothetical decay chains originating from the 3n and 4n compound nucleus (CN) evaporation channels following complete fusion of ^{50}Ti + ^{249}Bk leading to $^{299}119^{*}$ (red bordered boxes), ^{50}Ti + ^{249}Cf leading to $^{299}120^{*}$ (light blue bordered boxes), and ^{54}Cr + ^{248}Cm, ^{58}Fe + ^{244}Pu, and ^{64}Ni + ^{238}U, which all lead to $^{302}120^{*}$ (dark blue bordered boxes). Decay modes of unknown isotopes were assumed based on [6–8].

2 The search for new elements beyond Og

For the heaviest currently known elements with Z=112-118, employing fusion-evaporation reactions induced by doubly-magic neutron-rich ^{48}Ca on targets of $_{92}$U-$_{98}$Cf [10] is well established, having been confirmed in multiple laboratories (e.g., [11–16]) and would, therefore, also appear to be a natural choice for attempts to synthesize yet heavier elements. Cross sections stay relatively constant at a level of 0.5-10 pb over an extended region, suggesting that the accumulated data set provide a good basis for theoretical models to predict values for elements 119 and 120. The one caveat is that the corresponding target materials would have to be $_{99}$Es and $_{100}$Fm, which are not available in quantities that suffice to make targets for such synthesis experiments. Current technology for the production of transuranium isotopes, e.g., at the ORNL's High Flux Isotope Reactor (HFIR) is capable of producing μg and pg amounts of long-lived $_{99}$Es and $_{100}$Fm isotopes, respectively [17], which is orders of magnitude less than the experiments discovering and studying the heaviest known elements required. Still, fusion reactions with actinide targets are generally expected to provide the highest cross sections for the synthesis of new elements with Z≥119. Thus, reactions induced by beams heavier than ^{48}Ca will be needed. Five of the most commonly suggested reactions giving access to Z≥119 have been studied experimentally and are listed in Table 1. None of these experiments have led to the conclusive discovery of a new element, though.

A total of more than one year of accelerator beamtime proved insufficient for the discovery of a new element, whereas any of the known elements would have been synthesized in significant amounts given the current sensitivity levels of such experiments. This suggests that either cross sections are significantly lower for all of the actinide-based reactions using beams heavier than ^{48}Ca, or that some aspects relevant for performing a successful search experiment are not understood well enough. Likely, it is a combination of both. One such crucial aspect is

⋆e-mail: duellmann@uni-mainz.de

 DOI: 10.1051/epjconf/201716300015

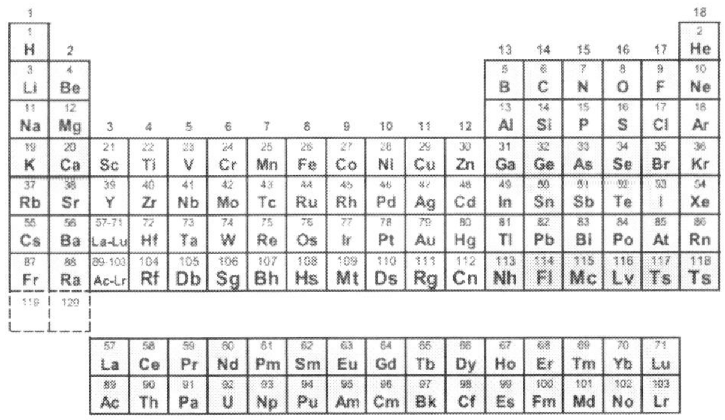

Figure 1. Periodic Table of the Elements. The known superheavy elements with Z=104-118 are highlighted. The positions of the elements with Z=119 and 120, which are yet unknown and would form the beginning of the 8th period, are also indicated.

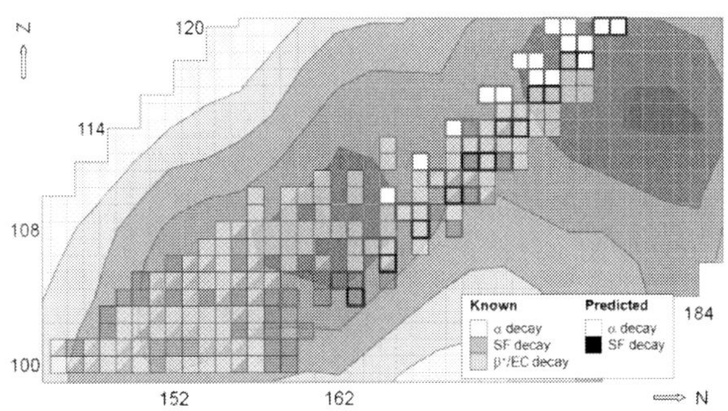

Figure 2. Chart of nuclei of the heaviest elements. The underlying contour plot shows calculated shell correction energies [5]. Known nuclei are indicated by colored boxes, new nuclei occurring in the hypothetical 3n/4n decay chains produced in complete fusion of different target/projectile combinations leading to Z=119 and Z=120 are indicated by white and black boxes. See text for more details. Taken from [9].

Table 1. Fusion reactions employed to search for elements beyond Z = 118.

	Compound nucleus	Beamtime invested	Cross section limit^{a}	Ref.
^{50}Ti + ^{249}Bk	^{299}119*	4.5 months	^{b}	[18]
^{64}Ni + ^{238}U	^{302}120*	4 months	≥90 fb	[19]
^{58}Fe + ^{244}Pu	^{302}120*	2 months	≥400 fb	[20]
^{54}Cr + ^{248}Cm	^{302}120*	>1 month	≥580 fb	[21]
^{50}Ti + ^{249}Cf	^{302}120*	>1 month	^{b}	[18]

^{a} 63.2% confidence level ("one-event limit")
^{b} under final analysis

that of the optimum beam energy, which indeed seems to be understood insufficiently, as is detailed elsewhere [18]. One has to bear in mind that some of the predictions of the nuclear half-lives of the produced isotopes, especially in the case of Z=120, suggest these to be in the order of microseconds, i.e., comparable to the flight-time through the used separators. This may imply that losses due to decay-in-flight might have negatively affected experiments. Still, the predominant reason for the lack of finding any new element seems to be insufficient sensitivity. Preparations for improved experiments are ongoing at several laboratories. Thus, it appears appropriate to reflect on which aspects should get most attention to perform the best possible experiments, thus maximizing the likelihood of success. Obvious, purely technical aspects include higher primary beam intensities, targets that can withstand such intensities over extended periods [22, 23], more efficient separators, more sensitive detection systems with dead-time free data acquisition [24, 25] that allow safe identification of single decay chains from new elements, and longer irradiation times. As detailed in [18], where the system ^{50}Ti+^{249}Cf is employed to illustrate the current situation, optimum beam energies suggested by different theoretical frameworks differ by significantly larger amounts than can be covered in a single experiment. This highlights the need for a well-informed choice of the irradiation energy. Performing searches that are significantly more sensitive than those summarized in Tab 1 at multiple beam energies (which were all obtained at one single energy each) would consume years of beamtime.

In this situation, performing complementary experiments studying these fusion reactions, which yield information that guides future new element search efforts, appears well justified. For this, we recall that the formation of a superheavy nucleus is parametrized as the product of three individual steps, each characterized with its own probability: $\sigma_{ER} = \sigma_{cap} \cdot P_{CN} \cdot W_{sur}$ with σ_{ER} the evaporation residue (ER) cross section, σ_{cap} the capture cross section, P_{CN} the compound nucleus formation probability, and W_{sur} the exit channel survival probability. The formation of superheavy elements is significantly suppressed by the quasifission (QF) process, in which two colliding nuclei reseparate on a fast time-scale before coalescing into a compound nucleus, see [26], and Fig. 4 for a schematic of the relevant processes.

With QF being the complementary process to compound nucleus formation, P_{QF} = 1-P_{CN}, where P_{QF} can reach large values. The QF process thus profits from high rates for systems where P_{CN} is small, e.g., SHE formation. Its study thus gives access to phenomena relevant for SHE formation, but in comparatively very short experiments. Observables include mass angle distributions [27] or mass energy distributions [28]. Combined with the corresponding experimental data, which can be recorded in short experiments, this will allow for the testing of different models, thus providing guidance concerning which ones deliver the most trustworthy predictions. Better informed selection of, e.g., the optimum irradiation energy will be invaluable. First steps in this direction have been made recently [26, 29]. A combined effort encompassing theoretical as well as experimental work will provide an improved basis for the certainly coming future efforts to discover new elements with Z>118 and to thus open a new row in the Periodic Table of the Elements.

2.1 R&D activities towards a new cw linac

Superheavy element research is a pillar of the GSI research program for many decades and has led to high-visibility results. These include the discovery of six new elements [30] and chemical investigations for the first classification of new elements in the periodic table [31]. Technical advancements recently facilitated further milestone successes, including synthesis of elements up to Ts [12–14, 32], extensive systematic nuclear structure studies of odd-mass even-Z isotones (N constant) and odd-mass odd-Z isotopes (constant Z) in the "high-cross section" region Z=102-106 around N=152 (see [33] for a recent example) as well first such studies in Mc decay chains [13], high-precision mass measurements of single atoms of the heaviest elements [34, 35], laser spectroscopic studies of their electron shell [36], as well as advances in chemical studies up to the heaviest studied elements [37], resulting in a worldwide unique breadth and depth of the research program. Beams for all these studies are provided by the UNILAC linear accelerator, a pulsed machine that provides up to three types of projectiles from different ion-sources to several experiments in quasi-parallel operation. The maximum duty cycle is 25%. This offers unique possibilities to significantly enhance beam availability by the installation of a continuous-wave (cw) heavy-ion linac for highly intense beams, as has been suggested, e.g., in [38]. With the foundation of the Helmholtz Institute in Mainz, Germany, the R&D towards such a machine has started [39]. With its continuous beam, a cw machine would be ideal for a steady irradiation at significantly higher intensities than are presently available and will thus allow advancing the GSI programs in superheavy element research, materials research and biophysics. In the field of SHE, this would provide for experiments at higher sensitivity, as well as for in-depth studies of nuclei produced with higher cross sections thanks to higher production rates, Currently, a first "crossed H-bar"(CH) type cavity has been built to demonstrate the performance of this novel accelerating structure and is in the commissioning phase. Fig. 5 shows a photograph of the cavity embedded by two sc solenoids mounted on a common support structure, which will be placed inside the cryostat.

Details on the cw linac demonstrator as well as a possible layout of the full machine are given in [40]. Next steps, which have started, include the installation of a string of cavities.

References

[1] P. J. Karol et al., Pure Appl. Chem. 88, 139 (2016).
[2] P. J. Karol et al., Pure Appl. Chem. 88, 155 (2016).
[3] L. Öhrström, and J. Reedijk, Pure Appl. Chem. 88, 1225 (2016).
[4] U. Forsberg et al., Nucl. Phys. A 953, 117 (2016).
[5] Z. Patyk, R. Smolanczuk, and A. Sobiczewski, Nucl. Phys. A 626, 337c (1997).
[6] I. Muntian, Z. Patyk, and A. Sobiczewski, Phys. At. Nucl. 66, 1015 (2003).
[7] I. Muntian et al., Acta Phys. Pol. B 34, 2073 (2003).
[8] R. Smolańczuk, J. Skalski, and A. Sobiczewski, Phys. Rev. C 52, 1871 (1995).
[9] Ch. E. Düllmann, EPJ Web Conf. 131, 08004 (2016).
[10] Y. T. Oganessian, and V. K. Utyonkov, Nucl. Phys. A 944, 62 (2015).
[11] S. Hofmann et al., Eur. Phys. J. A 32, 251 (2007).
[12] S. Hofmann et al., Eur. Phys. J. A 48, 62 (2012).
[13] D. Rudolph et al., Phys. Rev. Lett. 111, 112502 (2013).

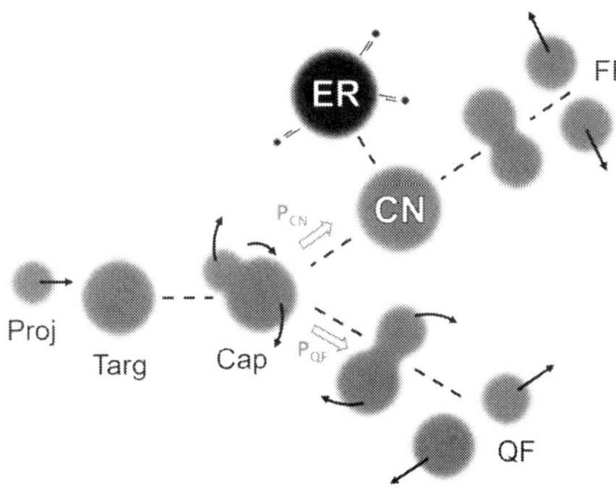

Figure 3. Schematic of the relevant processes in nucleus-nucleus collisions aimed at synthesizing the heaviest elements. Proj: projectile; Targ: target; Cap: captured system; CN: compound nucleus; ER: evaporation residue; FF: fusion-fission; QF: quasifission; P_{CN}: probability for the captured system to form a compound nucleus; P_{QF}: probability for the captured system to fission. Obviously, $P_{QF}=1-P_{CN}$.

Figure 4. Superconducting cw linac demonstrator comprising a CH cavity embedded by two sc solenoids on a support frame (taken from [40]).

[14] J. Khuyagbaatar et al., Phys. Rev. Lett. 112, 172501 (2014).
[15] L. Stavsetra et al., Phys. Rev. Lett. 103, 132502 (2009).
[16] D. Kaji et al., J. Phys. Soc. Jpn. 86, 034201 (2017).
[17] J. B. Roberto et al., Nucl. Phys. A 944, 99 (2015).
[18] Ch. E. Düllmann et al., (in preparation) (2016).
[19] S. Hofmann, J. Phys. G 42, 114001 (2015).
[20] Y. T. Oganessian et al., Phys. Rev. C 79, 024603 (2009).
[21] S. Hofmann et al., Eur. Phys. J. A 52, 180 (2016).
[22] S. N. Dmitriev, and A. G. Popeko, J. Radioanal. Nucl. Chem. 305, 927 (2015).
[23] E. Jäger et al., J. Radioanal. Nucl. Chem. 299, 1073 (2014).
[24] I. H. Lazarus et al., IEEE Trans. Nucl. Sci. 48, 567 (2001).
[25] J. Khuyagbaatar et al., Phys. Rev. Lett. 115, 242502 (2015).
[26] D. J. Hinde et al., EPJ Web Conf. 131, 04004 (2016).
[27] R. du Rietz et al., Phys. Rev. C 88, 054618 (2013).
[28] M. G. Itkis et al., Nucl. Phys. A 944, 204 (2015).
[29] H. M. David et al., (in preparation) (2016).

[30] G. Münzenberg, Nucl. Phys. A 944, 5 (2015).
[31] M. Schädel, Eur. Phys. J. D 45, 67 (2007).
[32] Ch. E. Düllmann et al., Phys. Rev. Lett. 104, 252701 (2010).
[33] S. Antalic et al., Eur. Phys. J. A 51, 41 (2015).
[34] M. Block et al., Nature 463, 785 (2010).
[35] E. Minaya Ramirez et al., Science 337, 1207 (2012).
[36] M. Laatiaoui et al., Nature 538, 495 (2016).
[37] A. Yakushev et al., Inorg. Chem. 53, 1624 (2014).
[38] S. Hofmann et al., in EXON 2004, edited by Y. E. Penionzhkevich, and E. A. Cherepanov (World Scientific, Singapore, Peterhof, Russia, 2004), pp. 157.
[39] S. Minaev et al., Phys. Rev. ST Accel. Beams 12, 120101 (2009).
[40] W. Barth et al., EPJ Web Conf. 138, 01026 (2017).

Pre-equilibrium emission to study clustering in nuclei

D. Fabris[1,⋆], F. Gramegna[2], M. Cicerchia[2,3], T. Marchi[4], S. Barlini[5], S. Piantelli[5], M. Bini[5], M. Bruno[6], G. Casini[5], M. Cinausero[2], M. D'Agostino[6], M. Degerlier[7], N. Gelli[5], G. Mantovani[8], L. Morelli[6], J. Mabiala[2], A. Olmi[5], G. Pasquali[5], G. Poggi[5], S. Valdré[9], E. Vardaci[10], O.V. Fotina[11], V.L. Kravchuk[12], M. Colonna[13], and A. Ono[14]

[1] *INFN Sezione di Padova, Padova, Italy*
[2] *INFN Laboratori Nazionali di Legnaro, Legnaro (PD), Italy*
[3] *Dipartimento di Fisica e Astronomia, Universitá di Padova, Padova, Italy*
[4] *KU Leuven, Department of Physics and Astronomy Instituut voor Kern- en Stralingsfysica, 3001 Leuven, Belgium*
[5] *INFN Sezione di Firenze and Dipartimento di Fisica e Astronomia, Universitá di Firenze, Firenze, Italy*
[6] *INFN Sezione di Bologna and Dipartimento di Fisica e Astronomia, Universitá di Bologna, Bologna, Italy*
[7] *Science and Art Faculty, Physics Department, Nevsehir Haci Bektas Veli University, Nevsehir, Turkey*
[8] *INFN Sezione di Ferrara and Dipartimento di Chimica, Universitá di Ferrara, Ferrara, Italy*
[9] *Grand Accelerateur National d'Ions Lourds, 14076 Caen, France*
[10] *INFN Sezione di Napoli and Dipartimento di Fisica, Universitá di Napoli, Napoli, Italy*
[11] *Skobeltsyn Institute of Niclear Physics, Lomonosov Moscow St. University, Moscow, Russia*
[12] *National Research Center "Kurchatov Institute", Moscow, Russia*
[13] *Laboratori Nazionali del Sud, Catania, Italy*
[14] *Tohoku University, Sendai, Japan*

Abstract. Clustering structural properties in medium-mass systems have been investigated studying the competition between the evaporation and pre-equilibrium particle emission in central collisions. At variance with light nuclei, in the case of heavier nuclei there are no experimental evidences of such structure effects, since the determination of pre-formed clusters within nuclear matter is less obvious.
The two systems $^{16}O + ^{65}Cu$ and $^{19}F + ^{62}Ni$, leading to the same compound system $^{81}Rb^*$, have been studied at the same beam velocity (16 AMeV). The experiments have been performed using the GARFIELD + RCo multi-detection system at the Legnaro National Laboratories.
Recent results of the data analysis and of the comparison of the experimental data with different statistical and dynamical model calculations are reported in this contribution.

1 Introduction

The concept that cluster of nucleons might be pre-formed prior to emission from nuclei has a history of more than 50 years and, actually, the α-cluster model is one of the oldest models which was used to describe the nucleus. The original idea was introduced by Hafstad and Teller in 1938 [1]. Examining the binding energy per nucleon of light nuclei as a function of the mass number, they hypothesized that nuclei, with even and equal number of protons and neutrons, are particularly stable and their ground state could be described in terms of geometric arrangments of α-particles [2]. These so called α-conjugate nuclei are typically observed as excited states close to the decay threshold into clusters. This has been summarized, in the late sixties, in the Ikeda diagram [3] which links the energy required to liberate the cluster constituents to the excitation energy at which the cluster structure prevail in the host nucleus. Moreover, in neutron-rich nuclei there is the possibility that additional neutrons may act as valence particles and can be exchanged between the α-particle cores. These covalent neutrons stabilize the unstable multi-cluster states, giving rise to nuclear structure which can be described by molecular concepts. The extended Ikeda diagram is a new threshold diagram which describes the structure of these non-alpha conjugate nuclei [4].

Clustering effects can show up in several ways along the nuclear chart, from the stability to the drip-lines. While for light nuclei several links between cluster emission and its connection with nuclear structure and dynamics have been pointed out [4] [5], this is less obvious moving towards heavier systems. In fact, in reactions involving mediun-mass nuclei the determination of pre-formed clusters in nuclear matter is more complicated. These preformed clusters have been observed especially close to the nuclear surface, making strong the link between pre-equilibrium emission and cluster structure.

An interesting way to investigate the structural properties of medium-mass systems is to study the competition between evaporation and pre-equilibrium particles emis-

⋆e-mail: fabris@pd.infn.it

sion in central collision, as a function of different entrance channel parameters. Pre-equilibrium light charged particles and/or neutrons are fast, forward focused particles emitted during the very early stages of the collision before the attainment of full statistical equilibrium of the compoud system. Comparing such particles with those emitted after thermal equilibration it is possible to derive information on the interplay between equilibrium and non equilibrium processes. Moreover, structural properties, like cluster pre-formation probabilities, may be evidenced both from the experimental comparison between different entrance channel leading to the same compound system and from the comparison of experimental data with model predictions. Two opposite mechanisms have been suggested for cluster emission in pre-equilibrium reactions: on one side, the α-particle is assumed to be pre-formed inside the nucleus and it can be treated as a single strongly coupled object. On the other side, the coalesence models assume that clusters (not only α-particle) are formed, in a dynamical way, during the course of the reaction [5].

Many attempts have been done to shed light on these structures, but there is still a lack of experimental data capable to give a direct feedback to the theoretical predictions.

2 Previous studies

In the past, the system ^{16}O + ^{116}Sn has been investigated with the aim of identifying the amount of pre-equilibrium emission in asymmetric entrance channel reactions, in an energy range (E/A = 8$\div$16 AMeV) where pre-equilibrium emission is expected. Experimental spectra of protons and α-particles have been compared with the predictions of Moscow Pre-equilibrium Model (MPM) [6]. This model is a modified version of the statistical code PACE2 where a non-equilibrium stage before the complete thermalization of the compoud nucleus has been inserted. The relaxation processes occurring during the fusion reaction is accounted for by the exciton model, based on the Griffin model [7], in which the description of the angular distribution of the fast emitted particles is still an intricate question [8]. The main parameter to be set is the initial number of excitons (n_o=$n_{particles}$+n_{holes}) in the projectile, which can be estimated from the empirical trend obtained in the work of N. Cindro et al. [9]. In the present case of ^{16}O this number is n_o= 17 (16p + 1h).

The proton spectra have been quite well reproduced by the model at all the incident energies considered over the measured angular range. On the contrary, an enhanced fast α-particles production was observed, especially at the most forward measured angles, indicating that pre-equilibrium emission contribution is not sufficient to explain the increased prodution of α-particles. A possible esplanation might be related to the α-cluster structure of the ^{16}O projectile.

The effect of the projectile clustering structure has been taken into account introducing in the model a preformation cluster probability. A second starting configuration has been considered in which the ^{16}O projectile is supposed to be divided into ^{12}C plus an α-particle. The probability of occurence of this configuration with respect to the original one (i.e. ^{16}O projectile as a whole) is a free parameter to be determined from the comparison with the experiment. The comparison of the data with the model predictions with different percentages of α-clustering probability has shown that the proton distributions seem to be described with no-clustering in the projectile nucleus. For alpha-particles, on the contrary, a quite large clustering probability has to be considered to try to describe the spectral shape of the energy distribution at forward angles.

3 The experiment

To further investigate, in a model independent way, the possible effects of the α-cluster structure of the projectile, two different entrance channel reactions have been studied in an energy range where fast particle emission was predicted. The decay of ^{81}Rb* compund nucleus populated by the two fusion reactions ^{16}O + ^{65}Cu and ^{19}F + ^{62}Ni have been studied at 16 AMeV incident energy. The same projectile velocity has been chosen since the pre-equilibrium emission is expected to mostly depend on this parameter [10] As a consequence, the non-equilibrium processes are predicted to be almost the same for the two systems, while some little differences may appear in the evaporative part of the emitted particle spectra due to the slightly different initial excitation energies of the compound nucleus (E*=209 MeV and E*=240 MeV respectively for ^{16}O and ^{19}F induced reactions). The observation of any difference of fast α-particles in the experimental spectra between the two reactions could be interpreted, in a model independent way, as possible influence of the projectile α-structure effect.

The ^{16}O and ^{19}F beams have been provided by the ALPI-TANDEM XTU accelerator complex at the Laboratori Nazionali di Legnaro (Italy). The experimental set-up used is the GARFIELD detection array implemented with the Ring Counter (RCo), at the most forward angles, fully equipped with digital electronics [11].

The GARFIELD apparatus consists of two large volume cylindrical drift chambers, each equipped with Micro-Strip Gas Chambers (MSGC) as amplified ΔE stage followed by CsI(Tl) scintillators residual energy detectors. Intermediate mass fragments and light charged particles are detected in an angular range from $\theta = 29^o$ to 151°. The Ring Counter is a three stage annular detector, covering the $\theta = 5^o$- 17° angular range, with an Ionization Chamber (IC) as first stage, followed by reverse mounted nTD Silicon Strip detectors (Si) and CsI(Tl) scintillators.

The GARFIELD plus RCo apparatus can perform complete high quality charged particle identification (both Z and A) and energy determination in a nearly 4π coverage ($\theta = 5^o$- 151°) for light charged particles and, in the most forward direction ($\theta = 5^o$- 17°), also for fragments with charge up to Z=14.

Light charged particles, detected in GARFIELD and RCo, have been measured in coincidence with Evaporation Residues (ER) collected in the first two stages (IC-Si) of the RCo within the angulare range $\theta = 8.6^o$- 17°, just beyond the grazing angle. The ERs have been selected setting proper gates in the reconstructed Z versus Energy distributions.

4 Experimental results and preliminary analysis

The double differential proton and alpha energy spectra, in coincidence with ERs, have been sorted out and the spectra obtained from the two systems have been compared. An example is shown in Fig 1, where the comparison between proton and α-particles spectra, normalized to the maximum, at the most backward and forward angle of the GARFIELD angular range are reported. The two reactions show very similar proton spectra except for a small difference at the most forward angles. Effect that can be ascribed to the slightly larger excitation energy in the ^{19}F induced reaction. A much larger difference is, on the contrary, observed in the α-particles spectra. The predicted emission spectra performed with an evaporative code like PACE2 (or PACE4), which takes into account the difference in the compound nucleus excitation energies, confirm that the purely statistical emission spectra should be very similar for the two systems, supporting the idea that a second fast emission source for both systems is needed when comparing with experimental data.

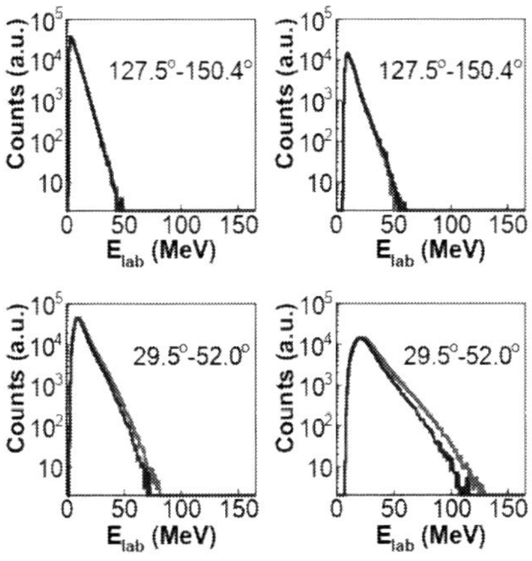

Figure 1. Double differential energy spectra in the laboratory frame (normalized to the maximum) for protons (left panels) and α-particles (right panels) for the two reactions 256 MeV ^{16}O + ^{65}Cu (blue line) and 304 MeV ^{19}F + ^{62}Ni (red line) at different detected angles (θ = 139°- 151°, θ = 29°- 41°).

In the preliminary analysis, a first estimate of the expected amount of fast emission in the two cases has been performed comparing the data with the predictions of the Moscow Pre-equilibrium Model (MPM). The calculations, done with an initial exciton number of n_o= 17 (16p + 1h) for the ^{16}O + ^{65}Cu case, reasonably describe the shape of the α-particles spectra, except for an understimation at the forward angles in GARFIELD, but, with the same initial n_o, the model strongly overstimates the pre-equilibrium protons emission. Performing the same comparison in the case of ^{19}F + ^{62}Ni reaction, with an initial n_o= 20 (19p + 1h), a quite similar results have been obtained, with an overproduction of fast α-particles even larger than in the ^{16}O case, while the fast protons are largely overstimated. A possible explanation for the extra yield of fast α-particles in both systems, may be due to the fact that even the ^{19}F can have an alpha structure and, in particular, that its α+^{15}N state is characterized by an energy (4.01 MeV) even smaller than the α+^{12}C (7.2 MeV) of the ^{16}O. An example of these comparisons, at the most forward angles in GARFIELD θ = 29°- 41°, for both systems are shown in Fig.2.

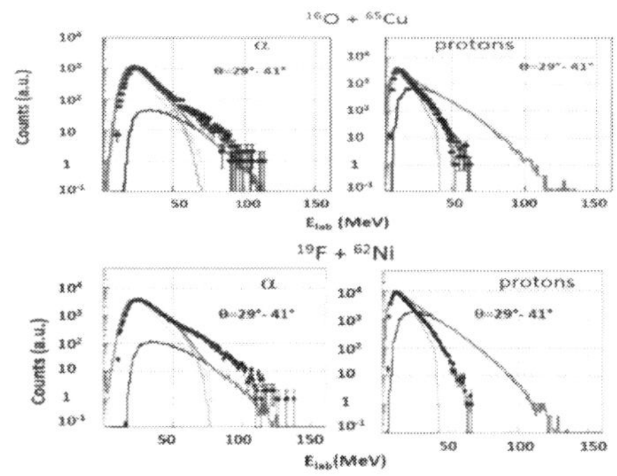

Figure 2. Comparison of laboratory energy spectra (normalized to the area) (black dots) for α-particles and protons for the two reactions ^{16}O + ^{65}Cu (upper panels) and ^{19}F + ^{62}Ni (lower panels) at θ = 29°- 41° with preliminary calculations from MPM (red line). The model evaporative PACE2 (green line) and pre-equilibrium (blue line) are shown. Evaporative PACE4 calculations (orange dots) are also reported.

By changing the initial configuration parameters (i.e. diminishing the exciton number) the description ot the α-particles sligtly improves, but, on the contrary, the proton spectra description becomes worse. A unique set of initial parameters of the MPM model seems not to be able to describe both protons and α-particles decay channels (like in the ^{16}O + ^{116}Sn case) indicating that some implementations to the model has to be introduced. For example in the used calculation no light charged particle isotopes are considered and this may strongly influence the relative proton and alpha decay probability [12] [13].

For a further analysis other theoretical approaches have been considered in order to follow the evolution of the reaction in an event-by-event base. Only in this way it is possible, then, to filtrate the model prediction through a software replica of the experimental set-up, taking into account the geometry of the apparatus, the energy threshols, the energy resolutions and the solid angle of each detector, for a realistic comparison with the experimental data.

First, simulations have been performed with the statisical-model code GEMINI++ [14], which describes the decay of hot nuclei formed in fusion reactions, using a Monte Carlo code, and generates light charged particles distribution emitted after the thermal equilibrium is reached.

Moreover, to take into account the dynamical part of the reaction, two models have been considered: the first one is the Stochastic Mean Field (SMF) [15], implemented in the TWINGO code [16], which simulates the time evolution of the nuclear matter through transport equations of one-body distribution function. In this model each nucleon is considered to be composed of many test particles which are subject to a mean field. The second model is the Antisymmetrized Molecular Dynamics (AMD) [17] in which the dynamics of the reaction is treated through the equation of motion of single nucleons, considered as gaussian wave packet, and the clustering effects of the colliding partners can be taken into account through the nucleon-nucleon correlations term.

Dynamical models generate primary excited fragments distributions, which then undergo to a statistical de-excitation. To predict the secondary fragments distributions the statistical code GEMINI++ has been applied, as afterburner, to the results of TWINGO and AMD, and the final results have been directly compared to the experimental data.

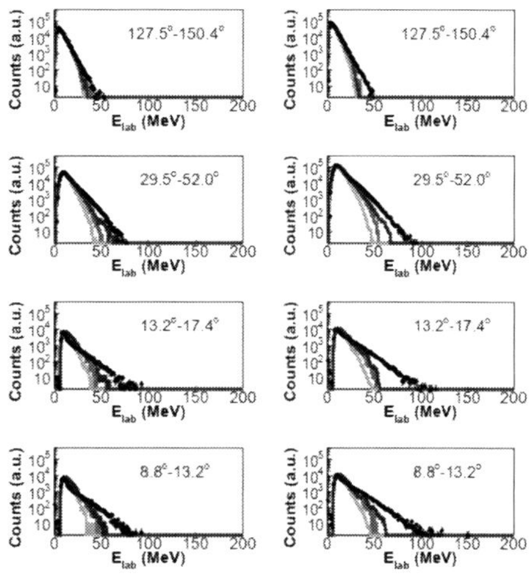

Figure 3. Comparison of protons laboratory energy spectra (normalized to the maximum) (black dots) for the two reactions $^{16}O + ^{65}Cu$ (left panels) and $^{19}F + ^{62}Ni$ (right panels) with GEMINI++ (red line), TWINGO coupled GEMINI++ (green line) and AMD coupled GEMINI++ (blue line), at different angular range in GARFIELD and RCo.

The experimental energy spectra of protons and α-particles, in coincidence of Evaporation Residues, have been first compared with the predictions of GEMINI++ alone and then with TWINGO and AMD coupled with GEMINI++ for both systems. GEMINI++ and TWINGO give similar information, apart from the temperature of the emitting thermalized source. In fact, the SMF approach does not reconstruct properly the nucleons (due to the test particle method) and in particular the clusters (due to the mean field approach) but it only defines the fragment size, number and excitation at a certain collision time. Therefore, only those particles emitted from the produced excited fragments, the decay of which is established by the after-burner, are finally implemented in the spectra. No pre-equilibrium spectra can therefore be provided. On the contrary, AMD is able to reconstruct fast nucleons and clusters, like it can be observed in Fig. 3 for the proton spectra. In this case also the most forward spectra from the Ring Counter have been sorted out.

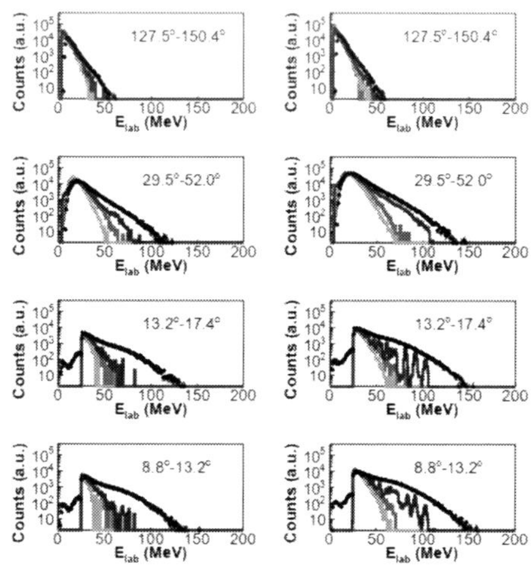

Figure 4. Comparison of α-particles laboratory energy spectra (normalized to the maximum) (black dots) for the two reactions $^{16}O + ^{65}Cu$ (left panels) and $^{19}F + ^{62}Ni$ (right panels) with GEMINI++ (red line), TWINGO coupled GEMINI++ (green line) and AMD coupled GEMINI++ (blue line), at different angular range in GARFIELD and RCo.

In the case of α-particles, at forward angles, the component of fast emission is more evident in the experimental spectra. This effect is partially described by the predictions from the AMD+GEMINI code, which include a possible influence of alpha clustering structure in the projectile, as shown in Fig. 4. However, the statistics of simulated events with AMD has still to be incremented to better describe the experimental spectra, avoiding unphysical statistical fluctuations. Moreover, further calculations can be provided, varying the input parameters in the code related to the clusterization effects to look for an optimization in the description of the data.

As a further step, the total multiplicity distributions of light charged particles, in coincidence of Evaporation Residues, have been obtained. The comparison between the two systems has evidenced that the ^{19}F induced reaction emits, in general, more protons and α-particles than the ^{16}O induced one, this can be understood because of the slightly higher excitation energy of the $^{19}F + ^{62}Ni$ system. Moreover, in Fig. 5 are shown the comparison between the experimental proton and α-particle multeplicity distributions with the different models predictions for the two systems, normalized to the number of ER. From these first results, one can notice that the AMD predictions overestimate, in general, the experimental protons and α's multiplicities, while TWINGO calculations predict always less particles emission. Moreover, the measured proton multeplicities are always lower then the predictions of GEMINI++. This observation is also true for the α-particles emitted in the ^{16}O + ^{65}Cu reaction, while in the ^{19}F induced reaction the α's are correctly reproduced by GEMINI++ calculations. The difference between the two systems is very interesting and further analysis is ongoing, looking to specific decay channels.

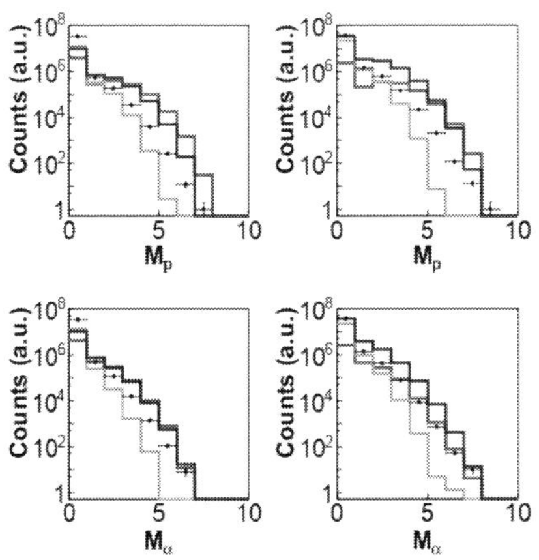

Figure 5. Protons (upper panels) and α-particles (lower panels) multeplicitiy distributions, in coincidence with ER, (black dots) for the two reactions in comparison with GEMINI++ (red line), TWINGO coupled GEMINI++ (green line) and AMD coupled GEMINI++ (blue line).

5 Conclusions

Secondary particle emission from ^{16}O + ^{65}Cu and ^{19}F + ^{62}Ni reactions at the same beam velocity 16 AMeV have been investigated to probe possible α-clustering effects in medium-mass systems. From the first comparison between the two systems a difference in the fast α-decay channel has been evidenced, which can be related to the difference in the projectile structure. The comparison of proton and α-particle energy spectra, in coincidence with evaporation residues, with the predictions of Moscow Pre-equilibrium Model, which takes into account both the pre-equilibrium and evaporation emission, shows that the model, with a unique set of initial parameters, is not able to describe protons and α-particles at the same time. The model has to be upgraded paying more attention to the clustering structure effects, introducing a correct filter in the Evaporation Residues distribution and including all possible decay channels.

Preliminary results show that experimental spectra cannot be described by both GEMINI++ and TWINGO predictions: the first simulation only considers complete fusion processes, while the second one, even if it is considers the dynamics of the reaction, is not able to reconstruct properly the fast emitted nucleons and clusters. The AMD model, which includes cluster structure effects in the projectile, in turns, seems to be more in agreement with the experimental data with respect to SMF model, giving a better description of the fast α-particles production. However, the AMD input parameters need to be further adjusted to have a more complete description of the data.

The data analysis is still undergoing to study more exclusive observables, like for example particle-particle correlations in events with multiplicities of light charged particles greater than two, and to compare them to different model simulations.

References

[1] L.R. Hafstad, E. Teller, *Phys. Rev.* **54**, 681 (1938).
[2] M. Freer, *Rep. Prog. Phys.* **70**, 2149 (2007).
[3] K. Ikeda et al., *Prog. Theor. Phys.* Suppl. **464**, (1968).
[4] W. von Oertzen et al., *Phys. Rep.* **432**, 43 (2006).
[5] P.E. Hodgson, E. Běták, *Phys. Rep.* **374**, 1 (2003).
[6] O.V. Fotina et al., *Int. Jour. Mod. Phys.* **E 19**, 1134 (2010).
[7] J.J. Griffin, *Phys. Rev. Lett.* **17**, 478 (1966).
[8] M. Blann, M.B. Chadwick, *Phys. Rev.* **C 62**, 034604 (2000).
[9] N. Cindro et al., *Phys. Rev. Lett.* **66**, 868 (1991).
[10] J. Cabrera et al., *Phys. Rev.* **C 68**, 034613 (2003).
[11] M. Bruno et al., *EPJA* **49**, 128 (2013)and ref. therein.
[12] F. Gramegna et al., *Jour. Phys. Conf. Series* **580**, 012011 (2015).
[13] D. Fabris et al., *Acta Phys. Polonica* **B 46**, 447 (2015).
[14] R.J. Charity, *Phys. Rev.* **C 82**, 014610 (2010).
[15] M. Colonna et al., *Nucl. Phys.* **A 642**, 449 (1998).
[16] A. Guarnera et al., *Phys. Lett.* **B 373**, 267 (1996).
[17] A. Ono et al., *Phys. Rev.* **C 59**, 853 (1999).

Study of the population of neutron-rich heavy nuclei in the A~200 mass region via multinucleon transfer reactions

E. Fioretto[1,*], L. Corradi[1], F. Galtarossa[1,2], S. Szilner[3], D. Montanari[4], T. Mijatović[3], G. Pollarolo[5], H.M. Jia[6], D. Ackermann[7], D. Bourgin[4], G. Colucci[8], S. Courtin[4], G. Fruet[4], A. Goasduff[1], J. Grebosz[9], F. Haas[4], D. Jelavić Malenica[3], S.C. Jeong[10], P.R. John[8], M. Milin[11], G. Montagnoli[8], N. Skukan[3], F. Scarlassara[8], N. Soić[3], A.M. Stefanini[1], E. Strano[8], V. Tokić[3], C.A Ur[12], J.J. Valiente-Dobón[1], and Y.X. Watanabe[13]

[1]INFN, Laboratori Nazionali di Legnaro, I-35020 Legnaro (Padova), Italy
[2]Dipartimento di Fisica e Scienze della Terra, Università di Ferrara, I-44121 Ferrara, Italy
[3]Ruđer Bošković Institute, HR-10002 Zagreb, Croatia
[4]Institut Pluridisciplinaire Hubert Curien, CNRS-IN2P3, Université de Strasbourg, F-67037 Strasbourg Cedex 2, France
[5]Dipartimento di Fisica Teorica, Università di Torino, and INFN Sezione di Torino , I-10125 Torino, Italy
[6]China Institute of Atomic Energy, P. O. Box 275(10), Beijing 102413, P. R. China
[7]Grand Accélérateur National d'Ions Lourds (GANIL), Boulevard Henry Becquerel, F-14076 Caen Cedex 5, France
[8]Dipartimento di Fisica dell'Università and INFN Sezione di Padova, I-35131 Padova, Italy
[9]The Henryk Niewodniczański Institute of Nuclear Physics, Polish Academy of Sciences, 31-342 Kraków, Poland
[10]Institute for Basic Science, Daejeon 305-811, Korea
[11]Department of Physics, Faculty of Science, University of Zagreb, HR-10002 Zagreb, Croatia
[12]Horia Hulubei National Institute of Physics and Nuclear Engineering, RO-077125 Bucharest-Magurele, Romania
[13]Wako Nuclear Science Center, Institute of Particle and Nuclear Studies, High Energy Accelerator Research Organization (KEK), Wako, Saitama 351-0198, Japan

Abstract. Multineutron and multiproton transfer channels, populated in the inverse kinematics reaction ^{197}Au+^{130}Te at E_{lab}=1.07 GeV, were measured at Laboratori Nazionali di Legnaro using the presently heaviest ion beam delivered by the PIAVE-ALPI accelerator complex and detecting both projectile-like and target-like ions. To this end the large solid angle magnetic spectrometer PRISMA was coupled to a second arm for the detection of the heavy fragments in kinematic coincidence with the light ones selected and identified with the spectrometer. The data analysis is still in progress and will allow to compare the yields of both light and heavy partner with theoretical predictions performed with the GRAZING code to get quantitative information on transfer channels and the effect of evaporation and fission on the production rate of primary fragments. The mass integrated Z distribution, extracted from the experimental data, evidenced the population of proton pick-up channels that, in conjunction with the neutron stripping ones from the ^{130}Te, open the path for the production of neutron-rich heavy nuclei. In the following, we will present some preliminary results as well as details on the experimental configuration and perspectives for future investigations in the neutron-rich heavy region.

1 Introduction

On the basis of theoretical predictions, multinucleon transfer (MNT) processes in heavy-ion collisions around the Coulomb barrier have been proposed as a possible mechanism for the population of neutron-rich heavy nuclei [1], complementary to the fragmentation [2], opening new fields of research in nuclear physics such as the production and the study of the nuclear structure of unexplored exotic nuclei. In particular, there is presently a significant interest in the population of neutron-rich nuclei (in the mass region A~200) located along the neutron closed shell N = 126 (the last waiting point in the r-path), having the largest impact on the astrophysical r-process [3, 4] and being crucial for understanding the shell evolution far from the stability, and in the actinide region where nuclear structure information is important being the benchmark for theoretical models that provide predictions for the heaviest nuclei [5].

The feasibility of using MNT reactions around the Coulomb barrier for the production of neutron-rich heavy nuclei has been so far poorly investigated but recently benefited from the construction of new generation tracking spectrometers, based on the trajectory reconstruction and the use of the state-of-art large area particle detectors. The large acceptance of these spectrometers and the high resolving power of their detection systems gave a boost to this kind of studies. Different groups have recently started with or proposed measurements, focused on the determination of the heavy partner yield, in connection with the preparatory phases of facilities, such as KISS (KEK Iso-

*e-mail: fioretto@lnl.infn.it

tope Separator System) [6] at the Radiactive Isotope Beam Factory (RIBF) in RIKEN and GALS (Gas cell and Laser ion Separation) at the Joint Institute for Nuclear Research in Dubna [7], aiming to the study of β-decay spectroscopy of neutron-rich heavy nuclei in the A∼200 and N∼126 region populated via MNT. This information can be obtained either through γ-particle coincidence experiments [8], gating the light partner identified with high-resolution tracking spectrometers and measuring the yields of heavy partners via the associated γ-rays spectra Doppler corrected for the heavy partner, or via integral measurements of delayed β-γ coincidences [9]. As alternative technique, aiming to estimate the production rate of neutron-rich heavy nuclei, one can use a high-resolution kinematical coincidence. This technique is a powerful tool widely used for studying heavy-ion induced reactions from elastic and quasi-elastic to multinucleon transfer processes [10–13]. The method is mainly based on the simultaneous detection of the reaction fragments using two different set-ups placed at the kinematically correlated angles. The use of this technique in binary reactions allows to measure the yields of both partners getting information on the influence of secondary processes on the production rate of primary fragments.

In this framework, a benchmark experiment was recently carried out at LNL, aiming to measure the yields of light and heavy fragments populated in the inverse kinematics reaction ^{197}Au+^{130}Te at E_{lab}=1.07 GeV via MNT processes, by using the kinematical coincidence technique. The inverse kinematics method allowed to get a high detection efficiency and good mass and nuclear charge resolutions due to forward focusing of the reaction products and, for the present reaction, the higher kinetic energy of both binary partners at forward angles. This method has been already successfully applied in recent measurements performed with heavy-ion magnetic spectrometer PRISMA for the ^{96}Zr+^{40}Ca [14] and ^{116}Sn+^{60}Ni [15] systems aiming to the study of nucleon-nucleon correlation in the sub-barrier regime. The use of a ^{130}Te target (the most neutron-rich stable Te isotope) in conjunction with a ^{197}Au beam opens up both neutron stripping and proton pick-up channels, allowing the production of neutron-rich heavy nuclei in the A∼ 200 mass region, and proton stripping and neutron pick-up channels leading to the population of the benchmark nucleus ^{132}Sn. Because of the presence of fission and/or neutron evaporation that mainly affect the final yields of the heavy partner, it is extremely important to compare the experimental yields with theoretical predictions in order to estimate their effects in the regions of interest.

2 Experimental set-up

Correlated fragments produced in the interaction of a ^{197}Au beam with a thin ^{130}Te target were detected by using a kinematical coincidence set-up consisting of the large solid angle magnetic spectrometer PRISMA [16–18] (acting as main arm of the coincidence) coupled to an ancillary detection system acting as second arm, as displayed in figure 1.

PRISMA is the large acceptance magnetic spectrometer of LNL designed to be used for heavy-ion induced reactions at energies up to 10 AMeV. Its optical design consists of a magnetic quadrupole followed by a dipole with a bending angle of 60° and a curvature radius of 120 cm. The maximum magnetic rigidity amounts to 1.2 Tm. PRISMA is equipped with a position sensitive Micro-Channel Plate detector (MCP) [19] at the entrance of the spectrometer followed (at ∼6 m of distance) by a large area focal plane detector [20]. The MCP, placed at 25 cm from the target, provides a timing signal for time-of-flight (TOF) measurements and X and Y position information. The focal plane detector is composed of a large area Multi-Wire Parallel Plate Avalanche Counter (MWPPAC), providing a timing signal and X and Y position information, followed by a transverse field segmented ionization chambers (IC), which allows the nuclear charge Z identification (based on ΔE-E technique) and the measurement of the energy loss ΔE_{IC} and total energy E_{IC} of the detected fragments. The timing signal of the PPAC is routinely used for time-of-flight (TOF) measurements with the MCP acting as start detector. The complete ion identification in PRISMA is performed through an event-by-event reconstruction of the trajectory inside the magnetic elements and the determination of the nuclear charge Z, the mass number A, the atomic charge state q, and the absolute velocity vector of each ion that reaches the focal plane detector.

The second arm consists of a Parallel Plate Avalanche Counter (PPAC) followed by an axial field ionization chamber, also known as Bragg chamber (BC) [21, 22]. This kind of ionization chamber allows to perform a Bragg curve spectroscopy, first proposed by Gruhn [23], providing simultaneously and with intrinsically high resolution many details about the nature of the fragments stopped in the active volume, such as the nuclear charge Z (due to the nearly linear dependence of the Bragg peak (BP) amplitude on Z), the range R (the length of the ion track) and the energy E of the fragments stopped in the active volume. The PPAC is composed of a central cathode and two anode planes orthogonally oriented on each side at a distance of 2 mm. The cathode is made of a double aluminized (20 μg/cm^2 + 20 μg/cm^2) Mylar foil with a thickness of 1.5 μm. The anode planes consist of 10 μm diameter gold-plated tungsten wires 1 mm apart. The PPAC provides a fast timing signal and X and Y information with timing and position resolutions of the order of few hundreds ps and 1 mm, respectively. The BC has an active depth of 33 cm. The uniformity of the electric field in the active volume is obtained by means of 114 guard rings. The active volumes of PPAC and BC are separated by the entrance window of the ionization chamber, also acting as cathode, made of an aluminized (20 μg/cm^2) Mylar foil (1.5 μm). The BC was used to get simultaneously the nuclear charge Z_{BC} and the total energy E_{BC} of the fragments. The corresponding signals have been obtained by sending the output signal of the charge preamplifier to two amplifiers with shaping times of 0.25 and 4 μs for the amplitude of Bragg peak and the energy, respectively. The detection system was placed at ∼ 90 cm from the target and had a geometrical solid angle quite similar to PRISMA (i.e. 80 msr).

3 Experiment details

The ^{197}Au beam was delivered by the PIAVE-ALPI superconducting accelerator complex of LNL with an intensity of 1.5 pnA and at an energy of E_{lab}=1.07 GeV. The target, isotopically enriched to 99.6% in ^{130}Te, with a thickness 200 μg/cm^2 (2 mm strip) was sandwiched between two 20 μg/cm^2 films of C in order to reduce the sputtering process on the target surfaces. Figure 1 shows the layout of the experimental set-up. PRISMA was placed around the grazing angle corresponding to θ_{lab}=37° for the chosen bombarding energy. The coincident Au-like ions were detected with the second arm symmetrically placed with respect the beam axis at θ_{lab}=-37°. PRISMA and the second arm were connected to a sliding seal scattering chamber of 1 m diameter. Two Si detectors were mounted inside the scattering chamber at θ_{lab}=61° and θ_{lab}=71°, respectively, in order to measure Rutherford scattered Te ions for the normalization of different runs and to monitor the beam quality during the experiment. A specific Ta beam stopper was designed to minimize the background in the entrance detector of PRISMA due to secondary electrons produced by the beam.

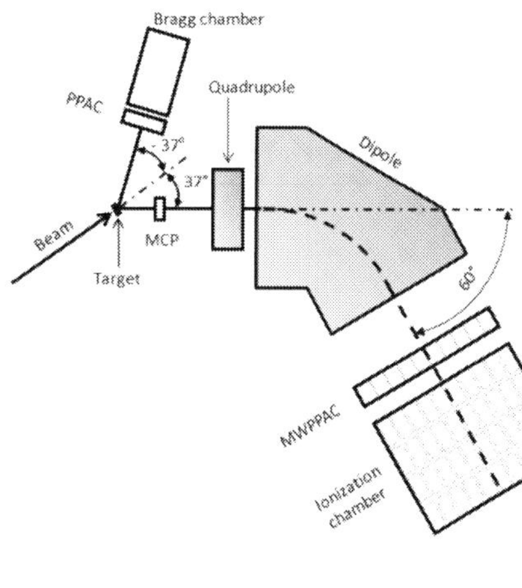

Figure 1. Sketch of the kinematic coincidence set-up used for the ^{197}Au+^{130}Te inverse kinematics reaction.

4 Light partner: PRISMA data

The magnetic spectrometer PRISMA and its focal plane detector were set in order to select and identify Te-like ions. The IC was operated with CH_4 as filling gas at a pressure of 93 hPa to stop target-like fragments in its active volume. In order to get a good Z resolving power, one has to properly take into account the direction followed by the different reaction products reaching the IC. In particular, the ion range R can be estimated from the signal of each IC section and the position information from MCP and MW-PPAC. Figure 2 display the two-dimensional plot energy E_{IC} vs range R measured with PRISMA detectors. The intense band which extends up to higher energies in the matrix corresponds to Te-like ions (Z=52) populated in binary reactions ranging from the quasi-elastic regime to more dissipative processes with increasing total kinetic energy losses (TKEL). Events located below this band in the same

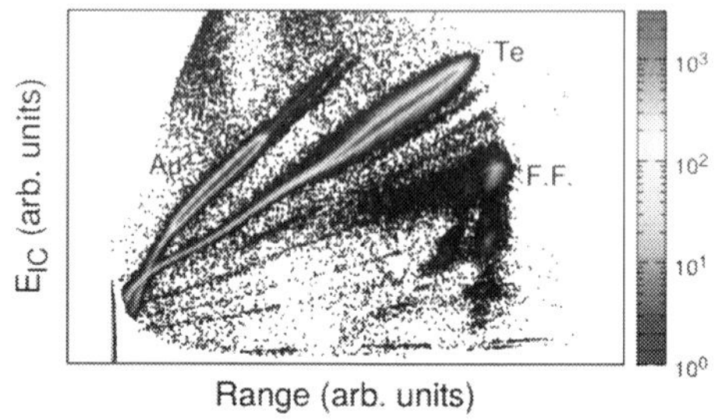

Figure 2. Energy E vs total range R of the ions detected at the focal plane of PRISMA for the ^{197}Au+^{130}Te reaction at E_{lab}=1.07 GeV.

figure have been associated to fission fragments (F.F.) produced in the reaction. They are not stopped in the active volume of the IC and, therefore show the characteristic punch-through effect in the matrix. Populated Au-like ions were almost stopped in the ΔE section and have an energy well below the threshold at which the Bragg peak occurs. The corresponding events are located in the intense band above the Te-like one. Due to the PRISMA settings, Au-like ions and fission fragments detected at the focal plane of the spectrometer correspond only to the ones having the correct p/q ratio (as the selected Te-like ions). As one can see in figure 2, the separation of the channels involving the transfer (stripping and pick-up) of few protons turned out to be very difficult due to the overwhelming of the elastic component. It is worthwhile to mention that in this experiment we chose a rather low bombarding energy (around the Coulomb barrier), in order to reduce the effect of secondary processes having in mind that the transfer yields, especially for proton transfer channels, may be modest. Figure 3 shows the mass integrated Z distribution corresponding to the nuclear species populated in the reaction. The estimated Z resolving power in this nuclear charge region was $\Delta Z/Z \sim 1/65$ that should be sufficient to distinguish different nuclear charges around Z=52, where the Te-like ions are located, confirming the good operation of the PRISMA IC. One observes the population of transfer channels involving both stripping and pick-up processes up to four protons. As expected from optimum Q-value considerations, the produced proton pick-up channels, are in the path of the population of neutron-rich heavy nuclei. One can also notice that the proton pick-up channels have yields larger than the corresponding proton stripping ones. The yields for the different channels will be extracted via a multigaussian fit of the experimental distribution.

For heavy ions around $A\sim$120-140 the mass discrimination becomes difficult due to the decrease of the mass

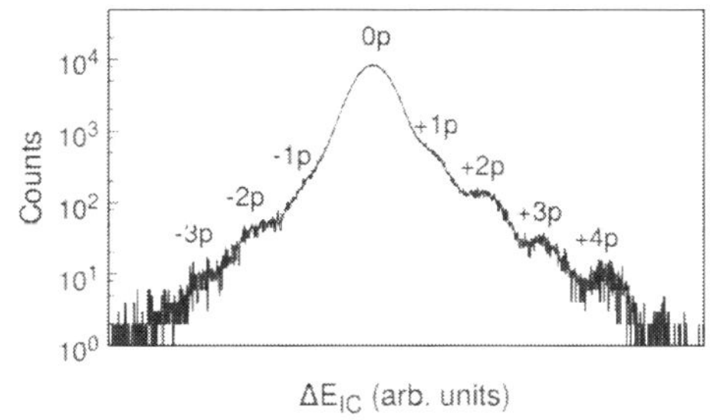

Figure 3. Mass integrated Z distribution of the fragments populated in the ^{197}Au+^{130}Te reaction at E_{lab}=1.07 GeV.

separation with increasing mass number. In addition, the mass resolving power of magnetic spectrometers may be affected by the presence of optical aberrations, especially for non-central trajectories, which have to be properly taken into account in the ion trajectory reconstruction, particularly for heavy masses. In order to reduce these effects and to improve the matching of the different sections of the focal plane detector, a dedicated off-line analysis was applied. This implied, in a first step, to tune the quadrupole length and the distance between target and quadrupole in order to take into account the effect of the fringing fields. Then, an iterative procedure was employed making use of the position information at the entrance and at focal plane of the spectrometer and performing a least-square fit of the A/q distributions with a n-order polynomial. From the resulting A/q distributions we extracted the mass spectrum summing over all atomic charge states. The outcome of this procedure is displayed in the mass A vs X position matrix of figure 4 showing the quality of the separation obtained for Te isotopes. This new procedure gave a mass resolving power of about $\Delta A/A \sim 1/240$. Even if both neutron pick-up and stripping were observed, the dominant flux was in the neutron stripping side, which was measured down to the (-6n) channel. A similar situation

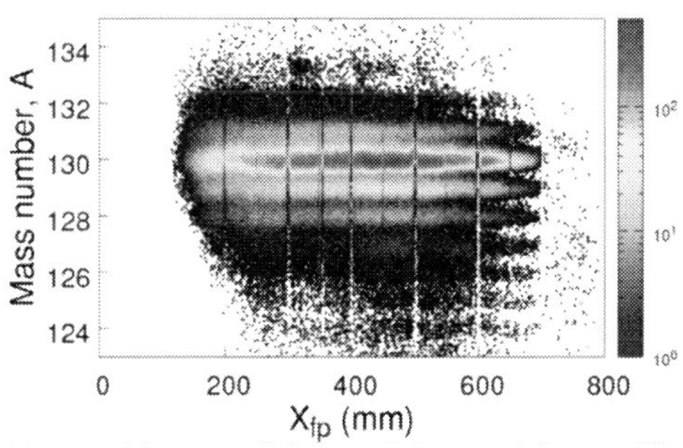

Figure 4. Mass number A vs X position scatter-plot obtained for Te isotopes.

was also observed in the recently measured ^{136}Xe+^{198}Pt [8] and ^{136}Xe+^{238}U [24] systems, making use of the most neutron-rich stable ^{136}Xe isotope, and the ^{40}Ar+^{208}Pb reaction [25] . To learn more about the character of the final populated channels we have to compare them with theory. The on-going analysis will allow to extract the experimental yields and compare them with the theoretical predictions obtained with the GRAZING code [26, 27]. The model takes into account on the same footing surface degrees of freedom and particle transfer and the exchange of many nucleons proceeds via a multi-step mechanism of single nucleons (both, protons and neutrons, via stripping, and pick-up processes). The model has been so far successfully applied in the description of multinucleon transfer reactions [28–30] as well as of fusion reactions and barrier distributions [31].

In a dedicated run PRISMA was also set with correct magnetic fields and gas pressure in the IC in order to select and detect Au-like ions at the same angle θ_{lab}=37°. The main goal was to evaluate the channel identification also for very heavy ions [32]. Even if a good A/q resolution was obtained, the energy of Au-like ions was not sufficient to separate the different atomic charge states.

5 Heavy partner: kinematical coincidence data

In the kinematical coincidence measurement, the focal plane detector of PRISMA always acts as master trigger for the data acquisition system that has been implemented in order to include experimental information coming from the detectors of the second arm. These latter were optimized for the detection of the heavy partners. In particular, the BC was operated with CF_4 gas at a pressure of 53 hPa in order to stop Au-like ions into its active volume. It is worthwhile to stress that in the present experimental configuration light and heavy partners enter into the detectors of both PRISMA and the second arm. This can be observed in figure 5 that displays the E_{BC} (total energy measured with the Bragg chamber) vs ΔTOF (difference in TOFs between the PRISMA MCP and the PPAC of the second arm). One can see the excellent separation between Te-like and Au-like events identified imposing the coincidence condition in PRISMA. After several cross

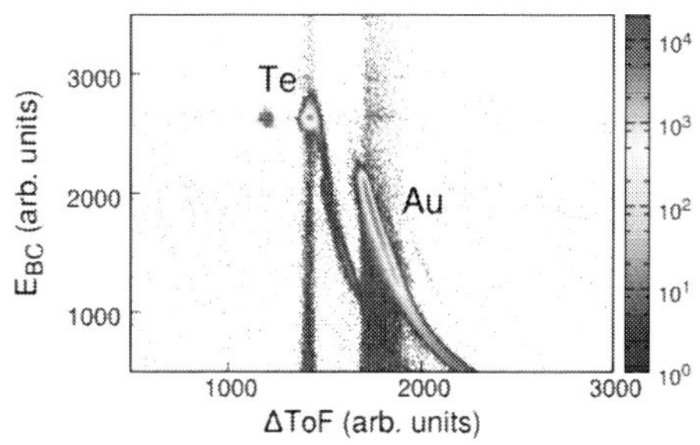

Figure 5. Scatter-plot energy E_{BC}, measured with the BC, vs ΔTOF between the PRISMA MCP and the PPAC in the second arm.

checks on the experimental spectra the correct operation of the kinematical coincidence was confirmed. We proceeded with the construction of the mass distribution of the heavy transfer products assuming a pure binary process and making use of the information on the correlated in-plane scattering angles as well as the difference in TOFs of the coincident fragments entering in the PRISMA MCP and the PPAC of the second arm. The obtained mass distribution is shown in figure 6, where the peaks are well centered at $A=197$ and $A=130$, corresponding to Au-like and Te-like ions, respectively. The mass resolution turned out to be $\Delta A/A \sim 1/43$. The construction of the mass-correlation matrix and its comparison with simulations, in order to estimate the effect of secondary processes, is in progress. The extracted cross sections for the heavy partner will be compared with theoretical calculations done with the GRAZING code.

6 Perspectives

The use of neutron-rich radioactive ion beams (RIB) that should be available, with reasonable intensity, in the forthcoming facility SPES [33] will provide important data to evaluate the possibility to access more exotic neutron-rich nuclei. As well known, transfer processes are mainly governed by form factors and optimum Q-value considerations. On the basis of theoretical predictions, transfer processes induced by neutron-rich RIBs on heavy targets will allow to populate neutron-rich heavy nuclei in the Pt-Os mass region with higher production cross sections. As an example, figure 7 displays the comparison of the estimated two-dimensional *Z-A* distributions for the heavy fragments populated in the reactions ^{136}Xe+^{198}Pt (upper panel) and ^{140}Xe+^{198}Pt (lower panel) at E_{lab} = 850 MeV. One can see the change in the population pattern from ^{136}Xe (the most neutron-rich Xe stable beam) to ^{140}Xe (neutron-rich RIB). The calculated production yield is not corrected by neutron evaporation, which of course limits the final yield of neutron-rich nuclei very far from stability.

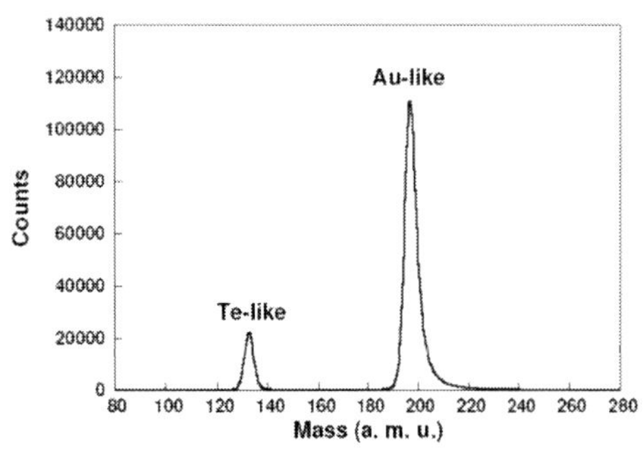

Figure 6. Mass distribution reconstructed with the kinematical coincidence. Light and heavy reaction products can be clearly distinguished.

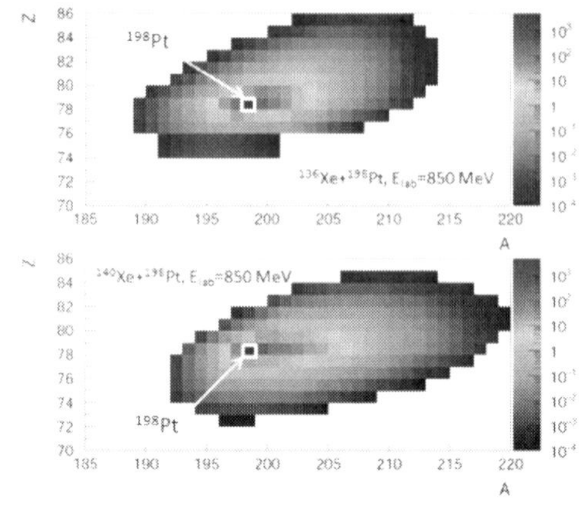

Figure 7. Total cross sections, calculated with the GRAZING code, for the heavy fragments populated in the ^{136}Xe+^{198}Pt (upper panel) and ^{140}Xe+^{198}Pt (lower panel) reactions at 850 MeV.

7 Summary

Recently the production of neutron-rich heavy nuclei around neutron magic number $N=126$ via MNT reactions has been focused theoretically and experimentally. In this framework, transfer channels populated in the inverse kinematics reaction ^{197}Au+^{130}Te were studied in a benchmark experiment by using the large solid angle magnetic spectrometer PRISMA coupled to a new gas detection system acting as second arm in a kinematical coincidence measurement. The population of neutron stripping and proton pick-up channels from ^{130}Te was evidenced. These processes open up the path leading to the production of neutron-rich heavy nuclei. The ongoing analysis will proceed with the extraction of the production cross sections for both light and heavy partner and their comparison with theoretical calculation performed with the GRAZING code. The mass spectrum of the heavy transfer products has been reconstructed by using the kinematical coincidence. This will allow to construct the mass-mass correlation matrix and compare it with simulations in order to estimate the effects of secondary processes, such as evaporation and fission. The extracted experimental yields for the heavy partner will also be compared with theoretical predictions obtained with the GRAZING code.

8 Acknowledgements

Grateful thanks are due to the staff of the Tandem/PIAVE-ALPI accelerator complex and to M. Loriggiola for preparing ^{130}Te strip targets of excellent quality. Funds for performing the research leading to the present results were received from the European Union Seventh Framework Programme FP7/2007- 2013 under Grant Agreement No. 262010 - ENSAR. This work was also partly supported by the Croatian Science Foundation under the project 7194.

References

[1] C.H. Dasso, G. Pollarolo, A. Winther, Phys. Rev. Lett. **73**, 1907 (1994)

[2] T. Kurtukian-Nieto *et al.*, Phys. Rev. C **89**, 024616 (2014)

[3] K.-L. Kratz *et al.*, Astrophys. J. 403, 216 (1994)

[4] H. Grawe *et al.*, Rep. Prog. Phys. 70, 1525 (2007)

[5] V. Zagrebaev and W. Greiner, Phys. Rev. Lett. **101**, 122701 (2008)

[6] Y. Hirayama *et al.*, Nucl. Instr. and Meth. B **352**, 4 (2015)

[7] S. Zemlyanoy *et al.*, EPJ Web of Conferences **86**, 00067 (2015)

[8] Y.X. Watanabe *et al.*, Phys. Rev. Lett. **115**, 172503 (2015)

[9] K. Novikov *et al.*, J. Phys.: Conference Series **515**, 012016 (2014)

[10] I. Tserruya *et al.*, Nucl. Instr. and Meth. **196**, 225 (1982)

[11] G. Casini *et al.*, Nucl. Instr. and Meth. A **277**, 445 (1989)

[12] H. Emling *et al.*, Nucl. Phys. A **211**, 600 (1973)

[13] F.W.N. de Boer *et al.*, Z. Phys. A **325**, 457 (1986)

[14] L. Corradi *et al.*, Phys. Rev. C **84**, 034603 (2011)

[15] D. Montanari *et al.*, Phys. Rev. Lett. **113**, 052501 (2014)

[16] A.M. Stefanini *et al.*, Nucl. Phys. A **701**, 217c (2002)

[17] S. Szilner *et al.*, Phys. Rev. C **76**, 024604 (2007)

[18] L. Corradi *et al.*, Nucl. Instr. and Meth. in Phys. Res. B **317**, 743 (2013)

[19] G. Montagnoli *et al.*, Nucl. Instr. and Meth. in Phys. Res. A **547**, 455 (2005)

[20] S. Beghini *et al.*, Nucl. Instr. and Meth. in Phys. Res. A **551**, 364 (2005)

[21] E. Fioretto *et al.*, J. Phys: Conference Series **533**, 012006 (2014)

[22] E. Fioretto *et al.*, EPJ Web Conf. **117**, 01004 (2016)

[23] C.R. Gruhn *et al.*, Nucl. Instr. and Meth. 196, 33 (1982)

[24] A. Vogt *et al.*, Phys. Rev. C **92**, 024619 (2015)

[25] T. Mijatović *et al.*, Phys. Rev. C **94**, 064616 (2016)

[26] A. Winther, Nucl. Phys. A **572**, 191 (1994)

[27] GRAZING code, http:/www.to.infn.it/~nanni/grazing

[28] L. Corradi, G. Pollarolo and S. Szilner, J. Phys. **G36**, 113101 (2009)

[29] L. Corradi *et al.*, Phys. Rev. C **66**, 024606 (2002)

[30] S. Szilner *et al.*, Phys. Rev. C **71**, 044610(2005)

[31] G. Pollarolo and A. Winther, Phys. Rev. C **62**, 054611 (2000)

[32] L. Corradi *et al.*, EPJ Web of Conferences **86**, 00007 (2015)

[33] SPES: Selective Production of Exotic Species; https://web.infn.it/spes/

 DOI: 10.1051/epjconf/201716300018

Fusion cross section measurements of astrophysical interest for light heavy ions systems within the STELLA project

Guillaume Fruet[1,*], *Sandrine* Courtin[1,2], *David G.* Jenkins[3], *Marcel* Heine[1], *Daniele* Montanari[1,2], *Luke G.* Morris[3], *Philip* Adsley[3,4], *Christian* Beck[1], *Serge* Della Negra[4], *Florent* Haas[1], *Fairouz* Hammache[4], *Oliver S.* Kirsebom[5], *Anne* Meyer[4], *Patrick H.* Regan[6,7], *Matthias* Rudiger[6], *Nicolas* de Séréville[4], and *Christelle* Stodel[8]

[1] *Université de Strasbourg, CNRS, IPHC UMR 7178, F-67000 Strasbourg, France*
[2] *USIAS, Université de Strasbourg, F-67000 Strasbourg, France*
[3] *Department of Physics, University of York, YO10 5DD York, United Kingdom*
[4] *Institut de Physique Nucléaire d'Orsay, UMR8608, CNRS/IN2P3, Université Paris-Sud 11, F-91406 Orsay, France*
[5] *Department of Physics and Astronomy, Aarhus University, DK-8000 Århus C, Denmark*
[6] *Department of Physics, University of Surrey, GU2 7XH Guildford, United Kingdom*
[7] *AIR Division, National Physical Laboratory, TW11 0LW Teddington, United Kingdom*
[8] *Grand Accélérateur National d'Ions Lourds, CNRS/IN2P3, F-14076 Caen, France*

Abstract. This contribution is focused on the STELLA project (STELlar LAboratory), which aims at the measurement of fusion cross sections between light heavy ions like $^{12}C+^{12}C$, $^{12}C+^{16}O$ or $^{16}O+^{16}O$ at deep sub-barrier energies. The gamma-particle coincidence technique is used in order to reduce background contributions that become dominant for measurements in the nanobarn regime.
The experimental setup composed of an ultra high vacuum reaction chamber, a set of 3 silicon strip detectors, up to 36 $LaBr_3(Ce)$ scintillators from the UK FATIMA collaboration, and a fast rotating target system will be described. The $^{12}C+^{12}C$ fusion reaction has been studied from E_{lab} = 11 to 5.6 MeV using STELLA at the Andromède facility in Orsay, France. Preliminary commissioning results are presented in this article.

1 Introduction

The measurement of fusion cross sections between light heavy ions like ^{12}C, ^{16}O or ^{20}Ne is at the center of interest to both nuclear physics and astrophysics. In particular, the $^{12}C+^{12}C$ fusion is a key reaction in the nucleosynthesis process during the C burning phase in the core of massive stars, in type Ia supernovae, and in superbursts from accreting neutron stars [1].

For reactions involving light ions, structural effects may seriously affect the fusion process. For instance, the ^{12}C-^{12}C system exhibits strong resonances from a few MeV/A down to the sub-barrier regime, suggesting possible cluster formation in the compound nucleus [2], while an alternative explanation involves relatively large spacings and narrow widths of ^{24}Mg compound nucleus energy levels [3].

The fusion reaction mechanism has been widely studied in the past [4–6], and despite considerable efforts in both experimental and theoretical studies questions concerning the behaviour of such reactions at very low colliding energies remain open. Indeed, it can be seen in figure 1 that data show discrepancies among the various experiments with large uncertainties at the lowest measured energies. Moreover, extrapolations towards the astrophysical region differ by orders of magnitude [7].

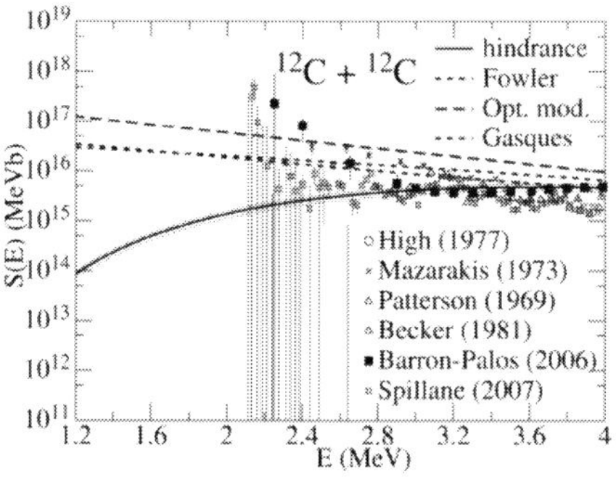

Figure 1. S-factor as a function of center of mass energy for the $^{12}C+^{12}C$ reaction. Theoretical calculations differ by orders of magnitude in the Gamow window $E_G = 1.5\pm0.3$ MeV at T = 5×10^8 K for this system.

The measurement technique will be presented in Sect. 2. The dedicated STELLA setup as well as technical developments will then be presented in Sect. 3. Fi-

*e-mail: guillaume.fruet@iphc.cnrs.fr

DOI: 10.1051/epjconf/201716300018

nally, preliminary results from the commissioning phase of STELLA will be discussed in Sect. 4.

2 Experimental approach

The $^{12}C+^{12}C$ fusion reaction has been studied at several beam energies around and below the Coulomb barrier ($E_C \sim 6.6$ MeV). In this regime, the three main exit channels are:

$$^{12}C + ^{12}C \rightarrow ^{20}Ne + \alpha \quad (Q = 4.62 \text{ MeV}), \quad (1)$$

$$^{12}C + ^{12}C \rightarrow ^{23}Na + p \quad (Q = 2.24 \text{ MeV}), \quad (2)$$

$$^{12}C + ^{12}C \rightarrow ^{23}Mg + n \quad (Q = -2.6 \text{ MeV}). \quad (3)$$

At stellar temperature $T = 5 \times 10^8$ K, the corresponding Gamow energy for $^{12}C+^{12}C$ is $E_G = 1.5 \pm 0.3$ MeV, below the neutron emission threshold. Thus, the exit channel (3) has not been studied within this work.

The detection of either light charged particles or γ's from the deexcitation of both residual nuclei ^{20}Ne and ^{23}Na has been widely used in the past to identify fusion events (see [4, 6, 8]).

Recently, it has been shown in Ref. [9] that the detection of the light charged particles in coincidence with the γ-ray from the heavy partner allows a very reliable selection of fusion events, due to a drastical rejection of background coming essentially from target contaminants (^{13}C, ^1H, ^2H, ...) or room background radiation.

For the present measurement, ^{12}C targets of 20, 30 and 50 $\mu g/cm^2$ are used. It is of great importance to avoid carbon build-up on them, a process that will affect precise determination of the reaction energy and probability by increasing the target thickness [10].

The differential cross section is then given by:

$$\left(\frac{d\sigma}{d\Omega}\right)_{lab} = \frac{N}{I \cdot \epsilon_\gamma \cdot \Delta\Omega \cdot \Delta t \cdot N_t}, \quad (4)$$

where N is the number of coincidence events, I the beam intensity, Δt the acquisition time, N_t the number of atoms in the target, $\Delta\Omega$ the solid angle covered by the particle detector, and ϵ_γ the γ efficiency.

To clarify the evolution of the S-factor towards the Gamow window, fusion cross sections below the nanobarn range should be measured. One possibility to obtain sufficient statistics is to increase the beam intensity, typically at values ≥ 1 pμA. Under such conditions, the usual thin fixed targets have a relatively short lifetime of few hours and an improved target system would be useful.

3 Technical developments

3.1 Reaction chamber and particle detectors

A reaction chamber has been designed and machined at IPHC-CNRS Strasbourg, France. The bottom part consists of a stainless steel cylinder used to accomodate the particle detector support, feedthrough and the pumping system. The upper part is a 2.5 mm thick aluminium dome of 20 cm diameter around which are placed $LaBr_3(Ce)$ scintillators for gamma detection.

Light charged particles like alphas and protons are detected using annular double-sided silicon strip detectors (DSSD), S1 and S3 design by Micron Semiconductor. These detectors are divided into 16 and 24 rings, respectively.

A PCB to support the silicon chips has been developed and built using R00403C material, which is known for low outgazing to prevent carbon build-up in targets.

Along with this development, a new pin connection system has been conceived. It allows to read signals of detectors from below the target plane, thus reducing material budget inside the reaction chamber. In the arrangment configuration seen in figure 2, the total solid angle coverage is $\sim 24\%$ of 4π.

Figure 2. Mechanical drawing of the reaction chamber and the arrangment of annular Si-strip detectors around the target which is at the center of the chamber. The beam is going from right to left.

The resolution of these detectors is about 35-40 keV (FWHM) for particle energies around 5.5 MeV giving sufficient resolution to separate the different energies associated to protons and alphas coming from the fusion reaction. Aluminium foils of 0.8 μm thickness protect detectors at backward angles from low energy electrons emitted from the target, while a 10 μm foil prevents intense elastically scattered ^{12}C beam to reach the forward detector.

To obtain an absolute normalisation for the fusion cross section two Si surface barrier detectors are used to measure scattered beam. They are placed symetrically at an angle $\theta_{lab} = 45°$ with respect to the beam axis, where the Mott cross section has a rather flat and energy-independent local maximum decreasing systematic uncertainties associated with the precise location of the monitors. A Faraday cup after the chamber was also used to control the beam intensity.

The signals from the Si detectors are processed using a data acquisition system based on the μTCA technology. It is a digital time stamp acquisition system supporting 96

DOI: 10.1051/epjconf/201716300018

channels, and a dedicated card distributes an external 125 MHz frequency signal used to synchronise the clocks of particle and γ cards to reconstruct coincident events offline.

3.2 Gamma detectors

At bombarding energies below the Coulomb barrier, the states that can be populated in ^{20}Ne and ^{23}Na exhibit a deexcitation pattern which cascades essentially through their first excited state [11, 12]. The energies of the corresponding γ transitions are E_γ = 1634 keV and E_γ = 440 keV for exit channel (1) and (2), respectively.

Novel generation scintillators are widely used in nuclear physics as they provide sufficient energy resolution combined with a high detection efficiency. In this project, $LaBr_3$(Ce) detectors from the UK FATIMA collaboration have been combined with our system. Details concerning these detectors can be found in Ref. [13].

Comprehensive Geant4 simulations have been carried out to retrieve the highest photopeak efficiency for the detection of γ lines of interest. A cylindrical setup where all the detectors are facing the beam line has been constructed. A mechanical drawing of the supporting structure is shown figure 3 and a detailed description of the efficiency studies can be found in Ref. [14].

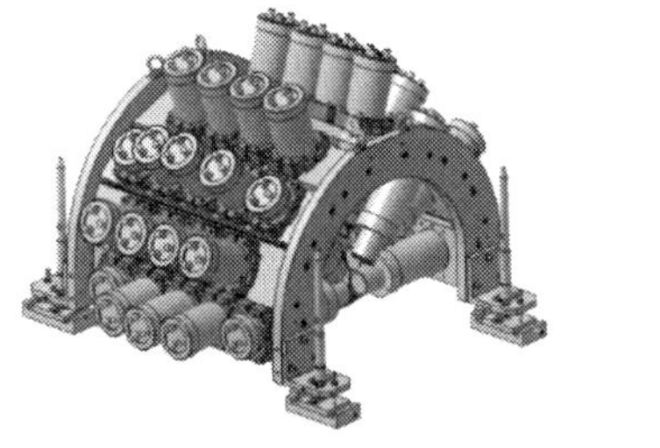

Figure 3. Cylindrical supporting structure of the 36 $LaBr_3$(Ce) crystals. The detectors face the beam line and are arranged in the most compact configuration.

In the optimal configuration where 36 $LaBr_3$(Ce) are used the photopeak efficiency is $\epsilon \sim 3\%$ for E_γ = 1634 keV, and $\epsilon \sim 8\%$ for E_γ = 440 keV.

The detectors self activity, which originates from the decay of the ^{138}La isotope as well as the ^{224}Ac contaminant, produces a constant background in γ spectra that is irrelevant in coincidence analysis. This can be used to perform an online calibration of the spectra allowing to correct for the gain drift of the PMTs that may occur with change of room temperature. A preliminary framework for this procedure is also presented in Ref. [14].

3.3 Target developments

In order to avoid carbon build-up on targets, several efforts have been made to guarantee an ultra high vacuum in the reaction chamber. Thus, only compatible materials have been used inside STELLA, like suited high vacuum connectors and PCB for DSSDs.

The pumping system is composed of a primary dry pump and a cryopump cooled down to a temperature of about 15 K. The latter has a sufficient diameter to cover the entire volume of the cylindrical part of the chamber, and a pressure of $\sim 3 \times 10^{-8}$ mbar is obtained close to the target position.

A serious limitation while using thin carbon targets is their lifetime when beam intensity is higher than 1 pμA. One possibility to overcome this difficulty is to use a fast rotating target system where the beam spot location is distributed along a path on the target surface. This enables better heat dissipation which is one of the cause of the target breaking process.

In collaboration with GANIL, a dedicated rotating target system has then been developed for the STELLA project and tested under beam at the end of the commissioning phase. Details of the apparatus will be given in a forthcoming technical paper.

3.4 The Andromède facility

For the first campaign, the STELLA station has been installed to the Andromède facility [15] in Orsay, France. Andromède is a 4 MV Pelletron accelerator which can provide various beams from light ions such as ^{4}He, ^{12}C, ^{16}O, up to heavy Au-cluster or CH_4 molecules.

STELLA was placed at the dedicated 90° line where the presence of a magnetic dipole ensures the delivery of a high purity ^{12}C beam. The accelerator ran in very stable energy and intensity conditions during the whole data taking.

During the experiment, we used $^{12}C^{3+}$ and $^{12}C^{2+}$ beams from E_{lab} = 11 to 5.6 MeV impinging onto fix targets of 20, 30 and 50 $\mu g/cm^2$ with beam intensity varying from 40 pnA to 450 pnA. Preliminary results are presented in the next section.

4 Preliminary results

A typical matrix of the particle energy as a function of ring number of a DSSD located at backward angles ($148° < \theta_{lab} \leq 168°$) is shown in figure 4. This spectrum was obtained at a beam energy of 11 MeV and the solid lines correspond to kinematics calculations for the different protons and alphas (p_i and α_i) associated to excited states in the heavy reaction ejectile. No contaminant contribution is present in the spectrum at this energy, where the total fusion cross section is $\sim$ 20 mb.

The corresponding single γ spectrum is shown in figure 5. This is the sum over the 28 detectors used during

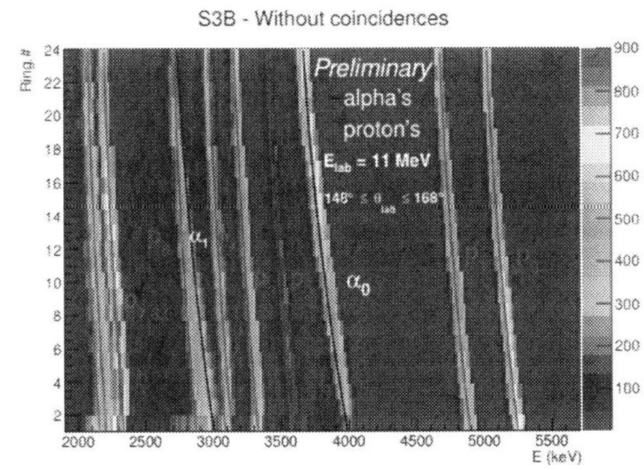

Figure 4. Angular distribution of particles obtained with a DSSD located at backward angles. The solid lines are kinematics calculations at E_{lab} = 11 MeV for the proton (red) and alpha (black) exit channels.

the first phase of the experiment.

As mentioned in Sect. 3.2, a continuous background is visible in the single γ spectrum mainly due to $LaBr_3$ self activity and gamma transitions originating from the first excited states of ^{20}Ne and ^{23}Na are masked by this contribution.

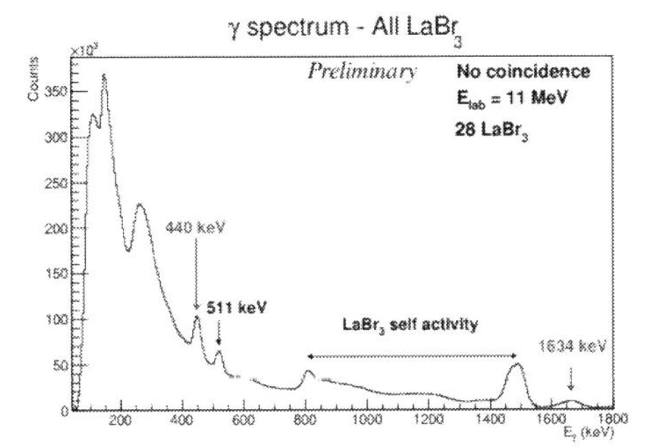

Figure 5. Gamma spectrum summed over 28 $LaBr_3$ obtained at E_{lab} = 11 MeV. The self activity as well as the two important γ lines at 440 and 1634 keV from the C+C fusion are indicated by arrows.

Nevertheless, this background in the single γ spectrum can be suppressed when analysing coincident events. A typical spectrum of the detection time difference between γ's and particles is shown in figure 6 where a well defined peak around 350 ns with few random coincidences spread all over the spectrum is visible. This allows a reliable selection of coincident events.

As an example, the effect of requiring a γ coincidence in the particle spectrum from a DSSD located at backward angles when gating around E_γ = 440 keV is depicted in figure 7, where single and coincident spectra are shown

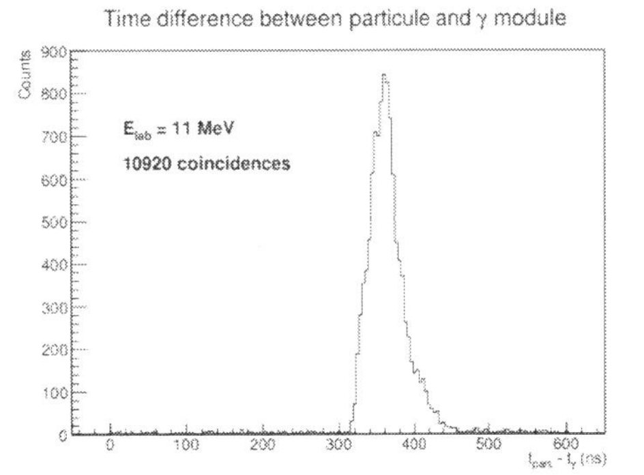

Figure 6. Time difference between γ and particle detection. A background free coincident peak is clearly visible.

together. The p_0 and α_0 exit channels are associated to

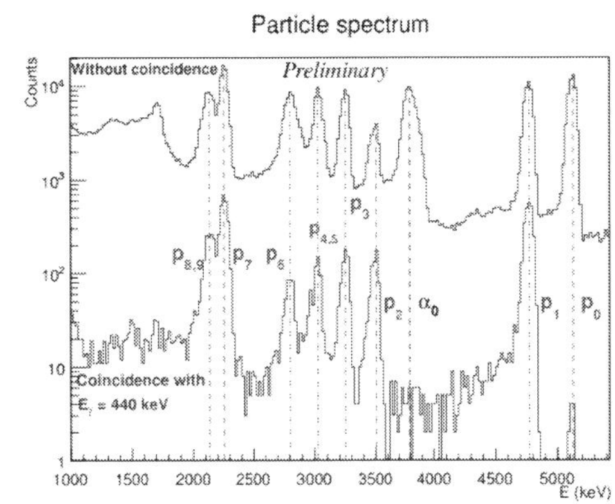

Figure 7. Single particle spectrum (upper part) and the corresponding one obtained by gating around E_γ = 440 keV (lower part) for a DSSD located at backward angles. The p_0 and α_0 exit channels, which have no true coincidence with γ's, are effectively suppressed in the coincident spectrum.

the population of the ground state of the heavy partner ^{23}Na and ^{20}Ne, respectively. No coincident gammas are expected and their contributions are effectively suppressed in the coincident spectrum.

In figure 7, the protons associated to the different excited states in ^{23}Na are separated enough to obtain partial fusion cross sections after correcting for the decay branching ratios tabulated in Ref. [12].

From the spectra displayed in figure 7, a first experimental value can be obtained for the detection efficiency of E_γ = 440 keV by taking the ratio of integrated coincident and single p_1 peaks. This gives a result of about 6%, which is in agreement with Geant4 simulation of the 28 detectors setup used during the experiment. Further checks

DOI: 10.1051/epjconf/201716300018

will be done by comparing simulation results to γ source runs where a ^{152}Eu source has been utilized.

5 Summary

A new dedicated experimental station (STELLA) has been developed and built in IPHC-CNRS, Strasbourg, to measure fusion cross sections of light heavy ions. It makes use of the γ-particle coincidence technique to ensure reliable selection of fusion events at deep sub-barrier energies.

During the first campaign, STELLA has been coupled to the $LaBr_3$ detectors from the UK FATIMA collaboration and installed at the Andromède facility in Orsay, France, to study the $^{12}C+^{12}C$ fusion reaction from E_{lab} = 11 to 5.6 MeV.

Identification of the various exit channels of the reaction has been proven possible using both single and coincident particle spectra. This allows to take into account in the total fusion cross section the feeding of the ground and excited states in the ^{23}Na and ^{20}Ne exit channels.

In collaboration with GANIL, a fast rotating target system has been developed and tested under beam. With this apparatus, a beam intensity of ~ 5 pμA may be used in the future. Then, considering the measured resonance of Spillane *et al.* at E_{cm} = 2.14 MeV [16] a statistical uncertainty of 30% can be obtained in about 3 days, whereas several weeks are needed following Gasques *et al.* extrapolation [17].

Acknowledgement

We wish to thank G. Frémont (GANIL) and M. Loriggiola (LNL, Legnaro) for preparation of ^{12}C targets for this commissioning run.

References

[1] L.R. Gasques *et al.*, PRC **76**, 035802 (2007)
[2] R.G. Stokstad *et al.*, PRL **37**, 888891 (1976)
[3] C.L. Jiang *et al.*, PRL **110**, 072701 (2013)
[4] L.J. Patterson *et al.*, The Astr. J. **157**, 367373 (1969)
[5] B. Cujec *et al.*, NPA **266**, 461493 (1976)
[6] H.W. Becker *et al.*, ZPA **303**, 305312 (1981)
[7] C.L. Jiang *et al.*, PRC **75**, 015803 (2007)
[8] K.U. Kettner *et al.*, PRL **38**, 337340 (1977)
[9] C.L. Jiang *et al.*, NIMA **682**, 1215 (2012)
[10] E.F. Aguilera *et al.*, PRC **73**, 064601 (2006)
[11] D.R. Tilley *et al.*, NPA **636**, 249 (1998)
[12] R.B. Firestone, Nuclear Data Sheets **108**, 1 (2007)
[13] O.J. Roberts *et al.*, NIMA **748**, 9195 (2014)
[14] M. Heine *et al.*, JoP: Conf. Ser. **763**, 012005 (2016)
[15] M.J. Eller *et al.*, NIMB **365**, 367370 (2015)
[16] T. Spillane *et al.*, PRL **98**, 122501 (2007)
[17] L.R. Gasques *et al.*, PRC **72**, 025806 (2005)

Fusion hindrance for the positive Q-value system $^{12}C+^{30}Si$

F. Galtarossa[1,*], A. M. Stefanini[1,**], G. Montagnoli[2], C. L. Jiang[3], G. Colucci[2], S. Bottoni[3], C. Broggini[2], A. Caciolli[2], P. Colović[4], L. Corradi[1], S. Courtin[5], R. Depalo[2], E. Fioretto[1], G. Fruet[5], A. Gal[5], A. Goasduff[1], M. Heine[5], S. P. Hu[6], M. Kaur[7], T. Mijatović[4], D. Montanari[5], F. Scarlassara[2], E. Strano[2], S. Szilner[4], and G. X. Zhang[8]

[1] INFN, Laboratori Nazionali di Legnaro, I-35020 Legnaro (PD), Italy
[2] Dip. di Fisica e Astronomia, Univ. di Padova, and INFN-PD, I-35131 PD, Italy
[3] Physics Division, Argonne National Laboratory, Argonne, IL 60439, USA
[4] Ruđer Bošković Institute Zagreb, Croatia
[5] IPHC, CNRS-IN2P3, Université de Strasbourg, F-67037 Strasbourg Cedex 2, France
[6] College of Phys. Sci. and Tech., Shenzhen Univ., Shenzeng, 518060, China
[7] Department of Physics, Punjab Technical University Jalandhar, Kapurthala, India
[8] School of Phys. and Nucl. Energy Eng., Beihang Univ., Beijing, 100191, China

Abstract. The fusion excitation function for the positive-Q-value system $^{12}C+^{30}Si$ (Q_{fus} = +14.1 MeV) has been measured in inverse kinematics down to the μb level and compared with standard coupled-channel calculations. The appearance of the fusion hindrance phenomenon and the evidence of a S-factor maximum have been observed. This result can be significant to extrapolate the behavior of lighter astrophysical relevant systems at deep sub-barrier energies, where existing experimental data are still contradicting and affected by large errors.

1 Introduction

Fusion reactions between light heavy ions have a prominent role in the dynamical stellar evolution. Reactions such as $^{12}C+^{12}C$, $^{12}C+^{16}O$ and $^{16}O+^{16}O$ are the main processes during the carbon and oxygen burning stages of massive stars [1, 2]. Existing data for these systems do not extend in the energy range of typical astrophysical scenarios because at the involved temperatures (T ~ 10^9 K) the associated Gamow energies are well below the Coulomb barrier V_C. For this reason the available data sets are still contradicting and affected by large errors, due to the significant experimental difficulties in measuring cross sections in the subnanobarn range. An example is reported in figure 1 where the S factors (see Sect. 2) for fusion between ^{12}C and $^{12,13}C$ are plotted as a function of E/V_C. No clear conclusion about the reaction rates in stars can be safely drawn from these data.

In order to obtain the astrophysical reaction rates one has, therefore, to rely on phenomenological extrapolation methods [3]. In particular, the possible presence of fusion hindrance and the trend of the S factor at very low energies for medium-light systems may assume a relevant role in the extrapolation procedure. If the S factor develops a maximum, the fusion cross sections and consequently the reaction rates in stars at the lower energies important for astrophysics will be substantially lower than what predicted by simple extrapolations of the high-energy trends, where sets of measurements already exist.

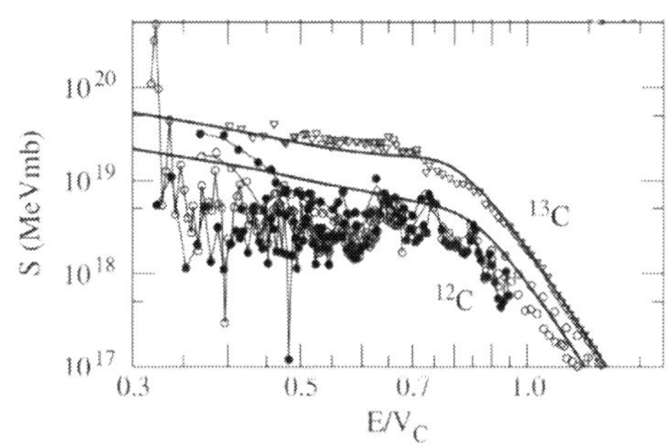

Figure 1. Measured S factors of fusion between ^{12}C and $^{12,13}C$ as a function of E/V_C. The two solid lines are theoretical predictions from coupled-channels calculations. Taken from Ref. [1].

In this contribution we will present the preliminary results of a measurement performed at the Laboratori Nazionali of Legnaro of INFN where the excitation function for the system $^{12}C+^{30}Si$ was measured down to the deep sub-barrier energy regime. The measurement was done by separating the evaporation residues from the beam by means of an electrostatic deflector and then detecting them in the E-ΔE-ToF spectrometer widely used in recent years for similar experiments. The aim of the experiment was to look for an indication of fusion hindrance in this positive Q-value system and the possible presence of a S-

* e-mail: franco.galtarossa@lnl.infn.it
** e-mail: alberto.stefanini@lnl.infn.it

factor maximum, which would be a relevant clue to deduce the behavior of astrophysical relevant systems.

2 Fusion hindrance and S factor

Since its discovery about 15 years ago [4], the phenomenon of fusion hindrance at low energies has been observed in many medium-heavy systems [1]. Its signature is a steeper decrease of the fusion cross section in the deep sub-barrier regime with respect to coupled-channels (CC) calculations employing a standard Woods-Saxon (WS) potential. Experimental data can be reproduced by using different potentials, as for instance the M3Y with the addition of a repulsive term [5, 6].

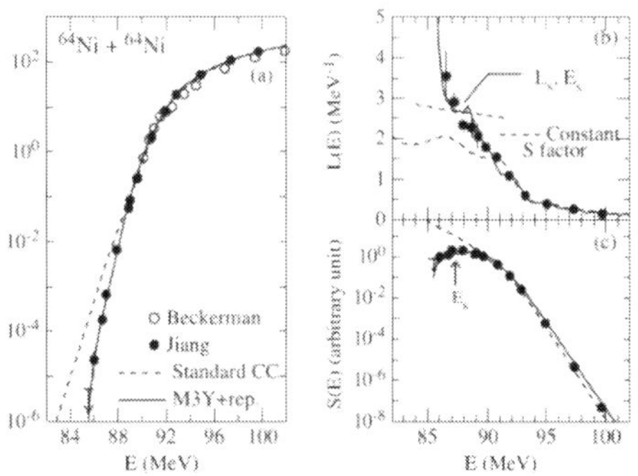

Figure 2. Experimental fusion cross section (a), logarithmic derivative (b) and S factor (c) for the system $^{64}Ni+^{64}Ni$. Solid circles are from Ref. [7], open circles from Ref. [8]. The dash and solid curves represent calculations with a standard Woods-Saxon potential and with the M3Y+rep. potential, respectively. Taken from Ref. [1].

An alternative way to make evident the presence of hindrance, without invoking model calculations, is to introduce the logarithmic derivative of the energy-weighted cross section, L(E), and the S factor, S(E), which are defined as:

$$L(E) = \frac{d[ln(E\sigma)]}{dE} = \frac{1}{E\sigma}\frac{d(E\sigma)}{dE} \tag{1}$$

$$S(E) = E\sigma(E)e^{2\pi\eta} \tag{2}$$

where $\eta = Z_1Z_2e^2/\hbar v$ is the Sommerfeld parameter. The onset of hindrance can usually be visualized with S(E) reaching a maximum and L(E) crossing the logarithmic derivative for a costant S factor, $L_{CS}(E) = \pi\eta/E$, obtained by imposing the maximum condition for S(E). This is shown in figure 2 for the benchmark case $^{64}Ni+^{64}Ni$ [7]. There is no fundamental principle underlying the constant S factor nor the intersection between L(E) and $L_{CS}(E)$ but they represent a simple and direct way to observe the discrepancy between data and standard CC calculations. The energy E_s at which the maximum appears is usually taken as the threshold energy for hindrance.

Medium-heavy systems, namely those with $\zeta = Z_1Z_2\sqrt{\frac{A_1A_2}{A_1+A_2}} > 1500$, are usually associated with negative ground-state Q values for fusion, $Q_{fus} < 0$. For these systems the S factor always shows a maximum, due to energy conservation considerations: when the center-of-mass energy E_{cm} tends to $-Q_{fus}$, the fusion cross section, and then S, must tend to 0. Figure 3 shows the experimental S factors as a function of the center-of-mass energy for different negative Q-value systems (except for $^{40}Ca+^{48}Ca$) where the fusion hindrance was observed. They all develop a maximum.

The same is not true for systems with $Q_{fus} > 0$: when E_{cm} tends to 0, the fusion cross section does not necessarily tend to 0 and neither does S. Of course the extrapolation of the behavior of these systems to lower energies largely depends on whether or not the S factor develops a maximum.

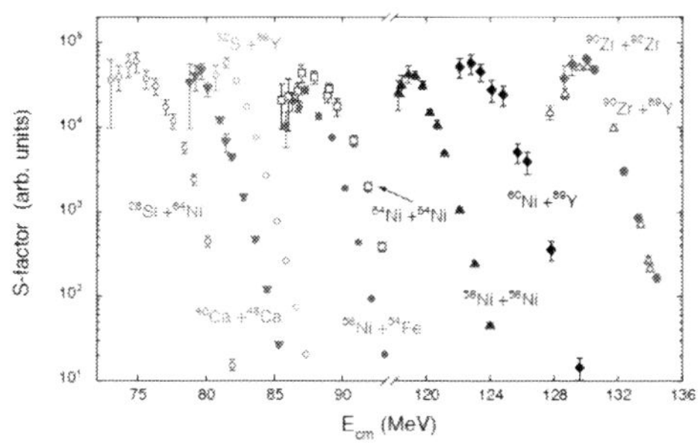

Figure 3. Experimental S factors as a function of the center-of-mass energy for different negative-Q-value systems (except for $^{40}Ca+^{48}Ca$ which has Q_{fus} = +4.56 MeV).

Recent experiments investigated the presence of hindrance in systems with positive Q_{fus}. Fusion excitation functions were measured for the systems $^{28}Si+^{30}Si$ (Q_{fus} = +14.3 MeV) [9, 10], $^{36}S+^{48}Ca$ (Q_{fus} = +7.55 MeV) [11] and $^{27}Al+^{45}Sc$ (Q_{fus} = +9.63 MeV) [12] down to ~ 4 μb, 600 nb and 300 nb, respectively. By comparing data with standard CC calculations, in all the systems indications of fusion hindrance were observed. On the other hand, no clear S-factor maxima showed up in the energy ranges covered by the experiments.

More recently indications of S-factor maxima were observed in the system $^{40}Ca+^{48}Ca$ (Q_{fus} = +4.56 MeV) [13] and more clearly in $^{24}Mg+^{30}Si$ (Q_{fus} = +17.9 MeV) [14]. This observation suggests that the cross section falls off very steeply at even lower energies than the measured ones and this behavior might be common to other positive Q-value systems, such as those involved in the carbon burning of stars.

In this context the $^{12}C+^{30}Si$ system (Q_{fus} = +14.1 MeV), which is not itself a relevant case in astrophysics, may represent a critical case to predict the behavior of lighter systems like $^{12}C+^{12}C$, $^{12}C+^{16}O$ and $^{16}O+^{16}O$. Figure 4 (top) shows the systematics for several medium-light systems with $Q_{fus} > 0$ (in red) and $Q_{fus} < 0$ (in black) where E_s is plotted as a function of the parameter ζ previously defined (μ is the reduced mass of the system). The points for $^{12}C+^{12}C$ and $^{16}O+^{16}O$ are not directly measured and rely on extrapolations. Figure 5 shows the similar systematics for the measured (red) and extrapolated (black) threshold cross section, σ_s, for some medium-light systems. One can clearly see that the measurement of E_s and σ_s for $^{12}C+^{30}Si$, whose position in ζ is indicated by the blue arrows in the figures, could be relevant for a more reliable extrapolation.

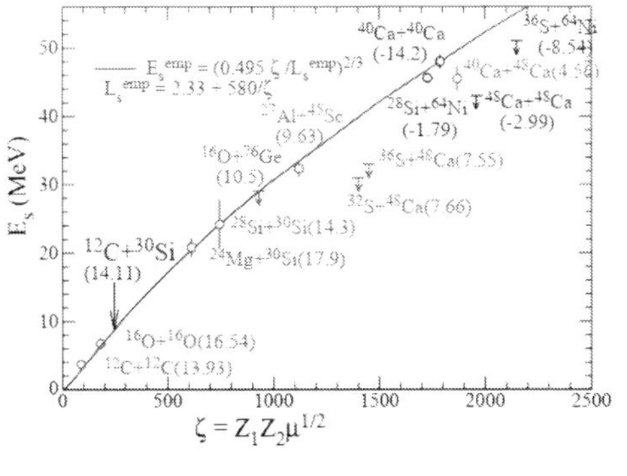

Figure 4. Systematics of threshold energies for hindrance, E_s, for various medium-light systems with positive (red) and negative (black) Q values for fusion. The points for $^{12}C+^{12}C$ and $^{16}O+^{16}O$ are determined from extrapolation procedures. The blue line is the predicted threshold for hindrance on the basis of the phenomenological systematics of Jiang et al. [18]. The predicted location for $^{12}C+^{30}Si$ is indicated by the blue arrow.

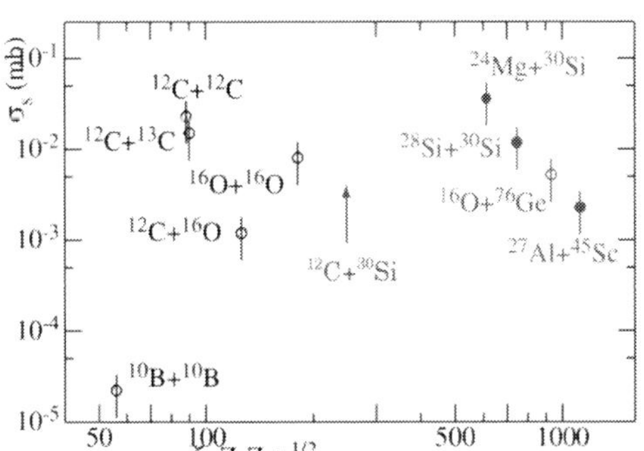

Figure 5. Systematics of threshold cross sections for hindrance, σ_s, for various medium-light systems determined by measurements (red) or by extrapolations (black). The predicted location for $^{12}C+^{30}Si$ is indicated by the blue arrow.

Existing data for this system [15], and for the nearby $^{12}C+^{28}Si$ [16, 17], extended down to ~ 200 mb, too far from the range in which the hindrance could start to play a role. Moreover, these data exhibit large error bars. The aim of our experiment was to refine and extend the measurement of the excitation function down in energy as much as possible to search for a possible indication of hindrance.

3 Experiment

The experiment was performed at the Laboratori Nazionali of Legnaro of INFN (Padova, Italy). The ^{30}Si beam was provided by the XTU Tandem accelerator in the energy range ~ 33-80 MeV, with an intensity of ~ 10 pnA. The 50 $\mu g/cm^2$ ^{12}C targets, enriched to 99.9% in mass 12 to reduce as much as possible the contamination coming from the higher-mass ^{13}C isotope, were installed in a sliding-seal scattering chamber. The choice of the inverse kinematics allowed to produce the compound nucleus (CN) with enough kinetic energy to reach the detector telescope, and still produce a signal in the Si detector of sufficient amplitude to start the Time-of-Flights (ToF) and trigger the DAQ. On the other hand, since the electric rigidities of CN and scattered beam are more similar compared to direct kinematics conditions, the contribution of background components might be more significant in this case.

The evaporation residues (ER), separated from the beam in the electrostatic deflector [20], were detected at $\theta_{lab} = 3°$ with a E-ΔE-ToF telescope schematically depicted in figure 6 and described in more detail in Ref. [19].

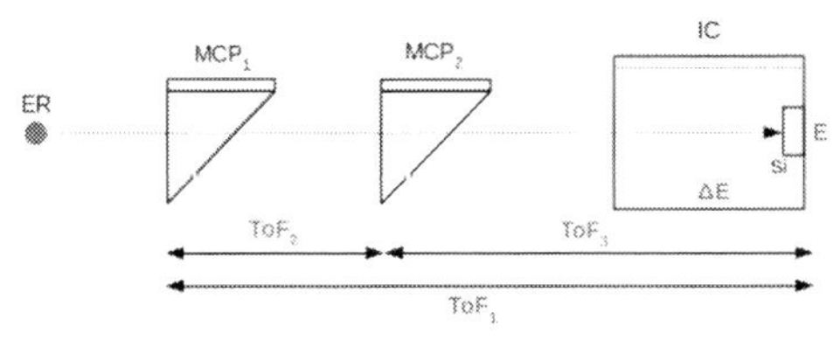

Figure 6. The E-ΔE-ToF telescope following the electrostatic deflector. Measured quantities are labeled in red.

The measured quantities are three ToFs, an energy loss ΔE in the ionisation chamber (IC) and a residual energy E in the silicon detector (Si). Four silicon detectors, placed around the target at $\theta_{lab} = 16°$ with respect to the beam direction, in the up, down, right and left position, were used for beam control and normalization between the different runs by measuring the Rutherford scattering from the target. Three angular distributions were measured at E_{lab} = 45, 59 and 80 MeV in the range $-6° \leq \theta_{lab} \leq +10°$. We report in figure 7 the angular distribution obtained at E_{lab} = 59 MeV (points), together with the gaussian curve which well fits the data.

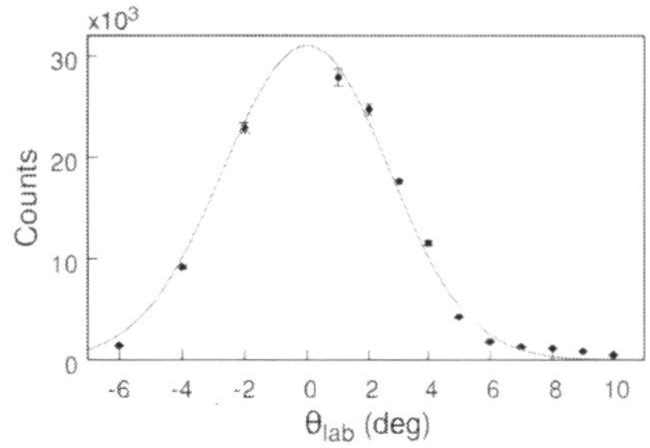

Figure 7. Angular distribution of the ER at E_{lab} = 59 MeV (points). The curve represents a gaussian fit. The plotted errors are only statistical uncertainties.

The evaporation residues can be separated from the scattered beam in the online analysis by combining E-ToF and E-ΔE matrices. Figure 8 shows an example of two-dimensional spectra where the residual energy detected in the Si detector is plotted versus the ToF between the first MCP detector and the Si detector at two different bombarding energies, one above (top panel) and the other below (bottom panel) the barrier. The ER are in both cases well separated from the beam.

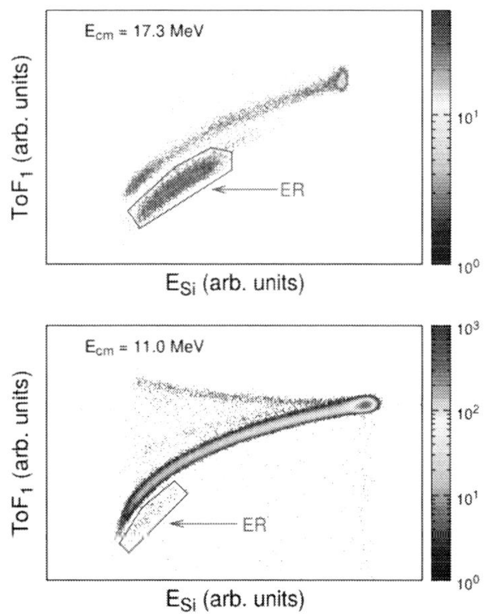

Figure 8. Two-dimensional spectra ToF-E for the present experiment at the indicated bombarding energies, above (top) and below (bottom) the Coulomb barrier. Evaporation residues are in both cases well separated from the scattered beam.

4 Results

Figure 9 shows the resulting excitation function. To account for a possible contamination from higher-mass carbon isotopes, some points were measured with a ^{13}C target and used to correct the plotted cross sections. The energy loss in the target was taken into account as well in the analysis. The new data well agree with those of Ref. [15] but extend down to a cross section almost 5 orders of magnitude lower.

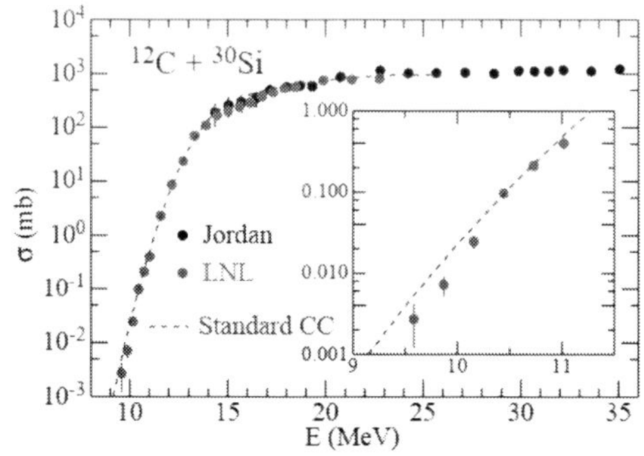

Figure 9. Experimental excitation functions measured in Ref. [15] (black points) and in the present experiment (red points), together with standard CC calculations (dashed blue line). In the insert a zoom in the low-energy region is shown.

Experimental data were compared to preliminary CC calculations employing an Akyüz-Winther (AW) potential [21]. Since the 2^+ and 3^- states of ^{12}C are quite high in energy (4.44 and 9.64 MeV, respectively), in the calculations this nucleus was considered inert. It is well known that the effect of coupling to high-lying states in CC calculations mainly consists in an overall shift of the calculated cross section to slightly lower energies, without significant influence on the shape of the excitation function [22, 23]. This coupling can then be accounted for by an adiabatic renormalization of the ion-ion potential in order to reproduce the experimental cross sections in the vicinity of the barrier. For this reason the depth of the AW potential here used was slightly modified and its value set to V_0 = 48.24 MeV. The parameter r_0, which accounts for the radii of the interacting nuclei, and the diffuseness a were set to 1.10 fm and 0.61 fm, respectively. This potential produces a barrier V_c = 13.6 MeV.

For the ^{30}Si the CC calculations included the coupling to the 2^+ and 3^- states, whose energies E and deformation parameters β_λ are reported in table 1. We remind that the deformation parameter is related to the reduced transition probability of the electromagnetic decay from the excited state λ^π to the ground state by [1]:

$$\beta_\lambda^2 = \left(\frac{4\pi}{3Ze^2R^\lambda}\right)^2 4\pi B(E\lambda) \ , \qquad (3)$$

with Z and R the nuclear charge and radius, respectively.

Table 1. Parameters for the low-lying excited states of ^{30}Si.

λ^{π}	E (MeV)	β_{λ}
2^+	2.235	0.33
3^-	5.488	0.28

In the insert of figure 9 one can see how the CC calculations start to overpredict the experimental cross sections from E ~ 10.5 MeV, an indication that the fusion hindrance is present in the measured energy range.

To confirm this hypothesis we report in figure 10 the experimental and calculated logarithmic derivative L(E) and the experimental S factor S(E).

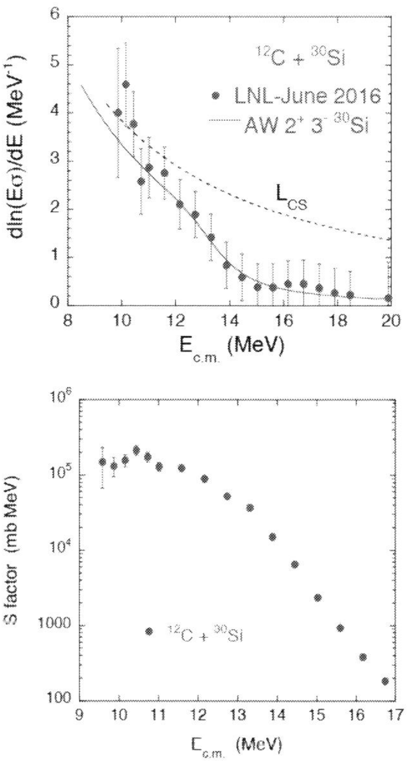

Figure 10. Experimental logarithmic derivative (top) and S factor (bottom) for the present experiment. For L(E) preliminary standard CC calculations are plotted as well.

The plotted values $L_i(E)$ were obtained by calculating the difference quotient:

$$L_i(E) = \frac{[ln(E\sigma)]_{i+1} - [ln(E\sigma)]_i}{E_{i+1} - E_i} \quad . \qquad (4)$$

Though the quite large error bars, one can clearly notice a crossing of the $L_{CS}(E)$ function, which was introduced in Sect. 2, as well as the appearance of a maximum of the S factor around E = 10.5 MeV. This can be taken as a confirmation of the onset of hindrance for this positive Q-value system.

5 Conclusions

In this contribution we have described the measurement of the fusion excitation function for the positive Q-value system ^{12}C+^{30}Si in inverse kinematics, performed at the INFN Laboratori Nazionali of Legnaro. The aim of the experiment was to search for a possible indication of fusion hindrance and presence of a S-factor maximum in this system, which could positively lead the extrapolation to lower mass systems that are relevant in an astrophysical context. The evaporation residues, separated from the beam by means of an electrostatic deflector, were detected with a E-ΔE-ToF telescope. The excitation function has been extended down to ~ 3 μb, nearly 5 orders of magnitude lower than existing measurements for this system. We could observe the presence of fusion hindrance by comparing the experimental cross sections with preliminary standard CC calculations which took into account the coupling to the 2^+ and 3^- states of ^{30}Si and consider the ^{12}C as inert. The observation of the crossing between the logarithmic derivative L(E) and L_{CS} and the apperance of a maximum of the S factor give a further confirmation of the presence of hindrance in the measured energy range and are relevant for the extrapolation procedure to lower energies.

The analysis is still ongoing. We expect that more accurate calculations, employing for example a "shallow" potential, will allow to determine more reliably the threshold energy and cross section for hindrance, adding an important point in the systematics of figures 4 and 5.

References

[1] B. B. Back, H. Esbensen, C. L. Jiang and K. E. Rehm, Rev. Mod. Phys. **86**, 317 (2014).
[2] C. J. Horowitz, H. Dussan and D. K. Berry, Phys. Rev. C **77**, 045807 (2008).
[3] C. L. Jiang *et al.*, Phys. Rev. C **75**, 015803 (2007).
[4] C. L. Jiang *et al.*, Phys. Rev. Lett. **89**, 052701 (2002).
[5] S. Misicu and H. Esbensen, Phys. Rev. Lett. **96**, 112701 (2006).
[6] H. Esbensen, C. L. Jiang, and A. M. Stefanini, Phys. Rev. C **82**, 054621 (2010).
[7] C. L. Jiang *et al.*, Phys. Rev. Lett. **93**, 012701 (2004).
[8] M. Beckerman *et al.*, Phys. Rev. Lett. **45**, 1472 (1980).
[9] C. L. Jiang *et al.*, Phys. Rev. C **78**, 017601 (2008).
[10] G. Montagnoli *et al.*, Phys. Rev. C **90**, 044608 (2014).
[11] A. M. Stefanini *et al.*, Phys. Rev. C **78**, 044607 (2008).
[12] C. L. Jiang *et al.*, Phys. Rev. C **81**, 024611 (2010).
[13] C. L. Jiang *et al.*, Phys. Rev. C **82**, 041601(R) (2010).
[14] C. L. Jiang *et al.*, Phys. Rev. Lett. **113**, 022701 (2014).
[15] W. J. Jordan, J. V. Maher and J. C. Peng, Phys. Lett. B **87**, 38 (1979).

[16] S. Gary and C. Volant, Phys. Rev. C. **25**, 1877 (1982).

[17] Y. Nagashima *et al.*, Phys. Rev. C. **26**, 2661 (1982).

[18] C. L. Jiang, B. B. Back, R. V. F. Janssens and K. E. Rehm, Phys. Rev. C **75**, 057604 (2007).

[19] A. M. Stefanini *et al.*, Phys. Rev. C **82**, 014614 (2010).

[20] S. Beghini *et al.*, Nucl. Instr. Meth. A **239**, (1985) 585.

[21] Ö. Akyüz and A. Winther, "*Nuclear structure and heavy-ion physics*", Proc. Int. School of Physics Enrico Fermi, Varenna, edited by R. A. Broglia and R. A. Ricci (North Holland, Amsterdam, 1981) (1981): 492.

[22] N. Takigawa, K. Hagino, M. Abe and A. B. Balantekin, Phys. Rev. C **49**, 2630 (1994).

[23] K. Hagino and N. Takigawa, Prog. Theor. Phys. **128**, 1001 (2012).

Clustering in light nuclei and their effects on fusion and pre–equilibrium processes.

Fabiana Gramegna[1,*], *Magda* Cicerchia[1,2,**], *Daniela* Fabris[3], *Tommaso* Marchi[4], *Marco* Cinausero[1], *Meltem* Degerlier[5], *Justin* Mabiala[1], *Giorgia* Mantovani[6], *Luca* Morelli[7], *Michela* D'Agostino[7], *Mauro* Bruno[7], *Sandro* Barlini[8], *Maurizio* Bini[8], *Gabriele* Pasquali[8], *Silvia* Piantelli[9], *Giovanni* Casini[9], *Giuseppe* Pastore[8], *Diego* Gruyer[10], *Pietro* Ottanelli[8], *Simone* Valdré[10], *Nicla* Gelli[9], *Alessandro* Olmi[9], *Giacomo* Poggi[8], *Emanuele* Vardaci[11], *Ivano* Lombardo[11], *Daniele* Dell'Aquila[11], *Silvia* Leoni[12], *Natalya* Cieplicka-Orynczak[13], *Bogdan* Fornal[13], *Daniele* Mengoni[2,3], *Gianmaria* Collazuol[2,3], *Antonio* Caciolli[2,3], *Maria* Colonna[14], *Akira* Ono[15], and *Giorgio* Baiocco[16]

[1]*Laboratori Nazionali di Legnaro, 35020 Legnaro (Pd), Italy*
[2]*University of Padova, Dep. of Physics and Astronomy, 35100 Padova, Italy*
[3]*INFN Padova, 35100 Padova, Italy*
[4]*KU Leuven, Dep. of Phys. and Astron. Instituut voor Kern- en Stralingsfysica, 3001 Leuven, Belgium*
[5]*Science and Art Faculty, Phys. Dep., Nevsehir Haci Bektas Veli Univ., Nevsehir, Turkey*
[6]*INFN Ferrara and University of Ferrara, Dep. of Chemistry, Ferrara, Italy*
[7]*INFN Bologna and University of Bologna, Dep. of Physics and Astronomy, Bologna, Italy*
[8]*INFN Firenze and University of Firenze, Dep. of Physics and Astronomy, Firenze, Italy*
[9]*INFN Firenze, Firenze, Italy*
[10]*GANIL, Caen, France*
[11]*INFN Napoli and University of Napoli, Dep. of Physics and Astronomy, Napoli, Italy*
[12]*INFN Milano and University of Milano, Dep. of Physics and Astronomy, Milano, Italy*
[13]*Institute of Nuclear Physics, Polish Academy of Sciences Krakow, Poland*
[14]*Laboratori Nazionali dl Sud, Catania, Italy*
[15]*Tohoku University, Sendai, Japan*
[16]*University of Pavia, Dep. of Physics, Pavia, Italy*

Abstract. The study of nuclear cluster states bound by valence neutrons is a field of recent large interest. In particular, it is important to study the pre-formation of α-clusters in α-conjugate nuclei and the dynamical condensation of clusters during nuclear reactions [1–5]. The NUCL–EX collaboration has recently initiated an experimental campaign of exclusive measurements of fusion–evaporation reactions with light nuclei as interacting partners. In collisions involving light systems, the low expected multiplicity of fragments increases the probability of achieving a quasi complete reconstruction of the event. In particular the formation and decay modes of an excited ^{24}Mg system have been studied through two different reactions, ^{12}C (95 MeV)+ ^{12}C and ^{14}N (80.7 MeV)+ ^{10}B, which have been used to produce fused systems with nearly the same mass and excitation energy (~60 MeV). In particular, even the de-excitation of the Hoyle state in ^{12}C have been studied, both in peripheral (projectiles de-excitation) and in central collisions (six α-particles channel). Moreover, a research campaign studying pre-equilibrium emission of light charged particles and cluster properties of light and medium-mass nuclei has been carried out. For this purpose, a comparative study of the three nuclear systems ^{18}O+^{28}Si, ^{16}O+^{30}Si and ^{19}F+^{27}Al has been recently studied using the GARFIELD+RCo 4π setup [6]. The experimental data are compared with the predictions of simulated events generated with the statistical models (GEMINI^{++} and HFℓ) and through dynamical models like Stochastic Mean Field (SMF) and Antisymmetrized Molecular Dynamics (AMD) and filtered with a software replica of our apparatus in order to take into account the experimental conditions.

1 Introduction

Nuclear reactions involving heavy-ions permit the formation of transient nuclear systems, which are excited in the continuum, typically well above the particle and fragment emission threshold. This can be considered quite an important and unique opportunity to probe nuclear systems in extreme conditions of density, temperature and isospin ratio, using experimental techniques. The properties of these excited nuclei, however, cannot be measured directly, since extremely short reaction times in the femtosecond range are involved. The information can, therefore, be accessed only through the decay products of the reaction. However, the experimental analyses of such ex-

*e-mail: fabiana.gramegna@lnl.infn.it
**e-mail: magda.cicerchia@lnl.infn.it

periments, which are performed on an event−by−event basis, are very challenging since the products in the exit channels are a large number of particles and fragments, that cannot be fully detected due to the finite size and resolution of the experimental apparatus and the dead layers. However, on the experimental side, complex detectors based on a ~ 4π geometry, with high granularity and low detection and energy thresholds, have been built: in particular, the GARFIELD-RCo multidetector [6] have been used in the experiments described in the present contribution.

Moreover, in the data analysis, particular care has to be devoted to the observables which have to be considered. When dealing with complex nuclear reactions, in fact, it has been demonstrated that inclusive observables can be typically reproduced by different models, even based on very different physical constraints, which may bring to contradictory interpretations. On the contrary, the comparison to exclusive observables, like the particle-particle and particle-fragment correlations becomes an essential tool to study the excited nuclei decay properties in order to extract information on both the properties of the excited sources, the decay mechanisms, and the occurring competing mechanisms. In this contribution, we will present results which were obtained in different measurement campaigns, performed at the Tandem XTU accelerator of the Legnaro National Laboratory.

2 Experimental Details

The experiments have been performed at the Laboratori Nazionali di Legnaro, with respectively the ^{12}C and the ^{14}N plus the ^{16,18}O and ^{19}F beams provided by the TANDEM−XTU acceleration system. The GARFIELD plus RCo experimental setup, fully equipped with digital electronics, was used [6, 7]. The apparatus has 488 detecting cells with a geometrical coverage of the order of 80% of 4π. A sketch of the layout is shown in Figure 1. In particular, GARFIELD consists of a couple of drift chambers with gas microstrip read−out for the ΔE signals, opportunately coupled to CsI(Tl) scintillation detectors for the residual energy information: the two drift chambers are filled with CF_4 gas at a pressure of 50 mbar and cover the polar angles θ=30° − 150°. They are divided in subsectors, each of which covers ~ 7° in azimuthal angle; moreover, in each sector four polar angle regions are defined by four CsI(Tl) scintillators, each of which covers ~ 14° in θ. For both chambers the ΔE−E technique allows charge identification of particles and fragments. Light particles (up to Z=3) are identified in both charge and mass through the Fast vs. Slow pulse shape analysis (PSA) technique of the CsI(Tl) signals. The energies of particles and fragments are determined with an accuracy of the order of few percent with a 0.8÷1AMeV threshold.

The forward polar angles (7°− 17°) are covered by a three−stage device, the Ring Counter (RCo), consisting of an ionization chamber (IC) (filled with CF_4 gas at a pressure of 50 mbar) divided in eight azimuthal sectors, each one followed by an eight strip 300 μm thick silicon detector (Si) and 6 CsI(Tl) scintillators. This allows for an angular resolution of about 1.4° for the polar angle and 22° for the azimuthal one. The particle and fragment iden-

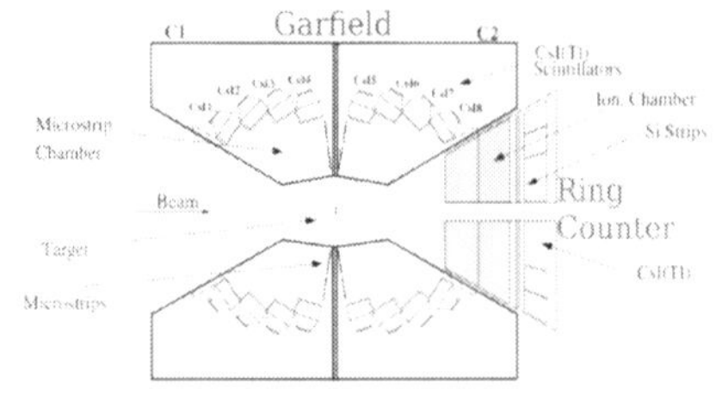

Figure 1: Layout of the GARFIELD + RCo array.

tification is obtained through ΔE−E technique in IC−Si or Si−CsI(Tl) and/or with CsI(Tl) scintillator fast-slow PSA technique. Using the gas section, particles and fragments are identified in charge with energy thresholds as low as 0.8 − 1 AMeV. Moreover, reaction products characterized by $1 \leq Z \leq 12$ and energies sufficient to reach the CsI scintillators can be identified in mass with an energy threshold of ~ 6 AMeV. In some cases, mass determination for light fragments stopped in the silicon is also possible, with a quite lower threshold (~ 2-3 AMeV), using the PSA technique. A percent accuracy in the determination of the particle and fragment energy is possible in the RCo detector. Due to the high count rate of elastic scattering, the small angles were shielded through an appropriate collimation system. The Evaporation Residues (ER) have been collected in an angular range θ=8.6° and 17° (just beyond the grazing angles). Coincidences with ER have been selected setting proper gates in the IC vs. Si ΔE-E energy spectra of the RCo and asking for any coincident particle in the entire apparatus.

3 The fusion−evaporation case: decay of excited ^{24}Mg

In this paragraph, we will consider the experiments where the following reactions were studied: a ^{12}C beam at 95 MeV onto a ^{12}C target and a ^{14}N beam at 80.7 MeV onto a ^{10}B target. The aim of these measurements consisted in investigating both the reaction mechanisms and the decay of excited nuclei formed in nuclear collisions, by detecting charged particle and studying their correlations and characteristics.

3.1 Events Selection

In order to carry out a dedicated analysis, two different event selections have been performed. The first one, related to *central events*, was fulfilled asking for the detection of the evaporation residue (ER) in the forward direction (RCo) and one or more light particles in coincidence, detected at large θ angles (GARFIELD). The ER are characterized by a velocity close to the centre of mass velocity of the colliding system [2]. A distribution which correlates the total detected charge vs. the total longitudinal momentum was built and the resulting plot is shown in Figure 2 for the ^{12}C+ ^{12}C system, where the almost complete fusion-evaporation events are located in the upper right region. Since a large number of events was collected, for which the total charge is completely detected, a very strict selection was performed for the subsequent analysis. In particular, only those events were considered, for which both the total incoming charge is Z_{tot} =12 and at least 95% of the incoming momentum have been measured. It was, however, verified that the less stringent selection Z_{tot} ≥10 did not actually bias the characteristics of the events. This second selection was, therefore, used when dealing with specific decay channels, for which more statistics was needed for a complete analysis.

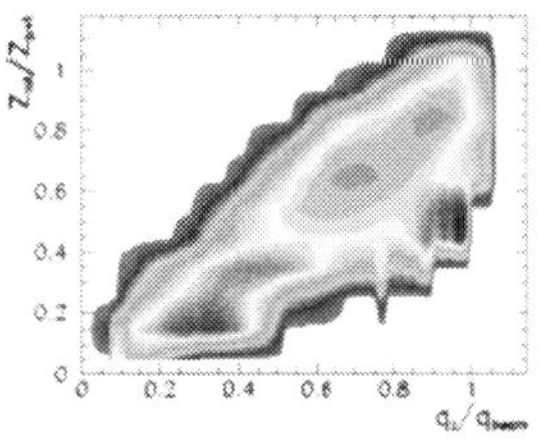

Figure 2: Correlation between the total detected charge and the total detected momentum along the beam axes for the ^{12}C+^{12}C at 95 MeV.

The second event selection was, on the contrary, dedicated to a subset of peripheral events, corresponding to inelastic reactions ^{12}C+ ^{12}C → ^{12}C* + ^{12}C, which was defined by the detection of three α−particles in the forward cone and no charged particles in the remaining solid angle. In this way, the selected channels were related to well defined quantum states of the quasi−projectile and of the quasi−target.

3.2 Central collisions

Experimental *central events* have been compared to Monte Carlo statistical model prediction and, in particular, with the Monte Carlo code HF*l*, opportunely developed to study the decay of light Compound Systems [1, 8]. For the ^{12}C+ ^{12}C case, a maximum value J_0^{max} =18$\hbar$ and a diffuseness parameter ΔJ =2 was used for the Angular Momentum in the code. Code prediction are filtered through a software replica of the apparatus, taking into account experimental energy thresholds, dead layers, angular restrictions, energy resolution and solid angle for each detector. The selected sample of experimental data is, in general, compatible with the expected behavior of a complete fusion evaporation reaction. However, some exceptions have been observed, when studying specific decay channels: in particular a larger population have been observed when two or three α−particles are emitted in coincidence with an O or a C residue, respectively. This is shown in table 1, where, for the ^{12}C + ^{12}C case, for each measured residue, the most probable experimental channel and its branching ratio are compared to the HF*l* model predictions. In

Table 1: In the ^{12}C +^{12}C case, the most probable experimental channel and its branching ratio are compared to the HF*l* model predictions for each measured residue. In the case of ^{14}N+^{10}B system, the results for the selected channels with C+3α, O+2α, Ne+α respectively are shown for comparison. Errors on the experimental values (~ 5%) are estimated, considering a sum of the statistical errors plus the indetermination due to possible α−particle contamination from badly detected ^{3}He.

	System ^{12}C+^{12}C		
Z_{res}	Channel	BR_{HFl}(%)	BR_{EXP}(%)
5	$^{11-xn}B + xn + p + 3\alpha$	100	99
6	$^{12-xn}C + xn + 3\alpha$	78	98
7	$^{15-xn}N + xn + p + 2\alpha$	95	91
8	$^{16-xn}O + xn + 2\alpha$	15	63
9	$^{19-xn}F + xn + p + \alpha$	88	92
10	$^{22-xn}Ne + xn + 2p$	82	55
10	$^{22-xn}Ne + xn + \alpha$	3	26
	System ^{14}N+^{10}B		
Z_{res}	Channel	BR_{HFl}(%)	BR_{EXP}(%)
6	$^{12-xn}C + xn + 3\alpha$	77	96
8	$^{16-xn}O + xn + 2\alpha$	15	56
10	$^{22-xn}Ne + xn + \alpha$	3	47

particular, both the α energy spectra and the angular distributions in the (O+2α) channel are not compatible with statistical model calculations, so that either a contamination from direct reactions or a persistence of α-correlations in the ^{24}Mg compound are supposed. However, the total discrepancy with HF*l* calculations, cannot be completely ascribed to entrance channel effects, since some discrepancies are also observed when looking at the energy spectrum of α−particles in coincidence with O for the ^{14}N induced reaction. Even for the (C + 3α) and the (Ne +α) channels an anomalously high branching ratio can be observed for both entrance channel systems and they seem to be solely attributed to a possible persistence at high excitation energy of α structure correlations in the ^{12}C+ ^{12}C molecular state and/or in the ^{24}Mg compound [2]. This can be observed looking at the results in Table 1.

3.3 Hoyle State in ^{12}C*

The data collected in the measurement of ^{12}C+^{12}C at 95MeV have been also analyzed in order to investigate the decay of the Hoyle state of ^{12}C*. Two different data selections have been chosen: a first one, which corresponds to peripheral binary collisions, where the quasi−projectile is excited to the Hoyle state and the target has been left in the ground state; a second selection, which is dedicated to the central events, where the formation of a ^{24}Mg* is considered and the Hoyle state is obtained as a step of the decay chain. A first selection of central events was performed looking for 6 α−particles events detected by the apparatus, with a complete reconstruction (Z_{tot} =12) of the ^{24}Mg* decay. To exclude events deriving from different channels than the pure 6 α−particles, like, for instance, when also a neutron could have been emitted, the dissipated energy was reconstructed through the quantity Q_{kin}, defined as: $Q_{kin} = \sum_{i=1}^{6} E_i - E_{beam}$. A selection of events

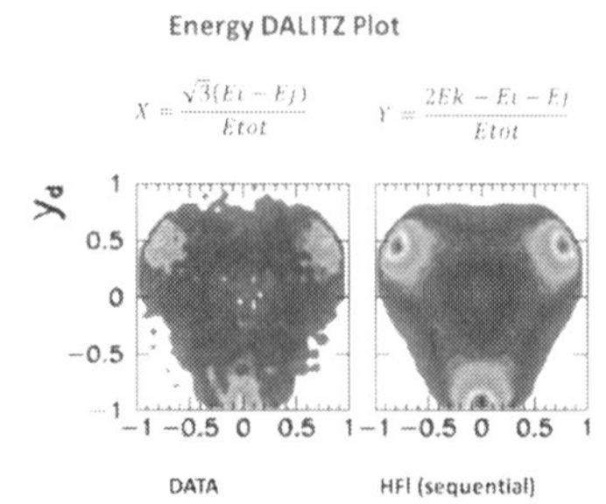

Figure 3: Energy Dalitz Plots: correlations within three α−particles. Left panel: Experimental data; Right panel: simulated data from HF*l*.

have been, therefore, carried out by selecting the opportune Q_{kin} region and by analyzing the correlation of three (over six) α−particles in each event, to finally choose the minimum excitation energy of the reconstructed $^{12}C^{\star}$ excitation spectrum (see [9] for details). The characteristics of the Hoyle state decay are very similar in the two samples of data: for example, for the semi-peripheral case, the energy Dalitz plot (see Figure 3) is shown for both the experimental and simulated data, this last derived from the prediction of the statistical evaporation code HF*l*. Both semi-peripheral and central data sets give similar results and mainly point to a sequential decay through the population of an intermediate $^{8}B_{gs}$, with a very small contribution (~1.1%) from a simultaneous three α−particle process. More details on the particular analysis performed have been presented in the work by L. Morelli et al. [9].

4 Fusion−Evaporation vs. Pre-equilibrium case: ^{18}O+^{28}Si, ^{16}O+^{30}Si and ^{19}F+^{27}Al

In order to continue the extensive experimental program, the comparative study of the four reactions: ^{18}O+^{28}Si, ^{16}O+^{30}Si and ^{19}F+^{27}Al at 7 AMeV and ^{16}O+^{30}Si at 8 AMeV has been performed. This is both related to the study of light compound system decay and, furthermore, to the study of pre−equilibrium emission and its possible link to the internal structure of the interacting partners, especially for what it concerns clusters. The main goal of the experiment is to measure and compare the pre−equilibrium component of the particle emission spectra for these systems, especially the α−particle emission, since it may be related either to the pre−formation of clusters in nuclei or to their formation during the dynamics of the collision. For central impact parameters, and in the case of complete fusion, the studied reactions all lead to the same compound nucleus, $^{46}Ti^{\star}$, even if with slightly different excitation energies as shown in Table 2, for which a small difference in the de-excitation chain is expected. On the contrary, choosing the same beam velocity (7 AMeV), the non-equilibrium processes are predicted to be almost the same [10]. From the comparison between statistical and pre-equilibrium light charged particle multiplicities and studying the difference between specific exit

Table 2: Reactions main characteristics.

Projectile	Target	E_{beam} (AMeV)	Mass Asymm.	CN Lab. Vel.	E* (MeV)
^{16}O	^{30}Si	7	0.30	1.28	88.0
^{18}O	^{28}Si	7	0.22	1.44	98.5
^{16}O	^{30}Si	8	0.30	1.37	98.4
^{19}F	^{27}Al	7	0.17	1.52	103.5

channels in all the studied systems, even through particle-particle correlations, it will be possible to extract information on the fast emission mechanism and its competition with statistical evaporation from a thermalized system. The influence of possible structure effects, especially those related to the $\alpha-$ clustering in these medium-mass systems, can derive from the comparison of the different entrance channels, where the neutron and proton numbers of the colliding partners have been mutually exchanged. The ^{16}O+^{30}Si at 8 AMeV have been considered in order to have the same compound nucleus excitation energy of the ^{18}O+^{28}Si at 7 AMeV, hence the same statistical component.

4.1 Events Selection

Even in this case, different groups of events have been selected. The first selection, which will be discussed in this paper, is related to *central events* and it was performed asking for the detection of the ER in the forward direction (RCo) in coincidence with one or more light charged particles in the remaining apparatus. The ER are characterized by a velocity close to the centre of mass velocity of the reaction (see Table 2). A correlation between the detected charge vs. the energy distribution in the Laboratory system for the reaction ^{18}O+^{28}Si at 7 AMeV is shown in Figure 4 for the total events (left panel) and for the selected central events (right panel) for the experimental data set and for simulated data from SMF+GEMINI [12] and AMD+ GEMINI [13]. See the following paragraph for more details on theoretical calculations. A further selection on a subset of peripheral events is under study, corresponding to inelastic reactions, which can be defined by the detection of four α−particles in the forward cone and no charged particles in the remaining solid angle. In this way, we can select channels where the quasi−projectile and the quasi−target are in a well defined single quantum state. This event selection will not be discussed further in the present paper, since the analysis is still too preliminary.

4.2 Central events: preliminary results

The energy distributions of the light charged particles detected in coincidence with the residue have been studied and compared to those obtained by model predictions performed either with the statistical model GEMINI^{++} [11], which is based on a complete fusion hyphotesis, or by a dynamical code followed by GEMINI^{++} as afterburner. In particular, the comparison to two dynamical

 DOI: 10.1051/epjconf/201716300020

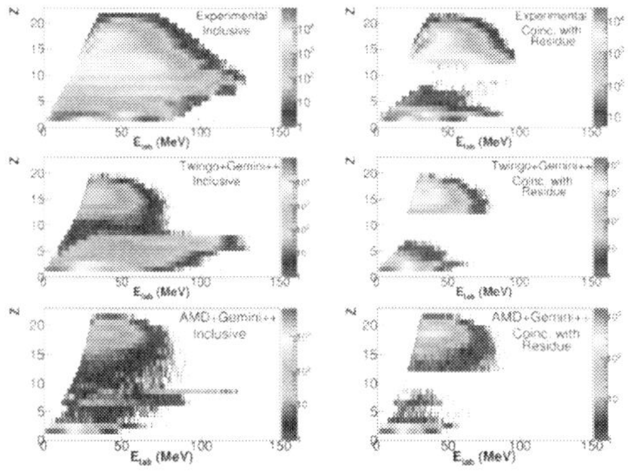

Figure 4: Correlation between the detected charge Z and the energy E for total events (left upper panel) and for events where the coincidence between at least one light charged particle in the whole apparatus and one heavy fragment in the RCo (Evaporation Residue) have been selected (right upper panel) for the ^{18}O+^{28}Si at 7 AMeV. The same for the TWINGO+GEMINI (Stochastic Mean Field approach) simulated events in the middle left and right panels. The same for the AMD+GEMINI (Antisymmetric Molecular Dynamics approach) simulated events in the lower left and right panels. Model predictions are filtered through a software replica of the apparatus.

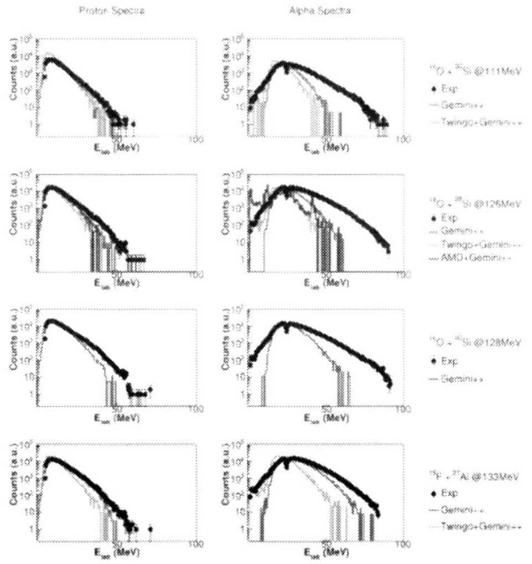

Figure 5: Experimental energy distributions (black full dots) collected at 8.6° $\leq \theta \leq$ 13.2° for protons (left panels) and α–particles (right panels), for the four studied reactions, respectively: ^{16}O+^{30}Si at 7 AMeV (upper panels); ^{18}O+^{28}Si at 7 AMeV (second upper panels); ^{16}O+^{30}Si at 8 AMeV (third upper panels); ^{19}F+^{27}Al at 7 AMeV (bottom panels). In all panels comparison to model predictions are provided: GEMINI statistical model (red line) and TWINGO + GEMINI dynamical plus statistical model (green lines). For the case ^{16}O+^{30}Si at 8 AMeV no TWINGO +GEMINI simulation have still been run, while for the ^{18}O+^{28}Si at 7 AMeV case even few events simulated with AMD+GEMINI (blue lines) are compared to experimental data. Model prediction are filtered through a software replica of the apparatus.

codes are under study: the TWINGO code [12], based on the Stochastic Mean Field approximation, which uses the test particle method to describe each nucleons, and the Antisymmetrized Molecular Dynamics (AMD) [13], which is based on gaussian wave–packets and consider particle–particle correlations with the possibility of describing the cluster structure of the interacting partners. From the preliminary results, one can observe a difference in the four systems when compared to the complete fusion scenario. While, in fact, for p, d, t spectra the complete fusion seems to account for almost all the central collision available cross section, with a minor discrepancy at forward angles expected by the systematics for pre-equilibrium emission at these low energies, this seems not the case for the α–particles. In fact, as it is shown as an example in the Figure 5 for a defined angular region 8.6° $\leq \theta \leq$ 13.2°, the results from the GEMINI complete fusion predictions are not in agreement with the data for the α–particles. This difference is observed in a large angular range and the gap is growing more and more in the forward detection region (RCo). This result, which could be accounted for by a quite strong and unexpected pre-equilibrium emission, is essentially related to α–particles, since, as already said, a very small amount of fast emission can be attributed to Z=1 particles. Moreover, from the comparison of the different systems with their own model predictions, one can infer that the amount of fast emission seems to be dependent on the system/entrance channel, as one can observe, for example, looking at the ^{19}F+^{27}Al at 7 AMeV case, where the difference between data and GEMINI are relatively smaller.

For what it concerns the dynamical models and, in particular the TWINGO code, unfortunately, some limitations in describing the pre-equilibrium phenomenon are present since, due to the particle-test method, all nucleons emitted before the equilibration cannot be properly reconstructed in the final event. Moreover, being TWINGO based on Stochastic Mean Field approximation, it is not even able to build up clusters like α-particles and, therefore, also these fast particles are missing in the final spectrum. Therefore, only the nucleons and α–particles recontructed by the after-burner GEMINI are present in the final distributions: in other words, only the statistical decay of the excited systems produced at 300 fm/c (selected collision time in our case at which the after-burner is applied) can be accounted for. Since the system has anyway lost some nucleons and clusters in the early stage of the reaction, the TWINGO+GEMINI case results in a colder system population with respect to GEMINI alone: predictions on the fast emission can then be obtained only as a difference to the experimental spectra. The AMD+GEMINI simulation can give different information, as we discussed previously, being able to describe nucleon -nucleon correlations. For the case ^{18}O+^{28}Si at 7 AMeV we have at present few events simulated with AMD+GEMINI (blue lines in Figure 5, which have been compared to experimental data. However, the statistics is still too low and subject to statistical fluctuations to permit to draw any conclusion at present. Another observable is the Light Charged Particles multiplicity distribution, which are shown in Figure 6 for protons (upper panels) and for α–particles (lower panels). The multiplicity distributions for both protons and α−particles for all systems are different from simulations. In general, for protons GEMINI seems to overestimate the experimental results for all systems, especially in specific channels (2p,3p,4p). TWINGO predictions are closer to the data for the $^{19}F+^{27}Al$ at 7 AMeV system. For α−particles the situation is even more complex and the situation seems to evolve in a very different way from one system to another. This, anyway, is an indication that further and more detailed studies have to be performed on the specific decay channels, to really understand the underlying phenomena, in a similar way like it was performed in the previous experiment. It is, therefore, necessary to use

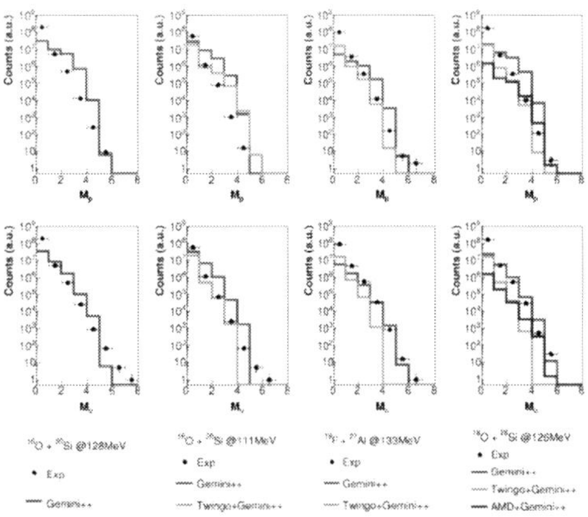

Figure 6: Experimental Light Charged Particle multiplicities (black full dots) in coincidence with ER: proton multiplicity (upper panels) and α−particle multiplicity (bottom panels), for the four studied reactions, respectively: $^{16}O+^{30}Si$ at 8 AMeV (left panels); $^{16}O+^{30}Si$ at 7 AMeV (second left panels); $^{19}F+^{27}Al$ at 7 AMeV (second right panels); $^{18}O+^{30}Si$ at 7 AMeV (right panels). In all panels comparison to model predictions are provided: GEMINI statistical model (red line) and TWINGO + GEMINI dynamical plus statistical model (blue lines). Model predictions are filtered through a software replica of the apparatus.

other theoretical models to account for this. On one side, calculations for the different systems are undergoing with the Moscow Pre−equilibrium Model (MPM) [5, 14, 15], which was already used to compare its predictions with previous measured data [3]. This code is based on the Hybrid Exciton Model coupled to a Statistical Decay Code (PACE2). Moreover, the Antisymmetric Molecular Dynamics (AMD) calculations are under study, since they can properly take into account the possible α−cluster structure in the colliding systems. AMD is based on a Stochastic Equation of Motion for the Gaussian Wave Packets representing the colliding nucleons: unfortunately the time needed to run the code for a sufficient number of events to compare to exclusive data is long and up to now only global observable have been obtained like the one shown in Figure 4. More time is needed to calculate all the studied systems and to compare the simulated events to exclusive observables in the experimental data. Several calculations may also be needed to change and check the optimized input parameters related to the clusterization factor.

5 Conclusions

We have presented complete results for the ^{24}Mg compound system, formed through two different entrance channels. In the case of this light system, the complete charge detection allows a precise identification of the emission source as ^{24}Mg, at an excitation energy $E^*=60$ MeV. We especially focussed on multiple α−particle emission channels, since they are believed to bear information on exotic α−cluster states, which could be populated at high excitation energy. For the exotic decay channel of a highly excited ^{24}Mg into six α−particles, the complex analysis performed, even through correlation function technique, shows that the decay is essentially sequential, due to an important contribution of intermediate Be states. For slightly heavier systems and higher involved energies, the $^{16}O+^{30}Si$ at 7 AMeV, $^{18}O+^{28}Si$ at 7 AMeV, $^{16}O+^{30}Si$ at 8 AMeV and $^{19}F+^{27}Al$ at 7 AMeV have been studied. The analysis is still very preliminary, even though particular results seem to come out looking at the α−decay channels, where already in a quasi-inclusive mode (ER coincidence - no multiple channel selection) unexpected differences are evidenced between the different studied systems for what it concerns the competition between evaporative and fast emission channels. Comparison to specific simulation models like the AMD are under study, to evidence the possible influence of α−structure effects in the dynamics of the reactions. Study of more exclusive decay channels, together with pre-equilibrium predictions will be performed in order to get a better insight on the reaction interplay.

References

[1] L. Morelli et al., Journ. of Phys. G **41**, 075107 (2014)
[2] L. Morelli et al., Journ. of Phys. G **41**, 075108 (2014)
[3] D. Fabris et al., PoS(X LANSPA , 061 (2013)
[4] V.L. Kravchuk, et al. EPJ WoCs, **2** 10006 (2010)
[5] O. V. Fotina et al., Int. Journ. Mod. Phys. E **19** 1134 (2010)
[6] M. Bruno, F. Gramegna et al., Eur. Phys. Journ. A **49** 128 (2013)
[7] Gramegna F. et al., Nucl. Instr. And Meth. A, **389** (1997) 474; Gramegna F. et al., 2004 IEEE Nucl. Science Symposium, Rome, 16−22 October 2004
[8] Baiocco G., PhD Thesis (2012) University of Bologna and University of Caen, Basse−Normandie
[9] L. Morelli et al., J. Phys. G: Nucl. Part. Phys. 43 (2016) 045110
[10] P.E. Hodgson, E. Běták, Phys. Rep. 374 (2003) 1-89.
[11] R. J. Charity, Phys. Rev. C82 (2010) 014610.
[12] M. Colonna et al., Nucl. Phys. A 642, 449 (1998).
[13] A. Ono, Phys. Rev. C59 (1999) 853.
[14] D.O. Eremenko et al. Phys Atom. Nucl. 65 (2002) 18
[15] O.V. Fotina et al. Phys. Atom. Nucl. 73 (2010) 1317c

Skyrme tensor force in the collision $^{16}O+^{40}Ca$

Lu Guo[1,*], *Chong* Yu[1], *Long* Shi[1], and *Cédric* Simenel[2]

[1]*School of Physics, University of Chinese Academy of Sciences, Beijing 100049, China*
[2]*Department of Nuclear Physics, RSPE, Australian National University, Canberra, Australia*

Abstract. The role of tensor force is investigated by using the time-dependent Hartree-Fock (TDHF) theory in the collision $^{16}O+^{40}Ca$. The full tensor force is incorporated in our TDHF implementation. The calculations are performed in three-dimensional Cartesian coordinate without any symmetry restrictions. We study the effect of tensor force on Coulomb barrier, upper fusion threshold energy, and energy contribution of the time-odd and tensor terms in Skyrme energy functional. The Coulomb barrier obtained from the energy functional with frozen density approximation is compared with the available experimental data. We find that the tensor force may change the upper fusion threshold energy in the order of a few MeV for the collision $^{16}O+^{40}Ca$. The tensor force has a non-negligible effect in heavy-ion collisions.

1 Introduction

The microscopic description of heavy-ion collisions such as time-dependent Hartree-Fock (TDHF) theory provides a useful foundation for a fully microscopic many-body theory. The dynamical and quantum effects have been automatically taken into account in TDHF theory. It has wide applications in fusion [1–10], fission [11–16], transfer reaction [17–21], deep inelastic collisions [22–28], quasifission [21, 29–33], and giant resonances [34–41]. For a review, see Refs. [42, 43].

TDHF theory was proposed by Dirac in 1930 [44] and was first applied in nuclear physics in 1976 [45]. After the first application, many groups in the late 70s and 80s performed more extensive calculations in nuclear large amplitude collective motion [46]. However, at that time, limited computer capacity restricted most calculations with many approximations. For instance, the reaction was assumed to be in an axial symmetric plane and a simplified Skyrme force with the omission of spin-orbit coupling was used. These approximations turned out to be a hindrance for the theoretical development. For example, TDHF calculations predicted that for $^{16}O+^{16}O$ reaction at $E_{c.m.} = 34$ MeV the partial waves $L \leq 6$ do not lead to fusion and the corresponding deep inelastic cross section was expected to be 132 mb [47]. According to TDHF prediction, an experiment to search for a fusion L window was carried out. However, experimentally there is no evidence for the occurrence of such phenomenon predicted by TDHF calculations [47]. This conflict between TDHF prediction and experimental observations is called the puzzle of small fusion window, and promotes the theoretical development.

A few years later in 1986, Umar *et. al* included the time-even terms of spin-orbit force in TDHF calculations [48], and found the upper fusion threshold energy was increased by more than two times. This indicates that earlier TDHF calculations without spin-orbit coupling underestimated the energy dissipation from the collective kinetic energy into the internal excitations so that the energy window of fusion reactions was too small in comparison with the experiments. After including the spin-orbit coupling, more intrinsic degrees of freedom is accessable and the dissipation is enhanced. The inclusion of time-even spin-orbit force in TDHF solved the puzzle of small fusion window. However, a meaningful collision theory should satisfy Galilean invariance which guarantees the calculation results will not depend on the choice of the frame of reference. This invariance is particularly important for reaction dynamics so that both the time-even and time-odd terms of spin-orbit force should be included simultaneously. In recent years, TDHF calculations with the full spin-orbit force become possible thanks to the development of computational power. The strong spin-excitation was shown in spin-saturated system $^{16}O+^{16}O$ [22]. The full spin-orbit force was found to contribute about 40%~65% of the total dissipation depending on the different Skyrme parameters and bombarding energies [26].

These studies indicate the nucleon-nucleon interaction plays a significant role in heavy-ion collisions. The most obviously missing component of nuclear force is tensor force, which is well known to be crucial to explain the properties of the deuteron. In nuclear structure, the tensor force plays an important role in the shell evolution of exotic nuclei [49], spin-orbit splitting [50, 51], deformation [52], rotation [53], Gamow-Teller and spin-dipole excitations [54]. However, in heavy-ion collisions the full tensor force has been neglected in most calculations due to the complexity of collision dynamics. In Ref. [55], the

*e-mail: luguo@ucas.ac.cn

time-even spin-current tensor was shown to become important as the increase of the mass of colliding systems. The role of time-even tensor force in the dissipation dynamics in deep-inelastic collisions has been explored in Ref. [26]. Recently the full tensor force was shown to play a non-negligible effect in $^{16}O+^{16}O$ inelastic collisions [9, 28].

The article is organized as follows. Section 2 will present the TDHF theory with the full version of Skyrme interaction and energy density functional including the tensor force. In Sec. 3 we illustrate the role of tensor force in the asymmetric reaction $^{16}O+^{40}Ca$. A summary is given in Sec. 4.

2 Theoretical method

Starting from the time-dependent action

$$S = \int_{t_1}^{t_2} dt \langle \Psi(\mathbf{r},t)|H - i\hbar\partial_t|\Psi(\mathbf{r},t)\rangle, \tag{1}$$

and applying the variational principal $\delta S=0$ with respect to the many-body wave-function $\Psi(\mathbf{r},t)$, one may obtain the time evolution of mean-field

$$i\hbar\partial_t\phi_\lambda = \hat{h}\phi_\lambda. \tag{2}$$

In the above TDHF equation, the many-body wave-function has been approximated as the Slater determinant composed by the single-particle states ϕ_λ

$$\Psi(\mathbf{r},t) = \frac{1}{\sqrt{N!}}\det\{\phi_\lambda(\mathbf{r},t)\}. \tag{3}$$

The initial wave-functions in dynamical evolution employ the nuclear ground state obtained from HF equation

$$\hat{h}\varphi_\lambda(\mathbf{r}) = \epsilon_\lambda\varphi_\lambda(\mathbf{r}). \tag{4}$$

The time evolution of single-particle states is expressed as

$$\phi_\lambda(\mathbf{r},t+\Delta t) = e^{-i\hat{h}\Delta t/\hbar}\phi_\lambda(\mathbf{r},t), \tag{5}$$

with $\hat{h}$ the single-particle Hamiltonian. Here $\mathbf{r}$ denotes the three-dimensional Cartesian coordinate and the spin of nucleon.

We employ the Skyrme effective interaction in our TDHF calculation

$$\begin{aligned}
v(\mathbf{r}_1,\mathbf{r}_2) = {} & t_0(1+x_0\hat{P}_\sigma)\delta(\mathbf{r}_1-\mathbf{r}_2) \\
& + \frac{1}{2}t_1(1+x_1\hat{P}_\sigma)[\mathbf{k}'^2\delta(\mathbf{r}_1-\mathbf{r}_2)+\delta(\mathbf{r}_1-\mathbf{r}_2)\mathbf{k}^2] \\
& + t_2(1+x_2\hat{P}_\sigma)\mathbf{k}'\delta(\mathbf{r}_1-\mathbf{r}_2)\mathbf{k} \\
& + \frac{1}{6}t_3(1+x_3\hat{P}_\sigma)\rho^\alpha(\frac{\mathbf{r}_1+\mathbf{r}_2}{2})\delta(\mathbf{r}_1-\mathbf{r}_2) \\
& + iW_0(\sigma_1+\sigma_2)\cdot\mathbf{k}'\times\delta(\mathbf{r}_1-\mathbf{r}_2)\mathbf{k} \\
& + \frac{t_e}{2}\Big\{[3(\sigma_1\cdot\mathbf{k}')(\sigma_2\cdot\mathbf{k}')-(\sigma_1\cdot\sigma_2)\mathbf{k}'^2]\delta(\mathbf{r}_1-\mathbf{r}_2) \\
& + \delta(\mathbf{r}_1-\mathbf{r}_2)[3(\sigma_1\cdot\mathbf{k})(\sigma_2\cdot\mathbf{k})-(\sigma_1\cdot\sigma_2)\mathbf{k}^2]\Big\} \\
& + t_o\Big\{3(\sigma_1\cdot\mathbf{k}')\delta(\mathbf{r}_1-\mathbf{r}_2)(\sigma_2\cdot\mathbf{k})-(\sigma_1\cdot\sigma_2)\mathbf{k}' \\
& \delta(\mathbf{r}_1-\mathbf{r}_2)\mathbf{k}\Big\},
\end{aligned} \tag{6}$$

where t_i, x_j $(i,j=0,...,3)$, W_0, α, t_e, and t_o are the Skyrme parameters. In the above equation, from the sixth to ninth line represents the tensor force, in which the coupling constants t_e and t_o denote the strength of the triplet-even and triplet-odd tensor interactions, respectively. The operator $\mathbf{k}=\frac{1}{2i}(\nabla_1-\nabla_2)$ acts on the right, $\mathbf{k}'=-\frac{1}{2i}(\nabla'_1-\nabla'_2)$ acts on the left. The spin exchange operator is $P_\sigma=\frac{1}{2}(1+\sigma_1\sigma_2)$.

It is natural to express the Skyrme interaction with the energy density functional (EDF)

$$E = \int d^3r H(\rho,\tau,\mathbf{j},\mathbf{s},\mathbf{T},\mathbf{J};\mathbf{r}). \tag{7}$$

In above equation, the number density ρ, kinetic density τ, current density $\mathbf{j}$, spin density $\mathbf{s}$, spin-kinetic density $\mathbf{T}$, and spin-current pseudotensor density $\mathbf{J}$ are defined as

$$\begin{aligned}
\rho_q(\mathbf{r}) &= \rho_q(\mathbf{r},\mathbf{r}')|_{\mathbf{r}=\mathbf{r}'} = \sum_\sigma \rho_q(\mathbf{r}\sigma,\mathbf{r}'\sigma'), \\
s_q(\mathbf{r}) &= s_q(\mathbf{r},\mathbf{r}')|_{\mathbf{r}=\mathbf{r}'} = \sum_{\sigma\sigma'} \rho_q(\mathbf{r}\sigma,\mathbf{r}'\sigma')\langle\sigma'|\hat{\sigma}|\sigma\rangle, \\
\tau_q(\mathbf{r}) &= \nabla\cdot\nabla'\rho_q(\mathbf{r},\mathbf{r}')|_{\mathbf{r}=\mathbf{r}'}, \\
j_q(\mathbf{r}) &= -\frac{i}{2}(\nabla-\nabla')\rho_q(\mathbf{r},\mathbf{r}')|_{\mathbf{r}=\mathbf{r}'}, \\
J_{q,\mu\nu}(\mathbf{r}) &= -\frac{i}{2}(\nabla_\mu-\nabla'_\mu)s_{q,\nu}(\mathbf{r},\mathbf{r}')|_{\mathbf{r}=\mathbf{r}'}, \\
T_{q,\mu}(\mathbf{r}) &= \nabla\cdot\nabla' s_{q,\mu}(\mathbf{r},\mathbf{r}')|_{\mathbf{r}=\mathbf{r}'}, \\
F_{q,\mu}(\mathbf{r}) &= \frac{1}{2}\sum_{\nu=x}^{z}(\nabla_\mu\nabla'_\nu+\nabla'_\mu\nabla_\nu)s_\nu(\mathbf{r},\mathbf{r}')|_{\mathbf{r}=\mathbf{r}'}.
\end{aligned} \tag{8}$$

Here, $q=n(p)$ stands for neutron (proton).

From these densities, one can define the isoscalar ($t=0$) and isovector ($t=1$) densities and currents as

$$\begin{aligned}
\rho_0(\mathbf{r}) &= \rho_n(\mathbf{r})+\rho_p(\mathbf{r}), \\
\rho_1(\mathbf{r}) &= \rho_n(\mathbf{r})-\rho_p(\mathbf{r}).
\end{aligned} \tag{9}$$

With the above definitions, the full version of Skyrme energy functional is expressed as

$$\begin{aligned}
\mathcal{H} = \mathcal{H}_0 + \sum_{t=0,1}\Big\{ & A_t^s\mathbf{s}_t^2 + \left(A_t^{\Delta s}+B_t^{\Delta s}\right)\mathbf{s}_t\cdot\Delta\mathbf{s}_t + B_t^{\nabla s}(\nabla\cdot\mathbf{s}_t)^2 \\
& + (A_t^T+B_t^T)\Big(\mathbf{s}_t\cdot\mathbf{T}_t - \sum_{\mu,\nu=x}^{z} J_{t,\mu\nu}J_{t,\mu\nu}\Big) \\
& + B_t^F\Big[\mathbf{s}_t\cdot\mathbf{F}_t - \frac{1}{2}\Big(\sum_{\mu=x}^{z} J_{t,\mu\mu}\Big)^2 - \frac{1}{2}\sum_{\mu,\nu=x}^{z} J_{t,\mu\nu}J_{t,\nu\mu}\Big]\Big\},
\end{aligned} \tag{10}$$

where $\mathcal{H}_0$ is the basic Skyrme functional used in Sky3D code [56] and most TDHF calculations. In our code, we incorporated the full version of Skyrme energy functional as shown in Eq. (10), including all the terms from central, spin-orbit and tensor forces. These new terms have been clarified to be especially important in the studies of nuclear structure. The basic functional $\mathcal{H}_0$ used in Sky3D code [56] is expressed as

$$\mathcal{H}_0 = \sum_{t=0,1}\Big\{A_t^{\rho}\rho_t^2 + A_t^{\Delta\rho}\rho_t\Delta\rho_t + A_t^{\tau}\big(\rho_t\tau_t - \mathbf{j}_t^2\big) + A_t^{\nabla J}\rho_t\nabla\cdot\mathbf{J}_t + A_t^{\nabla J}\mathbf{s}_t\cdot\nabla\times\mathbf{j}_t\Big\}. \quad (11)$$

The coupling constants A and B appearing in Eqs. (10) and (11) have been defined in Refs. [51, 57]. Some authors used the coupling constants C which is the sum of parameters A and B, $C = A + B$. The terms with the coupling constants A come from the Skyrme central and spin-orbit forces, while those with the B parameters are from the tensor force. In the calculations, we set $C_t^{\nabla S} = C^{\Delta S} = 0$ because the terms containing the gradient of spin-density may cause the spin instability both in nuclear structure and reaction studies as pointed out in Refs. [28, 51].

In the energy functional Eq. (10), the spin-current pseudotensor density $\mathbf{J}$ is expressed in its Cartesian components $J_{\mu\nu}$. In Ref. [58] the spin-current density has been decomposed into pseudoscalar, (antisymmetric) vector, and (symmetric) traceless pseudotensor components as

$$J_{\mu\nu}(r) = \frac{1}{3}\delta_{\mu\nu}J^{(0)}(r) + \frac{1}{2}\sum_{k=x}^{z}\varepsilon_{\mu\nu k}J_k^{(1)}(r) + J_{\mu\nu}^{(2)}(r). \quad (12)$$

Here $\delta_{\mu\nu}$ is the Kronecker symbol and $\varepsilon_{\mu\nu k}$ is the Levi-Civita tensor. The pseudoscalar $J^{(0)}$, vector $J^{(1)}$, and pseudotensor $J^{(2)}$ components are given in terms of the Cartesian form

$$J^{(0)}(\mathrm{r}) = \sum_{\mu=x}^{z}J_{\mu\mu}(\mathrm{r}),$$

$$J_k^{(1)}(\mathrm{r}) = \sum_{\mu\nu=x}^{z}\varepsilon_{k\mu\nu}J_{\mu\nu}(\mathrm{r}),$$

$$J_{\mu\nu}^{(2)}(\mathrm{r}) = \frac{1}{2}[J_{\mu\nu}(r) + J_{\nu\mu}(\mathrm{r})] - \frac{1}{3}\delta_{\mu\nu}\sum_{k=x}^{z}\mathrm{J_{kk}}(\mathrm{r}). \quad (13)$$

The vector spin-current density $J^{(1)}(\mathbf{r}) \equiv \mathbf{J}(\mathbf{r})$ is often called spin-orbit current, as it enters the spin-orbit functional in Eq. (11). The terms of energy functional involving the spin-current density in Eq. (10) can be instead expressed as

$$\sum_{\mu,\nu=x}^{z}J_{t,\mu\nu}J_{t,\mu\nu} = \frac{1}{3}J_t^{(0)2} + \frac{1}{2}J_t^{\,2} + \sum_{\mu,\nu=x}^{z}J_{t,\mu\nu}^{(2)}J_{t,\mu\nu}^{(2)}, \quad (14)$$

$$\frac{1}{2}\Big(\sum_{\mu=x}^{z}J_{t,\mu\mu}\Big)^2 + \frac{1}{2}\sum_{\mu,\nu=x}^{z}J_{t,\mu\nu}J_{t,\nu\mu} = \frac{2}{3}J_t^{(0)2} - \frac{1}{4}J_t^{\,2} + \frac{1}{2}\sum_{\mu,\nu=x}^{z}J_{t,\mu\nu}^{(2)}J_{t,\mu\nu}^{(2)}. \quad (15)$$

To test the accuracy of numerical calculations, the energy functional involving the spin-current density has been implemented by using the above two approaches in our code.

Table 1. Isoscalar and isovector tensor coupling constants $C_0^J = A_0^J + B_0^J$ and $C_1^J = A_1^J + B_1^J$ (in MeV·fm^5) for the five sets of Skyrme parametrizations.

Force	C_0^J	C_1^J
SLy5	0.0	60.0
SLy5t	-20.0	-65.0
T22	0.0	0.0
T26	120.0	120.0
T44	120.0	0.0

In the next section, we will show the energy contributions with this two approaches are identical, as they should be.

The set of nonlinear TDHF equations has been solved on a three-dimensional Cartesian coordinate and without any symmetry restrictions. We calculate the ground state of ^{16}O and ^{40}Ca in a numerical box of 24×24×24 fm^3. For the dynamical evolution of reaction ^{16}O+^{40}Ca, we used a box with $32 \times 24 \times 32$ fm^3 grid points and grid spacing 1.0 fm. The initial distance between the projectile and target is taken to be 20 fm. From infinity to initial distance the nucleus was assumed to move on a pure Rutherford trajectory so that the initial boost are properly treated in TDHF evolution. We expand the Taylor expansion up to the sixth order and employ a time step $\Delta t = 0.2$ fm/c in the dynamical evolution. The choice of these parameters guarantees a good numerical accuracy during the dynamical evolution for all the cases studied here. The total TDHF energy and particle number have been well conserved and shifted less than 0.1 MeV and 0.01, respectively.

3 Results and discussions

We employ the full version of Skyrme energy functional as shown in Eq. (10) in our calculations. The static properties and reaction dynamics are treated with the same energy density functional and a unified theoretical framework. To examine the accuracy of our code, we compared our results with those obtained by other code in three aspects. First, we have reproduced the upper fusion threshold energy in the collision ^{16}O+^{16}O reported in Ref. [28] within the accuracy of 1 MeV by our code. Second, we calculated the energy contribution from the new terms as a function of time, and our results reproduced those reported in Ref. [28] within a negligible discrepancy. Third, we implemented the energy functional involving the spin-current density in two approaches as shown in Eqs. (14) and (15) in our code, and found that the results with two approaches are identical as they should be and also reproduced those shown in Ref. [28].

In present work, we employ the five sets of Skyrme parametrizations SLy5 [60, 61], SLy5t [50], and T22,T26 and T44 [51] to study the effect of tensor force in the collision dynamics of ^{16}O+^{40}Ca reaction. The isoscalar and isovector tensor coupling constants $C_0^J = A_0^J + B_0^J$ and $C_1^J = A_1^J + B_1^J$ are shown in Tab. 1 for this five sets parameters with a wide range of coupling constants. Note that the parameters of Skyrme tensor force have been fitted in two ways. One is to fit the parameters of tensor force based

DOI: 10.1051/epjconf/201716300021

Table 2. Energy (in MeV) and radii (in fm) of the Coulomb barrier obtained with the frozen density approximation for the five sets of Skyrme parametrizations and experimental data [59] in the collisions $^{16}O+^{40}Ca$.

Force	V(MeV)	R(fm)
SLy5	23.36	9.20
SLy5t	23.36	9.20
T22	23.34	9.21
T26	23.33	9.21
T44	23.21	9.25
Expt.	23.06	9.21

Table 3. Upper fusion threshold energies (in MeV) for the five sets of Skyrme parametrizations in the collisions $^{16}O+^{40}Ca$.

Force	Threshold (MeV)
SLy5	142
SLy5t	145
T22	138
T26	143
T44	140

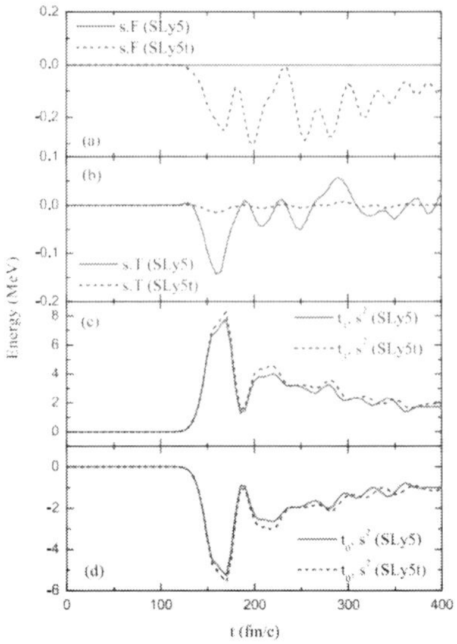

Figure 1. Energy contributions from the terms involving time-odd densities and currents for head-on collisions $^{16}O+^{40}Ca$ with SLy5 and SLy5t parametrizations at $E_{c.m.} = 170$ MeV.

on the existing Skyrme force, e.g., SLy5 plus tensor force denoted as SLy5t. The other is to fit all the Skyrme parameters on the same footing, e.g., the set of TIJ parametrizations with a wide range of isoscalar and isovector tensor couplings. With such fitting procedure, one may draw a different physical scenario. The effect of pure tensor force can be clarified by comparing the results with SLy5 and SLy5t. For the TIJ parametrizations, the calculations can figure out the role of isoscalar and isovector tensor terms in collision dynamics.

The energy and radii of Coulomb barrier are listed in Tab. 2 for $^{16}O+^{40}Ca$ with the five sets of forces and the experimental data [59]. Here the Coulomb barrier is obtained by the frozen density (FD) approximation within the EDF theory [62–64]. The interaction potential in the approaching phase can be expressed as

$$V^{FD}(R) = E[\rho_{P+T}](R) - E[\rho_T] - E[\rho_P]. \qquad (16)$$

$\rho_{P+T} = \rho_P + \rho_T$ is the sum of ground state density of projectile and target at the relative distance R, and $E[\rho_{P+T}](R)$ is the Skyrme EDF as shown in Eq (10). Note that the Pauli principle and the coupling between the collective motion and intrinsic states have been neglected in FD approximation. When the overlap of two densities is small, e.g., at the position of Coulomb barrier, EDF with FD approximation is a good tool to estimate the Coulomb barrier. However, at the smaller relative distance, since the Pauli effect is strong, FD approximation will not properly account for the interaction potential [65].

The energy and radii of Coulomb barrier with SLy5t are observed to be exactly same as those with SLy5 for $^{16}O+^{40}Ca$, as expected, because the tensor force has no contribution to the ground state EDF for the spin-saturated nuclei ^{16}O and ^{40}Ca. There exist some differences among the T22,T26 and T44 parameter sets. These differences, for the spin-saturated system, come from the rearrangement of other terms of Skyrme EDF in the fit of the parametrizations. Note that the structure of ^{16}O and ^{40}Ca, for instance, matter and charge radii, has small difference for T22,T26 and T44 parameters, which may cause a slight change of the barrier height. The Coulomb barrier with FD-EDF overestimates the experimental data due to its omission of the coupling between the collective motion and single particle degrees of freedom [66].

The upper fusion threshold is quite sensitive to the details of Skyrme EDF as reported in Ref. [3]. The inclusion of pure tensor force SLy5t increased the upper threshold by 3 MeV. The SLy5 and T22 have the same isoscalar coupling constants, and the upper fusion threshold is found to reduce as the decrease of isovector coupling. The same trend is also observed for T26 and T44. By comparing T22 and T44 with the same isovector coupling, the increase of isoscalar coupling increases the upper fusion threshold. Our results in $^{16}O+^{40}Ca$ are consistent with the findings in Ref. [28] for $^{16}O+^{16}O$.

The energy contributions from the terms involving time-odd densities and currents (a) $s \cdot F$, (b) $s \cdot T$, (c) s^2 with t_0 parameter, and (d) s^2 with t_3 parameter are examined for $^{16}O+^{40}Ca$ head-on collisions with SLy5 and SLy5t parametrizations at $E_{c.m.} = 170$ MeV. The results are shown in Fig. 1. This is deep-inelastic collisions as seen from Tab. 3. These terms were not included in Sky3D code [56]. At the initial time, the energy contribution from these terms both for SLy5 and SLy5t are zero, as expected, because these time-odd terms contribute zero for the ground state of even-even nuclei. Since the term $s \cdot F$ comes from the pure tensor force, its energy with SLy5 remains zero during the time evolution. For SLy5t because

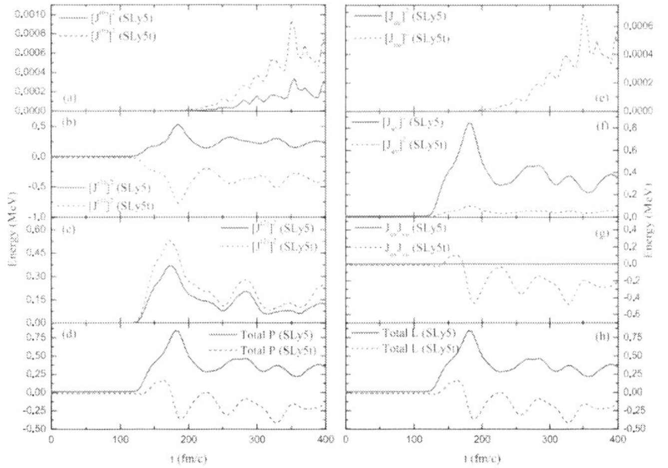

Figure 2. The energy contribution from the time-even terms in head-on collisions $^{16}O+^{40}Ca$ at $E_{c.m.}$ = 170 MeV.

the inclusion of tensor force changes the time-dependent mean-field and hence the densities s and F itself, the energy contribution keeps to be zero in the early stage and then starts to oscillate. Since $s \cdot T$ term comes from both the central and tensor forces, its energy with SLy5 evolves with time, while for SLy5t the more pronounced effect appears compared with SLy5. The s^2 terms from the central force are found to have similar trends.

In order to test the numerical realization of our code and understand the contribution of tensor force in energy functional, we calculate the energy contribution from the J^2 terms both in its coupled form and Cartesian form as given in Eqs. (14) and (15). The left column in Fig. 2(a-d) shows the energy contribution of scalar, vector, and tensor components of J^2, and the summation of these three terms, while the right column in Fig. 2(e-h) is the contribution from the diagonal, symetric and anti-symetric components and also the corresponding summation. The total contribution calculated with two approaches are exactly identical shown in Fig. 2(d) and (h), as they should be.

We also check the final kinetic energy of the fragments for the case of $E_{c.m.}$=170 MeV, which is 28.1 MeV and 26.3 MeV for SLy5 and SLy5t, respectively. This energy provides the amount of energy transferred from relative motion to internal degrees of freedom, and could be a direct measure of the energy dissipation.

The contribution to the total energy from J^2 term is shown in Fig. 3 for the five sets of Skyrme parameters in the collisions $^{16}O+^{40}Ca$ at $E_{c.m.}$ = 46.5 MeV and b = 5.0 fm. In the early stage of dynamical evolution, the energy from J^2 keeps constant as that in the ground state. As time evolves, these terms are highly excited in the dynamic process and present an evident effect in heavy-ion collisions of spin-saturated system $^{16}O+^{40}Ca$, while they have negligible effect in the ground state of spin-saturated nucleus. The perturbative addition of tensor terms with SLy5t has an opposite sign with respect to the other four forces. The TIJ forces contribute to the energy up to a few MeV.

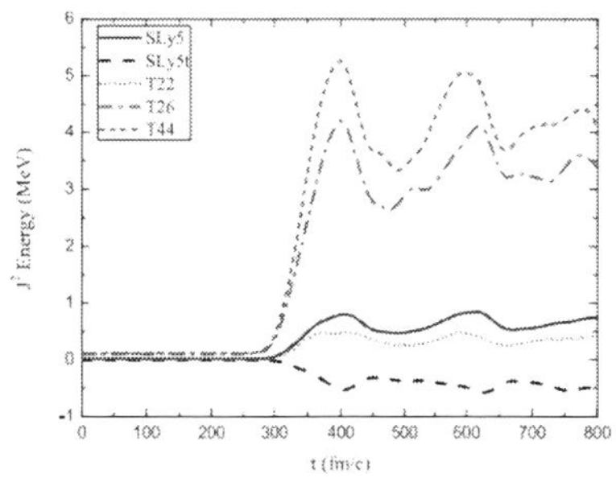

Figure 3. The energy contribution from the J^2 terms in the collisions $^{16}O+^{40}Ca$ at $E_{c.m.}$ = 46.5 MeV and b = 5.0 fm.

4 Conclusion

We study the role of tensor force within the TDHF theory for the collision $^{16}O+^{40}Ca$. The full tensor force is incorporated in our TDHF implementation. The calculations are carried out in three-dimensional and symmetry unrestricted Cartesian coordinate. We employ the five sets of Skyrme parametrizations SLy5, SLy5t, T22, T26 and T44 with a wide range of isoscalar and isovector tensor coupling to study the role of tensor force in heavy-ion collision dynamics. The tensor force is found to change the upper fusion threshold energy in the order of a few MeV in the spin-saturated system $^{16}O+^{40}Ca$. The time-odd and tensor terms in Skyrme energy functional show an evident effect in the dynamical evolution. The tensor force plays a non-negligible effect in heavy-ion collisions.

DOI: 10.1051/epjconf/201716300021

5 Acknowledgments

This work is partly supported by the Natural Science Foundation of China under Grants Nos. 11175252 and 11575189, President Fund of UCAS, the NSFC-JSPS international cooperation program under Grant No. 11711540016, and the Australian Research Council Grants Nos. FT120100760 and DP160101254. The computation of this work was performed in the Tianhe-1A supercomputer located in the Chinese National Supercomputer Center.

References

[1] C. Simenel *et al.*, Phys. Rev. Lett. **86**, 2971 (2001)
[2] C. Simene *et al.*, Phys. Rev. Lett. **93**, 102701 (2004)
[3] A.S. Umar, V.E. Oberacker, Phys. Rev. C **73**, 054607 (2006)
[4] K. Washiyama, D. Lacroix, Phys. Rev. C **78**, 024610 (2008)
[5] L. Guo, T. Nakatsukasa, EPJ WEB of CONFERENCE **38**, 09003 (2012)
[6] C. Simenel *et al.*, Phys. Rev. C **88**, 024617 (2013)
[7] K. Washiyama, Phys. Rev. C **91**, 064607 (2015)
[8] M. Tohyama, A.S. Umar, Phys. Rev. C **93**, 034607 (2016)
[9] G.L. SHI Long, Nuclear Physics Review **34**, 41 (4) (2017)
[10] K. Godbey *et al.*, Phys. Rev. C **95**, 011601(R) (2017)
[11] C. Simenel, A.S. Umar, Phys. Rev. C **89**, 031601 (2014)
[12] G. Scamps *et al.*, Phys. Rev. C **92**, 011602 (2015)
[13] P. Goddard *et al.*, Phys. Rev. C **92**, 054610 (2015)
[14] P.M. Goddard *et al.*, Phys. Rev. C **93**, 014620 (2016)
[15] A. Bulgac *et al.*, Phys. Rev. Lett. **116**, 122504 (2016)
[16] Y. Tanimura *et al.*, Phys. Rev. Lett. **118**, 152501 (2017)
[17] C. Simenel, Phys. Rev. Lett. **105**, 192701 (2010)
[18] G. Scamps, D. Lacroix, Phys. Rev. C **87**, 014605 (2013)
[19] K. Sekizawa, K. Yabana, Phys. Rev. C **88**, 014614 (2013)
[20] N. Wang, L. Guo, Physics Letters B **760**, 236 (2016)
[21] K. Sekizawa, K. Yabana, Phys. Rev. C **93**, 054616 (2016)
[22] J.A. Maruhn *et al.*, Phys. Rev. C **74**, 027601 (2006)
[23] L. Guo *et al.*, Phys. Rev. C **76**, 014601 (2007)
[24] L. Guo *et al.*, Phys. Rev. C **77**, 041301(R) (2008)
[25] C. Simenel, Phys. Rev. Lett. **106**, 112502 (2011)
[26] G.F. Dai *et al.*, Phys. Rev. C **90**, 044609 (2014)
[27] G.F. Dai *et al.*, Sci. China-Phys. Mech. Astron. **57**, 1618 (2014)
[28] P.D. Stevenson *et al.*, Phys. Rev. C **93**, 054617 (2016)
[29] C. Golabek, C. Simenel, Phys. Rev. Lett. **103**, 042701 (2009)
[30] A. Wakhle *et al.*, Phys. Rev. Lett. **113**, 182502 (2014)
[31] V.E. Oberacker *et al.*, Phys. Rev. C **90**, 054605 (2014)
[32] A.S. Umar *et al.*, Phys. Rev. C **92**, 024621 (2015)
[33] A.S. Umar *et al.*, Phys. Rev. C **94**, 024605 (2016)
[34] C. Simenel, P. Chomaz, Phys. Rev. C **68**, 024302 (2003)
[35] J.A. Maruhn *et al.*, Phys. Rev. C **71**, 064328 (2005)
[36] T. Nakatsukasa, K. Yabana, Phys. Rev. C **71**, 024301 (2005)
[37] A.S. Umar, V.E. Oberacker, Phys. Rev. C **71**, 034314 (2005)
[38] P.G. Reinhard *et al.*, Eur. Phys. J. A **32**, 19 (2007)
[39] C. Simenel, P. Chomaz, Phys. Rev. C **80**, 064309 (2009)
[40] B. Avez, C. Simenel, Eur. Phys. J. A **49**, 76 (2013)
[41] S. Fracasso *et al.*, Phys. Rev. C **86**, 044303 (2012)
[42] C. Simenel, Eur. Phys. J. A **48**, 152 (2012)
[43] T. Nakatsukasa *et al.*, Rev. Mod. Phys. **88**, 045004 (2016)
[44] P.A.M. Dirac, Proc. Cambridge Phil. Soc. **26**, 376 (1930)
[45] P. Bonche *et al.*, Phys. Rev. C **13**, 1226 (1976)
[46] J.W. Negele, Rev. Mod. Phys. **54**, 913 (1982)
[47] A. Lazzarini *et al.*, Phys. Rev. C **24**, 309 (1981)
[48] A.S. Umar *et al.*, Phys. Rev. Lett. **56**, 2793 (1986)
[49] T. Otsuka *et al.*, Phys. Rev. Lett. **95**, 232502 (2005)
[50] G. Colò *et al.*, Phys. Lett. B **646**, 227 (2007)
[51] T. Lesinski *et al.*, Phys. Rev. C **76**, 014312 (2007)
[52] M. Bender *et al.*, Phys. Rev. C **80**, 064302 (2009)
[53] V. Hellemans *et al.*, Phys. Rev. C **85**, 014326 (2012)
[54] C.L. Bai *et al.*, Phys. Rev. Lett. **105**, 072501 (2010)
[55] Y. Iwata, J.A. Maruhn, Phys. Rev. C **84**, 014616 (2011)
[56] J. Maruhn *et al.*, Comput. Phys. Commun. **185**, 2195 (2014)
[57] D. Davesne *et al.*, Phys. Rev. C **80**, 024314 (2009)
[58] E. Perlińska *et al.*, Phys. Rev. C **69**, 014316 (2004)
[59] L.C. Vaz *et al.*, Physics Reports **69**, 373 (1981)
[60] E. Chabanat *et al.*, Nucl. Phys. A **635**, 231 (1998)
[61] E. Chabanat *et al.*, Nucl. Phys. A **643**, 441 (1998)
[62] K.A. Brueckner *et al.*, Phys. Rev. **173**, 944 (1968)
[63] C. Simenel, B. Avez, Intl. J. Mod. Phys. E **17**, 31 (2008)
[64] K. Washiyama, D. Lacroix, Phys. Rev. C **78**, 024610 (2008)
[65] C. Simenel, *et al.*, Phys. Rev. C **95**, 031601(R) (2017)
[66] C. Simenel *et al.*, Phys. Rev. C **88**, 064604 (2013)

EPJ Web of Conferences **163**, 00022 (2017)
FUSION17 DOI: 10.1051/epjconf/201716300022

Evolving theoretical descriptions of heavy-ion fusion: from phenomenological to microscopic approaches

K. Hagino[1,2,3]

[1]*Department of Physics, Tohoku University, Sendai 980-8578, Japan*
[2]*Research Center for Electron Photon Science, Tohoku University, 1-2-1 Mikamine, Sendai 982-0826, Japan*
[3]*National Astronomical Observatory of Japan, 2-21-1 Osawa, Mitaka, Tokyo 181-8588, Japan*

Abstract. We overview the current status of theoretical approaches for heavy-ion fusion reactions at subbarrier energies. We particularly discuss theoretical challenges in the coupled-channels approach, that include i) a description of deep subbarrier hindrance of fusion cross sections, ii) the role of nuclear dissipation, iii) fusion of unstable nuclei, and iv) an interplay between fusion and multi-nucleon transfer processes. We also present results of a semi-microscopic approach to heavy-ion fusion reactions, that combines the coupled-channels approach with state-of-the-art microscopic nuclear structure calculations.

1 Introduction

Fusion is a reaction to form a compound nucleus. It plays an important role in several phenomena in nuclear physics and nuclear astrophysics, such as synthesis of superheavy elements and the energy production in stars. In the potential between two colliding nuclei, the so called Coulomb barrier is formed by a strong cancellation between the Coulomb and a nuclear interactions. This Coulomb barrier has to be overcome in order for fusion to take place, and thus the height of the Coulomb barrier defines the energy scale of the reaction. In this contribution, we mainly consider fusion reactions in heavy-ion systems at energies around the Coulomb barrier, that is, heavy-ion subbarrier fusion reactions.

Why are we interested in subbarrier fusion reactions? One obvious reason for this is that subbarrier fusion is relevant to superheavy elements and nuclear astrophysics. That is, many superheavy elements have been synthesized by heavy-ion fusion reactions at energies slightly above the Coulomb barrier. Also, it is crucially important to understand the dynamics of subbarrier fusion reactions in order to make a reliable extrapolation of experimental data down to extremely low energies at which fusion reaction takes place in stars.

Besides this obvious reason, there are many other good reasons why subbarrier fusion reactions are interesting to study. Firstly, there is a strong interplay between the nuclear reaction and the nuclear structure in there. This strong interplay has been well realized in a large enhancement of fusion cross sections at subbarrier energies as compared to a prediction of a one-dimensional potential model [1–5]. This is in marked contrast to high energy nuclear reactions, in which the couplings play a much less important role and thus the reaction dynamics is much simpler.

Secondly, fusion offers a unique opportunity to study quantum tunneling with many degrees of freedom. That is, fusion takes place only by quantum tunneling at energies below the Coulomb barrier, and the subbarrier enhancement of fusion cross sections can be viewed as a result of coupling assisted tunneling. Heavy-ion fusion reactions are unique in this respect because a variety of intrinsic degrees of freedom are involved, such as static and dynamical nuclear deformations with several multipolarities, as well as several types of particle transfer processes with several values of a transfer Q-value, which can be both negative and positive. Also, in heavy-ion fusion reactions, the incident energy can be easily varied in order to study the energy dependence of the tunneling probability, whereas the energy is basically fixed in many other tunneling phenomena in nuclear physics, such as alpha decays.

In order to analyze heavy-ion subbarrier fusion reactions, the coupled-channels approach has been developed [1, 6], which has enjoyed a lot of success in reproducing experimental data in many reaction systems. In this contribution, we shall first overview the present status of this approach and then discuss several remaining theoretical challenges.

2 Coupled-channels approach

The field of heavy-ion subbarrier fusion started at the late 70's, when a large subbarrier enhancement of fusion cross sections was discovered e.g., in the ^{16}O+^{154}Sm system [7]. For this particular system, the enhancement of fusion cross sections has been well understood in terms of the deformation of the target nucleus, ^{154}Sm. This nucleus is a typical deformed nucleus with a quadrupole deformation parameter of $\beta_2 \sim 0.3$. When the target nucleus is deformed, the potential between the projectile and the target nuclei depends upon the orientation angle, θ, of the deformed target. The fusion cross sections are then computed as,

$$\sigma_{\rm fus}(E) = \int_0^1 d(\cos\theta)\,\sigma_{\rm fus}(E;\theta), \qquad (1)$$

where $\sigma_{\rm fus}(E;\theta)$ is the fusion cross section at the incident energy E for a fixed orientation angle θ. This formula well reproduces experimental fusion cross sections for many systems with a deformed target nucleus, including the ^{16}O+^{154}Sm system [1, 8].

Eq. (1) is valid only when the excitation energy of a rotational excitation can be neglected as compared to the curvature of the Coulomb barrier [1]. In more general cases, one needs to solve the coupled-channels equations,

$$\left[-\frac{\hbar^2}{2\mu}\boldsymbol{\nabla}^2 + \epsilon_k - E\right]\psi_k(\boldsymbol{r}) + \sum_{k'}\langle\phi_k|V(\boldsymbol{r},\xi)|\phi_{k'}\rangle\,\psi_{k'}(\boldsymbol{r}) = 0, \qquad (2)$$

where μ is the reduced mass for the relative motion between the colliding nuclei, ϕ_k is the intrinsic wave function, for which ϵ_k is the excitation energy. $V(\boldsymbol{r},\xi)$ is the total potential, which includes both the bare and the coupling potentials, ξ denoting the intrinsic coordinate. Notice that the coupled-channels equations, Eq. (2), are derived by expanding the total wave function, Ψ, with the basis functions ϕ_k, as,

$$\Psi(\boldsymbol{r},\xi) = \sum_k \psi_k(\boldsymbol{r})\,\phi_k(\xi). \qquad (3)$$

Important ingredients for the coupled-channels approach are the internuclear potential, V, as well as the nature of the intrinsic degrees of freedom, ξ. For the latter, one often employs the macroscopic collective model [1]. Such approach has successfully accounted for experimental data for many systems, and has been a standard tool in analyzing experimental data for heavy-ion subbarrier fusion reactions [6]. The coupled-channels approach also offers a natural explanation for the barrier distribution [9], which is intimately related to the eigenchannel representation of the coupled-channels equations [1].

3 Remaining challenges in the coupled-channels approach

3.1 Deep subbarrier hindrance of fusion cross sections

Despite its success, there are still several theoretical challenges in the coupled-channels approach to subbarrier fusion. In this section, we discuss four main challenges.

The first problem which we consider is the hindrance phenomena of fusion cross sections at deep subbarrier energies [4]. That is, even when a standard coupled-channels calculation well reproduces experimental fusion cross sections in the vicinity of the Coulomb barrier, it appears that

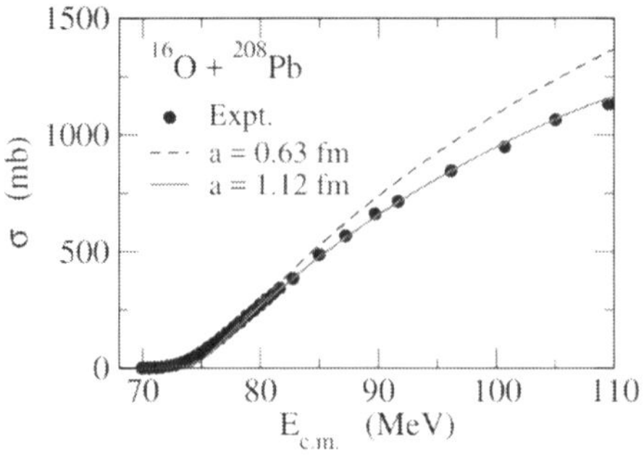

Figure 1. Fusion cross sections for the ^{16}O+^{208}Pb system. The dashed and the solid lines denote the results of single-channel calculations with a Woods-Saxon internuclear potential with the surface diffuseness parameter of a = 0.63 and 1.12 fm, respectively. The experimental data are taken from Ref. [15].

such calculation tends to overestimate fusion cross sections at deep subbarrier energies. Even though the exact origin for this phenomenon has not yet been clarified, so far there are mainly two models which account for the deep subbarrier hindrance of fusion cross sections. One is based on the sudden approach, in which fusion reaction is assumed to take place so fast that the density distribution of each colliding nucleus is frozen during fusion [4, 10]. The frozen density approximation leads to a repulsive core in an internucleus potential, which results in a shallow pocket. High angular momenta are cut-off when a potential is shallow, and this angular momentum cut-off is the main cause of deep subbarrier hindrance in this model. The second model, on the other hand, is based on the adiabatic approach, in which reaction is assumed to take place so slowly that the density distribution is optimized at every instant [11, 12]. This results in a deep and thick internucleus potential, and a tunneling of such thick potential has a responsibility to hinder fusion cross sections.

So far, both models have been equally successful in reproducing the observed fusion hindrance phenomenon. In order to disentangle them, it will be necessary to properly model the dynamics, such as the energy dissipation, around and after the touching point of the colliding nuclei [13], for which most of the current approaches resort either to the incoming wave boundary condition [1] or to a short range imaginary potential. Such modeling will be important also to understand the dynamics of quasi-fission and fusion reactions relevant to superheavy nuclei.

3.2 Fusion above the barrier

We next discuss fusion cross sections at energies above the Coulomb barrier. It has been a long standing problem that a Woods-Saxon internuclear potential with a standard value of surface diffuseness parameter, that is, $a \sim$ 0.63 fm, systematically overestimates fusion cross sections above the Coulomb barrier [14]. If the surface diffuseness parameter is phenomenologically increased, experimental data appear to be reproduced (see Fig. 1 for a

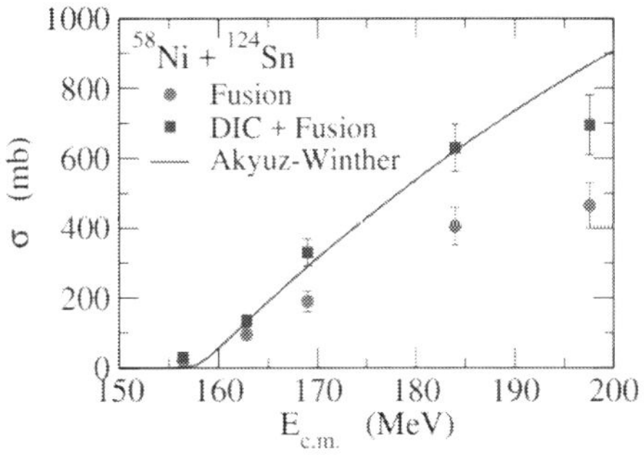

Figure 2. The experimental data for fusion cross sections (the filled circles) and a sum of fusion and deep-inelastic collision cross sections (the filled squares) for the $^{58}Ni+^{124}Sn$ system. These data are taken from Ref. [17]. The solid line shows the result of a single-channel calculation with the Akyüz-Winter potential [18].

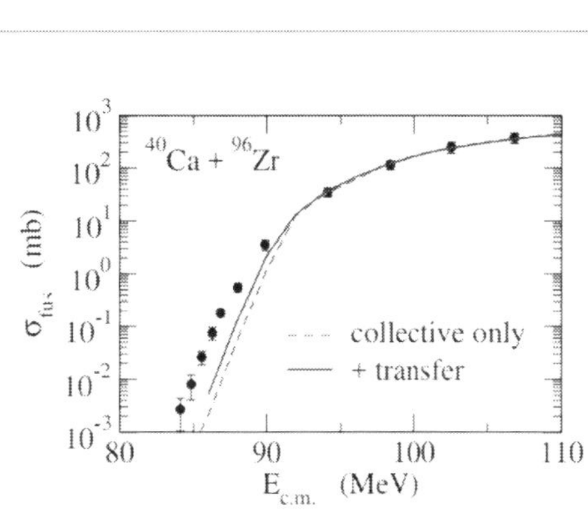

Figure 3. Fusion cross sections for the $^{40}Ca+^{96}Zr$ system. The dashed line is obtained by including only the collective octupole phonon excitations in the projectile and the target nuclei, while the solid line is obtained by including in addition the multi-neutron transfer channels. The strengths of the transfer couplings are determined so as to reproduce the experimental transfer cross sections for this system. The experimental data for fusion cross sections are taken from Ref. [34].

typical example of the situation, that is, the fusion cross sections for the $^{16}O+^{208}Pb$ system), but the exact origin of such surface diffuseness anomaly has not yet been understood.

The situation shown in Fig. 1 appears similar to the one in heavier systems where deep-inelastic collision well competes with fusion. In Ref. [16], Reisdorf argued that a barrier passing calculation well accounts for a sum of fusion and deep-inelastic cross sections e.g., for the $^{58}Ni+^{124}Sn$ system, even though it overestimates fusion cross sections themselves (see Fig. 2). The overestimate of fusion cross sections shown in Fig. 1 may have a similar origin [14].

It has been known that energy and angular momentum dissipations play an important role in deep inelastic collisions. Therefore, in order to resolve the surface diffuseness anomaly, it is crucially important to understand the nuclear dissipation in heavy-ion reactions. So far, current fusion models are all "friction free", in a sense that friction is taken into account only as a strong absorption well inside the barrier. It will be important to extend it by including the effect of dissipations at larger distances. Such model will provide a quantal model for deep inelastic collisions, and at the same time it will also describe dissipative tunneling in heavy-ion fusion reactions [19].

3.3 Fusion of unstable nuclei

Fusion of unstable nuclei has been discussed for some time [5, 20–22], but its underlying dynamics has not yet been completely understood. For unstable nuclei, the breakup of projectile nucleus becomes important because of its weakly bound property. This process complicates the whole fusion process, especially when one intends to separate between complete and incomplete fusion cross sections [5, 20]. Furthermore, transfer processes may also significantly affect the dynamics of fusion reactions of weakly bound nuclei [23–25]. It is therefore important for any theoretical calculation to take into account fusion, breakup, and transfer processes simultaneously in order to model fusion of unstable nuclei. Such calculation would require a large scale computing, and has been rather scarce. An exception is a time-dependent wave packet approach [26], which however has been limited only to total fusion cross sections. Because of this, the enhancement of fusion cross sections observed e.g., in the $^{15}C+^{232}Th$ system [27], as compared to fusion cross sections for the $^{12,13,14}C+^{232}Th$ systems, has remained theoretically unexplained.

3.4 Interplay between fusion and transfer

The nucleon transfer processes, especially the two-neutron transfer process, are important subjects not only in subbarrier fusion but also in connection to the pairing correlation in neutron-rich nuclei [28–30]. The multi-neutron transfer process [31] would play an important role also in fusion of neutron-rich skin nuclei, in order to reach the island of stability in the superhevy region.

An important theoretical issue here is to reproduce experimental fusion and transfer cross sections simultaneously. Recently, we have investigated this problem for the $^{40}Ca+^{96}Zr$ system [32]. By including the multi-neutron transfer process in the coupled-channels approach, we have found that the coupling strengths for the transfer couplings, which reproduce the transfer cross sections, largely underestimate fusion cross sections (see Fig. 3). To this end, we have included one octupole phonon excitation in ^{40}Ca, the octupole phonon excitations in ^{96}Zr up to the three-phonon levels, and the multi-nucleon transfer process up to three-neutron transfer, with both simultaneous and direct two-neutron transfer couplings [32]. We have assumed a transfer to a single effective channel for each transfer partition, and set its energy to be the same as the optimum Q-value, $Q = 0$. Following Ref. [33], we have assumed that the properties of the collective excitations do

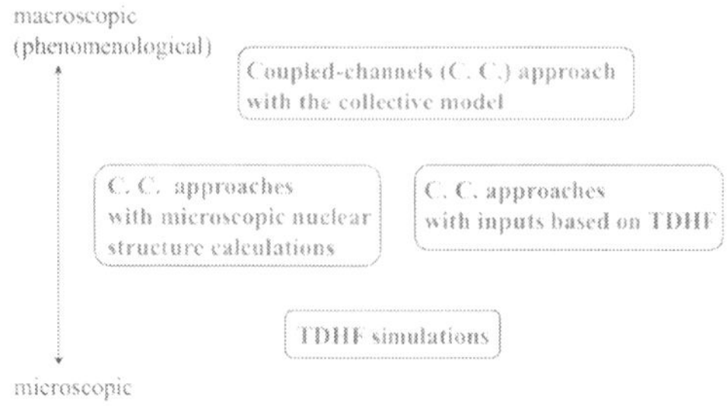

Figure 4. The current status of theoretical approaches to heavy-ion subbarrier fusion reactions.

not change even after the transfer. A striking fact is that the transfer process enhances fusion cross sections only a little if the coupling strengths are chosen so as to reproduce the transfer data. This indicates that the coupling scheme which has been employed in this calculation needs a further extension, by taking into account e.g., a distribution of transfer Q-value as well as changes in collective states after transfer. We have reached a similar conclusion also for the ^{40}Ca+^{58,64}Ni systems [35].

Very recently, it has been found experimentally that the multi-nucleon transfer processes in ^{16,18}O, ^{19}F + ^{208}Pb reactions populate highly excited states in the target-like nuclei [36]. A population of high excited states would lead to energy dissipation. It would be another challenging problem to describe such process within the framework of coupled-channels approach [19].

4 Semi-microscopic approach to heavy-ion fusion reactions

4.1 Current status of fusion modelings

Figure 4 summarizes the current status of theoretical approaches to heavy-ion subbarrier fusion reactions. The coupled-channels approach with the collective model is categorized as a macroscopic and phenomenological approach, which needs empirical information on the coupling strengths and an internuclear potential. On the other hand, a full time-dependent Hartree-Fock (TDHF) simulation, which has rapidly been developed in the past few years [37–39], does not require any empirical input once the energy functional is specified. This is a microscopic approach, starting from the nucleon degree of freedom. A big challenge in this approach, however, is a well known fact that TDHF cannot describe quantum tunneling, and thus heavy-ion fusion reactions at energies below the Coulomb barrier. One would have to go beyond the mean-field approximation in order to resolve this problem, e.g., by using the time-dependent generator coordinate method (TDGCM) [40]. No realistic calculation, however, has been done so far because such calculation is still quite numerically expensive.

In between the macroscopic and the full microscopic approaches, one can also think about semi-microscopic approaches. These are coupled-channels calculations with inputs either from a microscopic nuclear structure calculation [41, 42] or from a TDHF simulation [43–45]. The density-constrained TDHF (DC-TDHF) method [43] is categorized into this group. These semi-microscopic approaches can be applied to fusion at energies below the Coulomb barrier, as one solves a quantum mechanical equation once it is set up using microscopic inputs.

4.2 Coupled-channels calculations with microscopic nuclear structure calculations

As an example of the semi-microscopic approaches to heavy-ion subbarrier fusion reactions, we present in this subsection the approach which combines the coupled-channels calculations with the multi-reference covariant density functional theory (MR-CDFT) for nuclear collective excitations [41]. In this approach, the nuclear potential in the coupled-channels equations, Eq. (2), is assumed to be a deformed Woods-Saxon type with a microscopic multipole operator, $Q_{\lambda\mu} = \sum_i r_i^\lambda Y_{\lambda\mu}(\hat{\boldsymbol{r}}_i)$. That is,

$$V(r,\xi) = \frac{-V_0}{1+\exp\left(\frac{r-R_0-\sqrt{\frac{2\lambda+1}{4\pi}}R_T\alpha_{\lambda0}}{a}\right)} \tag{4}$$

with

$$\alpha_{\lambda\mu} = \frac{4\pi}{3e}\frac{1}{Z_T R_T^\lambda}Q_{\lambda\mu}, \tag{5}$$

where R_T and Z_T are the radius and the atomic number of the target nucleus, respectively (we here consider excitations in the target nucleus). In writing Eq. (4), we have used the isocentrifugal approximation [1]. The Coulomb coupling is also taken into account in a similar fashion. The wave functions for the collective states, $\phi_k(\xi) = \phi_{IM}(\xi)$, in Eq. (3) are many-body wave functions generated from microscopic nuclear structure calculations, such as the multi-reference covariant density functional theory. These microscopic wave functions also yield all the matrix elements of the multipole operator, $\langle\varphi_{I0}|Q_{\lambda0}|\varphi_{I'0}\rangle$. The coupled-channels equations can then be constructed in a similar way as in the macroscopic approach for a given internuclear potential.

In heavy-ion fusion reactions, the coupling between the ground state and the lowest-lying collective state play the most important role, even though the couplings from the lowest-lying state to higher states are often important as well. Since the strength for the coupling between the ground state and the lowest-lying state can often be estimated from an experimental transition probability, we introduce an overall scaling factor to all the matrix elements so that the transition from the lowest-lying collective state to the ground state is consistent with experimental data. The MR-CDFT calculation then provides the relative strengths among collective levels, which are often not available experimentally. The excitation energy, on the other hand, is often known well for many levels, and we simply use them in the calculations whenever they are available.

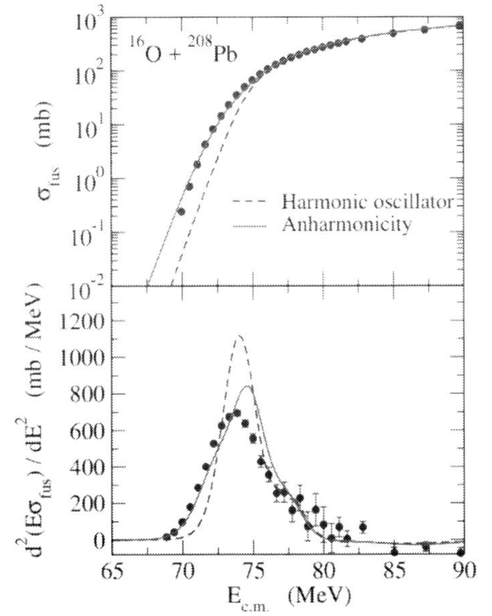

Figure 5. The fusion cross sections (upper panel) and the fusion barrier distribution (lower panel) for the $^{16}O+^{208}Pb$ system obtained with the semi-microscopic coupled-channels calculation (the solid line). For this purpose, the multi-reference covariant density functional theory is employed, and states up to "two-phonon" configurations are included in the calculation. The dashed line shows the results of the two-phonon coupling in the harmonic oscillator limit. The experimental data are taken from Ref. [15].

Figures 5 shows the fusion cross section $\sigma_{fus}(E)$ and the fusion barrier distribution $D_{fus}(E) = d^2(E\sigma_{fus})/dE^2$ [2, 9] for the $^{16}O+^{208}Pb$ reaction so obtained. The dashed line shows the result of the coupled-channels calculations including up to the double quadrupole and octupole phonon states (that is, 3^-_1, 2^+_1, $(3^-_1)^2$, $(2^+_1)^2$, and $3^-_1 \otimes 2^+_1$ states) in ^{208}Pb in the harmonic oscillator limit. As shown in the figure, this calculation overestimates the height of the main peak in the barrier distribution, as in many previous calculations [15, 46, 47]. The result of the semi-microscopic coupled-channels calculations is shown by the solid line. For this purpose, we generate the collective states with the MR-CDFT approach with the PC-PK1 interaction [48]. In the coupled-channels calculation, in addition to the entrance channel, we include the one-octupole phonon state, 3^-_1, at 2.615 MeV, the "one-quadrupole" phonon state, 2^+_1, and several states which are strongly coupled to those 3^-_1 and 2^+_1 states by the octupole and the quadrupole couplings. The whole two-octupole-phonon candidate states are included in this model space. It is remarkable that the semi-microscopic calculation yields a much lower main peak in the fusion barrier distribution, and the agreement with the experimental data is considerably improved both for the fusion cross sections and for the barrier distribution.

For this good reproduction, we find that the coupling between the 3^-_1 and the 2^+_1 states play an important role. The couplings between the two-octupole-phonon states and the excited negative parity states also play a role. In the calculations in the harmonic oscillator limit, the 3^-, 2^+_1 and the 5^-_1 states are treated as independent phonon states, and the couplings among those states are absent. In contrast, in the present semi-microscopic calculation, the 2^+_1 state has in part the two octupole phonon character, $(3^-)^2$. Likewise, the 1^- and 5^-_1 states have both the $(3^-)^3$ and the $3^- \otimes 2^+$ characters. Apparently those anharmonicity effects in the transition strengths lead to the strong couplings between the ground state and those states via multiple octupole excitations, significantly improving the previous coupled-channels calculations.

5 Summary

Heavy-ion subbarrier fusion reactions show a strong interplay between nuclear reaction and nuclear structure, and contain a variety of rich physics, such as coupling assisted tunneling and energy dissipation. The coupled-channels approach with a macroscopic and phenomenological description for nuclear structure has been developed in order to understand the dynamics of subbarier fusion. We have argued in this paper that a theoretical description of subbarrier fusion is now gradually being shifted from the phenomenological approach to more microscopic modelings. A full time-dependent Hartree-Fock simulation is a good example for this, even though this approach still has a difficulty in applying to fusion at energies below the Coulomb barrier. We have also developed the semi-microscopic coupled-channels approach, which uses inputs generated from microscopic nuclear structure calculations. We have shown the result of such calculation for the $^{16}O+^{208}Pb$ system, which has considerably improved the previous coupled-channels calculations in the harmonic oscillator limit. This approach is more flexible than the conventional approach for the coupled-channels calculations, because it can be applied also to transitional nuclei, which show neither the vibrational nor the rotational characters.

There are still many theoretical challenges in heavy-ion subbarrier fusion reactions, for which dissipation and multi-nucleon transfer make two important keywords. These are: i) deep subbarrier hindrance of fusion cross sections, ii) fusion above the barrier and the role of nuclear dissipation, iii) fusion of unstable nuclei, and iv) an interplay between fusion and transfer. Microscopic and semi-microscopic approaches to subbarrier fusion will hopefully resolve some of these challenges in a near future.

We thank J.M. Yao and G. Scamps for collaborations and for useful discussions.

References

[1] K. Hagino and N. Takigawa, Prog. Theor. Phys. **128**, 1061 (2012).

[2] M. Dasgupta, D.J. Hinde, N. Rowley and A.M. Stefanini, Annu. Rev. Nucl. Part. Sci. **48**, 401 (1998).

[3] A.B. Balantekin and N. Takigawa, Rev. Mod. Phys. **70**, 77(1998).

[4] B.B. Back, H. Esbensen, C.L. Jiang and K.E. Rehm, Rev. Mod. Phys. **86**, 317 (2014), and references therein.

[5] L.F. Canto, P.R.S. Gomes, R. Donangelo, J. Lubian and M.S. Hussein, Phys. Rep. **596**, 1 (2015).

[6] K. Hagino, N. Rowley, and A.T. Kruppa, Comp. Phys. Comm. **123**, 143 (1999).

[7] R.G. Stokstad *et al.*, Phys. Rev. Lett. **41**, 465 (1978).

[8] J.R. Leigh *et al.*, Phys. Rev. C**52**, 3151 (1995).

[9] N. Rowley, G.R. Satchler, and P.H. Stelson, Phys. Lett. B**254**, 25 (1991).

[10] S. Misicu and H. Esbensen, Phys. Rev. Lett. **96**, 112701 (2006).

[11] T. Ichikawa, K. Hagino, and A. Iwamoto, Phys. Rev. Lett. **103**, 202701 (2009).

[12] T. Ichikawa, Phys. Rev. C**92**, 064604 (2015).

[13] T. Ichikawa, K. Hagino, and A. Iwamoto, Phys. Rev. C**75**, 064612 (2007).

[14] J.O. Newton *et al.*, Phys. Lett. **B586**, 219 (2004); Phys. Rev. C**70**, 024605 (2004).

[15] C.R. Morton *et al.*, Phys. Rev. C**60**, 44608 (1999).

[16] W. Reisdorf, J. of Phys. **G20**, 1297 (1994).

[17] F.L.H. Wolfs, Phys. Rev. C**36**, 1379 (1987).

[18] O. Akyüz and A. Winther, *Proc. Int. School of Physics Enrico Fermi, Course LXXVII*, ed. by R.A. Broglia, R.A. Ricci, and C.H. Dasso (North-Holland, Amsterdam, 1981).

[19] M. Tokieda and K. Hagino, arXiv:1704.00254.

[20] L.F. Canto, P.R.S. Gomes, R. Donangelo, and M.S. Hussein, Phys. Rep. **424**, 1 (2006).

[21] K. Hagino, A. Vitturi, C.H. Dasso, and S.M. Lenzi, Phys. Rev. C**61**, 037602 (2000).

[22] A. Diaz-Torres and I. Thompson, Phys. Rev. C**65**, 024606 (2002).

[23] R. Raabe *et al.,* Nature **431**, 823 (2004).

[24] A. Lemasson *et al.*, Phys. Rev. Lett. **103**, 232701 (2009).

[25] D.H. Luong *et al.*, Phys. Rev. C**88**, 034609 (2013).

[26] M. Ito, K. Yabana, T. Nakatsukasa, and M. Ueda, Phys. Lett. **B637**, 53 (2006).

[27] M. Alcorta *et al.*, Phys. Rev. Lett. **106**, 172701 (2011).

[28] W. von Oertzen and A. Vitturi, Rep. Prog. Phys. **64**, 1247 (2001).

[29] K. Hagino and H. Sagawa, Phys. Rev. C**72**, 044321 (2005).

[30] G. Potel *et al.*, Phys. Rev. Lett. **107**, 092501 (2011).

[31] L. Corradi, G. Pollarolo, and S. Szilner, J. of Phys. **G36**, 113101 (2009).

[32] G. Scamps and K. Hagino, Phys. Rev. C**92**, 054614 (2015).

[33] H. Esbensen and S. Landowne, Nucl. Phys. **A492**, 473 (1989).

[34] A.M. Stefanini *et al.*, Phys. Lett. **B728**, 639 (2014).

[35] G. Scamps, D. Bourgin, K. Hagino, F. Haas, and S. Courtin, arXiv:1609.06428.

[36] D.C. Rafferty *et al.*, Phys. Rev. C**94**, 024607 (2016).

[37] C. Simenel, Euro. Phys. J. **A48**, 152 (2011).

[38] K. Sekizawa and K. Yabana, Phys. Rev. C**88**, 014614 (2013); C**93**, 054616 (2016).

[39] G. Scamps and D. Lacroix, Phys. Rev. C**87**, 014605 (2012).

[40] P.-G. Reinhard, R.Y. Cusson, and K. Goeke, Nucl. Phys. **A398**, 141 (1983).

[41] K. Hagino and J.M. Yao, Phys. Rev. C**91**, 064606 (2015); J.M. Yao and K. Hagino, Phys. Rev. C**94**, 011303(R) (2016).

[42] T. Ichikawa and K. Matsuyanagi, Phys. Rev. C**88**, 011602(R) (2013); Phys. Rev. C**92**, 021602(R) (2015).

[43] A.S. Umar and V.E. Oberacker, Phys. Rev. C**77**, 064605 (2008); Euro. Phys. J. A**39**, 243 (2009).

[44] K. Washiyama and D. Lacroix, Phys. Rev. C**78**, 024610 (2008); Phys. Rev. C**79**, 024609 (2009).

[45] C. Simenel, M. Dasgupta, D.J. Hinde, and E. Williams, Phys. Rev. C**88**, 064604 (2013).

[46] H. Esbensen and S. Misicu, Phys. Rev. C **76**, 054609 (2007).

[47] S. Yusa, K. Hagino, and N. Rowley, Phys. Rev. C**85**, 054601 (2012).

[48] P.W. Zhao *et al.*, Phys. Rev. C**82**, 054319 (2010).

Quasifission Dynamics in the Formation of Superheavy Elements

D.J. Hinde[1,*], *M.* Dasgupta[1], *D.Y.* Jeung[1], *G.* Mohanto[1,**], *E.* Prasad[1,***], *C.* Simenel[1], *E.* Williams[1], *I.P.* Carter[1], *K.J.* Cook[1], *Sunil* Kalkal[1], *D.C.* Rafferty[1], *E.C.* Simpson[1], *H.M.* David[2], *Ch.E.* Düllmann[2,3,4], and *J.* Khuyagbaatar[2,3]

[1]*Department of Nuclear Physics, Research School of Physics and Engineering, Australian National University, Canberra, ACT 2601, Australia*
[2]*GSI Helmholtzzentrum für Schwerionenforschung, D-64291 Darmstadt, Germany*
[3]*Helmholtz Institute Mainz, D-55099 Mainz, Germany*
[4]*Johannes Gutenberg Universität Mainz, D-55099 Mainz, Germany*

Abstract. Superheavy elements are created through the fusion of two heavy nuclei. The large Coulomb energy that makes superheavy elements unstable also makes fusion forming a compact compound nucleus very unlikely. Instead, after sticking together for a short time, the two nuclei usually come apart, in a process called quasifission. Mass-angle distributions give the most direct information on the characteristics and time scales of quasifission. A systematic study of carefully chosen mass-angle distributions has provided information on the global trends of quasifission. Large deviations from these systematics at beam energies near the capture barrier reveal the major role played by the nuclear structure of the two colliding nuclei in determining the reaction outcome, and thus implicitly in hindering or favouring superheavy element synthesis.

1 Introduction

Superheavy elements (SHE) are formed by heavy-ion fusion reactions. Fusion cross sections can be considerably suppressed [1] by quasifission [2]. This non-equilibrium process results when the combined di-nuclear system, formed as the two nuclear surfaces stick together, subsequently separates into two (fission-like) fragments, with the initial kinetic energy largely or completely damped. Quasifission can occur very rapidly, typically in less than 10^{-20}s, before a compact compound nucleus can be reached [2–5]. The probability of quasifission (P_{QF}) can be very large, thus the complementary probability of compound nucleus formation ($P_{CN} = 1 - P_{QF}$) can be small, quite likely lower than 10^{-3} in reactions forming superheavy elements. Understanding the competition between quasifission and fusion is thus very important in predicting the optimal fusion reactions to use to form new elements and isotopes in the superheavy mass region.

A key characteristic, important for superheavy element formation, is the "sticking time" following contact of the two nuclear surfaces [6]. It is expected that the sticking time is correlated with P_{CN}: where the sticking time is longer, then P_{CN} would be expected to be larger (more favourable for SHE synthesis). The average sticking time can be extracted from measurements of quasifission angular distributions. The two colliding nuclei always approach each other along the beam axis, and after contact rotate with angular velocities that can be calculated. Measurement of the rotation angle thus allows estimation of the sticking time. As the system rotates, mass flow also occurs between the two nuclei. Measurement of the velocity vectors of both fragments gives direct information on the centre-of-mass angle and mass-ratio of the fragments at scission, defined as $M_R = M_1/(M_1 + M_2)$, as well as providing excellent discrimination against fission events resulting from peripheral (transfer-induced) processes [4, 7]. The mass-ratio (or mass) plotted as a function of the centre-of-mass angle is referred to as a mass-angle distribution, or MAD. This gives direct information on the dynamical time scales, as long as the system undergoes less than a full rotation (taking $\sim 10^{-20}$s). This is usually the case for collisions of heavy nuclei, as shown first by measurements at GSI [2, 8], and by later results from ANU [3–5, 9–15]. Inter-relationships of different measurements, and background to the results presented here are given in recent conference proceedings [16–20].

2 Experimental Setup

The detector configuration used at the Australian National University (ANU) to measure MAD is shown in Fig.1. It consists of two or three large area multi-wire proportional counters (MWPC). They detect reaction products from the interaction of the pulsed beam with targets typically 50-200 μg/cm^2 in thickness, oriented with the normal to the target face at 60° to the beam. This eliminates shadowing of the detectors by the target frame. The average en-

* e-mail: david.hinde@anu.edu.au
** Current address: BARC, Mumbai, India
*** Permanent address: Department of Physics, School of Mathematical and Physical Sciences, Central University of Kerala, Kasaragod 671314, India.

 DOI: 10.1051/epjconf/201716300023

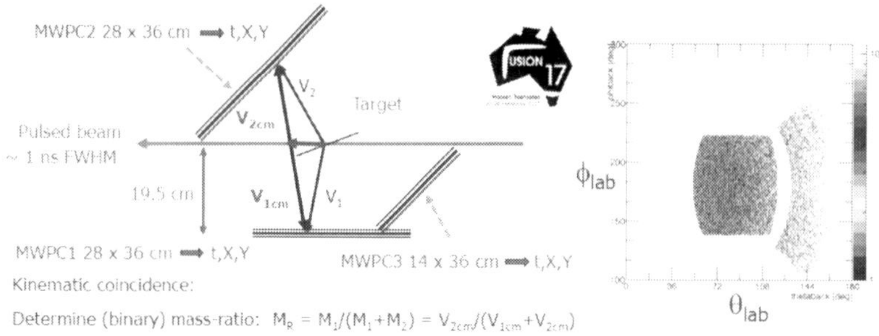

Figure 1. Enhanced ANU experimental setup to measure fission mass-angle distributions over the widest achievable angular range. The angular coverage of the MWPC1 and MWPC3 detectors, for events in coincidence with MWPC2, are shown in scattering angle θ and azimuthal angle ϕ.

ergy loss of the beam and the fragments in the target material is corrected for in the analysis. The wide coverage of MWPC1 and MWPC3 in scattering angle (θ) and azimuthal angle (ϕ) for coincidence events with MWPC2 is demonstrated in Fig.1. From the good resolution of the detectors in position (~1 mm) and time (< 500 ps), and knowledge of the interaction time of the beam pulse with the target, the velocity vectors of the detected particles are determined. From the deduced velocity vectors in the centre-of-mass frame, for each binary event the centre-of-mass angle and the mass-ratio are determined, thus generating the MAD.

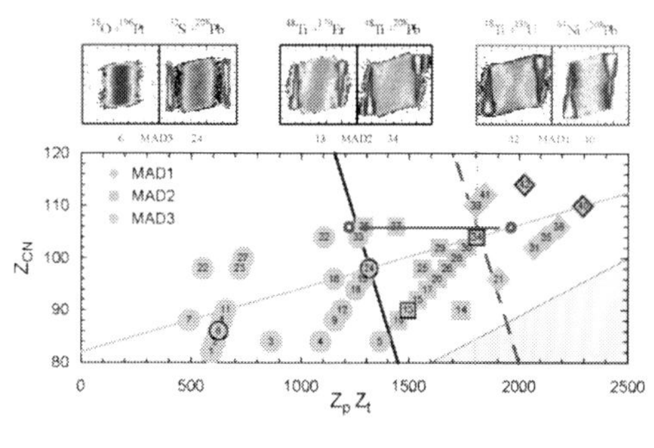

Figure 2. The symbol indicates the classification of MAD observed, shown as a function of the charge product of the colliding nuclei Z_pZ_t and the atomic number of the compound nucleus Z_{CN}=Z_p+Z_t. The numbers refer to the specific reaction in Ref.[4]. The diagonal full blue line represents the empirical boundary between reactions with no mass-angle correlation (left) and those that have (right). The diagonal red dashed line indicates the boundary of reactions which no longer exhibit a peak at symmetry in the angle-integrated fission mass distribution. The thin purple line represents the locus of reactions with Pb. Examples of each type of MAD are shown in the panels above, with their reaction number. The purple circles and arrow refer to Cr+Pb measurements discussed towards the end of the paper.

3 MAD systematics

Experimental MAD have been divided into three categories [4] having: (i) a mass-angle correlation with a minimum yield at mass-symmetry - associated with short sticking times (MAD1); (ii) a mass-angle correlation with peak yield at mass-symmetry - resulting from intermediate sticking times (MAD2); and (iii) no significant mass-angle correlation and a narrow mass-distribution – associated with long sticking times, including fission following fusion (MAD3). Examples of each type of MAD are shown in the upper panels of Fig.2. The systematic trends of MAD characteristics with the identity of the two colliding nuclei was studied [duRietz13], to determine global trends of quasifission dynamics. This is in analogy with the evaluation of the smooth liquid drop model dependence of nuclear masses on N and Z, where deviations highlight the effects of nuclear structure. Choosing bombarding energies E well-above the mean capture barrier B (around E/B=1.08), nuclear structure effects were minimised. It was found [duRietz13] at these bombarding energies that the MADs are indeed strongly correlated with global variables. The simplest variables are the Coulomb repulsion in the entrance channel (related to the product of the proton numbers of the projectile and target nuclei Z_pZ_t), and the compound nucleus atomic number Z_{CN}, as illustrated in Fig.2. However, for particular cases, it has been found that the nuclear structure of the nuclei in the entrance channel is extremely important in determining the sticking times and MAD characteristics. This relates to doubly-magic neutron-rich nuclei such as ^{48}Ca and ^{208}Pb, and prolate deformed actinide nuclei, all used in SHE formation reactions, as described below.

4 Effect of spherical closed shells

To investigate in detail the effect of closed shells in the entrance channel on quasifission probabilities and characteristics, measurements [13] of MADs were made for ^{40,44,48}Ca projectiles bombarding targets of ^{208,204}Pb (forming ^{248,252,256}No with Z_{CN} =102), and for ^{48}Ti bombarding ^{200}Hg (^{248}No) and ^{208}Pb (forming ^{256}Db with Z_{CN}

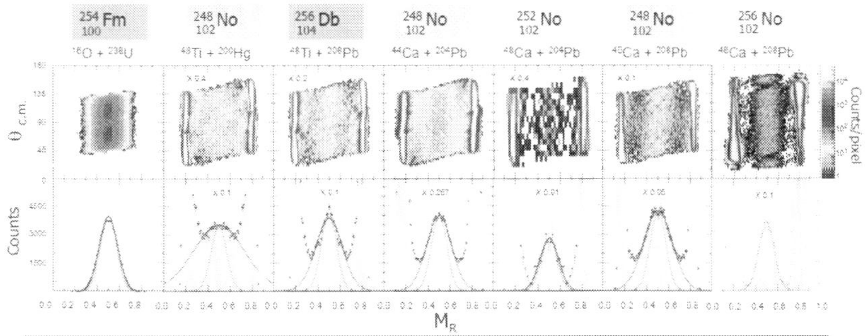

σ_{MR}	0.081	0.237	0.120	0.114	0.084	0.126	0.068
N_{Magic}	2	0	2	2	3	4	4
Δ(N/Z)	0.57	0.32	0.35	0.29	0.09	0.54	0.14

Figure 3. Measured mass-angle distributions for the indicated reactions at E/B~0.98 (upper panels), compared to ^{16}O+^{238}U (left) measured at an above-barrier energy, where fast quasifission would be expected to be negligible. The MAD for ^{48}Ca+^{208}Pb shows a dip in coverage at 30° and 150° due to the gap between detectors shown in Fig.1. In the projected mass ratio spectra for $45° < \theta_{c.m.} < 135°$ (lower panels) the scale factor multiplies the counts scale on the left. Gaussian functions with σ_{MR} = 0.07 (thin red lines) are shown for reference. Gaussian fits to the region around M_R=0.5 are shown (turquoise lines), whose standard deviations σ_{MR} are tabulated, along with the total number of magic numbers of the projectile and target nuclei N_{Magic}, and the difference Δ(N/Z) between the projectile and target nucleus N/Z ratios. (Adapted from Ref.[13])

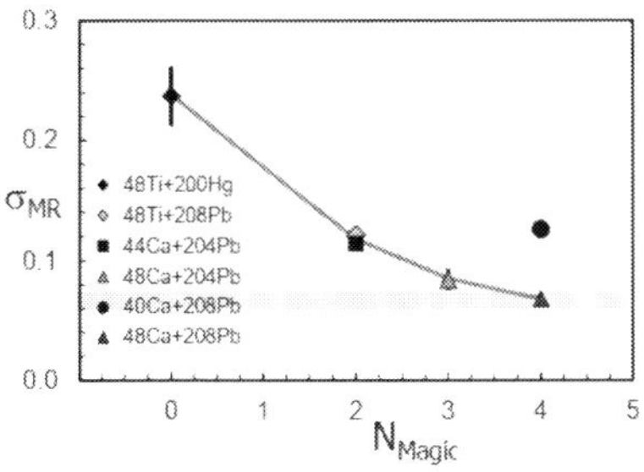

Figure 4. Fitted standard deviations σ_{MR} of the fission-like mass distributions as a function of the number N_m of magic numbers in the entrance channel, for the reactions shown in Fig.3. The line guides the eye. The expectations of σ_{MR} for fusion-fission are in the range 0.06 to 0.08. The value of σ_{MR} for the ^{40}Ca+^{208}Pb reaction is large, and the MAD in Fig.3 shows a mass-angle correlation. This was attributed [13] to the N/Z mismatch between the projectile and target nuclei causing transfer reactions early in the collision, which attenuate the effect of the four magic numbers in the entrance channel.

=104). Measurements were made a few percent below the average fusion barrier energy to give "gentle" collisions with minimal relative velocity at contact. The MAD and projected M_R spectra are shown in Fig.3, together with reference data for the ^{16}O + ^{238}U reaction at an above-barrier energy, forming ^{254}Fm. The standard deviations σ_{MR} of the Gaussian fits are tabulated.

Despite having similar or identical Z_pZ_t, and forming similar or identical compound nuclei, values of σ_{MR} differ by more than a factor of three, indicating a significant variation in the characteristics and probability of quasifission. Indeed, for the most recent (unpublished) MAD data for the ^{48}Ca+^{208}Pb reaction, the characteristics appear consistent with a fusion-fission mechanism. To reveal the systematic behaviour, the variation of σ_{MR} with the total number of magic numbers N_{Magic} in the entrance channel [13] is shown in Fig.4. Large values of σ_{MR} are found for no magic numbers, reducing to values expected for fusion-fission for maximal N_{Magic} (for ^{48}Ca+^{208}Pb). This suggests that reactions involving nuclei having several magic numbers form a compact compound nucleus with higher probability. It seems likely that this is associated with reduced energy dissipation as the two nuclei interpenetrate, allowing more compact shapes to be reached. The exception is the reaction ^{40}Ca+^{208}Pb, with 4 magic numbers, where the mass-width and MAD are consistent with quasifission being much more likely. It was suggested [13] that transfer reactions before contact can be expected to attenuate the entrance-channel magicity in this reaction, driven by the extreme mismatch of the N/Z ratios of projectile and target nuclei. This was supported by TDHF calculations [13]. Supporting the "magic number" hypothesis, systematic analysis of heavy element cross sections in many reactions forming Th isotopes indicated somewhat enhanced fusion probabilities with two magic numbers in the entrance channel [21].

More recent ANU quasifission results, for the reaction of isotopes of Cr with Pb, support these conclusions regarding magicity and N/Z matching. MAD and mass-ratio projections at a sub-barrier energy for each reaction studied are shown in Fig.5. The three reactions on the left all form the same compound nucleus ^{258}Sg. The panels are ordered from left to right first by the num-

 DOI: 10.1051/epjconf/201716300023

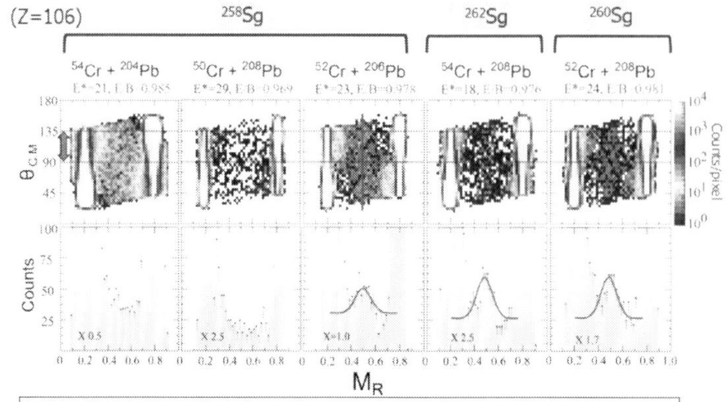

N_{Magic}	1	2	2	2	3
$\Delta(N/Z)$	0.24	0.45	0.35	0.29	0.37

Figure 5. MAD and projected mass-ratio distributions for backward angles from 90° to 135° (as indicated in the MAD plots), for reactions of ^{50,52,54}Cr isotopes with ^{204,206,208}Pb. Sub-barrier beam energies (denoted by E/B), resulted in the low excitation energies E*. As in Fig.4, the total number of magic numbers of the projectile and target nuclei N_{Magic}, and the difference $\Delta(N/Z)$ between the projectile and target nucleus N/Z ratios are shown for each reaction. The reaction outcome changes from a minimum in yield at mass-symmetry (left) to a narrow peak at symmetry, which shows no evidence of a mass-angle correlation.

ber of magic numbers in the entrance channel N_{Magic}, and then by the difference $\Delta(N/Z)$ between the N/Z values of the target and projectile nuclei. The left-most reaction has only a single magic number in the entrance channel, and shows a U-shaped mass distribution, consistent with MAD1, as expected from systematics (right-hand purple circle in Fig.2). With two magic numbers, the reactions better matched in N/Z (smaller values of $\Delta(N/Z)$) show a peak at mass-symmetry, associated with an angle-independent ridge in the MAD. With three magic numbers, but less favourable $\Delta(N/Z)$, a similar result is observed.

Fits were performed to the mass-ratio spectra, including an asymmetric U-shaped background from fast quasifission (derived from the reactions without a mass-symmetric peak) and a mass-symmetric Gaussian peak. As the beam energy for the Cr+Pb reactions is increased, Fig.6(a) shows that the mass-symmetric peak progressively becomes a smaller fraction of the total fission yield within the range 0.3< M_R <0.7. This might be expected, as the relative velocity at contact increases, as does the contribution from high angular momenta, for which the potential energy surface will be more repulsive. From the symmetric yield fraction, and preliminary absolute cross section normalization, estimates of the cross sections for the symmetric yield were extracted, as shown in Fig.6(b). Also shown are the xn evaporation residue cross sections

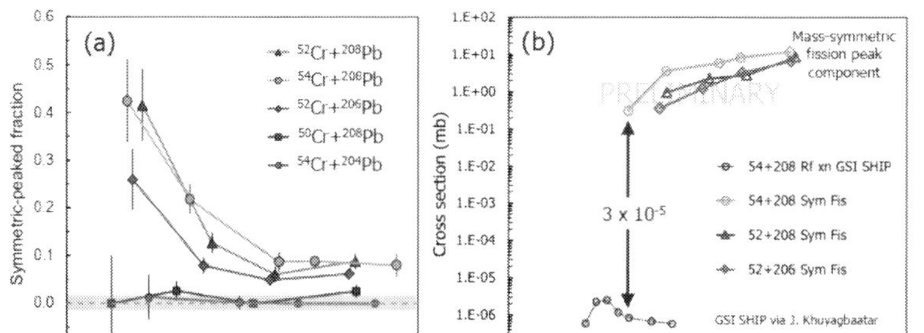

Figure 6. (a) The extracted fraction of symmetric peaked fission events to all fission and quasifission events in the range 0.3< M_R <0.7, for the indicated Cr+Pb reactions, as a function of beam energy with respect to the capture barrier energy (E/B). The smooth fall with increasing E/B might be expected. (b) Preliminary cross sections for the symmetric-peaked fission component for each reaction (where present) as a function of compound nucleus excitation energy. The summed xn evaporation residue cross sections for the ^{54}Cr+^{208}Pb reaction are also shown.

for the $^{54}Cr+^{208}Pb$ reaction [22]. If the symmetric yield originated only from fusion-fission, this would correspond to a probability of surviving fission decay of 3×10^{-5} at ~18 MeV excitation energy. The suppression of SHE cross sections from fission competition in this reaction is much larger than from fast quasifission, though it is still not clear whether the mass-symmetric peak might have a contribution from slow quasifission.

These Cr+Pb reactions show qualitatively similar changes in MAD as the $^{40,44,48}Ca+^{204,208}Pb$ reactions. However, the transition from a U-shaped mass distribution to a narrow peak at mass-symmetry is an even more drastic change in reaction outcome, and indicates a definite bifurcation in reaction trajectories that is not so apparent in the Ca+Pb MADs, since the quasifission yield is also shows a mass-symmetric peak. This is consistent with bifurcations in reaction trajectories between quasifission and fusion for a given projectile-target combination, evidenced by the observation of both quasifission and ER cross sections.

In reactions forming Sg (Z=106), Fig.2 indicates that a MAD of class 3 (no mass-angle correlation) would be expected only for Z_pZ_t<1200 (arrowed purple circle), in the absence of the favourable effects of magic numbers. This would correspond to a reaction with an Al projectile or lighter, with around half the atomic number of Cr. The dramatic change in reaction outcome with a small change in neutron number in the Cr+Pb reactions indicates that the observed systematic behaviour observed at higher beam energies does not necessarily allow prediction of behaviour in near-barrier reactions, where the nuclear structure of the system can play a very significant role. It is notable that this characteristic is also the case in comparing spontaneous and low energy nuclear fission with fission at high excitation energies.

5 Effect of deformation alignment

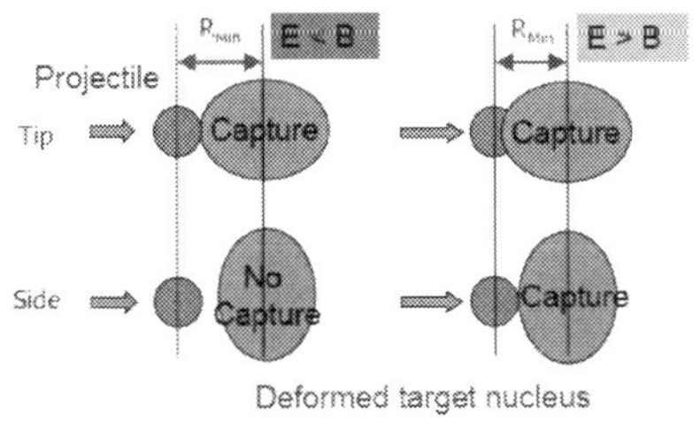

Figure 7. At beam energies below B, only deformation-aligned collisions result in contact and capture, whilst at E/B > 1, the distance of closest approach (R_{Min}) is smaller (red arrow), thus capture also occurs for more compact equatorial configurations, giving longer sticking times, expected to be correlated with larger P_{CN}.

Static deformation of the heavy reaction partner affects nuclear collisions because the capture barrier height depends on the relative orientation of the projectile nucleus and the deformation axis of the statically deformed heavy reaction partner [6]. For beam energies below the average barrier energy *B*, in collisions with prolate-deformed nuclei, it is only collisions in the "deformation aligned" configuration, giving axial or tip collisions (illustrated on the left in Fig.7) that result in capture. This is because the distance of closest approach at energies below the average barrier does not pass inside the smaller barrier radius corresponding to anti-aligned (side or equatorial) collisions. At energies well-above *B*, the distance of closest approach is reduced, and is inside the anti-aligned capture barrier. Thus, all orientations contribute to capture [7, 23].

An important question relevant to SHE synthesis is whether the elongated contact configuration (axial collision) or compact configuration (equatorial collision) affects the subsequent reaction dynamics. This has been most clearly demonstrated by the dependence on beam energy of the MAD characteristics [5, 10, 11, 24] and also mass-distributions [6, 25], for measurements from below to above the average capture barrier. Already in 1995, interpretation of measurements of fission angular distributions from below to above-barrier [7, 23, 26] led to the suggestion [23] that deformation alignment should be important in SHE synthesis. Specifically, as a result of the elongated contact configuration, the formation of a compact compound nucleus, and thus SHE cross sections, was suggested to be suppressed at E/B<1. Experimental evaporation residue cross sections [27, 28] have confirmed the correlation between fission characteristics and heavy element yields from below- to above-barrier. Indeed, superheavy element yields from reactions with prolate deformed actinide target nuclei have only been found at beam energies above predicted average capture barrier energies.

Let us now look at experimental results for collisions with prolate actinide nuclei. Fig.8 shows mass-angle distributions (with projected mass-ratio spectra below) for four different measurements. Panels (a) and (b) present the "reference" reaction $^{48}Ca+^{208}Pb$ with Z_pZ_t=1640 at a sub-barrier energy (already presented). This exhibits no mass-angle correlation and a narrow mass-width. As discussed previously, this is now believed to result from the magic numbers and the N/Z matching in the entrance channel. Fig.8(c) and (d) show data for the reaction $^{34}S+^{232}Th$ [5], measured at a below-barrier energy. This reaction has a smaller Z_pZ_t of 1440, but has no magic numbers. Higher statistics in this measurement than obtained for the similar $^{34}S+^{232}Th$ reaction [11] show clearly that the dominant process is fast quasifission, with limited mass flow from the projectile/target mass partition, and less than ~90° of angular rotation. On the basis of Z_pZ_t and Z_{CN}, the systematic shown in Fig.2 would predict a mass-angle correlation, but with symmetric peaked fission. Thus the deformation aligned collisions give MAD characteristics of a collision of heavier nuclei. The vertical dashed lines show the mass-ratio expected if one of the fragments was ^{208}Pb - the measured mass-splits closely follow these lines, a feature addressed in detail in Ref. [5]. There also appears to be a distinct component giving events peaked at mass-symmetry, but with lower probability. This can almost certainly be associated with a component having a significantly longer sticking time.

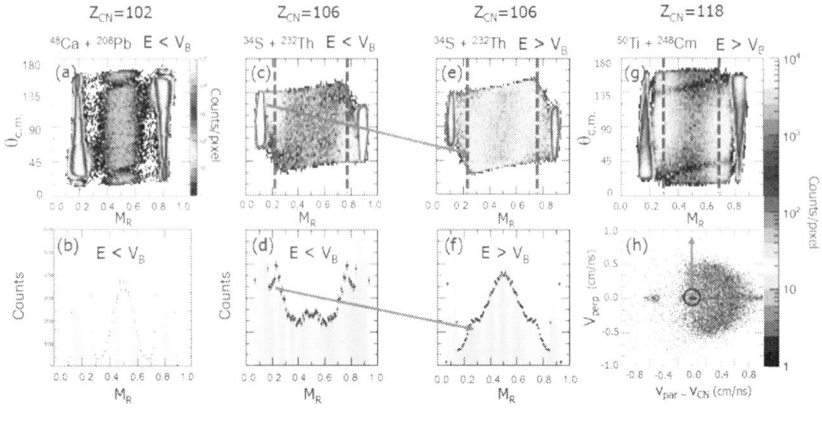

Figure 8. MAD and projected mass distributions for the four reactions indicated (see text). For the ^{50}Ti+^{248}Cm reaction, rather than the mass distribution, panel (h) shows the source velocity deduced assuming binary kinematics. The x-axis shows the component parallel to the beam (expressed as the difference between the measured velocity and the calculated centre-of-mass velocity), whilst the y-axis shows the component perpendicular both to the fission plane and the beam axis [7]. For a binary event originating from capture (corresponding to full momentum transfer from the projectile), the events should be close to the centre of the plot, around (0,0). The group of events at (-V_{CN},0) correspond to spontaneous fission of the ^{248}Cm target. The MAD for ^{50}Ti+^{248}Cm, gated as shown in (h), was measured at an energy above the expected barrier energy. It most closely resembles that for the ^{34}S+^{232}Th reaction at the sub-barrier energy shown in (c).

Fig.8(e) and (f) show data for the same reaction at an above-barrier energy. The mass-asymmetric component now makes up a small fraction of the total fission-like events. Its mass-centroid is essentially unchanged [5], as indicated by the dashed lines, but the group has moved from backward to more forward angles (both characteristics indicated by the arrows). This can be associated with increased mean angular momentum at this above-barrier energy, allowing a greater rotation angle (in a given time) before scission. The predominance of the group of events around mass-symmetry (but having a clear mass-angle correlation) is associated with the higher probability of equatorial collisions. The three dimensional geometry (resulting in a *sinθ* probability weighting [7]) means that for prolate nuclei, the probability of equatorial collisions at energies well-above the average barrier energy is much higher than that of axial (deformation aligned) collisions. The mass-angle correlation shows, despite the compact contact configuration in equatorial collisions, that these collisions predominantly result in quasifission, in contrast with the sub-barrier ^{48}Ca+^{208}Pb data, which shows no mass-angle correlation. Although equatorial collisions with a prolate deformed nucleus do seem to result in longer sticking times and increased P_{CN} compared with an equivalent spherical nucleus, the comparison of MADs (a) and (e) suggests that the enhancement is not as great as the effect of colliding two doubly-magic neutron-rich nuclei at a sub-barrier energy. More detailed analysis of a larger number of measurements is needed to reach a more quantitative conclusion about this factor in SHE formation reactions.

For all reactions with actinide target nuclei, there is a non-negligible probability of sequential fission of the target-like nucleus after the transfer of nucleons between the colliding nuclei early in the reaction. The analysis method allowing clean separation of the full momentum transfer fission following capture is illustrated for the ^{50}Ti+^{248}Cm reaction in Fig.8(h). The tightly grouped events in the centre of the graph (0,0) correspond to binary events [4, 7], originating from a source with the velocity of the centre-of-mass in the collision, with only a small momentum spread due to neutron evaporation. The surrounding broad semi-circular distribution of events corresponds to three-body events, most likely of fission of various target-like nuclei resulting from nucleon transfers, with the projectile-like nuclei recoiling with various angles and energies giving a broad distribution of velocities to the target-like nuclei. The tight group of events at (-V_{CN},0) correspond to spontaneous fission of the ^{248}Cm target nuclei, stationary in the laboratory frame. These show the expected asymmetric-peaked fission mass distribution. A typical gate on FMT events, as used to generate the MAD in Fig.8(g), is shown by the black circle.

The ^{50}Ti+^{248}Cm reaction forms the compound nucleus ^{298}Og (Z=118). The MAD shown in Fig.8(h) was measured at the Australian National University, using a target from Mainz/GSI, at a beam energy above the predicted [29] average capture barrier energy, where equatorial collisions having the longest sticking times would be expected to be dominant. However, the MAD more closely resembles the sub-barrier measurement for ^{34}S+^{232}Th shown in Fig.8(c), rather than the above-barrier measurement shown in Fig.8(e). The relative yield of mass-symmetric events is smaller than for the sub-barrier ^{34}S+^{232}Th reaction, but the fast quasifission events still show a considerable mass flow towards mass-symmetry. As in panels (c) and (e), the dashed lines show the expected mass-ratio if one of the quasifission fragments were ^{208}Pb.

The role of ^{208}Pb in quasifission mass distributions was investigated in the ^{40}Ca+^{238}U reaction, both experimentally and through TDHF calculations [15]. It was concluded that the closed shells centred on ^{208}Pb play a strong role in determining the mass-splits, but for that reaction, only for axial (tip) collisions. In the ^{50}Ti+^{248}Cm reaction, the yield around ^{208}Pb seems to be associated with all orientations. Various possible contributions to the observed behaviour need to be investigated. These include the possible effect of shells in the lighter fragment, sequential fission of heavy target-like nuclei, deviation of the orientation-dependent effective capture barriers from expectations, and changes in dynamics resulting from the heavier nuclei involved in the collision.

A program of measurements of quasifission (including MAD over a wide angular range) for projectiles from ^{48}Ca to ^{64}Ni, bombarding targets from ^{208}Pb to ^{249}Cf, is in progress at the ANU. Analysis and interpretation of this body of data will give more quantitative insights into SHE synthesis reactions.

6 Conclusions

Mass-angle distributions give the most direct information on the characteristics and time scales of quasifission. A systematic study of carefully chosen mass-angle distributions has provided information on the global trends of quasifission. Large deviations from these systematics reveal the major role played by the nuclear structure of the two colliding nuclei in determining the quasifission probability and time scale, and thus implicitly in hindering or favouring superheavy element synthesis.

The observation of rapid changes in quasifission outcomes, depending on magicity, neutron number, and beam energy is a severe challenge for models of quasifission and SHE formation to reproduce. And yet this level of sensitivity of reaction dynamics to nuclear structure is what models must attempt to reproduce, in order to map out the optimum experimental opportunities to create new superheavy elements and isotopes in the future.

References

[1] D.J. Hinde, M. Dasgupta, A. Mukherjee, Phys. Rev. Lett. **89**, 282701 (2002)

[2] J. Tõke, B. Bock, G.X. Dai, A. Gobbi, S. Gralla, K.D. Hildenbrand, J. Kuzminski, W. Müller, A. Olmi, H. Stelzer, Nucl. Phys. A **440**, 327 (1985)

[3] R. du Rietz, D.J. Hinde, M. Dasgupta, R.G. Thomas, L.R. Gasques, M. Evers, N. Lobanov, A. Wakhle, Phys. Rev. Lett. **106**, 052701 (2011)

[4] R. du Rietz, E. Williams, D.J. Hinde, M. Dasgupta, M. Evers, C.J. Lin, D.H. Luong, C. Simenel, A. Wakhle, Phys. Rev. C **88**, 054618 (2013)

[5] E. Prasad, A. Wakhle, D.J. Hinde, E. Williams, M. Dasgupta, M. Evers, D.H. Luong, G. Mohanto, C. Simenel, K. Vo-Phuoc, Phys. Rev. C **93**, 024607 (2016)

[6] M. Dasgupta, D.J. Hinde, N. Rowley, A.M. Stefanini, Annu. Rev. Nucl. Part. Sci. **48**, 401 (1998)

[7] D.J. Hinde, M. Dasgupta, J.R. Leigh, J.C. Mein, C.R. Morton, J.O. Newton, H. Timmers, Phys. Rev. C **53**, 1290 (1996)

[8] W.Q. Shen, J. Albinski, A. Gobbi, S. Gralla, K.D. Hildenbrand, N. Herrmann, J. Kuzminski, W.F.J. Müller, H. Stelzer, J. Tõke et al., Phys. Rev. C **36**, 115 (1987)

[9] R.G. Thomas, D.J. Hinde, D. Duniec, F. Zenke, M. Dasgupta, M.L. Brown, M. Evers, L.R. Gasques, M.D. Rodriguez, A. Diaz-Torres, Phys. Rev. C **77**, 034610 (2008)

[10] R. Rafiei, R.G. Thomas, D.J. Hinde, M. Dasgupta, C.R. Morton, L.R. Gasques, M.L. Brown, M.D. Rodriguez, Phys. Rev. C **77**, 024606 (2008)

[11] D.J. Hinde, R. du Rietz, M. Dasgupta, R.G. Thomas, L.R. Gasques, Phys. Rev. Lett. **101**, 092701 (2008)

[12] D.J. Hinde, R.G. Thomas, R. du Rietz, A. Diaz-Torres, M. Dasgupta, M.L. Brown, M. Evers, L.R. Gasques, R. Rafiei, M.D. Rodriguez, Phys. Rev. Lett. **100**, 202701 (2008)

[13] C. Simenel, D.J. Hinde, R. du Rietz, M. Dasgupta, M. Evers, C.J. Lin, D.H. Luong, A. Wakhle, Phys. Lett. B **710**, 607 (2012)

[14] E. Williams, D.J. Hinde, M. Dasgupta, R. du Rietz, I.P. Carter, M. Evers, D.H. Luong, S.D. McNeil, D.C. Rafferty, K. Ramachandran et al., Phys. Rev. C **88**, 034611 (2013)

[15] A. Wakhle, C. Simenel, D.J. Hinde, M. Dasgupta, M. Evers, D.H. Luong, R. du Rietz, E. Williams, Phys. Rev. Lett. **113**, 182502 (2014)

[16] Hinde, D.J., du Rietz, R., Williams, E., Simenel, C., Lin, C.J., Wakhle, A., Cook, K.J., Dasgupta, M., Evers, M., Luong, D.H., EPJ Web of Conferences **66**, 03037 (2014)

[17] Hinde, D.J., Williams, E., du Rietz, R., Dasgupta, M., Wakhle, A., Simenel, C., Luong, D.H., Cook, K.J., EPJ Web of Conferences **86**, 00015 (2015)

[18] Hinde, D.J., Williams, E., Mohanto, G., Simenel, C., Dasgupta, M., Wakhle, A., Carter, I.P., Cook, K.J., Jeung, D.Y., Luong, D.H. et al., EPJ Web of Conferences **117**, 08006 (2016)

[19] Hinde, D. J., Williams, E., Mohanto, G., Simenel, C., Jeung, D. Y., Dasgupta, M., Prasad, E., Wakhle, A., Vo-Phuoc, K., Carter, I. P. et al., EPJ Web Conf. **123**, 03005 (2016)

[20] Hinde, D.J., Dasgupta, M., Jeung, D.Y., Mohanto, G., Prasad, E., Simenel, C., Walshe, J., Wahkle, A., Williams, E., Carter, I.P. et al., EPJ Web Conf. **131**, 04004 (2016)

[21] D.J. Hinde, M. Dasgupta, Phys. Lett. B **622**, 23 (2005)

[22] S. Hofmann, Progress in Particle and Nuclear Physics **62**, 337 (2009), heavy-Ion Collisions from the Coulomb Barrier to the Quark-Gluon Plasma

[23] D.J. Hinde, M. Dasgupta, J.R. Leigh, J.P. Lestone, J.C. Mein, C.R. Morton, J.O. Newton, H. Timmers,

Phys. Rev. Lett. **74**, 1295 (1995)

[24] D.J. Hinde, R. du Rietz, M. Dasgupta, R.G. Thomas, L.R. Gasques, Phys. Rev. Lett. **101**, 092701 (2008)

[25] K. Nishio, H. Ikezoe, S. Mitsuoka, I. Nishinaka, Y. Nagame, Y. Watanabe, T. Ohtsuki, K. Hirose, S. Hofmann, Phys. Rev. C **77**, 064607 (2008)

[26] J.C. Mein, D.J. Hinde, M. Dasgupta, J.R. Leigh, J.O. Newton, H. Timmers, Phys. Rev. C **55**, R995 (1997)

[27] S. Mitsuoka, H. Ikezoe, K. Nishio, K. Satou, J. Lu, Phys. Rev. C **65**, 054608 (2002)

[28] K. Nishio, H. Ikezoe, I. Nishinaka, S. Mitsuoka, K. Hirose, T. Ohtsuki, Y. Watanabe, Y. Aritomo, S. Hofmann, Phys. Rev. C **82**, 044604 (2010)

[29] W.J. Swiatecki, K. Siwek-Wilczyńska, J. Wilczyński, International Journal of Modern Physics E **Vol. 13, No 1**, 261 (2004)

Inclusive Breakup Theory of Three-Body Halos

Mahir S. Hussein[1,2,3,⋆], *Lucas A.* Souza[1,⋆⋆], *Emanuel* Chimanski[1,⋆⋆⋆], *Brett* Carlson[1,⋆⋆⋆⋆], and *Tobias* Frederico[1,†]

[1] *Instituto Tecnológico de Aeronáutica, DCTA, 12.228-900 São José dos Campos, SP, Brazil*
[2] *Instituto de Estudos Avançados, Universidade de São Paulo C. P. 72012, 05508-970 São Paulo-SP, Brazil*
[3] *Instituto de Física, Universidade de São Paulo, C. P. 66318, 05314-970 São Paulo,-SP, Brazil*

Abstract. We present a recently developed theory for the inclusive breakup of three-fragment projectiles within a four-body spectator model [1], for the treatment of the elastic and inclusive non-elastic break up reactions involving weakly bound three-cluster nuclei in $A(a,b)X$ / $a = x_1 + x_2 + b$ collisions. The four-body theory is an extension of the three-body approaches developed in the 80's by Ichimura, Autern and Vincent (IAV) [2], Udagawa and Tamura (UT) [3] and Hussein and McVoy (HM) [4]. We expect that experimentalists shall be encouraged to search for more information about the $x_1 + x_2$ system in the elastic breakup cross section and that also further developments and extensions of the surrogate method will be pursued, based on the inclusive non-elastic breakup part of the b spectrum.

1 Introduction

We report a recently developed theory to treat the inclusive breakup of three-fragment weakly bound nuclei [1]. The natural three fragment candidates for projectiles are Borromean, two-nucleon and unstable three-fragment halo nuclei. Our theory is an extension of the inclusive breakup models used for incomplete fusion reactions and in the surrogate method with two fragment projectiles. The three-body approach was developed in the 80's by Ichimura, Autern and Vincent (IAV) [2], Udagawa and Tamura (UT) [3] and Hussein and McVoy (HM) [4]. These three-body theories were extended to obtain the fragment yield in the reaction $A(a,b)X$, where the projectile is $a = x_1 + x_2 + b$. The inclusive breakup cross section is a sum of the four-body elastic breakup cross section plus the inclusive non-elastic breakup cross section that involves the absorption cross sections of the participant fragments, x_1 and x_2, and which generalizes the three-body formula reviewed in Austern, et al. [5]. The connection between IAV, UT and HM theories was shown in [6].

The new formula contains the four-body dynamics both in the elastic breakup cross section and in the inclusive non-elastic breakup ones. It can be applied to treat reactions with stable/unstable projectiles composed of three-fragments, like weakly bound Borromean and two-nucleon halo nuclei, where fingerprints of Efimov physics [7] and universality [8] could be revealed through the appearance of long-range correlations between the three-fragments. In addition one can seek the generalization of the surrogate method applied to reactions like (d,p) and (d,n) to the reactions (t,d) and (^{3}He,d), among others. Of particular interest is also the two-fragment correlation contribution to the elastic breakup cross section and the inclusive non-elastic breakup cross section through a three-body absorption interaction, which appears naturally in the four-body formulation. One could examine, for example, the relation between the pairing, in the case of two-neutron halos, and the three-body formulation of the projectile and its importance to the reaction mechanism underlying the inclusive breakup cross-sections.

We expect that these developments could stimulate experimentalists to search for more information about the $x_1 + x_2$ system in the elastic breakup cross section, and that also further theoretical developments and extensions of the surrogate method will be pursued, based on the inclusive non-elastic breakup part of the b spectrum.

⋆e-mail: hussein@if.usp.br
⋆⋆e-mail: lucasufsj@gmail.com
⋆⋆⋆e-mail: evchimanski@gmail.com
⋆⋆⋆⋆e-mail: brett@ita.br
†e-mail: tobias@ita.br

2 Inclusive Break-up theory

The many-body Hamiltonian for the $b + x_1 + x_2 + A$ system to be applied to derive the scattering dynamics of the three-fragment projectile is

$$H_{(b,x_1,x_2,A)} = T_b + T_{x_1} + T_{x_2} + V_{b,x_1} + V_{b,x_2} + V_{x_1,x_2} + h_A + T_A + V_{b,A} + V_{x_1,A} + V_{x_2,A} \quad (1)$$

where the kinetic energies are given by the T's and the microscopic Hamiltonian of the target nucleus is h_A. In the spectator approximation the microscopic potential $V_{b,A}$ is associated by standard methods in reaction theory to the optical potential U_b, and the target is considered infinitely massive, namely $T_A = 0$.

The inclusive breakup cross-section is an integral over the spectator fragment position $\mathbf{r}_b$ and a sum over the x_1 +

 DOI: 10.1051/epjconf/201716300024

$x_2 + A$ bound and scattering states. This leads to the b-spectrum and angular distribution

$$\frac{d^2\sigma_b}{dE_b d\Omega_b} = \frac{2\pi}{\hbar v}\rho_b(E_b)\sum_c \delta(E - E_b - E_c) \times \left|\left\langle \chi_b^{(-)}\Psi^c_{x_1 x_2 A}\left|V_{b,x_1} + V_{b,x_2} + V_{x_1,x_2}\right|\Xi\right\rangle\right|^2 \quad (2)$$

where $\Xi(\mathbf{r}_b, \mathbf{r}_{x_1}, \mathbf{r}_{x_2}, A)$ is the exact eigenstate of the $A + a$ many body Hamiltonian. The final state wave function is $\chi_b^{(-)}(\mathbf{k}_b, \mathbf{r}_b)\Psi^c_{(x_1 x_2 A)}$, where $\Psi^c_{(x_1 x_2 A)}$ runs over bound and continuum states. The density of b continuum states is $\rho_b(E_b) \equiv [d\mathbf{k}_b/(2\pi)^3]/[dE_b d\Omega_b] = \mu_b k_b/[(2\pi)^3\hbar^3]$, with μ_b the reduced mass of the $b + A$ system.

The connection with the four-body (4B) scattering problem is obtained by eliminating the target internal degrees of freedom, and by using the product approximation, $\Xi = \Psi_0^{4B(+)}\Phi_A$, where $\Psi_0^{4B(+)}$ is the exact 4B scattering wave function in the incident channel, and Φ_A is the ground state wave function of the target nucleus. Together with the 4B approximation, the energy conservation δ in Eq. (2) is associated with the imaginary part of an optical model Green's function operator

$$\mathrm{Im}G_X^{(+)} = -\pi\,\Omega_X^{(-)}\,\delta(E_x - H_0)\,(\Omega_X^{(-)})^\dagger - (G_X^{(+)})^\dagger\,W_X\,G_X^{(+)} \quad (3)$$

where the Möller operator is $\Omega_X^{(-)} = [1 + G_X^{(-)}(V_X)^\dagger]$. The imaginary part of the optical potential for particles x_1 and x_2 contains single fragment terms, W_{x_i}, and a three-body term. In addition, closure has to be used to perform the sum over c and general nuclear reaction theory to transform the microscopic interactions $V_{x_1,A}$ and $V_{x_2,A}$ into complex optical potentials U_{x_1} and U_{x_2} (see e.g. [5]).

2.1 Inclusive breakup cross-section

Following the above steps, the inclusive breakup cross section is reduced to a sum of two distinct terms, the elastic breakup and the non-elastic breakup cross sections

$$\frac{d^2\sigma_b}{dE_b d\Omega_b} = \frac{d^2\sigma_b^{EB}}{dE_b d\Omega_b} + \frac{d^2\sigma_b^{INEB}}{dE_b d\Omega_b}$$

with the 4B elastic breakup cross section contribution being

$$\frac{d^2\sigma_b^{EB}}{dE_b d\Omega_b} = \frac{2\pi}{\hbar v_a}\rho_b(E_b)\int \frac{dk_{x_1}}{(2\pi)^3}\frac{dk_{x_2}}{(2\pi)^3} \times |\langle\chi^{3B(-)}_{x_1,x_2}\chi_b^{(-)}|V_{bx_1} + V_{bx_2}|\Psi_0^{4B(+)}\rangle|^2 \times \delta(E - E_b - E_{(\mathbf{k}_{x_1},\mathbf{k}_{x_2})}) \quad (4)$$

where $\chi^{3B(-)}_{x_1,x_2}$ is the full scattering wave function of the two unobserved fragments in the final channel.

The inclusive cross-section for the inelastic breakup contains the optical potentials U_{x_1}, U_{x_2} and the fragment-fragment interaction V_{x_1,x_2} to all orders:

$$\frac{d^2\sigma_b^{INEB}}{dE_b d\Omega_b} = \frac{2}{\hbar v_a}\rho_b(E_b)\langle\hat\rho_{x_1,x_2}|W_{x_1} + W_{x_2} + W_{3B}|\hat\rho_{x_1,x_2}\rangle \quad (5)$$

where the source function

$$\hat\rho_X(\mathbf{r}_{x_1}, \mathbf{r}_{x_2}) = (\chi_b^{(-)}|\Psi_0^{4B(+)}\rangle = \int d\mathbf{r}_b\left[\chi_b^{(-)}(\mathbf{r}_b)\right]^\dagger \Psi_0^{4B(+)}(\mathbf{r}_b, \mathbf{r}_{x_1}, \mathbf{r}_{x_2}) \quad (6)$$

carries the full 4B dynamics in the optical model description. The imaginary parts of the optical potentials U_{x_1} and U_{x_2} are W_{x_1} and W_{x_2}, respectively. Notice the presence of the imaginary part of a three-body optical potential (W_{3B}), associated with inelastic excitations that are irreducible to single fragment inelastic processes. This will be discussed in more detail in a later sections. We point out that the 4B inclusive non-elastic breakup cross section differs significantly from the 3B Austern formula [5].

The inelastic breakup cross-section is a sum of three terms,

$$\frac{d^2\sigma_b^{INEB}}{dE_b d\Omega_b} = \rho_b(E_b)\frac{k_a}{E_a}\left[\frac{E_{x_1}}{k_{x_1}}\sigma_R^{x_1} + \frac{E_{x_2}}{k_{x_2}}\sigma_R^{x_2} + \frac{E_{CM}(x_1, x_2)}{(k_{x_1} + k_{x_2})}\sigma_R^{3B}\right] \quad (7)$$

where the approximation of a weakly bound projectile is used to write the kinetic energy: $E_{x_i,Lab} = E_{a,Lab}(M_{x_i}/M_a)$ with M_a and M_{x_i} being the mass numbers of the projectile and fragment, respectively. The single fragment inclusive cross-sections are

$$\sigma_R^{x_1} = \frac{k_{x_1}}{E_{x_1}}\langle\hat\rho_{x_1,x_2}|W_{x_1}|\hat\rho_{x_1,x_2}\rangle, \quad \sigma_R^{x_2} = \frac{k_{x_2}}{E_{x_2}}\langle\hat\rho_{x_1,x_2}|W_{x_2}|\hat\rho_{x_1,x_2}\rangle, \quad (8)$$

and the double fragment inclusive cross-section is

$$\sigma_R^{3B} = \frac{(k_{x_1} + k_{x_2})}{E_{CM}(x_1, x_2)}\langle\hat\rho_{x_1,x_2}|W_{3B}|\hat\rho_{x_1,x_2}\rangle \quad (9)$$

which represents the two-fragment irreducible inelastic processes.

We elaborate here on the physical interpretation of the different contributions to the inclusive inelastic cross-section. The absorption of the fragment x_i by the target is given by $\sigma_R^{x_i}$ $(i = 1, 2)$, in the case in which the other fragment x_j just scatters off the target through the optical potential $U_{x_j A}$. It is noteworthy to stress that $\sigma_R^{x_i}$ is different from the one in the 3B theory of the $b - x - A$ system. The cross-sections $\sigma_R^{x_1}$, $\sigma_R^{x_2}$ are related to the three-body IAV cross section through a convolution of the latter with the distorted wave densities $|\chi_{x_2}^{(+)}(\mathbf{r}_{x_2})|^2$, and $|\chi_{x_1}^{(+)}(\mathbf{r}_{x_1})|^2$ of the spectator fragments x_1 and x_2, respectively. If an eikonal-type approximation of the projectile distorted wave $\chi_a^{(+)} \times \Phi_a(r_b, r_{x_1}, r_{x_2})$ $(a \equiv x_1 + x_2 + b)$ is used in the 4B theory, the difference with the 3B formulation is clearly exposed. Thus, it is expected that the $\sigma_R^{x_1}$ in a (t,p) reaction will differ from the σ_R^{x} extracted in a (d,p) reaction. Furthermore, at low energies $\sigma_R^{x_1}$, $\sigma_R^{x_2}$, and σ_R^{3B} are associated with the formation of compound nuclei $A + x_1$, $A + x_2$ and $A + (x_1 + x_2)$.

2.2 Examples

The first example to which we apply the 4B formulation of the inelastic breakup cross-sections is the case of the reaction ^{9}Be ($= \alpha + \alpha + n$) + ^{208}Pb, where 4B-CDCC (four-body Continuum Discretized Coupled Channels) calculations were performed by Descouvemont and collaborators

[9] for the elastic scattering, elastic breakup, and total fusion cross-sections. However so far, CDCC cannot obtain the partial or incomplete fusion cross-sections. Despite this fact, we can point out that compound nuclei formation with α detected, namely $\alpha + {}^{208}$Pb = ^{210}Po, n + ^{208}Pb = ^{209}Pb, and α + n + ^{208}Pb = ^{211}Po, would be very interesting to observe, in order to gather information about the spectrum of α particles in the analysis of the inclusive cross sections. It would be interesting to investigate the properties of these compound nuclei experimentally, as they are formed in such a hybrid reaction.

A second example is a two-neutron Borromean nuclei. The inclusive α spectrum for the ^{6}He projectile accounts for the formation of n + A and 2n + A compound nuclei. However it would be experimentally difficult to distinguish these in CN decay.

The third example is a two-proton Borromean nuclei. One might consider the inclusive proton spectra in the breakup of ^{20}Mg + ^{208}Pb with the formation of ^{209}Bi, ^{226}U and ^{227}Np at different excitation energies. However, ^{20}Mg presents a very short lifetime and low intensity as a secondary beam.

3 IB Cross Sections: HM Source function

The source function computed with the Hussein and McVoy (HM) model [4] can be written as:

$$\langle \mathbf{r}_{x_1}, \mathbf{r}_{x_2} | \hat{\rho}^{4B}_{HM} \rangle = \hat{S}_b(\mathbf{r}_{x_1}, \mathbf{r}_{x_2}) \chi^{(+)}_{x_1}(\mathbf{r}_{x_1}) \chi^{(+)}_{x_2}(\mathbf{r}_{x_2}) \tag{10}$$

where the four-body scattering state is approximated by the product of the distorted waves of the three fragments and the projectile bound state wave function. The internal motion modifies the S-matrix of the b fragment as

$$\hat{S}_b(\mathbf{r}_{x_1}, \mathbf{r}_{x_2}) \equiv \int d\mathbf{r}_b \Phi_a(\mathbf{r}_{x_1}, \mathbf{r}_{x_2}, \mathbf{r}_b) \langle \chi^{(-)}_b | \chi^{(+)}_b \rangle (\mathbf{r}_b), \tag{11}$$

which should be compared to the 2B S-matrix element of b given by $S_{\mathbf{k}'_b, \mathbf{k}_b} = \int d\mathbf{r}_b \langle \chi^{(-)}_b | \chi^{(+)}_b \rangle (\mathbf{r}_b)$. In this approximation the inclusive cross-section in the 4B theory with $a = x_1 + x_2 + b$ is

$$\frac{E_{x_1}}{k_{x_1}} \sigma^{x_1}_R = \int d\mathbf{r}_{x_1} \int d\mathbf{r}_{x_2} |\hat{S}_b(\mathbf{r}_{x_1}, \mathbf{r}_{x_2})|^2 \times |\chi^{(+)}_{x_2}(\mathbf{r}_{x_2})|^2 \, W(\mathbf{r}_{x_1}) |\chi^{(+)}_{x_1}(\mathbf{r}_{x_1})|^2, \tag{12}$$

which should be contrasted with the cross-section in the 3B theory ($a = x + b$)

$$\frac{E_x}{k_x} \sigma^x_R = \int d\mathbf{r}_x |\hat{S}_b(\mathbf{r}_x)|^2 W(\mathbf{r}_x) |\chi^{(+)}_x(\mathbf{r}_x)|^2, \tag{13}$$

where $\hat{S}_b(\mathbf{r}_x) \equiv \int d\mathbf{r}_b \langle \chi^{(-)}_b | \chi^{(+)}_b \rangle (\mathbf{r}_b) \Phi_a(\mathbf{r}_b, \mathbf{r}_x)$. Comparing the cross-sections in Eqs. (12) and (13), one singles out the distorted wave $|\chi^{(+)}_{x_2}(\mathbf{r}_{x_2})|^2$, which damps the 4B cross-section with respect to the 3B one.

4 HM and 3B Glauber theory

We begin with the Glauber phase [10]

$$[\psi^{(-)}_{\vec{p}'_b}(\vec{r}'_b)]^* \psi^{(+)}_{\vec{p}_b}(\vec{r}_b) = \exp\left[-i \frac{m}{\hbar p'_b} \int_{-\infty}^{z'_b} dz' V\left(\sqrt{b'^2_b + z'^2}\right) - i \frac{m}{\hbar p_b} \int_{z_b}^{\infty} dz' V\left(\sqrt{b^2_b + z'^2}\right) - i \vec{q} \cdot \vec{r}_b \right], \tag{14}$$

where $\hbar \vec{q} = \vec{p}'_b - \vec{p}_b$, and $\vec{p}_b$ ($\vec{p}'_b$) is the incoming (outgoing) momentum of the spectator. We take the initial value of the momentum to be $\vec{p}_b = p_b \hat{k}$ and the final value to be $\vec{p}'_b = p'_b \sin\theta \hat{i} + p'_b \cos\theta \hat{k}$. The coordinates (b'_b, z'_b) can be written in terms of the coordinates (b_b, z_b) as

$$b'_b = b_b \cos\theta - z_b \sin\theta \tag{15}$$

$$z'_b = b_b \sin\theta + z_b \cos\theta . \tag{16}$$

Figure 1 illustrates the coordinates (b_b, z_b) and (b'_b, z'_b).

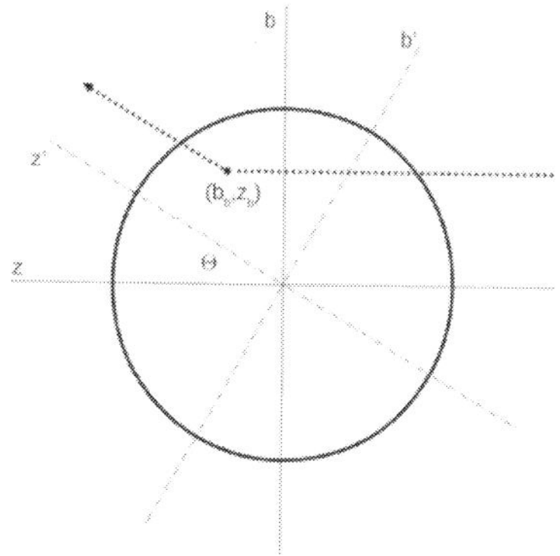

Figure 1. Appropriate kinematics and definitions of reference frame to compute the Glauber formula for the spectator distorted waves.

In order to get insight into the structure of the source function, we approximate the potential by a square well

$$V(r) = (V_0 - i W_0) \Theta(R - r) . \tag{17}$$

The spectator distorted density then becomes

$$[\psi^{(-)}_{\vec{p}'_b}(\vec{r}'_b)]^* \psi^{(+)}_{\vec{p}_b}(\vec{r}_b) = \exp\left[-i \frac{m}{\hbar^2} (V_0 + iW_0) \times \left(\frac{1}{p'_b} \left(\sqrt{R^2 - b'^2_b} - z'_b \right) + \frac{1}{p_b} \left(z_b + \sqrt{R^2 - b^2_b} \right) \right) - i \vec{q} \cdot \vec{r}_b \right] \tag{18}$$

for $b_b < R$ and $\exp[-i \vec{q} \cdot \vec{r}_b]$ for $b_b \geq R$.

The formula simplifies for the case of strong absorption, where the spectator distorted wave density becomes

$$[\psi^{(-)}_{\vec{p}'_b}(\vec{r}'_b)]^* \psi^{(+)}_{\vec{p}_b}(\vec{r}_b) = \Theta(b_b - R) \, \Theta(b'_b - R) \, e^{-i \vec{q} \cdot \vec{r}_b} . \tag{19}$$

Standard eikonal calculations ignore the difference $(b_b, z_b) \neq (b'_b, z'_b)$, inwhich case the distorted wave density can be simply written as

$$[\psi^{(-)}_{\vec{p}'_b}(\vec{r}'_b)]^* \psi^{(+)}_{\vec{p}_b}(\vec{r}_b) = \Theta(b_b - R) \, e^{-i \vec{q} \cdot \vec{r}_b} \tag{20}$$

DOI: 10.1051/epjconf/201716300024

The internal motion modified S-matrix of the b fragment within the approximations above is given by

$$\hat{S}_b(\mathbf{r}_{x_1}, \mathbf{r}_{x_2}) = \int d\mathbf{r}_b \Phi_a(\mathbf{r}_{x_1}, \mathbf{r}_{x_2}, \mathbf{r}_b) \Theta(b_b - R)\, e^{-i\vec{q}\cdot\vec{r}_b} \tag{21}$$

which produces a long range correlation between the fragments x_1 and x_2 for weakly bound projectiles, due to the long tail of the wave function penetrating in the classically forbidden region. This give us a taste of the possible rôle of Efimov physics, for example when the projectile is a two-neutron halo s-wave state such as ^{11}Li, in creating a long range correlation between the two neutrons. The actual set of coordinates to compute (21) can be visualized in Fig. 2.

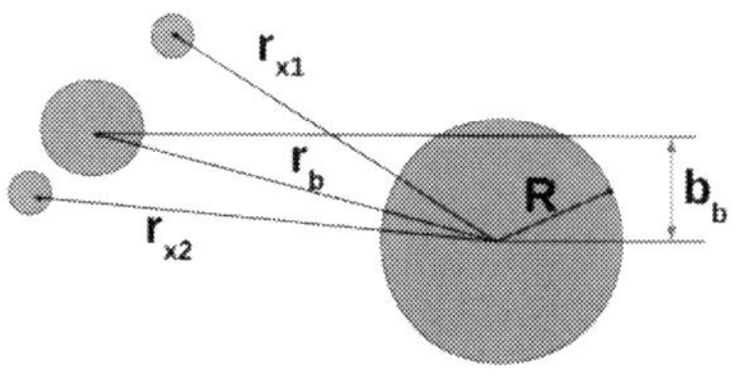

Figure 2. Definition of the coordinates of the fragments in configuration space. The target is represented by the largest circle, the unobserved fragments x_1 and x_2 are represented by the smallest circles, and the remaining circle represents the detected particle b.

Figure 2 also supplies a glimpse of the effect of an extended weakly bound system on the correlation between the two fragments x_1 and x_2. The formula (21) restricts b to be outside the target absorptive region. However the tail of the projectile wave function can provide a long range correlation between the two fragments and the target.

It is instructive to calculate $\hat{S}_b(\mathbf{r}_{x_1}, \mathbf{r}_{x_2})$ in the case of two-fragment projectiles, such as the deuteron in the (d, p) reaction. Here $b = p$ and $x = n$, and we obtain, within the same strong absorption (black disk) eikonal approximation,

$$\hat{S}_p(\mathbf{r}_n) = \int d\mathbf{r}_p \Phi_d(\mathbf{r}_n, \mathbf{r}_p) \Theta\left(b_p - R\right) e^{-i\vec{q}\cdot\vec{r}_p} \tag{22}$$

which corresponds to an incomplete Fourier transform involving $\mathbf{r}_p$ and q.

In the high energy regime, the cross section can be represented as an integral over impact parameter. Since Eq. (13),

$$\frac{E_x}{k_x}\sigma_R^x = \int d\mathbf{r}_x |\hat{S}_b(\mathbf{r}_x)|^2 \left[W(\mathbf{r}_x)|\chi_x^{(+)}(\mathbf{r}_x)|^2\right], \tag{23}$$

contains the integrand of a b-integral of the reaction cross section of x, $[W(\mathbf{r}_x)|\chi_x^{(+)}(\mathbf{r}_x)|^2]$, which can be replaced by $[1 - |S_x(b_x)|^2]$, and the factor $|\hat{S}_b(\mathbf{r}_x)|^2$ is basically the survival probability of the observed fragment, b, one can write, after rearranging terms,

$$\sigma_R^x = 2\pi \int db \langle\Phi_a||S_b(b)|^2 \left[1 - |S_x(b)|^2\right] \Phi_a\rangle \tag{24}$$

One can say that the above equation is the high energy eikonal limit of IAV or the HM formulae. At high energy, the source function Eq.(6) for two-fragment projectiles reduces to $\hat{\rho}_x^{IAV} = \hat{\rho}_x^{HM} = (\chi_b^{(-)}|\chi_b^{(+)}\chi_x^{(+)}\Phi_a\rangle = \langle\chi_b^{(-)}|\chi_b^{(+)}\rangle|\chi_x^{(+)}\Phi_a\rangle$. When used in the expression for the cross section Eq. (13) or in its original form,

$$\frac{d^2\sigma_b^{INEB}}{dE_b d\Omega_b} = \frac{2}{\hbar v_a}\rho_b(E_b)\langle\hat{\rho}_x^{HM}|W_x|\hat{\rho}_x^{HM}\rangle, \tag{25}$$

it is a simple exercise to reduce the expression to the Glauber cross section above. The above form of the cross section has been extensively used by [11, 12] to calculate one nucleon stripping and pickup reactions. These authors, take $\hat{R}_1 = |\hat{S}_b(\mathbf{r}_x)|^2 \left[W(\mathbf{r}_x)|\chi_x^{(+)}(\mathbf{r}_x)|^2\right]$ as an operator and calculate the expectation value in the state from which the nucleon is removed or added, $\langle\Phi_i|\hat{R}_1|\Phi_i\rangle$. This is then used to represent the cross section in a particularly transparent form. The partial cross section for removal of a nucleon, from a single-particle configuration j^π populating the residue final state α with excitation energy $E_\alpha^\star$, is calculated as

$$\sigma_\alpha = \left(\frac{A}{A-1}\right)^N C^2 S(\alpha, j^\pi)\langle\Phi_{j^\pi}|\hat{R}_1|\Phi_{j^\pi}\rangle \tag{26}$$

where $S_\alpha^\star = S_{n,p} + E_\alpha^\star$ is the effective separation energy for the final state α and $S_{n,p}$ is the ground-state to ground-state nucleon separation energy. Here the factor N, in the A-dependent center-of-mass correction factor that multiplies the shell-model spectroscopic factors $C^2S(\alpha, j^\pi)$, is the number of oscillator quanta associated with the major shell of the removed particle [11]. The cross section for the removal of a nucleon from the j^π single particle shell is then the sum over all bound final states, α of the residual nucleus $A - 1$.

Here, the radioactive projectile is now denoted by A. In this manner, useful nuclear structure information contained in the spectroscopic factor of the fragment in the projectile A, $C^2S(\alpha, j^\pi)$, can be extracted. Further, within the four-body theory, the practitioners of the eikonal theory extend their application to two nucleon removal or pickup. They use the following form of the reaction factor $\langle\Phi_i|\hat{R}_2|\Phi_i\rangle = \langle\Phi_i||S_b|^2(1 - |S_{x_1}|^2)(1 - |S_{x_2}|^2)|\Phi_i\rangle$. This formula should come out from σ_R^{3B} of Eq. (9) of our theory in the high energy limit. However, there are several shortcomings in the above model, as there is no reference to the individual absorptions of x_1 and x_2. Further, the model above misses the correlation between the two interacting fragments. Currently we are investigating this point.

5 Double fragment inelastic processes

The inclusive cross-section expressed by σ_R^{3B} given by Eq. (9) is new and corresponds to genuine three-body absorption processes to inelastic channels. This part of the cross-section is associated with a three-body optical potential U_{3B} depending on the relative coordinates of the fragments x_1, x_2 and the target, and cannot be reduced to connected terms of the 3B transition matrix with two-body potentials in different subsystems and the target in the ground state.

DOI: 10.1051/epjconf/201716300024

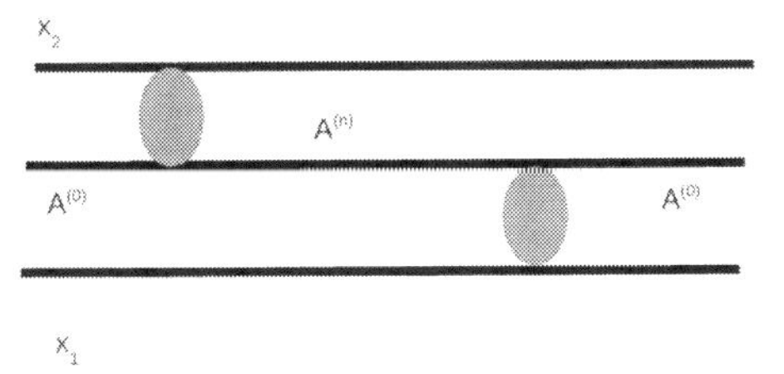

Figure 3. Three-body optical potential U_{3B}. Excitation of the target by particle x_1 and de-excitation by x_2.

The structure of $W_{3B} = \mathrm{Im}[U_{3B}]$ results from conventional nuclear reaction theory. It includes, for example, processes like the virtual excitation of the target by one of the fragments and its virtual de-excitation by the other, as well as other 3B processes irreducible to the re-scattering terms with the target in the ground state and with the final result being the full capture, or complete fusion, of both fragments. This is illustrated schematically in figure 3. If a resonant process exists, such as the excitation of giant pairing vibrations from the transfer of two neutrons from the projectile, it should furnish a large contribution to the optical three-body potential at the resonance energy.

5.1 Hint on the 3B optical potential U_{3B}

The schematic figure 3 furnishes a path to formally build the the three-body optical potential by using projection operators out of the target ground state,

$$U_{3B} = PV_{x_1A}Q(QG_{xA}(E_x)Q)QV_{x_2A}P + PV_{x_2A}Q(QG_{xA}(E_x)Q)QV_{x_1A}P, \quad (27)$$

where $E_x = E_{x_1} + E_{x_2}$ and the Q-projected 3B Green's function of the $x_1 + x_2 + A$ system is

$$QG_{xA}Q = [E_x - QH_0Q + Q\,V_{xA}\,PG_0P\,V_{xA}\,Q + i\varepsilon]^{-1} \quad (28)$$

with $V_{xA} \equiv V_{x_1A} + V_{x_2A}$. The imaginary part of the Q-projected Green's function of the $x_1x_2A \equiv xA$ system includes the virtual propagation in all states except for the elastic channel,

$$Im[QG_{xA}Q] = -\pi\Omega_Q^{(-)}Q\delta(E_x - H_0)Q(\Omega_Q^{(-)})^\dagger + (QG_{xA}Q)^\dagger Q\,V_{xA}\,P\delta(E_x - H_0)P\,V_{xA}\,QG_{xA}Q \quad (29)$$

Then, again using standard techniques, the imaginary part of U_{3B} can be isolated in (27) and can be written as

$$\begin{aligned} W_{3B} &= U_{3B}^\dagger - U_{3B} = \\ &= \pi[PV_{x_1A}Q\Omega_Q^{(-)}Q\delta(E_x - H_0)Q(\Omega_Q^{(-)})^\dagger QV_{x_2A}P \\ &\quad + PV_{x_1A}Q(QG_{xA}Q)^\dagger Q\,V_{xA}\,P\,\delta(E_x - H_0)\,P \\ &\quad \times P\,V_{xA}\,Q(QG_{xA}Q)QV_{xA}P] + (x_1 \leftrightarrow x_2) \end{aligned} \quad (30)$$

where one can identify the virtual propagation of the x_1x_2A system through the inelastic channels in the Q-space.

6 Beyond HM, UT, and IAV formulas

The key points for going beyond the IAV formula for the inclusive inelastic cross-section are to include in the source term the full four-body dynamics and the three-body optical potential, by computing $|\Psi_0^{4B(+)}\rangle$ without approximating the four-body problem. For the time being we will be content to generalize the three-body IAV formula to the four-body case and present the CFH formula given in [1].

The generalization given by CFH of the IAV formula to determine the inclusive cross-section observed by the detection of the spectator particle b, contains HM and UT like terms. The steps to derive a 4B version of the IAV formula follow [6] and we write

$$|\Psi_0^{4B(+)}\rangle = G_{b,x_1,x_2,A}^{(+)}V_x|\Psi_0^{4B(+)}\rangle \approx G_{b,x_1,x_2,A}^{(+)}\,V_x\,|\chi_a^{(+)}, \Phi_a\rangle \quad (31)$$

where $V_x = V_{b\,x_1} + V_{b\,x_2}$ is the interaction between the detected fragment and the other two, $|\Phi_a\rangle$ the projectile internal wave function and $\chi_a^{(+)}$ the distorted wave of the center of mass. The four body Green's function is

$$G_{b,x_1,x_2,A}^{(+)} = [E - T_b - T_{x_1} - T_{x_2} - V_{x_1,x_2} - U_b - U_{x_1} - U_{x_2} + i\varepsilon]^{-1}. \quad (32)$$

where we have not explicitly included U_{3B}, which should also contribute to the Green's function.

The generalization of the IAV three-body source function for the four-body case is:

$$\hat{\rho}_{x_1,x_2}^{CFH} = \langle\chi_b^{(-)}|\Psi_0^{4B(+)}\rangle = \langle\chi_b^{(-)}|G_{b,x_1,x_2,A}^{(+)}\,V_x\,|\chi_a^{(+)}\Phi_a\rangle \quad (33)$$

By manipulating the above formula in analogy to the development provided in Ref. [6], one gets the 4B extension of the 3B Ichimura-Austern-Vincent as

$$\hat{\rho}_{x_1,x_2}^{CFH} = \hat{\rho}_{x_1,x_2}^{UT} + \hat{\rho}_{x_1,x_2}^{HM} \quad (34)$$

where the 4B Hussein-McVoy (HM) term is

$$\hat{\rho}_{x_1,x_2}^{HM} = \langle\chi_b^{(-)}|\chi_a^{(+)}\Phi_a\rangle \quad (35)$$

and the 4B Udagawa-Tamura (UT) term is

$$\hat{\rho}_{x_1,x_2}^{UT} \equiv G_{x_1,x_2,A}^{(+)}\langle\chi_b^{(-)}|\,[U_b + U_{x_1} + U_{x_2} - U_a]\,|\chi_a^{(+)}\Phi_a\rangle\,. \quad (36)$$

where U_a is the optical potential of the projectile which generates its distorted wave, $\chi_a^{(+)}$.
The Green's function

$$G_{x_1,x_2,A}^{(+)} = [E - E_b - T_{x_1} - T_{x_2} - V_{x_1,x_2} - U_{x_1} - U_{x_2} + i\varepsilon]^{-1} \quad (37)$$

should also contain in principle the three-body optical potential. Further studies of this particular point will be developed.

7 Inclusive breakup and Efimov Physics

The inclusive cross-section formulae for the elastic breakup (4) and for the inelastic process (5) applied to a three-body halo nuclei, e.g., a weakly bound neutron-neutron-core nuclei close to the drip-line, contains a long-range correlation between the two neutrons and the core

DOI: 10.1051/epjconf/201716300024

b, through the extended wave function of the halo and the scattering wave $\chi_{x_1}^{(+)}(\mathbf{r}_{x_1})\chi_{x_2}^{(+)}(\mathbf{r}_{x_2})\chi_{x_b}^{(+)}(\mathbf{r}_{x_b})$, which appears both in (4) and in the source function in (5), when such an approximation is made.

However, we expect that interesting physics remains beyond such an approximation if the 4B scattering wave function, $\Psi_0^{4B(+)}(\mathbf{r}_b, \mathbf{r}_{x_1}, \mathbf{r}_{x_2})$ is taken in full. It should contain dynamics of the continuum of the neutron-neutron-core system beyond the bound halo state. The reaction mechanism now includes Efimov physics [7] and with that the long range correlation between the halo fragments, which could be influenced by the presence of Efimov bound, virtual or resonant states [8] (see also [13] for a discussion of observing Efimov states in halo reactions). Indeed, the existence of Borromean Efimov states was observed about a decade ago [14] in the resonant three-body recombination of cold cesium atoms in magneto-optical traps.

8 Digression on two-fragment projectile inclusive breakup cross section

It is important to remind the reader of the advances made in the application of the inclusive breakup theory of two-fragment projectiles, as was originally developed in [2–5]. Quite recently this theory was applied to the (d,p) reaction [19–21, 23] as a mean to test the validity of the Surrogate Method [24–26], employed to extract neutron capture and fission cross sections of actinide target nuclei, such as ^{238}U, ^{232}Th, of importance for the development of next generation fast breeder reactors. These neutron capture reactions on other targets are also important for the study of element formation following supernova explosion through the astrophysical s-process. The general conclusion of [19–21, 23] was that the Surrogate Method, the extraction of the cross section for the formation and decay of the compound nucleus of the $(n + A)$ subsystem, as well as the total capture cross section (the cross section for the formation of the compound nucleus), is justified as long as the direct part of the cross section is calculated and subtracted from the inclusive breakup total reaction cross section $\sigma_R^{(n+A)}$. Another topic of importance is the Trojan Horse Method [27–30] used to extract from the inclusive breakup data a direct reaction of interest to nuclear astrophysics following nova explosion, which otherwise would be difficult to measure in the laboratory. The THM uses the advantage that the surrogate charged fragment x is a bare nucleus with no electrons surrounding it. This avoids the problem of electron screening. Further, since this fragment is brought by the primary projectile to the region of the target, the Coulomb barrier is already surmounted, and accordingly no hindrance due to barrier penetration and tunneling is present in the x induced reaction. The two methods, SM, and THM, are two pieces of the same quantity, namely $\sigma_R^{(x+A)}$ in the breakup reaction, $a + A \rightarrow b + (x + A)$. Thus the THM can be easily justified within the theory alluded to above as being a process contributing to the direct part of $\sigma_R^{(x+A)}$. As a final remark, if the surrogate fragment x is a neutron then the SM and the THM are the same!

It would certainly be important to extend these studies to the case of the three-fragment projectiles discussed in this contribution. In particular, it would be particularly interesting to extend the THM to cases involving three-fragment projectiles, where three types of reactions can be extracted, $x_1 + A \rightarrow y_1 + B_1$, $x_2 + A \rightarrow y_2 + B_2$, and $x_1 + x_2 + A \rightarrow y_3 + B_3$. Work along these lines is in progress.

9 Conclusions and Perspectives

We have reported in this contribution that the general structure of the CFH cross section in the DWBA limit is similar in structure to the 3B one , with the full post form (or four-body IAV), which can be written as the sum of the prior four-body UT cross section plus the four-body HM one plus an interference term when the 4B source term detailed in Eq. (33) is used to compute the inelastic part of the inclusive cross-section. The major difference between the 4B and 3B cases resides in the structure of the reaction cross sections for the absorption of one of the interacting fragments, which we find to be damped by the absorption effect of the other fragment. It is expected that in a (t,p) reaction, an absorption cross section of the $n + A$ subsystem would be smaller than the corresponding one in a (d,p) reaction. Another important new feature is the 3B absorption from a three-body optical potential, the formal structure of which we have sketched.

The perspectives of our study are in the use of the Faddeev-Yakubovski equations [15, 16] to develop an expansion of the four-body wave function, to attempt to go beyond the CFH/IAV formulas. This effort could also be accomplished by building the four-body CDCC wave function to compute the source term. In addition, we can ask about the dynamics that are built in the three-body optical potential when giant pairing vibrations (GPV) are possible [17, 18]. These collective states involve the coherent excitation of particle-particle pairs, in complete analogy to the coherent excitation of particle-hole pairs that constitutes the microscopic foundation of multipole giant resonances. For example, the GPV opens interesting prospects for nuclear structure studies in reactions of the type (t, p), 2n Borromean cases such as (^{6}He, ^{4}He), (^{11}Li, ^{9}Li), (^{14}Be, ^{12}Be), (^{22}C, ^{20}C), and the 2p halo cases, (^{17}Ne, ^{15}O), and (^{20}Mg, ^{18}Ne). The theory we have developed is an appropriate framework to study these types of collective nuclear excitations, associated with pairing correlations in the target.

From the point of view of a two-neutron s-wave dominated halo target our theory naturally leads to an inquiry about the rôle of Efimov physics in inclusive breakup cross-section, both in the elastic and inelastic contributions, which should have its place beyond the CFH formula.

Further, the theory can be extended to allow for inclusive reactions where two or more fragments are detected. We expect that the developments discussed in our recent work should stimulate experimental and theoretical works to seek more information about the $x_1 + x_2$ system in the elastic and inelastic breakup cross sections, and in particular to theorists to extend the

Surrogate Method and the Trojan Horse Method, both based on the inclusive non-elastic breakup part of the *b* spectrum in a two-fragment projectile induced reaction, to three-fragment projectiles.

Acknowledgements. This work was partly supported by the Brazilian agencies, Fundação de Amparo à Pesquisa do Estado de São Paulo (FAPESP), the Conselho Nacional de Desenvolvimento Científico e Tecnológico (CNPq). MSH also acknowledges a Senior Visiting Professorship granted by the Coordenação de Aperfeiçoamento de Pessoal de Nível Superior (CAPES), through the CAPES/ITA-PVS program.

References

[1] B. V. Carlson, T. Frederico, M. S. Hussein, Phys. Lett. B **767**, 53 (2017).

[2] M. Ichimura, N. Austern, and C. M. Vincent, Phys. Rev. C **32**, 431 (1985).

[3] T. Udagawa and T. Tamura, Phys. Rev. C **24**, 1348 (1981).

[4] M. S. Hussein and K. W. McVoy, Nucl. Phys. A **445**, 124 (1985)

[5] N. Austern, Y. Iseri, M. Kamimura, M. Kawai, G. Rawitscher, and M. Yahiro, Phys. Rep. **154**, 125 (1987).

[6] M. S. Hussein, T. Frederico, and R. C. Mastroleo, Nucl. Phys. A **511**, 269 (1990).

[7] V. Efimov, Phys. Lett. B **33**, 563 (1970) 563; Few-Body Syst. **51**, 79 (2011).

[8] T. Frederico, M.T. Yamashita, A. Delfino, L. Tomio, Prog. Part. Nucl. Phys. **67**, 939 (2012).

[9] P. Descouvemont, T. Druet, L. F. Canto, and M. S. Hussein Phys. Rev. C **91**, 024606 (2015).

[10] L. F. Canto and M. S. Hussein, *Scattering Theory of Molecules, Atoms and Nuclei*, World Scientific , 2013

[11] P. G. Hansen and J. A. Tostevin, Annu. Rev. Nucl. Part. Sci. **53**, 219 (2003).

[12] H. Esbensen and G. F. Bertsch Phys. Rev. C **64**, 014608 (2001)

[13] A. O. Macchiavelli, Few-Body Syst. **56**, 773 (2015)

[14] T. Kraemer, et al., Nature **440**, 315 (2006).

[15] L.D.Faddeev, Zh. Eksp. Teor. Fiz. **39**, 1459 (1960) [Sov. Phys. JETP **12**, 1014 (1961)].

[16] O. A. Yakubovsky, Yad. Fiz. **5**, 1312 (1967) [Sov. J. Nucl. Phys. **5**, 1312 (1967)].

[17] R. A. Broglia and D. R. Bes, Phys. Lett. B **69**, 129 (1977).

[18] F. Cappuzzello, D. Carbone, M. Cavallaro, M. Bond, C. Agodi, F. Azaiez, A. Bonaccorso, A. Cunsolo, L. Fortunato, A. Foti, S. Franchoo, E. Khan, R. Linares, J. Lubian, J. A. Scarpaci and A. Vitturi, Nature Communications **6**, 6743 (2015).

[19] Q. Ducasse et al., arXiv:1512.06334 [nucl-ex]

[20] G. Potel, F. M. Nunes, and I. J. Thompson, Phys. Rev. C **92**, 034611 (2015).

[21] J. Lei and A. M. Moro, Phys. Rev. C **92**, 044616 (2015); J. Lei and A. M. Moro, C **92**, 061602(R) (2015).

[22] G. Potel et al. To be published in the European Physical Journal A (2017)

[23] B.V. Carlson, R. Capote, M. Sin, Few-Body Syst. **57**, 307 (2016).

[24] Jutta E. Escher and Frank S. Dietrich, Phys. Rev. C **74**, 054601 (2006)

[25] Jutta E. Escher, Jason T. Burke, Frank S. Dietrich, Nicholas D. Scielzo, Ian J. Thompson, and Walid Younes, Rev. Mod. Phys., **84**, 353 (2012)

[26] J. E. Escher, A. P. Tonchev, J. T. Burke, P. Bedrossian, R. J. Casperson, N. Cooper,, R. O. Hughes, P. Humby, R. S. Ilieva,, S. Ota1,, N. Pietralla,, N. D. Scielzo, and V.Werner, EPJ Web of Conferences **122**, 12001 (2016)

[27] S. Typel and G. Baur, Ann. Phys. (NY) **305**, 228 (2003)

[28] C. Spitaleri et al., Phys. of Atomic Nuclei **74**, 1725 (2011)

[29] A. Tumino et al., Few Body Systems, (2012) DOI 10.1007/s00601-012-0407-1

[30] R. Tribble et al., Rep. Prog. Pays. **77**, 106901(2014)

DOI: 10.1051/epjconf/201716300025

Universal damping mechanism of quantum vibrations in deep sub-barrier fusion reactions

Takatoshi Ichikawa[1,*]

[1] Center for Nuclear Study, the University of Tokyo, Hongo, Tokyo 113-0033, Japan

Abstract. I demonstrate that the coordinate dependence of the coupling strengths of colliding nuclei is important for describing the deep sub-barrier fusion hindrance. To this end, I firstly show the performance of an extended coupled-channel model by phenomenologically introducing a damping factor in the coupling potential. The damping factor stimulates the damping of quantum vibrations occurring near the touching point of colliding nuclei and introduce a coordinate dependence in the coupling strengths. Next, I directly show the coordinate dependence of the transition strengths of colliding nuclei by microscopically calculating excited states using the random phase approximation method. The obtained transition strengths of colliding nuclei as a function of the center-of-mass distance strongly correlate with the damping factor that reproduces very well the fusion hindrance. This is a direct justification for the concept of the coordinate-dependent coupling strengths. Finally, I conclude that the damping of quantum vibrations near the touching point is the universal mechanism for the deep sub-barrier fusion hindrance.

1 Introduction

At extremely low incident energies, called the deep sub-barrier energies, the steep fall-offs of fusion cross sections compared to the estimations of the standard coupled-channel (CC) model have been observed in a wide range of mass systems (see Ref. [1] for details). These steep falloff phenomena are often referred to as the fusion hindrance. An important observation for understanding this fusion hindrance is that the potential energy at the touching point of the colliding nuclei strongly correlates with the threshold incident energy for the emergence of the fusion hindrance. That is, the fusion hindrance would be associated with dynamics in the overlap region of the two colliding nuclei (see Fig. 1 in Ref. [2]).

To describe these fusion hindrance phenomena, recent theoretical works have been mainly devoted to how to extend the standard CC model in this overlap region. Two different models based on assumptions opposite to each other have been proposed [1]. One is the sudden approach proposed by Mişicu and Esbensen [4, 5]. They constructed a heavy ion-ion potential to stimulate the effect of nuclear incompressibility, which gave a shallow potential pocket. Recently, Simenel *et al.* used the density constrained Hartree-Fock method to demonstrate that Pauli repulsion reduces the tunneling probabilities at deep-sub-barrier energies [3].

The other is the adiabatic approach proposed by Ichikawa *et al.* [6]. In this approach, neck formations between the colliding nuclei are taken into account in the overlap region. Based on this picture, the sudden and adiabatic processes were smoothly jointed by phenomenologically introducing the damping factor in the coupling form factor [7]. That is, a coordinate dependence was introduced in the coupling potential. Later, I showed that the physical origin of the damping factor is the damping of quantum vibrations of the target and the projectile near the touching point using the random-phase approximation (RPA) method for not only the light mass-symmetric ^{16}O + ^{16}O and ^{40}Ca + ^{40}Ca systems [8] but also the mass-asymmetric ^{16}O + ^{208}Pb system [9].

In this paper, I demonstrate the importance of the coordinate dependence of the coupling strength using a phenomenological model and examine this assumption by a microscopic RPA calculation. I show that the damping of the quantum vibrations near the touching point occurs, which is a universal mechanism in the deep sub-barrier fusions and is responsible for the fusion hindrance.

In the standard CC model (and the sudden model), the total wave functions are expanded by the basis of isolated nuclei. Then, important input parameters are the excitation energy and the coupling strength for each excitation. The excitation energy is usually taken from the experimental data and the coupling strength is estimated from measured transition probability. However, in the CC calculations, it is usually assumed that these values are unchanged during the fusion process, even when the two nuclei strongly overlap with each other. This seems to be an unrealistic assumption. As shown in Ref. [8], the single-particle wave functions are drastically changed by level repulsions, which are associated with the neck formations.

So far, many theoretical works for deep sub-barrier fusions have paid attention only how to construct the inter-

* e-mail: ichikawa@cns.s.u-tokyo.ac.jp

DOI: 10.1051/epjconf/201716300025

nuclear potential in the overlap region. However, in this paper, I would like to stress that the treatment of the coupling potential in the overlap region is also much important for describing the deep sub-barrier fusion hindrance.

2 Coordinate-dependent coupling strength phenomenologically introduced by the damping factor

Before introducing the coordinate dependence in the coupling strength in the standard CC model, I first briefly describe the standard CC model (for details see Refs. [10, 11, 14]).

For heavy-ion fusion reactions, the no-Coriolis approximation is often used [11, 14]. Then, the CC equations [11, 14] are given by

$$\left[-\frac{\hbar^2}{2\mu}\frac{d^2}{dr^2}+\frac{J(J+1)\hbar^2}{2\mu r^2}+V^{(0)}(r)+\epsilon_n-E\right]u_n(r)+\sum_m V_{nm}(r)u_m(r)=0, \quad (1)$$

where r is the radial component of the relative motion coordinate, μ is the reduced mass, E is the incident energy in the center-of-mass frame, J is the total angular momentum, and ϵ_n is the excitation energy of the n-th channel. A bare nuclear potential $V^{(0)}$ consisting of the Coulomb and nuclear interactions is given by $V^{(0)}(r) = Z_T Z_P e^2/r + V_N^{(0)}(r)$, where Z_T and Z_P denote the proton number of the target and projectile, respectively. The matrix elements of the coupling Hamiltonian V_{nm} are calculated with the collective model including the Coulomb and nuclear components.

In coupling matrix elements, I consider only vibrational couplings in this paper. The nuclear coupling Hamiltonian can be generated by changing the target radius in the nuclear potential of $V_N^{(0)}$ to a dynamical operator $R_0 \to R_0 + \hat{O}_\lambda$. Therefore, the nuclear coupling Hamiltonian is given by $V_N(r,\hat{O}_\lambda) = V_N^{(0)}(r-\hat{O}_\lambda)$ [11, 14]. For the vibrational coupling, the operator $\hat{O}_\lambda$ is given by $\hat{O}_\lambda = \beta_\lambda/\sqrt{4\pi}\cdot R_T(\alpha_{\lambda 0}^\dagger + \alpha_{\lambda 0})$, where $\alpha_{\lambda 0}^\dagger$ and $\alpha_{\lambda 0}$ are the creation and annihilation operators of the phonons, respectively, and R_T is the radius of the target nucleus. Here, the eigenvalues λ_α and the eigenvectors $|\alpha\rangle$ of $\hat{O}_\lambda$ are given by $\hat{O}_\lambda|\alpha\rangle = \lambda_\alpha|\alpha\rangle$. The deformation parameter β_λ is an input parameter and can be estimated from an experimental transition probability $B(E\lambda)$ which is given by

$$\beta_\lambda = \frac{4\pi}{3Z_T R_T^\lambda}\sqrt{\frac{B(E\lambda)\uparrow}{e^2}}, \quad (2)$$

where e is the elementary charge. The matrix elements of the nuclear coupling Hamiltonian are expanded by the eigenvalues and eigenvectors [10, 11, 14], and thus are defined as

$$\begin{aligned} V_{nm}^{(N)} &= \langle n|V_N(r,\hat{O}_\lambda)|m\rangle - C_0(r) \\ &= \sum_\alpha \langle n|\alpha\rangle\langle\alpha|m\rangle V_N(r,\lambda_\alpha) - C_0(r), \end{aligned} \quad (3)$$

where C_0 is the coupling constant given by $C_0(r) = \langle 0|V_N(r,\hat{O}_\lambda)|0\rangle$.

A main idea to include the coordinate dependence in the coupling strength is the introduction of the damping factor $\Phi(r,\lambda_\alpha)$ in the coupling form factor. I employ the following form for the nuclear coupling potential with respect to the eigen channel α,

$$V_N(r,\lambda_\alpha) = V_N^{(0)}(r) + \left[-\frac{dV_N^{(0)}}{dr}\lambda_\alpha + \frac{1}{2}\frac{d^2V_N^{(0)}}{dr^2}\lambda_\alpha^2\right]\Phi(r,\lambda_\alpha). \quad (4)$$

Here,

$$\Phi(r,\lambda_\alpha) = \begin{cases} 1 & r > R_d + \lambda_\alpha \\ & \text{(Two-body region),} \\ e^{-(r-R_d-\lambda_\alpha)^2/2a_d^2} & \text{Otherwise} \\ & \text{(Overlap region),} \end{cases} \quad (5)$$

where R_d is the damping dependent touching distance between the target and the projectile defined by $R_d = r_d(A_T^{1/3} + A_P^{1/3})$. The coefficients r_d and a_d are the damping radius and diffuseness parameters. The damping factor Φ represents the physical process for gradually transitioning from sudden to adiabatic approximations by diminishing the excitation strengths of the target and/or projectile vibrational states after the two colliding nuclei overlap one another. An important modification is that the touching point in the damping factor depends on λ_α, namely, the excitation strength begins to reduce at different distances in each eigen channel.

Although I attempted to apply several functional forms to the damping factor, I found that the form of Eq. (5) can well reproduce various experimental data. Later, I will show the justification of this functional form by a direct calculation of the transition strength using the RPA method.

After these modifications, in the two-body region ($r > R_d + \lambda_\alpha$), the standard CC equations of Eq. (1) work well because $\Phi = 1$. Conversely, in the overlapping region ($r < R_d + \lambda_\alpha$), the coupling matrix elements become $V_{nm} \to 0$ because $\Phi \to 0$. Then, the standard CC equations of Eq. (1) are close to the one-dimensional Schrödinger-like equations given by

$$\left[-\frac{\hbar^2}{2\mu}\frac{d^2}{dr^2}+\frac{J(J+1)\hbar^2}{2\mu r^2}+V^{(0)}(r)+\epsilon_n-E\right]u_n(r)=0 \quad (r \ll R_d+\lambda_\alpha). \quad (6)$$

If an adiabatic one-body potential $V_{\rm 1bd}^{(0)}$ is substituted to $V^{(0)}$ in Eq. (6), one can avoid double counting of the CC effects and correctly estimate the tunneling probability in the one-body process. Subsequently, all the physical quantities are smoothly joined.

An example of the application for this model is a calculation for the ^{16}O+^{208}Pb system. I incorporate the damping factor in the computer code `ccfull` [11]. For the ^{16}O+^{208}Pb system, I include the coupling to the low-lying 3^- phonon states and the double-octupole phonon excitations for both the ^{16}O and ^{208}Pb nuclei. In the calculations,

I adopt the Yukawa-plus-exponential (YPE) potential as a basic heavy ion-ion potential $V^{(0)}(r)$. In order to fit the experimental fusion cross sections, the radius parameter r_0 is adjusted to be 1.2 fm for ^{16}O+^{208}Pb system. For the damping factor, I use r_d = 1.255 fm and a_d = 1.14 fm. The details of all calculations and parameters are given in Ref. [2].

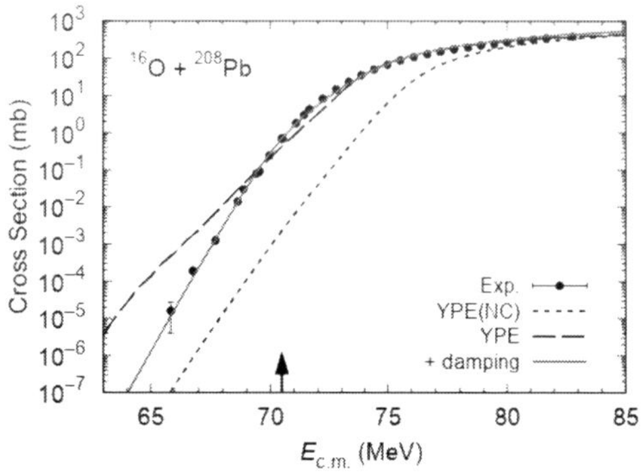

Figure 1. Calculated and experimental fusion cross sections versus incident energies for the ^{16}O + ^{208}Pb system. The solid circles denote the experimental data [12, 13]. The solid and long-dashed lines indicate the calculated results with and without the damping factor, respectively. The short-dashed lines indicate the calculated results of no coupling. The arrow indicates the potential energy at the touching point V_{Touch}.

Figure 1 shows the fusion cross sections thus obtained. The fusion cross sections obtained with the damping factor are in good agreement with the experimental data (see the solid line). We see that drastic improvement has been achieved by taking into account the damping of the CC form factors, as compared to the result without the damping factor (the dashed line).

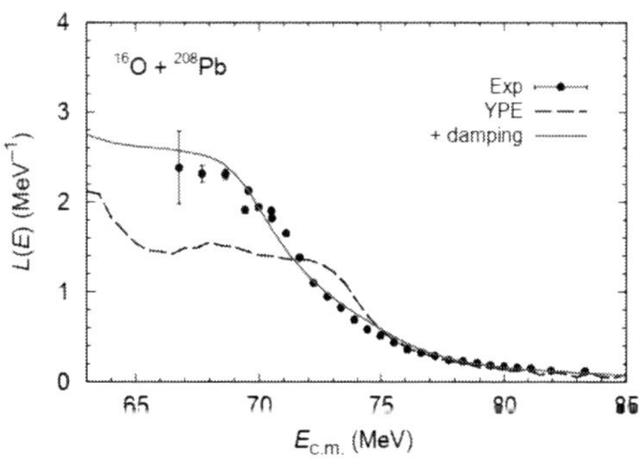

Figure 2. Logarithmic derivative representations of the fusion cross sections versus incident energies for the ^{16}O + ^{208}Pb system. The solid and dashed lines indicate the results calculated with and without the damping factor, respectively. The solid circles are obtained from the experimental fusion cross sections of Refs. [12, 13].

Figure 2 compares the logarithmic derivatives $d\ln(E_{c.m.}\sigma_{fus})/dE_{c.m.}$ of the experimental fusion cross sections with the calculated results. It is again remarkable that only the result with the damping factor achieves nice reproduction of the experimental data. The result with the damping factor becomes saturated below $E_{c.m.}$ = 68 MeV. The measurement at further lower incident energies for this system will thus provide a strong test for the present adiabatic model.

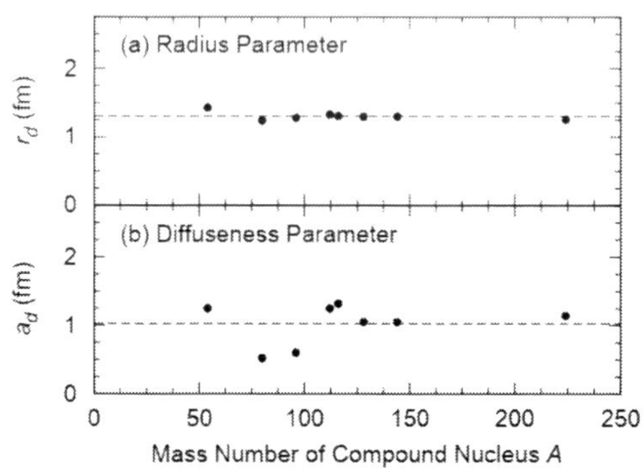

Figure 3. Systematic trends of the (a) radius and (b) diffuseness parameters in the damping factor versus the mass numbers of the compound nuclei for various systems. The solid circles denote the obtained values fitted to the experimental data. The dashed lines indicate the average value of all the obtained values in individual radius and diffuseness parameters.

I have also tested this model in various systems and investigated the systematic trends of the obtained radius and diffuseness parameters in the damping factor. I perform calculations for the medium-heavy mass systems of ^{64}Ni + ^{64}Ni, ^{58}Ni + ^{58}Ni, and ^{58}Ni + ^{54}Fe, the medium-light mass systems of ^{40}Ca + ^{40}Ca, ^{48}Ca + ^{48}Ca, and ^{24}Mg + ^{30}Si, and the mass-asymmetric systems of ^{48}Ca + ^{96}Zr and ^{16}O + ^{208}Pb. The obtained results are shown in Fig. 3. Clearly, the values of r_d are almost constant at around an average value of 1.31 fm. Except for a few points, the values of a_d are also distributed around an average value of 1.02 fm.

Thus, the damping factor strongly correlates with the damping of the transition strength of individual colliding nuclei when they approach one another. In this respect, the damping of the transition strengths would start at the overlapping between the tails of the density distributions for colliding nuclei. That is, the radius parameter of the damping factor would correlate with a range of interactions between the colliding nuclei. This would result in an almost constant value of r_d in all the systems.

3 Direct calculation of coordinate-dependent transition strength using the RPA method

As shown in the previous section, the damping factor phenomenologically introduced works very well for describing the deep sub-barrier fusion hindrance in various systems. The damping factor tells us the importance of the coordinate dependence of the coupling strength. To justify this assumption, I directly calculate the transition strength for ^{16}O + ^{208}Pb as a function of the center-of-mass distance R using the RPA method.

I calculate the mean-field potential for the ^{16}O + ^{208}Pb system using the folding procedure with the single Yukawa function [15]. Before the touching point, I assume the spherical shape for both nuclei. After the touching point, I describe the nuclear shapes with the reflection-asymmetric lemniscatoids parametrization [16]. (The parametrization dependence is negligible, because in this paper I do not discuss the strongly overlapping region.) Based on these densities, I also calculate the Coulomb potential. I use the radius for the proton and neutron potentials, R_0, with $R_0 = 1.27A^{1/3}$ fm, where A is the total nucleon number. The depths of the neutron and proton potentials for individual ^{16}O and ^{208}Pb nuclei, V_T and V_P, are taken from Ref. [17].

Using the obtained mean-field potentials, I solve the axially-symmetric Schrödinger equation with the spin-orbit force and obtain the single-particle wave functions $\phi_i(R)$ and the eigen energies $\epsilon_i(R)$. The details of the model and the parameters are given in Ref. [9]. Then, the z component of the total angular momentum, Ω, is the good quantum number. Note that the parity is not a good quantum number because the mean-field potential for the whole system breaks the space-reflection symmetry. I expand the single-particle wave functions in terms of the deformed harmonic-oscillator bases in the cylindrical coordinate representation. The deformation parameter of the basis functions is determined so as to cover the target and the projectile. The basis functions with energies lower than 26 $\hbar\omega$ are taken into account.

The obtained single-particle energies as a function of the center-of-mass distance R is shown in Fig. 1 of Ref. [9]. As shown in the figure, the single-particle wave functions are drastically changed by level repulsions, which are associated with the neck formations. We can also see extremely strong Coulomb effect of ^{208}Pb on ^{16}O. The single-particle energies of the $p_{1/2}$ and $p_{3/2}$ states in ^{16}O remarkably increase with decreasing R due to the increasing Coulomb effect from ^{208}Pb. Then, many level crossings and repulsion between the energy levels of ^{16}O and ^{208}Pb occur. Below R = 13 fm, the Coulomb barrier becomes lower due to the attractive nuclear mean-field potential. Then, the strong mixture of the single-particle states between ^{16}O and ^{208}Pb starts in many levels, which causes many level splittings seen in the Nilsson diagram.

I now solve the RPA equation at each R for the mass-asymmetric ^{16}O + ^{208}Pb system. The RPA equation then reads

$$\begin{pmatrix} A & B \\ B^* & A^* \end{pmatrix}\begin{pmatrix} X^\nu(R) \\ Y^\nu(R) \end{pmatrix} = \hbar\Omega_\nu(R)\begin{pmatrix} 1 & 0 \\ 0 & -1 \end{pmatrix}\begin{pmatrix} X^\nu(R) \\ Y^\nu(R) \end{pmatrix}, \quad (7)$$

where

$$A_{minj} = \langle\phi(R)|\left[a_i^\dagger a_m\left[H, a_n^\dagger a_j\right]\right]|\phi(R)\rangle \quad (8)$$

$$B_{minj} = \langle\phi(R)|\left[a_i^\dagger a_m\left[H, a_j^\dagger a_n\right]\right]|\phi(R)\rangle. \quad (9)$$

The symbols X and Y are the forward and backward amplitudes, respectively, $\hbar\Omega$ is the excitation energy, $a^\dagger$ (a) indicates creation (annihilation) operator, and H is the Hatree-Fock Hamiltonian. I calculate the first excited 3^- (octupole vibrational) states of ^{16}O and ^{208}Pb, which give the main contributions in the standard CC calculations. I can easily apply the RPA method to the di-nuclear system, because its wave function is described with a one-center Slater determinant. I take the single-particle levels for each neutron and proton from first to 200th and the coherent superposition of all one-particle one-hole states with excitation energies below 30 MeV. I follow the diabatic single-particle configuration corresponding to the ground state of ^{16}O. I use the density-dependent residual interaction taken from Ref. [18] and tune it so that the energy of the spurious center-of-mass motion becomes zero. I calculate B(E3) values for the RPA solutions with $\Omega = 0$ in individual nuclei using the shifted octupole operator, $\widehat{Q}_{30}(R - R_0')$, where R_0' is the center-of-mass position of the projectile or target nucleus.

At R = 20 fm, the obtained first excited octupole modes for ^{16}O and ^{208}Pb are isolated. When the two nuclei approach each other, however, these modes start to fragment into several states. To evaluate the octupole collective strengths carried by low-energy excitations, I then calculate the energy-weighted sum of B(E3) strengths. By checking the spectrum of all obtained RPA modes as a function of R, I determined to take the sum for octupole excitations with $E \leq 4$ MeV and $E \leq 6$ MeV for ^{208}Pb and ^{16}O, respectively.

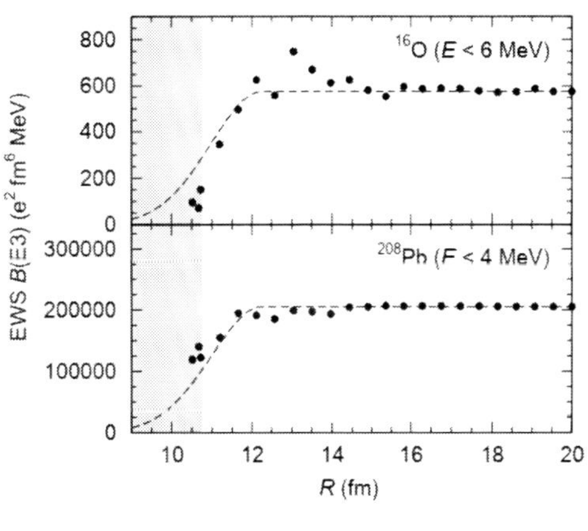

Figure 4. Energy-weighted sums of B(E3) for (a) ^{16}O and (b) ^{208}Pb as functions of R. The solid circles show the results of the RPA calculations. The dashed curves represent the damping factor which well reproduces the experimental data of the fusion cross section for ^{16}O + ^{208}Pb. The gray area indicates the overlap region of the colliding nuclei.

Figure 4 shows the B(E3) strengths for (a) ^{16}O and (b) ^{208}Pb as functions of R. The calculated values (the solid circles) drastically decrease near the touching point (the boundary between the white and gray areas) in both nuclei. The transition densities and currents for the RPA modes with the maximum B(E3) at the touching point are depicted in Figs. 2 (b) and (d) of Ref. [9]. These figures indicate that the octupole collectivities of both ^{16}O and ^{208}Pb are considerably diminished by each colliding partner.

To see the correlation with the damping factor phenomenologically introduced in the CC calculation, I compare the calculated results with the damping factor that well reproduced the experimental data of the fusion cross section for ^{16}O + ^{208}Pb shown in the previous section. In this case, the largest eigenvalue of λ_α is 1.46 fm. In Figs. 4 (a) and (b), the dashed curves represent the damping factor with these parameters normalized at R = 20 fm. We can see that the damping factor strongly correlates with the calculated energy-weighted sums of B(E3) in the low-energy region, which clearly indicates that the damping of the quantum vibrations indeed occurs when the colliding nuclei approach each other. The assumed functional form of Eq. (5) is also justified. These results clearly show the importance of the coordinate dependence of the coupling strength.

This assumption has been also examined in various systems. In Ref. [8], the damping of quantum vibrations was obtained in the ^{16}O + ^{16}O and ^{40}Ca + ^{40}Ca systems. This causes by the drastic change of single-particle wave functions constituting the low-energy collective excitations. Therefore, the damping of quantum vibrations would be the universal mechanism in all deep sub-barrier fusion reactions.

4 Summary

In summary, I have demonstrated the damping of the quantum octupole vibrations of both ^{16}O and ^{208}Pb, when they approach each other. That is, the coordinate dependence of the coupling strength plays an important role in describing the deep sub-barrier fusion hindrance. I have examined the coordinate-dependent coupling strength by introducing the damping factor in the coupling potential in the standard CC model. The damping factor suppresses the transition strength near the touching point of colliding nuclei. The CC calculations including the damping factor reproduce well the fusion hindrance in various systems. The obtained parameter r_d, which is the radius parameter in the damping factor, is almost constant in all systems.

I have also examined the coordinate dependence of the transition strength by calculating the excited states of ^{16}O and ^{208}Pb as a function of R using the RPA method. The obtained B(E3) strengths are substantially damped near the touching point of the colliding nuclei. The obtained energy-weighted sum of B(E3) in the low-energy region exhibits a strong correlation with the damping factor that reproduces well the experimental data of the fusion cross section for ^{16}O + ^{208}Pb. This is a clear evidence that the damping of the quantum octupole vibrations indeed occur near the touching point in the deep sub-barrier fusion reactions.

The drastic change of single-particle wave functions constituting the low-energy collective excitations would commonly occur in all deep sub-barrier reactions. Therefore, the damping of quantum vibrations, that is, the coordinate-dependent coupling strengths, in both the target and projectile near the touching point seems to be the universal mechanism causing the fusion hindrance, which should be taken into account in the standard CC model.

I thank K. Matsuyanagi, K. Hagino, and A. Iwamoto for collaborations and useful discussions.

References

[1] B.B. Back *et al.*, Rev. Mod. Phys. **86**, 317 (2014).
[2] T. Ichikawa, Phys. Rev. C **92**, 064604 (2015).
[3] C. Simenel, A. S. Umar, K. Godbey, M. Dasgupta, and D. J. Hinde Phys. Rev. C **95**, 031601(R)
[4] Ş Mişicu and H. Esbensen, Phys. Rev. Lett. **96**, 112701 (2006); Phys. Rev. C **75**, 034606 (2007).
[5] H. Esbensen and Ş Mişicu, Phys. Rev. C **76**, 054609 (2007).
[6] T. Ichikawa, K. Hagino, and A. Iwamoto, Phys. Rev. C **75**, 057603 (2007).
[7] T. Ichikawa, K. Hagino, A. Iwamoto, Phys. Rev. Lett. **103**, 202701 (2009); EPJ Web Conf. **17**, 07001 (2011).
[8] T. Ichikawa, K. Matsuyanagi, Phys. Rev. C **88**, 011602(R) (2013).
[9] T. Ichikawa, K. Matsuyanagi, Phys. Rev. C **92**, 021602(R) (2015).
[10] H. Esbensen and S. Landowne, Phys. Rev. C **35**, 2090 (1987).
[11] K. Hagino, N. Rowley, and A.T. Kruppa, Comp. Phys. Comm. **123**, 143 (1999).
[12] M. Dasgupta et al., Phys. Rev. Lett. **99**, 192701 (2007).
[13] C. R. Morton, A. C. Berriman, M. Dasgupta, D. J. Hinde, and J. O. Newton, K. Hagino, and I. J. Thompson, Phys. Rev. C**60**, 044608 (1999).
[14] K. Hagino and N. Takigawa, Prog. Theor. Phys. **128**, 1061 (2012).
[15] M. Bolsterli *et al.*, Phys. Rev. C **5** 1050 (1972).
[16] G. Royer and B. Remaud, Nucl. Phys. **A444**, 477 (1985).
[17] P. Möller *et al.*, At. Data Nucl. Data Tables **59** (1995) 185.
[18] S. Shlomo and G. Bertsch, Nucl. Phys. **A 243**, 507 (1975).

Evidence of enhanced three α radius in α + ^{12}C inelastic scattering

Makoto Ito[1,*] and *Makoto* Nakao[1]

[1] *Department of Pure and Applied Physics, Kansai University, 3-3-35 Yamatecho, Suita, Japan*

Abstract. The microscopic coupled-channels calculations (MCC), which is based on precise internal wave functions and a realistic nucleon-nucleon interaction, is performed for the α + ^{12}C inelastic scattering in the energy range of E_α = 80 to 400 MeV. The MCC calculations nicely reproduce the observed differential cross sections for the elastic and inelastic scattering, which goes to the 2_1^+, 3_1^-, 0_2^+ states. The partial wave analysis for the differential cross sections has also performed. From the partial wave analysis, the nuclear radius of three α rotational state in ^{12}C with a life time of 10^{-21} second, which has been expected to have much more extended radius than the ground ^{12}C nucleus, is speculated. Present analysis predicts about 1.0 fm enhancement in the matter radius of the three α rotational state in comparison to the normal radius of the ground state, which is known to be proportional to the mass number to the one third. The spatial extension of the three α rotational state is comparable to the extended radius observed in the neutron halo phenomena. Constraint on the recent ab-initio calculation for the 3α states in ^{12}C is also discussed.

1 Introduction

Cluster structures, in which a nucleus is decomposed into several sub-units, are realized in the low excited states of light mass nuclei [1]. In the cluster structure, the sub-units are weakly coupled each other, and their matter radius is prominently extended in comparison to the radius of the ground state, which obeys the law of $\propto A^{1/3}$. In particular, such the enhancement of the matter radius is extensive in the 3α structure of the ^{12}C nucleus [2, 3]. The ground state of ^{12}C has a radius of 2.40 fm [2] with the spin-parity of 0^+ (0_1^+), while the excited 0_2^+ state at E_x = 7.65 MeV, which is called the Hoyle state, is considered to have a radius of 3.47 fm with the well developed 3α cluster structure [2]. The radius of the Hoyle state was evaluated by the bound state approximation, in which the resonant wave function is expanded by the finite basis set although the Hoyle state is a resonant state, which has a infinite extension in the wave function. The extension of the radius in 0_2^+ is predicted to be about 1 fm, which is comparable to the extension of the neutron halos discussed in ^{11}Li [4]. In this article, we propose a new method to obtain a signature of the enhanced matter radius of the 3α cluster state.

Unfortunately, a direct measurement of the radius of the Hoyle 0_2^+ states with the developed 3α cluster structure is impossible due to its short life time, but there are several attempts to get an evidence of the enhanced matter radius of the 3α Hoyle states from the inelastic scattering of ^{12}C, which excites the Hoyle 0_2^+ state as a final state. In the inelastic scattering of light ions by ^{12}C, which excites the ^{12}C(0_2^+) state in a final state, an oscillating pattern in the differential cross section of the scattered ion is discussed in connection to the enhanced radius of the final 0_2^+ state with the 3α structure [5–9]. However, the relation of the extended radius and the oscillating pattern of the cross section of the ^{12}C inelastic scattering still remains unclear [10, 11].

In the coupled-channel approaches [7–9], for example, nuclear interactions for the system of the ^{12}C target plus a projectile are constructed from the double folding model, which employs the internal wave function of ^{12}C, obtained from the 3α resonating group method (3αRGM) [2] and from an effective density-dependent nucleon-nucleon interaction. This coupled-channel calculation is called microscopic coupled-channels (MCC) [14]. Since the 3α RGM wave function can precisely describe the 3α structure in 0_2^+, the constructed nuclear interaction in the 0_2^+ channel is more attractive at a surface region than the interaction in the ground channel.

The MCC calculations employing the long-range 3α folding interaction have been applied to the α+^{12}C [7] and ^{3}He+^{12}C [8, 9] systems, and the calculation nicely reproduces the angular distributions of various exit channels. In the differential cross section for the 0_2^+ channel, a scattering angle for the first Airy minimum is shifted to the larger angle region, and the number of the Airly minima increases in a comparison to other inelastic channels, going to the rotational and vibrational states such as the 2_1^+ and 3_1^- states. From this result, the authors have claimed that the evolution of the Airy structure originated from the spatial extension of the nuclear interaction in the final 0_2^+ channel [7–9].

In the nuclear reaction, the final 3α states can be assessed through the transition from the incident ground channel, and the 3α state is observed as the final state in the reaction process. Therefore, a coupling potential, which

*e-mail: itomk@kansai-u.ac.jp

induces a transition from the initial channel to the final channel, is expected as a main ingredient for the angular distribution of the inelastic scattering, and the distortion potential or the density distribution in the exit channel may be not so effective in comparison to the effect of the coupling potential. In fact, a dominance of the coupling potential in the inelastic scattering has been already pointed out by Takashina et al. [10, 11]. By employing the MCC calculation, which is the same method as in Refs. [7–9], the authors have clearly demonstrated that the Airy structure in the angular distribution is dominated by not the size of the 0_2^+ state, but by a spatial distribution of the coupling potential for the $0_1^+ \to 0_2^+$ transition.

There is another problem in the relation of the monopole $0_1^+ \to 0_2^+$ transition and the spatial size of the final 0_2^+ state. If an anomalous feature or an enhancement factor appears in the differential cross section in the final 0_2^+ channel, such the anomalous feature should be discussed in comparison to a cross section of a "reference channel", which should reveal normal behavior without the extension of the spatial size. However, there is no appropriate reference channels available to compare with the 0_2^+ channel in other inelastic channels. For example, the ground 0_1^+ channel with a normal radius corresponds to the elastic scattering, which is completely different from the inelastic scattering going to the 0_2^+ channel. Furthermore, other collective channels, such as the 2_1^+ and 3_1^- channels, are also inappropriate to compare with the 0_2^+ channel because of the finite spins in the 2_1^+ and 3_1^- channels. The finite spin generate the reorientation coupling, in which the direction of the intrinsic spins rotates with respect to the orbital angular momentum in the relative motion of the colliding nuclei. Such the reorientation coupling strongly affects on the shape of the differential cross sections. Therefore, the collective compact channels, such as 0_1^+, 2_1^+ and 3_1^-, are considered to be invalid to compare with the 0_2^+ channel.

In order to overcome these problems, we introduce a new viewpoint in considering the relation of the enhanced matter radius of the 3α state and the inelastic scattering. In the present analysis, we focus on the inelastic scattering, going to two kinds of the 2^+ states: the 2_1^+ and 2_2^+ states. The former state is a normal rotational state of the ground 0_1^+ state, while the latter state corresponds to the 3α rotational state of the Hoyle 0_2^+ state [2, 3] with a life of 10^{-21} sec. (width of $\Gamma \sim 1$ MeV). In a modern theory, the Hoyle rotational 2_2^+ state, which has been recently identified in experiment [12], is interpreted in terms of an α halo state with a dilute 3α structure [3]. We compare the differential cross section of α + ^{12}C(0_1^+) $\to$ α + ^{12}C(2_2^+) with the reference reaction of α + ^{12}C(0_1^+) $\to$ α + ^{12}C(2_1^+). Both of these $2_{1,2}^+$states have the same spin-parity and are excited as a final state in the inelastic scattering. The reaction process is completely common, and only the difference is the spatial size: a compact size in 2_1^+ and an extended size in 2_2^+. Therefore, we consider that the comparison of the differential cross section going to these two kinds of 2^+ states gives a clear sign relevant to the difference of the nuclear radius.

As pointed out in Refs [10, 11], the potential or density distribution in the final 3α channel itself may play a minor role in inelastic scattering but a coupling potential contains an important information about the matter radius of the final 3α state. This is because the coupling potential is determined by an overlap of the wave function between the initial ground state and the final 3α state. Therefore, an extended structure of the 3α state is indirectly reflected in the inelastic scattering to the 3α final state through a size of the coupling potential. We believe that it is still important to consider a relation between an inelastic scattering and the size of the coupling potential. Recently, we have formulated the method of scattering radius, which can characterize a spatial size of the reaction area or a size of the coupling potential for the individual exit channels in a general exclusive reactions [13]. In this method, the angle-integrated cross section decomposed into each of the incident partial wave are used to measure the spatial size for the production area of the final channel. By applying the method of the scattering radius to the inelastic scattering going to the $2_{1,2}^+$ channels, we will demonstrate that a direct evidence of the enhanced matter radius in the 2_2^+ state clearly appears in the α + ^{12}C inelastic scattering.

Organization of this article is the following. The framework of MCC is explained in Sec. 2. Nucmerical calculations are compared with the experiment in Sec. 3. In this section, the anomalous feature in 2_2^+ in the differential cross section is clearly pointed out. From the partial wave analysis, the lower bound of the matter radius of 2_2^+ is speculated. The final section is devoted to summary and discussion.

2 Theoretical framework

In this section, we explain the framework of the microscopic coupled-channels (MCC) calculation. In the MCC framework, the nuclear interactions of colliding nuclei are constructed by the double folding (DF) model, in which the reliable nucleon-nucleon interaction and the internal wave functions calculated from the microscopic cluster model are employed.

2.1 Coupled-channels equations

We calculate the differential cross sections of an α particle scattered by ^{12}C in the formulation of MCC [7, 8, 14]. In MCC framework, we solve a set of the coupled-channels equation for the α–^{12}C system, which is given in the symbolic form

$$\Big(T_f(\mathbf{R}) + V_{f,f}(\mathbf{R}) - E_f\Big)\chi_f(\mathbf{R}) = -\sum_{i\neq f} V_{f,i}(\mathbf{R})\chi_i(\mathbf{R}) . \tag{1}$$

Here the subscripts of f or i design a channel. In Eq. (1), $T_f(\mathbf{R})$ represents the kinetic energy of the relative motion of the α–^{12}C system with a relative coordinate $\mathbf{R}$, while $V_{f,i}(\mathbf{R})$ denotes the coupling potential for the transition from channel i to channel f. The total energy in the channel f, E_f, is given by the relation of $E_f = E - \epsilon_f$ with the α incident energy E and the ^{12}C internal energy of ϵ_f.

$\chi_f(\mathbf{R})$ is the α–^{12}C relative wave function for the channel f, which should be solved in the coupled-channels equation.

In the pragmatic calculation of MCC, the equation is expanded in terms of the total spin J, and the individual equations for a fixed J are solved. Thus, the cross sections are calculated for each of J, which are called the partial cross section. The distribution of the partial cross sections are used to characterize the spatial size for the final state production [13].

2.2 Double folding potential

In Eq. (1), $V_{f,i}(\mathbf{R})$ denotes the coupling potential for the transition of $i \to f$ in ^{12}C, which is calculated from the double folding model. The coupling potential calculated from the double folding procedure is symbolically written as

$$V_{f,i}(\mathbf{R}) = N_R \int\int \rho_{f,i}^{(^{12}C)}(\mathbf{r}_1)\rho^{(\alpha)}(\mathbf{r}_2)v_{NN}^{DDM3Y}(\mathbf{s})\,d\mathbf{r}_1 d\mathbf{r}_2 \quad (2)$$

with $\mathbf{s} = \mathbf{r}_2 - \mathbf{r}_1 - \mathbf{R}$. Here $\mathbf{r}_1$ ($\mathbf{r}_2$) denotes a coordinate measured from the center of mass in the ^{12}C (α) nucleus. $\rho_{f,i}(\mathbf{r})$ represents the diagonal ($f = i$) or transition ($f \neq i$) densities of ^{12}C, which are calculated by the microscopic 3α cluster model, resonating group method (RGM) [2], while $\rho^{(\alpha)}(\mathbf{r})$ denotes the density of α particle, which reproduce the charge form factor of the electron scattering. In Eq. (2), v_{NN}^{DDM3Y} represents the effective nucleon-nucleon (NN) interaction which acts between a pair of nucleon contained in the ^{12}C nucleus and the α particle. In the present calculation, we adopt the DDM3Y (Density Dependent Michigan 3-range Yukawa) interaction [15].

The normalization factor N_R is introduced because the folding potential contains ambiguity in its strength. This ambiguity arises from the neglection of the anti-symmetrization effect between the projectile and the target except for the knockon exchange [16]. Here this factor is set to $N_R = 1.42$ over all the α incident energy, which is consistent to the previous MCC calculation in Ref. [7, 8]. In the α scattering, the double folding potential needs a considerably large N_R, say about 1.3 as pointed out in the first application of DDM3Y to the α scattering [15]. This modification in the potential strength is originated from the violation of density saturation of an α particle. The improvement of the double folding potential for the α scattering has recently been discussed in Ref. [17].

In addition to the folding potential, we introduce the absorptive potential with the Saxon-Woods form factor in the diagonal ($f = i$) transition in order to simulate other reaction process, and the parameter set of Saxon-Woods is tuned so as to reproduce all of the observed differential cross sections as much as possible. As for the internal excitation of ^{12}C, we include the low-lying collective states (2_1^+ and 3_1^-) and the 3α cluster states (0_2^+, 0_3^+, 2_2^+) in addition to the ground 0_1^+ state. The coupling of all the excited states is included, and the multi-polarity $\lambda = |\mathbf{I}_f - \mathbf{I}_i|$ for the internal transition of ^{12}C($I_i \to I_f$) is considered up to $\lambda = 3$

3 Results

3.1 Calculation of differential cross sections

We have solved the coupled-channels equation, which is given in Eq. (1), for the α + ^{12}C scattering at E_α = 386 MeV [12] with the nuclear interaction in Eq. (2). The MCC calculations reproduce the differential cross sections of the scattering to 0_1^+, 2_1^+, 3_1^- states. Furthermore, the theoretical calculation nicely reproduce the inelastic scattering to the excited state observed at $E_x \sim 10$ MeV by taking the summation of the differential cross sections of the 0_3^+ and 2_2^+ states.

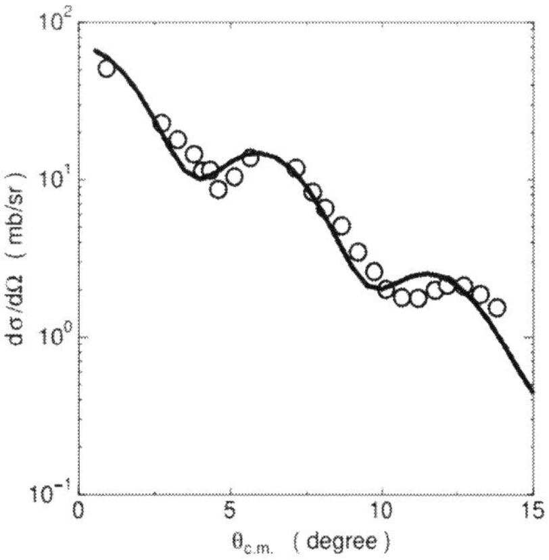

Figure 1. Comparison of observed differential cross section at $E_x \sim 10$ MeV (open circles) with theoretical calculation (solid curve). The data is taken from Ref. [12].

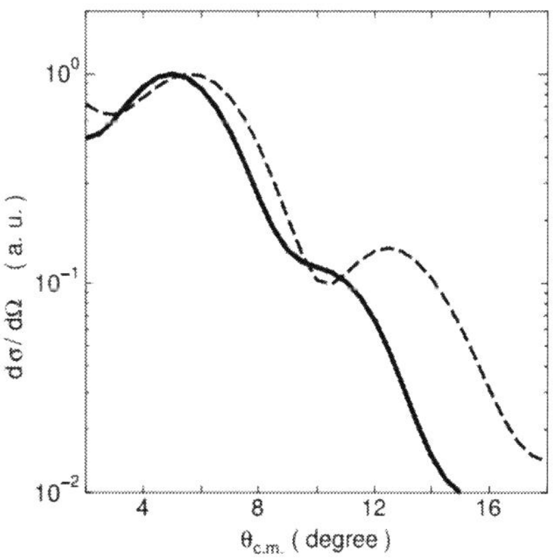

Figure 2. Theoretical differential cross sections of the final state of 2_1^+ (dashed curve) and 2_2^+ (solid curve) in the α + ^{12}C inelastic scattering. The magnitudes of both the cross sections are normalized by the maximum value around $\theta_{c.m.} \sim 5°$.

The comparison of the observed cross section of $E_x \sim 10$ MeV and the theoretical calculation is shown in Fig. 1.

 DOI: 10.1051/epjconf/201716300026

The experimental strength is considered to be the incoherent mixture of the 2_2^+ and 0_3^+ states because these two resonant states are overlapped each other, which have a broad width with a close energy spacing. The theoretical calculation (solid curve), which is obtained by the the incoherent summation of these two states, nicely reproduce the global behavior of the distribution, such as the gradual oscillation with the shallow valleys. This behavior in the experiment can never be reproduced by either the strength of 2_2^+ or that of 0_3^+. The incoherent summation of these two states is essential for the reproduction of the differential cross section.

In Fig. 2, the component of the 2_2^+ differential cross section in Fig. 1 is compared with the cross section of the 2_1^+ state, which also reproduces the observed cross section. The peak position of the 2_2^+ cross section (solid curve) shifts to the forward angular region in comparison to the 2_1^+ cross section (dashed curve), and its angular distribution has the shrunk and rapid fall-down structure. The shift and shrinkage features are completely consistent to the result of the multi-pole decomposition analysis (MDA) of the experimental cross section, which is performed in the range of $\theta_{c.m.} < 15°$ [12].

3.2 Partial wave analysis

In order to analyze the difference of the differential cross sections shown in Fig. 2 more deeply, we have performed the partial wave analysis, in which the angle-integrated cross sections in Fig. 2 are decomposed into the individual components of the total spin J. The partial wave expansion for the transition of $i \to f$ is defined as

$$\int d\Omega \frac{d\sigma}{d\Omega}(i \to f) = \sum_J \sigma(J, f). \tag{3}$$

The individual components of the angle integrated cross sections for the transition from the initial 0_1^+ state to the final state (f) are called the partial cross section, $\sigma(J, f)$. Since the incident channel of α + $^{12}C(0_1^+)$ is spinless, the total spin J is equal to the incident orbital spin L between α and ^{12}C and hence, $\sigma(J = L, f)$.

The results of the partial wave decomposition are shown in Fig. 3. The crosses show the partial wave distribution (L-distribution) of the angle-integrated 2_1^+ cross section ($\sigma(L, 2_1^+)$), while the L-distribution of 2_2^+ is plotted by the solid circles ($\sigma(L, 2_2^+)$). In the comparison of $\sigma(L, 2_1^+)$ with $\sigma(L, 2_2^+)$, we can clearly see the extended L-distribution of 2_2^+ to the higher L region. This extension of 2_2^+ in the L-space (Fig. 3) is just opposite to the shrinkage in the θ-space, the differential cross section (Fig. 2). This is nothing but the uncertainty relation of $\Delta L \cdot \Delta\theta \sim 1$, which can be hold in the diffraction scattering. According to the classical relation of $L = kb$, where k and b denote the incident wave number and the impact parameter, respectively, the extension of the L-distribution in Fig. 3 clearly means that the production area of the 2_2^+ state is more extended than the area of the 2_1^+ production.

The distribution in the L-space can be transformed into a size of the production area of the final state by applying

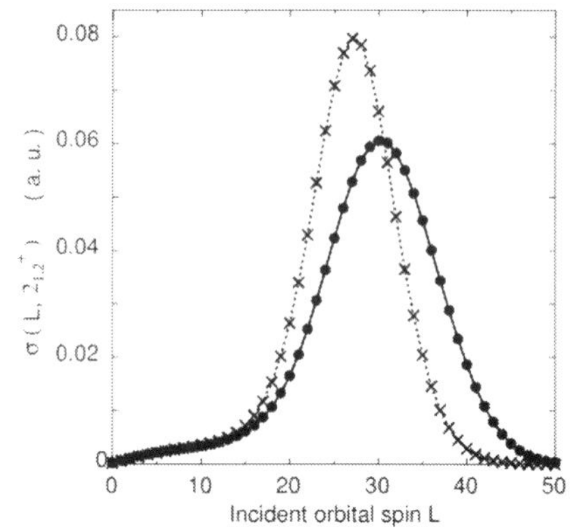

Figure 3. Partial wave decomposition of the angle integrated cross sections calculated as the function of the incident orbital spin L (L-distribution). The crosses with the dotted curve and the solid circles with the solid curve show the partial cross sections of the 2_1^+ and 2_2^+ final states, respectively. The magnitudes of all the partical cross section is normalized by the total cross section, which is the summations of the partial cross sections.

the method of the scattering radius [13]. In this method, the effective orbital spin $\bar{L}$ for the transition of $0^+ \to 2_{1,2}^+$ is derived according to the following expression:

$$\bar{L}(2_i^+) = \sqrt{\frac{\sum_L \hat{L}^4 \sigma(L, 2_i^+)}{\sum_L \hat{L}^2 \sigma(L, 2^+)}} \quad (i = 1, 2) \tag{4}$$

with a definition of $\hat{L} = \sqrt{L(L+1)}$. The radius of the final-state production, which is called the scattering radius (R_{SC}), is simply obtained according to the classical relation of $\bar{L} = kR_{SC}$. R_{SC} naturally goes to the matter radius of a target nucleus in the high energy limit of the proton elastic scattering [6, 13, 18].

Table 1. Effective orbital spins $\bar{L}$ and the scattering radius R_{SC} calculated from the definition in Eq. (4). The theoretical mean radius of the density distribution ($\bar{r}$) for the 2_1^+ (E_x = 4.44 MeV) and 2_2^+ (E_x = 10.30 MeV) states are also shown in the right-most column [2].

State	E_x (MeV)	$\bar{L}$	R_{SC} (fm)	$\bar{r}$ (fm)
2_1^+	4.44	29.67	4.58	2.38
2_2^+	10.3	33.69	5.20	4.00

The results of $\bar{L}$ and R_{SC} for the 2_1^+ and 2_2^+ states are summarized in table 1. The calculated effective orbital spin $\bar{L}$ is slightly shifted to the higher value than the peak position of the L-distribution in Fig. 3 because of the finite width in their distribution. The resultant scattering radius for 2_1^+ is $R_{SC}(2_1^+)$ = 4.58 fm, while the radius for 2_2^+ is $R_{SC}(2_2^+)$ = 5.20 fm. The obtained R_{SC} are larger than the root-mean-squared radius ($\bar{r}$) of the 2_1^+ and 2_2^+ states (2.38 fm and 4.00 fm) [2]. The magnitude correlation in R_{SC} seems to correspond to that in $\bar{r}$ but the correspondence of R_{SC} to $\bar{r}$ is non trivial because the magnitude of R_{SC} strongly depends on the definition of $\bar{L}$ in Eq. (4). Therefore, we should focus on the relative difference of R_{SC} between 2_1^+ and 2_2^+ in the present analysis.

The difference of the scattering radius is $\Delta R_{SC} = R_{SC}(2_2^+) - R_{SC}(2_1^+) = 0.62$ fm, while the difference of $\bar{r}$ in the structure calculation is $\Delta r = \bar{r}(2_2^+) - \bar{r}(2_1^+) = 1.62$ fm. The enhancement in R_{SC} is smaller than that in $\bar{r}$, and this reduction of ΔR_{SC} is attributed to the property of the inelastic scattering, which is induced by the coupling potential from the initial ground state to the final 3α states. This point will be discussed in the last section.

3.3 Energy systematics and radius of 2_2^+ state

We have extended the MCC calculation with the method of the scattering radius to the lower α incident energy of $E_\alpha \leq 240$ MeV. In the lower energy region, unfortunately, there is no experimental data on the excited state at $E_x \sim 10$ MeV, which contains the 2_2^+ component. Therefore, the parameters of the absorptive potential of the 3α cluster states (0_2^+, 0_3^+, 2_2^+) are set to a common value, and the parameters are tuned so as to reproduce the cross section of 0_2^+ (and 0_3^+) as much as possible. The energy systematics of the $R_{SC}(2_2^+)$ and $R_{SC}(2_1^+)$ is shown in Fig. 4. In all the energy region considered, $R_{SC}(2_2^+)$ (solid line with solid circles) is larger than $R_{SC}(2_1^+)$ (dotted line with crosses) by about 1 fm over the entire incident energy. The enhancement of $\Delta R_{SC} = R_{SC}(2_2^+) - R_{SC}(2_1^+) \sim 1$ fm is equivalent to the fact that the differential cross section of the 2_2^+ state is more shrunk than that of the 2_1^+ state. The reason why the ΔR_{SC} in the lower energy region is more enhanced than that in the higher energy, $E_\alpha = 386$ MeV, is expected to be the distortion of the final state. In the lower energy region, the α particle can feel the extended distortion potential in the final 2_2^+ channel, which leads to the shrinkage of the angular distribution and the enhancement of R_{SC}.

From the systematic values of ΔR_{SC}, we can speculate the lower bound of the matter radius of the 2_2^+ state. In a naive consideration, we can image the relation of ΔR_{SC} and the matter radius $\bar{r}$, such as

$$\bar{r}(2_2^+) \geq \bar{r}(2_1^+) + \Delta R_{SC} \; . \tag{5}$$

In this relation, the inequality means the fact that ΔR_{SC} speculated from the inelastic scattering does not necessarily identify a true enhancement of the matter radius, $\Delta\bar{r}$. In the present analysis, $\Delta R_{SC} \sim 1$ fm is predicted in the lower energy region of $E_\alpha \leq 240$ MeV. The matter radius of 2_1^+ is still unknown but we can safely assume $\bar{r}(2_1^+) \sim \bar{r}(0_1^+)$. This assumption is because the 2_1^+ state is the first excited state of the ground 0_1^+ state and hence, we can expect that the spatial size of 2_1^+ is not drastically changed from the size of the ground 0_1^+. This assumption is also supported by the 3α RGM calculation [2]. Therefore, the matter radius of the Hoyle rotational state is speculated to $\bar{r}(2_2^+) \geq 3.3$ fm if we employ the experimental value of $\bar{r}(0_1^+) = 2.3$ fm [19].

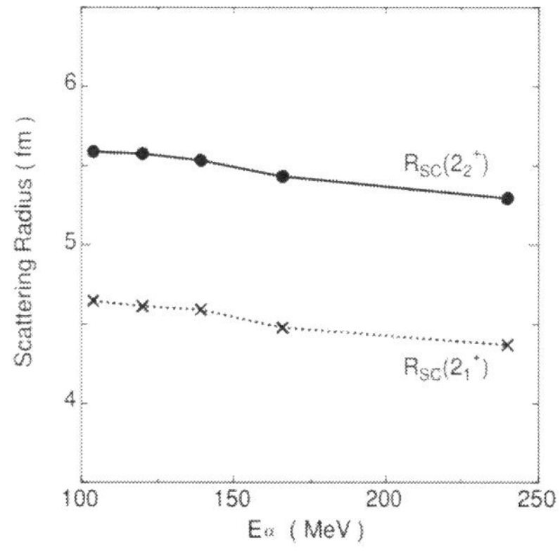

Figure 4. Energy systematics of the scattering radius of $R_{SC}(2_2^+)$ (solid line with solid circles) and $R_{SC}(2_1^+)$ (dotted line with crosses). The abscissa represent the incident α energy in the laboratory system.

4 Summary and discussion

In summary, we have performed the MCC calculation with 3α RGM + DDM3Y and shown that the enhanced radius of the 2_2^+ state can be probed via the differential cross section by comparing with the respective cross section of the 2_1^+ state. The inelastic differential cross section of the 2_2^+ state is more shrunk than that of the 2_1^+ state, and this shrinkage is a first evidence of the extended matter radius of the final 2_2^+ state in the inelastic scattering of α + $^{12}C(0_1^+) \to \alpha$ + $^{12}C(2_2^+)$. According to the prescription of the partial wave decomposition, the partial wave components of the cross sections are converted into the scattering radius, which characterizes a size of the reaction area for the final state production. According to the evaluation of the scattering radius, the shrunk structure of the differential cross section corresponds to about 0.6 ~ 1 fm enhancement in the size of the production area of the final 2_2^+ state.

Enhancement of R_{SC} ($\Delta R_{SC} \sim 1$ fm) is smaller than the difference of the respective root-mean-squared radius ($\Delta\bar{r} \sim 1.6$ fm), which is predicted by the structure calculation [2]. This is because of the feature peculiar in the inelastic scattering, in which the size of the final wave function is filtered by the size of the initial wave function. As pointed out in Refs [10, 11], the inelastic scattering going to the excited states are dominated by the size of the coupling potential, which induces the transition from the incident 0_1^+ state to the final $2_{1,2}^+$ state. Since the coupling potential is determined by the overlap of the initial and final wave function, the size of the coupling potential is restricted by the distribution of the initial wave function. Thus, the inelastic scattering cannot necessarily probe a whole size of the density (or potential) in the final state, and the ΔR_{SC} values obtained from the inelastic scattering is considered to underestimate the true enhancement of the matter radius in the final state. However, we believe

DOI: 10.1051/epjconf/201716300026

that about 1 fm in the difference of R_{SC} is still meaningful enhancement.

Furthermore, the identification of 1 fm enhancement imposes a strong constraint on the recent ab-initio calculation which seems to reproduce the excitation energy of the Hoyle 0_2^+ and 2_2^+ states [20]. In this structure calculation, the matter radius of the Hoyle rotational 2_2^+ state is almost same as the radius of the ground state (2.4 fm) [20], which is much smaller than the prediction by the 3α cluster model ($\sim$ 4.0 fm) [2, 3]. In previous studies, there was no information about the matter radius of the excited states that should be compared with the theoretical calculation. Therefore, the speculation of the lower bound of the matter radius of the Hoyle rotational 2_2^+ state from the experimental observables is quite important in future studies. Since the experimental information of the differential cross section of the 2_2^+ state is still insufficient, the measurement of the differential cross section of the excited state at $E_x \sim$ 10 MeV and careful MDA to separate the 2_2^+ component should be extended over a wide angle and energy region.

It is quite meaningful to reproduce the differential cross section by the MCC calculation because MCC is based on the internal wave function obtained from the microscopic 3α cluster model. The reproduction of the experimental data strongly support that the size enhancement induced by the transition to the 3α structure is a realistic picture in the reaction process. In the more stringent check of the present result, however, we should apply another kind of the reaction calculation without the presupposition of the size difference of the 2_1^+ and 2_2^+ state. For example, one should assume a standard form factor for the coupling potential, such as the normal rotational or vibrational excitation, and the optimal range of the coupling potential should be searched so as to reproduce the observed cross section. The comparison to the calculation without the cluster assumption is important in future studies.

Finally, let us stress the new insight in the present approach. The comparison of the yrast 2_1^+ state and the Hoyle rotational 2_2^+ state is a new insight in the discussion of the inelastic scattering of ^{12}C, which was absent in the previous studies about relation of the matter radius and the inelastic scattering to the Hoyle 0_2^+ state [5–8, 10, 11]. In the case of 0_2^+ excitation, the enhancement of the matter radius of the final 0_2^+ state may be reflected in the differential cross section but it is difficult to extract an evidence only from the cross section of the 0_2^+ state. The anomalous feature in the 0_2^+ cross section should be identified from the comparison with a certain kind of "reference reactions". Such the reference reaction should be set to the other 0^+ state, which has non α cluster structure. In the vicinity of the 0_2^+ state, there is no such the excited 0^+ state with the spatially compact structure. On the contrary, in the case of the inelastic scattering to the 2^+ state, there is a pair of the 2^+ channels, such as the compact 2_1^+ state and the excited 2_2^+ one, which can be compared directly.

The comparison of the same (finite) spin-parity states excited by the inelastic scattering is essential in the present analysis. The similar comparison of the yrast state and the cluster state can be generalized to other nuclei because a cluster state with a finite spin always appear above the respective yrast state with a compact shell model like structure. One of such the interesting applications is the ^{10}Be nucleus. In this nucleus, the intrinsic structure can be nicely described by the covalent (or molecular) orbit structure of two valence neutron around the two center cores of ^{8}Be = $\alpha + \alpha$ [21]. The yrast 2_1^+ and the exited 2_2^+ state have the spatially compact structure, while the 2_3^+ state is considered to have the well developed structure of $\alpha + \alpha + N + N$, in which two neutrons form the covalent σ^+ orbital [21]. σ^+ orbital. The calculation of the α + ^{10}Be inelastic scattering is now underway.

Acknowledgment

The author is sincerely thankful to Profs. H. Horiuchi, K. Iida, M. Itoh, K. Ogata, T. Wada, and Dr. K. Minomo for their useful comments and discussions. He is deeply thankful to Prof. A. Kohama for valuable comments and encouragement. This work has been done in collaboration with Mr. M. Tomita, M. Iwasaki, and R. Otani and supported by Grants-in-Aid for Scientific Research in Japan (No 26400284).

References

[1] K. Ikeda et al., Prog. Theor. Phys. **68**, 1 (1980).

[2] Y. Fukushima and M. Kamimura, *Proceedings of the International Conference Nuclear Structure*, J. Phys. Soc. Jpn. **44**, 225 (1977); M. Kamimura, Nucl. Phys. A **351**,456 (1981).

[3] Y. Funaki, A. Thosaki, H. Horiuchi, P. Schuck, and G. Reöpke, Eur. Phys. J. A **24**, 321-342 (2005), and references therein.

[4] I. Tanihata, H. Savajolos, R. Kanungo, Prog. Part. Nucl. Phys. **68**, 215 (2013).

[5] A. N. Danilov, T. L. Belyaeva, A. S. Demyanova, S. A. Ganchalov, and A. A. Ogloblin, Phys. Rev. C **80**, 054603 (2009).

[6] K. Iida, S. Koide, A. Kohama, and K. Oyamatsu, Mod. Phys. Lett. A **27**, 1250020 (2012).

[7] S. Ohkubo and Y. Hirabayashi, Phys. Rev. C**70**, 041602(R) (2004).

[8] S. Ohkubo and Y. Hirabayashi, Phys. Rev. C**75**, 044609 (2007).

[9] Sh. Hamada, Y. Hirabayashi, N. Burtebayev, and S. Ohkubo, Phys. Rev. C **87**, 024311 (2013).

[10] M. Takashina and Y. Sakuragi, Phys. Rev. C **74**, 054606 (2006).

[11] M. Takashina, Phys. Rev. C **78**, 014602 (2008).

[12] M. Itoh et al., Phys. Rev. C**84**, 054308 (2011).

[13] M. Tomita, M. Iwasaki, R. Otani, M. Ito, Phys. Rev. C **89**, 034619 (2014).

[14] M. Ito, Y. Hirabayashi, and Y. Sakuragi, Phys. Rev. C **66**, 034307 (2002), and references therein.

[15] A.M. Kobos, B.A. Brown, P.E. Hodgson, G.R. Satchler, A. Budzanowski, Nucl. Phys. A **384**, 65 (1982).

[16] T. Wada, H Horiuchi, Prog. Theor. Phys. **80**, 502 (1988), and references therein.

DOI: 10.1051/epjconf/201716300026

[17] K. Egashira, K. Minomo, M. Toyokoawa, T. Matsumoto, M. Yahiro, Phys. Rev. C**89**, 064611 (2014).
[18] A. Kohama, K. Iida, and K. Oyamatsu, J. Phys. Soc. Jpn. **85**, 094201 (2016), and references therein.
[19] A. Ozawa et al., Nucl. Phys. A**691**, 599-617 (2001).
[20] E. Epelbaum, H. Krebs, T. A. Lähde, D. Lee, Ulf-G. Meiβner, Phys. Rev. Lett. **109**, 252501 (2012), and references therein.
[21] M. Ito and K. Ikeda, Rep. Prog. Phys. **77**, 096301 (2014), and references therein.

EPJ Web of Conferences 163, 00027 (2017)
FUSION17
DOI: 10.1051/epjconf/201716300027

Production of neutron-rich nuclides in the vicinity of $N = 126$ shell closure in multinucleon transfer reactions

Alexander Karpov[1,2,*] and *Vyacheslav* Saiko[1,2,**]

[1] *Flerov Laboratory of Nuclear Reactions, JINR, 141980 Dubna, Russia*
[2] *Dubna State University, 141980 Dubna, Russia*

Abstract. Multinucleon transfer in low-energy nucleus-nucleus collisions is widely discussed as a method of production of yet-unknown neutron-rich nuclei hardly accessible (or inaccessible) by other methods. Modeling of complicated dynamics of nuclear reactions induced by heavy ions is done within a multidimensional dynamical model of nucleus-nucleus collisions based on the Langevin equations. The model gives a continuous description of the system evolution starting from the well-separated target and projectile in the entrance channel of the reaction up to the formation of final reaction products. In this paper, rather recent sets of experimental data for the ^{136}Xe+^{198}Pt,^{208}Pb reactions are analyzed together with the production cross sections for neutron-rich nuclei in the vicinity of the $N = 126$ magic shell.

1 Motivation

Nowadays, production and study of neutron-rich nuclei is one of the main trends in nuclear physics. A number of facilities over the world are already working on this subject. These studies, in part, are motivated by a quantitative understanding of the astrophysical r-process, which goes through neutron-rich nuclei far away from the beta-stability line. Such nuclides located in the vicinity of the neutron closed shells form the so-called r-process waiting points. Knowing the properties of these nuclides plays a key role in the modeling of the r-process.

One of the least explored neutron-rich areas of the nuclear chart is that close to the $N = 126$ shell closure. That is caused by the low values of the fragmentation cross sections – the only method of production of neutron-rich nuclei in this area used so far. During the last decade, 3-4 new nuclides have been added to each of the isotonic chains in the vicinity of $N = 126$ by ^{238}U and ^{208}Pb fragmentation. Thus, the most neutron-enriched nucleus with $N = 126$ known at present, ^{202}Os, was produced [1, 2] with the cross section 4.4 ± 2.0 pb [2]. Moreover, only the lower limit of the ^{202}Os half-life was identified. Since the fragmentation cross sections will continue their rapid decrease with each step in the direction of neutron excess, alternative efficient methods of production of these nuclei are still of great value.

Multinucleon transfer in the processes of deep inelastic (DI) collisions of heavy ions at near-barrier energies has been considered to be such an appropriate method since the discovery of this type of nuclear reactions in 1966 by V.V. Volkov with collaborators [3]. During several years after the discovery, a number of light neutron-rich nuclei were produced for the first time in DI collisions of light ions with heavy targets (see [4] and references therein).

A recently renewed interest in the DI processes is caused, in part, by a widely discussed possibility of synthesizing the unknown neutron-rich isotopes of medium-mass, heavy, and superheavy nuclei [5–7, 13, 14]. A number of experimental studies have already been performed [8–12], aiming at the validation of this idea and studying the peculiarities of the multinucleon transfer reactions in this mass region.

Semiclassical approaches based on the Langevin equations are proved to be powerful tools for analysis of low-energy nucleus-nucleus collisions, allowing one to model low-energy collisions of heavy ions in their complexity of intersecting and competing reaction channels. In this work we analyze new experimental data on the ^{136}Xe+^{198}Pt,^{208}Pt [8–10] reactions within a recently developed multidimensional dynamical model [15] which is based on the same principles as the one in Ref. [13]. The model is briefly described in the next section. Finally, the optimal conditions for production of neutron-rich nuclei with $N = 126$ in multinucleon transfer reactions are discussed.

2 Model

A near-barrier collision of two nuclei may be described as consisting of three main stages: approaching stage, mononucleus formation, and separation into final fragments. It is necessary to realistically describe the system evolution at all these stages. The first step on this way is the choice of the most important degrees of freedom of the interacting system. Their number should be large enough to repro-

*e-mail: karpov@jinr.ru
**e-mail: saiko@jinr.ru

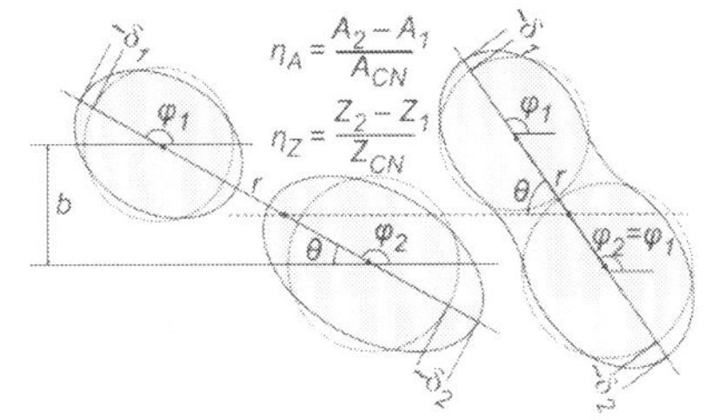

Figure 1. Schematic view of the nuclear degrees of freedom.

duce the main features of the process. On the other hand, we are limited by the present computational possibilities. The developed model has eight degrees of freedom: distance r between the centers of two nuclei, two ellipsoidal surface deformations δ_i ($i = 1, 2$), mass and charge asymmetries η_A, η_Z, two angles φ_i of rotation of the nuclei, and angle θ between the symmetry axis and beam direction. They are schematically shown in Fig. 1.

Potential energy, inertia and friction tensors are the main quantities regulating the nuclear dynamics. We calculate them on a grid, before starting the dynamical calculations. The potential energy takes into account the transition from the diabatic regime to adiabatic regime of nuclear interaction as a relaxation process with the characteristic time τ_{DA} [16]:

$$V(r,\delta_1,\delta_2,\eta_A,\eta_Z;t) = V_{\rm diab}\exp\left(-\frac{t}{\tau_{DA}}\right) + +V_{\rm adiab}\left[1-\exp\left(-\frac{t}{\tau_{DA}}\right)\right]. \quad (1)$$

The diabatic potential is calculated using the double-folding method with the Migdal nucleon-nucleon forces. The adiabatic potential energy is calculated within the macro-microscopic approach

$$V_{\rm adiab}(r,\delta_1,\delta_2,\eta_A,\eta_Z) = V_{\rm mac}(r,\delta_1,\delta_2,\eta_Z,\eta_A) + +\delta E(r,\delta_1,\delta_2,\eta_Z,\eta_A), \quad (2)$$

where $V_{\rm mac}$ is the macroscopic part of the potential calculated within the finite-range liquid-drop model [17], δE is the shell correction calculated of the basis of the two-center shell model [16, 18]. The details on the potential energy calculations may be found in Refs. [16, 19].

A rather traditional approach based on the Langevin equations was applied for the modeling of the time evolution of the nuclear system. The corresponding system of equations reads:

$$\left\{\begin{array}{ll} \dot{q}_i & = \mu_{ij}p_j, \\ \dot{p}_i & = F_i^{\rm driving} + F_i^{\rm friction} + F_i^{\rm random}, \end{array}\right. \quad (3)$$

where q_i and p_i are the collective degrees of freedom and their conjugate momenta, respectively, m_{ij} is the mass tensor $\parallel \mu_{ij} \parallel = \parallel m_{ij} \parallel^{-1}$.f The driving, friction, and random forces are calculated by the following expressions:

$$\begin{aligned} F_i^{\rm driving} &= T\frac{\partial S}{\partial q_i}, \quad S = 2\sqrt{aE^*}, \\ F_i^{\rm friction} &= -\sum_{j,k}\gamma_{ij}\mu_{jk}p_k, \\ F_i^{\rm random} &= \sum_j \theta_{ij}\xi_j(t). \end{aligned} \quad (4)$$

Here S is the entropy of the excited nuclear system, a is the level density parameter, $E^* = E_{\rm tot} - V - E_{\rm kin}$ is the excitation energy, $E_{\rm tot}$ is the total energy of the system, $E_{\rm kin}$ is the kinetic energy stored in all collective degrees of freedom, $\xi_i(t)$ are the normalized random variables with Gaussian distribution $\langle\xi_i(t)\rangle = 0$, $\langle\xi_i(t),\xi_j(t')\rangle = 2\delta_{ij}\delta(t-t')$, and θ_{ij} are the amplitudes of the random force determined from the Einstein equation $\theta_{ik}\theta_{kj} = \gamma_{ij}T$, $T = \sqrt{E^*/a}$ is the nuclear temperature, and γ_{ij} is the friction tensor.

Since the deformations, charge and mass asymmetries are finite collective modes, one should use an effective temperature $T_i^{\rm eff}$, which takes into account quantum fluctuations, instead of the heat-bath temperature T for amplitudes of the random force $\theta_{\delta_i\delta_i}$, $\theta_{\eta_Z\eta_Z}$, and $\theta_{\eta_A\eta_A}$ [20]

$$T_i^{\rm eff} = \frac{\hbar\omega_i}{2}\coth\left(\frac{\hbar\omega_i}{2T}\right), \quad (5)$$

where ω_i is the frequency with respect to the mode $i = \{\delta_1,\delta_2,\eta_Z,\eta_A\}$. Asymptotically, $T_i^{\rm eff} \simeq T$ when $T \gg \hbar\omega_i$ and $T_i^{\rm eff} = \hbar\omega_i/2$ when $T = 0$. It is clear that the use of the effective temperature instead of the thermodynamic one is extremely important at the first, approaching, stage of nucleus-nucleus collisions, when the temperature T is close to zero. This, in particular, simulates the zero-vibrations of the target and projectile, as well as mass and charge transfer between the not yet excited reaction partners. The difference in the use of $T_i^{\rm eff}$ and T gradually disappears, when the energy dissipates and the system gets excited. For the charge and mass asymmetry degrees of freedon, the use of the effective temperature is important even for well-excited systems due to large values of $\hbar\omega_{\eta_i}$ (up to several MeV).

The inertia tensor components are calculated within the Werner-Wheeler approach for incompressible irrotational flow [21] as functions of the collective degrees of freedom: m_{ij} where $\{i, j\} = \{r,\delta_1,\delta_2,\eta_A,\eta_Z\}$. The inertia coefficients for the angular degrees of freedom are the corresponding inertia moments. The one-body dissipation mechanism [22, 23] of nuclear viscosity is used for the friction tensor γ_{ij} calculation. The friction coefficient γ_{rr} calculated within the one-body friction model goes to zero for separated nuclei. However, the energy dissipation may start even before the contact. In order to account for this effect, we introduce a phenomenological nuclear friction force with a Woods-Saxon-type form factor (the friction strength depends on the amplitude γ_r^0). Thus, the friction force in the r direction is the weighted sum of these two friction forces (see, e.g., Ref. [13]). Similarly, we introduce the dissipation and inertia coefficients for the mass and charge asymmetries, which are necessary for modeling nucleon transfer between the colliding (not yet touching) nuclei.

Numerical solution of the Langevin equations (3) starts from the approaching stage of collision, when the target and projectile are separated by 50 fm, and finishes, when two reaction products are formed and separated again by an approximately 50 fm distance. Each solution of these equations is a trajectory in the space of collective coordinates. It provides full information about a single collision, such as: charges and masses of primary (excited) reaction products, their kinetic energies, scattering angles, reaction time, etc. In order to study the characteristics of final fragments we use the statistical model of de-excitation of an excited rotating nucleus [19].

In this approach, the differential cross sections are calculated as follows: (i) a large number of trajectories at different impact parameters $0 < b < b_{max}$ are simulated; (ii) additional limitations on energies, angles, etc. are imposed according to the experimental measurement conditions; (iii) the differential cross sections are then calculated in a standard way, for example:

$$\frac{d^2\sigma}{dEd\Omega}(E,\theta) = \int_0^{b_{max}} \frac{\Delta N(b,E,\theta)}{N_{tot}(b)} \frac{b\,db}{\sin\theta\Delta\theta\Delta E}, \quad (6)$$

where ΔN is a number of trajectories in specific energy and angle bins and N_{tot} is the total number of simulated trajectories for each impact parameter.

3 Results and discussions

The ^{136}Xe + ^{208}Pb reaction has been studied recently in Ref. [8] at several energies. The covered angular range was $25° \leq \theta_{lab} \leq 70°$. Both reaction products were detected in coincidence using the time-of-flight method. This allows one to determine the characteristics of the primary (excited) reaction fragments that have survived against sequential fission. The experimental resolution (FWHM) is 7 units for the fragment mass and 25 units for each of the fragment energies. These experimental conditions and uncertainties are taken into account in the calculations.

Figure 2 shows the comparison of the experimental [8] and calculated energy, mass, and angular distributions for the primary fragments of the ^{136}Xe + ^{208}Pb reaction at two incident energies $E_{c.m.}$ = 526 and 617 MeV. In addition, the total mass yields of final fragments (including the events of sequential fission) are calculated without any restrictions and assuming ideal mass resolution (thin histograms). A good overall agreement with the data should be mentioned. An evident but not strong underestimation of the maximum value of the TKEL for higher collision energy should be the subject for further model tune.

It is also interesting to compare the calculations with the data not only for the integral mass yields but also for the independent isotopic yields. Such a comparison is shown in Fig. 3 for the isotopes of Po, Rn, and Ra determined in two independent experiments [8, 11] at two rather close collision energies. The data in Ref. [8] were accumulated in the $45° \leq \theta_{lab} \leq 55°$ angular range, while the entire solid angle was covered in Ref. [11]. The dashed histograms show the primary fragments yields. Since the

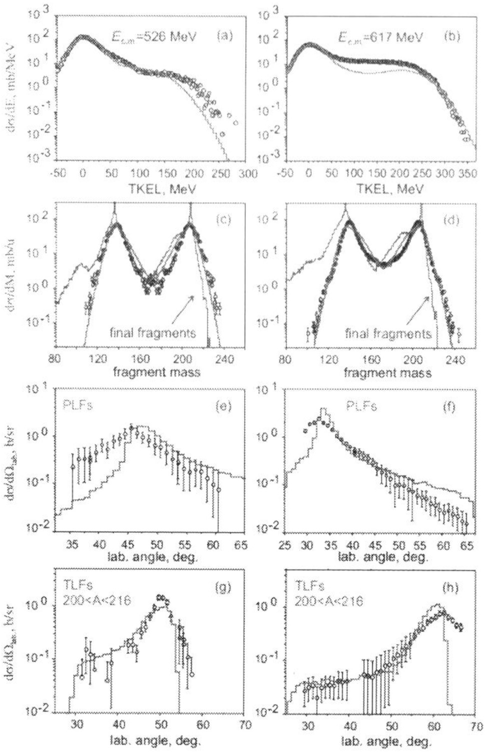

Figure 2. Energy, mass, and angular distributions of primary fragments for the ^{136}Xe + ^{208}Pb reaction for two collision energies $E_{c.m.}$ = 526 and 617 MeV. Experimental data (symbols) are from Ref. [8]. The angular distributions are shown for light reaction fragments ($A \leq 172$) and heavy fragments with masses $200 < A < 216$. The thin histograms are the calculated mass distributions of final fragments including the sequential fission events.

de-excitation process for $E_{c.m.}$ = 450 MeV goes predominantly by neutron evaporation, the final fragment yields (solid histograms) represent the nearly-parallel-shifted primary ones. For higher collision energy, the excitation is large enough to have a substantial probability of sequential fission for isotopes of radon and radium. The calculations are in a good agreement with the data for proton-enriched nuclei, while they distinctly deviate from the data on the neutron-rich side for $E_{c.m.}$ = 450 MeV and reproduce the data for $E_{c.m.}$ = 514 MeV.

An even more interesting for this particular paper comparison can be made for the production of osmium isotopes in the same projectile-target combination. This domain of nuclides is located in the vicinity of N = 126. It is this domain that we intend to study. The corresponding experiments were performed at $E_{c.m.}$ = 450 MeV in Ref. [11] and at $E_{c.m.}$ = 496 MeV in Ref. [9]. In the later case, osmium isotopes having kinetic energies in the $400 < E(\text{Os}) < 600$ MeV range were registered for

 DOI: 10.1051/epjconf/201716300027

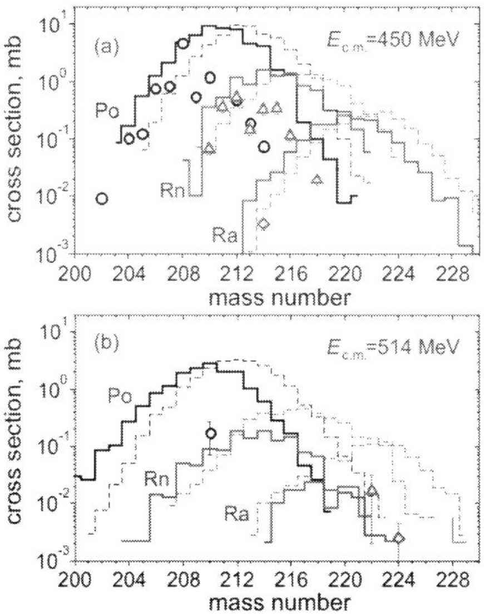

Figure 3. Calculated primary (dashed histograms) and final (solid histograms) yields of Po, Rn, and Ra isotopes for the ^{136}Xe + ^{208}Pb reaction in comparison with experimental data (circles, triangles, and diamonds for Po, Rn, and Ra, respectively) taken from Ref. [11] for $E_{c.m.}$ = 450 MeV and Ref. [8] for $E_{c.m.}$ = 514 MeV.

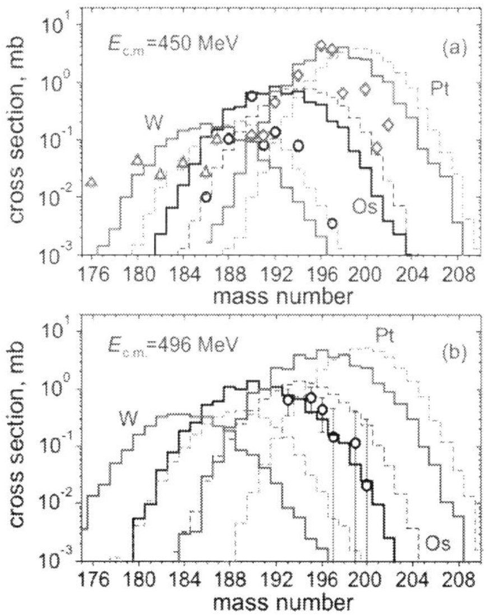

Figure 4. Same as in Fig. 3 but for W, Os, and Pt isotopes. The data (triangles, circles, and diamonds are for W, Os, and Pt, respectively) are taken from Ref. [11] for $E_{c.m.}$ = 450 MeV and Ref. [9] for $E_{c.m.}$ = 496 MeV. The data of Ref. [9] are multiplied by ten (see the text).

$31° \leq \theta_{lab} \leq 54°$. The calculated cross sections are compared with the data in Fig. 4. The data shown in Fig. 9 of Ref. [9] are not absolute cross sections but relative yields, which have been normalized to the theoretical predictions of Ref. [24]. We have normalized these data to the present calculations (the data of Ref. [9] have been multiplied by ten). Thus, one may compare only the slopes of the calculated and measured cross sections. A good agreement of the calculations with the data for $E_{c.m.}$ = 496 MeV should be noticed. As in the case of Po, Rn, and Ra isotopes, overestimation of the cross sections for production of neutron-rich nuclides is seen for the data of Ref. [11]. This suggests that additional studies (both theoretical and experimental) are required.

In order to further verify the model and following the suggestion of L. Corradi made during the FUSION17 conference, the experimental data for the ^{136}Xe+^{198}Pt system at $E_{c.m.}$ = 643 MeV [10] have been analyzed. The results are summarized in Fig. 5 for the projectile-like fragments (PLF) and in Fig. 6 for the osmium and mercury isotopes. In the later case, the data accumulated in different TKEL gates are shown as well. The cross sections are integrated over the experimentally covered angles from 24° to 34°. The calculations reproduce all the data for this system quite well. A shift (maximum of three mass units) of the calculations with respect to the data for the PLF towards neutron-richness is however seen for larger proton transfers.

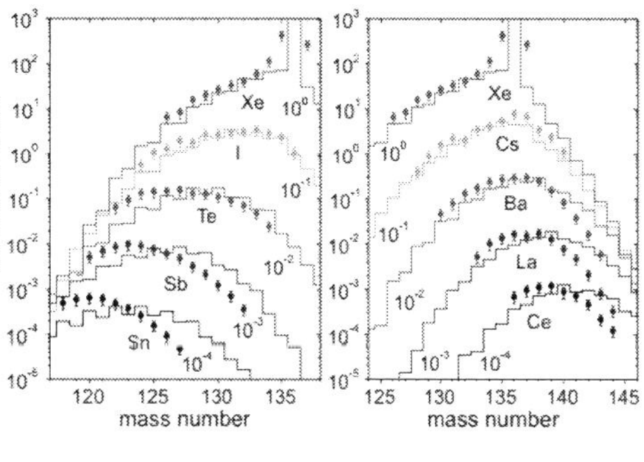

Figure 5. Calculated (histograms) and experimental [10] (symbols) cross sections of PLF for the ^{136}Xe + ^{198}Pt reaction at $E_{c.m.}$ = 643 MeV.

Finally, the data on the production cross sections of the N = 126 nuclides in the ^{136}Xe+^{198}Pt,^{208}Pb reactions are compared with the calculations in Fig. 7 (a). Regarding the comparison with the experimental data, one may see a good agrement with the data of Ref. [10] and proton-rich nuclei from Ref. [11]. The calculated cross sections for production of neutron-rich nuclei in the ^{136}Xe+^{208}Pb reaction strongly overestimate the data. According to the calculations, the cross sections are larger for the ^{136}Xe+^{198}Pt combination than for the ^{136}Xe+^{208}Pb one. The difference increases with the increasing neutron-richness of a nuclide. The cross sections calculated at two different in-

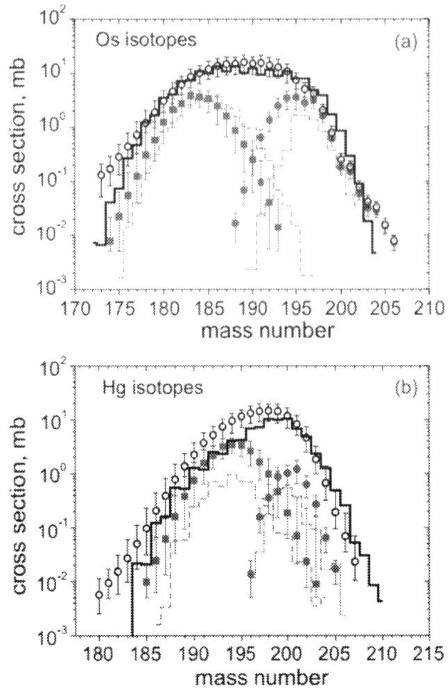

Figure 6. Calculated (solid histograms) and experimentally deduced in Ref. [10] (open circles) cross sections for Hg and Os isotopes integrated over the angular range from 24° to 34°. Isotopic distributions for different ranges of TKEL from -25 to 25 MeV (dash-dotted histograms and filled circles) and from 175 to 225 MeV (dashed histograms and filled squares) are shown.

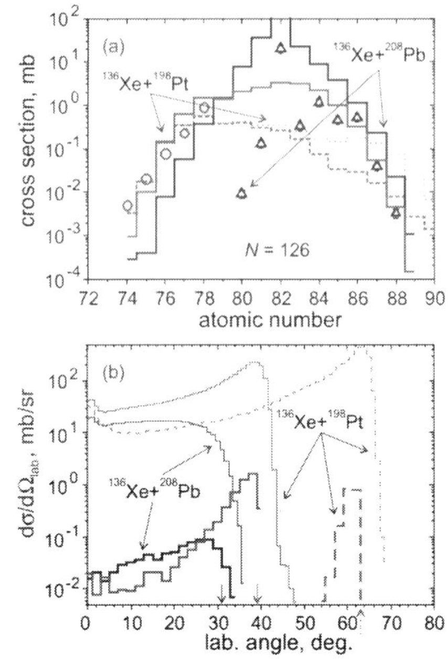

Figure 7. (a): Calculated (histograms) and experimental (symbols) cross sections for production of isotopes with N = 126 in the ^{136}Xe+^{198}Pt,^{208}Pb reactions. The solid and dashed histograms are for $E_{c.m.}$ = 450 and 643 MeV, respectively. The thin dashed curve is integrated over all angles and the thick one – over the experimentally covered angles from 24° to 34°. The experimentally deduced cross sections for the ^{136}Xe+^{198}Pt system is from Ref. [10] and for ^{136}Xe+^{208}Pb is from Ref. [11]. (b): Calculated angular distributions in the laboratory frame for the Ir isotopes for the ^{136}Xe+^{198}Pt,^{208}Pb reactions. The thick histograms are for the ^{203}Ir$_{126}$ nucleus and the thin histograms are integrated over all iridium isotopes. The solid and dashed histograms are for $E_{c.m.}$ = 450 and 643 MeV, respectively. The vertical arrows indicate the corresponding grazing angles.

cident energies for ^{136}Xe+^{198}Pt are quite close. Thus, an optimal energy should be searched for in future studies.

Rather important conclusions can be made regarding the optimal angular ranges for production of neutron-rich nuclei in multinucleon transfer reactions [see Fig. 7 (b)]. While the angular distributions are broad being integrated over an isotope mass (iridium in our case), they are much more localized for a specific neutron-rich isotope (the ^{203}Ir$_{126}$ case is shown). The maximum for all the three studied reactions are close to the corresponding angles of grazing collisions. Obviously, the degree of localization depends on the system and incident energy. The sharpest angular distribution is predicted for the ^{136}Xe+^{198}Pt system at $E_{c.m.}$ = 643 MeV. It is consistent with the calculations of cross sections [compare thick and thin histograms for this system in Fig. 7 (a)], as well as the data and calculations in Fig. 6. This figure shows that the cross sections for production of neutron-rich isotopes are almost exhausted by events having low TKEL that corresponds to peripheral collisions. Highly excited transfer products formed in central collisions do not survive. Low values of neutron binding energies amplify this effect. At lower near-barrier energies, even central collisions lead to low excitation energies of primary products (see difference of the primary and final cross sections shown in Fig. 4 for two collision energies). Moreover, the interaction time increases with the decreasing impact parameter that stimulates nucleon transfer. Thus, the angular distributions for neutron-rich nuclides become broader extending towards the forward angles. The most uniform distribution is obtained for the ^{136}Xe+^{208}Pb combination at $E_{c.m.}$ = 450 MeV that can be explained by the lowest energy above the barrier (contact point potential). This peculiarities should be taken into account, while constructing experimental setups for studying multinucleon transfer reactions and planning the corresponding experiments.

Figure 8 shows the contour map of the production cross sections for ^{136}Xe+^{208}Pb ($E_{c.m.}$ = 450 MeV). It is seen that the last shown line corresponding to 100 nb follows the border of the known neutron-rich nuclei in a wide range of atomic numbers. The most neutron-rich nuclides in the N = 126 area can be synthesized in this reaction with the cross sections exceeding 100 nb, that is more than five orders of magnitude higher than one may expect for the fragmentation reactions used so far.

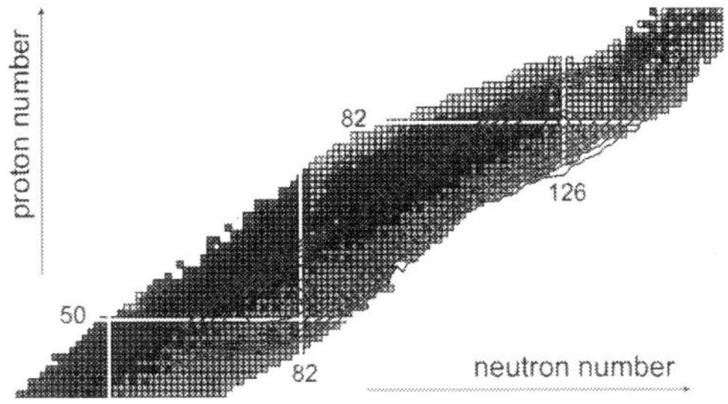

Figure 8. Upper part of the chart of nuclides. The contour lines are the multinucleon transfer cross sections for the ^{136}Xe + ^{208}Pb system at $E_{c.m.}$ = 450 MeV. The contour lines are drawn over each order of the cross section down to 100 nb.

To conclude, the developed dynamical model of nucleus-nucleus collisions has shown its applicability for deep inelastic reactions. The performed calculations allows us to make rather optimistic conclusions about the perspectives of using the multinucleon transfer reactions for production of neutron-rich nuclei in the vicinity of the N = 126 shell closure.

References

[1] S.J. Steer, Zs. Podolyák, S. Pietri, M. Górska, G.F. Farrelly, P.H. Regan, D. Rudolph, A.B. Garnsworthy, R. Hoischen, J. Gerl, H.J. Wollersheim, H. Grawe, K.H. Maier, F. Becker, P. Bednarczyk, L. Cáceres, P. Doornenbal, H. Geissel, J. Grebosz, A. Kelic, I. Kojouharov, N. Kurz, F. Montes, W. Prokopowicz, T. Saito, H. Schaffner, S. Tashenov, A. Heinz, T. Kurtukian-Nieto, G. Benzoni, M. Pfützner, A. Jungclaus, D.L. Balabanski, C. Brandau, A. Brown, A.M. Bruce, W.N. Catford, I.J. Cullen, Zs. Dombrádi, M.E. Estevez, W. Gelletly, G. Ilie, J. Jolie, G.A. Jones, M. Kmiecik, F.G. Kondev, R. Krucken, S. Lalkovski, Z. Liu, A. Maj, S. Myalski, S. Schwertel, T. Shizuma, P.M. Walker, E. Werner-Malento, and O. Wieland, Int. J. Mod. Phys. E **18**, 1002 (2009).

[2] J. Kurcewicz, F. Farinon, H. Geissel, S. Pietri, C. Nociforo, A. Prochazka, H. Weick, J.S. Winfield, A. Estradé, P.R.P. Allegro, A. Bail, G. Bélier, J. Benlliure, G. Benzoni, M. Bunce, M. Bowry, R. Caballero-Folch, I. Dillmann, A. Evdokimov, J. Gerl, A. Gottardo, E. Gregor, R. Janik, A. Kelić-Heil, R. Knöbel, T. Kubo, Yu.A. Litvinov, E. Merchan, I. Mukha, F. Naqvi, M. Pfützner, M. Pomorski, Zs. Podolyák, P.H. Regan, B. Riese, M.V. Ricciardi, C. Scheidenberger, B. Sitar, P. Spiller, J. Stadlmann, P. Strmen, B. Sun, I. Szarka, J. Taïeb, S. Terashima, J.J. Valiente-Dobón, M. Winkler, and Ph. Woods, Phys. Lett. B **717**, 371 (2012).

[3] A.G. Artukh, G.F. Gridnev, V.L. Mikheev, V.V. Volkov, and J. Wilczynski, Nucl. Phys. A **215**, 91 (1973).

[4] V.V. Volkov, Phys. Rep. **44**, 93 (1978); V.V. Volkov, Nuclear Reactions of High-Inelastic Transfers, Energoizdat, Moscow, 1982 (in Russian).

[5] C.H. Dasso, G. Pollarolo, and A. Winther, Phys. Rev. Lett. **73**, 1907 (1994).

[6] R. Broda, J. Phys. G **32**, R151 (2006).

[7] L. Corradi, G. Pollarolo, and S. Szilner, J. of Phys. G **36**, 113101 (2009).

[8] E.M. Kozulin, E. Vardaci, G.N. Knyazheva, A.A. Bogachev, S.N. Dmitriev, I.M. Itkis, M.G. Itkis, A.G. Knyazev, T.A. Loktev, K.V. Novikov, E.A. Razinkov, O.V. Rudakov, S.V. Smirnov, W. Trzaska, and V.I. Zagrebaev, Phys. Rev. C **86**, 044611 (2012).

[9] K. Novikov, E.M. Kozulin, I.M. Harca, S. Dmitriev, A. Flueras, P. Greenlees, F. Hanappe, S.V. Khlebnikov, T. Loktev, J. Maurer, A. Di Nitto, J. Pakarinen, P. Ruotsalainen, M. Sandzelius, J. Sorri, M. Sin, W.H. Trzaska, E. Vardaci, and V.I. Zagrebaev, J. Phys.: Conf. Ser. **515**, 012016 (2014).

[10] Y.X. Watanabe, Y.H. Kim, S.C. Jeong, Y. Hirayama, N. Imai, H. Ishiyama, H.S. Jung, H. Miyatake, S. Choi, J.S. Song, E. Clement, G. de France, A. Navin, M. Rejmund, C. Schmitt, G. Pollarolo, L. Corradi, E. Fioretto, D. Montanari, M. Niikura, D. Suzuki, H. Nishibata, and J. Takatsu, Phys. Rev. Lett. **115**, 172503 (2015).

[11] J.S. Barrett, W. Loveland, R. Yanez, S. Zhu, A.D. Ayangeakaa, M.P. Carpenter, J.P. Greene, R.V.F. Janssens, T. Lauritsen, E.A. McCutchan, A.A. Sonzogni, C.J. Chiara, J.L. Harker, and W.B. Walters, Phys. Rev. C **91**, 064615 (2015).

[12] S. Heinz, O. Beliuskina, V. Comas, H.M. Devaraja, C. Heinz, S. Hofmann, E. Kozulin, F. Morherr, G. Münzenberg, D. Ackermann, F.P. Heßberger, B. Kindler, B. Lommel, R. Mann, and J. Maurer, Eur. Phys. J. A **51**, 140 (2015).

[13] V.I. Zagrebaev and W. Greiner, J. Phys. G **34**, 1 (2007).

[14] V.I. Zagrebaev and W. Greiner, J. Phys. G **35**, 125103 (2007).

[15] A.V. Karpov and V.V. Saiko, Phys. Part. Nucl. Lett. (2017) in press.

[16] V. Zagrebaev, A. Karpov, Y. Aritomo, M. Naumenko, and W. Greiner, Phys. Part. Nucl. **38**, 469 (2007).

[17] P. Möller, J.R. Nix, W.D. Myers, and W.J. Swiatecki, At. Data Nucl. Data Tables **59**, 185 (1995).

[18] J. Maruhn and W. Greiner, Z. Phys. A **251**, 431 (1972).

[19] Nuclear Reaction Video project, http://nrv.jinr.ru.

[20] L.D. Landau and E.M. Lifshitz, Statistical Physics. Part 1, Pergamon Press, Oxford, 1980.

[21] K.T.R. Davies, A.J. Sierk, and J.R. Nix, Phys. Rev. C **13** 2385 (1976).

[22] A.J. Sierk and J.R. Nix, Phys. Rev. C **21**, 982 (1980).

[23] A.V. Karpov and G.D. Adeev, Phys. At. Nucl. **65**, 1596 (2002); Eur. Phys. J. A **14**, 169 (2002).

[24] V.I. Zagrebaev and W. Greiner, Phys. Rev. Lett. **101**, 122701 (2008).

DOI: 10.1051/epjconf/201716300028

Direct measurement of the ^{7}Be(n, α)^{4}He reaction cross sections for the cosmological Li problem

Takahiro Kawabata[1,⋆], *Yuki* Fujikawa[2], *Tatsuya* Furuno[1], *Tatsuya* Goto[2], *Toshikazu* Hashimoto[1], *Masaya* Ichikawa[1], *Makoto* Itoh[1], *Naohito* Iwasa[3], *Yoshiko* Kanada-En'yo[1], *Ami* Koshikawa[1,2], *Shigeru* Kubono[4], *Eisuke* Miyawaki[2], *Masatoshi* Mizuno[2], *Keigo* Mizutani[1], *Takahiro* Morimoto[1,2], *Motoki* Murata[1], *Takuya* Nanamura[1,2], *Shunji* Nishimura[4], *Takuya* Nanamura[1,2], *Shintaro* Okamoto[2], *Yuichi* Sakaguchi[2], *Itsushi* Sakata[2], *Akane* Sakaue[1], *Ryo* Sawada[2], *Yuki* Shikata[2], *Yu* Takahashi[2], *Daiki* Takechi[2], *Tomoya* Takeda[2], *Chisato* Takimoto[2], *Miho* Tsumura[1], *Ken* Watanabe[2], and *Sota* Yoshida[2]

[1]*Department of Physics, Kyoto University, Kitashirakawa-Oiwake, Sakyo, Kyoto 606-8502, Japan*
[2]*Faculty of Science, Kyoto University, Kitashirakawa-Oiwake, Sakyo, Kyoto 606-8502, Japan*
[3]*Department of Physics, Tohoku University, Aoba, Sendai, Miyagi 980-8578, Japan*
[4]*RIKEN Nishina Center, RIKEN, 2-1 Hirosawa, Wako, Saitama 351-0198, Japan*

Abstract. The cross sections of the ^{7}Be(n, α)^{4}He reaction for p-wave neutrons were experimentally determined at $E_{c.m.}$ = 0.20–0.81 MeV close to the Big Bang nucleosynthesis (BBN) energy window for the first time on the basis of the detailed balance principle by measuring the time-reverse reaction. The obtained cross sections are much larger than the cross sections for s-wave neutrons inferred from the recent measurement at the n_TOF facility in CERN, but significantly smaller than the theoretical estimation widely used in the BBN calculations. The present results suggest the ^{7}Be(n,α)^{4}He reaction rate is not large enough to solve the cosmological lithium problem.

1 Introduction

The primordial abundances of the light elements produced in the Big Bang nucleosynthesis (BBN) provide important insights into the early universe. Accurate estimation of the primordial abundances is crucial to test the cosmological theories by comparing the predicted values with the observations.

The BBN theory relies on nuclear reactions among the primordial light elements and their electroweak decays. This theory has only one free parameter "baryon density", which is related to the baryon-to-photon ratio. A precise value for the baryon density was reported as $\Omega_b h^2 = 0.0223 \pm 0.0002$ from the anisotropy analysis of the cosmic microwave background by the Planck satellite [1].

The theoretical predictions of the primordial abundances reasonably well agree with the observations for the helium and deuterium. However, there remains a serious problem that the ^{7}Li abundance does not agree with any theoretical BBN calculations. The primordial abundance of ^{7}Li inferred from observations of metal-poor stars is ^{7}Li/H = $1.6 \pm 0.3 \times 10^{-10}$ [2], whereas the calculated values by the BBN theory is as large as ^{7}Li/H = $4.68 \pm 0.67 \times 10^{-10}$ [3]. This discrepancy is known as the cosmological lithium problem, and has been of great interest in recent years.

Several ideas have been proposed to solve this problem [4]. One idea is to improve the current understanding of the stellar processes that exhaust lithium in metal-poor stars. Other ideas are to find new physics beyond the standard BBN model, *e.g.*, cosmological variation of fundamental constants [5], decay of supersymmetric particles [6], and so on. However, there is still no experimental evidence to confirm these models.

From a view of nuclear physics, nuclear-reaction rates involved in the BBN theory should be examined. The main process of the ^{7}Li production in the BBN is the electron-capture decay of ^{7}Be, which is synthesized in the ^{3}He(^{4}He,γ)^{7}Be reaction. Direct measurements of the cross section for the ^{3}He(^{4}He,γ)^{7}Be reaction were extensively carried out in the past by several groups, and uncertainties in this thermonuclear reaction rate are now very small. There is no room to modify the ^{7}Be production rate to solve the lithium problem [7].

Recently, it was pointed out that the ^{7}Li abundance will be greatly reduced in the BBN calculation if the destruction rate of ^{7}Be is enhanced. One of the candidate channels to destruct ^{7}Be is the ^{7}Be(n,α)^{4}He reaction.

The effective energy windows for the neutron induced reactions are given in Ref. [8]. For s-wave neutrons at low energies, the cross sections follow the $1/v$ rule, *i.e.*, the cross sections are inversely proportional to the velocity of neutrons [9]. Therefore, the reaction rate has a peak at lower energies, whereas the centrifugal barrier shifts

⋆e-mail: kawabata@scphys.kyoto-u.ac.jp

the effective energy to higher energies for higher partial waves. At the BBN temperature around $T_9 \sim 0.7$, the effective energy window is $E = 30 \pm 24$ and 90 ± 42 keV for the s- and p-wave neutrons, respectively. Unfortunately, the cross section for the ^{7}Be(n,α)^{4}He reaction at the cosmological energy has been scarcely measured.

Since ^{7}Be is a short-lived nucleus, it is not easy to directly measure the cross section for the ^{7}Be(n,α)^{4}He reaction. This reaction cross sections were previously measured at the thermal energy much lower than the BBN energy [10]. Very recently, the cross section for this reaction was measured using a radioactive ^{7}Be target irradiated by a low-energy neutron beam at 10 meV–10 keV [11], although these reaction energies are still lower than the BBN energy. In this measurement, two α particles emitted from the ^{7}Be(n,α)^{4}He reaction were detected.

Such low-energy reactions proceed via the s wave only. Actually, the energy dependence of the measured cross section in Ref. [11] agrees with the $1/v$ rule, and therefore, no sizable p-wave component was observed at $E_n < 10$ keV. This result suggests the p-wave cross section is less than a few hundreds μb at $E_n \sim 10$ keV. In the s-wave reaction, negative-parity states at $E_x \sim 19$ MeV near the neutron-decay threshold in ^{8}Be are formed at first. Since the direct 2α decay of these negative-parity states is forbidden due to the parity conservation, the two α particles must be emitted after the γ decay populating low-lying positive natural-parity states. The branching ratio of the electromagnetic decay above the particle emission threshold is generally small and thus, the cross section for the ^{7}Be destruction by s-wave neutrons is suppressed.

Among various positive natural-parity states in ^{8}Be populated by the γ decay, the dominant decay channels proceed through the ground and first excited states because the $E1$ decay probability is proportional to the cube of the decay energy. However, these dominant channels were not measured in Ref. [11] due to difficulties in the low-energy α-particle detection, but the minor decay channels emitting the high-energy α-particles via the positive natural-parity states at $E_x > 8$ MeV in ^{8}Be were measured. The branching ratio between the dominant and minor decay channels was inferred by the theoretical calculation assuming the direct radiative capture mechanism [12]. It should be noted that the total ^{7}Be(n,α)^{4}He cross section for the s-wave neutrons was estimated to be as small as a few mb in the BBN energy window, and the s-wave reaction does not contribute to solve the cosmological Li problem.

Contrary to the s-wave reaction where the cross section decreases with the increasing neutron energy according to the $1/v$ rule, the p-wave reaction might become dominant in the BBN energy window. The p-wave cross sections can be determined by measuring the time-reverse ^{4}He(α, n)^{7}Be reaction on the basis of the detailed balance principle. Since resonant states in ^{8}Be formed by the α+α collision must be positive natural-parity states, these states decay to the ground and first excited states in ^{7}Be by emitting p-wave neutrons.

The cross section for the ^{4}He(α, n)^{7}Be reaction at the lowest energy ever measured was reported in Ref. [13]. Residual ^{7}Be nuclei were implanted in aluminum foil, and the number of ^{7}Be in the foil was determined by the off-line measurement. Although the cross sections leading to the ground and first excited states in ^{7}Be could not be separated, one can estimate the upper limit of the cross section for the ^{7}Be(n,α)^{4}He reaction from the previous data. However, the lowest reaction energy of $E_{c.m.} = 0.7$ MeV is much higher than the BBN energy window, and this result is not directly related to the cosmological lithium problem.

The cross sections for the ^{7}Be(n,α)^{4}He reaction at $E_{c.m.} = 0.0113$–5.754 MeV were calculated from the cross section for the mirror reactions ^{7}Li(p,α)^{4}He and ^{4}He(α,p)^{7}Li by assuming the charge symmetry [14]. However, it is not trivial whether the charge symmetry is still good approximation at low reaction energies near the threshold. Therefore, the direct measurement of the ^{4}He(α,n)^{7}Be reaction at the cosmological energies was strongly desired.

In the present work, we have measured the cross section for the ^{4}He(α,n)^{7}Be reaction at low reaction energies $E_\alpha = 38.50$–39.64 MeV. The final states in the residual ^{7}Be nucleus were unambiguously identified by the on-line spectroscopy of the scattered neutrons. On the basis of the detailed balance principle, we have successfully obtained the cross sections for the ^{7}Be(n,α)^{4}He reaction at $E_{c.m.} = 0.20$–0.81 MeV close to the BBN energy window for the first time.

2 Experiment

The experiment was carried out at the neutron time-of-flight facility in Research Center for Nuclear Physics (RCNP), Osaka University [15]. An α beam accelerated up to 39.89 or 39.56 MeV was transported to a He gas target in the beam swinger magnet. During the measurement using the α beam at 39.56 MeV, aluminum plates with the thicknesses of 12 and 24 μm were installed in the beam line to degrade the beam energy to 39.15 and 38.76 MeV. Background particles caused by the beam degraders were swept away with the bending magnets and movable collimator system before the target. Taking the energy loss in the target into account, the beam energies at the center of the He gas target were determined to be 39.64, 39.30, 38.90, and 38.50 MeV as the nominal beam energies in the present work. These beam energies are close to the threshold energies at $E_\alpha = 37.98$ and 38.84 MeV for the ^{4}He(α,n)^{7}Be reaction leading to the ground state and the first excited state at $E_x = 0.429$ MeV in ^{7}Be, respectively. The uncertainties of the beam energies were estimated at 40 keV, while the reaction energies fluctuated about ± 100 keV depending on the position where the reaction occurred in the target due to the difference in the energy loss of α particles.

The scattered neutrons at $E_n = 2$–7 MeV were detected by a BC-501A liquid scintillation detector, which was located at 13-m away from the target. The sensitive volume of the scintillation detector was a cylindrical shape with the diameter of 200 mm and the depth of 50 mm along the neutron trajectory. The detection threshold of the light output from the BC-501A liquid scintillator was set at 0.153 MeVee. A conventional pulse-shape discrimination

 DOI: 10.1051/epjconf/201716300028

technique was applied to distinguish neutrons from γ rays. The detection efficiency of neutrons by the BC-501A liquid scintillation detector was estimated by using the computer code SCINFUL-CG [16]. We also measured the neutron detection efficiency at E_n = 2.5, 4.9, and 8.1 MeV using the tagged neutrons emitted from the $d + d \rightarrow {}^3$He + n reaction. The measured efficiency was 0.323 ± 0.014, 0.302 ± 0.020, and 0.201 ± 0.017 at E_n = 2.5, 4.9, and 8.1 MeV, respectively. The calculated efficiency agrees with the measurement within the measurement uncertainties.

The He gas was filled at 1 atm in the target cell with the effective length of 6.3 cm. The target cell has the entrance and exit windows with the diameter of 12 mm, which are made of the 6-μm Aramid films. The window material was carefully chosen from three candidates (tantalum, Havar alloy, and Aramid) through the background measurements, and its thickness was minimized to sustain the gas pressure by the mechanical consideration of the breaking strength.

The mass thicknesses of the He gas and Aramid films were 1.0 and 1.7 mg/cm^2, respectively, and the energy loss of the α beam across 6.3 cm in the He gas was about 0.2 MeV. During the measurement, the temperature and pressure of the He gas were monitored using the Pt resistor and diaphragm gauge.

The measurements were carried out at 5 scattering angles between 0° and 20° in the laboratory frame. A typical neutron-energy spectrum in the ^{4}He(α,n)^{7}Be reaction measured at E_α = 39.30 MeV and θ_{lab} = 0° is shown in Fig. 1. The two prominent peaks due to the ground ($3/2^-_1$) and first excited ($1/2^-_1$) states are clearly observed on the continuous background due to the window films in Fig. 1(a). The background-free spectra were successfully obtained by subtracting background spectra taken from the empty-cell measurement as seen in Fig. 1(b). The neutron-energy resolution was 250–570 keV at the full width at half maximum. The energy spread of the neutrons was mainly due to the fluctuation of the reaction energy and kinematical effects.

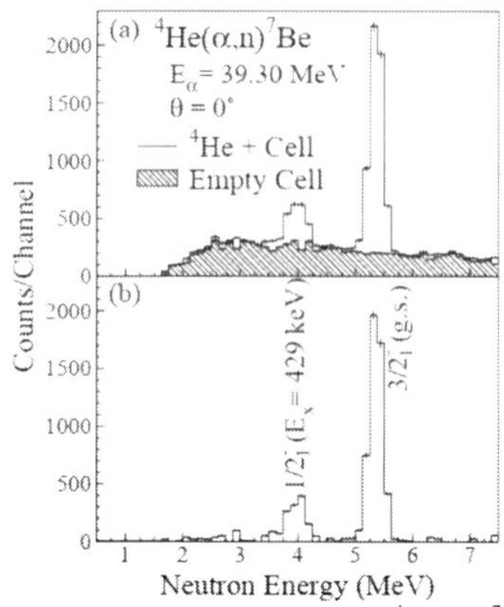

Figure 1. Neutron energy spectra in the ^{4}He(α,n)^{7}Be reaction measured at E_α = 39.30 MeV and θ_{lab} = 0°. (a) Open and hatched spectra were taken using the He gas-filled and empty cells, respectively. (b) Background-free spectrum obtained by subtracting the background spectrum taken from the empty-cell measurement.

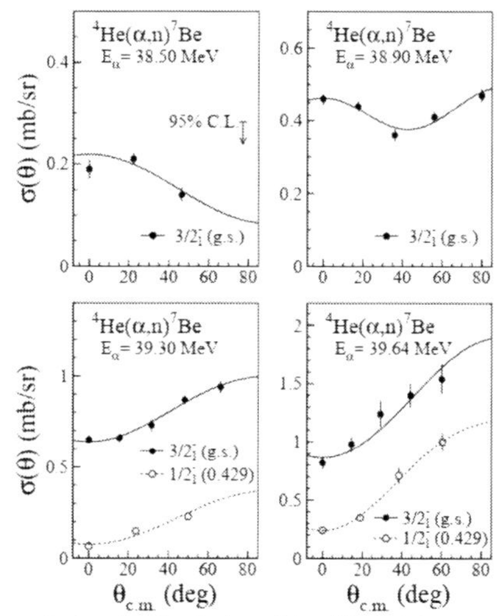

Figure 2. Angular distribution of the differential cross sections for the ^{4}He(α,n)^{7}Be reaction. The solid and open circles show the differential cross sections for the ground and first excited states in ^{7}Be, respectively. The solid and dashed lines are series of the even-order Legendre polynomials to fit the experimental data.

3 Results and discussion

The angular distributions of the differential cross sections are shown in Fig. 2. The solid and open circles show the differential cross sections for the ground and first excited states in ^{7}Be, respectively. The vertical bars present the measurement uncertainties including both the statistical and systematic uncertainties. The cross sections for the first excited state were not measured at E_α = 38.90 and 38.50 MeV since these beam energies are below or very close to the threshold energy for the first excited state at E_α = 38.84 MeV. Unfortunately, the ^{4}He(α,n)^{7}Be events could not be reliably separated from the background events at E_α = 38.50 MeV and $\theta_{c.m.}$ = 77.3°, because the energy resolution was poor due to the kinematical effects. Therefore, the upper limit for the differential cross section was given at the 95% confidence level.

Since the angular distribution of the cross section for the two-body scattering of the identical particles must be symmetric with respect to $\theta_{c.m.}$ = 90°, the measured cross section can be fitted by a series of the even-order Legendre polynomials

$$\sigma(\theta) = \frac{\sigma}{4\pi}\left(1 + \sum_{\substack{l=2\\ l:\,\mathrm{even}}}^{l_{\max}} \alpha_l P_l(\cos\theta)\right),$$

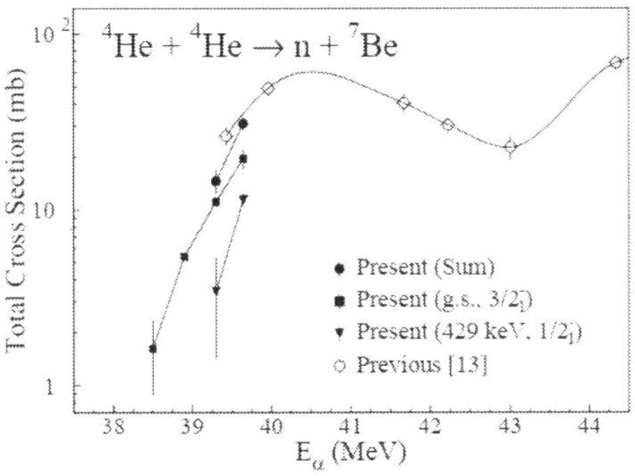

Figure 3. Measured total cross sections of the ^{4}He(α,n)^{7}Be reaction for the ground (solid squares) and first excited (solid triangles) states. The sum of the two states is shown by solid circles. Previous results from Ref. [13] are shown by the open circles. The solid lines interpolating the data points are drawn for eye-guide.

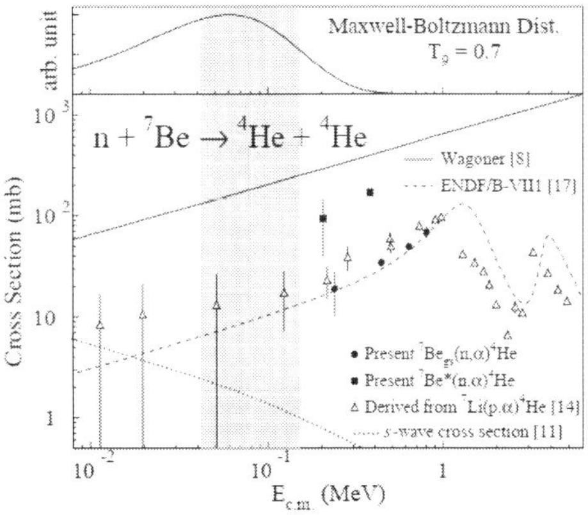

Figure 4. Measured total cross sections for the ^{7}Be(n,α)^{4}He reaction on the ground (solid circles) and first excited (solid squares) states. The cross sections for the first excited state are larger than those for the ground state, but those for the first excited state make no sizable contribution to the destruction rate of ^{7}Be at the BBN temperature. The open triangles, solid line, dashed line, and dotted line in the bottom panel are the previous evaluations from Refs. [14], [8], [17], and [11], respectively. The solid line in the top panel shows the Maxwell-Boltzmann distribution at T_9 = 0.7. The shaded area presents the effective-energy window for the p-wave reaction at T_9 = 0.6–0.8.

and the total cross section σ was obtained. Because the present measurement was carried out at beam energies close to the n + ^{7}Be threshold energy, the Legendre polynomial expansion should involve a few low-order terms only. We decided l_{max} = 4 for the ground state at $E_\alpha \geq$ 38.90 MeV and 2 for the other states. The fit functions are shown by the solid and dashed lines in Fig. 2.

The total cross sections were evaluated by using the different l_{max} values in order to estimate the uncertainties stem from the assumption about the l_{max} values, and found the variation of the total cross sections was less than 10% in most cases except the low energy cases where the number of the measured angles is 3. The uncertainties are, therefore, as large as 45% and 60% for the ground state at E_α = 38.50 MeV and the first excited state at E_α = 39.30 MeV, respectively.

The obtained total cross sections are shown in Fig. 3. The solid squares and triangles are the present results for the ground and first excited states. The sum of the two states are shown by the solid circles to directly compare the present results with the previous results [13] presented by the open circles. The solid lines interpolating the data points are drawn for eye-guide. The present and previous results agree within the measurement uncertainties around E_α = 39.4 MeV. It should be noted that the present measurement extends the experimental data toward lower energies down to E_α = 38.50 MeV for the first time.

The cross section for the first excited state decreases with the beam energy more rapidly than that for the ground state. This fact is naturally understood from the threshold effect, *i.e.*, the neutron penetrability for the p-wave centrifugal barrier and the phase space volume for the first excited state are strongly suppressed as approaching to the threshold energy at E_α = 38.84 MeV.

From a view of the cosmological lithium problem, the total cross section for the ^{7}Be(n,α)^{4}He reaction is important. The cross section for the reaction ^{7}Be + n → ^{4}He + ^{4}He is related to that for the time-reverse reaction ^{4}He + ^{4}He → ^{7}Be + n by the detailed balance principle:

$$\frac{\sigma(^7\text{Be} + n \rightarrow {}^4\text{He} + {}^4\text{He})}{\sigma(^4\text{He} + {}^4\text{He} \rightarrow {}^7\text{Be} + n)} = \frac{\hat{S}_{^4\text{He}}\hat{S}_{^4\text{He}}k^2_{^4\text{He}+^4\text{He}}}{2\hat{S}_{^7\text{Be}}\hat{S}_n k^2_{^7\text{Be}n}}, \quad (1)$$

where $\hat{S}_A = 2S_A + 1$. S_A is the spin of A, and $\hbar k_{AB}$ is the relative momentum in the $A + B$ channel. The measured ^{4}He(α,n)^{7}Be cross sections for the ground and first excited states in ^{7}Be are separately converted to the cross section of the time-reverse reactions for p-wave neutrons. The solid circles and squares in Fig. 4 show the total cross sections of the (n,α) reaction on the ground and first excited states in ^{7}Be.

The shaded area presents the effective-energy window for the p-wave reaction at T_9 = 0.6–0.8, and the peak of the thermal-energy distribution at T_9 = 0.7 given by the Maxwell-Boltzmann distribution is located in the effective-energy window as shown in the top panel of Fig. 4.

The cross sections evaluated by the indirect methods are compared with the present results in Fig. 4. The estimation from p + ^{7}Li scattering [14] is plotted by the open triangles, whereas the cross section from the evaluated nuclear data library ENDF/B-VII.1 [17] based on the R-matrix analysis of several indirect reactions is shown by the dashed line. It was found that these evaluated cross sections are very close to the present data for the ^{7}Be$_{g.s.}$(n,α)^{4}He reaction but inconsistent with the measurement in Ref. [11] around $E_n \sim$ 10 keV. ENDF/B-VII.1

suggests the p-wave cross section is about 3 mb, whereas no sizable p-wave component was observed in Ref. [11].

The cross section for s-wave neutrons inferred in Ref. [11] is shown by the dotted line. It is smaller than the cross section for p-wave neutrons around the thermal energy at $T_9 \sim 0.7$, and thus the p-wave reaction is dominant compared to the s-wave reaction in the BBN energy window.

The cross section for the $^7Be(n,\alpha)^4He$ reaction was first estimated by Wagoner [8] as shown by the solid line in the bottom panel of Fig. 4. Currently, this evaluation is widely used in the BBN calculations. The present values of the $^7Be(n,\alpha)^4He$ cross sections are much smaller than the Wagoner's calculation. The cross sections of the first excited state at E_x = 492 keV in ^7Be are larger than those of the ground state owing to the kinematical factors in Eq. (1). Although low-lying excited states can be thermally populated in the high-temperature environment, the excitation energy of the first excited state is too high to make a sizable contribution to the destruction rate of ^7Be at the BBN temperature. Moreover, the cross sections for the $^7Be^*(n,\alpha)^4He$ reaction are still smaller than the Wagoner's estimation. As a conclusion, the present results suggest that the $^7Be(n,\alpha)^4He$ reaction does not solve the cosmological lithium problem.

4 Summary

The cross sections for the $^4He(\alpha,n)^7Be$ reaction were measured at low energies between E_α = 38.50 and 39.64 MeV. The final states in the residual ^7Be nucleus were unambiguously identified by the on-line spectroscopy of the scattered neutrons. On the basis of the detailed balance principle, the cross sections for the $^7Be(n,\alpha)^4He$ reaction were obtained at $E_{c.m.}$ = 0.20–0.81 MeV slightly above the BBN energy window for the first time. It was found that the p-wave reaction is dominant compared to the s-wave reaction in the BBN energy window, however the obtained cross sections are significantly smaller than the theoretical estimation widely used in the BBN calculations. The present results suggest the $^7Be(n,\alpha)^4He$ reaction rate is not large enough to solve the cosmological lithium problem. This conclusion agrees with the recent result from the direct measurement of the s-wave cross sections using a low-energy neutron beam [11] and the evaluated nuclear data library ENDF/B-VII.1 [17].

The present work was performed as a graduation research by the undergraduate senior students at Kyoto University (KADAIKENKYU P4) over the three school years. All the processes of the research project, *i.e.*, planning, development, measurement, and analysis were done by the undergraduate students under the supervision of the faculties. The authors are grateful to RCNP for the deep understanding about the importance of the undergraduate education through the hands-on training at the large accelerator facility. The authors also thank the cyclotron crews at RCNP for their efforts to provide a clean and stable beam, and Prof. T. Wakasa from Kyushu University for his kind support to carry out the present measurement at the neutron time-of-flight facility. This work was supported by JSPS KAKENHI Grant Numbers JP26287058 and JP15H02091.

References

[1] P.A.R. Ade et al. (Planck Collaboration), Astron. Astrophys. **571**, A16 (2014)

[2] L. Sbordone et al., Astron. Astrophys. **522**, 26 (2010)

[3] R.H. Cyburt, B.D. Fields, K.A. Olive, T.H. Yeh, Rev. Mod. Phys. **88**, 015004 (2016)

[4] B.D. Fields, Ann. Rev. Nucl. Part. Sci. **61**, 47 (2011)

[5] A. Coc, N.J. Nunes, K.A. Olive, J.P. Uzan, E. Vangioni, Phys. Rev. D **76**, 023511 (2007)

[6] D.G. Yamazaki, M. Kusakabe, T. Kajino, G.J. Mathews, M.K. Cheoun, Phys. Rev. D **90**, 023001 (2014)

[7] R.H. Cyburt, B. Davids, Phys. Rev. C **78**, 064614 (2008)

[8] R.V. Wagoner, Astrophys. J. Suppl. Ser. **18**, 247 (1969)

[9] C.E. Rolfs, W.S. Rodney, *Cauldrons in the Cosmos: Nuclear Astrophysics* (University of Chicago Press, 1988)

[10] P. Bassi et al., Il Nuovo Cimento **28**, 1049 (1963)

[11] M. Barbagallo et al., Phys. Rev. Lett. **117**, 152701 (2016)

[12] A. Mengoni, T. Otsuka, M. Ishihara, Phys. Rev. C **52**, R2334 (1995)

[13] C.H. King, S.M. Austin, H.H. Rossner, W.S. Chien, Phys. Rev. C **16**, 1712 (1977)

[14] S.Q. Hou, J.J. He, S. Kubono, Y.S. Chen, Phys. Rev. C **91**, 055802 (2015)

[15] H. Sakai et al., Nucl. Instrum. Methods Phys. Res. A **369**, 120 (1996)

[16] E. Kim, A. Endo, Y. Yamaguchi, J. Nucl. Sci. and Tech. **39**, 693 (2002)

[17] M.B. Chadwick, M. Herman, P. Oblozinsky et al., Nucl. Data Sheets **112**, 2887 (2011)

DOI: 10.1051/epjconf/201716300029

Influence of neutron transfer channels on fusion enhancement in sub-barrier region

Khushboo[1,*], S. Mandal[1], N. Madhavan[2], S. Muralithar[2], J. J. Das[2,**], S. Nath[2], A. Jhingan[2], J. Gehlot[2], B. Behera[3], S. Verma[1], H. Singh[4], S. Kalkal[1,***], and R. Singh[1]

[1]Department of Physics and Astrophysics, University of Delhi, Delhi-110007, India
[2]Inter University Accelerator Centre, Aruna Asaf Ali Marg, New Delhi-110067, India
[3]Department of Physics, Panjab University, Chandigarh-160014, India
[4]Department of Physics, Kurukshetra University, Kurukshetra-136119, India

Abstract. Fusion cross-section measurements were performed for system ^{40}Ca + ^{70}Zn around the Coulomb barrier energies using Heavy Ion Reaction Analyzer (HIRA). The observed enhancement in experimental fusion cross-sections was investigated via coupled-channels formalism. The coupling of inelastic excitations alone could not reproduce the experimental data, however, the effect of octupole state of the projectile was observed to be significant. The multi-neutron positive Q-value transfer channels were also included in the calculations using semi-classical model. It was observed that two neutrons pick-up channel gave a major contribution to the fusion enhancement and successfully reproduce the experimental data at above as well as below barrier energies. The coupling of more than two neutrons transfer could not give any significant enhancement to sub-barrier fusion.

1 Introduction

The fusion reaction is an extensively studied phenomenon at and near the Coulomb barrier. A numerous investigations were performed in past years with various systems which showed a large enhancement of fusion cross-sections in comparison with theoretical predictions at sub-barrier energies [1]. It was demonstrated under the framework of coupled-channels (CC) formalism that fusion is influenced by inelastic excitations of colliding nuclei and transfer of neutrons [2]. The effect of inelastic excitations were well understood with the available theoretical models. However, due to the complicated mechanism involved in coupling the transfer channels in theoretical models, the role of multi-neutron transfer on fusion was not explored in detail.

The influence of transfer on fusion was noticed in systems having a positive Q-value for neutron transfer channels [3–9]. However, in few recent work, in spite of positive Q-values, no effect of the transfer channel on fusion enhancement was observed [10–12]. Apart from this, it was demonstrated in a literature that only outermost neutrons (up to two neutrons) of nuclei give a significant contribution to the fusion enhancement [13]. The importance of two neutrons pick-up channel was also highlighted for system ^{40}Ca + ^{64}Ni whose experimental data was well reproduced by coupling with two neutron transfer. This system has positive Q-value for up to six neutrons pick-up channels [14]. Further, the effect of pair transfer on fusion enhancement was reported recently by Stefanini *et al.* [15, 16] for ^{40}Ca + ^{96}Zr system where the fusion cross-sections were measured at deep sub-barrier energies. No fusion hindrance was observed in this system. It was pointed out that strong transfer couplings may be the reason for this observation. Beside these observations, the importance of multi-neutron transfer on sub-barrier fusion was also indicated in an article [5].

Considering all the above observations, in order to explore the effect of transfer channels on fusion enhancement, ^{40}Ca + ^{70}Zn system was selected for the present study. ^{40}Ca is a doubly magic nucleus, however, octupole state was shown to be an important state for ^{40}Ca [17]. The system has positive Q-value for seven neutrons pick-up channels (2n-8n), therefore, importance of multi-neutron transfer on fusion can be examined with this system. The Q-values for 2n to 8n pick-up channels are 4.14, 1.88, 5.96, 2.31, 4.73, 0.14 and 0.98 MeV respectively.

2 Experiment Details

The fusion cross-sections for ^{40}Ca + ^{70}Zn were measured at Inter University Accelerator Centre (IUAC), New Delhi. A pulsed beam of ^{40}Ca was used to bombard a self-supporting ^{70}Zn target (95% isotopic enrichment) of 670 μg/cm^2 thickness. The measurements were carried out using the Heavy Ion Reaction Analyzer (HIRA) [18] which

*e-mail: s12khushboo@gmail.com
**Present address: National Superconducting Cyclotron Laboratory (NSCL) 640 South Shaw Lane, Michigan State University, East Lansing, Michigan-48824, USA
***Present address: School of Physics and Materials Science, Thapar University, Patiala-147004, India

was kept at zero degree with respect to beam direction with 5 mSr of solid angle acceptance. At the focal plane of HIRA, a Multi-Wire proportional Counter (MWPC) of area 15 × 5 cm^2 was placed followed by segmented Ionization Chamber (IC) of 7.0 × 3.5 cm^2 area and active lengths of 3 cm-5.8 cm-13 cm for detection of evaporation residues (ERs) arriving at the focal plane. During the experiment, MWPC and IC were operated at a pressure of 2 mbar and 30 mbar of isobutane gas respectively. In addition to these detectors, two Silicon Surface Barrier Detectors (SSBD) were kept symmetrically at an angle of 25° on both the sides of incoming beam inside the target chamber. These detectors were used to monitor the beam during experiment and for normalization of fusion cross-sections. A carbon foil of thickness 40 $\mu g/cm^2$ was placed 10 cm downstream from the target for equilibration of charge states of ERs. For estimation of HIRA transmission efficiency for ERs, a High Purity Germanium (HPGe) detector was mounted on top of the target chamber at 90° to the beam direction.

The cross-section measurements were performed around the Coulomb barrier from laboratory energies 142 to 106 MeV. The fusion cross-sections were calculated using the yield of evaporation residues (ERs) which were separated from the beam-like particles at the focal plane using energy loss in MWPC versus time of flight (TOF) spectrum [5]. For the cross-section calculations, transmission efficiency of HIRA was estimated experimentally by gamma ray coincidence technique [5] and it was found to be 7.8% at 130 MeV laboratory energy. It was also calculated as 7.6% using Monte-Carlo code TERS [19].

3 Results and Discussion

The experimental fusion cross-sections were extracted and compared with the theoretical calculated cross-sections based on coupled-channels formalism as shown in figure 1. The experimental error in fusion cross-sections includes statistics related error and the error in transmission efficiency. The reported energies are corrected for loss in half thickness of the target. For initial estimation of theoretical fusion cross-sections, Akyuz-Winther (AW) parameterization was employed for Woods-Saxon potential. The AW parameters for the ^{40}Ca + ^{70}Zn system were V_0 = 69.87 MeV, r_0 = 1.18 fm and a_0 = 0.67 fm, which gave barrier parameters as V_b = 76.76 MeV, R_b = 10.48 fm and $\hbar\omega$ = 3.70 fm. The calculations were performed using CCFULL code [20] by considering projectile and target nuclei as a vibrator. To begin with, the coupling was switched off and no coupling fusion excitation function was obtained which underpredicted the experimental cross-sections. Therefore, in order to obtain a good fit to the above barrier experimental data, potential parameters were varied. The new set of parameters gave a slightly lower potential barrier as compared to the barrier obtained with AW parameters (V_b). Using these new set of potential parameters, calculations were performed to estimate the fusion cross-sections with state 3^- (E_{3^-} = 3.74 MeV, β_3 = 0.33) of the projectile. Coupling to this state gave a considerable enhancement to the cross-sections in the above barrier region and overpredicted the data in near barrier region. The new set of potential parameters could not give an appropriate fit to the above and near barrier data with projectile coupling hence, it was decided to use AW parameters for further calculations. After including the projectile octupole (3^-) state with AW parameters, the cross-sections were enhanced significantly and a reasonable fit to the above and near barrier fusion cross-sections was obtained as can be seen in figure 1. Whereas, the state 2^+ (E_{2^+} = 3.90 MeV, β_2 = 0.123) of ^{40}Ca gave a negligible contribution to the fusion cross-sections. In projectile excitations, 3^- state of ^{40}Ca seems to be an important state.

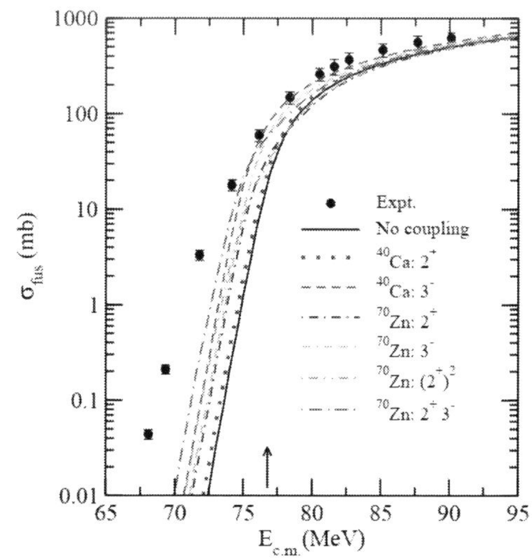

Figure 1. Experimental fusion cross-sections for ^{40}Ca + ^{70}Zn system along with coupled-channels calculations performed with CCFULL code. The arrow at the bottom indicates the position of the barrier.

In case of target excitations, both 2^+ (E_{2^+} = 0.88 MeV, β_2 = 0.23) as well as 3^- (E_{3^-} = 2.86 MeV, β_3 = 0.22) states enhanced the cross-sections by a similar amount in below barrier region however, 3^- state gave an additional enhancement in the above barrier region. The multi-phonon and mutual excitations were also taken into consideration in the calculations. The two phonon 2^+ state of target further enhanced the cross-sections slightly and the excitation 2^+ 3^- gave a considerable enhancement to the cross-sections. When both projectile as well target excitations were considered simultaneously in the calculations, then excitation ^{40}Ca: 3^- and ^{70}Zn: 2^+ 3^- overpredicted the data in near barrier region while not explaining the data in sub-barrier region. Whereas, the excitation ^{40}Ca: 3^- and ^{70}Zn: two phonons of 2^+ reproduced the data reasonably well in above and near barrier energy region. This is depicted in figure 2.

The inelastic excitations alone could not reproduce the experimental data. This indicates that apart from inelastic excitations, transfer channels may play an important role for ^{40}Ca + ^{70}Zn as Q-value is positive for multi-neutron

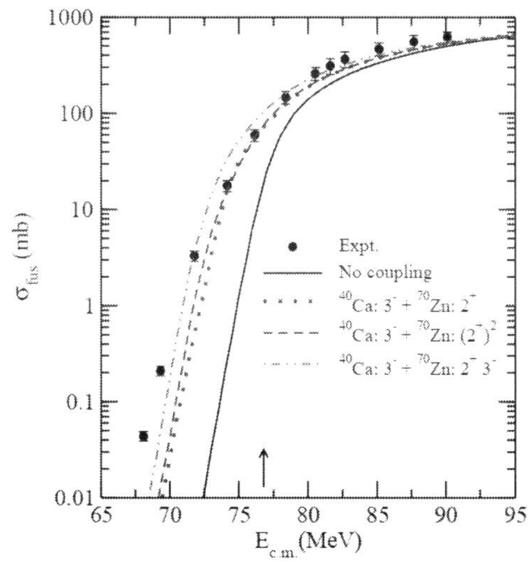

Figure 2. Experimental fusion cross-sections for $^{40}Ca + ^{70}Zn$ system along with coupled-channels calculations performed with CCFULL code. The combined effect of projectile and target inelastic states are shown with different colored lines (See text). The arrow at the bottom indicates the position of the barrier.

pick-up channels. The code CCFULL considers only one pair of transfer channel between ground states. In order to examine the role of multi-neutron transfer, semi-classical model (empirical coupled-channels, ECC) of Zagrabaev *et al.* [21] was undertaken to perform the calculations through which one can incorporate up to four neutron transfer channels in the calculations on the basis of Q-value. Initially, the fusion excitation function was obtained without coupling of transfer channels as shown in figure 3. After that, neutron pick-up channels were successively added in the calculations. It can be seen that one neutron pick-up channel has negligible effect on fusion enhancement. The two neutron pick-up channel enhanced the sub-barrier cross-sections to a large extent and give an appropriate fit to the entire experimental data. The transfer channels 3n and 4n pick-up could not give any significant enhancement to the cross-sections which implies that the sub-barrier fusion enhancement was mainly due to the two neutron transfer despite the presence of a large number of positive Q-value transfer channels.

A detail investigation is still required to clarify the impact of two or multi-neutron transfer on fusion. The explicit coupling of multi-neutron transfer channels in the theoretical calculations will be advantageous to disentangle the neutron transfer and structure effects on fusion enhancement. The structure of the colliding nuclei may also affect the transfer coupling itself and thereby fusion. The excitation process involved during the rearrangement of nucleons between colliding partners may influence the dynamics of the sub-barrier fusion process. The presently available exotic ions from radioactive beam facilities opens up a possibility to explore the dynamics of the fusion reactions far away from the stability line where transfer and break up mechanisms may strongly influence the fusion cross-sections around the barrier and in sub-barrier region.

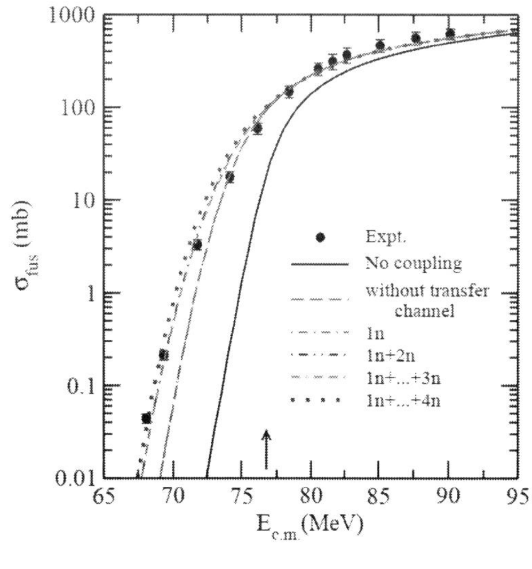

Figure 3. Experimental fusion cross-sections for $^{40}Ca + ^{70}Zn$ system compared with the theoretical cross-sections obtained using ECC model. The arrow at the bottom indicates the position of the barrier.

4 Summary

Fusion excitation function was measured for the system $^{40}Ca + ^{70}Zn$ to explore the role of multi-neutron transfer on sub-barrier fusion enhancement. The cross-sections were analyzed within coupled-channels formalism using CCFULL and ECC model. The effect of octupole vibration of ^{40}Ca was found to be major which gave a noticeable enhancement to the cross-sections in the entire energy range. The present results show the importance of transfer channels for explaining the sub-barrier fusion enhancement, however, transfer of only two neutrons was sufficient to reproduce the fusion cross-sections in above and below barrier energy region. It was observed that transfer channels beyond two neutrons have a negligible effect on sub-barrier fusion enhancement indicating the less importance of large number of neutron transfer on sub-barrier fusion.

5 Acknowledgements

We would like to thank Pelletron group of IUAC for providing a good quality of beam throughout the experiment. The authors are extremely thankful to target laboratory of GSI and H. J. Wollersheim for providing the target. We would like to acknowledge the help received from Ranjeet during the experiment. One of the author (Khushboo) would like to thank SERB, Goverment of India for providing international travel grant to attend FUSION17 conference. The financial support of the IUAC research project under UFR-51314 is gratefully acknowledged.

References

[1] M. Dasgupta, D. J. Hinde, N. Rowley, and A. M. Stefanini, Annu. Rev. Nucl. Part. Sci. **48**, 401 (1998)

[2] A. M. Stefanini, D. Ackerman, L. Corradi, J. H. He, G. Montagnoli, S. Beghini, F. Scarlassara, and G. F. Segato, Phys. Rev. C **52**, R1727 (1995)

[3] R. A. Broglia, C. H. Dasso, S. Landowne, and A. Winther, Phys. Rev. C **27**, 2433 (1983)

[4] H. Timmers, D. Ackermann, S. Beghini, L. Corradi, J. He, G. Montagnoli, E. Scarlassara, A. Stefanini, and N. Rowley, Nucl. Phys. A **633**, 421 (1998)

[5] S. Kalkal, S. Mandal, N. Madhavan, E. Prasad, S. Verma, A. Jhingan, R. Sandal, S. Nath, J. Gehlot, B. R. Behera, M. Saxena, S. Goyal, D. Siwal, R. Garg, U. D. Pramanik, S. Kumar, T. Varughese, K. S. Golda, S. Muralithar, A. K. Sinha, and R. Singh, Phys. Rev. C **81**, 044610 (2010)

[6] H. Q. Zhang, C. J. Lin, F. Yang, H. M. Jia, X. X. Xu, Z. D. Wu, F. Jia, S. T. Zhang, Z. H. Liu, A. Richard, and C. Beck, Phys. Rev. C **82**, 054609 (2010)

[7] G. Montagnoli, A. M. Stefanini, H. Esbensen, C. L. Jiang, L. Corradi, S. Courtin, E. Fioretto, A. Goasduff, J. Grebosz, F. Haas, M. Mazzocco, C. Michelagnoli, T. Mijatovic, D. Montanari, C. Parascandolo, K. E. Rehm, F. Scarlassara, S. Szilner, X. D. Tang, and C. A. Ur, Phys. Rev. C **87**, 014611 (2013)

[8] M. Trotta, A. M. Stefanini, L. Corradi, A. Gadea, F. Scarlassara, S. Beghini, and G. Montagnoli, Phys. Rev. C **65**, 011601(R) (2001)

[9] G. Montagnoli, A. M. Stefanini, C. L. Jiang, H. Esbensen, L. Corradi, S. Courtin, E. Fioretto, A. Goasduff, F. Haas, A. F. Kifle, C. Michelagnoli, D. Montanari, T. Mijatovic, K. E. Rehm, R. Silvestri, P. P. Singh, F. Scarlassara, S. Szilner, X. D. Tang, and C. A. Ur, Phys. Rev. C **85**, 024607 (2012)

[10] A. M. Stefanini, G. Montagnoli, F. Scarlassara, C. Jiang, H. Esbensen, E. Fioretto, L. Corradi, B. Back, C. Deibel, B. D. Giovine, J. Greene, H. Henderson, S. Marley, M. Notani, N. Patel, K. Rehm, D. Sewerinyak, X. Tang, C.Ugalde, and S. Zhu, Eur. Phys. J. A **49**, 63 (2013)

[11] Z. Kohley, J. F. Liang, D. Shapira, R. L. Varner, C. J. Gross, J. M. Allmond, A. L. Caraley, E. A. Coello, F. Favela, K. Lagergren, and P. E. Mueller, Phys. Rev. Lett. **107**, 202701 (2011)

[12] H. M. Jia, C. J. Lin, F. Yang, X. X. Xu, H. Q. Zhang, Z. H. Liu, L. Yang, S. T. Zhang, P. F. Bao, and L. J. Sun, Phys. Rev. C **86**, 044621 (2012)

[13] V. A. Rachkov, A. V. Karpov, A. S. Denikin, and V. I. Zagrebaev, Phys. Rev. C **90**, 014614 (2014)

[14] D. Bourgin, S. Courtin, F. Haas, A. M. Stefanini, G. Montagnoli, A. Goasduff, D. Montanari, L. Corradi, E. Fioretto, J. Huiming, F. Scarlassara, N. Rowley, S. Szilner, and T. Mijatovic, Phys. Rev. C **90**, 044610 (2014)

[15] A. M. Stefanini, G. Montagnoli, H. Esbensen, L. Corradi, S. Courtin, E. Fioretto, A. Goasduff, J. Grebosz, F. Haas, M. Mazzocco, C. Michelagnoli, T. Mijatovic, D. Montanari, G. Pasqualato, C. Parascandolo, F. Scarlassara, E. Strano, S. Szilner, D. Torresi, Phys. Lett. B **728**, 639 (2014)

[16] A. M. Stefanini, G. Montagnoli, H. Esbensen, L. Corradi, S. Courtin, E. Fioretto, A. Goasduff, J. Grebosz, F. Haas, M. Mazzocco, C. Michelagnoli, T. Mijatovic, D. Montanari, G. Pasqualato, C. Parascandolo, F. Scarlassara, E. Strano, S. Szilner, N.Toniolo, D. Torresi, EPJ Web of Conferences **86**, 00056 (2015)

[17] J. F. Liang, J. M. Allmond, C. J. Gross, P. E. Mueller, D. Shapira, R. L. Varner, M. Dasgupta, D. J. Hinde, C. Simenel, E. Williams, K. Vo-Phuoc, M. L. Brown, I. P. Carter, M. Evers, D. H. Luong, T. Ebadi, and A. Wakhle, Phys. Rev. C **94**, 024616 (2016)

[18] A. K. Sinha, N. Madhavan, J. . Das, P. Sugathan, D. Kataria, A. Patro, and G. Mehta, Nucl. Inst. Meth. A **339**, 543 (1994)

[19] S. Nath, Comput. Phys. Commun. 180, 2392 (2009)

[20] K. Hagino, N. Rowley, and A. T. Kruppa, Comput. Phys. Commun. **123**, 143 (1999)

[21] V. I. Zagrebaev, A. S. Denikin, A. V. Karpov, A. P. Alekseev, M. A. Naumenko, V. A. Rachkov, V. V. Samarin, V. V. Saiko, NRV web knowledge base on low-energy nuclear physics, http://nrv.jinr.ru/.

The cross sections of fusion-evaporation reactions: the most promising route to superheavy elements beyond Z=118

Khuyagbaatar Jadambaa[1,2,⋆]

[1]*Helmholtz Institute Mainz, 55099 Mainz, Germany*
[2]*GSI Helmholtzzentrum für Schwerionenforschung, 64291 Darmstadt, Germany*

Abstract. The synthesis of superheavy elements beyond oganesson (Og), which has atomic number $Z = 118$, is currently one of the main topics in nuclear physics. An absence of sufficient amounts of target material with atomic numbers heavier than californium ($Z = 98$) forces the use of projectiles heavier than ^{48}Ca ($Z = 20$), which has been successfully used for the discoveries of elements with $Z = 114 - 118$ in complete fusion reactions. Experimental cross sections of ^{48}Ca with actinide targets behave very differently to "cold" and "hot" fusion-evaporation reactions, where doubly-magic lead and deformed actinides are used as targets, respectively. The known cross sections of these reactions have been analysed compared to calculated fission barriers. It has been suggested that observed discrepancies between the cross sections of ^{48}Ca-induced and other fusion-evaporation reactions originate from the shell structure of the compound nucleus, which lies in the island of the stability. Besides scarcely known data on other reactions involving heavier projectiles, the most promising projectile for the synthesis of the elements beyond Og seems to be ^{50}Ti. However, detailed studies of ^{50}Ti, ^{54}Cr, ^{58}Fe and ^{64}Ni-induced reactions are necessary to be performed in order to fully understand the complexities of superheavy element formation.

1 Introduction

The search for "hypothetical" nuclei from the island of stability, referred to as superheavy nuclei (SHN), which are predicted to be located at around proton number $Z = 114$ and neutron number $N = 184$ according to most theoretical calculations [1–3], is the one of the main goals of fundamental nuclear physics research. To date, isotopes of elements up to $Z = 118$ and $N = 177$ have already been discovered [4]. The heaviest elements ($Z = 114 - 118$) have been directly produced in ^{48}Ca-induced reactions with actinide targets, first at Dubna Gas-Filled Recoil Separator (DGFRS) at the Flerov Laboratory of Nuclear Reactions in Dubna, Russia. The findings of these works were confirmed world-wide by different laboratories using the same reactions [5–15]. Recently, four elements with $Z = 113, 115, 117$ and 118 have officially been named as nihonium (Nh), moscovium (Mc), tennessine (Ts) and oganesson (Og), respectively [16].

Presently, the synthesis of elements beyond Og is the hot topic in the research field, despite that the predicted magic $Z = 114$ has already been reached. However, an absence of sufficient amounts of target material with atomic numbers heavier than californium ($Z = 98$) currently prevents the use of ^{48}Ca for synthesis of elements beyond Og. The predicted magic $N = 184$ also remains out of reach due to limited access to both radioactive beams and appropriate target materials.

Of special interest is the element with $Z = 120$, wherein some theories predict a stronger shell closure compared to the "classical" magic $Z = 114$ i.e. a shift in the location of the center of the island of stability. Several attempts to synthesize the element with $Z = 120$ have already been carried out using ^{64}Ni+^{238}U [17], ^{58}Fe+^{244}Pu [18], ^{54}Cr+^{248}Cm [19], and ^{50}Ti+^{249}Cf [20] reactions. However, none of these studies reported observation of the element with $Z = 120$, despite reaching sensitivity levels that would be sufficient to detect at least one event from any of the known ^{48}Ca-induced reactions. This non-observation can firstly be due to entrance-channel effects [21], as the coulomb repulsion forces of these reactions are significantly greater than those for ^{48}Ca-induced reactions. Secondly, the survival probabilities of compound nuclei could also be smaller when compared with Og, according to most theoretical predictions [1–3]. Also, impacts of the double magicity ($Z = 20$ and $N = 28$) of ^{48}Ca on capture, fusion and survival probabilities are still not yet fully understood [22] despite its effect on the reaction Q-value.

However, knowledge of the effect of these two processes remains poor, not only for elements beyond Og but also for lighter ones. For the observation of a single atom of a superheavy element, experiments must often be performed for several days, weeks or months [4–15, 17–20, 23]. Therefore, choosing a suitable projectile-target combination and optimum beam energy for a reaction are of upmost importance to maximize the countable yield of heavy nuclei.

⋆e-mail: J.Khuyagbaatar@gsi.de

The cross sections of ^{48}Ca-induced fusion-evaporations reactions with actinide targets remain at the level of pbarns as a function of Z of the compound nucleus (Z_{CN}), as opposed to those of "cold"- and "hot"-fusion reactions (where doubly-magic lead and deformed actinides are used as targets, respectively), which decrease exponentially. It has already been suggested that such behaviour is associated with the survival probability of the compound nucleus [4].

In the present paper, a compilation of the known data on ER cross sections of the above mentioned reactions were performed and tested for a presence of any systematic trends. This would help to understand the discrepancy in the case of ^{48}Ca, and may also be useful for the selection of preferable projectile and target combinations leading to elements beyond Og.

2 The cross sections of the fusion-evaporation reactions

Since all of the SHN have been exclusively produced in fusion-evaporation reactions, their experimental cross sections may provide an important information on the preferable choice of projectile and target combinations for the synthesis of the as yet unknown SHN. The evaporation residue (ER) cross section of fusion reactions passing through a formation of compound nucleus (CN) is often described by the three-term expression,

$$\sigma_{ER}(E^*, J) = \sigma_{cap}(E, J)[1 - P_{QF}(E, J)]W_{CN}(E^*, J), \quad (1)$$

where σ_{cap} is the cross section characterizing the formation of the di-nuclear system (capture) at collision energy E. P_{QF} is the fraction of di-nuclear systems not evolving via formation of CN with a compact shape i.e. splitting into two fragments (quasi-fission), and W_{CN} is the survival probability of CN against fission through particle evaporation at the excitation energy of $E^* = E - Q$, where Q is the energy necessary for the fusion. Terms σ_{cap} and W_{CN} describe independent processes and have been studied substantially, both experimentally and theoretically.

A theoretical description of P_{QF}, or its complementary P_{CN}=1-P_{QF}, which characterizes the fusion probability, is not yet fully developed [24, 25]. The main problem is the dynamical evolution of the multidimensional di-nuclear system leading to equilibration in all possible degrees of freedom, which is still not describable despite developments in modern theory and computing. Therefore, P_{CN} is often extrapolated from the experimental observations of the nuclear reaction studies [26]. In the earlier studies, where mostly the light particles/ions were used as a projectile, the quasi-fission probability was shown to be negligible. With more massive ion beams with $Z > 20$ that gave access to superheavy elements, a reduction of P_{CN} has been observed due to quasi-fission. In reactions with strong coulomb repulsion forces between the collision partners ($Z_pZ_t > 1600$), the presence of quasi-fission was predicted [27]. As a result, σ_{ER} for fusion reactions leading to the formation of SHN drastically decreases as function of Z_{CN} or Z_pZ_t.

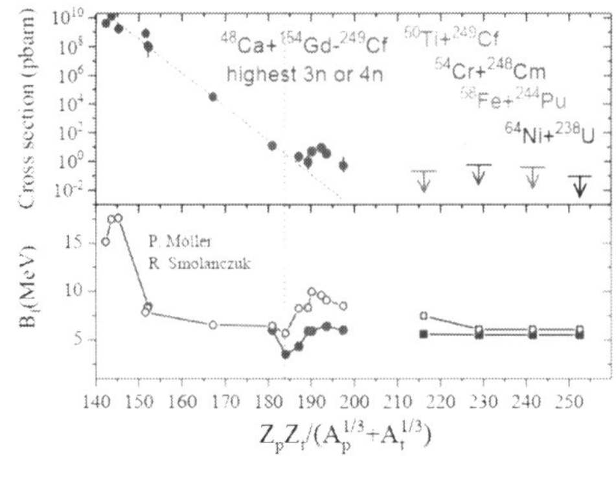

Figure 1. a) Compiled experimental ER cross sections of fusion-evaporation reactions with ^{208}Pb and ^{238}U targets, and ^{48}Ca projectiles [28–32]. Only maximum values of σ_{1n} and σ_{5n} are shown for ^{208}Pb and ^{238}U, respectively. Maximum values of either σ_{3n} or σ_{4n} from ^{48}Ca-induced reactions with deformed Gd-Cf targets are shown. Arrows mark upper limits. Dashed lines are drawn to guide an exponential descent. b) Theoretical $B_f - S_n$ calculated within the macro-microscopic FRDM approach are given [3, 37].

2.1 "Cold ", "hot"and ^{48}Ca-induced fusion reactions

Compiled ER cross sections of three types of fusion reactions are shown in Fig. 1 as a function of $Z_pZ_t/(A_p^{1/3}+A_t^{1/3})$ parameter which is equivalent to the Coulomb repulsion at contact between spherically shaped projectile and target nuclei. This or any other variables like effective entrance channel fissility are better for displaying the reaction entrance channel than Z_{CN} or Z_pZ_t. As representatives of "cold", and "hot" fusion, the maxima cross sections of 1 and 5 neutron evaporation channels, σ_{1n} and σ_{5n}, of reactions with ^{208}Pb and ^{238}U targets, respectively, were compiled [28, 29, 32]. In the case of the ^{48}Ca-induced reactions, only maximum of σ_{3n} or σ_{4n} with deformed targets were selected to test for any systematic trend. In this case the well-known orientation effect of the collision will be accounted for [24, 33, 34].

In the statistical model, the survival probability of the excited nucleus has an exponential dependence on the fission barrier, B_f, and neutron separation energy, S_n and is often expressed as

$$W_{CN}(E^*) \sim \prod_i^x e^{[(B_f - S_n)/T]_i},$$

where T and x are the temperature and number of emitted neutrons from the CN. The latter depends on available E* of the nucleus populated at each stage.

The fission probability of the highly-excited heavy nucleus strongly increases as function of E* [35]. Thus, the survival probability of the initial CN which has highest possible E*, against the fission (first-chance fission) is the most crucial process for residue formation [25, 36]. Accordingly, $B_f - S_n$ value of the initial CN can be applied for comparative analysis of W_{CN}. The calculated $B_f - S_n$

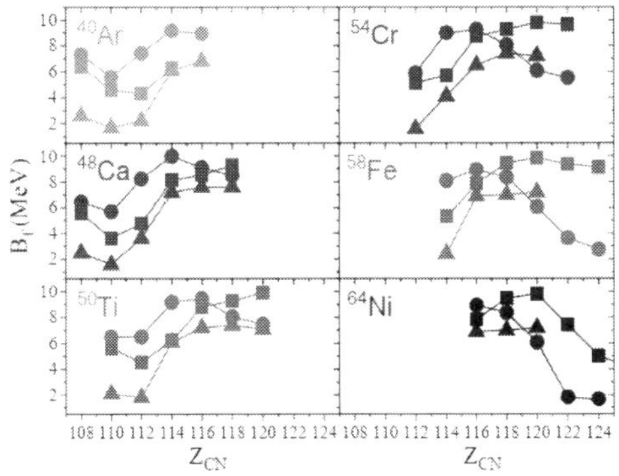

Figure 2. Calculated fission barriers of SHN, which can be formed as CN in ^{40}Ar, ^{48}Ca, ^{50}Ti, ^{54}Cr, ^{58}Fe and ^{64}Ni-induced reactions with available actinide targets, ^{226}Ra-^{249}Cf. The results from FRDM [37], HBF(Sk*) [40] and ETFSI [41] calculations are given by dots, squares and triangles, respectively.

of the initial CN corresponding to each reaction are shown in Fig. 1b) as a representative for the W_{CN}. Both variables have been extracted from the same mass FRDM-calculations [3, 37] to exclude systematic uncertainties due to the theoretical approaches.

Firstly, all three types of reactions have a decreasing trend of their σ_{ER} as a function of the coulomb parameter i.e. Z_{CN}, except in the case of ^{48}Ca leading to elements heavier than Hs ($Z = 108$). The decreasing slope is steeper in ^{48}Ca-induced reactions compared to other two, which have a similar slopes. As mentioned above, the fusion-evaporation reaction is the output of three consecutive physical processes. Accordingly, observed trends might have different origins. In "hot" fusion reactions where the coulomb repulsion force is smaller i.e., $Z_pZ_t < 1600$, QF was believed to be negligibly small. However this criteria has been shown to be obsolete i.e. experimental evidence for the presence of QF in reactions with $Z_pZ_t < 1600$ has been observed [25, 38]. In "cold" fusion reactions where $Z_pZ_t > 1600$, it is well known that QF dominates over fusion [27]. In addition, complied σ_{1n} of this type of reaction have been observed at energies below the fusion barrier, where σ_{cap} steeply decreases. Trends of $B_f - S_n$ for these two types of reactions do not show a significant difference (within the chosen scale of vertical axis) and vary smoothly as function of Z_{CN}. Therefore, the decreasing σ_{ER} trend of "cold" fusion reactions could mostly be driven by entrance channel effects which can modify both capture cross sections and fusion probabilities.

In the case of ^{48}Ca, $B_f - S_n$ values follow a U-shape as function of Z_{CN}, and initially decrease with increasing Z_{CN} up to Ds ($Z = 110$). This may explain the observed steeper decrease of σ_{ER} compared to other two reactions. (i) After Ds, $B_f - S_n$ rises again as B_f falls under the influence of the shell closure originating from the island of stability, thus impacting on W_{CN} and increasing the σ_{ER}. Consequently, σ_{ER} of SHN beyond Ds that are enhanced compared to the expected exponential descent could be due to W_{CN} [39].

To test the above-mentioned suggestion (i) initially proposed in Ref. [39], the results of other theoretical predictions of B_f are examined. Fission barriers calculated within three different theoretical models are shown in Fig. 2. The results of FRDM compared to two other theoretical models, both microscopic self-consistent calculations performed within the Hartree-Fock-Bogoliubov (HFB) approximation with the SkM* force [40] and Extended Thomas–Fermi plus Strutinsky Integral model (ETFSI) based on Skyrme SkSC4 functional [41], are shown.

Predicted fission barriers of CN formed in ^{40}Ar, ^{50}Ti, ^{54}Cr, ^{58}Fe and ^{64}Ni-induced reactions with ^{226}Ra-^{249}Cf actinide targets are also shown in Fig. 2, in addition to those formed in ^{48}Ca-induced reactions.

Overall, they all predict the existence of the island of stability despite their quantitative discrepancies. However, a precise prediction of an absolute B_f is presently still not yet established for heaviest nuclei and results often differ from experimental ones by up to several MeV [42]. In spite of this, they all predict an increasing tendency of B_f after passing Ds in the ^{48}Ca cases, thus supporting the arguments made in point (i) discussed above. One interesting feature can be noted in the case of CN formed in the ^{40}Ar+^{248}Cm reaction, where the calculations predict a similarly high B_f as for the ^{48}Ca+^{244}Pu and/or ^{48}Ca+^{242}Pu reactions where the σ_{ER} are known. One has to emphasize that it will be interesting to measure σ_{ER} of ^{40}Ar+^{248}Cm in order to help to understand the differences in the reaction mechanism.

Obvious differences in between theories can be found in cases where reactions lead to elements beyond Og because they predict differing values of Z to be magic. For example, FRDM calculations predict a magic proton number at $Z = 114$, compared to $Z = 120$ in HFB(Sk*), as can be clearly seen in Fig. 2. If the latter prediction is correct, then one may expect to observe a more enhanced W_{CN} of CN with $Z = 120$ compared to Cn-Og, which may also increase the σ_{ER}. Thus, the synthesis of the element with $Z = 120$ and measuring its production cross section will help to solve persistent discrepancies in the theoretical models.

2.2 The reactions leading to formation of elements beyond 118

For the synthesis of the element with $Z = 120$, ^{48}Ca is can not presently be used, thus one needs to use a heavier projectiles like ^{50}Ti, ^{54}Cr, ^{58}Fe, ^{64}Ni etc.. The experimental cross-section sensitivities of four different reactions leading to formation of element $Z = 120$, ^{64}Ni+^{238}U [17], ^{58}Fe+^{244}Pu [18], ^{54}Cr+^{248}Cm [19], and ^{50}Ti+^{249}Cf [20], are shown in Fig. 3 together with ^{48}Ca results. Values of calculated $B_f - S_n$ for CN calculated within the FRDM framework are also shown. To make a consistent comparison of σ_{ER} to those discussed in the previous section, the known data (maximum of either σ_{3n} or σ_{4n} on ^{50}Ti, ^{54}Cr,

 DOI: 10.1051/epjconf/201716300030

^{58}Fe and ^{64}Ni-induced reactions with deformed Gd-Cf targets are compiled.

As seen from Fig. 3 data on these reactions are scarce. A decreasing trend can only be drawn in the case of ^{50}Ti, which falls off much steeper than for ^{48}Ca (cf. Fig. 3a). However, it is probably not a realistic case for extracting a slope as the ER from the ^{50}Ti+^{176}Yb is an extremely neutron-deficient isotope of U (Z = 92) [43], and thus W_{CN} may be strongly reduced.

This is also the case for the ^{48}Ca+^{181}Ta reaction (see Fig. 3a), where recently measured upper limit [44] is not following the extrapolated exponential. Therefore, one can assume a similar slope for all reactions including ^{48}Ca data, although these need to be measured. Extrapolated exponentials, fitted to the known data of the particular reaction and assuming the same slope as for ^{48}Ca, are shown in Fig 3a.

In all four reactions leading to Z = 120 formation discussed here (^{64}Ni+^{238}U, ^{58}Fe+^{244}Pu, ^{54}Cr+^{248}Cm, and ^{50}Ti+^{249}Cf), one might expect a similar increase of W_{CN} due to shell effects originating from the island of stability as seen in ^{48}Ca-induced reactions (discussed in (i)). The CN is the same in all cases except for ^{50}Ti+^{249}Cf, where the CN has three fewer neutrons, but the B_f values are similar. From the known ^{48}Ca cross sections, leading to the formation of Fl-Og, this gain in cross section via the W_{CN} enhancement can be estimated to be roughly two orders of magnitude higher than an exponential descent. One might then tentatively expect a similar gain in σ_{ER} with respect to an exponential extrapolated, to be observed in reactions leading to Z = 120.

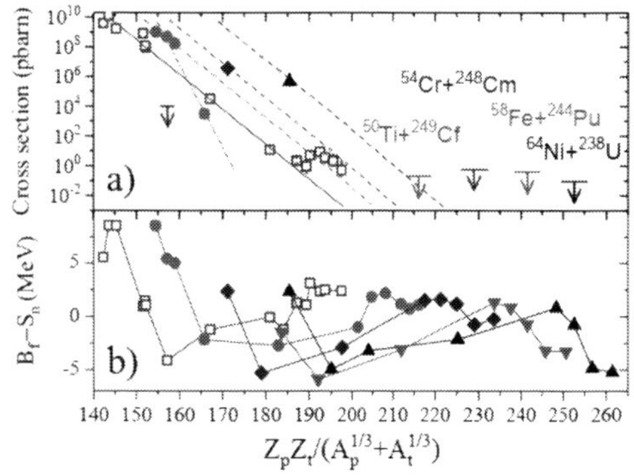

Figure 3. a) The compiled maximum of either σ_{3n} or σ_{4n} [28, 29, 32, 43, 45] and b) calculated CN's $B_f - S_n$ within the FRDM model [3, 37] are given for ^{48}Ca (open symbols), ^{50}Ti (dots), ^{54}Cr (rhombuses), ^{58}Fe (down-triangles) and ^{64}Ni (up-triangles) induced reactions with deformed Gd-Cf targets. a) Cross-section sensitivities are marked by arrows [17-20, 44]. Solid line is drawn to guide an exponential descent of σ_{ER} in ^{48}Ca-induced reactions up to formation of Hs and continued to higher $_{ZCN}$. Dashed-dotted line shows a trend of known σ_{ER} for ^{50}Ti-induced reactions. Dashes lines are the same as the solid one but shifted to fit known σ_{ER} of the particular reaction. See text for details

As the coulomb forces increase significantly when changing from ^{48}Ca to a heavier projectile, the extrapolated σ_{ER} decrease dramatically for a fixed Z_{CN}. Accordingly, reactions involving projectile and target combinations with the smallest coulomb forces are preferable for synthesizing element Z = 120. Although the resulting σ_{ER} estimate for the ^{50}Ti+^{249}Cf reaction of up to several fbarns (10^{-3} pbarns) is not very encouraging and is about two-three orders of magnitude smaller than those for ^{48}Ca-induced reactions, the true cross section could be quite different and therefore it is of great importance that the values be measured experimentally.

Note that the use of ^{50}Ti as a projectile also gives access to SHN with Z = 114 – 118, whereas the range of accessible SHN with heavier projectiles appears to be even more limited, within the current capabilities and level of understanding of the fusion-evaporation reaction.

3 Summary and outlook

The compiled ER cross sections of three different types of fusion-evaporation reaction were comparatively analysed relative to the calculated fission barriers (~ survival probability) from three different theoretical models. All three types of reaction (namely "cold", "hot" and ^{48}Ca-induced) show a similar decreasing trend of the ER cross sections as functions of the proton numbers in the compound nucleus. The known discrepancy in ER cross sections of ^{48}Ca with actinide targets were suggested to occur due to an enhanced survival probability of compound nuclei originating from shell closures associated with the island of stability, shown independently by the different theoretical frameworks discussed here.

A similarly enhanced survival probability for reactions leading to Z = 120 may also be expected, providing an important testing ground for theories which, although they interpret the behaviour of ^{48}Ca+actinide results well, may not produce an accurate description for heavier systems. Experimental data on ER cross sections of reactions involving heavier projectiles and deformed targets, which are necessary for reaching Z = 120, are very scarce. Thus, more measurements must be made.

Based on the present analysis, the main differences in the ER cross section of reactions ^{64}Ni+^{238}U, ^{58}Fe+^{244}Pu, ^{54}Cr+^{248}Cm, and ^{50}Ti+^{249}Cf leading to element with Z = 120 could be strongly due to their entrance channel effects. One can argue that the latter reaction with smallest Z_pZ_t might be the most preferable for a making the superheavy element with Z = 120. For a final conclusion to be made on which reactions are preferable, further investigations of fusion and quasi-fission processes must be performed.

References

[1] A. Sobiczewski et al., Phys. Lett. **22**, 500 (1966).

[2] W.D. Myers and W.J. Swiatecki, Nucl. Phys. **81**, 1 (1966).

[3] P. Möller et al., At. Data. Nucl. Data. Tab. **59**, 185 (1995).

[4] Yu.Ts. Oganessian and V.K. Utyonkov, Nucl. Phys. A **944**, 62 (2015)

[5] S. Hofmann et al., Eur. Phys. J. A **32**, 251 (2007).

[6] L. Stavsetra et al., Phys. Rev. Lett. **103**, 132502 (2009).

[7] R. Eichler et al., Radiochim. Acta **98**, 133 (2010).

[8] Ch.E. Düllmann et al., Phys. Rev. Lett. **104**, 252701 (2010).

[9] P.A. Ellison et al., Phys. Rev. Lett. **105**, 182701 (2010).

[10] J.M. Gates et al., Phys. Rev. C **83**, 054618 (2011).

[11] S. Hofmann et al., Eur. Phys. J. A **48**, 62 (2012).

[12] A. Yakushev et al., Inorg. Chem. **53**, 1624 (2014).

[13] D. Rudolph et al., Phys. Rev. Lett. **111**, 112502 (2013).

[14] J. Khuyagbaatar et al., Phys. Rev. Lett. **112**, 172501 (2014).

[15] D. Kaji et al., J. Phys. Soc. Jpn. **86**, 034201 (2017).

[16] L. Öhrström and J. Reedijk, Pure and Applied Chemistry **88**, 1225 (2016).

[17] S. Hofmann et al., GSI Scientific Report-2007, 131 (2008).

[18] Yu.Ts. Oganessian et al., Phys. Rev. C **79**, 024603 (2009).

[19] S. Hofmann et al., GSI Scientific Report-2011, 205 (2012).

[20] Ch.E. Düllmann et al., to be published.

[21] M.G. Itkis and A. Ya. Rusanov, Phys. Part. Nuclei **29**, 160 (1998).

[22] C. Simenel et al., Phys. Lett. B **710**, 607 (2012).

[23] J. Khuyagbaatar et al., GSI Scientific Report-2012, 131 (2013).

[24] K. Nishio et al.,Phys. Rev. C **82**, 024611 (2010).

[25] J. Khuyagbaatar et al., Phys. Rev. C **91**, 054608 (2015).

[26] M.G. Itkis et al., Nucl. Phys. A **787**, 150 (2007).

[27] W.J. Swiatecki, Phys. Scr. **24**, 113 (1981).

[28] http://nrv.jinr.ru/nrv/

[29] http://www.nndc.bnl.gov/

[30] K.E. Gregorich et al., Lawrence Berkeley National Laboratory, Berkeley, CA, USA, LBNL-Report 63617 (2007)

[31] S. Hofmann, Russ. Chem. Rev. **78**, 1123 (2009).

[32] J. Khuyagbaatar, unpublished data for 3n cross section of the ^{48}Ca+^{197}Au.

[33] D.J. Hinde et al., Phys. Rev. Lett. **74**, 1295 (1995).

[34] K. Nishio et al.,Phys. Rev. C **77**, 064607 (2008).

[35] R. Vandenbosch and J. R. Huizenga, Nuclear Fission, Academic, New York, (1973).

[36] R. Yanez et al., Phys. Rev. Lett. **112**, 152702 (2014).

[37] P. Möller et al., Phys. Rev. C **79**, 064304 (2009).

[38] A.C. Berriman et al., Nature (London) **413**, 144 (2001).

[39] Yu.Ts. Oganessian, J. Phys. G **34**, R165 (2007).

[40] A. Staszczak et al., Phys. Rev. C, **87**, 024320 (2013).

[41] A. Mamdouh et al., Nucl. Phys. A **679**, 337 (2001).

[42] A. Baran et al., Nucl. Phys. A, **944** 442 (2015).

[43] J. Khuyagbaatar et al., Phys. Rev. Lett. **115**, 242502 (2015).

[44] A. Mistry, private communication.

[45] D.A. Mayorov et al., Phys. Rev. C **92**, 054601 (2015).

 DOI: 10.1051/epjconf/201716300031

Near and sub-barrier reactions of ^{8}B

J. J. Kolata[1,*], Eli F. Aguilera[2,**], and Valdir Guimarães[3,***]

[1] Physics Department, University of Notre Dame, Notre Dame, Indiana, 46556-5670, USA
[2] Departamento de Aceleradores, Instituto Nacional de Investigaciones Nucleares, Apartado Postal 18-1027, Código Postal 11801, México, Distrito Federal, México
[3] Instituto de Fisica, Universidade de São Paulo, Rua do Matão Nr. 1371, 05508-900, São Paulo, SP, Brazil

Abstract. Four classes of measurements that are important for achieving a complete understanding of the reactions of exotic nuclei are identified and classified according to their degree of difficulty. Previously reported data for near- and sub-barrier fusion of ^{8}B+^{58}Ni are critically reviewed. The influence of breakup protons on the evaporation proton measurements for this system is shown to be small at all energies except for the lowest one measured, and corrections are made for this process. Model dependencies in the deduced fusion cross sections are assessed using three different evaporation codes. Data sets for targets of ^{58}Ni and ^{28}Si are shown to be consistent with each other, and with fusion enhancement up to energies that are greater than the Coulomb barrier V_b ($E_{c.m.} \lesssim V_b + 1.5\times\hbar\omega$). An important difference with the behavior of neutron-halo systems is thereby confirmed. Possible fusion suppression beyond this energy is suggested by the ^{28}Si data. A proposal to measure the ^{8}B+^{40}Ar fusion cross section using an active-target time projection chamber is discussed.

1 Introduction

The study of nuclear reactions of exotic, weakly-bound nuclei near the limits of stability has revealed a large number of interesting features. However, most of these studies have been carried out at energies far above the Coulomb barrier, due to the relative unavailability of these beams at Coulomb-barrier energies. As a result there is only a rather limited literature on reactions, such as complete and incomplete fusion, which turn out to be very important for achieving a complete understanding of the exotic structure of these nuclei (see Ref. [1] and references therein). Four different types of relevant experiments can be identified and ranked according to their relative degree of difficulty. The first of these is elastic scattering, the easiest reaction to study because of its large cross section, which leads to a determination of the total reaction cross section. As discussed below, this can be an important indicator of the effect of structure on nuclear reactions, and in some cases the effects of coupling to weakly-bound direct channels on elastic scattering can be dramatic, as illustrated in Fig. 1.

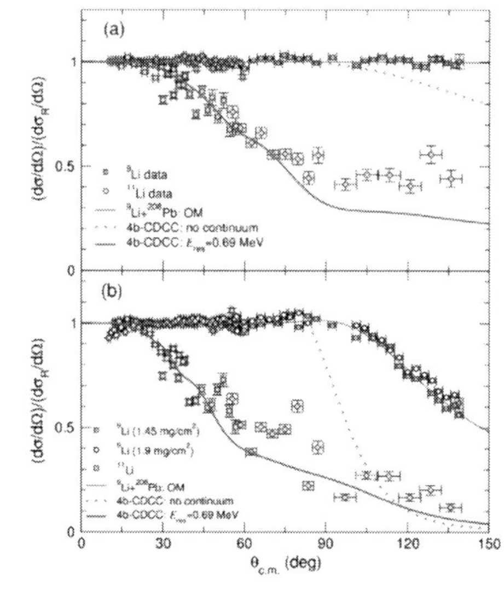

Figure 1. Elastic-scattering cross section (ratio-to-Rutherford) for ^{9}Li and ^{11}Li on a ^{208}Pb target at the same c.m. energies of 23.1 MeV (upper panel) and 28.3 MeV (lower panel). Figure reprinted with permission from Ref. [2]. ©2012, The American Physical Society.

The second category of experiments, measurement of transfer/breakup reactions, is also relatively easy but complicated by the fact that separation of various reaction types such as one- and two-particle transfer from direct breakup can be challenging. This separation was accomplished for the ^{6}He+^{209}Bi reaction in a series of experiments [3–5] carried out at the *TwinSol* facility [6]. Other examples are discussed in Ref. [1].

* e-mail: jkolata@nd.edu
** e-mail: eli.aguilera@inin.gob.mx
*** e-mail: valdirg@if.usp.br

The third category, most relevant for this conference, is fusion. Although fusion cross sections tend to be quite large above the Coulomb barrier, they rapidly grow smaller below the barrier where the most interesting effects are expected. Experimental techniques range from direct detection of the fusion products, to the light particles evaporated from them, to γ radiation emitted by the final systems. In the case of weakly-bound projectiles, it can be very difficult to separate complete fusion from incomplete fusion, i.e., the fusion of a part of the projectile with the target. As a result, only a few examples of the separation of these two types of fusion reactions, the fourth category of measurement, have been reported for exotic nuclei, and these are typically at above-barrier energies. Examples include Refs. [7, 8]. Another issue addressed by fusion studies of radioactive but not weakly-bound projectiles is the influence of neutron number on the fusion cross section, as for example in Refs. [9, 10] and a work reported at this conference [11].

2 Reactions of ^{8}B

2.1 Elastic scattering and breakup

The elastic scattering of ^{8}B on ^{58}Ni has been measured [12], and the extracted total reaction cross section was compared with that for other weakly- and strongly-bound projectiles on various targets in Ref. [13]. The reduced total reaction cross sections are illustrated in Fig. 2.

It can be seen in this figure that the weakly-bound, two-neutron-halo nucleus ^{6}He and the one-proton-halo nucleus ^{8}B both display approximately the same reduced total reaction cross sections. This is surprising since it is known that the two-neutron-transfer cross section for ^{6}He is remarkably large near to and below the Coulomb barrier [3], while the breakup cross section of ^{8}B is considerably smaller [14, 15]. The latter was extensively studied in the context of the coupled discretized continuum channels (CDCC) method [16]. The agreement with the data was excellent, so transfer/breakup is well understood for this system. It appeared, therefore, that the fusion cross section for ^{8}B might show some surprises.

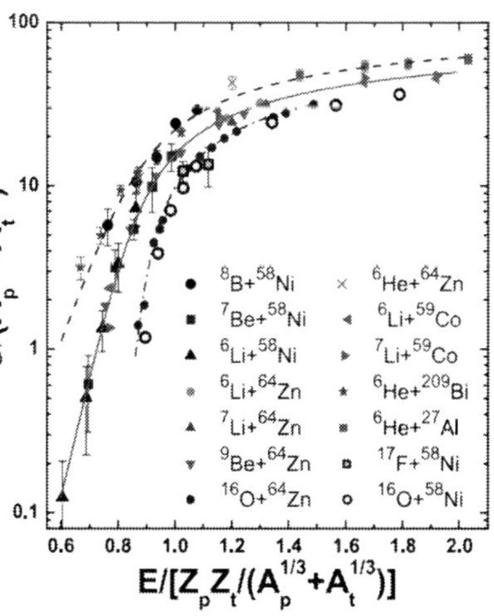

Figure 2. Reduced total reaction cross sections for a number of weakly- and strongly-bound projectiles (from Ref. [13]). ©2009, The American Physical Society.

2.2 Fusion

The fusion cross section for ^{8}B + ^{58}Ni was measured [17] by detecting evaporation protons emitted at backward angles. This method has an inherent advantage since the ^{66}As compound system is at the proton dripline, resulting in the emission of about 2.3 protons per fusion event and thereby increasing the statistical accuracy of the deduced cross section. However, it was suggested that the evaporation proton yield might have been contaminated by direct breakup protons even at backward angles [18]. Therefore, an intensive study of this possibility was carried out [19]. The calculations of Ref. [16] were extended to predict the direct proton yield at backward angles. It was found that the contamination was very small except at the lowest energy measured, and appropriate corrections were deduced. Furthermore, the evaporation α-particle yield was shown to agree with the proton yield. The statistics were much poorer since the α-particle multiplicity is only about 0.5. However, it was possible to recover the fusion cross section at the lowest energy since the α-particle yield is not expected to be contaminated by direct processes. In addition, possible systematic errors due to the calculation of the evaporated-particle multiplicities were determined. The same methodology was applied to the ^{8}B + ^{28}Si data [20] and the results are shown in Fig. 3.

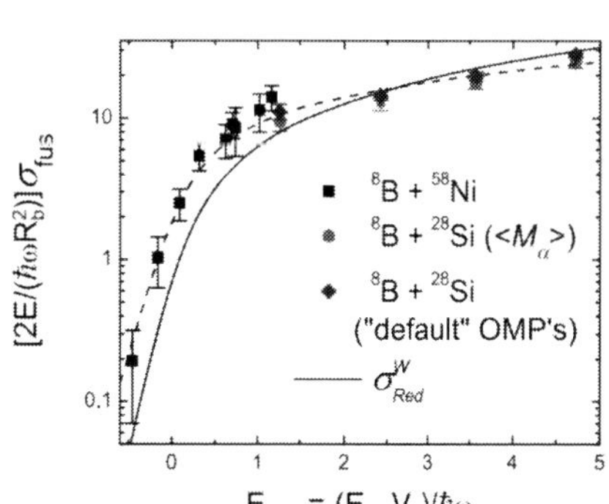

Figure 3. Reduced total fusion yields for ^{8}B+(^{58}Ni,^{28}Si) (from Ref. [13]). The solid curve represents the Universal Fusion Function (UFF) [21]. ©2016, The American Physical Society.

 DOI: 10.1051/epjconf/201716300031

The solid curve in this figure is the Universal Fusion Function (UFF) [21], which represents the expectation for fusion in "normal" circumstances. It can be seen that $^{8}B+^{58}Ni$ fusion appears to be enhanced up to about 1.5 times the reduced barrier, while the $^{8}B+^{28}Si$ fusion cross section indicates a possible fusion suppression at energies well above the barrier. The ^{58}Ni and ^{28}Si data agree reasonably well with each other at the one common (reduced) energy point. (It should be noted, however, that the ^{28}Si cross sections shown are the result of a reanalysis taking into account the fact that different evaporation codes predict different α-particle multiplicities. See Ref. [19] for a further discussion). In addition, the sum of the ^{58}Ni total fusion cross section plus the direct breakup yield equals the experimentally-measured total reaction cross section (Fig. 4). This is similar to the situation for ^{6}He reactions, in which the sum of the fusion and breakup/transfer yields equals the total reaction cross section. However, in the ^{6}He case, direct reactions, and particularly two-neutron transfer [3], dominate the total reaction cross section near the barrier, while in the case of ^{8}B the fusion and direct yields are more comparable.

Nevertheless, there are two important points to consider here. First of all, the cross sections shown in Fig. 3 depend on light-particle multiplicities calculated by various evaporation-model codes which have been shown to disagree with each other (though more-so for α-particle than proton evaporation). It would be helpful to compare with data from direct detection of evaporation residues, but this is a very difficult experiment at the available ^{8}B beam intensities. Secondly, it is very likely that the total fusion yield shown in this figure includes a large component of incomplete fusion (ICF) in which the ^{7}Be core fuses with the target while the weakly-bound proton continues on relatively undisturbed. The γ-ray detection methods that can resolve complete fusion (CF) from ICF processes [8] also require much greater beam intensities than are currently available. In the next section, we will discuss an experimental technique that can address both these issues.

3 Fusion Studies with a Time Projection Chamber

In a recent experiment [23] using the prototype active-target time projection chamber (pAT-TPC) [24] developed at Michigan State University, the fusion cross section for $^{10}Be+^{40}Ar$ was studied with extremely high efficiency. An entire excitation function (Fig. 5) was measured with an average beam intensity of only 100 particles per second in a counting period of 90 hours.

Furthermore, it was possible to directly image the evaporated charged particles (protons and α-particles), so that events in which only neutrons were emitted (NCP fusion) could be separated from the charged-particle (CP) events (Fig. 6). The individual cross sections for these types of events, as well as the summed fusion yield, agreed well with the predictions of the PACE4 fusion-evaporation code contained within the LISE++ package [25], as shown in Fig. 5. Note, however, that the target was actually P10 gas (90% Ar plus 10% CH_4) so that the C contaminant needed to be dealt with. As a result, the ^{40}Ar fusion yield at low energies was not well determined. See Ref. [23] for further details.

The results illustrated in Figs. 5 and 6 suggest that it will be possible to determine the $^{8}B + ^{40}Ar$ fusion yield by directly imaging the evaporation products, thus eliminating ambiguities due to the reliance on multiplicity calculations in the experiments discussed above. Furthermore, the ability to image light charged particles in this experiment suggests that it might be possible to distinguish complete fusion from incomplete fusion by looking for the associated higher-energy proton from ^{8}B breakup, as well as by measuring the kinetic energy of the recoiling evaporation residue. A calculation of the expected results for complete

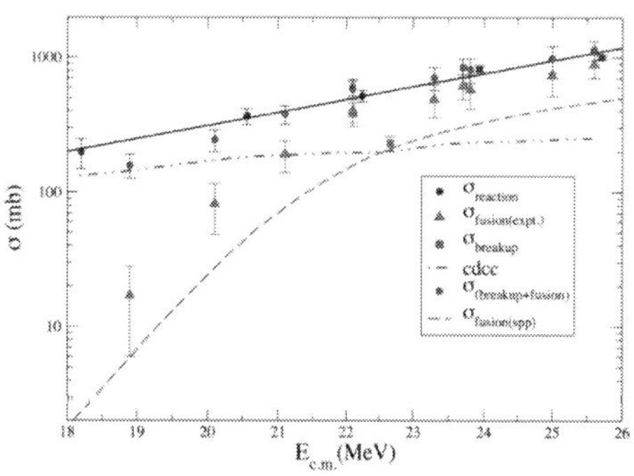

Figure 4. Cross sections for $^{8}B+^{58}Ni$. The total reaction cross section is from Refs. [12, 13], the total fusion cross section from Refs. [17, 19], and the breakup cross section from Refs. [14, 15] and also the CDCC calculation [16]. The fusion curve was calculated using the São Paulo potential [22].

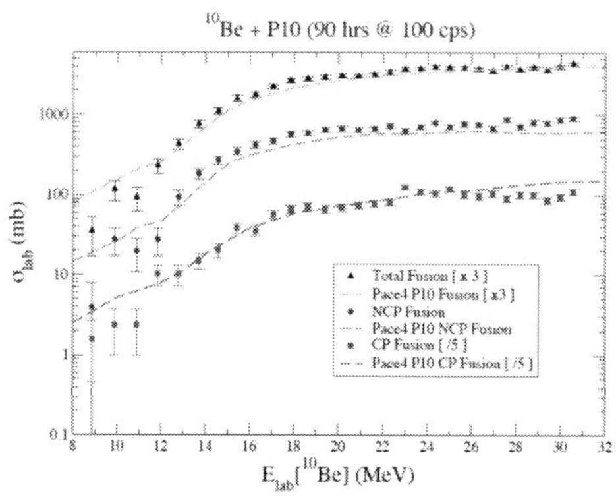

Figure 5. Total fusion cross section for $^{10}Be+P10$ (from Ref. [23]). The individual yields for neutron-only evaporation (NCP events) and charged-particle evaporation (CP events) are also shown. The curves are PACE4 calculations. ©2016, Elsevier.

 DOI: 10.1051/epjconf/201716300031

fusion is shown in Fig. 7 below. The cross sections were calculated using PACE4. The red points and associated curve are for a target of P10.

The blue points and curve assume a target of 90% Ar plus 10% H_2. This mixture worked well in a test chamber, but problems with the pAT-TPC (since repaired) prevented its use in the previous experiment. Incomplete fusion, if present, should act to increase the illustrated yield but hopefully could be identified as discussed above. This proposed experiment would then be an excellent test of the ability to separate CF from ICF for very-low-intensity exotic beams.

4 Modifications to *TwinSol*

One very important impediment to the proposed ^{8}B fusion studies with a TPC is the fact that this beam was heavily contaminated by ^{6}Li and ^{7}Be in the previous experiments at *TwinSol*, such that ^{8}B was only a very minor component. This was not a problem since ^{8}B could be unambiguously identified by its time of flight to the secondary target. However, TPC's have very definite rate limitations and therefore require a pure beam. As a result, extensive modifications of *TwinSol* are being carried out in order to provide such a beam. These are part of a major project to also provide higher-intensity and higher-Z radioactive beams.

The gas-cell primary target has been doubled in length to 5 cm, and 4 micron Ti windows have been installed (3 micron Ti is currently being tested). In addition, a drive mechanism that allows four different gas cells to be used before the chamber must be opened to replace damaged windows has been developed. To accomplish this, the primary target was moved upstream from the first solenoid by an additional 7 cm. Next, the fixed-aperture collimator at the crossover point between the two solenoids was replaced by a variable iris that can be adjusted without breaking vacuum. This collimator determines the momentum acceptance of the system and therefore can improve the purity of the transmitted beam. Finally, the focus of the secondary beam was moved about 5 m upstream. Due to space limitations, the pAT-TPC in previous experiments was installed at about 9 m from the primary target, while the designed focal point of *TwinSol* was at 5.6 m. As a result, the size of the beam was increased and therefore the transmission into the TPC was reduced. By moving the second solenoid 88 cm upstream, the optics were substantially improved and it was possible to achieve a reasonable beam image at 11 m. Another advantage of this is that the beam now travels through an aperture in a 1.25 m wide high-density concrete shielding wall, which greatly reduces the background levels for anticipated neutron and γ-ray experiments with radioactive beams.

The effects of all these modifications were investigated and optimized with Monte-Carlo calculations using LISE++ [25], which now includes calculation blocks for *TwinSol*. The ability to include angular distribution data

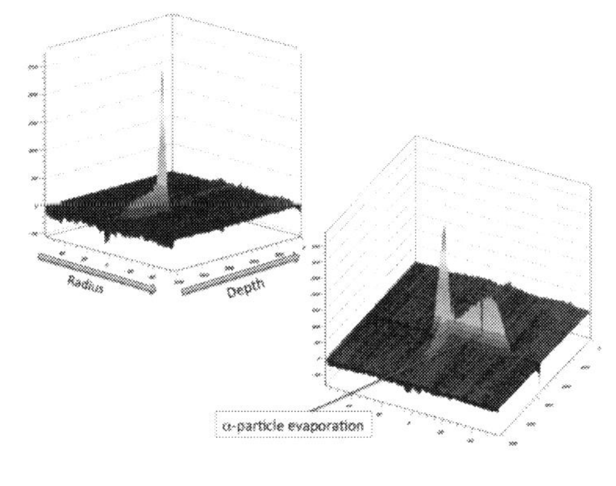

Figure 6. Ionization vs. radius and depth in the counter for two isolated fusion events. The leftmost plot shows a typical neutron-evaporation (NCP) event. The rightmost plot shows the effect of α-particle evaporation from the fused system. Note the deviation of the fusion-residue peak from the beam track, and its larger width, due to recoil of the evaporation product. (from Ref. [23]). ©2016, Elsevier.

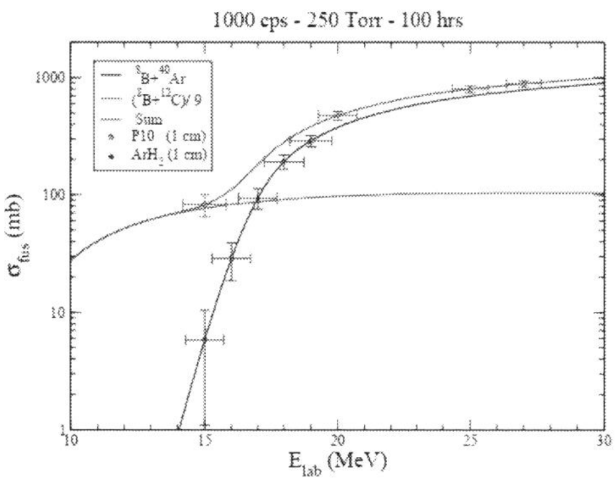

Figure 7. Calculated complete fusion cross sections for ^{8}B+^{40}Ar. The cross sections are averaged over a 1 cm path length of the incident ^{8}B, resulting in the horizontal and vertical error bars illustrated.

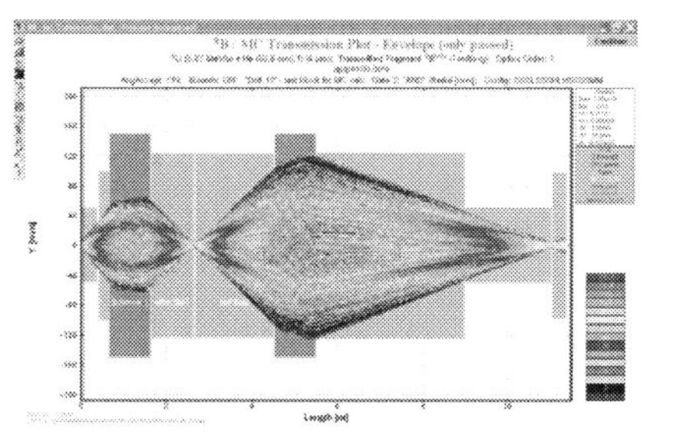

Figure 8. LISE++ calculation of the envelope of a 27-MeV ^{8}B beam focussed at 11 m from the primary target.

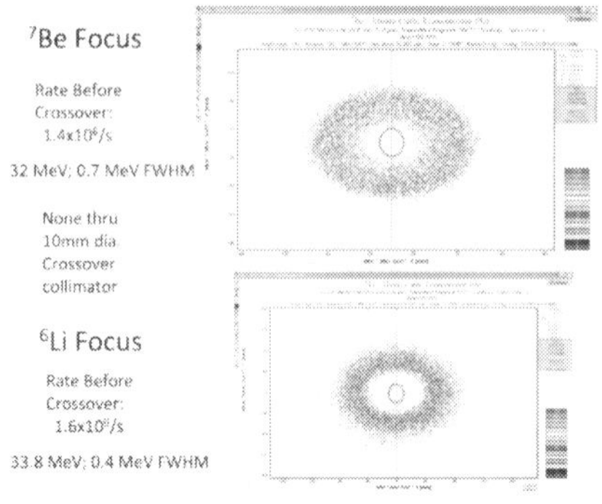

Figure 9. LISE++ calculation for the ^{7}Be and ^{6}Li images at the crossover collimator, including scattering in the gas-cell windows. The red circles illustrate 10 mm diameter collimators in each case. All the ^{8}B ions pass through a collimator of this diameter, leading to a rate of $3.6\text{x}10^4$/s through an 8 mm diameter collimator at the final focus (assuming a 1eμA ^{6}Li (3$^+$) primary beam).

for the production reaction in this code (either from experimental data or DWBA calculations) turned out to be very helpful both in estimating the final secondary beam rate and in adjusting the solenoid currents to maximize that rate. A sample calculation for a 27-MeV ^{8}B-ion envelope is illustrated in Fig. 8. The calculations suggest that the purity of the beam might be greater than 90% (see Fig. 9 below).

5 Summary and Conclusions

Elastic scattering, breakup, and fusion reactions of the exotic one-neutron-halo nucleus ^{8}B have been studied over a period of several years. The recent fusion data suggests that the total fusion cross section of this projectile is enhanced near and below the Coulomb barrier, while apparently being suppressed at energies well above the barrier. The sum of total fusion plus breakup yields equals the measured large total reaction cross section. This is similar to the situation for ^{6}He except that, in that case, the total reaction cross section near the barrier is dominated by neutron transfer, while the fusion yield is more similar to the direct-reaction yield for ^{8}B.

It seems likely that incomplete fusion (ICF) plays a substantial role in enhancing the ^{8}B total fusion cross section. A possible method to measure the ICF yield using a time projection chamber has been proposed. This method requires beams of high purity. Measures currently underway at *TwinSol* to improve the purity of the radioactive beams, as well as to provide higher-Z beams and a more background-free area for use in γ-ray and neutron detection experiments, have been described. The ^{8}B plus ^{40}Ar fusion experiment would be an excellent test of the ability of TPC's to detect and measure ICF reactions with very-low-intensity exotic beams.

Acknowledgements

This work has been partially supported by the US National Science Foundation under Grant No. PHY14-01343, and by the CONACYT (México). We also wish to thank Dr. Oleg Tarasov of Michigan State University for his help in using LISE++ and working to improve its *TwinSol* predictions.

References

[1] J.J. Kolata, V. Guimarães, and E. F. Aguilera, European Physical Journal A **52**, 123 (2016).
[2] M. Cubero, *et al.*, Phys. Rev. Lett. **109**, 262701 (2012).
[3] J. P. Bychowski, *et al.*, Phys. Lett. B **596**, 26 (2004).
[4] P. A. DeYoung, *et al.*, Phys. Rev. C **71**, 051601(R) (2005).
[5] J.J. Kolata, *et al.*, Phys. Rev. C **75**, 031302[R] (2007).
[6] M. Y. Lee *et al.*, Nucl. Instrum. Methods in Physics Research A **422**, 536 (1999).
[7] S.A. Krupko, *et al.*, AIP Conf. Proc. **1098**, 245 (2009).
[8] D.J. Hinde and M. Dasgupta, Phys. Rev. C **81**, 064611 (2010).
[9] J.J. Kolata, *et al.*, Phys. Rev. C **85**, 054603 (2012).
[10] P.F.F. Carnelli, *et al.*, Phys. Rev. Lett. **112**, 192701 (2014).
[11] R.T. deSouza, *et al.*, Phys. Rev. C **88**, 014602 (2013).
[12] E. F. Aguilera, *et al.*, Phys. Rev. C **79**, 021601[R] (2009).
[13] J.J. Kolata, *et al.*, Phys. Rev. C **79**, 027603 (2009).
[14] J.J. Kolata, *et al.*, Phys. Rev. C **63**, 024616 (2001).
[15] E. F. Aguilera, *et al.*, Physics of Atomic Nuclei **71**, 1163 (2008).
[16] J.A. Tostevin, *et al.*, Phys. Rev. C **63**, 024617 (2001).
[17] E. F. Aguilera, *et al.*, Phys. Rev. Lett. **107**, 092701 (2011).
[18] J. Rangel, *et al.*, Eur. Phys. J. A **49**, 57 (2013).
[19] E. F. Aguilera, P. Amador-Valenzuela, *et al.*, Phys. Rev. C **93**, 034613 (2016).
[20] A. Pakou, *et al.*, Phys. Rev. C **87**, 014619 (2013).
[21] L.F. Canto, *et al.*, Nucl. Phys. A **821**, 51 (2009).
[22] L.C. Chamon, *et al.*, Phys. Rev. C **66**, 014610 (2002).
[23] J.J. Kolata, *et al.*, Nucl. Instrum. Methods in Physics Research A **830**, 82 (2016).
[24] D. Suzuki, M. Ford, D. Bazin, W. Mittig, *et al.*, Nucl. Inst. Methods in Physics Research A **691**, 39 (2012).
[25] http://lise.nscl.msu.edu/lise.html

^{8}B + ^{208}Pb Elastic Scattering at Coulomb Barrier Energies

M. La Commara[1,*], M. Mazzocco[2], A. Boiano[1], C. Boiano[3], C. Manea[2], C. Parascandolo[1], D. Pierroutsakou[1], C. Signorini[2], E. Strano[2], D. Torresi[2], H. Yamaguchi[4], D. Kahl[4], P. Di Meo[1], J. Grebosz[5], N. Imai[4], Y. Hirayama[6], H. Ishiyama[6], N. Iwasa[7], S.C. Jeong[6], H.M. Jia[8], Y.H. Kim[6], S. Kimura[6], S. Kubono[9], C.J. Lin[8], H. Miyatake[6], M. Mukai[6], T. Nakao[10], M. Nicoletto[2], Y. Sakaguchi[4], A.M. Sánchez-Benítez[11], F. Soramel[2], T. Teranishi[12], Y. Wakabayashi[9], Y.X. Watanabe[6], L. Yang[8], and Y.Y. Yang[8]

[1] *Department of Physics "E.Pancini", University "Federico II" and INFN, Napoli, Italy*
[2] *Department of Physics and Astronomy, University of Padova and INFN, Padova, Italy*
[3] *INFN-Sezione di Milano, Milano, Italy*
[4] *CNS - The University of Tokyo, Wako-shi, Saitama, Japan*
[5] *IFJ PAN, Krakow, Poland*
[6] *KEK, Tsukuba, Ibaraki, Japan*
[7] *Department of Physics, Sendai, Miyagi, Japan*
[8] *China Institute of Atomic Energy, Beijing, China*
[9] *RIKEN Nishina Center, Wako, Saitama, Japan*
[10] *JAEA, Tokai, Ibaraki, Japan*
[11] *Department of Applied Physics, University of Huelva, Huelva, Spain*
[12] *Department of Physics, Kyushu University, Hakozaki, Fukuoka, Japan*

Abstract. The scattering process of weakly-bound nuclei at Coulomb barrier energies provides deep insights on the reaction dynamics induced by exotic nuclei. Within this framework, we measured for the first time the scattering process of the short-lived Radioactive Ion Beam (RIB) ^{8}B (S_p = 0.1375 MeV) from a ^{208}Pb target at 50 MeV beam energy. The ^{8}B RIB was produced by means of the in-flight facility CRIB (RIKEN, Japan) with an average intensity on target of 10 kHz and a purity about 25%. Elastically scattering ions were detected in the angular range $\theta_{c.m.}$ = 10°-160° by means of the detector array EXPADES. A preliminary optical model analysis indicates a total reaction cross section of about 1 b, a value, once reduced, 2-3 times larger than those obtained for the reactions induced by the stable weakly-bound projectiles ^{6,7}Li on a ^{208}Pb target in the energy range around the Coulomb barrier.

1 Introduction

Nuclei far from β−stability valley, as ^{6}He, ^{11}Be, ^{8}B, are very interesting because of their peculiarities as far as their structure and their way to interact with other nuclei are concerned, indeed during the last decades many theoretical and experimental efforts have been devoting to the investigation of halo nuclei in order to shed light on their unusual and intriguing features [1]. Nuclear collisions involving stable and unstable weakly-bound nuclei have attracted great interest nowadays. However, despite great efforts in experiments and theories over the past decades, the precise description on the influence of breakup channel on fusions is still elusive, due to the limited knowledge on the breakup process, especially for the unstable reaction systems. More knowledge in this framework can be obtained starting from the investigation of simple processes, as elastic scattering. It is well accepted by the nuclear physicists community that the presence of a proton (neutron) halo or skin may heavily influence the behavior of an exotic nucleus in terms of nuclear reaction mechanism as well as can strongly influence its structure [1] [2]. Elastic scattering, a simple process, is a very appropriate probe to study the size and surface diffuseness of exotic nuclei by comparing similarities and differences in reactions induced by weakly bound and tightly bound nuclei. Anyway, elastic scattering data for proton-halo isotopes around the Coulomb barrier are still not enough to fully understand the dynamics of this process. ^{8}B is a proton-drip-line β−decaying isotope, conveying pretty much interest because of the role it plays in the production of high-energy neutrinos in the sun and its quite unusual structure, being considered by now a one-proton halo nucleus [3]. It is a very interesting isotope since it has a very low proton separation energy, S_p = 0.1375 MeV, which makes it a very lightly bound nucleus being an ideal candidate for nuclear-reaction mechanism investigations with exotic nuclei. Furthermore, the studies on ^{8}B reaction-mechanism processes are rather exiguous due to the difficulty of producing the ^{8}B beam, being it produced only in very few facilities and anyhow with low intensities [4] [5] [6]. In

* e-mail: Marco.LaCommara@na.infn.it

this paper we report new experimental data for the elastic scattering of ^{8}B on a ^{208}Pb-target, at 50 MeV, which means just on the Coulomb barrier, showing very preliminary data results.

2 Experimental setup and data analysis

The radioactive beam ^{8}B at 50 MeV (6.25 A MeV) has been produced via the reaction ^{3}He(^{6}Li,^{8}B)n (Q_{value} = −1.97 MeV) at the facility CRIB [7], situated at the RIKEN laboratory in Japan. The primary beam of ^{6}Li had an intensity of 3eμA, and a cryogenic ^{3}He−target with pressure of 1 bar and a typical temperature of 90 K was used. The secondary beam had an intensity (on target) of 10^{4} pps and a purity of 20% consistent with the expected one, being the main contaminants ^{7}Be and ^{3}He, besides the ^{6}Li scattered beam. Anyhow all the contaminants had different times of flight through the separator and then it was possible to separate events coming from the various components of the beams by using proper gates in the ToF spectra. Charged products arising from the reaction of the ^{8}B beam impinging on a 2.2 mg/cm^{2} ^{208}Pb target (evaporated on a 1.5 μm thick mylar backing foil) were detected with the detector array EXPADES [8] utilized in a configuration consisting of 6 $\Delta E - E_{res}$ double-sided silicon-strip detectors, placed at 113 mm from the center of the ^{208}Pb target in a slightly asymmetric configuration such to ensure the whole angular-coverage ranging from 16° till 164° in the laboratory frame. In Fig.1 a very preliminary evaluation of the angular distribution for the elastic scattering process in the system ^{8}B+^{208}Pb is shown, where each of the four plotted points corresponds to the angular-range mean placement of the EXPADES detectors. The

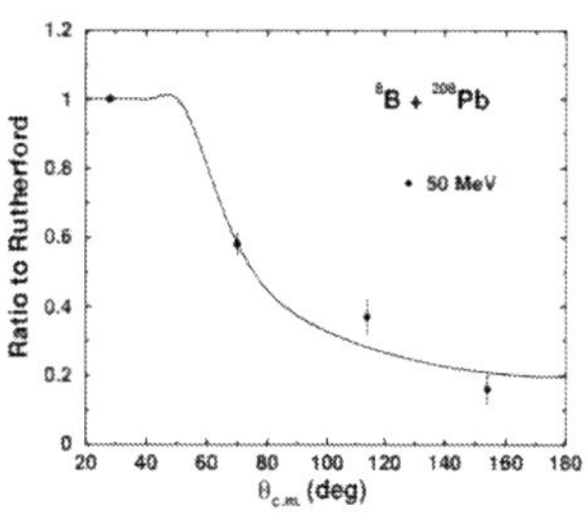

Figure 1. Preliminary elastic scattering angular distribution for the system ^{8}B+^{208}Pb at 50 MeV beam energy. The continuous line is the result of the optical model analysis of the collected data

experimental data have been normalized with a Monte-Carlo simulation accounting for the detector-geometry and the elastic-scattering of the process, whereas the simulated events have been generated according to a pure Rutherford cross-section. The continuous line represents the result of an optical model best-fit analysis (FRESCO code). In this way we could extract a −preliminary− evaluation of the reaction cross-section for the system ^{8}B+^{208}Pb at 50 MeV beam energy.

In Fig.2 the reaction X-sections for the systems ^{6,7}Li, ^{7}Be and ^{8}B +^{208}Pb in the energy range around the Coulomb barrier are shown [10], [11] and [12]. The data for the different systems have been reduced as reported in ref. [8] and multiplied by a normalization factor of the ^{7}Be+^{208}Pb reaction, in order to be comparable with each other. It seems, even if the data analysis is in the first stages of its route, that the reaction induced by the one-proton halo nucleus ^{8}B (S_{p} =0.1375 MeV) on the ^{208}Pb target has a total reaction X-section much larger than those measured for reactions induced by other weakly-bound isotopes on the same target.

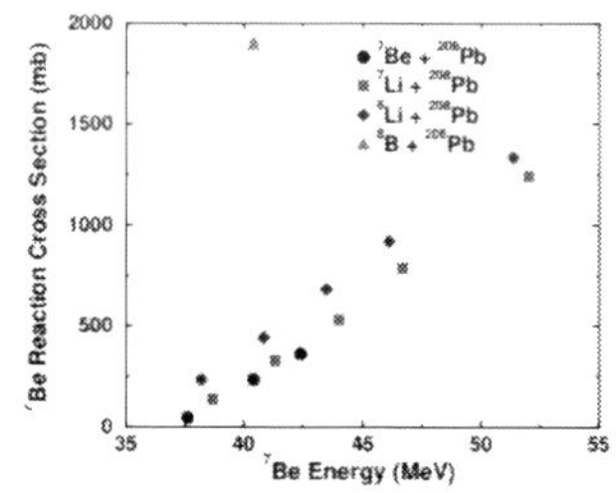

Figure 2. Reaction X-sections for the systems ^{6,7}Li, ^{7}Be and ^{8}B +^{208}Pb (see [10], [11] and [12]). The data were reduced according to the procedure recommended in [8] and then normalized to the system ^{7}Be +^{208}Pb

3 Discussion and outlook

This is to some extent consistent with the outcome for reactions induced by the same projectiles on ^{58}Ni [3]. If this will be confirmed, the data analysis will hopefully allow us to disentangle which mechanism, direct processes (such as break-up and/or transfer) or fusion, is responsible for accounting such large X-section value.

References

[1] N. Keeley *et al.*, *Prog. Part. Phys.* **59**, 579 (2007).
[2] P.R.S. Gomes *et al.*, *Phys. Rev.* **C71**, 034608 (2005).
[3] E.F. Aguilera *et al.*, *Phys. Rev.* **C79**, 021601(R) (2009).
[4] E.F. Aguilera and J.J. Kolata, *Phys. Rev.* **C85**, 014603 (2012).
[5] A. Pakou *et al.*, *Phys. Rev.* **C87**, 014619 (2013).
[6] Y.Y. Yang *et al.*, *Phys. Rev.* **C87**, 044613 (2013).
[7] Y. Yanagisawa *et al.*, *Nucl. Instr. and Meth. A* **539**, 74 (2005).
[8] D. Pierroutsakou *et al.*, *Nucl. Instr. and Meth. A* **834**, 46 (2016).
[9] P.R.S. Gomes *et al.*, *Phys. Rev.* **C71**, 017601 (2005).
[10] M. Mazzocco *et al.*, these Proceedings.
[11] N. Keeley *et al.*, *Nucl. Phys. A* **571**, 326 (1994).
[12] M. Mazzocco *et al.*, *Eur. Phys. Journal*, Web of Conferences, **117**, 06006 (2016).

 DOI: 10.1051/epjconf/201716300033

Survival Mediated Heavy Element Capture Cross Sections

Walter Loveland[1] and Larry Yao[1,]
[1]Dept. of Chemistry, Oregon State University, Corvallis, OR 97331 USA

Abstract. Formally, the cross section for producing a heavy evaporation residue, $\sigma_{\rm EVR}$, in a fusion reaction can be written as

$$\sigma_{\rm EVR}(E) = \frac{\pi\hbar^2}{2\mu E}\sum_{\ell=0}^{\infty}(2\ell+1)T(E,\ell)P_{\rm CN}(E,\ell)W_{\rm sur}(E,\ell), \quad (1)$$

where E is the center of mass energy, and T is the probability of the colliding nuclei to overcome the potential barrier in the entrance channel and reach the contact point. $P_{\rm CN}$ is the probability that the projectile-target system will evolve from the contact point to the compound nucleus. $W_{\rm sur}$ is the probability that the compound nucleus will decay to produce an evaporation residue rather than fissioning. However, one must remember that the $W_{\rm sur}$ term effectively sets the allowed values of the spin, which in turn, restricts the values of the capture and fusion cross sections. We point out the implications of this fact for capture cross sections for heavy element formation reactions.

1 Introduction

Formally, the cross section for producing a heavy evaporation residue, $\sigma_{\rm EVR}$, in a fusion reaction can be written as as the product of three factors, (see equation 1) that express the probability of bringing the colliding nuclei into contact, having that configuration evolve inside the fission saddle point and having the resulting nucleus survive against fission.

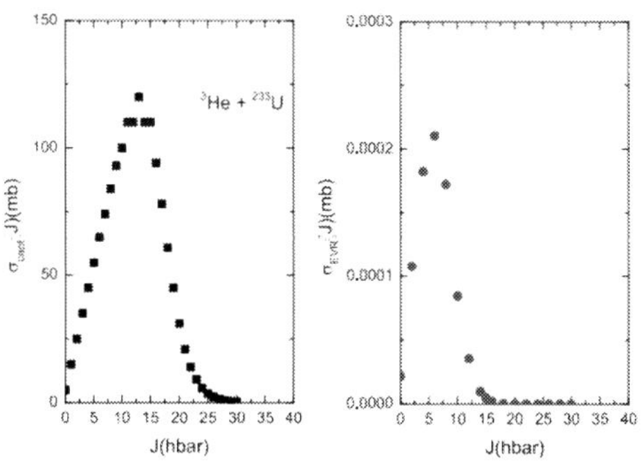

Figure 1. Spin dependence of capture and evaporation residue formation cross sections for ^{3}He + ^{233}U where the lab energy of the ^{3}He was 44.7 MeV.

Conventionally the EVR cross section is separated into three individual reaction stages (capture, fusion, survival) motivated, in part, by the different time scales of the processes. However, one must remember that the $W_{\rm sur}$ term effectively sets the allowed values of the spin, which in turn, restricts the values of the capture and fusion cross sections. To illustrate this point, we show, in Figure 1, the spin dependence of capture cross section for the reaction of ^{3}He + ^{233}U and the spin dependence of the evaporation residue cross section for the 5n channel [1]. Note the restrictions on the surviving spins due to the fission probability and the resulting differences between the capture and evaporation residue cross sections.

2 Survival Probabilities

For the most part, the formalism for calculating the survival, against fission, of a highly excited nucleus is understood [2]. One begins with a single particle model [3] of the level density in which one allows the level density parameter to be a function of the excitation energy. Masses and shell corrections are taken from [4]. The deformation dependent collective enhancement of the level density is taken from [5] for a particular xn outgoing reaction channel. Standard formulas are used to calculate the decay widths for decay by neutron, charged particle and γ-emission. The fission width is corrected for Kramers effects [6]. The fission barrier heights are calculated using liquid drop barriers and excitation energy dependent shell corrections.

What are the uncertainties in these calculations of W_{sur}? Lu and Boilley [7] found collective enhancement effects, Kramers effects, and the overall fission barrier height to have the biggest effect on the calculated survival probabilities. Loveland [8] concluded that, in general, fission barrier heights are known to within 0.5 - 1.0 MeV. For super-heavy nuclei, the change of fission barrier height by

1 MeV in each neutron evaporation step can cause an order of magnitude uncertainty in the 4n-channel. To minimize sensitivity to this factor, we have restricted our attention to nuclei where $Z \leq 110$ where the barrier heights are better known.

3 Outline of Calculations

We begin with the compilation of Duellmann of evaluated evaporation residue cross sections for reactions that produce heavy nuclei [9]. This compilation involves ~ 500 different reactions producing nuclei from Z=80 to Z=122, although we have limited our calculations to cases where $Z_{CN} \leq 110$. For each reaction (projectile, target and beam energy) we calculate the spin dependent capture cross sections using the Bass model [10]. The beam energies are chosen to correspond to the maximum of the excitation functions for the reaction channel being studied. For each value of the angular momentum, ℓ, the survival probability is calculated using the formalism(s) of [2]. If we assume P_{CN} is one, then, according to equation 1, we can straightforwardly estimate the evaporation residue formation cross section. If the calculated evaporation residue cross section is significantly greater then the measured cross section for this reaction, we have evidence, perhaps quantitative, that P_{CN} is less than 1.

As a "reality check" on these results, we calculated the EVR spin distribution for the reaction $^{176}Yb(^{48}Ca,4n)^{220}Th$ and compared it to the measured spin distribution of Henning et al. [11] (Figure 2) for the same reaction. The comparison of the results for the measured and calculated spin distributions indicates that the calculational procedure used in this work is appropriate. This same sort of agreement between the calculated spin distributions and the measured distributions was observed by [5] for the $^{208}Pb(^{48}Ca, 2n)^{254}No$ reaction.

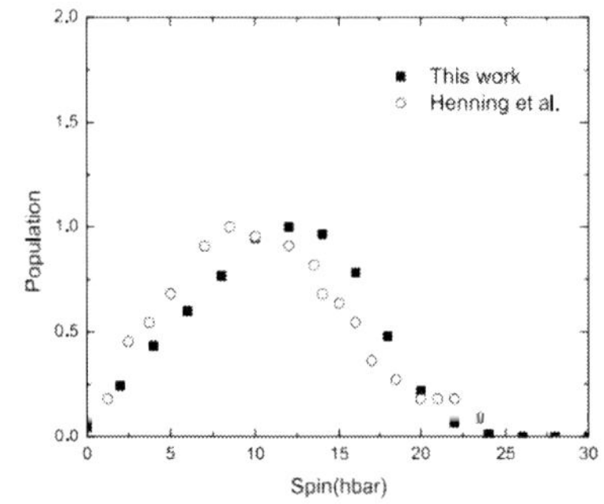

Figure 2. Spin dependence of calculated and measured evaporation residue formation cross sections for the $^{176}Yb(^{48}Ca,4n)^{220}Th$ reaction.

3.1 Limitations of Calculational Procedure

We have used relatively simple tools (the semi-empirical Bass model [10]) to simulate the dynamics of the capture process. Better tools might have included coupled channel calculations for the capture process, taking into account the deformations of the target and projectile nuclei and the excitations of collective modes. The procedures used to calculate the survival probabilities are thought to reflect all the important features of these processes. The success in calculating the EVR spin distributions (Fig. 2 and [5] support the methods chosen. The dramatic decrease in the survival probabilities for higher J values (as shown in Figures 3-4) demonstrate the important qualitative effect being studied in this work.

4 Results of Calculations

4.1 $Z_1Z_2 < 700$

There are 76 cases in this category. In each case, P_{CN} was calculated to be 1, i.e., the calculated evaporation residue cross section agreed with the measured cross section within experimental error. A sampling of some typical cases for Z_{CN} = 94-98, 101-103 and 104-105 is shown in Figure 3.

The most probable spins for the surviving evaporation residues are ~ 5 $\hbar$, while the capture cross sections can involve spins out to 30 $\hbar$ with average spins of 10-20 $\hbar$. Also note the dramatic reduction in cross section due to the fission of the completely fused system. Capture cross sections for these hot fusion reactions are typically 100-300 mb while the evaporation residue cross sections are ~ nb to μb.

For "convenience" it is sometimes assumed that the relevant capture cross sections and P_{CN} factors can be evaluated for J=0, i.e., drop the spin dependences. This extreme view is not supported by the data shown in Fig.3.

4.2 $700 < Z_1Z_2 < 1000$

There are 119 cases in this category. Many of these reactions are sub-barrier and the difference between the spin dependence of the capture cross section and the spin-mediated evaporation residue cross section is smaller. For all the cases studied, P_{CN} =1. A sampling of these cases is shown in Figure 4. Note not only the decrease in the mean spin of the evaporation residues but the loss of orders of magnitude in the cross sections due to fission de-excitation.

4.3 $Z_1Z_2 > 1000$

We can break this category into two groups, $1000 \leq Z_1Z_2 \leq 1600$ (136 cases) and $Z_1Z_2 \geq 1600$ (167 cases). Generally P_{CN} ~ 1 for the first group and $P_{CN} < 1$ for $Z_1Z_2 \geq 1600$. In Figure 5, we show some typical cases of "cold fusion" reactions with $Z_1Z_2 \geq 1600$. We plot the capture cross section and the "effective" capture cross section for each case where the "effective" capture cross section reflects the fraction of the capture cross section that contributes to evaporation residue formation.

In Figure 6, we show a similar sampling of the data for hot fusion reactions.

4.4 Effective capture cross sections

In Figure 7, we show the dependence of the "effective" capture cross sections calculated in this work as a function of the product of the atomic numbers of the colliding nuclei, Z_1 and Z_2. (Other scaling parameters could have been chosen such as $(Z_1Z_2/(A_1^{1/3}+A_2^{1/3})$ that might have been more effective but we have chosen the simplest parameterization). One notes that the effective capture cross sections quickly decrease to less than 10 mb for Z_1Z_2 values > 1000 and for the most "interesting" reactions (Z_1Z_2 > 1500) are only a few mb. In terms of impact parameters for heavy element synthesis reactions, this means that the interesting collisions are the near central collisions. Calculations of fusion and quasi-fission cross sections should focus on these collisions.

5 Conclusions

While measurements of capture, fusion and quasi-fission cross sections have scientific value, they must be "filtered' for relevance in considerations of heavy element synthesis. The "effective" capture cross sections are a fraction of the "real" capture capture cross sections due to the restrictions placed on these quantities due to the spin dependence of the survival probabilities. The "effective" capture cross sections decrease rather than increase with increasing Z_1Z_2 due to these survival probability effects. Only a few partial waves contribute to the effective capture cross sections for larger values of Z_1Z_2. Calculations of fusion cross sections for the synthesis of the heaviest nuclei should only consider small impact parameter collisions.

6 Acknowledgements

This work was supported in part by the U.S. Dept. of Energy, Office of Science, Office of Nuclear Physics under award number DE-SC0014380.

References

[1] C. Laue, et al., , Phys. Rev. C **59**, 3086 (1999)
[2] Nuclear Reactions Video Project (nrv.jinr.ru) - "Statistical Model of Decay of Excited Nuclei"
[3] A. V. Ignatyuk, IAEA report INDC(CCP)-233/L(1985)
[4] P. Moller, A.J. Sierk, T. Ichikawa, and H. Sagawa, At. Data and Nucl. Data Table **109-110**, 1 (2016).
[5] V.I. Zagrebaev, Y. Aritomo, M.G. Itkis, Y.T. Oganessian, and M. Ohta, Phys. Rev. C **65**, 014607 (2002)
[6] H.A. Kramers, Physica (Amsterdam) **7**, 284 (1940).
[7] H. Lu, D. Boilley, EPJ Web of Conferences **62**,03002 (2013).
[8] W. Loveland, Eur. J. Phys. A **51**, 120 (2015).
[9] Ch. E. Duellmann, private communication
[10] R. Bass, Phys. Rev. Lett. **39**, 265 (1977); R. Bass, Nuclear reactions with heavy ions (Springer-Verlag, Berlin, 1980).
[11] G. Henning, et al., Phys. Rev. Lett. **113**, 262505 (2014).
[12] Z. Qin et al. , Radiochemica Acta, **96**, 455 (2008)
[13] T. Sikkeland, J. Maly, and D.F. Lebeck, Phys. Rev. **169**, 1000 (1968)
[14] P. Eskola, Phys. Rev. C **7**, 280 (1973)
[15] K. Eskola, P. Eskola, M. Nurmia and A. Ghiorso, Phys. Rev. C **4**, 632 (1971)
[16] Yu. A. Lazarev et al. , Phys. Rev. C **62**, 064307 (2000)
[17] Y. Nagame, et al., J. Nucl. Radio Sciences **3**, 85 (2002).
[18] J. V. Kratz et al., Phys. Rev, C **45**, 1064 (1992).
[19] C. M. Folden III et al., Phys. Rev. C **79**, 027602 (2009)
[20] F.P. Hessberger et al., Eur. Phys. J. A **12**, 57 (2001)
[21] S. L. Nelson et al., Phys. Rev. C **78**, 024606 (2008)
[22] J.M. Gates et al., Phys. Rev. C **78**, 034604 (2008)
[23] K. Morita, et al., J. Phys. Soc. Japan **78**, 064201 (2009).
[24] J.M. Gates et al., Phys. Rev. C **77**, 034603 (2008)
[25] J. Dvorak et al., Phys. Rev. Lett. **100**, 132503 (2008)
[26] K. Nishio et al., Phys. Rev. C **82**, 024611 (2010)

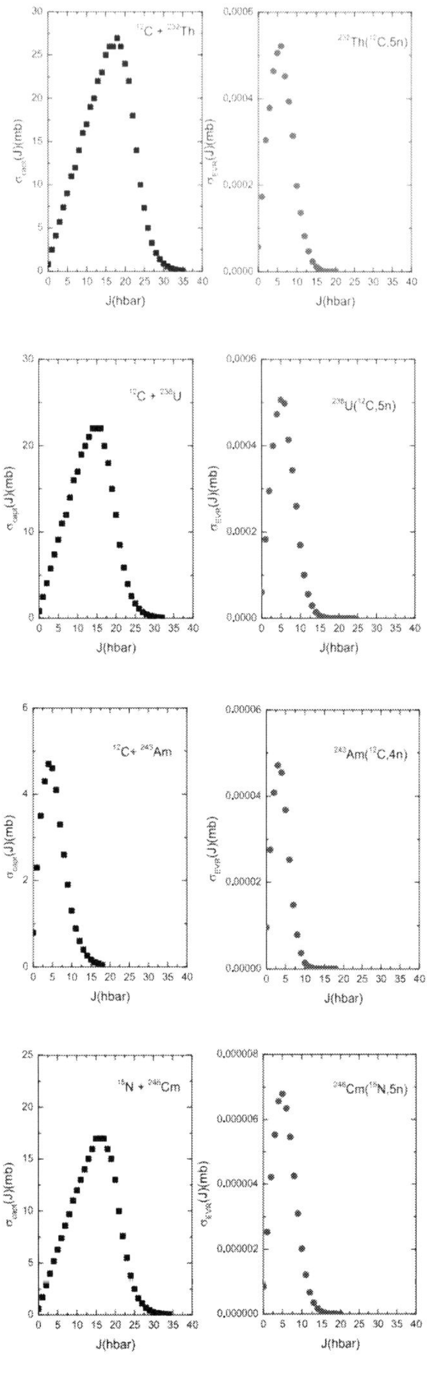

Figure 3. Spin dependence of calculated capture cross sections and the EVR cross sections for the reactions of ^{12}C + ^{232}Th, ^{243}Am, ^{249}Cf and the reaction of ^{15}N with ^{248}Cm. [12–15] where the lab frame beam energies were 70,73,70, and 86 MeV respectively.

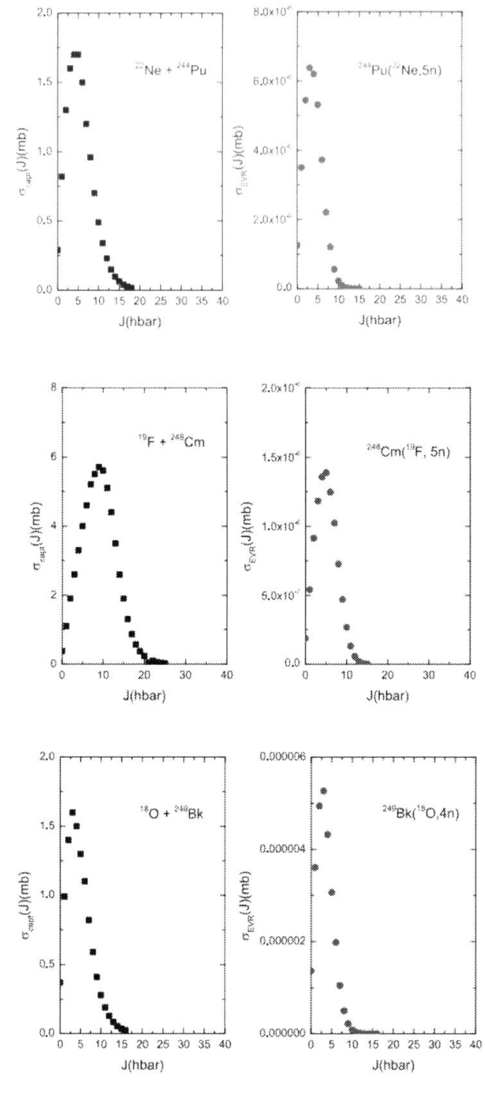

Figure 4. Spin dependence of calculated capture cross sections and the EVR cross sections for the reaction of ^{22}Ne + ^{244}Pu, ^{19}F + ^{248}Cm and ^{18}O + ^{249}Bk. [16–18] where the laboratory frame beam energies were 114,106, and 93 MeV, respectively.

 DOI: 10.1051/epjconf/201716300033

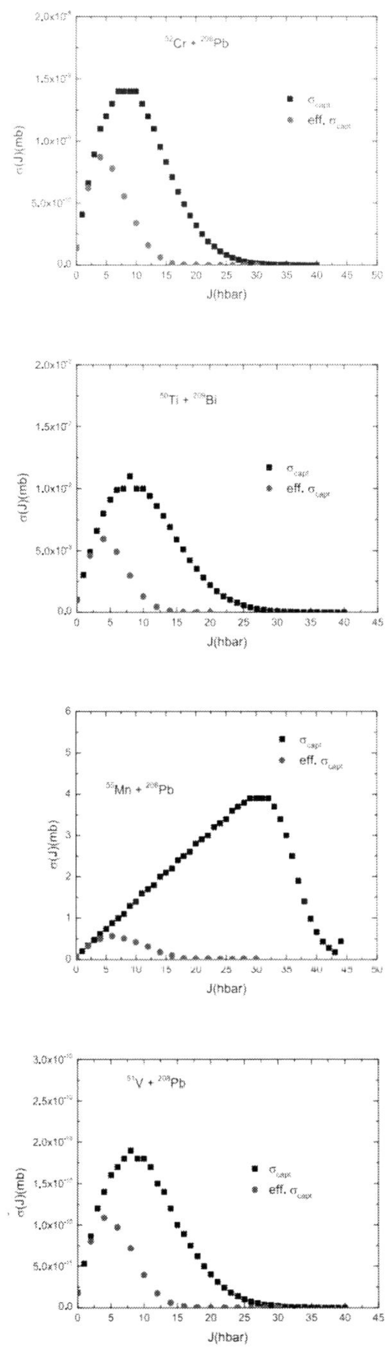

Figure 5. Spin dependence of calculated capture cross sections and the "effective" capture cross sections for the "cold fusion" reactions [19–22] where the lab frame beam energies were 253,241,283,and 240 MeV, respectively.

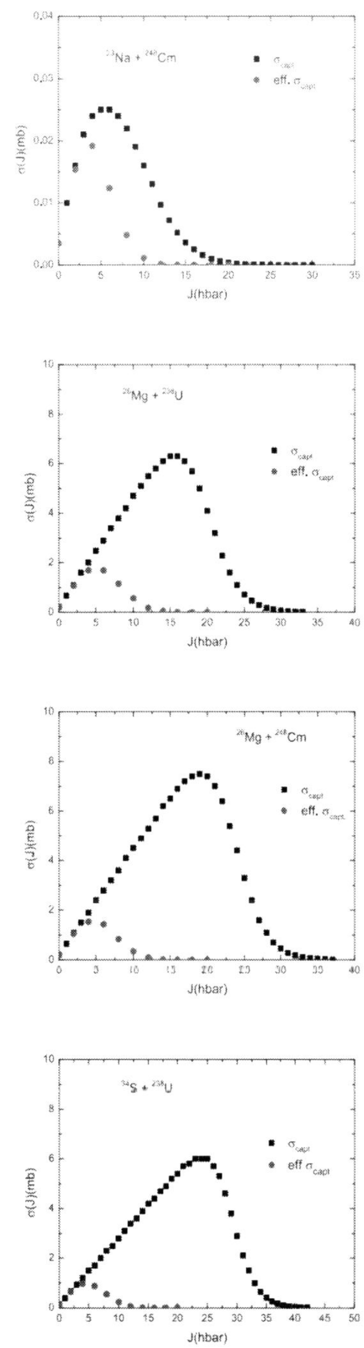

Figure 6. Spin dependence of calculated capture cross sections and the "effective" capture cross sections for the "hot fusion" reactions [23–26] where the lab frame beam energies were 130,139,144 and 188 MeV respectively.

 DOI: 10.1051/epjconf/201716300033

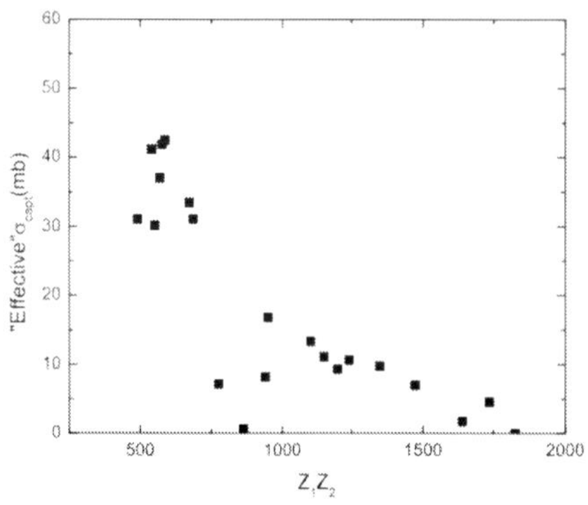

Figure 7. The dependence of the effective capture cross sections calculated in this work on the scaling parameter Z_1Z_2

Fission barriers from multidimensionally-constrained covariant density functional theories

Bing-Nan Lu[1,2], *Jie Zhao*[1,3], *En-Guang Zhao*[1,4], and *Shan-Gui Zhou*[1,4,5,6,★]

[1]*CAS Key Laboratory of Frontiers in Theoretical Physics, Institute of Theoretical Physics, Chinese Academy of Sciences, Beijing 100190, China*
[2]*Institut für Kernphysik (IKP-3) and Jülich Center for Hadron Physics, Forschungszentrum Jülich, D-52425 Jülich, Germany*
[3]*Microsystem and Terahertz Research Center, China Academy of Engineering Physics, Chengdu 610200, China*
[4]*Center of Theoretical Nuclear Physics, National Laboratory of Heavy Ion Accelerator, Lanzhou 730000, China*
[5]*School of Physical Sciences, University of Chinese Academy of Sciences, Beijing 100049, China*
[6]*Synergetic Innovation Center for Quantum Effects and Application, Hunan Normal University, Changsha, 410081, China*

Abstract. In recent years, we have developed the multidimensionally-constrained covariant density functional theories (MDC-CDFTs) in which both axial and spatial reflection symmetries are broken and all shape degrees of freedom described by $\beta_{\lambda\mu}$ with even μ, such as $\beta_{20}, \beta_{22}, \beta_{30}, \beta_{32}, \beta_{40}$, etc., are included self-consistently. The MDC-CDFTs have been applied to the investigation of potential energy surfaces and fission barriers of actinide nuclei, third minima in potential energy surfaces of light actinides, shapes and potential energy surfaces of superheavy nuclei, octupole correlations between multiple chiral doublet bands in ^{78}Br, octupole correlations in Ba isotopes, the Y_{32} correlations in $N = 150$ isotones and Zr isotopes, the spontaneous fission of Fm isotopes, and shapes of hypernuclei. In this contribution we present the formalism of MDC-CDFTs and the application of these theories to the study of fission barriers and potential energy surfaces of actinide nuclei.

1 Introduction

Many intrinsic nuclear shapes which break certain spatial symmetries play important roles in determining nuclear structure and fission properties [1–5]. The nuclear shape can be described by parametrizing the nuclear surface or the nucleon density distribution with the multipole expansion method. The deformation is then characterized by the expansion coefficient $\beta_{\lambda\mu}$.

Although the axial quadrupole (β_{20}) shape are prevalent in nuclear physics study, there have become more and more theoretical and experimental investigations on the β_{22} deformation, which manifests itself in the wobbling motion [2, 6], chiral doublet bands [7–11] and the termination of rotational bands [12], and the β_{30} deformation, which is characterized by parity doublet bands [13–15]. The impact of the non-axial octupole (β_{32}) deformation on the low-lying spectra have been also a hot topic in last decades [16–25].

Since the discovery of the fission phenomena [26], a quantitative description of the whole fission process has been very challenging. It was already known that the fission dynamics are mostly determined by the barriers prohibiting the dissolving of the nucleus [27]. Thus for solving the fission problem, one needs very accurate information on fission barriers, e.g., the heights.

Due to the shell effects, many actinide nuclei are found to be characterized by the two-humped fission barrier [28–37]. Consequently any models concerning the fission phenomena should be able to reproduce the barrier heights with certain assumptions on the nuclear shapes. For example, in early calculations of the fission barriers, the nuclei are usually assumed to be axially symmetric. Later it was found that the occurrence of the triaxial and the octupole deformations has non-negligible consequences in determining the heights of the inner and outer fission barriers, respectively. It has been known from the macroscopic-microscopic model calculations that the inner fission barrier is lowered when the triaxial deformation is allowed, so is the outer one by the reflection asymmetric distortion [30, 31]. Such lowering effects were also revealed in the non-relativistic [35] and relativistic [38, 39] density functional calculations, respectively. Recently, it is shown in Ref. [40] that the outer barriers are further lowered by the triaxial deformation compared with axially symmetric results. This lowering effect for the reflection-asymmetric outer barrier is of 0.5~1 MeV, accounting for 10%~20% of the barrier height. It is thus necessary to include these shape degrees of freedom properly in any systematic calculations.

To study nuclear ground states, shape isomers and PES's, it is desirable to have microscopic and self-consistent models in which all known important shape degrees of freedom are included. We have developed such

★e-mail: sgzhou@itp.ac.cn

a model, the so called multidimensionally-constrained covariant density functional theories (MDC-CDFTs), by breaking the reflection and axial symmetries simultaneously.

In MDC-CDFTs, the nuclear shape is assumed to be invariant under the reversion of x and y axes, i.e., the intrinsic symmetry group is V_4 and all shape degrees of freedom $\beta_{\lambda\mu}$ with even μ (β_{20}, β_{22}, β_{30}, β_{32}, β_{40}, $\cdots$) are included self-consistently. The MDC-CDFTs consist of two types of models: the multidimensionally-constrained relativistic mean field (MDC-RMF) model and the multidimensionally-constrained relativistic Hartree-Bogoliubov (MDC-RHB) model.

In the MDC-RMF model, the BCS approach has been implemented for the particle-particle (pp) channel. This model has been used to study potential energy surfaces and fission barriers of actinides [40–45], the spontaneous fission of several fermium isotopes [46], the Y_{32} correlations in $N = 150$ isotones [47], and shapes of hypernuclei [48, 49], see Refs. [50–52] for recent reviews.

In the MDC-RHB model, pairing correlations are treated by making the Bogoliubov transformation which generalizes the BCS quasi-particle concept and provides a unified description of particle-hole (ph) and pp correlations in a mean-field level. A separable pairing force of finite range [53–57] is adopted. The MDC-RHB model has been used to study the spontaneous fission of fermium isotopes [58], octupole correlations between multiple chiral doublet (MχD) bands in ^{78}Br [59], octupole correlations in Ba isotopes [15] and the Y_{32} correlations in neutron-rich Zr nuclei [60].

In this contribution, we present briefly the formalism of the MDC-CDFTs and some results of fission barriers. The formulae of the MDC-CDFTs is given in Sec. 2. The results and discussions are presented in Sec. 3. Finally we summarize in Sec. 4.

2 Formalism

In the CDFT [61–70], there are four types of covariant density functionals: the meson exchange or point-coupling nucleon interactions combined with the nonlinear or density dependent couplings [71–77] (see Ref. [78] for recent reviews). In this contribution we only show the CDFT with non-linear point coupling (NL-PC) interactions.

The starting point of the relativistic NL-PC density functional is the following Lagrangian:

$$\mathcal{L} = \bar{\psi}(i\gamma_\mu\partial^\mu - M_B)\psi - \mathcal{L}_{\rm lin} - \mathcal{L}_{\rm nl} - \mathcal{L}_{\rm der} - \mathcal{L}_{\rm Cou}, \quad (1)$$

where $M_{\rm B}$ is the nucleon mass and $\mathcal{L}_{\rm lin}$, $\mathcal{L}_{\rm nl}$, $\mathcal{L}_{\rm der}$ and $\mathcal{L}_{\rm Cou}$ are the linear coupling, non-linear coupling, derivative coupling and the Coulomb part, respectively, see Refs. [50–52] for details. Starting from the Lagrangian, we can deduce the relativistic mean field equations with the variational principle:

$$\{\boldsymbol{\alpha}\cdot\boldsymbol{p} + \beta[M_B + S(\boldsymbol{r})] + V(\boldsymbol{r})\}\psi_k(\boldsymbol{r}) = \epsilon_k\psi_k(\boldsymbol{r}), \quad (2)$$

where $S(\boldsymbol{r})$ and $V(\boldsymbol{r})$ are the scalar and vector potentials, respectively. As usual we suppose that the states are invariant under the time-reversal operation, which means that all time-odd components of the currents and the potentials vanish.

The pairing correlations play an important role in the fission process [79, 80]. We have implemented the BCS method or the Bogoliubov transformation for pairing in the MDC-RMF or MDC-RHB models.

The axially deformed harmonic oscillator (ADHO) basis [81, 82] was adopted in the development of MDC-CDFTs. In contrast to the traditional ADHO code, in this work the components with different quantum numbers are mixed, allowing us to treat triaxial and octupole nuclear shapes simultaneously.

The ADHO basis are defined as the eigensolutions of the Schrödinger equation with an ADHO potential,

$$\left[-\frac{\hbar^2\nabla^2}{2M} + \frac{1}{2}M(\omega_\rho^2\rho^2 + \omega_z^2 z^2)\right]\Phi_\alpha(\boldsymbol{r}\sigma) = E_\alpha\Phi_\alpha(\boldsymbol{r}\sigma), \quad (3)$$

where ω_z and ω_ρ are the oscillator frequencies along and perpendicular to the symmetry axis, respectively. The solution of Eq. (3) reads

$$\Phi_\alpha(\boldsymbol{r}\sigma) = C_\alpha\phi_{n_z}(z)R_{n_\rho}^{m_l}(\rho)\frac{1}{\sqrt{2\pi}}e^{im_l\varphi}\chi_{s_z}(\sigma), \quad (4)$$

where $\phi_{n_z}(z)$ and $R_{n_\rho}^{m_l}(\rho)$ are the HO wave functions, χ_{s_z} is a two component spinor and C_α is a complex number introduced for convenience. These bases are also eigenfunctions of the z component of the angular momentum j_z with eigenvalues $K = m_l + m_s$.

These bases form a complete set for expanding any two-component spinors. For a Dirac spinor with four components, the following expansion can be made,

$$\psi_i(\boldsymbol{r}\sigma) = \begin{pmatrix} \sum_\alpha f_i^\alpha\Phi_\alpha(\boldsymbol{r}\sigma) \\ \sum_\alpha g_i^\alpha\Phi_\alpha(\boldsymbol{r}\sigma) \end{pmatrix}, \quad (5)$$

where the sum runs over all the possible combination of the quantum numbers $\alpha = \{n_z, n_r, m_l, m_s\}$, f_i^α and g_i^α are the expansion coefficients.

In the general case that the axial symmetry as well as the space reflection symmetry are broken, different components with different m_j and π are mixed with each other. In this case it is convenient to choose the operator $\hat{S} = ie^{-i\pi j_z}$ with $S^2 = 1$ as the conserved quantity to classify the energy levels. S is a good quantum number for the basis with $\hat{S}\Phi_\alpha = S\Phi_\alpha = (-1)^{K_\alpha - \frac{1}{2}}\Phi_\alpha$, which means that the bases Φ_α with $K_\alpha = 1/2, -3/2, 5/2, -7/2, \cdots$ span the subspace with $S = 1$, while their time-reversed states span the one with $S = -1$. Note that now the block with $K = 1/2$ should be mixed with that with $K = -3/2$ instead of that with $K = 3/2$.

For deformed nuclei with the V_4 symmetry, we expand the potentials $V(\boldsymbol{r})$ and $S(\boldsymbol{r})$ and the densities in terms of the Fourier series,

$$f(\rho,\varphi,z) = f_0(\rho,z)\frac{1}{\sqrt{2\pi}} + \sum_{n=1}^{\infty} f_n(\rho,z)\frac{1}{\sqrt{\pi}}\cos(2n\varphi). \quad (6)$$

The matrix elements of the potentials can then be calculated from the expansion coefficients [42, 60].

In addition to the mean field energy, the centre of mass correction is calculated either phenomenologically or microscopically, depending on which effective interaction is used. The intrinsic multipole moments are calculated from the densities by

$$Q^{\tau}_{\lambda\mu} = \int d^3\boldsymbol{r}\rho^{\tau}_V(\boldsymbol{r})r^{\lambda}Y_{\lambda\mu}(\Omega), \qquad (7)$$

where $Y_{\lambda\mu}(\Omega)$ is the spherical harmonics and τ refers to the proton, neutron or nucleon. The deformation parameter $\beta_{\lambda\mu}$ is obtained from the corresponding multipole moment by

$$\beta_{\lambda\mu} = \frac{4\pi}{3NR^{\lambda}}Q_{\lambda\mu}, \qquad (8)$$

where $R = 1.2 \times A^{1/3}$ fm and N is the number of proton, neutron or nucleons.

3 Results

The double-humped fission barriers of actinide nuclei are often used to benchmark theoretical models for fission barriers. As the nucleus evolves from the ground state to the fission configurations, various shape degrees of freedom play important and different roles in determining the heights of the inner and outer barriers. As mentioned in the introduction, the inner barrier is lowered when the triaxial deformation is allowed, while for the outer barrier, the reflection asymmetric shape is favored. In Ref. [40] we have shown that, the reflection asymmetric outer barrier may be further lowered by the triaxial distortion with the MDC-RMF model.

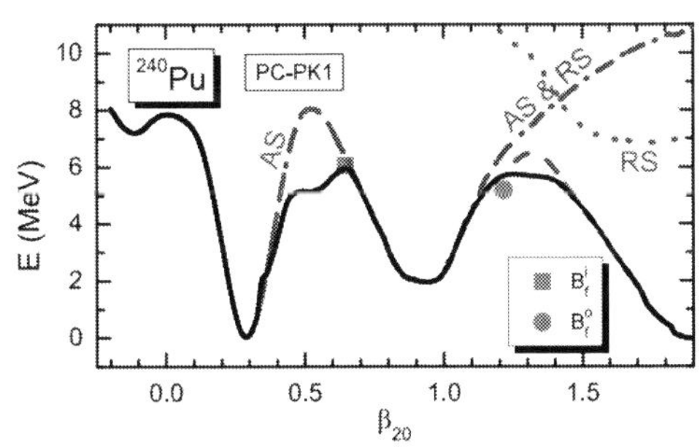

Figure 1. (Color online) Potential energy curves of ^{240}Pu with various self-consistent symmetries imposed. The solid black curve represents the calculated fission path with V_4 symmetry imposed; the red dashed curve that with axial symmetry (AS) imposed, the green dotted curve that with reflection symmetry (RS) imposed, the violet dot-dashed line that with both symmetries (AS & RS) imposed. The empirical inner (outer) barrier height [83] is denoted by the grey square (circle). The energy is normalized with respect to the binding energy of the ground state. The parameter set used is PC-PK1 [84]. Taken from Ref. [40].

In Fig. 1 we show the one-dimensional potential energy curves (PEC) from an oblate shape with β_{20} about −0.2 to the fission configuration with β_{20} beyond 2.0 which are obtained from calculations with different self-consistent symmetries imposed: the axial (AS) or triaxial (TS) symmetries combined with reflection symmetric (RS) or asymmetric (RA) cases. The importance of the triaxial deformation on the inner barrier and that of the octupole deformation on the outer barrier are clearly seen: The triaxial deformation reduces the inner barrier height by more than 2 MeV and results in a better agreement with the empirical value [83]; the RA shape is favored beyond the fission isomer and lowers very much the outer fission barrier. Besides these features, it was found in Ref. [40] that the outer barrier is also considerably lowered by about 1 MeV when the triaxial deformation is allowed. In addition, a better reproduction of the empirical barrier height can be seen for the outer barrier. It has been stressed that this feature can be revealed only when the axial and reflection symmetries are simultaneously broken [40].

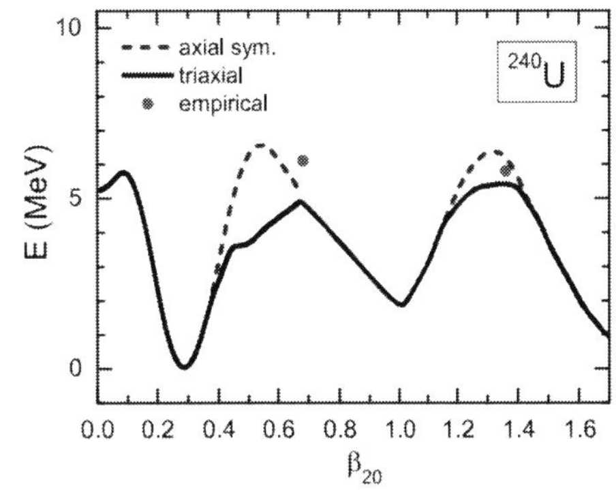

Figure 2. (Color online) Potential energy curves of ^{240}U. The solid black curve represents the calculated fission path with V_4 symmetry imposed, the blue dotted curve represents that with axial symmetry (AS) imposed. The empirical inner barrier height [85] is denoted by the dots. The energy is normalised with respect to the binding energy of the ground state. The parameter set used is PC-PK1 [84].

In Fig. 2 we show the calculated potential energy curves for ^{240}U. The results with and without non-axial deformations are presented. Near the second barrier both the axial and non-axial results are reflection asymmetric. The empirical values of the fission barriers are taken from Ref. [85] and depicted by red dots. The calculated inner and outer fission barrier heights are 4.96 and 5.43 MeV, respectively. This means that in ^{240}U, the outer fission barrier is higher than the inner one, which is in contrast to the empirical values. It's worthwhile to mention that the outer fission barrier height is very close to 5.5 MeV, a value which was recently obtained in JAEA [86]. It is clearly shown that the inclusion of non-axial shapes improve the agreement between the calculation and the experiment.

In Ref. [42] we applied the MDC-RMF model to a systematical study of fission barriers of the actinide nuclei with both axial and reflection symmetries broken. In Fig. 3 we show the calculated potential energy curves near

DOI: 10.1051/epjconf/201716300034

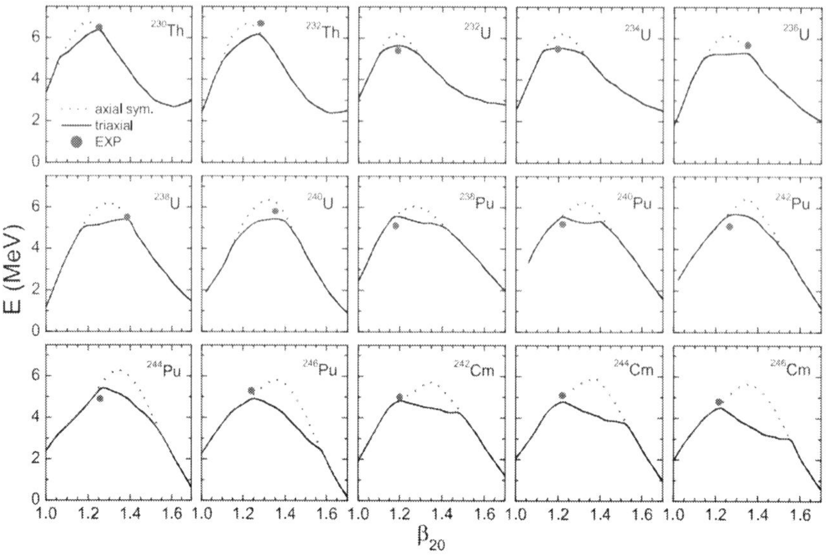

Figure 3. (Color online) Potential energy curves of even-even actinides nuclei in the outer barrier regions from the MDC-RMF calculations. The reflection asymmetric shapes are allowed. The triaxial results are displayed by solid lines, while the results from the axial symmetric calculations are shown by dashed lines. The binding energies are normalized with respect to the binding energies of the ground states of each nucleus. The empirical values of fission barrier heights are shown as red dots [83, 85]. Taken from Ref. [42].

the outer barriers of the actinide nuclei. For comparison the results with and without non-axial deformations are both presented. Note that near the second barrier all the results are reflection asymmetric. Because only the β_{20} deformation is constrained, the total binding energy is automatically minimized against other possible deformation parameters. The empirical values of the fission barriers are depicted by red dots. Clearly, all the axial symmetric calculations overestimate the barrier height. After including the triaxial deformation, the agreement between the calculated values and the empirical ones becomes better.

Besides the double-humped barriers of the PES's in the actinide region, the occurrence of a third barrier at large deformations beyond the second one was predicted by macroscopic-microscopic model calculations in the 1970s. But contradictory results about the existence of the third barrier were later obtained with different models. In Ref. [45] the MDC-RMF model was used to investigate the triple-humped barriers in light even-even actinides ^{232,234,236,238}U and ^{226,228,230,232}Th. The relativistic functionals PC-PK1 [84] and DD-ME2 [87] were adopted. Next we will mainly discuss the results from DD-ME2.

A third minimum was found on the PES's of ^{232,234,238}U: The pocket is rather shallow for ^{232}U and ^{234}U and very pronounced for ^{238}U. These results are quite similar to those obtained in recent calculations based on the macroscopic-microscopic model [88–91] and the Skyrme Hartree-Fock-Bogoliubov model [92].

In the case of Th isotopes, our calculation predicts a pronounced third minimum for ^{226}Th with a pocket depth of 2–3 MeV. As the neutron number N increases, both the energy of the third minimum and the height of the third barrier decrease, and the depth of the third well is also reduced. This trend has also been predicted in macroscopic-microscopic calculations [91]. For ^{230}Th only a shallow pocket of depth less than 1 MeV occurs around the third minimum. The pocket around the third minimum is very shallow for ^{232}Th.

4 Summary

In this contribution we present the formalism and some applications of the MDC-CDFTs in which all shape degrees of freedom $\beta_{\lambda\mu}$ with even μ are allowed. The potential energy surfaces (curves) of actinide nuclei and the effect of the triaxiality on the first and second fission barriers were investigated. It is found that besides the octupole deformation, the triaxiality also plays an important role upon the second fission barriers. The third minimum of actinide nuclei were also studied. A pronounced third minimum were predicted for ^{226}Th with a pocket depth of 2–3 MeV.

Acknowledgments

This work has been supported by the NSF of China (11120101005, 11275248, 11525524, 11621131001, 11647601 and 11711540016), the 973 Program of China (2013CB834400), the Key Research Program of Frontier Sciences of CAS, the HPC Cluster of SKLTP/ITP-CAS and the Supercomputing Center, CNIC of CAS.

References

[1] A. Bohr, B.R. Mottelson, *Nuclear Structure*, Vol. I (World Scientific, Singapore, 1998)

[2] A. Bohr, B.R. Mottelson, *Nuclear Structure*, Vol. II (World Scientific, Singapore, 1998)

[3] P. Ring, P. Schuck, *The Nuclear Many-Body Problem* (Springer-Verlag, Berlin/Heidelberg/New York, 1980)

[4] S. Frauendorf, Rev. Mod. Phys. **73**, 463 (2001)

[5] W. Nazarewicz, in *An Advanced Course in Modern Nuclear Physics (Lecture Notes in Physics)*, edited by J. Arias, M. Lozano (Springer, Berlin/Heidelberg, 2001), Vol. 581, pp. 102–140

[6] S.W. Odegard, G.B. Hagemann, D.R. Jensen, M. Bergstroem, B. Herskind, G. Sletten, S. Toermaenen, J.N. Wilson, P.O. Tjom, I. Hamamoto et al., Phys. Rev. Lett. **86**, 5866 (2001)

[7] S. Frauendorf, J. Meng, Nucl. Phys. A **617**, 131 (1997)

[8] K. Starosta, T. Koike, C.J. Chiara, D.B. Fossan, D.R. LaFosse, A.A. Hecht, C.W. Beausang, M.A. Caprio, J.R. Cooper, R. Krucken et al., Phys. Rev. Lett. **86**, 971 (2001)

[9] J. Meng, J. Peng, S.Q. Zhang, S.G. Zhou, Phys. Rev. C **73**, 037303 (2006)

[10] J. Meng, S.Q. Zhang, J. Phys. G: Nucl. Phys. **37**, 064025 (2010)

[11] J. Meng, S.Q. Zhang, P. Zhao, in *Relativistic Density Functional for Nuclear Structure*, edited by J. Meng (World Scientific Publishing Co. Pte. Ltd., 2016), Vol. 10 of *International Review of Nuclear Physics*, chap. 9. Novel rotational excitations, pp. 355–411

[12] A.V. Afanasjev, D.B. Fossan, G.J. Lane, I. Ragnarsson, Phys. Rep. **322**, 1 (1999)

[13] L.P. Gaffney, P.A. Butler, M. Scheck, A.B. Hayes, F. Wenander, M. Albers, B. Bastin, C. Bauer, A. Blazhev, S. Bonig et al., Nature **497**, 199 (2013)

[14] B. Bucher, S. Zhu, C.Y. Wu, R.V.F. Janssens, D. Cline, A.B. Hayes, M. Albers, A.D. Ayangeakaa, P.A. Butler, C.M. Campbell et al., Phys. Rev. Lett. **116**, 112503 (2016)

[15] X.C. Chen, J. Zhao, C. Xu, H. Hua, T.M. Shneidman, S.G. Zhou, X.G. Wu, X.Q. Li, S.Q. Zhang, Z.H. Li et al., Phys. Rev. C **94**, 021301(R) (2016)

[16] I. Hamamoto, B. Mottelson, H. Xie, X.Z. Zhang, Z. Phys. D **21**, 163 (1991)

[17] J. Skalski, Phys. Rev. C **43**, 140 (1991)

[18] X. Li, J. Dudek, Phys. Rev. C **49**, R1250 (1994)

[19] S. Takami, K. Yabana, M. Matsuo, Phys. Lett. B **431**, 242 (1998)

[20] M. Yamagami, K. Matsuyanagi, M. Matsuo, Nucl. Phys. A **693**, 579 (2001)

[21] J. Dudek, A. Gozdz, N. Schunck, M. Miskiewicz, Phys. Rev. Lett. **88**, 252502 (2002)

[22] J. Dudek, D. Curien, N. Dubray, J. Dobaczewski, V. Pangon, P. Olbratowski, N. Schunck, Phys. Rev. Lett. **97**, 072501 (2006)

[23] P. Olbratowski, J. Dobaczewski, P. Powalowski, M. Sadziak, K. Zberecki, Int. J. Mod. Phys. E **15**, 333 (2006)

[24] K. Zberecki, P. Magierski, P.H. Heenen, N. Schunck, Phys. Rev. C **74**, 051302(R) (2006)

[25] J. Dudek, A. Gozdz, K. Mazurek, H. Molique, J. Phys. G: Nucl. Phys. **37**, 064032 (2010)

[26] O. Hahn, F. Strassmann, Naturwissenschaften **27**, 11 (1939)

[27] N. Bohr, J.A. Wheeler, Phys. Rev. **56**, 426 (1939)

[28] V.M. Strutinsky, Nucl. Phys. A **95**, 420 (1967)

[29] V.M. Strutinsky, Nucl. Phys. A **122**, 1 (1968)

[30] V.V. Pashkevich, Nucl. Phys. A **133**, 400 (1969)

[31] P. Möller, S.G. Nilsson, Phys. Lett. B **31**, 283 (1970)

[32] M. Brack, J. Damgaard, A.S. Jensen, H.C. Pauli, V.M. Strutinsky, C.Y. Wong, Rev. Mod. Phys. **44**, 320 (1972)

[33] T. Burvenich, M. Bender, J.A. Maruhn, P.G. Reinhard, Phys. Rev. C **69**, 014307 (2004)

[34] M. Mirea, L. Tassan-Got, Cent. Eur. J. Phys. **9**, 116 (2011)

[35] M. Girod, B. Grammaticos, Phys. Rev. C **27**, 2317 (1983)

[36] V. Blum, J.A. Maruhn, P.G. Reinhard, W. Greiner, Phys. Lett. B **323**, 262 (1994)

[37] L. Csige, M. Csatlos, T. Faestermann, J. Gulyas, D. Habs, R. Hertenberger, M. Hunyadi, A. Krasznahorkay, H.J. Maier, P.G. Thirolf et al., Phys. Rev. C **85**, 054306 (2012)

[38] H. Abusara, A.V. Afanasjev, P. Ring, Phys. Rev. C **82**, 044303 (2010)

[39] K. Rutz, J.A. Maruhn, P.G. Reinhard, W. Greiner, Nucl. Phys. A **590**, 680 (1995)

[40] B.N. Lu, E.G. Zhao, S.G. Zhou, Phys. Rev. C **85**, 011301(R) (2012)

[41] B.N. Lu, J. Zhao, E.G. Zhao, S.G. Zhou, J. Phys: Conf. Ser. **492**, 012014 (2014)

[42] B.N. Lu, J. Zhao, E.G. Zhao, S.G. Zhou, Phys. Rev. C **89**, 014323 (2014)

[43] B.N. Lu, J. Zhao, E.G. Zhao, S.G. Zhou, Phys. Scr. **89**, 054028 (2014)

[44] B.N. Lu, J. Zhao, E.G. Zhao, S.G. Zhou, EPJ Web Conf. **38**, 05003 (2012)

[45] J. Zhao, B.N. Lu, D. Vretenar, E.G. Zhao, S.G. Zhou, Phys. Rev. C **91**, 014321 (2015)

[46] J. Zhao, B.N. Lu, T. Nikšić, D. Vretenar, S.G. Zhou, Phys. Rev. C **93**, 044315 (2016)

[47] J. Zhao, B.N. Lu, E.G. Zhao, S.G. Zhou, Phys. Rev. C **86**, 057304 (2012)

[48] B.N. Lu, E.G. Zhao, S.G. Zhou, Phys. Rev. C **84**, 014328 (2011)

[49] B.N. Lu, E. Hiyama, H. Sagawa, S.G. Zhou, Phys. Rev. C **89**, 044307 (2014)

[50] L.L. Li, B.N. Lu, N. Wang, K. Wen, C.J. Xia, Z.H. Zhang, J. Zhao, E.G. Zhao, S.G. Zhou, Nucl. Phys. Rev. **31**, 253 (2014), (in Chinese)

[51] B.N. Lu, J. Zhao, E.G. Zhao, S.G. Zhou, in *Relativistic Density Functional for Nuclear Structure*, edited by J. Meng (World Scientific Publishing Co. Pte. Ltd., 2016), Vol. 10 of *International Review of Nuclear Physics*, chap. 5. Superheavy nuclei and fission barriers, pp. 171–217
[52] S.G. Zhou, Phys. Scr. **91**, 063008 (2016)
[53] Y. Tian, Z.Y. Ma, Chin. Phys. Lett. **23**, 3226 (2006)
[54] Y. Tian, Z.Y. Ma, P. Ring, Phys. Lett. B **676**, 44 (2009)
[55] Y. Tian, Z.y. Ma, P. Ring, Phys. Rev. C **80**, 024313 (2009)
[56] T. Nikšić, D. Vretenar, P. Ring, Phys. Rev. C **78**, 034318 (2008)
[57] T. Nikšić, P. Ring, D. Vretenar, Y. Tian, Z.y. Ma, Phys. Rev. C **81**, 054318 (2010)
[58] J. Zhao, B.N. Lu, T. Nikšić, D. Vretenar, Phys. Rev. C **92**, 064315 (2015)
[59] C. Liu, S.Y. Wang, R.A. Bark, S.Q. Zhang, J. Meng, B. Qi, P. Jones, S.M. Wyngaardt, J. Zhao, C. Xu et al., Phys. Rev. Lett. **116**, 112501 (2016)
[60] J. Zhao, B.N. Lu, E.G. Zhao, S.G. Zhou, Phys. Rev. C **95**, 014320 (2017)
[61] B.D. Serot, J.D. Walecka, Adv. Nucl. Phys. **16**, 1 (1986)
[62] P.G. Reinhard, Rep. Prog. Phys. **52**, 439 (1989)
[63] P. Ring, Prog. Part. Nucl. Phys. **37**, 193 (1996)
[64] M. Bender, P.H. Heenen, P.G. Reinhard, Rev. Mod. Phys. **75**, 121 (2003)
[65] D. Vretenar, A.V. Afanasjev, G.A. Lalazissis, P. Ring, Phys. Rep. **409**, 101 (2005)
[66] J. Meng, H. Toki, S.G. Zhou, S.Q. Zhang, W.H. Long, L.S. Geng, Prog. Part. Nucl. Phys. **57**, 470 (2006)
[67] N. Paar, D. Vretenar, G. Colo, Rep. Prog. Phys. **70**, 691 (2007)
[68] T. Nikšić, D. Vretenar, P. Ring, Prog. Part. Nucl. Phys. **66**, 519 (2011)
[69] H. Liang, J. Meng, S.G. Zhou, Phys. Rep. **570**, 1 (2015)
[70] J. Meng, S.G. Zhou, J. Phys. G: Nucl. Part. Phys. **42**, 093101 (2015)
[71] J. Boguta, A.R. Bodmer, Nucl. Phys. A **292**, 413 (1977)
[72] R. Brockmann, H. Toki, Phys. Rev. Lett. **68**, 3408 (1992)
[73] Y. Sugahara, H. Toki, Nucl. Phys. A **579**, 557 (1994)
[74] B.A. Nikolaus, T. Hoch, D.G. Madland, Phys. Rev. C **46**, 1757 (1992)
[75] T. Burvenich, D.G. Madland, J.A. Maruhn, P.G. Reinhard, Phys. Rev. C **65**, 044308 (2002)
[76] C. Fuchs, H. Lenske, H.H. Wolter, Phys. Rev. C **52**, 3043 (1995)
[77] T. Nikšić, D. Vretenar, P. Finelli, P. Ring, Phys. Rev. C **66**, 024306 (2002)
[78] J. Meng, ed., *Relativistic Density Functional for Nuclear Structure*, Vol. 10 of *International Review of Nuclear Physics* (World Scientific Pub Co Pte Lt, 2016)
[79] S. Karatzikos, A.V. Afanasjev, G.A. Lalazissis, P. Ring, Phys. Lett. B **689**, 72 (2010)
[80] A. Bulgac, P. Magierski, K.J. Roche, I. Stetcu, Phys. Rev. Lett. **116**, 122504 (2016)
[81] H. Kucharek, P. Ring, Z. Phys. A **339**, 23 (1991)
[82] P. Ring, Y.K. Gambhir, G.A. Lalazissis, Comput. Phys. Commun. **105**, 77 (1997)
[83] R. Capote, M. Herman, P. Oblozinsky, P.G. Young, S. Goriely, T. Belgya, A.V. Ignatyuk, A.J. Koning, S. Hilaire, V.A. Plujko et al., Nucl. Data Sheets **110**, 3107 (2009)
[84] P.W. Zhao, Z.P. Li, J.M. Yao, J. Meng, Phys. Rev. C **82**, 054319 (2010)
[85] P. Möller, A. J. Sierk, T. Ichikawa, A. Iwamoto, R. Bengtsson, H. Uhrenholt, S. Aberg, Phys. Rev. C **79**, 064304 (2009)
[86] K. Nishio, private communications
[87] G.A. Lalazissis, T. Nikšić, D. Vretenar, P. Ring, Phys. Rev. C **71**, 024312 (2005)
[88] M. Kowal, J. Skalski, Phys. Rev. C **85**, 061302(R) (2012)
[89] P. Jachimowicz, M. Kowal, J. Skalski, Phys. Rev. C **87**, 044308 (2013)
[90] T. Ichikawa, A. Iwamoto, P. Möller, A.J. Sierk, Phys. Rev. C **86**, 024610 (2012)
[91] T. Ichikawa, P. Möller, A.J. Sierk, Phys. Rev. C **87**, 054326 (2013)
[92] J.D. McDonnell, W. Nazarewicz, J.A. Sheikh, Phys. Rev. C **87**, 054327 (2013)

EPJ Web of Conferences 163, 00035 (2017) DOI: 10.1051/epjconf/201716300035
FUSION17

Reaction dynamics studies for the system ^{7}Be + ^{208}Pb at Coulomb barrier energies

M. Mazzocco[1,2,*], A. Boiano[3], C. Boiano[4], M. La Commara[5,3], C. Manea[2], C. Parascandolo[3], D. Pierroutsakou[3], E. Strano[1,2], D. Torresi[1,2], L. Acosta[6], P. Di Meo[3], J.P. Fernandez-Garcia[7], T. Glodariu[8], J. Grebosz[9], A. Guglielmetti[10,4], G. Marquinez-Duran[6], I. Martel[6], M. Nicoletto[2], A. Pakou[11], A.M. Sánchez-Benítez[6], T. Sava[8], O. Sgouros[11], C. Signorini[1,2], F. Soramel[1,2], V. Soukeras[11], and L. Stroe[8]

[1]Dipartimento di Fisica e Astronomia, Università di Padova, Padova, Italy
[2]INFN-Sezione di Padova, Padova, Italy
[3]INFN-Sezione di Napoli, Napoli, Italy
[4]INFN-Sezione di Milano, Milano, Italy
[5]Dipartimento di Fisica, Università di Napoli, Napoli, Italy
[6]Departamento de Física Aplicada, Universidad de Huelva, Huelva, Spain
[7]INFN-LNS, Catania, Italy
[8]NINPE, Magurele, Romania
[9]IFJ-PAN, Krakow, Poland
[10]Dipartimento di Fisica, Università di Milano, Milano, Italy
[11]Department of Physics, University of Ioannina and HINP, Ioannina, Greece

Abstract. The scattering process of the Radioactive Ion Beam ^{7}Be from a ^{208}Pb target was measured at three near-barrier energies. The quasi-elastic angular distributions were analyzed within the framework of the optical model to extract the reaction cross sections. The results are compared with those obtained for the reactions induced by the mirror projectile ^{7}Li and by the lightest particle-stable lithium isotope ^{6}Li on the same target. The angular distributions for the production of the two ^{7}Be constituent clusters, namely ^{3}He and ^{4}He, were also measured. In agreement with what observed for the interaction of ^{7}Be with lighter targets, the production of the heavier helium isotope resulted to be much more abundant than that of its lighter counterpart.

1 Introduction

The interaction of light weakly-bound nuclei at energies around the Coulomb barrier has attracted the interest of the Nuclear Physics community for the last three decades at least (see, for instance, [1–8] and references therein). Early experiments were mainly aimed at measuring the fusion process, since theoretical calculations strongly disagreed on the predictions for the fusion cross section at sub-barrier energies. Being these projectiles rather weakly-bound, they can quite easily break up into smaller fragments while approaching a target nucleus and before a complete fusion process could occur. From a theoretical point of view, depending whether the breakup channel was treated as an additional open channel or as a removed flux from the entrance channel, models predicted either the enhancement or the hindrance of the sub-barrier fusion cross section. Discrepancies between different approaches were, in some cases, as large as four orders of magnitude.

Besides very first experiments, it was soon realized that breakup related effects enhanced the total reaction cross section at sub-barrier energies and that this increase was mainly due to direct processes rather than to compound nucleus reaction mechanisms. The attention was then turned towards the investigation of direct processes and it was observed that the large production of α particles at sub-barrier energies in ^{6}He- [9–11] and ^{8}He-induced [12, 13] reactions was essentially due to the 2n-stripping process. More recently, experiments performed with the 2n-halo ^{11}Li [14, 15] and the 1n-halo ^{11}Be [16] indicated that direct processes also have a quite strong influence on the elastic scattering, since large deviations from the standard Rutherford cross sections were observed already at very large impact parameters. On the other side, the data collected for the system ^{8}B + ^{58}Ni [17, 18] suggested that the reaction cross section enhancement at sub-barrier energies could be mainly due to the fusion process, establishing a sort of different behavior between n-rich and p-rich exotic nuclei.

The relevance of direct processes, especially of transfer channels, at near-barrier energies has been recently confirmed by the detailed studies performed at the Australian National University [19–23] for reactions induced by the stable weakly-bound nuclei ^{6,7}Li and ^{9}Be on a quite large variety of targets, from ^{58}Ni to ^{209}Bi.

* e-mail: marco.mazzocco@pd.infn.it

In this framework, we developed a small facility, named EXOTIC [24–26], at the Laboratori Nazionali di Legnaro (LNL) of the Istituto Nazionale di Fisica Nucleare (INFN) for the In-Flight production of light Radioactive Ion Beams (RIBs). So far, secondary beams of ^{17}F, ^{7}Be, ^{8}B, ^{15}O, ^{8}Li and ^{10,11}C have been delivered with intensities in the range 10^3-10^6 and energies in the range 2-6 AMeV. These RIBs have been employed for reaction dynamics studies at Coulomb barrier energies, for resonant scattering experiments and to investigate processes of astrophysical interest by means of the Trojan Horse Method.

In 2010 we undertook the study of ^{7}Be-induced reactions. This projectile was selected because is a weakly-bound p-rich nucleus with a very pronounced ^{3}He-^{4}He cluster configuration in its ground state. The ^{7}Be (S_α = 1.586 MeV) nuclear structure is quite similar to that of its mirror nucleus ^{7}Li (S_α = 2.468 MeV), whereas its binding energy is rather close to that of the lightest particle stable lithium isotope ^{6}Li (S_α = 1.475 MeV). A meaningful comparison between the ^{7}Be and ^{6,7}Li reaction dynamics at Coulomb barrier can easily be achieved owing to the tremendous amount of data available for the reactions induced by the stable projectiles on several different targets.

In 2014 at FUSION14 conference in New Delhi (India), the preliminary results concerning the measurement of the quasi-elastic angular distribution for the system ^{7}Be + ^{58}Ni at 22 MeV were presented, together with the differential cross sections for the ^{3}He and ^{4}He inclusive production. The final evaluation was published the following year [27]. The quasi-elastic data was in remarkable agreement with an earlier measurement performed at the University of Notre Dame (Indiana, USA) [17], while the ^{4}He production (measured for the first time) turned out to be 4-5 times larger than that for ^{3}He. This result clearly suggested that the near-barrier reaction dynamics was not dominated by the breakup process, in such a case, in fact, similar yields for the two helium isotopes would have been observed. Unfortunately, due to small geometrical efficiency and limited beam-time availability, no coincidences between projectile fragments were recorded. Therefore quite elaborated theoretical calculations were undertaken in order to determine the possible contributions of the n-stripping, n-pickup and exclusive breakup processes to the inclusive production cross section of ^{3}He and ^{4}He.

After the upgrade of the facility EXOTIC [28], we could deliver more energetic secondary beams and for the ^{7}Be RIB, in particular, it was possible to match the energy range around the barrier for the interaction with a ^{208}Pb target. The goals of the measurement were to measure for the first time the near-barrier quasi-elastic angular distribution and, with an improved detector setup, to detect (at least) a few coincidences between projectile fragments, in order to get a deeper insight on the reaction dynamics induced by this exotic projectile.

The paper is organized as follows: Sec. 2 will describe the secondary beam production and the experimental setup employed for the measurements. Sec. 3 will present the data analysis for the evaluation of the quasi-elastic scattering angular distributions, the theoretical analysis performed to extract the total reaction cross section, the differential cross sections for the inclusive production of ^{3}He and ^{4}He and the status of the analysis of coincidence events between projectile fragments. Some concluding remarks will finally be given in Sec. 4.

2 Experiment

The experiment was performed at INFN-LNL (Italy), where the ^{7}Be RIB was produced by means of the facility EXOTIC. A 48.8 MeV ^{7}Li^{+3} primary beam, delivered by the LNL XTU-Tandem accelerator, was impinging on a 5-cm long gas cell doubly-walled with 2.2 μm thick Havar windows. The gas pressure was about 1 bar and the target station was kept at liquid nitrogen temperature (90K). The RIB was produced through the charge exchange reaction ^{1}H(^{7}Li,^{7}Be)n (Q_{value} = -1.64 MeV) and was selected by a careful tuning of the quadrupole lenses, the dipole magnet and the Wien filter of the facility EXOTIC. Before reaching the 1 mg/cm^2 thick ^{208}Pb target, the ^{7}Be beam was crossing two Parallel Plate Avalanche Counters (PPACs) employed for beam tracking and trigger generation purposes. The outcoming secondary beam energy in these conditions was 40.5 MeV with an average intensity of about 2-3 × 10^5 pps and a 99%-purity. A second energy of 37.6 MeV was obtained by inserting in a suitable location along the beam-line between the two PPACs a 12.5 μm thick aluminum degrader. Furthermore, an additional measurement at 42.4 MeV was performed after warming up the target at room temperature and retuning all ion optical elements of the facility.

Charged particles produced by in the interaction between the RIB and the target were detected in the angular range [θ_{lab} = 55°-165°] by means of six modules of the EXPADES detector array [29, 30]. This was actually the first experiment (after the commissioning runs) performed with this setup, whose electronics was entirely developed by our collaboration. Each EXPADES module consisted of two Double Sided Silicon Strip Detectors (DSSSDs) arranged in a ΔE-E_{res} telescope configuration. Each DSSSD (design BB7 from Micron Semiconductor Ltd, UK) had an active area of 62.3 mm × 62.3 mm and both sides are segmented into approximately 2-mm wide strips. To contain the cost of the electronics, the signals from the strips were 2-by-2 short-circuited at the entrance of the preamplifier board. The thickness of the EXPADES inner and outer detector was 43-57 μm and 300 μm, respectively. The six modules were placed in symmetrical locations with respect to beam axis, covering the following angular ranges: θ_{lab} = ± [55°-84°], ± [96°-125°] and ± [138°-165°]. The average distance between the detectors and the central position of the target was about 115 mm. The trigger signal for the data acquisition system was provided by the "OR" of all the silicon detectors "AND" the cathode signals from both PPACs. Trigger rates of 5-10 Hz were routinely recorded during the experiment. The acquisition times were about 48 h, 24 h and 8 h for the measurements at 40.5, 37.6 and 42.4 MeV, respectively.

3 Data Analysis

3.1 Quasi-elastic scattering

Fig. 1 shows the angular distributions for the quasi-elastic scattering process in the reaction ^{7}Be + ^{208}Pb at three beam energies. Due to the target thickness and the energy resolution of the secondary beam, it was not possible to distinguish in the data analysis between events generated for a pure elastic scattering process and those leading to the population of the first excited state of the projectile at 0.429 MeV. Thus, hereafter we refer to these data as "quasi-elastic". To avoid large statistical fluctuations, practically unavoidable in present-day experiments involving low-intensity RIBs, each data point in Fig. 1 originates from the counting statistics collected in four adjacent vertical strips and to the weighted average between the evaluations obtained by groups of strips located at symmetrical positions with respect to the beam axis. The error bars, which include only statistical uncertainties, clearly show that the data set at the middle energy (filled circles in Fig. 1) is the most accurate one. The quasi-elastic differential cross sections decrease as both the scattering angle and the beam energy increase, since projectile and target classically get closer to each other and more nuclear reaction channels open up.

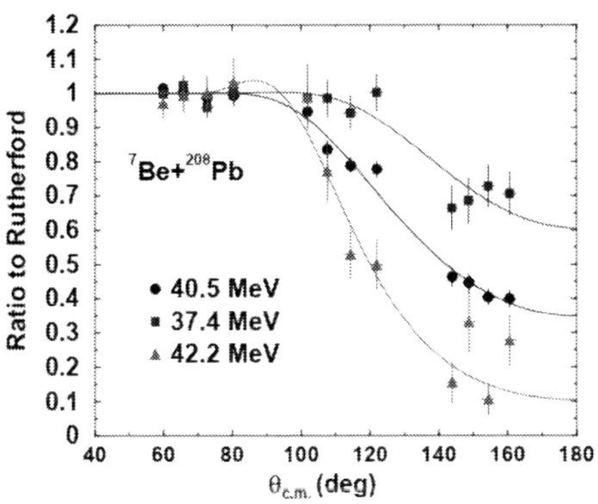

Figure 1. Quasi-elastic differential cross sections for the collision of the RIB ^{7}Be with a ^{208}Pb target at three beam energies: 40.5 MeV (circles), 37.4 MeV (squares) and 42.2 MeV (triangles). Lines are the results of a best-fit analysis of the experimental data within the framework of the optical model, as described in detail in the text.

The angular distributions plotted in Fig. 1 were analyzed within the framework of the optical model by using the code Fresco [31]. Both the real and the imaginary part of the nuclear interaction potential were described with Woods-Saxon wells. During the fitting procedure only the depths of the real (V_0) and imaginary (W_0) potential were let free to vary, whereas all other geometrical parameters were calculated according to systematics of Broglia and Winther [32] and kept fixed at the following values: $r_0 = r_i = 1.183$ fm and $a_0 = a_i = 0.63$ fm. Table 1 summarizes the results of the best-fit analysis of the experimental data

Table 1. Optical model best-fit parameters of the quasi-elastic angular distributions for the reaction ^{7}Be + ^{208}Pb at three beam energies. The radius and diffuseness of both real and imaginary Woods-Saxon wells were fixed at the following values: $r_0 = r_i = 1.183$ fm and $a_0 = a_i = 0.63$ fm.

E (MeV)	V_0 (MeV)	W_0 (MeV)	σ_R (mb)	χ^2
37.4	155.6	76.5	92.2	0.85
40.5	48.2	78.3	243.3	1.52
42.2	85.1	36.1	338.7	1.07

and the resulting curves are plotted with continuous lines in Fig. 1.

3.2 Total reaction cross section

The fourth column in Table I reports the total reaction cross sections extracted from the fitting procedure. These values are also plotted with filled circles in Fig. 2, where they are compared to the total reaction sections measured for the systems ^{6}Li + ^{208}Pb (diamonds) and ^{7}Li + ^{208}Pb (squared) [33] in a similar energy range. To make a meaningful comparison, the beam energies and the cross section values for lithium-induced reactions were normalized following the prescription given by P.R.S. Gomes *et al.* [34]. All values were than scaled to the Coulomb barrier and the geometrical size for the reaction ^{7}Be + ^{208}Pb.

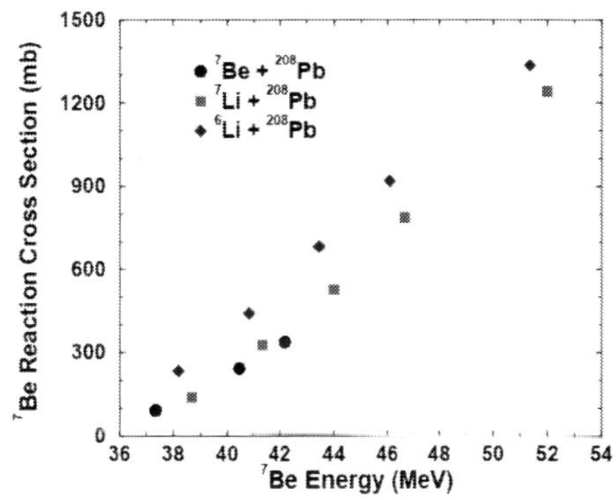

Figure 2. Total reaction cross sections for the systems ^{6,7}Li + ^{208}Pb and ^{7}Be + ^{208}Pb in the energy range around the Coulomb barrier. The beam energies and the cross section data for the reactions induced by lithium isotopes were scaled to account for the different Coulomb barrier and projectile geometrical size with respect to the reaction ^{7}Be + ^{208}Pb.

We can clearly appreciate from Fig. 2 that the ^{7}Be data follow essentially the trend drawn by the more tightly-bound (mirror) nucleus ^{7}Li rather than that given by the similarly weakly-bound projectile ^{6}Li, suggesting that, for the considered nuclei, nuclear structure has a larger influence on the reaction dynamics than the binding energy. However, a clearer picture of the whole scenario can be obtained only after measuring individually the cross sections for the most relevant reaction channels, especially transfer processes, non-capture breakup and (complete and incomplete) fusion. A first step in this direction can be done by measuring the energy spectra and the angular distributions for the inclusive production of the two ^{7}Be constituent cluster, ^{3}He and ^{4}He, as described in the next paragraph.

3.3 ^{3}He and ^{4}He inclusive production

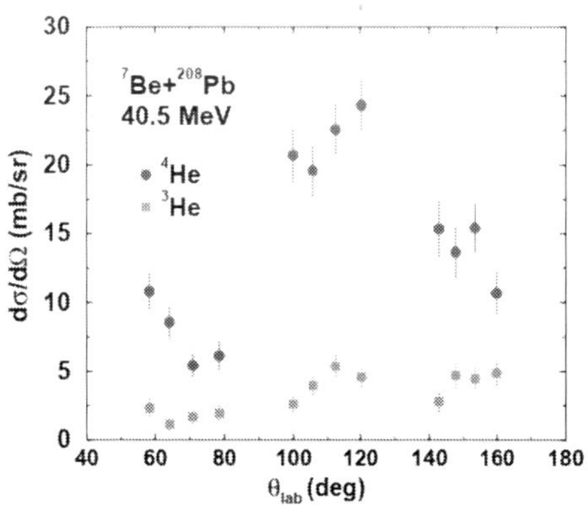

Figure 3. Angular distributions for the inclusive production of ^{3}He (squared) and ^{4}He (circles) in the reaction ^{7}Be + ^{208}Pb at 40.5 MeV beam energy.

Fig. 3 displays the angular distributions for the production of ^{3}He (squared) and ^{4}He (circles) in the interaction of the RIB ^{7}Be with a ^{208}Pb target at 40.5 MeV. The energy spectra for both helium isotopes evidence two distinctive components: the former mainly peaked at low-energy (2-6 MeV) with a large exponentially decreasing tail extending up to 10-12 MeV and the latter with a mostly Gaussian shape centered at about 16 and 18 MeV for ^{3}He and ^{4}He, respectively. The first structure should originate from particle evaporation after complete fusion reactions between beam particles and the aluminum target holder. In fact, the last collimator along the beam axis was placed too close to the second PPAC and did not prevent beam particles scattered from the PPAC exit window to hit the 60-mm wide aluminum target holder. This hypothesis is supported by three experimental evidences: (*i*) the low-energy contamination is mostly present at backward angles, being the detectors at forward angles essentially shielded by the target holder itself; (*ii*) the energy range where the maximum is located (2-6 MeV) and the exponentially decreasing behavior in the energy region up to about 12 MeV is compatible with the energy spectra of helium isotopes evaporated after a compound nucleus reaction between ^{7}Be and ^{27}Al [35], (*iii*) in more than 95% of coincidence events, either ^{4}He-^{4}He (clear signature of n-pickup process on a ^{208}Pb) or ^{4}He-^{3}He (non capture breakup) or ^{4}He-^{2}H (p-stripping), the energy of the less energetic helium fragment is larger than 12 MeV. Therefore the angular distributions plotted in Fig. 3 were obtained considering only helium isotopes whose total energy deposit was at least 12 MeV, thus minimizing possible contributions from the low-energy contamination.

Fig. 3 also illustrates that the ^{4}He production is more abundant by a factor 4-5 with respect to probability of producing the lighter helium isotope. This outcome qualitatively agrees with what we had previously observed for the reaction induced by the same RIB on a ^{58}Ni target. We can thus immediately conclude that, at near-barrier energies, the ^{7}Be reaction dynamics is not dominated by the breakup process even for the interaction with a heavy target. The ^{4}He angular distribution exhibits a maximum in the region around θ_{lab} = 120°, while that for ^{3}He has a flatter behavior with a much less pronounced peak approximately in the same angular range. A detailed analysis of the ^{3,4}He angular distributions and energy spectra will be the subject of a forthcoming investigation, maybe in the framework of the model recently developed by Lei and Moro [36].

3.4 Detection of coincidence events

As already anticipated, a few coincidence events between projectile fragments were detected at 40.5 MeV. In particular, we observed: 13 ^{4}He-^{2}H (originated from the p-stripping reaction: ^{7}Be + ^{208}Pb → ^{6}Li + ^{209}Bi, Q_{value} = -1.81 MeV), 19 ^{4}He-^{3}He (non capture breakup, S_α = 1.586 MeV) and 17 ^{4}He-^{4}He (n-pickup: ^{7}Be + ^{208}Pb → ^{8}Be + ^{207}Pb, Q_{value} = +11.53 MeV) coincidence events.

Detailed kinematic calculations are presently being performed with the code Kookaburra [21] in order to derive the geometrical efficiency of the coincidence detection system and to compare theoretical predictions with the relative energy spectra, Q_{value} distributions, opening and orientation angles reconstructed from the experimental data. The procedure will essentially follow the scheme described in Refs. [22, 23] and already successfully employed for the reactions: ^{6,7}Li + ^{58}Ni, ^{64}Zn, ^{208}Pb and ^{209}Bi and ^{9}Be + ^{144}Sm, ^{168}Er, ^{186}W, ^{196}Pt, ^{208}Pb and ^{209}Bi.

4 Summary

The interaction of the RIB ^{7}Be with a ^{208}Pb target has been studied for the first time at three near-barrier energies. The experiment benefited from the recent upgrade of the RIB In-Flight facility EXOTIC, which is now able to deliver ^{7}Be beams energetic enough to match the energy range around the Coulomb barrier also for reactions on heavy targets. We measured the quasi-elastic scattering differential cross section and extracted the total reaction cross section. The comparison with the reaction cross sections derived for the mirror projectile ^{7}Li and the similarly weakly-bound nucleus ^{6}Li interacting with the same target clearly indicate that the ^{7}Be data essentially follow the trend individuated by the heavier lithium isotopes. This outcome suggests that, for these projectiles, nuclear structure might play a more relevant role in the reaction dynamics rather then the projectile binding energy. In order to get a deeper insight on the dominant reaction mechanisms for the system ^{7}Be + ^{208}Pb, we started the investigation of the energy spectra and angular distributions for the inclusive production of ^{3}He and ^{4}He nuclei. We immediately observed that the ^{4}He production is a factor 4-5 higher than the ^{3}He one, confirming that also for the interaction with heavy targets the ^{7}Be reaction dynamics at near barrier is not dominated by the non capture breakup process. The detailed analysis of these events as well as of the coincidence events is presently in progress.

References

[1] L.F. Canto, P.R.S. Gomes, R. Donangelo, and M.S. Hussein, Phys. Rep. **424**, 1 (2006).
[2] J.F. Liang, and C. Signorini, Int. J. Mod. Phys. E **14**, 1121 (2005).
[3] N. Keeley, R. Raabe, N. Alamanos, and J.L. Sida, Prog. Part. Nucl. Phys. **59**, 579 (2007).
[4] N. Keeley, N. Alamanos, K.W. Kemper, and K.Rusek, Prog. Part. Nucl. Phys. **63**, 396 (2009).
[5] M. Mazzocco, Int. J. Mod. Phys. E **19**, 977 (2010).
[6] N. Keeley, K.W. Kemper and K. Rusek, Eur. Phys. J. A **50**, 145 (2014).
[7] L.F. Canto, P.R.S. Gomes, R. Donangelo, J. Lubian, and M.S. Hussein, Phys. Rep. **596**, 1 (2015).
[8] M. Dasgupta *et al.*, Phys. Rev. C **70**, 024606 (2004).
[9] A. Di Pietro *et al.*, Phys. Rev. C **69**, 044613 (2004).
[10] A. Navin *et al.*, Phys. Rev. C **70**, 044601 (2004).
[11] R. Raabe *et al.*, Nature **431**, 823 (2004).
[12] A. Lemasson *et al.*, Phys. Rev. Lett. **103**, 232701 (2009).
[13] A. Lemasson *et al.*, Phys. Rev. C **82**, 044617 (2010).
[14] M. Cubero *et al.*, Phys. Rev. Lett. **109**, 262701 (2012).
[15] J.P. Fernandez-Garcia *et al.*, Phys. Rev. Lett. **110**, 142701 (2013).
[16] A. Di Pietro *et al.*, Phys. Rev. Lett. **105**, 022701 (2010)
[17] E.F. Aguilera *et al.*, Phys. Rev. C. **79**, 021601(R) (2009).
[18] E.F. Aguilera *et al.*, Phys. Rev. Lett. **107**, 092701 (2011).
[19] R. Rafiei *et al.*, Phys. Rev. C **81**, 024601 (2010).
[20] D.H. Luong *et al.*, Phys. Rev. C **88**, 034609 (2013).
[21] E.C. Simpson *et al.*, Phys. Rev. C **93**, 024605 (2016).
[22] S. Kalkal *et al.*, Phys. Rev. C **93**, 044605 (2016).
[23] K.J. Cook *et al.*, Phys. Rev. C **93**, 064604 (2016).
[24] V.Z. Maidikov *et al.*, Nucl. Phys. A **746**, 389c (2004).
[25] F. Farinon *et al.*, Nucl. Instrum. Methods B **266**, 4097 (2008).
[26] M. Mazzocco *et al.*, Nucl. Instrum. Methods B **266**, 4665 (2008).
[27] M. Mazzocco *et al.*, Phys. Rev. C. **92**, 024615 (2015).
[28] M. Mazzocco *et al.*, Nucl. Instrum. Methods B **317**, 223 (2013).
[29] E. Strano *et al.*, Nucl. Instrum. Methods B **317**, 657 (2013).
[30] D. Pierroutsakou *et al.*, Nucl. Instrum. Methods A **834**, 46 (2016).
[31] I.J. Thompson, Comput. Phys. Rep. **2**, 167 (1988).
[32] R. Broglia and A. Winther, Heavy Ion Reactions (Addison-Wesley, Reading, MA, 1990), p. 113.
[33] N. Keeley *et al.*, Nucl. Phys. A **571** (1994) 326.
[34] P.R.S. Gomes, J. Lubian, I. Padron, R.M. Anjos, Phys. Rev. C **71** (2005) 017601.
[35] A. Gavron, Phys. Rev. C **21**, 230 (1980).
[36] J. Lei and A.M. Moro, Phys. Rev. C **95**, 044605 (2017).

Reactions of exotic nuclei with the quark-meson coupling model

E. McRae[1,*], C. Simenel[1], E. C. Simpson[1], and A. W. Thomas[2]

[1]Department of Nuclear Physics, Research School of Physics and Engineering, The Australian National University, Canberra, ACT 2601, Australia
[2]ARC Centre of Excellence for Particle Physics at the Terascale and CSSM, Department of Physics, The University of Adelaide, SA 5005, Australia

Abstract. The nucleon-nucleon interaction is an important requirement for investigations of nuclear structure and reactions, as well as for astrophysical models such as r-process nucleosynthesis and neutron stars. The traditional approach to low-energy nuclear physics is to treat nucleons as immutable objects interacting via phenomenological forces. The use of phenomenological interactions, rather than one derived from a microscopic theory, raises questions as to the reliability of predictions for exotic regions of the nuclear chart. The quark-meson coupling (QMC) model uses a relativistic mean-field approach to provide a microscopically derived nucleon-nucleon interaction, which takes into account the quark structure of the nucleon.

The Skyrme energy density functional is a popular phenomenological tool in studies of nuclear structure and reactions. In this work, the QMC density functional was used to produce a set of Skyrme parameterisations, in the hopes that they will give more reliable predictions for exotic nuclei. In conjunction with Hartree-Fock-Bogoliubov (HFB) calculations, the Skyrme-QMC (SQMC) parameterisations have been used to model the ground-state properties of individual nuclei and nucleus-nucleus potentials for Ca + Sn reactions. The SQMC parameterisation performs with an accuracy comparable to modern phenomenological functionals. From this, one can investigate the importance of the isovector terms of the nucleon-nucleon interaction, which are particularly significant for exotic, neutron-rich regions of the nuclear chart.

One of the notable successes of the QMC model is its derivation of nuclear spin-orbit coupling. The isovector dependence of the spin-orbit equation of state is remarkably similar to that of the modern UNEDF1 phenomenological density functional. HFB calculations along the Sn isotopic chain reveal that the isovector properties of the spin-orbit term impact binding energies to a level that will be significant for astrophysical r-process modelling.

1 Introduction

The behaviour of exotic nuclear systems is an important question for nuclear, particle and astrophysics. While new experimental facilities and techniques are constantly expanding the limits of known nuclei, many regions remain largely out of reach, including the astrophysical r-process path, superheavy elements, and the neutron dripline.

Theoretically, one is faced with the quantum many-body problem posed by a system of strongly interacting nucleons. One of the most successful tools currently used is the Skyrme energy density functional. It gives the energy density at each point in the nuclear system, as a function of the local nucleon densities. The functional can be derived from the Skyrme interaction [1], which is the most general two-body, zero-range nucleon-nucleon interaction up to second-order in derivatives. The downside of the Skyrme density functional is that it is phenomenological. Its set of 10–17 parameters are typically determined by a fit to experimental masses and radii of nuclides near stability, calling into question the predictions for very exotic systems. A possible solution to this is to obtain the interaction between nucleons from the more fundamental, high-energy degrees-of-freedom of the system.

The quark-meson coupling (QMC) model is a relativistic mean-field approach which self-consistently accounts for the in-medium modification of the quark structure of bound nucleons. Proposed by P. A. M. Guichon in Ref. [2], and since developed in Refs. [3–5], it confines three quarks to a nucleon bag and allows the quarks of different bags to interact by exchanging σ, ω and ρ mesons. This results in an energy density functional with only four free parameters (a coupling constant for each meson field and the mass of the σ meson). The reduced number of parameters should make the QMC model more predictive than standard phenomenological approaches.

Thanks to its inclusion of quark degrees-of-freedom, the QMC model offers a possible explanation to the famous "European Muon Collaboration" (EMC) effect [6–8] and also gives a remarkably successful description of hypernuclei [9, 10].

While the QMC model assumes that nucleons move slowly, the quarks are treated relativistically and the model

*e-mail: ellen.mcrae@anu.edu.au

naturally includes certain relativistic corrections. Most notably, the model derives nuclear spin-orbit coupling, a relativistic effect essential for shell structure of nuclei and dissipation in collisions.

2 Spin-orbit energy density functional

2.1 The QMC model

The spin-orbit energy density functional from the QMC model is given by [4]

$$\mathcal{H}_{\rm SO}^{\rm QMC} = \frac{-1}{4M_N^2}\Bigg[(G_\sigma + G_\omega(2\mu_s - 1))\rho\nabla\cdot\boldsymbol{J} + \left(\frac{G_\sigma}{2} + \frac{G_\omega}{2}(2\mu_s - 1) + \frac{3G_\rho}{8}(2\mu_v - 1)\right) \times \sum_{q=n,p}\rho_q\nabla\cdot\boldsymbol{J}_q\Bigg], \quad (1)$$

where ρ_n (ρ_p) is the neutron (proton) particle density and ρ is the total particle density, $\rho = \rho_n + \rho_p$; similarly for the spin-orbit density, $\boldsymbol{J}$. The coefficients are formed from the meson coupling constants (G_σ, G_ω, G_ρ), the isoscalar and isovector nucleon magnetic moments (μ_s and μ_v, respectively), and the free nucleon mass (M_N). Though not strictly necessary, the ω and ρ meson fields can be identified with real particles and their experimental masses used. This leaves only four free parameters in the QMC model, G_σ, G_ω, G_ρ and the σ meson mass, all of which are fixed by the central terms of the functional. Unlike non-relativistic approaches such as Skyrme, the QMC spin-orbit functional has no extra spin-orbit parameters to adjust. The small number of free parameters in the QMC model is expected to make it more predictive than a standard Skyrme functional, particularly for the spin-orbit term.

2.2 Skyrme

The most general form of the Skyrme density functional [11],

$$\mathcal{H}_{\rm SO}^{\rm Skyrme} = -\frac{1}{2}\Big[W_0\rho\nabla\cdot\boldsymbol{J} + W_0'\sum_{q=n,p}\rho_q\nabla\cdot\boldsymbol{J}_q\Big], \quad (2)$$

takes the same form of the nucleon densities but now with free coefficients W_0 and W_0'. As the Skyrme interaction is non-relativistic, obeying Galilean rather than Lorentz invariance, the coefficients of the spin-orbit term are completely unknown and must be adjusted to experiment.

2.3 Isovector dependence

The isovector dependence of the spin-orbit functionals quantifies how they are affected by differences between proton and neutron densities, which become increasingly important for studying very neutron rich nuclei. It is controlled by the second term of equations 1 and 2, as the proton and neutron densities appear individually. Therefore, the ratio of the two coefficients can be used to compare the isovector properties of different approaches, as done in Table 1.

Table 1. Ratio of the spin-orbit coupling constants illustrating the isovector dependence of spin-orbit energy density functionals.

W_0'/W_0	Model
1	Standard Skyrme
1.86	Modern Skyrme: UNEDF1 [12]
1.78	QMC
0.1	Standard relativistic mean-field (RMF) [13]
0.2	QMC, Hartree terms only, Dirac μ

Table 2. Parameters of the spin-orbit functional.

Model	W_0 (MeV fm^5)	W_0' (MeV fm^5)
QMC	82.8602	147.4411
UNEDF1	76.736144	142.63304

Traditionally, Skyrme functionals only had one spin-orbit parameter, W_0. The modern Universal Nuclear Energy Density Functional (UNEDF1) [12] released this constraint, and from its state-of-the-art fitting procedure, had actually derived a much stronger isovector dependence. The QMC model, with its more fundamental, relativistic basis gives a very similar value.

On the other hand, a standard relativistic mean-field (RMF) model of inert nucleons, neglecting exchange (Fock) terms, gives a substantially weaker isovector dependence. By keeping only the direct (Hartree) terms of the QMC spin-orbit functional and setting the nucleon magnetic moments to their Dirac ($\mu_s = \mu_v = 1$) rather than experimental ($\mu_s = 0.88$, $\mu_v = 4.7$) values, one would obtain a similarly weak isovector dependence. This implies that the weak isovector dependence of RMF is primarily due to these two approximations, neglect of exchange terms and use of Dirac magnetic moments. These are approximations which are not required in the QMC model.

There is a remarkable similarity between the spin-orbit functionals of UNEDF1 and the QMC model (see Tab. 2). However, the significance of this fact is dependent upon whether the isovector dependence of the spin-orbit term alone has any impact on Hartree-Fock calculations of nuclei. The following section investigates the use of a Skyrme parameterisation from the QMC model and the importance of the isovector properties of the spin-orbit term.

3 Skyrme-QMC parameterisation

There exist many codes based on Hartree-Fock approximations which can be used to compute properties of nuclear structure [14–16] and reactions [17–24]. The majority of them are based on Skyrme functionals. While the QMC functional has a similar form to the Skyrme functional, it is not identical.

We have constructed a Skyrme functional which is as close as possible to that of the QMC model, called Skyrme-QMC (SQMC) [25]. For the central part of the functionals there is a difference in the density dependence and they are matched around the saturation point of nuclear matter. But for the spin-orbit term, SQMC and the QMC model are identical, making SQMC a convenient tool to test the isovector dependence of the original QMC functional.

A similar SQMC was obtained in Ref. [26], however with $W_0 = W_0'$. Here, our SQMC has $W_0'/W_0 = 1.78$ as in the original QMC model.

4 Hartree-Fock calculations

Hartree-Fock-Bogoliubov codes are well-suited to the study of exotic nuclear structure as the treatment of pairing remains robust for weakly bound nuclei. Also, long isotopic chains provide a large range of neutron excesses, useful for revealing trends in the isovector properties of the functionals. The axial Hartree-Fock-Bogoliubov code, HFBTHO [15], was used to calculate the ground-state binding energies of tin ($Z = 50$) nuclei.

Relative to experimental data [27], as seen in Figure 1, the SQMC functional performs at a level comparable to UNEDF1, especially away from stability. This is particularly remarkable when one recalls that the stable tin masses were included in the fit of UNEDF1, while no experimental masses were used to derive the SQMC functional.

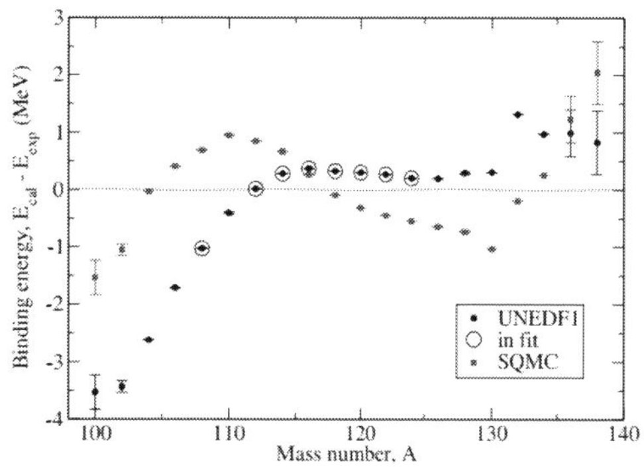

Figure 1. Difference between calculated and experimental binding energies for ^{A}Sn nuclei. The error bars are from the experimental data (E_{exp}).

It is possible to extract the impact of only the isovector dependence of the spin-orbit term by comparing a SQMC parameterisation with the isovector dependence of the QMC model ($W_0'/W_0 = 1.78$), against a baseline which has only one spin-orbit parameter ($W_0 = W_0' = 104.3872$ MeV fm^5) but is otherwise identical. Calculations along the Sn isotopic chain out to the dripline, as shown in the upper panel of Figure 2, reveal that the isovector dependence of the spin-orbit term leads to a change of approximately 1 MeV in the region of the r-process ($^{134-152}$Sn). A change half this size is expected to have a significant impact on r-process abundances [28]. This illustrates that it is crucial to properly account for the isovector properties of the spin-orbit functional, as predicted by the QMC model for example.

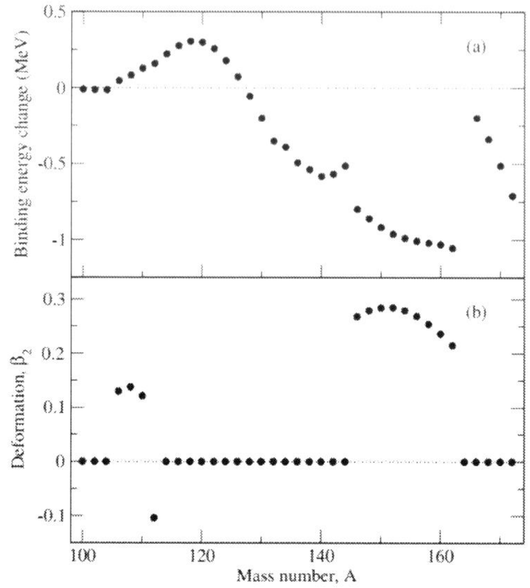

Figure 2. (a) Binding energy difference between SQMC with $W_0'/W_0 = 1.78$ and SQMC with $W_0'/W_0 = 1$, for ^{A}Sn nuclei. (b) Quadrupole deformation, β_2, predicted by the SQMC parameterisation.

The lower panel of Fig. 2 shows the deformation of the mean-field ground-states in the tin isotopes. We observe that the effect of the isovector contribution to the spin-orbit functional is clearly enhanced by the presence of prolate deformation in neutron-rich nuclei ($^{146-162}$Sn). A more detailed investigation of the interplay between deformation and the isovector contribution to the spin-orbit functional will be the focus of a future study.

5 Fusion

The SQMC parameterisation can also be used to investigate any effect the isovector dependence of the spin-orbit term has on nucleus-nucleus potentials and fusion cross-sections. Ref. [29] showed that isovector terms (central and spin-orbit) were significant for density-constrained time-dependent Hartree-Fock (DC-TDHF) calculations of the nucleus-nucleus potential of the ^{40}Ca + ^{132}Sn system. As a preliminary study, frozen Hartree-Fock (FHF) [30–33] was used with the SQMC parameterisation for the same system. FHF calculations provide the bare nucleus-nucleus potential from the Hartree-Fock ground-states, with no polarisation effects from dynamics.

Again we compare the two SQMC functionals with different spin-orbit isovector dependences ($W_0' = W_0 = 104.3872$ MeV fm^5 and $W_0'/W_0 = 1.78$), shown in Figure 3. While SQMC is in good agreement with the phenomenological UNEDF1 functional, the isovector dependence of the spin-orbit interaction has no impact on the fusion barrier of $^{40}Ca + ^{132}Sn$. Similar results were obtained for the $^{40}Ca + ^{100}Sn$ and $^{48}Ca + ^{100,132}Sn$ systems.

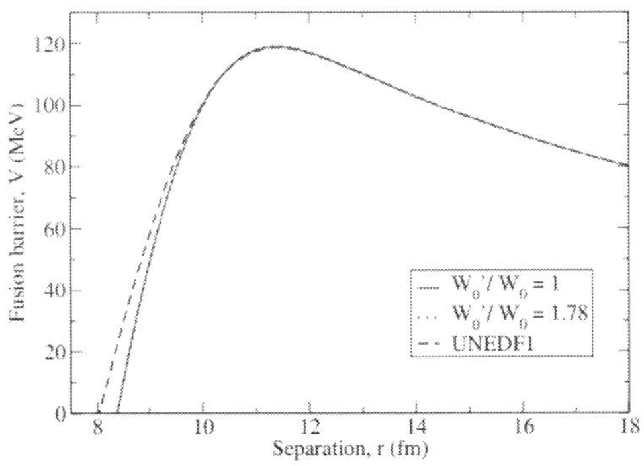

Figure 3. Nucleus-nucleus potential for $^{40}Ca + ^{132}Sn$.

Future work will involve calculations using TDHF [34, 35] and the DC-TDHF technique [36], which self-consistently treats dynamical effects and nuclear structure, to investigate any potential impact of the isovector dependence of the spin-orbit term coupling to the dynamics or deformation of the system.

6 Conclusions

The isovector dependence of the spin-orbit term from the QMC model, with its high-energy degrees of freedom, is remarkably similar to that of UNEDF1. This is significant because, while no change to bare fusion potentials was observed, this dependence is an important factor that should be accounted for in r-process calculations [25].

A Skyrme functional from the QMC model can successfully reproduce ground-state binding energies. In the future, the SQMC functional will be used to study further exotic structure effects or, using a time-dependent Hartree-Fock code, dynamic processes including reactions with exotic nuclei, fusion barriers, transfer and fission.

It will also be possible to perform similar investigations using the original QMC functional, rather than the SQMC parameterisation, by modifying Skyrme-based Hartree-Fock codes to accept the more complicated density dependence of the central terms. Stone *et al.* [5] modified the static SKYAX code [37] in this way and used nuclear data to fix the few free parameters of the QMC model. A similar implementation in a time-dependent code will allow for a study of the giant monopole resonance [38, 39] to explore the effects of the unique density dependence of the QMC model.

This research is supported by an Australian Government Research Training Program (RTP) Scholarship and ARC Grants FT120100760 and DP160101254.

References

[1] T.H.R. Skyrme, Nucl. Phys. **9**, 615 (1959)
[2] P.A.M. Guichon, Phys. Lett. B **200**, 235 (1988)
[3] P.A.M. Guichon, K. Saito, E. Rodionov, A.W. Thomas, Nucl. Phys. A **601**, 349 (1996)
[4] P.A.M. Guichon, H.H. Matevosyan, N. Sandulescu, A.W. Thomas, Nucl. Phys. A **772**, 1 (2006)
[5] J.R. Stone, P.A.M. Guichon, P.G. Reinhard, A.W. Thomas, Phys. Rev. Lett. **116**, 092501 (2016)
[6] A.W. Thomas, A. Michels, A.W. Schreiber, P.A.M. Guichon, Phys. Lett. B **233**, 43 (1989)
[7] I.C. Cloët, W. Bentz, A.W. Thomas, Phys. Lett. B **642**, 210 (2006)
[8] I.C. Cloët, W. Bentz, A.W. Thomas, Phys. Rev. Lett. **109**, 182301 (2012)
[9] K. Tsushima, K. Saito, J. Haidenbauer, A. Thomas, Nucl. Phys. A **630**, 691 (1998)
[10] P.A.M. Guichon, A.W. Thomas, K. Tsushima, Nucl. Phys. A **814**, 66 (2008)
[11] P.G. Reinhard, H. Flocard, Nucl. Phys. A **584**, 467 (1995)
[12] M. Kortelainen, J. McDonnell, W. Nazarewicz, P.G. Reinhard, J. Sarich, N. Schunck, M.V. Stoitsov, S.M. Wild, Phys. Rev. C **85**, 024304 (2012)
[13] M.M. Sharma, G. Lalazissis, J. König, P. Ring, Phys. Rev. Lett. **74**, 3744 (1995)
[14] K. Bennaceur, J. Dobaczewski, Comput. Phys. Comm. **168**, 96 (2005)
[15] M.V. Stoitsov, N. Schunck, M. Kortelainen, N. Michel, H. Nam, E. Olsen, J. Sarich, S. Wild, Comput. Phys. Comm. **184**, 1592 (2013)
[16] W. Ryssens, V. Hellemans, M. Bender, P.H. Heenen, Comput. Phys. Comm. **187**, 175 (2015)
[17] J.A. Maruhn, P.G. Reinhard, P.D. Stevenson, A.S. Umar, Comput. Phys. Comm. **185**, 2195 (2014)
[18] A.S. Umar, V.E. Oberacker, Phys. Rev. C **73**, 054607 (2006)
[19] C. Simenel, P. Chomaz, G. de France, Phys. Rev. Lett. **86**, 2971 (2001)
[20] C. Simenel, P. Chomaz, G. de France, Phys. Rev. Lett. **93**, 102701 (2004)
[21] K. Sekizawa, K. Yabana, Phys. Rev. C **88**, 014614 (2013)
[22] G. Scamps, D. Lacroix, Phys. Rev. C **87**, 014605 (2013)
[23] P.D. Stevenson, E.B. Suckling, S. Fracasso, E.D. Simmons, A.S. Umar, EPJ Web Conf. **86**, 00058 (2015)
[24] L. Guo, T. Nakatsukasa, EPJ Web Conf. **38**, 09003 (2012)
[25] E. McRae, C. Simenel, E.C. Simpson, A.W. Thomas, arXiv p. 1704.07991 (2017)
[26] X.B. Wang, C. Qi, F.R. Xu, Nucl. Phys. A **865**, 57 (2011)
[27] M. Wang, G. Audi, A.H. Wapstra, F.G. Kondev, M. MacCormick, X. Xu, B. Pfeiffer, Chin. Phys. C **36**, 1603 (2012)
[28] M. Mumpower, R. Surman, G. McLaughlin, A. Aprahamian, Prog. Part. Nucl. Phys. **86**, 86 (2016)

DOI: 10.1051/epjconf/201716300036

[29] K. Godbey, A.S. Umar, C. Simenel, Phys. Rev. C **95**, 011601 (2017)
[30] C. Simenel, B. Avez, Int. J. Mod. Phys. E **17**, 31 (2008)
[31] K. Washiyama, D. Lacroix, Phys. Rev. C **78**, 024610 (2008)
[32] C. Simenel, M. Dasgupta, D.J. Hinde, E. Williams, Phys. Rev. C **88**, 064604 (2013)
[33] C. Simenel, A.S. Umar, K. Godbey, M. Dasgupta, D.J. Hinde, Phys. Rev. C **95**, 031601 (2017)
[34] C. Simenel, Eur. Phys. J. A **48**, 152 (2012)
[35] T. Nakatsukasa, K. Matsuyanagi, M. Matsuo, K. Yabana, Rev. Mod. Phys. **88**, 045004 (2016)
[36] A.S. Umar, V.E. Oberacker, Phys. Rev. C **74**, 021601 (2006)
[37] P.G. Reinhard, unpublished
[38] P.D. Stevenson, D. Almehed, P.G. Reinhard, J.A. Maruhn, Nucl. Phys. A **788**, 343 (2007)
[39] Avez, B., Simenel, C., Eur. Phys. J. A **49**, 76 (2013)

 DOI: 10.1051/epjconf/201716300037

Interference effect between neutron direct and resonance capture reactions for neutron-rich nuclei

Futoshi Minato[1,*] and *Tokuro* Fukui[1,2]

[1] Nuclear Data Center, JAEA, Tokai, 319-1195, Japan
[2] INFN, Napoli, 80126, Italy

Abstract. Interference effect of neutron capture cross section between the compound and direct processes is investigated. The compound process is calculated by resonance parameters and the direct process by the potential model. The interference effect is tested for neutron-rich ^{82}Ge and ^{134}Sn nuclei relevant to r-process and light nucleus ^{13}C which is neutron poison in the s-process and produces long-lived radioactive nucleus ^{14}C ($T_{1/2}$ = 5700 y). The interference effects in those nuclei are significant around resonances, and low energy region if s-wave neutron direct capture is possible. Maxwellian averaged cross sections at kT = 30 and 300 keV are also calculated, and the interference effect changes the Maxwellian averaged capture cross section largely depending on resonance position.

1 Introduction

The neutron capture reaction is one of the simplest nuclear reactions, but its importance is recognized in various fields. For example, the heavy elements existing in the nature are considered to be produced by a successive neutron capture in a neutron-rich environment (the leading candidates are the s-process and r-process). In the field of accelerators and reactors, the neutron capture generates unstable nuclei in the structure materials used for confining neutrons and quantum beams in their long-term operation. The production of unstable nuclei by (secondary) neutrons cannot be ignored in terms of the radioactive protection and the decommissioning process. Vice versa, the neutron capture has an potential to transmute a radioactive unstable one to stable nucleus in terms of reducing radioactive wastes produced in various nuclear sites, especially in nuclear reactors. Moreover, using prompt γ-rays resulted from the neuron capture, non-destructive analysis in materials is also performed. It is therefore important to know accurate neutron capture cross sections systematically for nuclei in the nuclear chart, in particular at low energy region where its cross section becomes large.

The neutron capture at low energies can be divided into two parts, that is the compound process and the direct process. Both of the processes emit γ-rays with the same energy if the initial and final states are identical. However, the compound process stops by a compound state before γ-emission one hand, the direct process instantly goes to the final state by the electronic dipole radiation of incident neutron on the other hand. The compound process has a characteristic resonance structure in the cross section (see Fig. 2 , for instance). The strong peaks emerge when the incident neutron wave function smoothly connects with the interior wave function of $A + 1$ nucleus according to the R-matrix theory [1, 2]. For most nuclei, the compound process occupies its dominant part of the neutron capture reaction. This is because the number of resonances reaches to hundreds to millions at a capture state energy, $E^* = E + S_n$, where E and S_n are the incident neutron energy and the neuron threshold of $A + 1$ nucleus, respectively. As a consequence, the neutron capture cross section of the compound process ranges approximately from 10^{-1} to 10^5 barns at neutron resonance regions. On the other hand, typical neutron capture cross section of the direct process is the order of 0.1 to 1 mb.

However, it is not always the case that the compound process is larger than the direct one in case of light nuclei and magic nuclei, for which the number of resonances at the capture state energy is small. In this case, neutron capture cross section of the compound process becomes small and the direct process comes to play a comparable role to the compound one. One of the examples is ^{12}C [3] and ^{16}O [4]. The direct process of those nuclei is strong enough that it affects the evolution of stars and s-process. Besides light nuclei and magic nuclei, the direct process becomes significant in neutron-rich nuclei as well because the number of resonances available to form the compound state is expected to be small due to the small neutron threshold energy. For this reason, the direct process in neutron-rich nuclei has been actively investigated [5–8].

The total neutron capture cross section is described in general by summing the results of the compound and direct processes. However, it was discussed [9, 10] that there is an interference effect between these processes, similar to the potential scattering and compound one in the elastic channel. Such an interference effect was investigated

* e-mail: minato.futoshi@jaea.go.jp

in proton capture reaction [11], however, it has not been clarified if the interference really exists in the neutron capture reaction. The evaluated neutron nucleus reaction cross section library, JENDL [12], therefore did not consider the effect. However, several recent experiment works have shown unique data distributions in the cross sections of $^{9}Be(\gamma, n)$ and $^{16}O(n, \gamma)^{17}O$ that one cannot reproduce only by the compound and direct processes. Mengoni and Otsuka showed [13] that the cross sections can be reproduced theoretically if they take into account the interference effect.

It is then natural to think of whether the interference effect also becomes significant in neutron-rich nuclei. In fact, the compound and direct processes become comparable in the vicinity of the neutron magic number [7, 8], and therefore a clear interference effect is expected to be found around that mass region. We choose three nuclei as test ones, that is ^{13}C, ^{86}Ge, and ^{134}Sn. The first nucleus locates next to the stability line in the nuclear chart, but it is important for nuclear data evaluation because ^{13}C is neutron poison in *s*-process and can be a generator of long-lived nucleus ^{14}C ($T_{1/2} = 5700$ y). The latter two are typical neutron-rich nuclei which may be involved in *r*-process.

In addition to neutron direct capture reaction, there is a semi-direct process coming out of $E = 5 \sim 10$ MeV. But we do not discuss it because our interest in this work is less than $E = 1$ MeV.

This paper is organized as follows. In Sec. 2, we describe the formalism to calculate the neutron capture cross section briefly. Sec. 3 gives our result and Sec. 4 summarizes this work and gives the future perspectives.

2 Calculation

In neutron resonance region for the nuclei studied in this work, the exit channel is classified only to the elastic scattering and the capture reaction, and other channels do not take part in. We also assume that the target nucleus is spherical. The entrance and the capture channels are specified by $i = (J^\pi, j_i, l_i)$ and $f = (J_f^\pi, n_f, j_f, l_f)$, respectively. Here, l_i is the relative orbital angular momentum between the projectile and the target and $j_i = l_i \pm 1/2$; $\vec{J} = \vec{S} + \vec{l_i}$ is the total spin of the system, $\vec{S} = \vec{s} + \vec{s_n}$ the total spin, and s and s_n the spins of target and projectile, respectively; n_f is the principle quantum number, l_f the orbital angular momentum, $j_f = l_f \pm 1/2$ the total angular momentum of the bound state, and $J_f = J \pm 1$.

The cross section of neutron capture from entrance channel i with spin-parity J^π to capture channel f can be described by the *R*-matrix theory, which is given by [1, 2, 9, 10, 14, 15]

$$\sigma_{i\to f}(E) = \frac{\pi}{k^2} g_J \left| \sum_\lambda \frac{i e^{-i\phi_i} \Gamma_{\lambda ni}^{1/2} \Gamma_{\lambda\gamma f}^{1/2}}{(E_\lambda - E) - i\Gamma_\lambda/2} \right|^2, \tag{1}$$

where $\Gamma_\lambda = \Gamma_{\lambda ni} + \Gamma_{\lambda\gamma f}$ is the total decay width for resonance state λ, $\Gamma_{\lambda ni}$ the neutron width, $\Gamma_{\lambda\gamma f}$ the γ width, and E_λ the resonance energy; $g_J = (2J+1)/(2s+1)(2s_n+1)$ is the statistical factor. The hard-sphere phase shift in channel i is defined by ϕ_i. Incident neutron wave number is defined by $k = \sqrt{2mE}/\hbar$, where E and $m = A/(A+1)m_N$ are the incident neutron energy and the reduced mass, respectively.

If each resonance is isolated from others, Eq. (1) reads

$$\sigma_{BW,i\to f}(E) = \frac{\pi}{k^2} g_J \sum_\lambda \frac{\Gamma_{\lambda ni}\Gamma_{\lambda\gamma f}}{(E_\lambda - E)^2 + (\Gamma_\lambda/2)^2}. \tag{2}$$

Equation (2) is the so-called the multilevel Breit-Wigner formula. The cross section has a maximum value at $E = E_\lambda$.

The direct radiative neutron capture cross section can be calculated by

$$\sigma_{dir,i\to f}(E) = \frac{\pi}{k^2} g_J |U_{i\to f}^{dir}|^2, \tag{3}$$

where the collision matrix U_{dir} is written as [10, 15]

$$U_{i\to f}^{dir} = \frac{1}{\sqrt{2J+1}} \left(\frac{16}{9\hbar v}\right)^{1/2} \bar{e} \sum_i k_\gamma^{3/2} Q_{fi}. \tag{4}$$

The factor $v^{-1/2} = \sqrt{m/\hbar k}$ is coming from the assumption of a unit-flux incoming wave in the entrance channel. The effective charge of neutron is $\bar{e} = -eZ/A$, the emitted γ-ray wave number $k_\gamma = \epsilon_\gamma/\hbar c$, $\epsilon_\gamma = E - \epsilon_f$ the γ-ray energy, ϵ_f the single particle energy of the final state calculated by the Woods-Saxon potential, and $Q_{fi} = \mathcal{T}_{fi} B_{fi} A_{fi}$ the transition matrix [13], where B_{fi} is the parameter concerning the spectroscopic factor. The radial integral part, $\mathcal{T}_{fi}$, is given by [13, 16]

$$\mathcal{T}_{fi} = \int_0^\infty u(n_f j_f l_f; r)\, r \chi(j_i l_i; r)\, dr, \tag{5}$$

where the scattering wave function calculated by a neutron-nucleus optical potential is

$$\begin{aligned} \chi(j_i l_i; r) &= \sqrt{4\pi}\sqrt{2l_i+1}\, i^{l_i} \frac{i}{2}\left(I_{l_i}^{j_i}(r) - U_{l_i}^{j_i} O_{l_i}^{j_i}(r)\right) \\ &\xrightarrow[r\to\infty]{} \sqrt{4\pi}\sqrt{2l_i+1}\, i^{l_i} \frac{i}{2}\left(e^{i(kr-\frac{l\pi}{2})} - U_{l_i}^{j_i} e^{-i(kr-\frac{l\pi}{2})}\right) \end{aligned} \tag{6}$$

and $u(n_f j_f l_f; r)$ is the radial part of the single particle wave function of the final state. We here define the scattering matrix element $U_{l_i}^{j_i} = e^{-2i\phi_i}$. The angular momentum part, A_{fi} is written as

$$\begin{aligned} A_{fi} &= \hat{J}_f \hat{j}_f \hat{j}_i \hat{l}_i \sqrt{\frac{3}{4\pi}} \langle l_i 010 | l_f 0 \rangle \\ &\times \begin{Bmatrix} j_i & s & J \\ J_f & 1 & j_f \end{Bmatrix} \begin{Bmatrix} l_i & j_i & 1/2 \\ j_f & l_f & 1 \end{Bmatrix}. \end{aligned} \tag{7}$$

So far, we considered the compound process and the direct process separately. The neutron capture cross section taking into account both processes simultaneously is given by [9, 10]

$$\sigma_{i\to f}(E) = \frac{\pi}{k^2} g_J \left| \sum_\lambda \frac{i e^{-i\phi_i} \Gamma_{\lambda ni}^{1/2} \Gamma_{\lambda\gamma f}^{1/2}}{(E_\lambda - E) - i\Gamma_\lambda/2} + U_{i\to f}^{dir} \right|^2. \tag{8}$$

 DOI: 10.1051/epjconf/201716300037

If each resonance with the same capture state is isolated, Eq. (8) becomes as we have done in Eq. (2)

$$\sigma_{i\to f}(E) \simeq \sigma_{BW,i\to f} + \sigma_{dir,i\to f} + \frac{\pi}{k^2} g_J \sum_\lambda \frac{\Gamma_{\lambda n}^{1/2}\Gamma_{\lambda\gamma\mu}^{1/2} 2Re\left(-ie^{i\phi_i} U_{dir}(E_\lambda - E - i\Gamma_\lambda/2)\right)}{(E_\lambda - E)^2 + (\Gamma_\lambda/2)^2}. \tag{9}$$

The first and second terms are indentical to the Breit-Wigner formula and the direct capture cross section, respectively, and the third term is nothing but the one from the interference effect.

Single particle wave functions and levels are calculated by the Woods-Saxon potential, in which the parameters are $R_n = 1.236A^{1/3}$ fm, the diffuseness $a = 0.62$ fm, the potential depth $V = -50$ MeV, and the LS potential $V_{LS} = 19.4/r\frac{df(r)}{dr}\vec{l}\cdot\vec{s}$ MeV. In case of ^{13}C, we adjust V for each resonance to reproduce the neutron capture cross section at thermal energy and $E = 64$ keV, that is $V_{1p_{3/2}} = -38.6$, $V_{2s_{1/2}} = -51.4$, and $V_{1d_{5/2}} = -48.0$ MeV, retaining the other parameters. The scattering neutron wave function is calculated by the optical model with the Kunieda potential [17]. In order to take into account the non-local effect of the optical potential on the scattering wave function, the Perey and Buck method [18] with the non-local parameter $\beta = 0.85$ fm is adopted.

Because ^{82}Ge and ^{134}Sn are the short half-life nuclei, the information about the resonance parameters evaluated from experiments is not available. It is not a simple task to calculate those parameters by a theoretical method. Therefore, we adopted, in this work, $\Gamma_{\lambda ni}$, $\Gamma_{\lambda\gamma}$ and E_λ given in TENDL-2015 [19], in which their resonance parameters are calculated on the basis of statistical properties regarding the nuclear excitations. In case of ^{13}C, we used a newly evaluated resonance parameter of JENDL [20].

If resonance state is $J \neq 0$, the final state has two possibilities in case of $E1$ radiation, that is $J \pm 1$. In this case, we assume the equality in the γ decay widths, namely $\Gamma_{\lambda\gamma f} = \Gamma_{\lambda\gamma}/2$. Note that the interference effect between the compound process and direct process occurs when the entrance and exit channels of the two processes are exactly identical.

In order to calculate the interference effect given in Eq. (9), we are confronted by the sign problem in the term proportional to $Re(-ie^{\phi_i}U_{dir}(E_\lambda - E - i\Gamma_\lambda/2))$. In this work, the following prescription is adopted; the interference effect on the neutron capture cross section in the first resonance of ^{13}C ($J = 2$) and ^{134}Sn ($J = 1/2$) works constructively below the resonance and destructively above it as found in the experimental data of ^{16}O(n,γ) cross section [13]. Then, we apply the same sign to the interference term of the other resonances. In case of ^{82}Ge, the same sign as ^{13}C is adopted because the neutron capture reaction of ^{82}Ge is dominated by the p-wave neutron, which causes the interference effect to the first resonance of ^{13}C.

The sign of interference effect mentioned above is entirely arbitrary. As is obvious, it is not clear whether the interference effect works constructively below resonance and destructively above resonance as ^{16}O(n,γ). We do not discuss the inverse interference effect in this work, however it should be kept in mind that we may have an interference effect on neutron capture cross sections working oppositely to what we will show in the next section. One of the solutions to fix the interference effect would be to measure neutron capture cross section just below and above resonance as done for ^{16}O [13].

3 Result

Figure 1 shows the neutron capture cross sections of ^{13}C. The resultant ^{14}C nucleus has the excited states of 2^+ and 1^- below the neutron threshold energy, and the 0^+ ground state. Since the target has the initial spin of $J^\pi = 1/2^-$, the cross section is obtained by summing the results of total angular momentum, $J = 0$ to 3 based on the assumption that only E1 transition occurs. The top panel shows the cross section from $E = 10^{-8}$ to 1 MeV. The solid and dashed lines are the results of compound process + direct process + interference effect (Com+Dir+Int) and compound process + direct process (Com+Dir), respectively. The Woods-Saxon potential parameter V is determined so as to reproduce the most recent experimental data of the thermal neutron capture cross section, $\sigma_\gamma(th) = 1.496$ b [21] and at $E = 64$ keV [22] in case of Com+Dir+Int, as mentioned in the previous section. The interference effect plays a significant role from low energy region. This is because the s-wave neutron capture into the $1p_{1/2}$ bound state is concerned both with the compound and direct processes, and those cross sections are comparable. As incident neutron energy increases, the contribution from the s-wave neutron becomes weak, and the p-wave neutron comes to play a role from $E \sim 0.01$ MeV instead.

In the bottom panel of Fig. 1, the neutron capture cross section from $E = 0.01$ to 1 MeV is shown. The dotted line is the result of only the compound process. It is found that the direct process is essential to explain the experimental data [22] around 10^{-2} to 10^{-1} MeV. However, the interference effect around the resonance at $E = 152.4$ keV ($J = 2$) is not strong as studied in ^{16}O [16].

Figure 2 shows the neutron capture cross sections of ^{134}Sn and ^{82}Ge. We can see that the interference effect gives a significant change at the first resonance appearing at $E = 102.1$ keV for ^{134}Sn. The changes by the interference effect can be also observed in the other resonances distributing from $E = 0.5$ to 1.0 MeV although the modifications are not as strong as the first one. As seen in ^{13}C, the interference effect increases the cross section from low energy because the s-wave neutron capture into the $3p_{3/2}$ and $3p_{1/2}$ bound states is possible, and the compound and direct processes have a comparable cross section. In case of ^{82}Ge, because of the multilevel resonances composed of $J = 1/2$ and $3/2$ above $E = 100$ keV, the interference effect gives complicated changes.

In order to evaluate the effect of the interference effect in an astrophysical condition, we calculate the Maxwellian averaged cross section at $kT = 30$ and 300 keV. The results comparing Com+Dir+Int and Com+Dir at $kT = 30$ keV are shown in Table 1. By introducing the interference effect, the cross sections are increased for nuclei studied

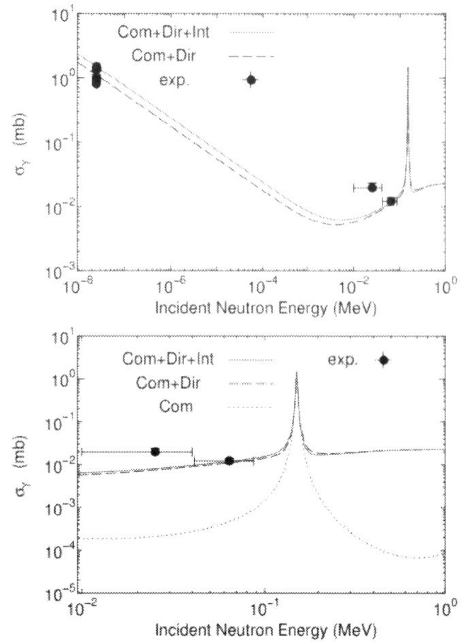

Figure 1. Neutron capture cross section for ^{13}C from $E = 10^{-8}$ to 1 MeV (top) and $E = 0.01$ to 1 MeV (bottom). The solid and dashed lines are the results of compound process + direct process + interference effect (Com+Dir+Int) and compound+direct process (Com+Dir), respectively. Note that the resonance parameter of ^{13}C is taken from Ref. [20]. The experimental data are taken from Ref. [21–23].

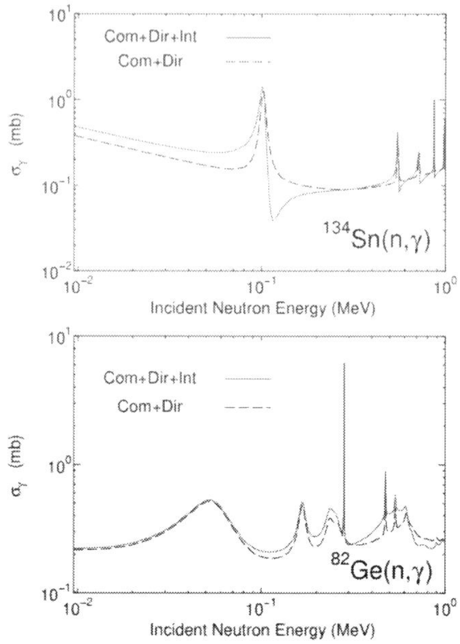

Figure 2. Neutron capture cross sections of ^{134}Sn (top) and ^{82}Ge (bottom) from $E = 0.01$ to 1 MeV. The results of compound process + direct process + interference effect (Com+Dir+Int), compound process + direct process (Com+Dir) are shown by the solid and dashed lines, respectively. The resonance parameters of ^{134}Sn and ^{82}Ge are taken from TENDL [19].

in this work. In particular, the neutron capture cross section for ^{134}Sn changes considerably as compared to the other two nuclei. The Maxwellian averaged cross section at $kT = 300$ keV is shown in Table 2. Contrary to the results obtained for $kT = 30$ keV, the interference effect decreases the neutron capture cross sections of ^{13}C and ^{134}Sn. On the other hand, the variation of ^{82}Ge by including the interference effect is not changed significantly from that for $kT = 30$ keV.

The difference of the interference effect at $kT = 30$ and 300 keV can be understood by the followings; Because the Maxwell-Boltzmann distribution has an arched curve with the peak at $E = kT$, the cross section is increased by the interference effect if the temperature is sufficiently smaller than the first resonance. However, if the Maxwell-Boltzmann distribution shifts to higher energy as the temperature increases, the cross section is affected not only by the constructive part but also by the destructive part of the interference effect. This case is realized in ^{13}C and ^{134}Sn. In case of ^{82}Ge, because there are a lot of resonances from $E = 0.01$ to 1 MeV, the interpretation of the difference between $kT = 30$ and 300 keV is not simple. However, the reduction of the capture cross section above $E = 700$ keV, as seen in Fig. 2, results in the smaller Maxwellian averaged cross section at $kT = 300$ keV than that at $kT = 30$ keV.

	^{13}C	^{82}Ge	^{134}Sn
Com+Dir+Int	0.0700	0.3574	0.3511
Com+Dir	0.0692	0.3412	0.2690
Variation (%)	+1.1	+4.7	+30

Table 1. The Maxwellian averaged cross sections at $kT = 30$ keV of ^{13}C, ^{82}Ge, and ^{134}Sn in unit of mb. The results of compound process + direct process + interference effect (Com+Dir+Int) and compound process + direct process (Com+Dir) are listed. The variation is calculated by $(\sigma(\mathrm{Com+Dir+Int})/\sigma(\mathrm{Com+Dir}) - 1) \times 100$.

	^{13}C	^{82}Ge	^{134}Sn
Com+Dir+Int	0.0776	0.6367	0.2566
Com+Dir	0.0780	0.6115	0.2591
Variation (%)	-0.5	+4.1	-1.0

Table 2. Same as Table 1 but at $kT = 300$ keV.

4 Summary

We have estimated the interference effect of the neutron capture cross sections on the neutron-rich nuclei ^{82}Ge and ^{134}Sn, and light nucleus ^{13}C. The neutron capture cross section of the compound process was calculated by the resonance parameters and that of the direct process was calculated by the potential model. The phase problem in the interference effect was chosen so that the interference effect became constructive below the resonance and destructive above it.

In case of ^{13}C, it was found that the interference effect increased the neutron capture cross section from $E = 10^{-8}$ to 10^{-2} MeV significantly. However, its effect was relatively small around the first resonance. In case of ^{134}Sn and ^{82}Ge, the interference effect was clearly observed around the resonances.

The Maxwellian averaged cross sections at $kT = 30$ and 300 keV were calculated. The changes of the neutron capture cross section by the interference effect were sensitive in case of $kT = 30$ keV for nuclei studied in this work. When the temperature was set to be $kT = 300$ keV, the interference effect became weak because of the constructive and destructive contribution of the interference effect caused by neighboring resonances.

This work suggests that the interference effect may have a large impact on neutron-rich nuclei as well as other light nuclei. However, in order to put forward the interference effect in the neutron capture reactions practically, we must solve the problems about the use of artificial resonance parameters, the phase problem, and the partial γ-decay width. Thus, new experimental measuring cross section and γ-ray energy and its strength around resonances is required and it will provide a further understanding of the interference effect.

Acknowledgment

This work is supported by YUGOKENKYU project at Japan Atomic Energy Agency.

References

[1] E. P. Wigner, Phys. Rev. **70**, 15 (1946); Phys. Rev. **70**, 606 (1946).
[2] E. P. Wigner and L. Eisenbud, Phys. Rev. **72**, 29 (1947).
[3] Y. Nagai, et al., Astrophys. J. **372**, 683 (1991).
[4] M. Igashira et al., Astrophys. J. **441**, L89 (1995).
[5] G. J. Matthews, A. Mengoni, F.-K. Thielemann, and W. A. Fowler, Astrophys. J. **270**, 740 (1983).
[6] L. Bonneau, T. Kawano, T. Watanabe, and S. Chiba, Phys. Rev. C **75**, 054618 (2007).
[7] T. Rauscher, *Nucl. Phys.* A**834**, 635 (2010).
[8] S. Chiba, *et al.*, *Phys. Rev. C* **77**, 015809 (2008).
[9] J. E. Lynn, *The theory of nuclear resonance reactions*, Oxford : Clarendon Press, (1968).
[10] I. Lovas, Sov. Phys. JETP, **14**, 840 (1962).
[11] C. Rolfs and R. E. Azuma, Nucl. Phys. **A227**, 291 (1974).
[12] K. Shibata et al., J. Nucl. Sci. Technol. **48**, 1 (2011).
[13] A. Mengoni and T. Otsuka, *AIP Conference Proceedings* (New York, 2000) **529**, 119 (2000).
[14] P. Descouvemont and D. Baye, Rep. Prog. Phys. **73**, 036301 (2010).
[15] A. M. Lane and J. E. Lynn, Nucl. Phys. **17**, 563 (1960).
[16] A. Mengoni, T. Otsuka, M. Ishihara, Phys. Rev. C **52**, R2334 (1995).
[17] S. Kunieda, et al., J. Nucl. Sci. Techno. **44**, 838 (2007).
[18] G. Perey and B. Buck, Nucl. Phys. **32**, 353 (1962).
[19] A. J. Koning and D. Rochman, *Nuclear Data Sheets* **113**, 2841 (2012).
[20] N. Iwamoto, private communication.
[21] R.B. Firestone, Zs. Revay, Phys. Rev. C **93**, 054306 (2016).
[22] T. Shima, T. Kii, T. Kikuchi, and F. Okazaki, JAERI Conference Proceedings, **4**, No.97, 131 (1996).
[23] O. Schwerer, editor. EXFOR formats description for users (EXFOR basics) (IAEA-NDS-206). Vienna: International Atomic Energy Agency (2008).

Effect of deformation and orientation on spin orbit density dependent nuclear potential

Rajni Mittal*, *Raj* Kumar, and *Manoj K.* Sharma
School of Physics and Materials Science, Thapar University, Patiala, Punjab, India.

Abstract. Role of deformation and orientation is investigated on spin-orbit density dependent part V_J of nuclear potential (V_N=V_P+V_J) obtained within semi-classical Thomas Fermi approach of Skyrme energy density formalism. Calculations are performed for $^{24-54}$Si+^{30}Si reactions, with spherical target ^{30}Si and projectiles $^{24-54}$Si having prolate and oblate shapes. The quadrupole deformation β_2 is varying within range of $0.023 \le \beta_2 \le 0.531$ for prolate and $-0.242 \le \beta_2 \le -0.592$ for oblate projectiles. The spin-orbit dependent potential gets influenced significantly with inclusion of deformation and orientation effect. The spin-orbit barrier and position gets significantly influenced by both the sign and magnitude of β_2-deformation. Si-nuclei with β_2<0 have higher spin-orbit barrier (compact spin-orbit configuration) in comparison to systems with β_2>0. The possible role of spin-orbit potential on barrier characteristics such as barrier height, barrier curvature and on the fusion pocket is also probed. In reference to prolate and oblate systems, the angular dependence of spin-orbit potential is further studied on fusion cross-sections.

1 Introduction

The aspect of nuclear deformations has been continually probed to analyze the fusion processes. A lot of work has been done by taking one of the target or projectile as deformed nucleus [1–4]. When the nuclear deformations are incorporated in theoretical calculations, it is inherently necessary to account for the relative orientation of the projectile and the target in the fusion dynamics. The orientation of the colliding nuclei has a significant effect not only on the fusion barrier height but also on the compactness of the touching configuration. In order to have better understanding of fusion barrier, the contribution of nuclear interaction potential plays a significant role. In the Skyrme energy density formalism (SEDF) [5–7], the spin-orbit dependent (V_J) and spin-orbit independent (V_P) potential add up to give the total nuclear potential. Therefore within this approach one can study effect of deformation and orientation on individual contribution of nuclear potential. In most of the previous studies [8, 9], the spin-orbit part was overlooked by either studying nuclei with the major shell closed for both protons and neutrons or taking its contribution to be small. Evidently, the contribution of spin-orbit potential in fusion dynamics is appreciable [4].

In the present work, the role of deformation (quadrupole deformed β_2) and orientation is investigated on spin-orbit density dependent potential V_J of nuclear potential. Here, an effort is made to analyse the possible role of orientation degree of freedom on spin-orbit part via different β_2-deformed shapes. To facilitate discussion on V_J due to shape and orientation of deformed nuclei, calculations are performed for $^{24-54}$Si+^{30}Si reactions having quadrupole deformation (β_2) within range of $0.023 \le \beta_2 \le 0.531$ (for prolate) and $-0.242 \le \beta_2 \le -0.592$ (for oblate). Calculations are done within the SEDF using SKRA force [10]. The Skyrme parameter set obtained from SKRA force seem to address the nuclear interactions adequately.

The above mentioned reactions are chosen because the target ^{30}Si is spherical, whereas projectiles are weakly and highly deformed with prolate (p) and oblate (o) shapes. The ground state deformation of nuclei are taken from the work of Moller and Nix [11]. It is manifested from the calculations that, independent of the magnitude of prolate (β_2>0) deformed nuclei, the spin-orbit barrier height (V_{JB}) (the spin-orbit barrier height (V_{JB}) is the maximum value of V_J and the corresponding position is spin-orbit barrier position (R_{JB}).) increases (corresponding spin orbit barrier position R_{JB} decreasing) from θ=0^0 to 90^0. For highly prolate nuclei, the spin-orbit barrier height and spin-orbit barrier position get influenced significantly in comparison to weakly prolate nuclei. In case of oblate shape nuclei (β_2<0), irrespective of magnitude of deformation, the optimum angle (highest spin-orbit barrier and lower interaction radius) is obtained at (0^0, 180^0). The effect of orientation on V_J is much prominent for nuclei with β_2<0 as compare to β_2> 0. Further the effect of spin-orbit term is explored on fusion cross-sections (σ_{fusion}) obtained by using Wong formula [12].

The purpose of this paper is two fold (i) to study the role of orientation on spin-orbit density part of nuclear potential. (ii) to see the effect of spin-orbit density interaction potential on fusion cross-section. In present study the role of quadrupole deformation is also studied via various prolate and oblate shape nuclei.

*e-mail: rajni.mittal1989@gmail.com

This work is organized as follows: Section 2 contains the methodology, which includes Skyrme Energy Density Formalism (SEDF) and Wong formula. Calculations and results are discussed in Section 3 and finally the outcome is summarized in Section 4.

2 Methodology

2.1 Semi-Classical Skyrme Energy Density Formalism (SEDF)

The SEDF in semiclassical Extended Thomas Fermi (ETF) method [6] provides a convenient way for calculating the interaction potential between two nuclei. In the Hamiltonian density, the kinetic energy density τ as well as the spin-orbit density $\vec{J}$ are functions of the nucleon density ρ_q, q = n, p. The nucleus-nucleus interaction potential in SEDF, based on semiclassical ETF model, is

$$V_N(R) = E(R)-E(\infty) = \int H(\vec{r})d\vec{r} - \int H_1(\vec{r})d\vec{r} - \int H_2(\vec{r})d\vec{r} \quad (1)$$

Above equation which is nuclear interaction potential calculated by using slab approximation method as discussed later in reference to Eq. (3).

The Skyrme Hamiltonian density in Eq. (1) reads as,

$$\begin{aligned}
H(\rho,\tau,\vec{J}) = & \frac{\hbar^2}{2m}\tau \\
& + \frac{1}{2}t_0\left[(1+\frac{1}{2}x_0)\rho^2 - (x_0+\frac{1}{2})(\rho_n^2+\rho_p^2)\right] \\
& + \frac{1}{12}t_3\rho^\alpha\left[(1+\frac{1}{2}x_3)\rho^2 - (x_3+\frac{1}{2})(\rho_n^2+\rho_p^2)\right] \\
& + \frac{1}{4}\left[t_1(1+\frac{1}{2}x_1)+t_2(1+\frac{1}{2}x_2)\right]\rho\tau \\
& - \frac{1}{4}\left[t_1(x_1+\frac{1}{2})-t_2(x_2+\frac{1}{2})\right](\rho_n\tau_n+\rho_p\tau_p) \\
& + \frac{1}{16}\left[3t_1(1+\frac{1}{2}x_1)-t_2(1+\frac{1}{2}x_2)\right](\vec{\nabla}\rho)^2 \\
& - \frac{1}{16}\left[3t_1(x_1+\frac{1}{2})+t_2(x_2+\frac{1}{2})\right] \\
& \times\left[(\vec{\nabla}\rho_n)^2+(\vec{\nabla}\rho_p)^2\right] \\
& - \frac{1}{2}W_0\left[\rho\vec{\nabla}\cdot\vec{J}+\rho_n\vec{\nabla}\cdot\vec{J_n}+\rho_p\vec{\nabla}\cdot\vec{J_p}\right] \\
& - \frac{1}{16}\left[(t_1x_1+t_2x_2)\vec{J}^2+(t_1-t_2)(J_p^2+J_n^2)\right]
\end{aligned} \quad (2)$$

Here, $\rho = \rho_n + \rho_p$, $\tau = \tau_n + \tau_p$, and $\vec{J} = \vec{J_n} + \vec{J_p}$ are the nuclear, kinetic energy and spin-orbit densities, respectively. m is the nucleon mass, and x_j, t_j (j=0,1,2,3), α and W_0 are the Skyrme force parameters. In Eq. (2), the last two terms of Hamiltonian corresponds to spin-orbit effect. Author of [10] has introduced new Skyrme parameter set known as SKRA force, whose parameters are t_0 = -2895.4 MeV fm^3, t_1 = 405.5 MeV fm^5, t_2 = -89.1 MeV fm^5, t_3 = 16660 MeV fm$^{3+3\alpha}$, x_0 = 0.08, x_1 = 0, x_2 = 0.2, x_3 = 0, α = 0.1422, W_0 = 129 MeV fm^5. This force is found to account better for nuclear interactions and satisfactory describes finite nuclei at low and medium energy. Since our present study involves the nucleon interaction at low energy, such force is expected to give reasonably good results. For the proximity potential, Gupta *et al.* [13] introduce the slab approximation of semi-infinite nuclear matter with surfaces parallel to xy plane, moving in z-direction, and separated by distance s having minimum value s_0. Then, the interaction potential $V_N(R)$ between two nuclei separated by R = R_1 + R_2 + s is given as (for more detail see Ref. [6, 13, 14]),

$$\begin{aligned}
V_N(R) &= 2\pi\bar{R}\int_{s_0}^{\infty} e(s)ds \\
&= 2\pi\bar{R}\int\{H(\rho,\tau,\vec{J}) - [H_1(\rho_1,\tau_1,\vec{J_1}) \\
&\quad +H_2(\rho_2,\tau_2,\vec{J_2})]\}dz \\
&= 2\pi\bar{R}\int\{(H(\rho)-[H_1(\rho_1)+H_2(\rho_2)]) \\
&\quad +(H(\vec{J})-[H_1(\vec{J_1})+H_2(\vec{J_2})])\}dz \\
&= V_P(R)+V_J(R).
\end{aligned} \quad (3)$$

Here, V_P(R) and V_J(R) are spin-orbit density independent and dependent parts of nuclear interaction potential. $\bar{R}$ is the mean curvature radius defining the geometry of the system. Hence nuclear potential derived within the SEDF approach is a sum of spin-orbit dependent (V_J) and independent (V_P) part. Then, adding the Coulomb (V_C) and centrifugal (V_ℓ) interactions to the nuclear interaction potential V_N(R), we get the total interaction potential for deformed and oriented nuclei [15], as

$$\begin{aligned}
V_T(R) &= V_N(R,A_i,\beta_{\lambda i},\theta_i,T) \\
&\quad +V_c(R,Z_i,\beta_{\lambda i},\theta_i,T) \\
&\quad +V_\ell(R,A_i,\beta_{\lambda i},\theta_i,T).
\end{aligned} \quad (4)$$

Here, $\beta_{\lambda i}$, λ = 2, 3, 4 are the static quadrupole, octupole, and hexadecapole deformations and i=1, 2 represents the projectile and target respectively. In the present work, the deformation effects up to quadrupole (β_{2i}) are included.

The deformation effect in Eq. (4) is introduced via the nuclear radii, as follows:

$$R_i(\alpha_i,T) = R_{0i}(T)[1+\sum_\lambda \beta_{\lambda i}Y_\lambda^{(0)}(\alpha_i)] \quad (5)$$

Here, R_{0i} is the half-density nuclear radius at temperature T=0 obtained by fitting the experimental data to respective polynomials in the nuclear mass region A = 4-238 [14, 16]. The T-dependence in the Eq. (5) is introduced as in [17]. In above expression, T is related to the incoming center-of-mass energy $E_{c.m.}$ or the compound nucleus (CN) excitation energy E^*_{CN} through semi-classical statistical relation [18] as

$$E^*_{CN} = \frac{A_{CN}}{8}T^2 - T \quad (6)$$

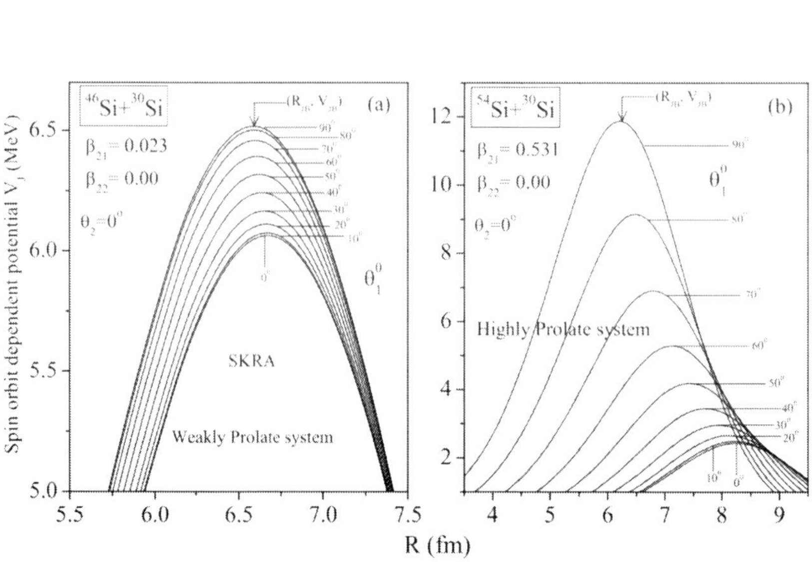

Figure 1. Figure shows the variation of spin-orbit dependent part $V_J(R)$ of interaction potential, plotted for different orientation degree of freedom θ (a) for weakly prolate and (b) for highly prolate deformed nuclei. V_{JB} represent the maximum value of V_J and R_{JB} is corresponding position, different for different orientations.

2.2 Wong formula

According to Wong [12], the fusion cross-section, in terms of angular-momentum ℓ partial waves, for deformed and oriented nuclei (with orientation angles θ_i), lying in same planes, and colliding with center-of-mass energy ($E_{c.m.}$), is

$$\sigma(E_{c.m.},\theta_i) = \frac{\pi}{k^2}\sum_{\ell}(2\ell+1)P_\ell(E_{c.m.},\theta_i), \quad (7)$$

with $k=\sqrt{\frac{2\mu E_{c.m}}{\hbar^2}}$, and μ as the reduced mass. Here, the penetrability P_ℓ is describes by the Hill-Wheeler [19, 20] approximation,

$$P_\ell = \left[1+\exp\left(\frac{2\pi(V_B^\ell(E_{c.m.},\theta_i)-E_{c.m.})}{\hbar\omega_\ell(E_{c.m.},\theta_i)}\right)\right]^{-1}. \quad (8)$$

with $\hbar\omega_\ell(E_{c.m.},\theta_i)$, evaluated at the barrier position $R=R_B^\ell$ corresponding to the maximum barrier height $V_B^\ell(E_{c.m.},\theta_i)$.

Wong has applied some approximations, and replaced the ℓ-summation in Eq.(7) by an integral, giving the fusion cross-section [12] as,

$$\sigma(E_{c.m.},\theta_i) = \frac{R_B^{0\,2}\hbar\omega_0}{2E_{c.m.}}\ln\left[1+\exp\left(\frac{2\pi}{\hbar\omega_0}(E_{c.m}-V_B^0)\right)\right], \quad (9)$$

Here, one may use the interaction between deformed nuclei of radii R_i with deformation parameters $\beta_{\lambda i}$ and making orientation angles θ_i with respect to the collision axis. Hence, the fusion cross-section integrated over the angles θ_i (i=1,2) is obtained as

$$\sigma(E_{c.m.}) = \int_{\theta_1,\theta_2=0}^{\pi/2}\sigma(E_{c.m.},\theta_1,\theta_2)sin\theta_1 d\theta_1 sin\theta_2 d\theta_2. \quad (10)$$

3 Calculation and Result

Within the framework of Skyrme Energy Density Formalism (SEDF), the role of deformation and orientation of colliding nuclei is investigated on spin-orbit density part V_J of nuclear potential (V_N=V_P+V_J) using SKRA force [10]. For the purpose of study, various Si-induced reactions *i.e.* ^{24}Si+^{30}Si, ^{26}Si+^{30}Si, ^{28}Si+^{30}Si, ^{38}Si+^{30}Si ^{40}Si+^{30}Si, ^{42}Si+^{30}Si, ^{44}Si+^{30}Si,^{46}Si+^{30}Si, ^{48}Si+^{30}Si, ^{50}Si+^{30}Si, ^{52}Si+^{30}Si and ^{54}Si+^{30}Si are considered. Above mentioned reactions are so chosen because, the target is spherical and projectiles corresponds to either prolate or oblate shapes. It is to be mentioned here that, for the nuclei which are not discovered experimentally, the relevant details are taken from the theoretical work of Moller and Nix [11].

First of all, we study the effect of orientation on spin-orbit potential by considering weakly and highly prolate systems i.e. ^{46}Si+^{30}Si and ^{54}Si+^{30}Si respectively as shown in left and right panels of figure 1. The variation of V_J is shown for different orientation degree of freedom θ *i.e.* from 0^0 to 90^0. It is observed in figure 1(a) that, for weakly prolate system with β_{21} =0.023 and β_{22} =0.00, V_J is increasing slowly with the change in orientation of projectile from 0^0 to 90^0. At an orientation θ=0^0, the height of spin-orbit barrier is ~ 6.0 MeV, which for orientation angle of θ=90^0, increases to ~ 6.5 MeV. Apparently, the spin-orbit barrier position R_{JB} is also affected by the orientation angles between the projectile and target as the maximum R_{JB} corresponds to θ_1=0^0 and the minimum to θ_1=90^0. On changing the orientation from θ_1 =0^0 to 90^0, the spin-orbit barrier position is shifting ~ 0.1 fm towards lower interaction radius, making system more compact.

 DOI: 10.1051/epjconf/201716300038

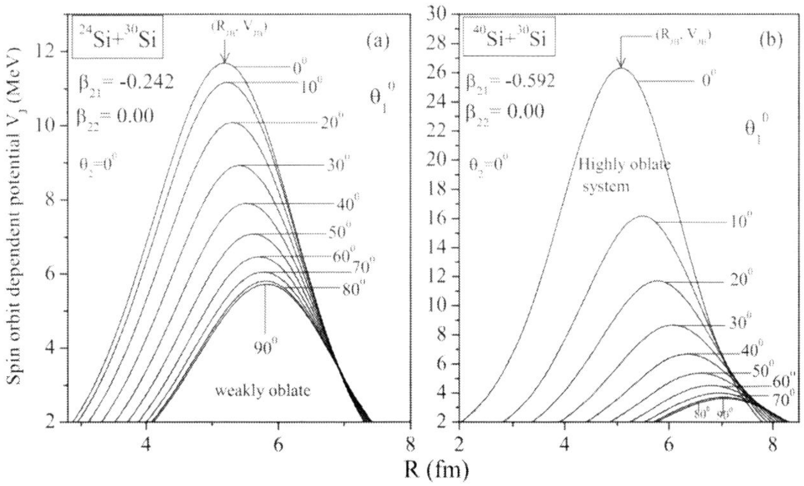

Figure 2. Same as Figure 1 but for oblate shape nuclei.

Table 1. The spin-orbit barrier height V_{JB} and spin-orbit barrier position R_{JB} of various prolate and oblate deformed systems with their optimum orientations [21]. The coulomb barrier V_{CB} is also shown in table below. β_2-deformations are taken from the theoretical work of Moller and Nix [11]. β_{21} and β_{22} represents the deformation of projectile and target respectively.

A_P (Si)	A_T (Si)	θ_1	θ_2	β_{21}	β_{22}	V_{CB} (MeV)	V_{JB} (MeV)	R_{JB} (fm)
Prolate deformed nuclei								
38	30	90	0	0.234	0	26.360	8.185	6.0
46	30	90	0	0.023	0	25.092	6.526	6.6
52	30	90	0	0.490	0	25.254	11.323	6.2
54	30	90	0	0.531	0	25.119	11.875	6.2
Oblate deformed nuclei								
24	30	180	0	-0.242	0	29.320	11.695	5.2
26	30	180	0	-0.353	0	29.266	14.548	5.1
28	30	180	0	-0.478	0	29.351	19.11	4.9
40	30	180	0	-0.592	0	29.975	26.321	5.1
42	30	180	0	-0.321	0	26.554	13.169	5.7
44	30	180	0	-0.263	0	26.110	11.502	5.9
48	30	180	0	-0.398	0	26.232	16.467	5.7
50	30	180	0	-0.422	0	26.154	17.794	5.7

Similar results are obtained for highly prolate system i.e. for $^{54}Si+^{30}Si$ reaction with β_{21} =0.531 and β_{22} =0.00 as depicted in figure 1(b). One may clearly see that, the enhancement in magnitude of V_J with respect to orientation degree of freedom is significantly high for highly prolate as compare to weakly prolate system. At an orientation angle of 90^0, V_J of highly prolate system is ~ 6 MeV higher than weakly prolate system.

Further, the effect of orientation on spin-orbit part is analyzed through weakly and highly oblate shape nuclei as shown in figure 2. It is evident from the figure that, regardless of magnitude of deformation, the decrement in magnitude of spin-orbit barrier height V_{JB} is observed from orientation 0^0 to 90^0 (see figure 2(a) and figure 2(b)). The higher spin-orbit barrier and lowest spin-orbit position is attained at $\theta_1=0^0$, $\theta_2=180^0$. On changing the orientation of projectile from 0^0 to 90^0, the V_J get affected as following: (i) the V_{JB} of weakly oblate system is decreased by a factor of ~ 6.0 MeV and shifted the spin-orbit barrier position ~ 0.6 fm towards higher interaction radius i.e. more elongated. (ii) for highly oblate system, the V_{JB} is decreased by ~ 23.0 MeV and R_{JB} moved ~ 2.0 fm towards higher interaction radius.

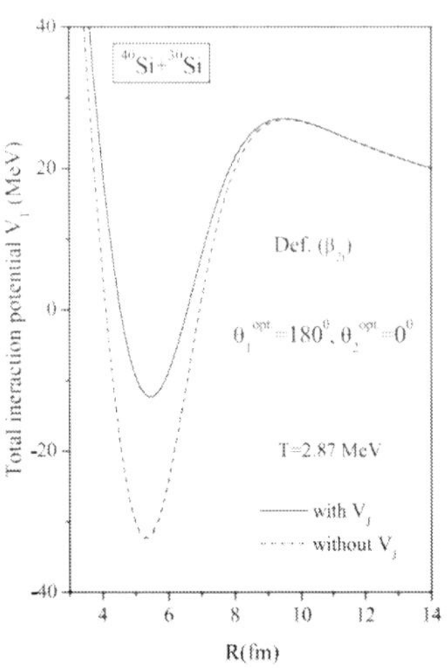

Figure 3. Total interaction potential V_T (MeV) of $^{40}Si+^{30}Si$ reaction as a function of inter nuclear distance R (fm) plotted with and without spin-orbit density effects at optimum orientations $\theta_1^{opt}=180^0$ and $\theta_2^{opt}=0^0$.

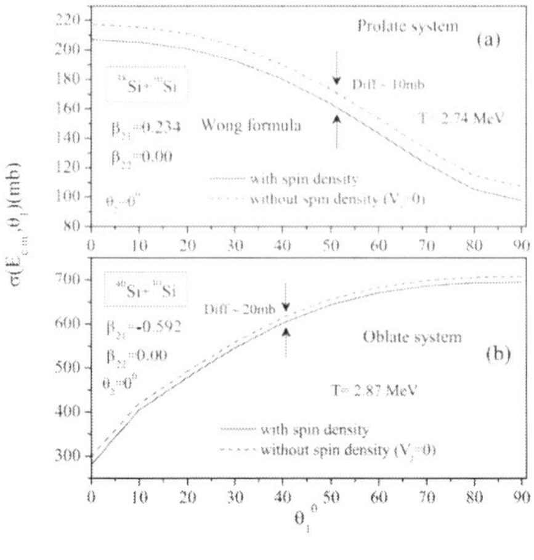

Figure 4. Variation of fusion cross-section σ_{fusion} at each angle with and without spin density effects (a) for prolate and (b) for oblate shape nuclei, plotted at their coulomb barrier energy (see Table 1).

Moreover, on comparing V_{JB} of weakly prolate (figure 1(a)) and weakly oblate system (figure 2(a)) at their respective optimum orientations, we find that, in comparison to weakly prolate system, V_{JB} of weakly oblate system is increased ~ 5.2 MeV. However, V_{JB} of highly oblate system is ~ 14.4 MeV higher than highly prolate system (see figure 1(b) and 2(b)). This analysis is further extended for other Si+Si reactions, where magnitude of deformation is in between $0.023 \leq \beta_2 \leq 0.531$ for prolate and $-0.242 \leq \beta_2 \leq -0.592$ for oblate systems. All prolate/oblate projectiles show similar result as we discussed above (therefore not shown here to avoid repetition). Table 1 summarizes the result for all considered reactions with their spin-orbit barrier height and position at the optimum orientations (θ_1, θ_2).

The present study suggest that, the magnitude of prolate/oblate deformations and corresponding orientation angle between the colliding nuclei, significantly influence the spin-orbit barrier height and position. Moreover, the compactness of the system also get affected within the sign of β_2-deformation.

Now in order to see the role of spin-orbit part in the total interaction potential ($V_T=V_P+V_J+V_C=V_N+V_C$), the spin-orbit independent potential (V_P) and Coulomb potential (V_C) parts are added to it. figure 3 shows the total interaction potential with and without the contribution of spin-orbit effect for $^{40}Si+^{30}Si$ reaction plotted at Coulomb barrier energy i.e. E_{CB}=29.98 MeV (temperature T=2.87 MeV). It is clear that, impact of spin-orbit part is significant on fusion pocket rather than on fusion barrier height. Pocket becomes more shallower with inclusion of V_J. Moreover, the barrier curvature ($\hbar\omega$) also gets reduced with the inclusion of spin-orbit potential. In view of this point, it would be of further interest to study the influence of spin-orbit part on fusion cross-sections. The fusion cross-sections are calculated with and without the inclusion of spin-orbit density part in the total interaction potential and plotted at different orientations as shown in figure 4. Calculations are performed for $^{38}Si+^{30}Si$ and $^{40}Si+^{30}Si$ reactions with respective prolate and oblate shapes. Since no experimental data is available for these reactions, therefore cross-sections are estimated at their respective Coulomb barrier energy (see Table 1). It is observed that, the spin-orbit density part decreases the fusion cross-sections to a small extent within the Wong formula. This fall off in cross-section is about ~ 10 mb for prolate system and ~ 20 mb for oblate system. As expected, for prolate system (as shown in figure 4(a)), independent of spin orbit effect, the fusion cross-sections are decreasing with increase in orientation angle of colliding nuclei with maxima at $\theta_2=0^0$. On the other hand, for oblate nuclei, σ_{fusion} increases with increase in orientation angle of colliding nuclei and become maximum at $\theta_2=90^0$ as shown in figure 4(b). For in between configurations the fusion cross-sections vary smoothly as a function of orientation angle.

4 Summary

A systematic study is carried out to see the influence of deformation and orientation on spin-orbit density part V_J by employing various prolate and oblate shape nuclei within the Skyrme Energy Density Formalism. The deformation is included upto β_2 lying within range of $0.023 \leq \beta_2 \leq 0.531$ for prolate and $-0.242 \leq \beta_2 \leq -0.592$ for oblate nuclei. We observed that, independent of magnitude of prolate deformed nuclei, higher spin-orbit barrier and lowest spin-orbit barrier position corresponds to orientation 90^0, which for oblate shape nuclei changes to orientation angle 0^0. The spin-orbit barrier and position is significantly influenced by both the sign and magnitude of deformation. The oblate shape nuclei (β_2<0) exhibit higher spin-orbit barrier (and lowest barrier position) in comparison to prolate systems (β_2>0). Also, the pocket of total interaction potential become more shallower with the inclusion of spin-orbit effect. In addition to this, the angular dependence of spin-orbit part is studied on fusion cross-section by taking both prolate and oblate systems. It is observed that, fusion cross-sections are maximum at θ_2=0^0 for prolate and θ_2=90^0 for oblate nuclei and their magnitude reduces slightly after inclusion of spin-density effect.

5 Acknowledgment

The financial support from the department of science and technology (DST), is gratefully acknowledged.

References

[1] A. Iwamoto and P. Moller, Nucl. Phys. A **605**, 334 (1996).

[2] D. Jain, R. Kumar and M. K. Sharma, Nucl. Phys. A **915**, 106-124 (2013).

[3] Rajni, R. Kumar and M. K. Sharma, Phys. Rev. C **90**, 044604 (2014).

[4] D. Jain, M. K. Sharma, Rajni, R. Kumar and R. K. Gupta, Eur. Phys. J. A **50**, 155 (2014).

[5] D. Vautherin and D. M. Brink, Phys. Rev. C **5**, 626 (1972).

[6] R. K. Gupta, D. Singh, and W. Greiner, Phys. Rev C **75**, 024603 (2007).

[7] J. Bartel and K. Bencheikh, Eur. Phys. J. A **14**, 179 (2002).

[8] C. Ngo *et al.*, Nucl. Phys. A **240**, 353 (1975); C. Ngo *et al.*, Nucl. Phys. A 252, 237 (1975).

[9] Fl. Stancu and D. M. Brink, Nucl. Phys. A **270**, 236 (1976).

[10] M. Rashdan, Mod. Phys. Lett. **15**, 1287 (2000).

[11] P. Moller and J. R. Nix, Nucl. Phys. A **361**, 117 (1981).

[12] C. Y. Wong, Phys. Rev. Lett. **31**, 766 (1973).

[13] P. Chattopadhyay and R. K. Gupta Phys. Rev. C **30**, 1191 (1984).

[14] R. K. Gupta, D. Singh, R. Kumar, and W. Greiner, J. Phys. G: Nucl. Part. Phys. **36**, 075104 (2009).

[15] R. K. Gupta, M. Balasubramaniam, R. Kumar, N. Singh, M. Manhas, and W. Greiner, J. Phys. G: Nucl. Part. Phys. **31**, 631 (2005).

[16] L. R. B. Elton, Nuclear Sizes (Oxford University Press, London, 1961); H. de Vries, C. W. de Jager, and C. de Vries, At. Data Nucl. Data Tables 36, 495 (1987).

[17] S. Shlomo and J. B. Natowitz, Phys. Rev. C **44**, 287 (1991).

[18] K. J. LeCouteur and D. W. Lang, Nucl. Phys. **13**, 32 (1959).

[19] D. L. Hill and J. A. Wheeler, Phys. Rev. **89**, 1102 (1953).

[20] T. D. Thomas, Phys. Rev. **116**, 703 (1959).

[21] R. K. Gupta, M. Balasubramanium, R. Kumar, N. Singh, M. Manhas, and W. Greiner, J. Phys. G: Nucl. Part. Phys. **31**, 631 (2005).

Study of Quasielastic scattering for ^{7}Li+^{159}Tb at around- barrier energies

A. Mukherjee[1,3,⋆], *Piyasi* Biswas[1,3], *Md. Moin* Shaikh[1], *Subinit* Roy[1,3], *A.* Goswami[1,3], *M.K.* Pradhan[1], *P.* Basu[1], *S.* Santra[2,3], *S.K.* Pandit[2,3], *K.* Mahata[2,3], and *A.* Shrivastava[2,3]

[1] *Saha Institute of Nuclear Physics, 1/AF Bidhannagar, Kolkata-700064, India*
[2] *Nuclear Physics Division, Bhabha Atomic Research Centre, Mumbai-400085, India*
[3] *Homi Bhabha National Institute, Anushaktinagar, Mumbai-400094, India*

Abstract. Quasielastic scattering cross sections for the reaction ^{7}Li+^{159}Tb have been measured at large back-angles, at energies around the Coulomb barrier. The quasielastic barrier distribution has been extracted from the measured quasielastic scattering excitation function, including and excluding α particle contribution. The peak of the quasielastic barrier distribution including α particle contribution shows a shift towards higher energy compared to the peak of the distribution without α particles. The quasielastic barrier distribution when compared to the calculated fusion barrier distribution, appears to show reasonable agreement for the system.

1 Introduction

Near barrier heavy ion fusion is strongly influenced by the internal structure of the colliding nuclei and coupling to the direct nuclear processes, like inelastic excitation and direct nucleon transfer. The coupling between the relative motion and intrinsic degrees of freedom of the interacting nuclei leads to an enhancement of fusion cross sections at energies below the average fusion barrier, relative to the cross sections obtained from the one-dimensional (1D) barrier penetration model calculations. The coupling essentially modifies the effective interaction potential and in turn splits the single, uncoupled fusion barrier into multiple barriers, thereby giving rise to a distribution of barriers.

It has been found that the nature of the couplings affecting the fusion process is not always apparent from the measured fusion cross sections. Rowley *et al.* proposed [1] a very sensitive approach for investigating the role of different couplings on the fusion cross sections at energies around the Coulomb barrier. The method consists in deriving the fusion barrier distribution (D_{fus}) from very precisely measured fusion cross sections, according to the prescription,

$$D_{fus}(E) = \frac{d^2}{dE^2}\left[E\sigma_{fus}(E)\right] \quad (1)$$

Over the past several years of research in heavy ion fusion, D_{fus} has been found to be a powerful tool to understand the effect of coupling of other channels on sub-barrier fusion and hence probe the reaction dynamics of nucleus-nucleus collision [2]. Since extraction of D_{fus} involves second derivative of $E\sigma_{fus}$, obtaining a meaningful barrier distribution requires very precisely measured fusion data.

⋆e-mail: anjali.mukherjee@saha.ac.in

Alternatively barrier distribution can also be obtained from large back-angle quasielastic (QEL) scattering excitation function [3]. QEL scattering is defined as the sum of elastic and inelastic scattering, transfer and other direct processes. Fusion is related to transmission through the barrier, whereas large back-angle QEL scattering is related to reflection at the barrier. Because of the conservation of reaction flux, these two processes may be considered as complementary to each other, and one may obtain information concerning one of them by investigating the other [3]. The barrier distribution of QEL scattering (D_{qel}) is obtained as [3],

$$D_{qel}(E) = -\frac{d}{dE}\left[\frac{d\sigma_{qel}}{d\sigma_{Ruth}}(E)\right] \quad (2)$$

As D_{qel} is derived from the first derivative, the uncertainty associated with D_{qel} extracted from back-angle QEL data is less than that associated with D_{fus}, which is extracted from the second derivative of fusion data. In view of the experimental benefits of D_{qel} over D_{fus}, the function D_{qel} may be considered to be a suitable alternative to D_{fus}.

Barrier distributions derived from the above two complementary approaches have been found to be similar for the systems where both the reactants are strongly bound [2, 4, 5]. In the past two decades, the influence of breakup of weakly bound projectiles, like ^{6,7}Li and ^{9}Be on fusion cross sections has been widely studied [6–9]. Some of these studies have been done with an aim to investigate the fusion barrier distributions for reactions with weakly bound projectiles on heavy targets [6]. Alongside, studies have also been carried out [10–16] to investigate the fusion barrier distributions derived from large back-angle QEL scattering cross sections for systems involving weakly bound nuclei. For such systems, the breakup channel is expected to play a significant role. For weakly bound systems, QEL scattering, including no-capture breakup (NCBU) and incomplete fusion (ICF) along with elastic, inelastic and transfer processes, may be considered to be complementary to the complete fusion (CF) of the projectile with the target. For such systems, a comparative investigation of barrier distributions extracted from fusion and back-angle QEL excitation functions appears to be very interesting.

Recently, large back-angle QEL scattering cross sections have been measured for the ^{7}Li + ^{159}Tb system, at near-barrier energies. A preliminary analysis of the data have been done and QEL barrier distribution has been extracted from the QEL excitation function. Results of the preliminary analysis will be presented here.

2 Experiment and Analysis

Beams of ^{7}Li in the energy range 17–34 MeV, in steps of 1 MeV, delivered by the 14UD BARC-TIFR Pelletron Accelerator, Mumbai, India were used to impinge a self-supporting ^{159}Tb target foil of thickness $\sim 1.1\ \mu g/cm^2$. To detect and identify the charged particles produced in the reaction a set of four ΔE-E telescopes of Si-surface barrier detectors were placed at $\pm 170^o$ and $\pm 160^o$, relative to the beam direction, inside a scattering chamber of diameter 1m. The thicknesses of the detectors were so chosen that the charged particles lose part of their kinetic energies in the first detector (ΔE) and stop in the second detector (E). However, the stop detectors used in the experiment were not thick enough to stop the Z=1 particles. Two Si-surface barrier detectors were placed at $\pm 20^o$ with respect to the beam direction for monitoring the beam and also for normalization purposes. In front of each telescope and monitor, a collimator was placed to define the solid angle. The energies of the incident beam were corrected for the loss of energy in the target material at half thickness of the target.

Figure 1(a) shows a typical two-dimensional ΔE-E_{tot} (where, E_{tot}=ΔE+E_{res}) spectrum obtained from the detector telescope taken at a scattering angle of $+170^o$. The events corresponding to (elastic + inelastic) scattering and various transfer or breakup products at E_{lab}=27 MeV are marked in the figure. The peak in the Z=3 band arises from contributions due to elastic scattering of ^{7}Li and inelastic scattering from the higher lying states of ^{159}Tb. The low lying levels of ^{159}Tb are very closely spaced and so the inelastic states of target could not be separated from the elastic events. The band may also contain contribution from the first inelastic state of ^{7}Li [14], if any. Besides, it may also contain the contribution from ^{6}Li$_{g.s.}$, produced via n-stripping of ^{7}Li [14, 17]. It needs to be mentioned here that the events corresponding to ^6Li could not be separated from those of ^{7}Li in the spectra. However, the Z=3 band predominantly consists of elastic and inelastic events. So for nomenclature purpose, here we refer the Z=3 band as (elastic + inelastic) band. The Z=1 band shows a fall back feature because the stop detectors were not thick enough to stop the Z=1 particles. So the events corresponding to Z=1 could not be used in the analysis. The x-axis of the two-dimensional ΔE-E_{tot} spectrum was energy calibrated using the elastic peaks of ^{7}Li beam scattered from ^{159}Tb target at different bombarding energies below the Coulomb barrier.

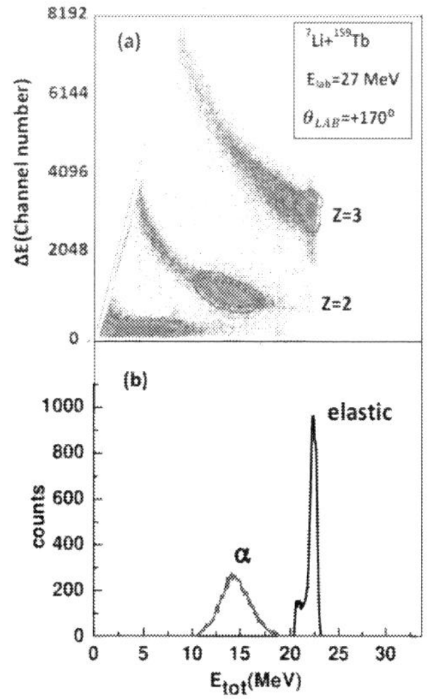

Figure 1. a) Typical ΔE-E_{tot} spectrum for the ^{7}Li+^{159}Tb reaction at E_{lab}=27 MeV and ΔE-E telescope angle of 170°. b) A portion of the Z=3 and Z=2 projections on the ΔE-E_{tot} axis.

The x-axis projection of Z=3 (elastic + inelastic) band and Z=2 (α-particle) band, is shown in Fig. 1(b). The projection of Z=2 band shows a broad continuous peak with the centroid at around 14.5 MeV which is nearly equal to 4/7 times the incident beam energy. The contribution of the α-particles, emitted mostly at energies corresponding to the beam velocity primarily originate from breakup related processes, like NCBU (^{7}Li→ $\alpha + t$) and ICF (where t is captured by ^{159}Tb, following the breakup of ^{7}Li into $\alpha + t$). The broad α-peak can also have contributions from single nucleon transfer channels resulting in α-particles.

The ratio of quasielastic to Rutherford cross-sections is given by the expression,

$$\frac{d\sigma_{qel}}{d\sigma_{Ruth}}(E,\theta_{tel}) = \left[\frac{N_{qel}(E,\theta_{tel})}{N_m(E,\theta_m)}\right]\left[\frac{(d\sigma_{Ruth}/d\Omega)(E,\theta_m)}{(d\sigma_{Ruth}/d\Omega)(E,\theta_{tel})}\right] X \left(\frac{\Delta\Omega_m}{\Delta\Omega_{tel}}\right) \quad (3)$$

where, $N_{qel}(N_m)$ is the average yield in telescope (monitor) detector, $\frac{d\sigma_{Ruth}}{d\Omega}(E,\theta_m(\theta_{tel}))$ is the calculated Rutherford scattering cross-section at the corresponding bombarding energy E, at monitor angle θ_m (telescope angle θ_{tel}), and $(\frac{\Delta\Omega_m}{\Delta\Omega_{tel}})$ is the solid angle ratio of monitor to telescope

DOI: 10.1051/epjconf/201716300039

detector.

The $\frac{\Delta\Omega_m}{\Delta\Omega_{tel}}$ ratio was determined from the measurements at the lowest bombarding energies of 17, 18 and 19 MeV, where the elastic scattering is purely Rutherford. The ratio was estimated to be 0.0050 ± 0.00004. The Rutherford scattering cross section in the centre of mass frame is calculated using the formula,

$$\frac{d\sigma_{Ruth}}{d\Omega} = \frac{1}{16}\left(\frac{(z_a z_A e^2)^2}{(4\pi\epsilon_0)^2}\right)\frac{1}{E_{cm}^2 sin^4(\theta/2)} \quad (4)$$

where a and A refer to the projectile and target respectively.

At first, the "partial" QEL counts $N_{qel}(E, \theta_{tel})$ at each bombarding energy was obtained by selecting the region, marked in Fig. 1(a) of the Z=3 (elastic + inelastic) band. The area under the peak, marked "elastic" in Fig. 1(b), gives the value of $N_{qel}(E, \theta_{tel})$. As the measurements were done at angles less than 180°, centrifugal correction was incorporated to obtain the effective c.m. energies. Figure 2(a) shows the "partial" QEL (elastic + inelastic) excitation function and the corresponding barrier distribution is shown in Fig. 2(b). Most of the error bars lie within the size of the data points.

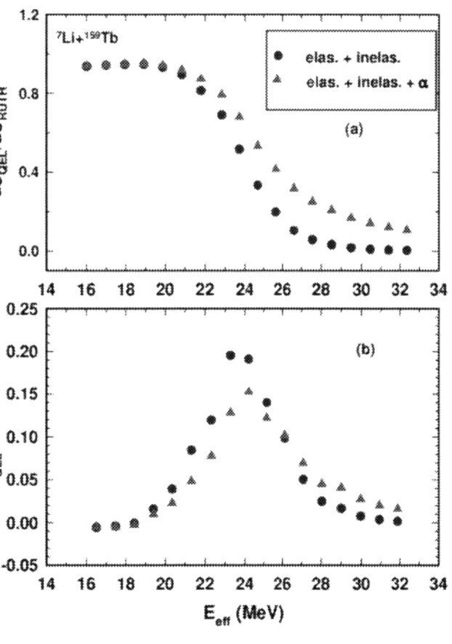

Figure 2. Comparison of a) quasielastic excitation function and b) quasielastic barrier distribution for ^{7}Li+^{159}Tb, excluding (•) and including (△) α particles.

Total QEL scattering cross sections is defined by adding the contribution of the α-particles to the above "partial" QEL scattering cross sections. For this, the α-particles shown in the marked region in Fig. 1(a) for the Z=2 band was considered. The area under the peak, marked "α" in Fig. 1(b), gives the contribution of the α particles. This when added to the Z=3 partial QEL scattered events gives the value of $N_{qel}(E, \theta_{tel})$, for determining the total QEL scattering cross sections. The experimental results of the total QEL scattering excitation function, including elastic, inelastic and α outgoing channels is shown in Fig. 2(a) and the corresponding QEL barrier distribution (D_{qel}) derived from the total QEL data is shown in Fig. 2(b). From Fig. 2(b) it can be observed that there is a shift in the peak of the barrier distribution when contribution of the α-particles is included in the QEL excitation function.

3 Discussion

It needs to be noted that this is an inclusive measurement and so the α-particle band will consist of α-particles produced in all possible ways in this reaction. In the ^{7}Li+^{159}Tb system, the compound nucleus evaporation α is expected to be negligible [7]. The processes that might contribute to the α- particle cross-sections for the ^{7}Li+^{159}Tb reaction are:

(1) NCBU: Breakup of ^{7}Li (B.U threshold for $\alpha + t$ is 2.46 MeV) into $\alpha + t$, either direct or sequential or both, where both the fragments escape without any being captured by the target

(2) Triton-ICF: t captured by the target following break up of ^{7}Li into $\alpha + t$ or a one step t-transfer to the target

(3) Neutron stripping: Single n-stripping from ^{7}Li will produce ^{6}Li which may break into $\alpha + d$ if excited above its breakup threshold 1.47 MeV

(4) Deuteron stripping: d-stripping from ^{7}Li will produce unbound ^{5}He that decays to α and a neutron

(5) Proton stripping: p-stripping from ^{7}Li will produce unbound ^{6}He which then decays to α and two neutrons

(6) Proton pickup: p-pickup by ^{7}Li will lead to ^{8}Be which immediately decays to two α-particles. Since this is an inclusive measurement, each ^{8}Be will contribute two α-particles to the total α-yield. But, the contribution of α-particles from the p-pickup may be expected to be very small compared to the total contribution of α-particles from other processes, like ICF (t+^{159}Tb) [17]. Hence, the extra α particle contribution arising from the double counting of the α-particles for the p-pickup channel may be neglected in comparison to the total α particle contribution for the reaction.

A comparison of the D_{qel}, including and excluding the α particle contribution, with the D_{fus} may shed some light on the importance of the α particle contribution in defining the QEL scattering events.

To compare D_{qel} with D_{fus}, an attempt was made to extract the D_{fus} from the measured fusion excitation function for ^{7}Li+^{159}Tb [7]. Unfortunately, D_{fus} could not be extracted from the reported fusion cross sections [7, 18], because only a few data points were available for differentiation. Therefore, a rough comparison of the experimental D_{qel} was made with the theoretical D_{fus}, extracted from the fusion cross sections calculated using the coupled channels code, CCFULL [19]. All parameters were taken from Ref. [7]. In addition to the coupling scheme used in Ref.[7], here coupling to the first excited state of the projectile ^{7}Li, having spin $1/2^-$ (E_{ex}=0.477 MeV) with B(E2↑)=21.8 $e^2 fm^4$ was also included using the rotational scheme [15], though it did not produce any appreciable change to the excitation function. It needs to be noted here that the CF cross sections for ^{7}Li+^{159}Tb at above barrier energies were seen to be suppressed by a factor of 0.74 [7]. The D_{fus} function in the present work was therefore extracted after multiplying the CC calculated CF cross sections by the factor of 0.74 [15]. To compare D_{qel} with D_{fus}, the D_{fus} values were normalized by $1/\pi R_b^2$, where the barrier radius R_b was taken from Ref. [7]. Figure 3 shows a comparison of the experimental D_{qel} with the theoretical D_{fus} (normalized). The dotted and the solid lines represent the D_{fus} obtained from the CC calculated fusion cross sections without and with coupling, respectively. The solid line can be seen to agree fairly well with the experimental D_{qel}, shown by the symbol △.

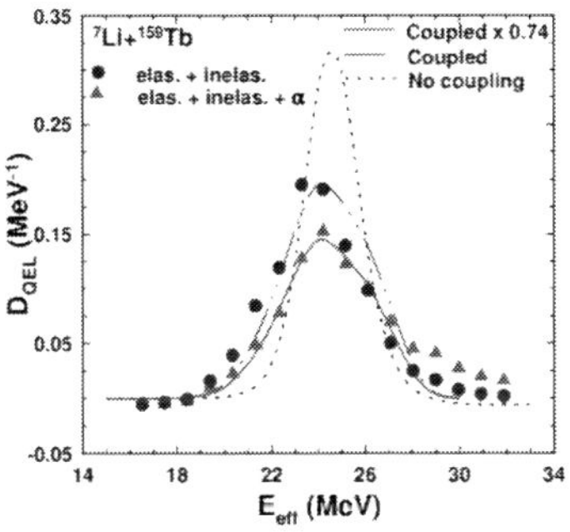

Figure 3. Comparison of the barrier distribution obtained from the quasielastic excitation function with and without inclusion of α particles for the system ^{7}Li+^{159}Tb. The dotted and the solid line represent the D_{fus} extracted from the CC calculated CF cross section without and with coupling, respectively. The calculated D_{fus} values have been normalized by the factor $1/(\pi R_b^2)$ to compare with D_{qel}. See text for details.

Thus, we see that the experimental D_{qel} including contribution of α particles agrees fairly well with the calculated D_{fus}. However, before reaching any definite conclusion, detailed analysis of the data need to be done and also the experimental D_{fus} for the system ^{7}Li+^{159}Tb has to be obtained.

4 Summary

The QEL excitation function for the system ^{7}Li+^{159}Tb has been measured at energies around the Coulomb barrier. The results of a preliminary analysis of the data have been presented at the conference. The D_{qel} functions have been extracted from the elastic + inelastic cross sections and also from the elastic + inelastic + α cross sections. Inclusion of α-particles shifts the D_{qel} peak towards the higher energy side. A rough comparison between the experimental D_{qel} and the calculated D_{fus} shows that the D_{qel} including the α-particle contribution agrees fairly well with the calculated D_{fus}. Detailed analysis of the data and measurement of the D_{fus} for the system ^{7}Li+^{159}Tb are necessary for a definite conclusion.

5 Acknowledgements

We sincerely thank the pelletron staff of BARC-TIFR, Mumbai for providing the beam. We would also like to thank Mr. Sujib Chandra Chatterjee of Saha Institute of Nuclear Physics for technical support during the experiment. One of the authors, Piyasi Biswas would like to thank the Council of Scientific & Industrial Research, India, for their financial support.

References

[1] N. Rowley, G.R. Satchler and P.H. Stelson Phys.A Lett. **B254**, 25 (1991)
[2] M. Dasgupta *et al.*, Annu. rev. Nucl. Part. Sci. **48**, 401 (1998) and references therein
[3] H. Timmers *et al.*, Nucl. Phys. **A584**, 190 (1995)
[4] H. Timmers *et al.*, Phys. Lett. **B399**, 35 (1997)
[5] H. Timmers *et al.*, Nucl. Phys. **A633**, 421 (1998)
[6] L.F. Canto *et al.*, Phys. Rep. **424**, 1 (2006), and references therein
[7] A. Mukherjee *et al.*, Phys. Lett. **B636**, 91 (2006), and
[8] M.K. Pradhan *et al.*, Phys. Rev. **C83**, 064606 (2011), and references therein
[9] Md. Moin Shaikh *et al.*, Phys. Rev. **C93**, 044616 (2016)
[10] C.J. Lin *et al.*, Nucl. Phys. **A787**, 281c (2007)
[11] J. Lubian *et al.*, Phys. Rev. **C78**, 064615 (2008)
[12] S. Mukherjee *et al.*, Phys. Rev. **C80**, 014607 (2009)
[13] D.S. Monteiro *et al.*, Phys. Rev. **C79**, 014601 (2009)
[14] M. Zadro *et al.*, Phys. Rev. **C87**, 054606 (2013), and references therein.
[15] C.S. Palshetkar *et al.*, Phys. Rev. **C89**, 024607 (2014)
[16] Md. Moin Shaikh *et al.*, Phys. Rev. **C91**, 034615 (2015)
[17] S.K. Pandit *et al.*, Phys. Rev. **C93**, 061602 (2016)
[18] R. Broda *et al.*, Nucl. Phys. **A248**, 356 (1975)
[19] K. Hagino, N. Rowley, and A.T. Kruppa, Comput. Phys. Commun. **123**, 143 (1999)

Cluster structure and Coulomb shift in two-center mirror systems

M. Nakao[1], H. Umehara[1], S. Sonoda[1], S. Ebata[2], and M. Ito[1,*]

[1]Department of Pure and Applied Physics, Kansai University, 3-3-35 Yamatecho, Suita, Osaka 564-8680, Japan.
[2]Nuclear Reaction Data Center, Faculty of Science, Hokkaido University, Sapporo 060-0810, Japan.

Abstract. The α + ^{14}C elastic scattering and the nuclear structure of its compound systems, ^{18}O = α + ^{14}C, are analyzed on the basis of the semi-microscopic model. The α + ^{14}C interaction potential is constructed from the double folding (DF) model with the effective nucleon-nucleon interaction of the density-dependent Michigan 3-range Yukawa. The DF potential is applied to the α + ^{14}C elastic scattering in the energy range of E_α/A_α = 5.5 ~ 8.8 MeV, and the observed differential cross sections are reasonably reproduced. The energy spectra of ^{18}O are calculated by employing the orthogonality condition model (OCM) plus the absorbing boundary condition (ABC). The OCM + ABC calculation predicts the formation of the 0^+ resonance around E = 3MeV with respect to the α threshold, which seems to correspond to the 0_4^+ resonance identified in the recent experiment. We also apply the OCM + ABC calculation to the mirror system, such as ^{18}Ne = α + ^{14}O, and the Coulomb shift of ^{18}O – ^{18}Ne is evaluated. We have found that the Coulomb shift is clearly reduced in the excited 0^+ state due to the development of the α cluster structure. This result strongly supports that the Coulomb shift is a candidate of new probe to identify the clustering phenomena.

1 Introduction

α particle is a building block in constructing the intrinsic structures of atomic nuclei because of its stable and inert structure. The so-called α cluster structure appears in the excited states of light mass system, and the structures have been extensively discussed in the $N = Z$ system along the stability line [1]. For example, the structure of ^{8}Be = 2α, ^{12}C = 3α, ^{16}O = α + ^{12}C and ^{20}Ne = α + ^{16}O are the typical examples of the α cluster structure. The studies of the α cluster structures in the $N = Z$ system is still developed. For example, in recent studies, the multi-α structures, such as ^{12}C = 3α [2], ^{16}O = 4α [2, 3] and ^{20}Ne = 5α [2], are extensively discussed, and they are interpreted in terms of the "nuclear α-particle condensation state" in a bosonic mean field [4].

The research on the α cluster structures are extended to nuclei which are not on the non-stability line with $N \neq Z$. In the $N \neq Z$ systems, the cluster degrees of freedom are also important, and fruitful structures are widely discussed. In particular, the α cluster structures are extensively investigated in neutron excess ($N > Z$) systems [5–7]. In neutron excess systems, the valence neutrons play a glue-like role among the cluster cores, which is associated with covalent electrons in molecular systems, and the various chemical-bonding structures are generated. We can see the typical example of such the chemical-bonding structure in Be isotopes, in which excess neutrons form the covalent orbit around the two center cores of ^{8}Be = α + α. In ^{10,12}Be, for example, the prominent structural changes among the chemical bonding structure, such as the covalent, atomic and ionic configurations, occurs in the unbound continuum region, and they are strongly excited by the α + ^{6,8}He resonant scattering [5].

An interesting viewpoint for the cluster structures in the $N \neq Z$ systems is the Coulomb shift for the mirror systems, which are obtained by replacing neutrons to protons. Generally speaking, the energy levels in the proton-rich side ($N < Z$) shift to the higher energy region than the levels in the neutron-rich side ($N > Z$) because of the difference in the Coulomb interaction. One of the famous phenomena relevant to the Coulomb shift is the so-called "Thomas-Ehrman Shift (TES)", which was discussed in the mirror system of ^{17}O = ^{16}O + n and ^{17}F = ^{16}O + p [8]. If the valence nucleon occupies the S-wave orbit, the wave function is spatially extended in comparison to the distribution of the non S-wave orbit because of lacking the centrifugal barrier in the S-wave single particle motion. Thus, the effect of the Coulomb interaction is more suppressed for the S-wave state than that for the non S-wave state. As a result of this suppressed Coulomb effect, the excitation energy of the $1/2^+$ state in the proton-rich ^{17}F seems to be lower than that in the neutron-rich ^{17}O [8].

TES is a kind of anomalous phenomena, which is induced by the S-wave configuration in the nucleon orbit. Such the S-wave configuration can also be seen in the cluster state, which often appears in the excited 0^+ state of the even-nuclear systems. In the cluster structure, generally speaking, the relative wave function of the α particle and the residual nuclei are weakly coupled, and they mainly perform the S-wave relative motion. Thus, the Coulomb interaction is not so effective in the case that the cluster

* e-mail: itomk@kansai-u.ac.jp

structures are strongly developed. In this situation, the reduction of the Coulomb shift, which is similar to the phenomenon discussed in ^{17}O – ^{17}F, is expected to be observed. Therefore, the reduced Coulomb shift is a new candidate to identify the cluster degrees of freedom in the excited 0^+ state. In fact, the reduction of the Coulomb shift due to the cluster formation is demonstrated in the mirror system of ^{10}Be – ^{10}C [9].

In the present report, we extend the analysis of the Coulomb shift studied in ^{10}Be – ^{10}C to the heavier systems, such as ^{18}O – ^{18}Ne, which are handled by the cluster models of α + ^{14}C and α + ^{14}O, respectively. In ^{18}O, the α + ^{14}C structure has already been discussed by the full microscopic model, in which the anti-symmetrization among all the nucleons are completely taken into account [10, 11]. However, the energy levels of the excited 0^+ states, which have the well developed α + ^{14}C structure, is much higher than the observed 0^+ levels and hence, the direct comparison of the theory with the experiment is still difficult. This overestimation of the excitation energies is a peculiar problem in the full microscopic models, which are difficult to reproduce the α threshold and the energy levels simultaneously.

In order to overcome the threshold problem, it is important to introduce the semi-microscopic model, which is free from the reproduction of the threshold energies theoretically. The orthogonality condition model (OCM) is very useful and powerful model as such the semi-microscopic approach [12]. In the OCM calculation, the functional space allowed by the Pauli's exclusion principle is constructed by solving the eigenvalue problem of the norm kernels, which are the overlap functions of the cluster wave function with the full anti-symmetrization. After constructing the Pauli allowed states, we can apply the simple potential model in solving the cluster relative motion.

In the present report, we apply the OCM to the mirror cluster-systems of ^{18}O = α + ^{14}C and ^{18}Ne = α + ^{14}O and discuss the Coulomb shift in connection to the cluster formation. The cluster interaction is constructed by the double folding (DF) model, and the validity of the DF potential is checked by the analysis of the α + ^{14}C elastic scattering. The DF potential, which is used for the scattering calculation, is applied to the OCM calculation with the α + ^{14}C and α + ^{14}O structures. The cluster states, which appear as the resonant state above the α threshold, are identified by imposing the absorbing boundary condition (ABC) [21]. In the calculation of OCM + ABC, we formulate a new computational technique to handle the norm kernel on the basis of the simple numerical calculation because the evaluation of the norm kernel in OCM is complicated for general cluster systems of $N \neq Z$.

The organization of this article is the following. In Sec. 2, the construction of the DF potential, and the new OCM framework plus the ABC method are briefly explained. The results of the numerical calculations are shown in Sec. 3. First, the application of the DF potential to the α + ^{14}C elastic scattering is presented. Secondly, the energy calculation for ^{18}O = α + ^{14}C by using OCM + ABC is performed. Finally, the Coulomb shift of ^{18}O and ^{18}Ne is discussed in connection to the development of the α clusters. The final section is devoted to the summary and discussion.

2 Framework

In our analyses, there are two steps in the calculations; first, we solve the scattering problem of the α + ^{14}C system by employing the double folding (DF) potential, and the validity of the DF potential is checked. Secondly, the energy spectra of the respective compound system, ^{18}O = α + ^{14}C, is calculated by the orthogonality condition model (OCM) under the absorbing boundary condition (ABC). In the following sections, we explain our models for solving the scattering and structure problems.

2.1 α – ^{14}C Double folding potential

We calculate the differential cross sections of an α particle scattered by ^{14}C in the formulation of the microscopic nuclear interaction. The nuclear potential of α and ^{14}C is calculated from the double folding (DF) model [13], which is symbolically written as a function of the α – ^{14}C relative coordinate **R**,

$$U_{\mathrm{DF}}(\mathbf{R}) = \iint \rho_\alpha(\mathbf{r}_\alpha)\rho_{14}(\mathbf{r}_{14}) \cdot v_{\mathrm{NN}}^{\mathrm{DDM3Y}}(s,\rho)d\mathbf{r}_\alpha d\mathbf{r}_{14} \quad (1)$$

with $s = |\mathbf{r}_{14} - \mathbf{r}_\alpha - \mathbf{R}|$. Here $\mathbf{r}_\alpha$ ($\mathbf{r}_{14}$) denotes a coordinate measured from the center of mass in α (^{14}C). $\rho_\alpha(\mathbf{r}_\alpha)$ denotes the density of α particle, which reproduce the charge form factor of the electron scattering, while $\rho_{14}(\mathbf{r}_{14})$ represents the density of ^{14}C, which is calculated by the Harmonic Oscillator (HO) model. The HO density of ^{14}C is designed to reproduce the observed matter radius, r_{rms} =2.30 fm, which is obtained from the analysis of the interaction cross section [14].

In Eq.(1), $v_{\mathrm{NN}}^{\mathrm{DDM3Y}}$ represents the effective nucleon-nucleon (NN) interaction which acts between a pair of nucleon contained in the α particle and the ^{14}C nucleus. In the present calculation, we adopt the DDM3Y (Density Dependent Michigan 3-range Yukawa) interaction [15–17], which is given by the functional form of

$$v_{\mathrm{NN}}^{\mathrm{DDM3Y}}(s,\rho) = v_{\mathrm{NN}}^{\mathrm{M3Y}}(s) \cdot C\{1 + \alpha\exp[-\beta\rho]\}, \quad (2)$$

where $v_{\mathrm{NN}}^{\mathrm{M3Y}}$ represents the Michigan 3-range Yukawa interaction [18]. The factors of C, α and β depend on the incident energy per nucleon, and they are determined so as to reproduce the strength of the nucleon potential inside of the nuclear matter [15]. In the previous application of DDM3Y, the so-called frozen density approximation (FDA), $\rho = \rho_\alpha + \rho_{14C}$, is employed [16]. However, this approximation is known to give a little shallow potential in the case of the α scattering [17]. Thus, we neglect the α density, ρ_α, according to the recent prescription called the target density approximation (TDA) [19].

In the realistic calculation of the scattering cross sections, we introduce the absorptive potential ($-iW$) and the

Coulomb potential (V_C) in addition to the DF potential. The final form of the interaction potential (U_{int}) is

$$U_{int} = N_r \cdot U_{DF} + V_C - iW. \tag{3}$$

Here we use the Saxon-Woods and its derivative form factor for the absorptive potential (W). In this interaction potential, we multiply the normalization factor of N_r to the DF potential because there is theoretical ambiguity in the strength of the DF potential. As for the Coulomb potential (V_C), we assume the uniform charge sphere with a radius of $R_C = r_0\left(4^{1/3} + 14^{1/3}\right)$. The six parameters in Saxon-Woods (depth, radius, diffuseness in each form factor) and N_r are tuned so as to reproduce the experimental cross sections as much as possible.

2.2 Orthogonality condition model

We calculate the energy spectra of $^{18}O = \alpha + {}^{14}C$ by employing the DF potential. In the energy calculation of the compound system, it is quite important to take into account the Pauli's exclusion principle between the α particle and ^{14}C nucleus. In the cluster systems, some spurious states, which is forbidden by the exclusion principle, arise in a simple potential model and hence, the forbidden states must be excluded from the computational space. The construction of the allowed functional space can be achieved by the orthogonality condition model (OCM) [12].

In OCM, the functional space that satisfies the exclusion principle can be produced by solving the eigenvalue problem of the norm kernel, which is symbolically given by

$$\int d\mathbf{r}'\hat{N}(\mathbf{r},\mathbf{r}')\chi_\mu(\mathbf{r}') = \lambda_\mu\,\chi_\mu(\mathbf{r}) \tag{4}$$

with the μ – th norm eigenvalue of λ_μ and the corresponding relative wave function of χ_μ. The solutions with $\lambda \neq 0$ are the Pauli allowed states for the α – ^{14}C relative motion, which should be included in the computational space, while the solution of $\lambda = 0$ are excluded. The explicit expression of the norm kernel $\hat{N}(\mathbf{r},\mathbf{r}')$ is

$$\hat{N}(\mathbf{r},\mathbf{r}') = \langle\delta(\mathbf{r}-\mathbf{a})\varphi_\alpha\phi_{14}|\mathcal{A}\{\delta(\mathbf{r}'-\mathbf{b})\varphi_\alpha\phi_{14}\}\rangle, \tag{5}$$

where φ_α and ϕ_{14} are the internal wave functions of α and ^{14}C, respectively. The intrinsic structure of the α particle is set to $(0s_{1/2})^4$, while (0, 2) representation in SU(3) is used for ^{14}C, which is projected to the $J^\pi = 0^+$ state. In Eq. (5), the bra-ket means the integration over the valuables of **a**, **b** and all the internal coordinate contained in α and ^{14}C.

In the previous studies, the wave function of χ_μ is expanded in terms of the Harmonic Oscillator (HO) basis, and the matrix elements of the norm kernel is derived according to the mathematical technique in RGM (resonating group method) [20]. However, the analytic treatment in RGM is quite tedious and complicated for general $N \neq Z$ systems. Therefore, we handle the norm kernel in a way of the numerical calculation based on the prescription of GCM (generator coordinate method) [20].

Here we briefly explain the computational step of the numerical treatment; first, we solve a simple HO problem with the shifted Gaussian basis, which is a basis function of the GCM norm kernel. Secondly, the GCM norm kernels corresponding to the shifted Gaussian used in solving HO problem are calculated. Finally, the GCM norm kernels are transformed into the matrix element of the norm kernels of the HO basis. In this computational process, only the GCM norm kernels, which are easily calculated in a numerical manner, is used to evaluate the matrix elements of the norm kernel in Eq. (5) with the HO basis. The details of the computational procedure will be published in a forthcoming article.

2.3 OCM under the absorbing boundary condition

After the construction of the Pauli allowed states, the α – ^{14}C relative wave function (Ψ) is expanded in terms of the Pauli allowed state, such as

$$\Psi = \sum_\mu{}' d_\mu \cdot \chi_\mu, \tag{6}$$

where the prime means that only the allowed states with $\lambda_\mu \neq 0$ is included in the summation. The energy spectra are calculated by solving the OCM equation, which is given by

$$\left(T + N_r \cdot U_{DF} + V_C - i\eta\tilde{W} - E\right)\sqrt{\hat{N}}\,\Psi = 0 \tag{7}$$

with the expansion of Eq. (6). In Eq. (7), T, U_{DF}, V_C denote the kinetic energy, the double folding potential, the Coulomb potential, respectively, while E and $\hat{N}$ are the energy eigenvalues, and the norm kernels, respectively. The normalization factor of N_r, which is fixed in the scattering calculation, is also used for the structure calculation, and only the real part of the interaction potential in Eq. (3) is included in the OCM equation.

In Eq. (7), $-i\eta\tilde{W}$ denotes the absorptive potential with the strength of η but it differs from the absorptive potential using in the scattering calculation. This absorptive potential is introduced to impose the absorbing boundary condition (ABC) [21] to identify the resonant states, which appears above the α threshold. In the present calculation, we employ the shifted linear function with a step function $\theta(x)$ like

$$\tilde{W}(r) = \theta(r - r_a)(r - r_a). \tag{8}$$

Here r_a represents a starting point of the absorber, which is set to $r_a = 8.0$ fm in the present calculation.

3 Results

3.1 Analysis of α + ^{14}C elastic scattering

From the comparison of the theoretical calculation with the experiment, we check the validity of the DF potential with TDA. Comparison of the scattering calculations with the observed differential cross sections is shown in Fig. 1. The scattering calculation is done by the computational code of Automatic Local Potential Search (ALPS) [22]. We have calculated the differential cross section in the energy range from $E_\alpha/A_\alpha = 5.5$MeV to $E_\alpha/A_\alpha = 8.8$MeV.

DOI: 10.1051/epjconf/201716300040

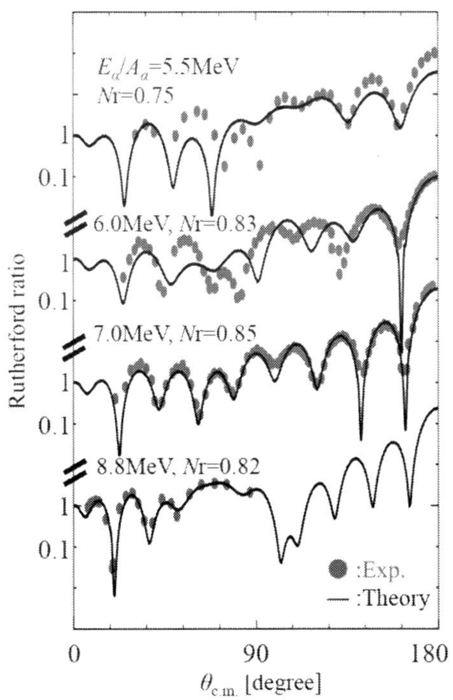

Figure 1. Compasiron of the theory and expermental data concering the α + ^{14}C elastic scattering. The theoretical calculations and the observed differential cross sections are plotted by the solid curves and the solid circles, respectively. The abscissa shows the scattering angle in the center of mass system, while the ordinate represents the Rutherford ratio. The N_r attached to the individual comparisons represents the normalization factor multiplied to the DF potential.

The theoretical calculation (solid curves in Fig. 1), reasonably reproduce the observed cross sections (solid circles) over a whole range of the collision energies. The parameters of the absorptive potential are energy dependent, while the normalization factors of the DF potential are almost the energy independent, say about $N_r \sim 0.8$. The reasonable reproductions of the observed cross section suggests that the employed DF potential with the TDA treatment is reliable interaction to describe the elastic scattering.

3.2 OCM + ABC calculation for $^{18}O = \alpha + {}^{14}C$

We solve the OCM + ABC equation, which is given in Eq. (7), for the $^{18}O = \alpha + {}^{14}C$ cluster system by employing the DF potential used in the elastic scattering at E_α/A_α = 5.5MeV. The 0^+ energy spectra plotted in a complex energy plane are shown in Fig. 2. In the case without absorbing potential, which is a calculation of $\eta = 0$, all the energy levels appears along the real axis (solid circles). This $\eta = 0$ calculation just corresponds to the simple bound-state-approximation. On the contrary, a series of the energy spectra are distributed in a complex energy plane (open circles) when the absorbing potential is switched on ($\eta = 0.3$) [21]. There are two kinds of the split continuum levels with a boundary of the sequence of the eigenvalues parallel to the real axis. The levels above the parallel line are the converged continuum state, in which the tail part of their wave function is smoothly damped in the asymptotic region, while the levels below the parallel line are the diverging state, where the wave function is artificially damped due to the finite range of the employed basis functions. On the contrary, the levels forming the line parallel to the real axis are called the indifferent state, and the wave functions in these states have a strong oscillating behavior over a whole spatial range.

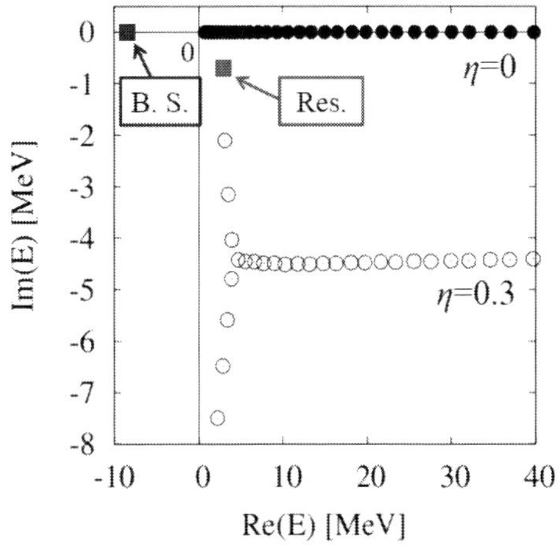

Figure 2. Complex energy spectra calculated from OCM + ABC in Eq. (7). The abscissa and ordinate represent the real and imaginary parts of the energy eigenvalues. The zero energy in the real axis is set to the α decay threshold. The spectra on the real axis (solid circles) are the solution without the absorber ($\eta = 0$), while those distributed in the complex energy plane (solid squares and open circles) are the results with the abosrbing potential ($\eta = 0.3$). The solid squares show the bound state (B.S.) and the resonant state (Res.), while the open circles mean the non-resonant continuum states.

In the finite absorber calculation ($\eta = 0.3$), the bound state (B.S.) and the resonant state (Res.), which are plotted by the solid square are clearly separated from the non-resonant continuum state (open circles) . The energy of the bound state is $E \sim -10$ MeV with respect to the α threshold, while the resonance energy is $E_R \sim 3$ MeV. The theoretical binding energy of the ground 0_1^+ state is over bound in comparison to the respective ground state in the experiment ($E_{exp.} = -6.2$ MeV). The absolute value of the binding energy is sensitive to the computational setting of the basis function in solving the eigenvalue problem of the norm kernel in Eq. (4). We have checked the several parameter sets of the shifted Gaussian in calculateing the GCM kernels and found the total binding energy considerably depends on the employed parameter sets, such as the width parameter or the mesh of the distant parameter in the shifted Gaussian. The basis dependence of the calculated energy will be discussed in future studies. However, the over binding feature of the ground state is invariant even if we change the parameter set of the shifted Gaussian. On the contrary, the energy of the excited 0_2^+ state ($E_R \sim 3$ MeV) obtained in the present calculation is not so sensitive to the choice of the parameter set in the shifted Gaussian. The energy of the calculated 0_2^+ state seems to correspond to the 0_4^+ state, which is assigned in the recent experiment [23] although the resonance energy is a little higher than the experimental observation ($E_{exp.} = 1.6$ MeV).

The deep binding feature of 0_1^+ is originated from the depth of the DF potential, which depends on the incident energy of the α particle. The parameters in DDM3Y (C, α, β in Eq. (2)) are energy dependent, and the DF potential gets deeper and deeper as the incident energy becomes the lower. In the present OCM + ABC calculation, we use the DF potential at the lowest energy, $E_\alpha/A_\alpha = 5.5$ MeV and hence, this incident energy gives the deeper potential than the realistic potential, which should give the correct binding energy. In order to eliminate the energy dependence, we need to average of the DF potential over a finite range of the α incident energies. The averaged potential is expected to improve the over bound feature of the ground 0_1^+ state.

In contrast to the 0_1^+ state, the excited 0_2^+ state in theory is expected to be insensitive to the depth of the DF potential because this resonance has the well developed $\alpha + {}^{14}$C structure, in which the main amplitude of the relative wave function is concentrated around the tail part of the nuclear potential. Once the α cluster structure is strongly developed, the relative wave function of $\alpha - {}^{14}$C becomes the higher nodal state involving a prominent extension. Thus, the relative wave function mainly feels the kinetic energy and the Coulomb interaction, and the nuclear interaction is not so effective in the cluster state.

3.3 Coulomb shift in ^{18}O – ^{18}Ne

We have performed the similar OCM + ABC calculation for $\alpha + {}^{14}$O, and the 0^+ energies are calculated for the ^{18}Ne nucleus. In this calculation, the nuclear potential of the $\alpha + {}^{14}$O system is set to the same DF potential as that used in the $\alpha + {}^{14}$C calculation, and only the charge number in the Coulomb potential of V_C in Eq. (7) is changed. The energy shift of ^{18}O – ^{18}Ne, which is defined by $\Delta E = |E({}^{18}\mathrm{O}) - E({}^{18}\mathrm{Ne})|$, is summairzed in table. 1.

Table 1. Energy shift $\Delta E = \left|E\left({}^{18}\mathrm{O}\right) - E\left({}^{18}\mathrm{Ne}\right)\right|$ for the 0^+ state. The energy shift for the ground 0_1^+ state is shown in the upper row, while the bottom row shows the energy shift for the excited 0^+ state.

	Exp. [MeV]	Theory [MeV]
ΔE (Ground 0^+)	1.1	1.6
ΔE (Excited 0^+)	–	0.8

The energy shift of ^{18}O – ^{18}Ne is insensitive to the choice of the parameter set in the shifted Gaussian because the variation of the energies with respect to the change of the basis set is common in these two systems. Theoretical energy shift of the 0_1^+ state, which is shown in the upper row, is about $\Delta E(0_1^+) = 1.6$ MeV. This energy shift is a little larger than the experimental observation of $\Delta E(0_1^+) = 1.1$ MeV. This overestimation of the present calculation is due to the deep binding feature of the 0_1^+ energies, which is calculated from the DF potential at the lowest α scattering energy. Thus, the energy average of the DF potential should be taken into account in the evaluation of the Coulomb shift of the ground state. On the contrary, the theoretical calculation predicts that the Coulomb shift for the excited 0^+ state is about $\Delta E(0_2^+) = 0.8$ MeV, which is reduced to be about half the ground shift of $\Delta E(0_1^+) = 1.6$ MeV. There is no experimental data of the energy shift, which is evaluated by $\Delta E = \left|E({}^{18}\mathrm{O}(0_4^+)) - E({}^{18}\mathrm{Ne}(0_4^+))\right|$. This is due to the lack of the experimental information of the resonant levels in ^{18}Ne above the $\alpha + {}^{14}$O threshold.

The reduction of the energy shift for the excited 0_2^+ state is induced by the development of the α clustering. If the compound system is excited to the cluster state, the spatial distance between the α particle and the residual nucleus (^{14}C or ^{14}O) is prominently extended. In fact, we have checked the relative wave functions of the ground 0_1^+ state and the excited 0_2^+ state and confirmed the root-mean-square (r.m.s.) radius of the wave function in the cluster 0_2^+ state is enhanced by about 3.5 fm in comparison to the r.m.s. radius of the 0_1^+ wave function. Therefore, the expectation value of the Coulomb interactions in the cluster state is weaker than the expectation value in the ground state, in which α and the residual nucleus is strongly overlapped each other. This suppression of the Coulomb effect leads to the reduction of the energy shift in the mirror systems, which can be clearly confirmed in table. 1. The present results strongly suggests that the reduced Coulomb shift is a new probe to identify the cluster degrees of freedom, which appears in the low excited state.

4 Summary and discussion

In summary, we have performed the scattering and structure calculations for the ${}^{18}\mathrm{O} = \alpha + {}^{14}$C system. In the scattering calculations, the $\alpha + {}^{14}$C interaction potential is constructed from the double folding (DF) model, which employs the effective nucleon-nucleon interaction of the density-dependent Michigan 3-range Yukawa (DDM3Y). In the derivation of the DF potential, only the density dependence of ^{14}C is considered in accordance to the recent prescription of target density approximation (TDA). The DF potential plus the absorptive potentials, which are determined in a phenomenological manner reasonably reproduces the observed differential cross section of the $\alpha + {}^{14}$C elastic scattering.

The DF potential is applied to the structure calculation based on the orthogonality condition model (OCM) under the absorbing boundary condition (ABC). In the previous applications of OCM, the norm kernel is evaluated analytically by employing the sophisticated technique of the resonating group method (RGM). Since the evaluation of the RGM norm kernels is quite tedious and complicated, we have formulated the simple computational method of the norm kernel on the basis of the generator coordinate method (GCM). In this treatment, only the GCM norm kernels, which can be evaluated from the numerical calculation, is used in handling the eigenvalue problem in OCM.

In the ABC + OCM calculation, we have obtained the two 0^+ states: the ground 0_1^+ state and the excited 0_2^+ state. The binding energy of the ground 0_1^+ state is a little over bound but the excitation energy of the excited 0_2^+ state seems to be coincident to the 0_4^+ state, which has been confirmed in the recent experiment. In order to improve the over binding problem of the ground state, we need to consider the energy average of the DF potential, which depends on the incident energy of the α particle. The energy average of the DF potential is now considering.

We have also performed the similar structure calculation for the mirror partner, the ^{18}Ne = α + ^{14}O system, by replacing the Coulomb potential of ^{18}O to that of ^{18}Ne. We compare the energy shift of these mirror systems, which is mainly originated from the difference of the Coulomb interaction. From this comparison, we have confirmed that the Coulomb shift of the excited 0_2^+ is prominently reduced in comparison to the shift of the ground 0_1^+ state. This difference of the Coulomb shift has a close connection to the intrinsic structure in the individual 0^+ states: the developed α cluster structure in the excited 0_2^+ state and the melted cluster structure in the ground 0_1^+ state. In the excited state, the spatial extension is prominent due to the α cluster formation, and this extension leads to the suppression of the Coulomb interaction between α and the residual nuclei (^{14}O and ^{14}C). On the contrary, in the ground state, the α particle and ^{14}C are strongly overlapped each other and hence, the Coulomb interaction is strong. Therefore, the Coulomb shift of the excited state is strongly suppressed in comparison to the ground state due to the spatial extension of the α cluster.

The present result strongly suggests a new possibility of the Coulomb shift as a probe of the cluster degrees of freedom. However, the present calculation is still insufficient to discuss the realistic Coulomb shift observed in the experiment because the present calculation does not reproduce the low-lying bound 0^+ states below the α threshold. In ^{18}O, for example, the 0_2^+ and 0_3^+ states appear at the excitation energy of 3.63 MeV and 5.34 MeV, respectively. These bound state cannot be reproduced by the present calculation, which includes the single cluster configuration of α + ^{14}C. Thus, we introduce other configurations to reproduce the bound levels. In order to improve the calculation, we need to include the two particle configuration around ^{16}O core and the excitation of ^{14}C in the α + ^{14}C configuration. As demonstrated in Ref. [24], the two particle configuration is essential to reproduce the bound 0^+ states, while the excitation of ^{14}C is expected to be important for the formation of the higher resonances. The extended calculation taking into account these configurations are now underway.

In addition to the theoretical improvement, the experimental investigation of ^{18}Ne is strongly desired. In the bound region, the 0^+ levels corresponding to those in ^{18}O have already been identified but the identification of the resonant levels above the α decay threshold is still insufficient in ^{18}Ne. In particular, the experimental investigation of the resonant 0^+ state is very interesting because the cluster degrees of freedom will be strongly activated. The development from both of the theory and experiment will be important subject in future studies.

References

[1] H. Horiuchi *et al.*, Suppl. Prog. Theor. Phys. **192**, 1 (2012), and references therein.
[2] A. Tohsaki, H. Horiuchi, P. Schuck and G. Reöpke, Phys. Rev. Lett. **87**, 192501 (2001).
[3] Y. Funaki *et al.*, Phys. Rev. Lett. **101**, 082502 (2008)
[4] Y. Funaki, H. Horiuchi, W. von Oertzen, G. Reöpke, P. Schuck, A. Tohsaki and T. Yamada, Phys. Rev. C **80**, 064326 (2009).
[5] M. Ito and K. Ikeda, Rep. Prog. Phys. **77**, 096301 (2014), and references therein.
[6] N. Itagaki, T. Otsuka, K. Ikeda, and S. Okabe, Phys. Rev. Lett. **92**, 142501 (2004).
[7] M. Kimura, Phys. Rev. C **75**, 034312 (2007).
[8] A. Bohr and B. R. Mottelson, *Nuclear Structure vol. I* (New York: Benjamin, 1969) p.320.
[9] M. Ito, EPJ Web of conf. **122** (2016).
[10] P. Descouvemont and D. Baye, Phys. Rev. C **31**, 2274 (1985).
[11] N. Furutachi *et al.*, Prog. Theor. Phys. **119**, 403 (2008).
[12] A. Saito, Suppl. Prog. Theor. Phys. **62**, 11 (1977).
[13] M. Ito, Y. Hirabayashi and Y. Sakuragi, Phys. Rev. C **66**, 034307 (2002), and references therein.
[14] A. Ozawa, T. Suzuki and I. Tanihata, Nucl. Phys. **A693**, 32 (2001).
[15] A. M. Kobos *et al.*, Nucl. Phys. **A384** (1982); A. M. Kobos, B. A. Brown, R. Lindsay and G. R. Satchler, Nucl. Phys. **A425**, 205 (1984).
[16] M. El-Azab Farid and G. R. Satchler, Nucl. Phys. **A438**, 525 (1985).
[17] M. E. Brandan and G. R. Satchler, Nucl. Phys. **A487**, 477 (1988).
[18] G. R. Satchler and W. G. Love, Phys. Rep. **55** (1979).
[19] K. Egashira *et al.*, Phys. Rev. C **89**,064611 (2014).
[20] H. Horiuchi, Suppl. Prog. Theor. Phys. **62**, 90 (1977).
[21] M. Iwasaki, R. Otani, Y. Takenaka and M. Ito, PTEP **2015**, 023D01.
[22] Y. Iseri (private communication).
[23] W. von Oertzen *et al.*, Eur. Phys. J. A **43**, 17 (2010).
[24] H. Furutani *et al.*, Suppl. Prog. Theor. Phys. **68**, 279 (1980).

DOI: 10.1051/epjconf/201716300041

STUDY OF FISSION USING MULTI-NUCLEON TRANSFER REACTIONS

Katsuhisa Nishio[1,★], *Kentaro* Hirose[1], *Vermeulen* Mark[1], *Hiroyuki* Makii[1], *Riccardo* Orlandi[1], *Kazuaki* Tsukada[1], *Masato* Asai[1], *Atsushi* Toyoshima[1], *Tetsuya K.* Sato[1], *Yuichiro* Nagame[1], *Satoshi* Chiba[2], *Yoshihiro* Aritomo[3], *Shouya* Tanaka[3], *Tsutomu* Ohtsuki[4], *Igor* Tsekhanovich[5], *Costel M.* Petrache[6], and *Andrei* Andreyev[7,1]

[1]*Advanced Science Research Center, Japan Atomic Energy Agency, 2-4 Shirakata, Tokai, Naka-gun, Ibaraki 319-1195, Japan*
[2]*Laboratory for Advanced Nuclear Energy, Tokyo Institute of Technology, 2-12-1, Ookayama, Meguro-ku, Tokyo, 152-8550 Japan*
[3]*Faculty of Science and Engineering, Kindai University, Higashi-Osaka, 577-8502, Japan*
[4]*Research Reactor Institute, Kyoto University, Kumatori-cho, Sennangun, Osaka 590-0494, Japan*
[5]*University of Bordeaux, 351 Cours de la Libration, 33405 Talence Cedex, France*
[6]*Centre des Sciences Nucléaire et des Sciences de la Matière, Université Paris-Saclay, CNRS/IN2P3, 91406 Orsay, France*
[7]*Department of Physics, University of York, Heslington, York, YO10 5DD, United Kingdom*

Abstract. It is shown that multi-nucleon transfer reaction is a powerful tool to study fission of exotic neutron-rich actinide nuclei, which cannot be accessed by particle-capture or heavy-ion fusion reactions. In this work, multi-nucleon transfer channels of the reactions of ^{18}O+^{232}Th, ^{18}O+^{238}U, ^{18}O+^{248}Cm, and ^{18}O+^{237}Np were used to measure fission-fragment mass distribution for each transfer channel. Predominantly asymmetric fission is observed at low excitation energies for all the studied cases, with an increase of the symmetric fission towards high excitation energies. Experimental data are compared with predictions of the fluctuation-dissipation model, where effects of multi-chance fission (neutron evaporation prior to fission) was introduced. It is shown that mass-asymmetric structure remaining at high excitation energies originates from low-excited and less neutron-rich excited nuclei due to higher-order chance fissions.

1 Introduction

Fission process is usually described as an evolution of a nuclear shape on a potential-energy surface which results from the subtle interplay of macroscopic nuclear properties and microscopic shell effects. Also dynamical effects should have an important role to explain various aspects in fission. New experimental techniques and associated new data are indispensable to further understand fission mechanism. Fission-fragment mass distributions (FFMDs) are one of the most fundamental data. Neutron- and charged particle capture reactions have been used to populate low-excited compound nuclei (CN) for fission studies [1, 2]. Spontaneous fission, starting from a ground state, is the extreme case in low energy fission. Around 2000, GSI in Darmstadt developed a Coulex-induced fission of relativistic RIBs in inverse kinematics, where comprehensive fission studies were performed for several tens of nuclei in the neutron-deficient Ac-U region [3]. The recent SOFIA experiment at GSI also followed the same approach but with a much improved technique [4, 5]. Recently, β/EC delayed fission was investigated for the very proton-rich nucleus using radioactive beams, and ^{180}Hg was found to show an asymmetric fission as a new region of mass-asymmetric fission [6].

Multi-nucleon transfer (MNT) reactions is another unique reaction which allow us to populate neutron-rich nuclei which cannot be accessed by other reactions such as particle capture and/or heavy-ion fusion reactions. Furthermore, excited states of CN range widely from under the fission barrier to higher energies, allowing us to measure the excitation energy dependence of FFMDs. The MNT reactions are further used for a surrogate reaction technique as a method to determine the neutron-induced fission cross sections [7]. Recently, an inverse kinematics technique was applied in the MNT channels of the ^{238}U+^{12}C reaction to study fission using the large-acceptance magnetic spectrometer VAMOS at GANIL [8–10]. In these experiments, sufficiently-high A and Z resolution for FFs was achieved due to their kinematic boost, allowing the simultaneous measurement of the complete mass- and atomic-number distributions of fission fragments (FFs).

At the tandem accelerator facility of the Japan Atomic Energy Agency (JAEA), we studied the MNT channels of the reactions ^{18}O + ^{232}Th,^{238}U,^{248}Cm,^{237}Np in normal kinematics to obtain FFMDs and their excitation-energy dependence for various isotopes (data for ^{18}O + ^{232}Th were published in [11]). An obvious advantage of this method is a relatively easy possibility to change the projectile and/or the target nuclei. In particular by using targets of the rarest highly-radioactive neutron-rich isotopes heavier than ^{238}U (e.g. Cm and Cf), nuclei to be studied can be extended to isotopes far heavier than uranium, which cannot be used at

★e-mail: nishio.katsuhisa@jaea.go.jp

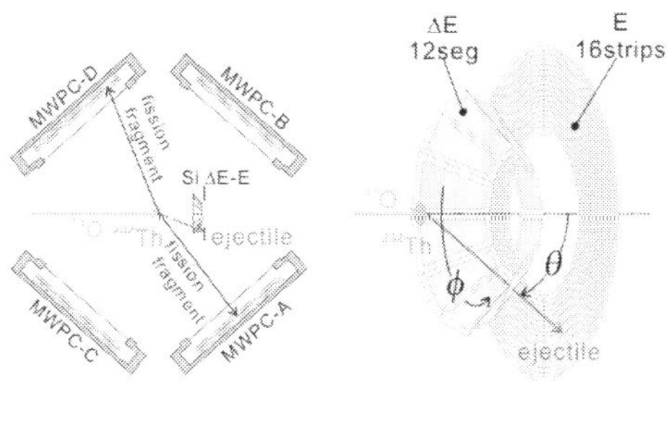

Figure 1. (Color online) Schematic detection set-up (left) and expanded view of the silicon ΔE-E detector telescope (right). See text for details.

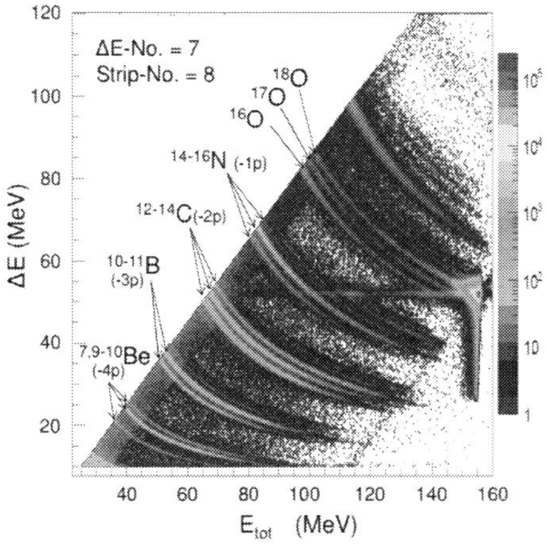

Figure 2. (Color online) ΔE-$E_{\rm tot}$ spectrum for ejectiles measured by one pair of the ΔE-E detectors obtained in the ^{18}O+^{232}Th reaction [11]. The curves corresponding to different ejectiles are marked with the respective isotopes. The scattered ^{18}O is also seen in the plot.

the accelerator facilities for the inverse kinematics experiments similar to VAMOS or SOFIA.

2 Experimental methods

An ^{18}O beam was supplied by the JAEA-tandem accelerator at a typical beam intensity of about 0.5 *p*nA. Beam energies were 157–162 MeV, depending on the different run. Targets were prepared by electrodeposition of oxide-target material on a Ni backing of about 90–300 μg/cm^2 thickness. Thickness of the target-material layer was around 35 – 148 μg/cm^2.

For the event-by-event identification of the transfer channel (thus, of the fissioning nucleus) and of respective coincident FFs, a detection system consisting of a ΔE - E silicon detector telescope and four multiwire proportional chambers (MWPC) were used, see Fig. 1. Specific transfer channels were identified by detecting projectile-like (ejectile) nuclei in twelve 75 μm-thick trapezoidal ΔE silicon detectors which were mounted in a cone around the beam axis, each with the azimuthal angle acceptance of $\Delta\phi = 22.5°$. After passing through the ΔE detector, the ejectiles impinged on the 300 μm-thick annular silicon strip detector (E-detector), divided in 16 annular strips, which allows determination of the scattering angle θ. The inner and outer radius of the detector are 24.0 mm and 48.0 mm, respectively, corresponding to the acceptance angle θ between 16.7° and 31.0° relative to the beam direction.

The energy calibration of the E-detectors was performed by removing two ΔE-detectors so that the elastically-scattered ^{18}O beam could hit the E-detector directly. The well-defined initial beam energy from the tandem and the measured scattering energy $E_{\rm elastic}(\theta)$ were then used to calibrate the strips of the E-detector. Elastic scattering was further used to calibrate the ΔE-detectors, by determining the energy deposition in the ΔE-detector as $E_{\rm elastic} - E_{\rm res}$, where $E_{\rm res}$ is the energy measured in the E-detector after passing through the ΔE-detector. From the peak of elastic scattering in the sum spectrum $E_{\rm tot}=\Delta E + E_{\rm res}$, the energy resolution was obtained to be ~1.0 MeV (FWHM), which also determines the precision for the excitation energies deduced in our study.

Figure 2 shows the ΔE - $E_{\rm tot}$ spectrum for ejectiles obtained in the ^{18}O+^{232}Th reaction, where the parabolic lines correspond to different transfer channels, including a clear separation of specific isotopes. Isotopic assignment was done in respect of the elastically-scattered peak of ^{18}O and the missing line of ^{8}Be. It was further confirmed with the energy-loss calculation using the program SRIM [12]. The identification of the ^{12}C line was also checked by accelerating a ^{12}C beam and measuring the elastic peak. The data from ΔE - $E_{\rm tot}$ spectra were also used to deduce the excitation energy of the respective fissioning nuclei, which were determined from reaction Q-value [13] and the measured (angle-dependent) ejectile energy $E_{\rm tot}$. In this procedure we assumed that no excitation energy is given to the ejectile, thus the quoted excitation energies should be considered as upper limits only.

The coincident FFs resulting from the fission of excited nuclei after the MNT reaction are detected by four 200 × 200 mm^2 position-sensitive MWPCs (see in Fig. 1). The MWPCs were operated with an isobutane gas of about 1.5 Torr [14]. The distance between the target and the center of the cathode was 224 mm, and each MWPC covers a solid angle of 0.67 sr. The positions of FFs's incidence on the MWPC were determined with a position resolution of 4.0 mm. Charge induced in the cathode of the MWPC was recorded to separate FFs from other reaction products. Typical rise time of the MWPC is 5 ns. Both FFs were detected in coincidence with a pair of MWPC facing both sides of the target, (+50.1°, −129.9°) or (−50.1°, +129.9°) relative to the beam direction. Fission-fragments time difference, ΔT, between two coincident MWPCs were measured to determine the masses of both FFs. Figure 3

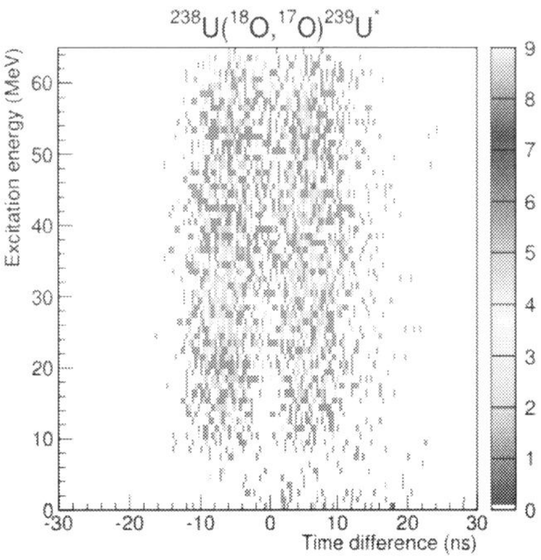

Figure 3. Fission events recorded on the time difference between the signals from coincided MWPCs and excitation energy obtained in one neutron transfer reaction $^{238}U(^{18}O,^{17}O)^{239}U^*$.

shows an example of recorded FFs on the time difference and excitation energy in the transfer channel of $^{238}U(^{18}O,$ $^{17}O)^{239}U^*$. Two regions are clearly observed in the low-excitation fissions, corresponding to the light- and heavy-fragment groups, which smear at high-excitation energies.

3 Results

FFs masses were determined event-by-event from the kinematic analysis, where the measured ΔT values and incident positions of both FFs were used. The momentum of the target-like fissioning recoil nucleus is determined by the measured momentum of ejectile under the assumption of a binary reaction process. To validate the calibration procedure, Figure 4 shows the comparison of FFMDs for $^{239}U^*$, populated in the $^{238}U(^{18}O,^{17}O)^{239}U^*$ reaction [15], with $n + ^{238}U$ [16]. The obtained FFMDs from MNT reactions agree well with the neutron-induced data, particularly the mass asymmetry at the peak positions at the lowest energy data and the increase of the symmetric fission with excitation energy are noteworthy. The result demonstrates that ^{18}O-induced neutron-transfer reaction can be a surrogate of neutron-induced fission to give FFMDs. The FFMD data for $^{233}Pa^*$, $^{233}Th^*$ and $^{236}U^*$ from the MNT reactions of $^{18}O+^{232}Th$ [11] agree with literature data obtained in proton- and neutron-induced fissions [17–20].

Figure 5 shows the FFMDs for nuclei of $^{238-240}U^*$, $^{239-241}Np^*$, $^{241-243}Pu^*$; selection from the MNT-channels of the $^{18}O + ^{238}U$ reaction [15]. The FFMDs of the $^{240}U^*$, $^{240,241}Np^*$ were obtained for the first time in this experiment. For the other nuclei, the known FFMD data were systematically extended to excitation energies as high as 60 MeV. It follows from Fig. 5 that mass-asymmetric fission dominates at low excitation energies for all the measured nuclei. The yield in the mass-symmetric fission region increases with excitation energy (see also Fig. 4) and

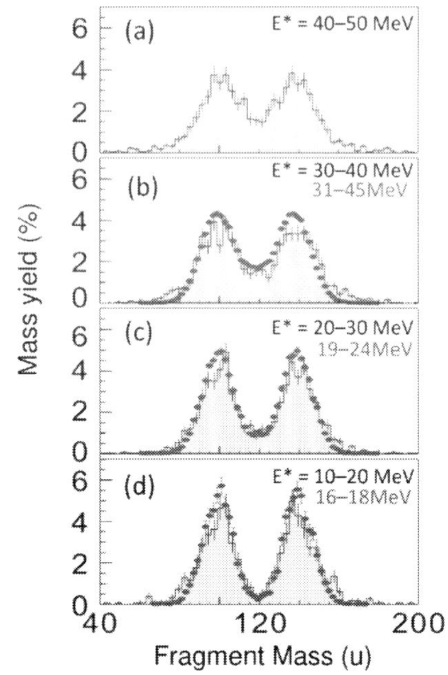

Figure 4. (Color online) Experimental FFMDs (black histogram with error bars) obtained in the $^{238}U(^{18}O,^{17}O)^{239}U^*$ reaction. Excitation energy ranges are indicated (black character). Data are compared with those from the $n+^{238}U$ [16] from the similar excitation energies (red character).

the double-peaked shapes tend to be washed out. However, even at the highest energies, $E^* = 50–60$ MeV, double-peak structure is preserved for all the studied nuclides in Fig. 5. It is also interesting to note that the measured spectra reveal smaller peak-to-valley ratio in the FFMDs for heavier elements as can be seen, for example, in the spectra of $E^* = 30–40$ MeV. All these features will be discussed in the following subsection in comparison with a fluctuation-dissipation fission model.

In the recent measurement of $^{18}O+^{248}Cm$, new FFMDs of eleven nuclei are further generated, $^{247,249}Cm$, $^{249,250,251,252}Bk$, $^{251,253}Cf$, $^{254,256}Es$, and ^{255}Fm [21]. In the similar experiments on the reaction of $^{18}O+^{237}Np$, we generated FFMDs for thirty nuclides [22] including new FFMD data for ^{240}Am, $^{245,247,248}Bk$, and $^{248,250}Cf$.

The evolution of the center of the light- and heavy-fragment groups ($\bar{A}_L$ and $\bar{A}_H$) with the mass of the CN in low energy fissions of $10 < E^* < 20$ MeV is shown in Fig. 6, where data obtained from the three MNT reactions, $^{18}O+^{232}Th$, $^{18}O+^{238}U$, and $^{18}O+^{248}Cm$ are used. It is found that the $\bar{A}_H$ values are kept constant around 141, whereas $\bar{A}_L$ increases linearly with mass of fissioning nucleus. The trend shows the dominant influence of the shell structure in heavy fragments [1].

 DOI: 10.1051/epjconf/201716300041

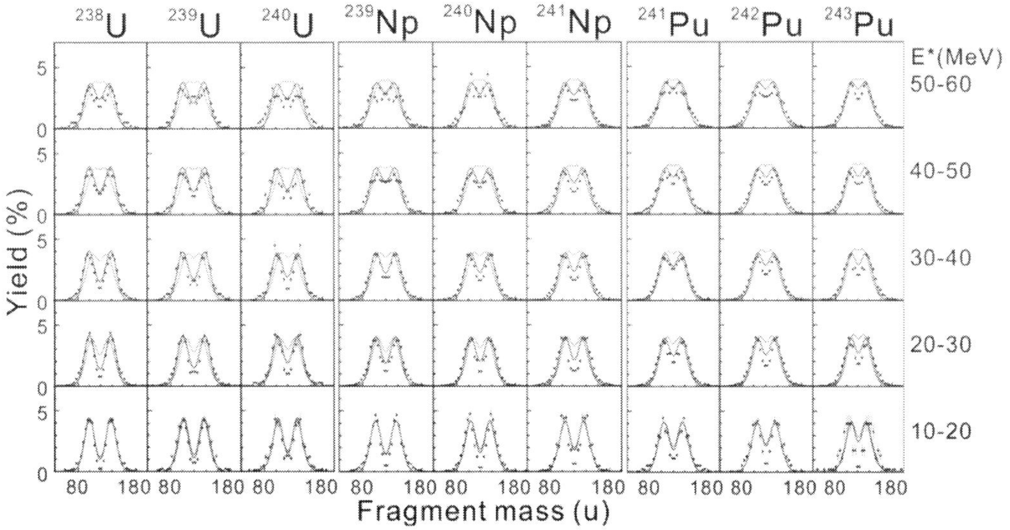

Figure 5. (Color online) Experimental FFMDs (points with error bars) of the U, Np and Pu isotopes and their dependence on the excitation energy in the range of E^* =10–60 MeV. The experimental FFMDs are compared with the Langevin calculations without (blue curves) and with (red curves) the inclusion of the multi-chance fission (see text).

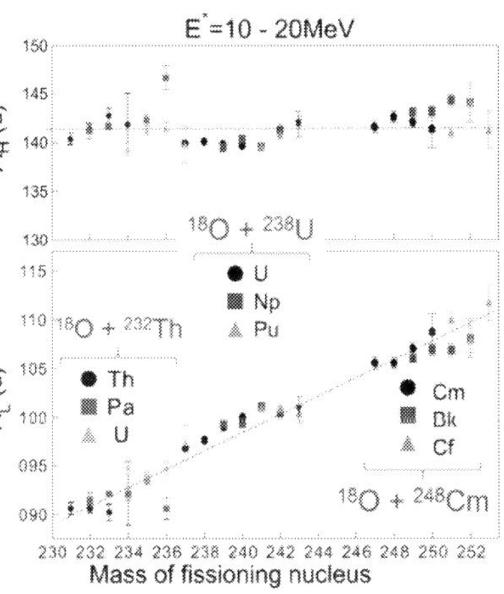

Figure 6. (Color online) Center of the light and heavy fragment groups ($\bar{A}_L$ and $\bar{A}_H$) as a function of mass of the fissioning nuclei in low-excitation fission of $10 < E^* < 20$ MeV. Data are obtained from the reactions of ^{18}O+^{232}Th, ^{238}U and ^{248}Cm using the same setup shown in Fig. 1.

4 Discussions

The measured FFMDs from the MNT reactions are compared with calculations based on the fluctuation-dissipation model developed in [23], where description of fission in Langevin equations from the low-excited state were attempted, and a good reproduction of the measured FFMDs for ^{234,236}U* and ^{240}Pu* from E^*=20 MeV was obtained. As described in [23], the nuclear shape and the corresponding energy is calculated by a two-center shell model [24]. The nuclear shape is defined by three parameters (distance between two potential centers, deformation of fragments, and mass-asymmetry), and the corresponding energy is given by a sum of the liquid-drop energy V_{LD} and the shell correction energy V_{shell}. The latter term is represented as $V_{\mathrm{shell}}(0)\exp(-E^*/E_{\mathrm{d}})$ using the shell correction energy at the zero temperature $V_{\mathrm{shell}}(0)$ and shell damping parameter E_{d}, where E_{d} =20 MeV was chosen as in [23]. For simplicity, we assumed that the total excitation energy of the system after the MNT reactions is given to the initial excitation energy (E^*) of the fissioning nucleus.

The results of Langevin calculation are shown in Fig. 5 by thin blue curves. Calculated FFMDs are broadened with the experimental mass resolution (~6.5 u at FWHM). Under this assumption, the mass asymmetry, i.e. the peak positions of the double-humped FFMD, for all isotopes are reproduced below E^* =20 MeV with clear deviations seen for higher energies. At the highest energy, the calculation shows structure-less symmetric fission in contrast to the measurement. It is seen that the peak-to-valley ratio is reproduced only for the uranium isotopes, $^{238-240}$U of E^* =10–20 MeV, as well as nuclei $^{231-234}$Th, $^{232-236}$Pa and $^{234-237}$U [11] studied in the ^{18}O+^{232}Th reaction. For the neptunium and plutonium isotopes of E^* =10–20 MeV, the calculated peak-to-valley ratio is smaller than the experiments. One of the possible reasons for the deviation could be in the treatment of the neck parameter ϵ $(0 < \epsilon < 1)$ [24] to define the shape of nucleus. In our work we adopted ϵ =0.35 derived as an optimal value in [23] to explain FFMDs of fissioning nucleus of mass of 234–240, which

 DOI: 10.1051/epjconf/201716300041

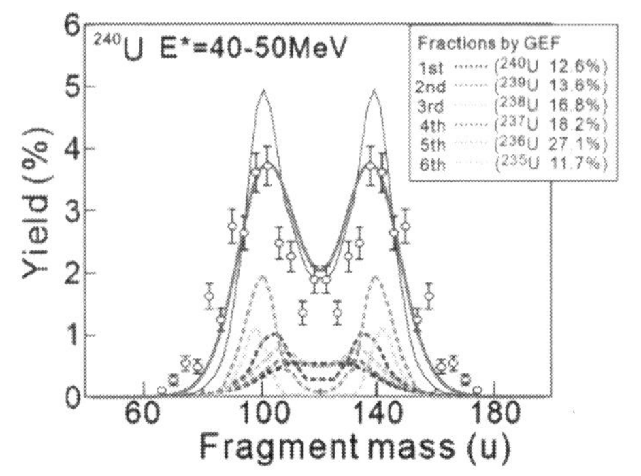

Figure 7. (Color online) Experimental FFMD of $^{240}U^*$ (blue symbols) measured at the initial excitation energy of 40–50 MeV obtained from the two neutron-transfer channel $^{238}U(^{18}O,^{16}O)^{240}U^*$, is compared with the Langevin calculation [23] taking into account multi-chance fission. Thin gray curve is obtained by summing all the fission-chances, which is then broadened with the experimental mass resolution (red curve).

could thus change for the heavier nuclei. Investigating the ϵ parameter for heavier nuclides will be a future subject.

In the above calculation we assumed that all the fission events originate from the initial excitation energy populated by MNT channel. As a next step we attempted to take into account the multi-chance fission (MCF). It is defined as a fission occurring after neutron emission from CN, thus FFs from low-excited and neutron-less excited residual nucleus can contribute (second chance fission). When the residual excited nucleus has enough high excitation energy, further competition between neutron-evaporation and fission (third chance fission) can take place. The higher chance fission successively occurs until the competition terminates. The experimentally observed FFMD is represented by a superposition of all the possible fission chances. These features are demonstrated by Fig. 7, which compares the experimental data for fission of $^{240}U^*$ at the initial excitation energy $E^* = 40$–50 MeV with the Langevin calculation taking into account the MCF. As the excitation energy for the calculation, the middle value 45 MeV for the bin-width (40-50 MeV) was used. Probabilities for each fission chances were calculated by the GEF code (Version 2015v2.2) [25], where spins of the compound nucleus were set at zero for simplicity. The reduction of the excitation energy of the compound nucleus due to the neutron emission was determined from neutron binding energies [26] and a mean energy of the emitted neutron, ~1.9 MeV, obtained by the PACE2 code [27]. At each step of MCF, the potential energy surface for the respective compound nucleus was adopted. The finally calculated FFMD shown by the thin gray line is the sum of the FFMDs over the possible chance fissions. It reproduces the observed peak positions of the experimental FFDM, but has narrower peaks than the measured ones. However, after introducing the experimental mass resolution (6.5 u in FWHM) as shown by the thick solid curve, the calculation well reproduces also the peak-to-valley ratio and the total width of the FFMD. It is seen that at this initial energy, the 1st- and 2nd-chance fission occur with somewhat lower probabilities, which exhibits more symmetric-like fission. On the contrary, the higher-fission chances, after emission of several neutrons (2–5, in this case), lead predominantly to an asymmetric mass split. It is evident that the mass-asymmetric fission observed in the data even at the high excitation energy originates from the lower-energy 4th-, 5th-, and 6th-chance fissions($^{235,236,237}U$) [15].

The same calculation procedure was applied for all the cases as displayed in Fig. 5, and the results are shown by the red thick curves. In contrast to the calculation without MCF (thin blue curves), the calculation with MCF well explains the variation of FFMDs with the excitation energies. Also mass-asymmetry and peak-to-valley ratio observed at the higher excitation energies are well reproduced. The calculation also demonstrates the decreasing peak-to-valley ratio of FFMDs for heavier elements (from uranium to plutonium), observed for example in the $E^* = 30$–40 MeV range, whereas the analysis without MCF predicts almost the same distributions through the isotopes. It should also be noted that the consideration of MCF validates that the shell effect responsible for mass-asymmetric fission disappears around $E^* = 30$–40 MeV (blue curves in Fig.5), resulting from the shell-damping energy $E_d = 20$ MeV entering in the excitation-energy dependence of the shell correction energy V_{shell}.

5 Summary and Outlook

It is shown that the multi-nucleon transfer reaction is a powerful tool to study fission for nuclei which cannot be accessed by particle-capture and/or heavy-ion fusion reactions. An advantage in the normal kinematics is that the nuclei to be studied can be significantly expanded by using available high-purity radioactive targets. Fission studies using the MNT reactions with other targets, such as ^{234}Am, ^{231}Pa, and ^{249}Cf, are planned at the JAEA tandem facility. Furthermore, a reaction using the ^{254}Es target will allow us to study low-energy fissions of fermium isotopes, where sharp transition from the mass-asymmetric fission (e.g. ^{256}Fm) to the sharp symmetric fission (e.g. ^{258}Fm) was observed in the spontaneous fission studies [28].

Other fission observables not described in this report are the fission barrier heights, determined by the threshold in the excitation function of a fission probability. We are also revealing that total spin brought to the fissioning system is nearly proportional to the number of transferred nucleons by measuring the center-of-mass FF angular distributions relative to the rotational axis of CN.

In addition to investigate the fission-fragment properties, a measurement of prompt neutrons in coincidence with FFs has stated to obtain neutron multiplicity $\bar{\nu}(A)$ from individual fragments with mass A and their excitation energy dependence, by mounting a neutron detector array around the present fission setup.

Acknowledgement

Special thanks are due to the crew of the JAEA tandem facility for their beam operation. Present study is supported by "Compre-hensive study of delayed-neutron yields for accurate evaluation of kinetics of high-burn up reactors" and "Development of prompt-neutron measurement in fission by surrogate reaction method and evaluation of neutron-energy spectra" by the Ministry of Education, Culture, Sports, Science and Technology of Japan (MEXT).

References

[1] *Nuclear Fission*, R.Vandenbosch and J.R. Huizenga (Academic Press, 1973).
[2] *Nuclear Fission Process*, edited by C. Wagemans (CRC Press, Boca Raton, FL, 1991).
[3] K.-H. Schmidt *et al.*, Nucl. Phys. A **665**, 221 (2000).
[4] G. Boutoux *et al.*, Physics Procedia **47**, 166 (2013).
[5] E. Pellereau *et al.*, Phys. Rev. C **95**, 054603 (2017).
[6] A.N. Andreyev *et al.*, Phys. Rev. Lett. **105**, 252502 (2010).
[7] J.E. Escher *et al.*, Rev. Mod. Phys. **84**, 353 (2012).
[8] M. Caamaño *et al.*, Phys. Rev. C **88**, 024605 (2013).
[9] M. Caamaño *et al.*, Phys. Rev. C **92**, 034606 (2015).
[10] C. Rodríguez-Tajes *et al.*, Phys. Rev. C **89**, 024614(2014).
[11] R. Léguillon *et al.*, Phys. Lett. B, **761**, 125 (2016).
[12] J.F. Siegler, http://www.srim.org/
[13] G. Audi *et al.*, Chinese Physics C, **36**, 1287, (2012). and *Ibid.*, M. Wang *et al.*, 1603.
[14] K. Nishio *et al.*, Phys. Rev. C **77**, 064607 (2008).
[15] K. Hirose *et al.*, submitted.
[16] V.D. Simutkin *et al.*, Nucl. Data Tables **119**, 331 (2014).
[17] S.I. Mulgin *et al.*, Nucl. Phys. **A824**, 1 (2009).
[18] I. Nishinaka, *et al.*, Phys. Rev. C, **70**, 014609 (2004)., and Proceedings of the Fourth International Conference on Fission and Properties of Neutron-rich Nuclei, Sanibel Island, Florida, USA, 11-17.Nov.2007. pp.206-211, World scientific.
[19] I.V. Ryzhov *et al.*, Phys. Rev. C, **83**, 054603 (2011).
[20] Ch. Straede, B. Jørgensen and H.-H. Knitter, Nucl. Phys. **A462**, 85 (1987).
[21] K. Nishio *et al.*, International Conference on Nuclear Data for Science and Technology (ND2016), 11–16 Sep.2016, Bruges Belgium,
[22] M. Vermeulen *et al.*, to be submitted.
[23] Y. Aritomo and S. Chiba, Phys. Rev. C **88**, 044614 (2013).
[24] K. Sato *et al.*, Z. Phys. A **288**, 383 (1978).
[25] K.-H. Schmidt and B. Jurado and C. Amouroux and C. Schmitt, Nuclear Data Sheets **131**, 107 (2016).
[26] P. Möller *et al.*, Atomic Data and Nuclear Data Tables. **109-110**, 1 (2016).
[27] A. Gavron, Phys. Rev. C **21**, 230 (1980).
[28] D.C. Hoffman, Nucl. Phys. A, **A502**, 21c (1989).

Investigating fusion dynamics at high angular momentum via fission cross sections

C.S. Palshetkar[1,*], D.J. Hinde[1,**], E. Williams[1], K. Ramachandran[1,***], M. Dasgupta[1], K.J. Cook[1], A. Wakhle[1], D.Y. Jeung[1], D.C. Rafferty[1], S.D. McNeil[1], I.P. Carter[1], and D.H. Luong[1]

[1]Department of Nuclear Physics, Research School of Physical Sciences and Engineering, The Australian National University, Canberra, ACT 2601, Australia

Abstract. A quantitative understanding of fusion dynamics at high angular momentum is attempted employing experimental fission cross sections as a probe and carrying out a simultaneous description of the fusion and fission cross sections at above barrier energies. For this, experimental fission fragment angular distributions for three systems: $^{16}O+^{148}Sm$, $^{28}Si+^{136}Ba$ and $^{40}Ca+^{124}Sn$, all forming the same compound nucleus ^{164}Yb at similar excitation energies, have been measured at four beam energies above their respective capture barriers. A simultaneous description of the angle integrated fission cross sections and evaporation residue/fusion cross sections available in literature for the systems is carried out using coupled-channels and statistical model calculations. Fission cross sections, which are most sensitive to the changes in angular momentum, provide very stringent constraints for model calculations thus indicating the need of precision evaporation residue as well as fission cross sections in such studies. A large diffuseness (a_o>0.65 fm) of the nuclear potential gives the best reproduction of the experimental data. In addition, different coupling schemes give very different angular momentum distributions, which, in turn, give very different fission cross section predictions. Both these observations hint at the explanation that depending on energy dissipation of the interacting nuclei occurring inside or outside the fusion pocket, very different fission cross sections can result due to heavily altered angular momentum and thus justifies the sensitivity of fission cross sections used as probes in the present work.

1 Introduction

Understanding fusion dynamics around the Coulomb barrier remains a topic of continued interest, particularly, due to a strong competition between compound and non-compound nuclear processes, altering the evolution of the compound nucleus (CN) formed by heavy interacting nuclei. This is especially important in heavy element (Z>100) synthesis where these competing processes strongly suppress CN formation [1]. Another observation challenging our understanding of the fusion process has come from a systematic study of fusion cross section measurements for a large variety of projectile-target combinations [2]. Here, a large value of the diffuseness parameter of the Woods-Saxon form of the nuclear potential, a_o ~0.7-1.4 fm, was found to be required to reproduce the experimental fusion cross sections. This is in contrast to the value of a_o=0.65 fm generally used for describing elastic scattering [3]. It was suggested in Ref. [2] that the large a_o values might be mocking up some dynamical effects not taken into account in model calculations for the fusion process.

A large diffuseness results in a barrier shifted to lower radius values and a shallow fusion pocket [4]. In model calculations, fusion is generally described to occur when the interacting nuclei dissipate energy inside the fusion pocket where they are trapped, thus forming a CN. However, the system might dissipate energy not just inside but also outside the barrier. This might alter the flux of trajectories reaching inside the barrier, leading to reduced fusion cross sections and, as a result, different angular momentum (l) distributions. Thus, the diffuseness parameter becomes an important parameter in such model calculations and the origin and implications of its large value should be investigated.

Dissipative effects are expected to be significant at higher bombarding energies and angular momenta. The effect of angular momentum in fusion could be probed by measuring the fission fragments, as fission is highly sensitive to the changes in angular momentum. This is because fission cross sections depend on the height of the fission barrier, which is angular momentum dependent. Previous works [5, 6] demonstrated a significant variation in the fission cross sections when different l distributions were used in the model calculations. In this work, we test the sensitivity of fission cross sections on angular momentum by a systematic measurement of fission cross sections for the ^{164}Yb CN populated through three different

*e-mail: asckj1@yahoo.co.in; Present Address: Department of Nuclear and Atomic Physics, Tata Institute of Fundamental Research, Mumbai- 400 005, India
**e-mail: David.Hinde@anu.edu.au
***Permanent Address: Nuclear Physics Division, Bhabha Atomic Research Center, Mumbai- 400 085, India

entrance channels, namely, $^{16}O+^{148}Sm$, $^{28}Si+^{136}Ba$ and $^{40}Ca+^{124}Sn$. Fission fragment angular distributions have been measured for these systems at four beam energies above their respective Coulomb barriers. Simultaneous descriptions of fission cross sections from the present measurement and the evaporation residue/fusion cross sections available in the literature have been carried out using coupled channels and statistical model calculations.

2 Experimental Details

Fission fragment angular distributions were measured separately for the three systems using the 14UD electrostatic tandem accelerator at the Heavy Ion Accelerator Facility at the Australian National University, Canberra. Pulsed beams, separated by 107 ns, of ^{16}O (laboratory beam energy, E_b^{lab}=81-110 MeV), ^{28}Si (E_b^{lab}=130-152 MeV) and ^{40}Ca (E_b^{lab}=166-191 MeV) were bombarded onto ^{148}Sm (30 μg/cm^2 + 40 μg/cm^2 ^{12}C backing), ^{136}Ba (40 μg/cm^2 + 12 μg/cm^2 ^{12}C backing) and ^{124}Sn (17 μg/cm^2 + 12 μg/cm^2 ^{12}C backing) targets respectively, oriented at ~30° or 45° with respect to the beam axis, at four beam energies above their respective Coulomb barriers. Charged particles arising from the reactions, namely, the fission fragments following fusion of the systems, elastically scattered beam particles and recoils were detected using the CUBE spectrometer [7] which consists of two large area Multi-Wire Proportional Counters (MWPC) with an active area of 27.9 cm x 37.5 cm each. For all three measurements, one detector (MWPC 2) was placed at forward angle with respect to the beam axis, corresponding to the detector center at polar angle of θ_2=45°, giving an angular coverage from 5° to 80° and azimuthal angle of ϕ_2=0°. The other detector (MWPC 1) was placed at backward angles with respect to the beam axis, the angle depending on the expected folding angle of the fission fragments from each reaction, and with ϕ_1=180°. For the $^{16}O+^{148}Sm$ and $^{28}Si+^{136}Ba$ measurements, MWPC 1 was centered at θ_1=135°, giving an angular coverage from 95° to 170°, while for the $^{40}Ca+^{124}Sn$ measurement, the detector center was at θ_1=90°, giving an angular coverage of 55° to 130°. Both MWPCs were placed 18 cm away from the center of the target. Two silicon surface barrier detectors were also placed out of the reaction plane at θ_m^{lab} ~23° to serve as monitors.

Energy loss and time of flight signals with respect to the beam pulse from each detector were recorded on an event by event basis. The timing signals were taken from the central cathode foils of each MWPC. A 2D plot showing the raw timing signals from each MWPC obtained in the $^{40}Ca+^{124}Sn$ measurement is shown in Fig. 1 where the fission events are clearly distinguishable from the elastics and recoils. For measurements of the other two systems, only the fission events were collected in the coincidence gate due to the thresholds set in the electronics for rejection of elastics. The position information from each MWPC was obtained from the delay line readouts from both ends of the X and Y anode wires (1 mm apart). Signals from MWPC 2 were always recorded in coincidence with MWPC 1 to reduce the dead-time of the acquisition system. A pulser signal fed through the scaler and data acquisition was used for the dead time corrections in the fission data. For the position calibration, events were recorded in singles mode, in which the entire active area of the detector gets illuminated. Thus, from the known positions, in mm, of the detector edges, position calibration was done. Time calibration of the delay lines was done by inputting pulses separated by 10 ns from an ORTEC time calibrator into the delay lines and recording the output pulses. The solid angle normalization of the fission detectors was done using elastic scattering measured at beam energies much below the Coulomb barrier at 74.3 MeV and 184.2 MeV for the $^{28}Si+^{197}Au$ and the $^{40}Ca+^{197}Au$ reactions respectively.

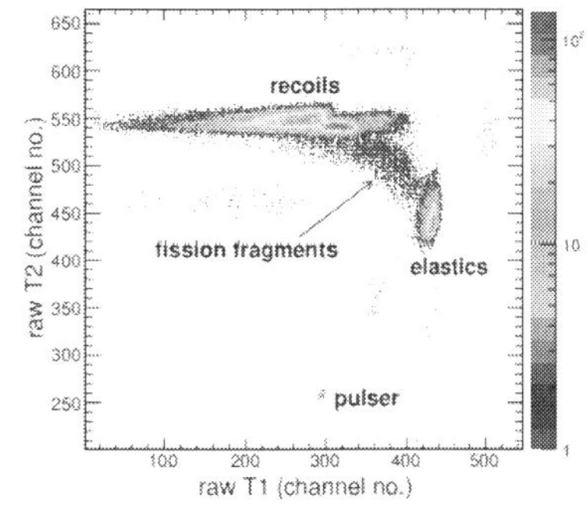

Figure 1. (Color online) Raw timing signals, from the central cathodes of the two MWPCs, recorded in the $^{40}Ca+^{124}Sn$ measurement.

3 Analysis and Results

The data analysis has three parts, (i) extracting the experimental fission cross sections from the measured angular distributions, (ii) coupled-channels calculations to get best reproduction of the experimental fusion cross sections and obtain the respective l-distributions (iii) using these l-distributions as input to statistical model calculations for fission cross section prediction.

3.1 Experimental fission cross sections

From the position information, the angles of the detected particles and from the timing information, their velocities were estimated. These velocities were then converted into masses using the kinematic coincidence technique described in the Appendix of Ref. [7]. By putting gates on the timing signals in Fig. 1, the fission events were clearly selected and yields corresponding to 5° angle bins were extracted. These were corrected for solid angle to get the differential fission cross sections. The cross sections were converted from laboratory frame to the center of mass frame assuming symmetric binary fission and taking the kinetic energy of each fragment to be half of the total kinetic energy (TKE), where TKE is estimated from Viola systematics [8]. The experimental differential fission cross sections thus obtained are shown in Fig. 2 by filled points.

The energies indicated in the figure which have been used while extracting the cross sections have been corrected for energy loss at the target center and converted to the center of mass frame. Due to limitations on the angular range covered with even such large area MWPCs, extrapolation of differential cross sections to 90° and 180° is required in order to extract the angle integrated fission cross sections. For this, the experimental differential cross sections were fitted with transition state model calculations for spin zero nuclei as done in Ref. [9] and are plotted in Fig. 2 as lines. The angle integrated fission cross sections thus obtained are given along with their errors in Fig. 2 and are plotted below in Fig. 4, 5 and 6 by filled squares.

3.2 Coupled-channels calculations

All coupled-channels (CC) calculations were carried out using the code CCFULL [10]. In this, it is necessary to constrain the potential parameters, the depth V_o, radius parameter r_o and the diffuseness parameter a_o. This was done as follows: (i) Starting with an initial set of V_o=200 MeV, r_o=1.0 fm and a_o=1.0 fm, the parameters were varied iteratively to fit the experimental fusion cross sections for the systems taken from literature. In doing so, the uncoupled barriers quoted in literature for the systems, namely, V_b=59.8 MeV [11] for ^{16}O+^{148}Sm and V_b= 118.6 MeV [12] for ^{40}Ca+^{124}Sn, were also reproduced. (ii) Next, the experimental fusion barrier distributions were extracted from the fusion cross sections available in literature for each system using the three point difference method detailed in Ref. [13]. (iii) Inelastic states of the projectile and target were added one by one and checked iteratively to see coupling of which state gives a substantial change in fusion cross section predictions and also simultaneously fits the experimental cross sections as well as barrier distributions. Once a satisfactory fit to both was obtained, these couplings were used for final calculations and the potential parameters fine tuned. The best fit potential parameters for all three systems required a_o>0.65 fm. Another set of calculations keeping the best fit V_o and a_o=0.65 fm was done where the r_o value was varied to get the same uncoupled barrier as the set with the larger a_o. Thus, in the following, predictions of the CC and statistical model calculations are always shown for two sets of diffuseness, one with the fitted larger a_o and the other with a_o=0.65 fm.

As a representative of the above described procedure, in Fig. 3, are shown the experimental fusion cross sections for the ^{16}O+^{148}Sm system (filled circles) taken from [11, 14–16] plotted in the linear (Fig. 3(a)) and logarithmic scale (Fig. 3(b)) while the barrier distribution extracted from the data in the present analysis is plotted in Fig. 3(c).

For CC calculations, ^{16}O was considered inert while for ^{148}Sm, a rotational coupling including static deformation with β_2=0.18 and β_4=0.05 was considered as done in [11]. In addition, a 3^- vibrational state in ^{148}Sm (see Table 1 for details) was also coupled. The best fit potential parameters which simultaneously reproduce the experimental fusion cross sections, barrier distribution as well as the uncoupled barrier of V_b=59.8 MeV [11] are

Table 1. Details of the vibrational states used in the CC calculations for the three systems studied in the present work. All the B(E2↑) have been taken from [17] and B(E3↑) values from [18].

Nucleus	J^π	E_x (MeV)	λ	B(Eλ↑) (e^2b^λ)	β_λ
^{148}Sm	3^-	1.161	3	0.2911	0.1418
^{28}Si	3^-	6.878	3	0.0042	0.4007
^{136}Ba	2^+	0.818	2	0.410	0.1258
	3^-	2.532	3	0.155	0.1253
^{40}Ca	3^-	3.737	3	0.184	0.411
^{124}Sn	2^+	1.132	2	0.1660	0.0953
	3^-	2.603	3	0.073	0.1056

V_o=294 MeV, r_o=0.898 fm and a_o=0.99 fm. For the set with a_o=0.65 fm using the same V_o=294 MeV, r_o=1.076 fm. From Fig. 3, it is clear that the larger diffuseness of a_o=0.99 fm (solid line) gives the best fit to the experimental data as well as barrier distribution. The better agreement is especially clear at above barrier energies. The *l*-distributions obtained from above procedure were used for the statistical model calculations.

For the ^{28}Si+^{136}Ba calculations, due to the unavailability of fusion cross sections for the system, those corresponding to a nearby system, ^{28}Si+^{142}Ce, taken from [19] were used. In the CC calculations, the 3^- state of ^{28}Si and the 2^+ and 3^- states in ^{136}Ba, with the deformation parameters given in Table 1, were coupled as vibrational states. While comparing the CC results for ^{28}Si+^{136}Ba with the experimental ^{28}Si+^{142}Ce fusion cross sections, the geometrical effects were effectively accounted for using the reduction procedure described in [20], where $\sigma_{fus} \rightarrow \sigma_{fus}/(A_p^{1/3}+A_t^{1/3})^2$ and $E_{c.m.} \rightarrow E_{c.m.}(A_p^{1/3}+A_t^{1/3})/(Z_pZ_t)$. Scaling factors for σ_{fus} of 0.0147 and for $E_{c.m.}$ of 0.0102 for the ^{28}Si+^{142}Ce and 0.0149 and 0.0104 respectively for ^{28}Si+^{136}Ba were thus obtained. The best fit potential parameters for this system are: V_o=194.5 MeV, r_o=0.971 fm, a_o=0.89 fm and for a_o=0.65 fm, V_o=194.5 MeV and r_o=1.064 fm.

A comparison of the CC results with the experimental fusion cross sections, showed that inelastic coupling alone was unable to reproduce the experimental cross sections, especially at below barrier energies (see Fig. 5 below). Thus, in addition to the inelastic states, transfer coupling was done for two different Q values, the 2n transfer with Q=3.002 MeV and an effective Q value, corresponding to the 2n+4n transfer with Q=5.133 MeV. These two sets of calculations were carried out to investigate which of them gave the biggest variation in the predicted cross sections. To account for the re-normalization of the barrier introduced due to transfer coupling, keeping V_o=194.5 MeV, the radius parameter was tweaked for the two different diffuseness. The strength parameter, F_t, was varied in steps and a final value of F_t=0.5 MeV was obtained which gave satisfactory reproduction of experimental fusion cross sections.

For the ^{40}Ca+^{124}Sn system, fusion evaporation cross sections were taken from [12, 21, 22]. Similar to the pro-

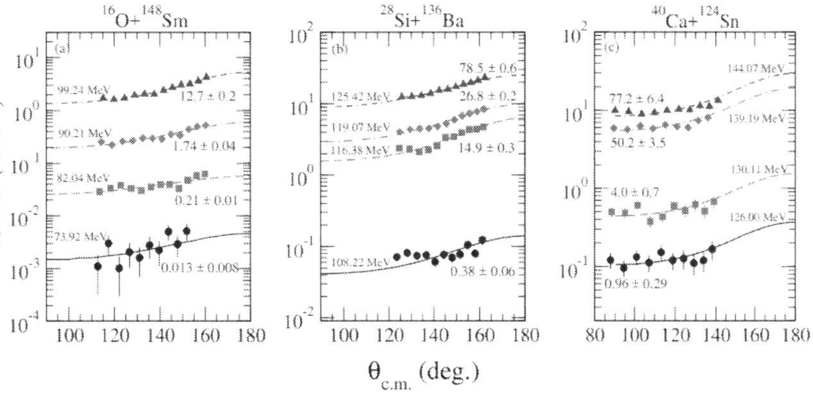

Figure 2. (Color online) Differential fission cross sections for (a) $^{16}O+^{148}Sm$, (b) $^{28}Si+^{136}Ba$ and (c) $^{40}Ca+^{124}Sn$ systems at four different energies in the center of mass frame. The values with errors are the angle integrated fission cross sections. The errors are statistical only.

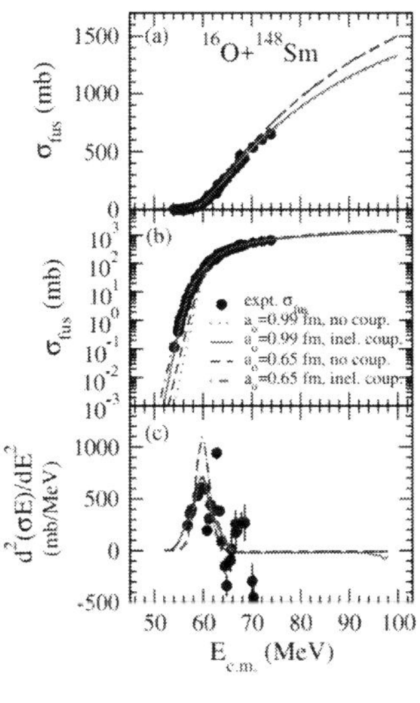

Figure 3. (Color online) Experimental fusion cross sections taken from [11, 14–16] (filled circles) plotted in the (a) linear and (b) logarithmic scale. The experimental barrier distribution extracted from the fusion data is plotted by the filled circles in (c). The dotted and dashed line are the results of the CC calculations without any coupling obtained by using a_o=0.99 fm and a_o=0.65 fm respectively. The solid line corresponds to the best fit obtained by including coupling of inelastic states and using a_o=0.99 fm while the dot-dash line are CC results for the same coupling scheme but using a_o=0.65 fm.

cedure done for the other two system studied in the present work, the CC calculations for $^{40}Ca+^{124}Sn$ were done initially by coupling only the inelastic states, namely, the 3^- state in ^{40}Ca and the 2^+ and 3^- states in ^{124}Sn, included in the vibrational coupling mode. The best fit potential parameters in this case were found to be with V_o=200 MeV, r_o=0.984 fm for a_o=1.0 fm and r_o=1.107 fm for a_o=0.65 fm. Similar to the observations for $^{28}Si+^{136}Ba$ system, coupling of inelastic states alone did not reproduce the experimental cross sections. This is in line with the results obtained in [12, 21, 22] where the importance of transfer coupling has been extensively shown. Hence, in the present analysis, calculations were performed with three different Q values: Q=3 MeV, F_t=0.5 MeV, Q=5.41 MeV (2n transfer), F_t=0.7 MeV and Q=9.5 MeV (4n transfer), F_t=0.8 MeV.

3.3 Statistical model fission calculations

Statistical model calculations for the three systems were carried out using the code PACE [23]. The *l*-distributions, obtained from the CC calculations done using different coupling schemes (inelastic coupling only or inelastic + transfer with different Q values), as described in sec 3.2, corresponding to the larger and smaller diffuseness of the potential, were given as input to PACE. The level density parameter at equilibrium deformation a_n was taken to be a_n=A/7.5, following the semi-empirical formulation given in Equation A9 of Ref. [24]. The two other important adjustable parameters in PACE are, the ratio, a_f/a_n, where a_f is the level density at the saddle point and the scaling factor, k, for the fission barrier. For estimating k, using a_f/a_n=1.00, for the $^{16}O+^{148}Sm$ system, k was varied in steps and for each k, the χ^2 of the fit of the resultant fission cross sections to the experimental values measured in the present work was determined. In this way, the k giving the minimum χ^2, k=0.997 for a_o=0.99 fm and k=1.084 for a_o=0.65 fm was obtained and used for the final calculations. This procedure was also carried out using a_f/a_n=1.04 and although different values of k were obtained, the resultant fission cross sections were very similar. Hence, in the following, results using only a_f/a_n=1.00 are shown.

The results of the simultaneous analysis of fusion and fission cross sections for the $^{16}O+^{148}Sm$ are shown in

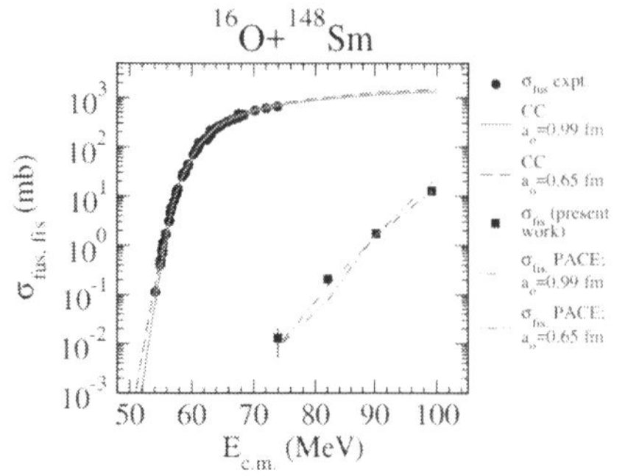

Figure 4. (Color online) Comparison of the experimental fusion [11, 14–16] (filled circles) and fission (filled squares) cross sections from the present work with the predictions of the statistical model calculations PACE. The solid and dashed line are the CC results obtained by coupling the inelastic states in the projectile and target (same inel. coup. results as plotted in Fig. 3 by solid and dot-dash lines) for the two different diffuseness parameters respectively. The corresponding fission cross sections estimated from PACE are plotted as dot-dash line for a_o=0.99 fm and dot-dot-dash lines for a_o=0.65 fm.

Fig. 4. The beam energy dependence of the fission calculations from PACE, corresponding to the larger diffuseness parameter, a_o=0.99 fm (dot-dash line) are in an overall better agreement with the experimental cross sections.

Taking the same k values as for $^{16}O+^{148}Sm$ system and a_f/a_n=1.00, PACE calculations were carried out for the other two systems as well. Results for the $^{28}Si+^{136}Ba$ are shown in Fig. 5. Fission predictions obtained by inputting the *l*-distributions corresponding to inelastic states only (Fig. 5(a)) is in disagreement with the experimental values, even though the experimental fusion cross sections at above barrier energies are reproduced satisfactorily. Results corresponding to inelastic + transfer, shown in Fig. 5(b) and (c) are well reproduced. Here again, though the difference in the predictions for the two diffuseness values is small, predictions corresponding to a_o=0.65 fm show a steeper fall with decreasing energy as compared to those for a_o=0.89 fm and poorly reproduce the experimental cross sections. Effect of different transfer Q values used for calculations is not significant.

Results for $^{40}Ca+^{124}Sn$ system are shown in Fig. 6. As for the other two systems shown above, the best fission predictions from PACE are obtained for inelastic + transfer coupling (Q=3 MeV) and once again the larger diffuseness gives better predictions. As the transfer Q value is increased, the corresponding *l*-distributions are also altered giving very different fission predictions. This shows the sensitivity of the fission cross sections to the type of couplings used for the calculations.

4 Summary and Conclusions

To summarize, this work focuses on investigating the sensitivity of fission cross sections in understanding the fusion

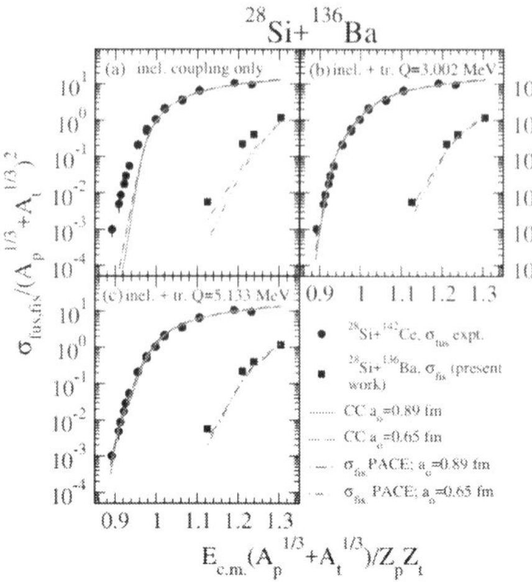

Figure 5. (Color online) Similar to Fig. 4 but for $^{28}Si+^{136}Ba$ system. The reduced cross sections have been plotted in order to show the experimental fusion cross sections for $^{28}Si+^{142}Ce$ [19] (filled circles) with predictions of model calculations for the $^{28}Si+^{136}Ba$ system. Results from PACE when *l*-distributions extracted from inelastic coupling only are used for the input in the calculations are plotted in (a). (b) shows the results for inelastic + transfer with Q=3.002 MeV while (c) shows the results for inelastic + transfer with Q=5.133 MeV.

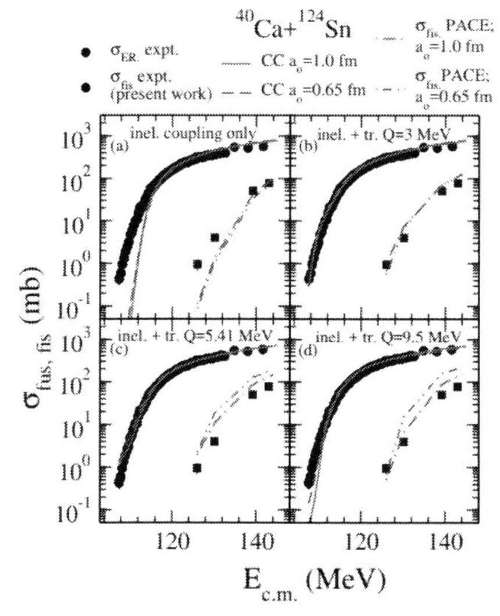

Figure 6. (Color online) Similar to Fig. 4 and 5 but for $^{40}Ca+^{124}Sn$ system. Results from PACE corresponding to *l*-distributions obtained from inelastic coupling only are plotted in (a) whereas those corresponding to results from inelastic + transfer coupling are plotted in (b) for transfer Q=3 MeV, in (c) for Q=5.41 MeV and in (d) for Q=9.5 MeV.

dynamics at high angular momentum. For this, measurement of fission cross sections for the CN ^{164}Yb, formed via different entrance channels has been carried out. A simultaneous description of the fusion as well as fission cross sections has been done using the CC and statistical model calculations.

Stringent constraints are put on model predictions by requiring a simultaneous reproduction of fusion as well as fission cross sections over the entire energy range of the available data. The fission cross sections are observed to be very sensitive to the shape of the *l*-distributions as well as the type of couplings used in the model calculations. With increasing projectile charge, the importance of transfer channels is seen, which again alters the *l*-distributions. A larger diffuseness of the nuclear potential used in the CC calculations and the *l*-distributions resulting from the same is required for the best reproduction of the experimental fission cross sections measured in the present work, in line with previous observations [2]. The present results provide support to the explanation for the requirement of the large diffuseness to be due to energy dissipation of the system possibly occurring outside the potential pocket thus altering the flux of trajectories and ultimately the fusion cross section. The present work emphasizes the requirement of precision absolute evaporation residue and as well as fission cross sections and the procedure of simultaneous analysis to understand the fusion process at high angular momentum.

5 Acknowledgments

The authors thank the accelerator and technical staff of the Heavy Ion Accelerator Facility at the Australian National University for their help during the experiment.

References

[1] D.J. Hinde and M. Dasgupta, Nucl. Phys. A **787**, 176c (2007).

[2] J.O. Newton, R.D. Butt, M. Dasgupta, D.J. Hinde, I.I. Gontchar, C.R. Morton and K. Hagino, Phys. Rev. C **70**, 024605 (2004).

[3] R.O. Akyuz, and A. Winther, *in Proc. Enrico Fermi. Int. School of Physics* edited by R.A. Broglia, C.H. Dasso and R. Richi North-Holland, Amsterdam, (1981).

[4] M. Dasgupta, D.J. Hinde, J.O. Newton and K. Hagino, Prog. Theo. Phys. Supp. **154**, 209 (2004).

[5] R.J. Charity, J.R. Leigh, J.J.M. Bokhorst, A. Chatterjee, G.S. Foote, D.J. Hinde, J.O. Newton, S. Ogaza and D. Ward, Nucl. Phys. A **457**, 441 (1986).

[6] Y. Kondo, B.A. Robson, J.J.M. Bokhorst, D.J. Hinde and J.R. Leigh, Phys. Rev. C **35**, 828 (1987).

[7] D.J. Hinde, M. Dasgupta, J.R. Leigh, J.C. Mein, C.R. Morton, J.O. Newton and H. Timmers, Phys. Rev. C **53**, 1290 (1996).

[8] V.E. Viola, K. Kwiatkowski and M. Walker, Phys. Rev. C **31**, 1550 (1985).

[9] B.B. Back, R.R. Betts, J.E. Gindler, B.D. Wilkins, S. Saini, M.B. Tsang, C.K. Gelbke, W.G. Lynch, M.A. McMahan and P.A. Baisden, Phys. Rev. C **32**, 195 (1985).

[10] K. Hagino, N. Rowley and A.T. Kruppa, Comp. Phys. Comm. **123**, 143 (1999).

[11] J.R. Leigh, M. Dasgupta, D.J. Hinde, J.C. Mein, C.R. Morton, R.C. Lemmon, J.P. Lestone, J.O. Newton, H. Timmers and N. Rowley, Phys. Rev. C **52**, 3151 (1995).

[12] A.M. Stefanini, L. Corradi, D. Ackermann, C.J. Lin, L.F. Zheng, S. Beghini, G. Montagnoli, F. Scarlassara and G.F. Segato, IL Nuovo Cimento **111**, 895 (1998).

[13] M. Dasgupta, D.J. Hinde, N. Rowley and A.M. Stefanini, Annu. Rev. Nucl. Part. Sci. **48**, 401 (1998).

[14] R.G. Stokstad, Y. Eisen, S. Kaplanis, D. Pelte, U. Smilansky and I. Tserruya, Phys. Rev. Lett. **41**, 465 (1978).

[15] R.G. Stokstad, Y. Eisen, S. Kaplanis, D. Pelte, U. Smilansky and I. Tserruya, Phys. Rev. C **21**, 2427 (1980).

[16] D.E. DiGregorio, M. diTada, D. Abriola, M. Elgue, A. Etchegoyen, M.C. Etchegoyen, J.O Fernández Niello, A.M.J. Ferrero, S. Gil, A.O. Macchiavelli, A.J. Pacheco, J.E. Testoni, P.R.Silveira Gomes, V.R. Vanin, R. Liguori Neto, E. Crema and R.G. Stokstad, Phys. Rev. C **39**, 516 (1989).

[17] S. Raman, C.W. Nestor, JR., and P. Tikkanen, Atomic Data and Nuclear Data Tables **78**, 1 (2001).

[18] T. Kibédi and R.H. Spear, Atomic Data and Nuclear Data Tables **80**, 35 (2002).

[19] S. Gil, F. Hasenbalg, J.E. Testoni, D. Abriola, M.C. Berisso, M. di Tada, A. Etchegoyen, J.O. Fernández Niello, A.J. Pacheco, A. Charlop, A.A. Sonzogni and R. Vandenbosch, Phys. Rev. C **51**, 1336 (1995).

[20] P.R.S. Gomes, J. Lubian, I. Padron and R.M. Anjos, Phys. Rev. C **71**, 017601 (2005).

[21] F. Scarlassara, S. Beghini, G. Montagnoli, G.F. Segato, D. Ackermann, L. Corradi, C.J. Lin, A.M. Stefanini and L.F. Zheng, Nucl. Phys. A **672**, 99 (2000).

[22] J.J. Kolata, A. Roberts, A.M. Howard, D. Shapira, J.F. Liang, C.J. Gross, R.L. Varner, Z. Kohley, A.N. Villano, H. Amro, W. Loveland and E. Chavez, Phys. Rev. C **85**, 054603 (2012).

[23] A. Gavron, Phys. Rev. C **21**, 230 (1980).

[24] J. Tõke and W.J. Świątecki, Nucl. Phys. A **372**, 141 (1981).

Dynamical Dipole mode in heavy-ion fusion reactions

C. Parascandolo[1,★], D. Pierroutsakou[1], R. Alba[2], A. Del Zoppo[2], C. Maiolino[2], D. Santonocito[2], C. Agodi[2], V. Baran[3,2], A. Boiano[1], M. Colonna[2], R. Coniglione[2], E. De Filippo[4], M. Di Toro[2,5], U. Emanuele[6], F. Farinon[7], A. Guglielmetti[8], M. La Commara[9,1], B. Martin[9,1], C. Mazzocchi[8], M. Mazzocco[10], C. Rizzo[2,5], M. Romoli[1], C. Signorini[10], R. Silvestri[9,1], F. Soramel[10], E. Strano[10], D. Torresi[10], A. Trifirò[6], and M. Trimarchi[6]

[1] INFN - Napoli, Napoli, Italy
[2] INFN - LNS, Catania, Italy
[3] University of Bucharest, Bucharest and NIPNE-HH, Magurele, Romania
[4] INFN - Catania, Catania, Italy
[5] Dip. di Fisica e Astronomia, Università di Catania, Catania, Italy
[6] INFN - Gr. Col. di Messina and Università di Messina, Messina, Italy
[7] GSI, Darmstadt, Germany
[8] Dip. di Fisica, Università di Milano and INFN - Milano, Milano, Italy
[9] Dip. di Fisica, Università di Napoli, Napoli, Italy
[10] Dip. di Fisica e Astronomia, Università di Padova and INFN - Padova, Padova, Italy

Abstract. The excitation of the Dynamical Dipole mode along the fusion path was investigated for the first time in the formation of a heavy compound nucleus in the $A \sim 190$ mass region. The compound nucleus was formed at identical conditions of excitation energy and spin from two entrance channels having different charge asymmetry: the charge asymmetric ^{40}Ca + ^{152}Sm and the nearly charge symmetric ^{48}Ca + ^{144}Sm at E_{lab}=11 and 10.1 MeV/nucleon, respectively. Both fusion–evaporation and fission events were studied simultaneously for the first time. The Dynamical Dipole mode excitation in the charge asymmetric channel was evidenced, in a model-independent way, by comparing the γ-ray multiplicity spectra and angular distributions of the two entrance channels with each other.

1 Introduction

The "Dynamical Dipole mode" (DD throughout the text), or pre-equilibrium Giant Dipole Resonance, is a collective oscillation of protons against neutrons with a dipole spatial pattern inside the atomic nucleus. It is a pre-equilibrium phenomenon, being excited along the fusion path of nuclei with a different ratio of neutrons and protons and decaying through emission of prompt γ-rays [1–5]. The DD radiation is characterized by the following characteristics: i) a centroid energy lower than that of the compound nucleus GDR in the same mass region indicating a high deformation of the emitting source [3, 4] ii) an anisotropic angular distribution with respect to the beam axis since the oscillation is confined in the reaction plane [6] and iii) a γ yield that should depend on both the reaction dynamics and the symmetry term of the nuclear matter Equation Of State [4].

From an experimental point of view, the DD excitation and subsequent γ decay has been observed in deep inelastic [5, 7–9] and fusion-evaporation heavy-ion collisions [9–16]. In some of the above-cited works [11–13], the first systematic study of its features as a function of the incident energy was performed in the ^{132}Ce mass region in fusion-evaporation experiments. In those measurements, the DD was evidenced by employing a model independent method, the so-called difference technique, that consists in: (a) forming the same compound nucleus at identical conditions of excitation energy and spin from two entrance channels, a nearly charge symmetric and a charge asymmetric one; (b) obtaining the difference between the γ-ray spectra and angular distributions of the two channels for fusion-evaporation events. The resulting difference showed an excess of yield in the more charge asymmetric channel, ascribed to a pre-equilibrium effect, namely the DD excitation. A second campaign of experiments [15, 16] was performed to probe the DD excitation in the same composite system with a different method: the DD radiation was extracted by subtracting the γ-ray multiplicity spectrum of the compound nucleus, calculated by means of the statistical model code CASCADE [18], from the charge asymmetric reaction experimental γ spectrum. The CASCADE calculation used in the comparison was tested with a near-symmetric reaction leading to the same compound nucleus ^{132}Ce.

The comparison of the two data sets [11–13, 15, 16] with each other and with the theoretical predictions, performed within a Boltzmann-Nordheim-Vlasov (BNV) transport model, based on a collective bremsstrahlung

★e-mail: concetta.parascandolo@na.infn.it

analysis of the entrance channel reaction dynamics [4], shows some discrepancies. On the one hand calculations, while being able to describe the phenomenon and absolute values of some observables, are not able to fully reproduce the existing experimental findings [14, 15]. On the other hand few experimental results exist reporting on the DD γ multiplicity and on its angular distribution, that can be directly compared with calculations. These elements called for more experimental efforts to disentagle the influence of each reaction parameter on the DD features and to provide severe constraints to the theoretical models.

The emission of DD γ-rays decreases the excitation energy and the initial temperature of the nucleus reaching the statistical phase. This cooling mechanism could be suitable to favour the compound nucleus survival against statistical fission and thus, the formation of a super heavy element in hot fusion processes [4, 19]. TDHF calculations [19] showed that the DD γ yield decreases as the mass of colliding ions increases since the reactions with small nuclei are less damped than those involving more nucleons. In order to understand the behavior of the DD in heavier systems than those studied before and to test its usefulness in super-heavy element production, we decided to study the DD in a composite system in the mass region A=190 [17].

2 The experiment: ^{40,48}Ca+^{152,144}Sm at 11 MeV/nucleon

The experiment was performed by using the ^{40}Ca (^{48}Ca) pulsed beam provided by the Superconducting Cyclotron of the Laboratori Nazionali del Sud (INFN-LNS, Italy), impinging on a 1 mg/cm^2 thick self-supporting ^{152}Sm (^{144}Sm) target enriched to 98.4%(93.8%) in ^{152}Sm (^{144}Sm) at E_{lab} = 440 (485) MeV. Both entrance channels populate the same compound nucleus through a quite different initial dipole moment, 30.6 fm for the ^{40}Ca + ^{152}Sm charge asymmetric reaction and 5.3 fm for the ^{48}Ca + ^{144}Sm more charge symmetric one. The mass asymmetry of the two entrance channels is very similar, namely 0.22 (0.18) for the ^{40}Ca + ^{152}Sm (^{48}Ca + ^{144}Sm) system. The formed compound nucleus had identical excitation energy in both reactions, E^* = (220±7) MeV, and identical spin distribution: L_{max} = 74$\hbar$ for fusion and L_{max} = 42$\hbar$ for fusion-evaporation, according to PACE2 calculations [20] with a level density parameter a = A/9.5 MeV^{-1}, A being the compound nucleus mass.

The γ-rays and the light charged particles were detected by using the 180 barium fluoride (BaF_2) modules of the MEDEA experimental apparatus [27] that covers the polar angular range between θ_{lab} = 30° and θ_{lab} = 170° and the full range in the azimuthal angle ϕ. The MEDEA ball has an inner radius of 22 cm and covers a total solid angle of 3.7π sr. The apparatus operates under vacuum inside a large scattering chamber to allow a simultaneous detection of γ-rays and light charged particles. The discrimination between γ-rays and light particles was performed by combining a shape analysis of the BaF_2 signal with a time of flight (TOF) measurement, between each BaF_2 detector and the radiofrequency signal of the Cyclotron.

The energy calibration of the BaF_2 detectors was obtained by using the composite sources of ^{241}Am + ^{9}Be (E_γ=4.43 MeV) and ^{238}Pu + ^{13}C (E_γ=6.13 MeV) and the 15.1 MeV γ-rays from the p + ^{12}C reaction at E_p =25 MeV. The charged particle calibration was deduced from the γ calibration as described elsewhere [21]. The time stability of the energy calibration was checked during the experiment by monitoring after each run the stability of the peak corresponding to a radioactive source.

The fusion-evaporation residues were detected by four Parallel Plate Avalanche Counters (PPACs) located symmetrically around the beam direction at 70 cm from the target in an annular configuration. The PPACs were centered at an angle θ_{lab} = 7° with respect to the beam direction, subtending 7° in θ and covering a total solid angle of 0.089 sr. The PPACs provided the energy loss ΔE and the TOF of the reaction products with respect to the radiofrequency signal of the accelerator. The fission events were selected by detecting the two kinematically coincident fission fragments with position sensitive PPACs, centered at θ = 52.5° symmetrically around the beam axis, at 16 cm from the target covering 22° in both θ and ϕ and allowing the study of γ-ray - fragment angular correlations. The PPACs gave information on the energy loss ΔE, the TOF and the x and y positions of the fragments that allowed to reconstruct angles, masses and velocity vectors of the fragments in the laboratory and the center-of-mass reference frame. Both evaporation residues and fission fragments were selected by applying appropriate contours in the relative bi-dimensional plots (ΔE, TOF) of the PPACs.

Down-scaled single PPAC events together with coincidence ones between a PPAC and at least one fired BaF_2 scintillator were collected during the experiment. A coincidence event was accepted if the deposited energy in a BaF_2 detector was greater than ~5.5 MeV for γ-rays. The coincidence request eliminated any cosmic ray contamination of the γ-ray spectra. By using the above trigger there are no normalization factors in the γ-ray (charged particle) spectra as the double differential γ (charged particle) multiplicity is obtained from the ratio of the number of coincidences between γ-rays (charged particles) and evaporation residues (fission fragments) and the number of single events of evaporation (fission).

3 Analysis and Results

The experiment was designed in such a way to form the same compound nucleus at identical excitation energy from the two entrance channels by taking into account incomplete fusion events [22]. These events are characterized by emission of pre-equilibrium light particles that reduces the compound nucleus average mass, average charge and average excitation energy and cannot be discarded in the TOF spectrum of the reaction products because they have overlapping velocity distributions with those of the complete fusion events [23].

Therefore, in the present work, the average excitation energy, the average mass and the average charge of the composite system, after pre-equilibrium particle emission, were evaluated by studying the energy spectra of the light

 DOI: 10.1051/epjconf/201716300043

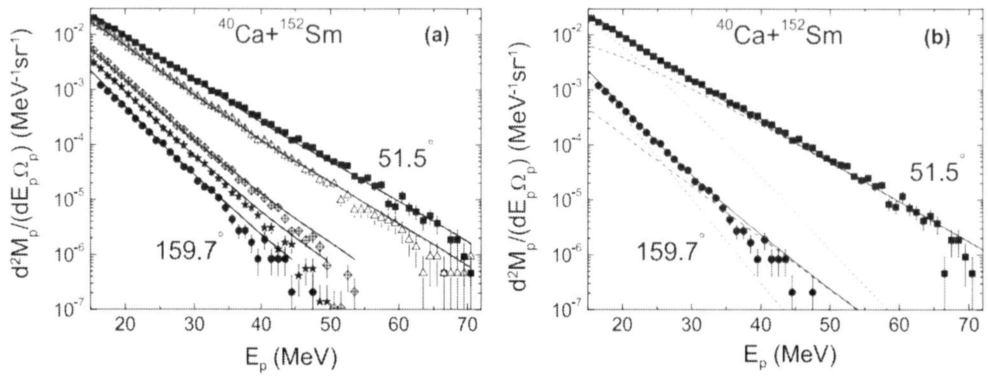

Figure 1. (a): Laboratory proton multiplicity spectra obtained at various angles in coincidence with the evaporation residues for the ^{40}Ca + ^{152}Sm reaction. The angles are from top to bottom: 51.5°, 68.1°, 111.9°, 128.5° and 159.7°. The solid lines show the result of the simultaneous fit with two sources described in the text. (b): Laboratory proton multiplicity spectra at two angles. The solid lines are the result of the fit with two sources. The dashed and dotted lines represent the *intermediate-velocity* source and the *slow* source components, respectively.

charged particles (α-particles and protons) detected in coincidence with evaporation residues and fission fragments while the pre-equilibrium neutron emission was estimated from our proton data and from existing neutron emission studies (see [17] and references therein). Here, we will present the light charged particle analysis for evaporation residues events.

3.1 Fusion–evaporation results

In Figure 1 the laboratory proton multiplicity spectra obtained with BaF_2 scintillators centered at polar angles $51.5° < \theta_{lab} < 159.7°$, in coincidence with the evaporation residues are reported for the ^{40}Ca + ^{152}Sm reaction. These spectra were analyzed by means of a moving source fit in which the particles were assumed to be emitted isotropically from two moving sources: a *slow* source describing the statistical evaporation from the hot compound nucleus and an *intermediate-velocity* (between the compound nucleus and the projectile velocity) source related to the pre-equilibrium particles emitted by the composite system before thermalization. The energy distribution of the evaporated particles was parameterized, in the source rest frame, adopting a surface-type Maxwellian distribution, while the distribution of the pre-equilibrium particles was taken to be that for volume emission from a thermal source (see [17] for more details). The same procedure was adopted for the α-particles.

The result of the simultaneous fit is shown with solid lines in panel (a) of Figure 1 for proton events and the relative contributions of the two sources are reported in panel (b) of the same figure for a backward and a forward angle, with the *slow* (*intermediate-velocity*) source component represented with a dotted (dashed) line.

To evaluate the average energy taken away by pre-equilibrium neutrons, not detected in the present experiment, we assumed that their energy spectra were very similar to the proton ones, apart from the Coulomb barrier. Then, the average kinetic energy of a pre-equilibrium neutron was taken to be that of a pre-equilibrium proton minus the Coulomb barrier while the pre-equilibrium neutron multiplicity was deduced by that of pre-equilibrium protons multiplied with the N/Z ratio of the compound nucleus. The adopted pre-equilibrium neutron multiplicity is in agreement within errors with neutron emission studies performed at similar center-of-mass incident energy above the Coulomb barrier [24].

By using the parameters extracted from the fitting procedure, the average excitation energy of the composite system after pre-equilibrium particle emission was deduced to be E^*=(219.6±6.8) MeV for the ^{40}Ca + ^{152}Sm and E^*=(220.5± 6.6) MeV for the ^{48}Ca + ^{144}Sm reaction. The average total mass and average total charge lost in each reaction were found to be: ΔA=3.2 ± 0.6 and ΔZ=1.5 ± 0.2 for the ^{40}Ca + ^{152}Sm system and ΔA=3.4 ± 0.5 and ΔZ=1.6 ± 0.2 for the ^{48}Ca + ^{144}Sm one. We considered, thus, that two units of charge and three units of mass were carried away from the initial composite system leading to the ^{189}Hg nucleus in both reactions. The average total momentum removed from the system was obtained by taking the product of the pre-equilibrium particle momentum with its multiplicity and then summing over particle types. This allowed us to derive the linear momentum transfer along the beam direction that was ~97% in both reactions.

This analysis ensured us that the two reactions lead to the formation of the same compound nucleus ^{189}Hg with the same average excitation energy, giving us confidence that any difference between the γ-ray spectra and γ-ray

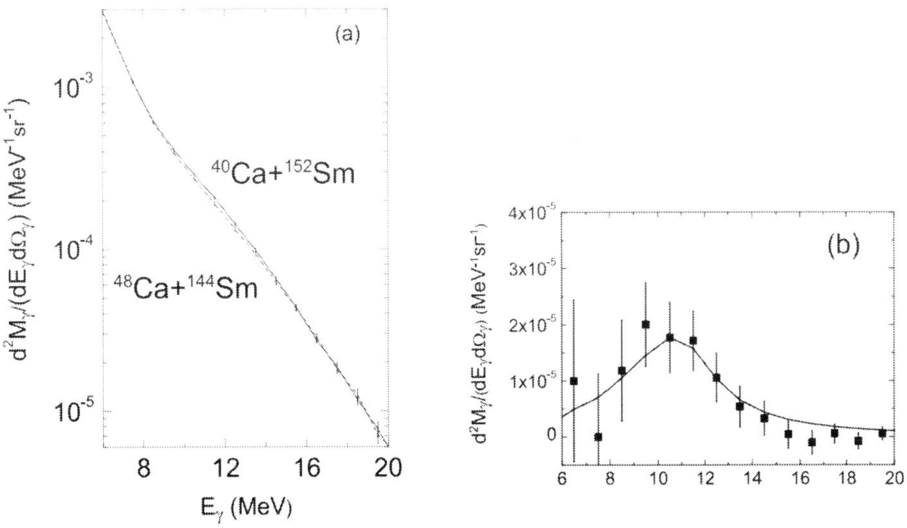

Figure 2. (a) Total center-of-mass bremsstrahlung-subtracted γ-ray multiplicity spectra obtained in coincidence with evaporation residues. The solid (dashed) line represents the charge asymmetric $^{40}Ca+^{152}Sm$ (nearly charge symmetric $^{48}Ca+^{144}Sm$) reaction; (b) the symbols correspond to the difference between the experimental spectra of panel (a), i.e., the DD γ-ray spectrum. The solid line is a Lorentzian curve folded by the experimental set up response function with characteristics described in the text.

angular distributions of the two reactions is related to an entrance channel effect.

After the evaluation of the excitation energy of both reactions, shown above, the incoherent bremsstrahlung component considered to originate primarily in neutron-proton (*np*) collisions and dominant for $E_\gamma > 30$ MeV, must be evaluated and subtracted from the data. An equal bremsstrahlung component is expected for the two reactions because of their very similar beam energy and size of the reaction partners and of the same temperature of the composite system [25, 26]. This is confirmed by the data of the two reactions, that are equal within errors for $E_\gamma > 20$ MeV. The *np* bremsstrahlung component was deduced by fitting simultaneously the center-of-mass fusion-evaporation γ-ray spectra of the two reactions at different polar angles in the energy range 30 MeV$<E_\gamma<$40 MeV. The fit was performed assuming an exponentially decreasing behavior in the *nn* center-of-mass system.

In the left-hand side panel of Figure 2 we display the center-of-mass bremsstrahlung-subtracted γ-ray multiplicity of the two reactions for fusion-evaporation events. The solid (dashed) lines represent the charge asymmetric $^{40}Ca+^{152}Sm$ (nearly charge symmetric $^{48}Ca+^{144}Sm$) reaction. The difference between the data of the two reactions, shown in the right-hand side panel of the figure, evidences an excess of γ-rays in the more charge asymmetric reaction, concentrated in the energy range E_γ=8-15 MeV. This excess can only be related to the DD excitation in the composite system of the $^{40}Ca+^{152}Sm$ reaction because of its larger charge asymmetry.

The DD γ spectrum in fusion-evaporation events is shown in Figure 2 (symbols in panel (b)) and can be reproduced well by means of a Lorentzian curve folded with the response function of the MEDEA apparatus [28] (solid line), with the DD centroid energy E_{DD} = 11 MeV and the width Γ_{DD} = 3.5 MeV. Since the DD centroid energy reflects the emitting source deformation, it is interesting to compare it with the centroid energy of the GDR, E_{GDR}, obtained by means of a statistical model calculation. We notice that $E_{DD,exp}$ is 3 MeV lower than E_{GDR}, indicating a high deformation of the emitting source during the DD γ emission. This result is in excellent agreement with expectations for a dipole oscillation along the symmetry axis of a deformed dinucleus during the early moments of the reaction [3, 4, 4, 5, 19, 29, 30] and with previous experimental works on lighter systems [10, 11, 13, 14, 19].

The DD γ-ray angular distribution provides important information, since it is a sensitive probe of the fusion dynamics and of the DD lifetime. Indeed, the amount of anisotropy, if present, is related to the interplay of the rotation angular velocity of the dinuclear system during the prompt DD emission and the instant at which this emission occurs [6, 13, 14].

Panels (a) and (b) of Figure 3 display the fusion-evaporation γ-rays angular distribution with respect to the beam axis for the two reactions (a) and for their difference (b), integrated over energy from 9 to 15 MeV and

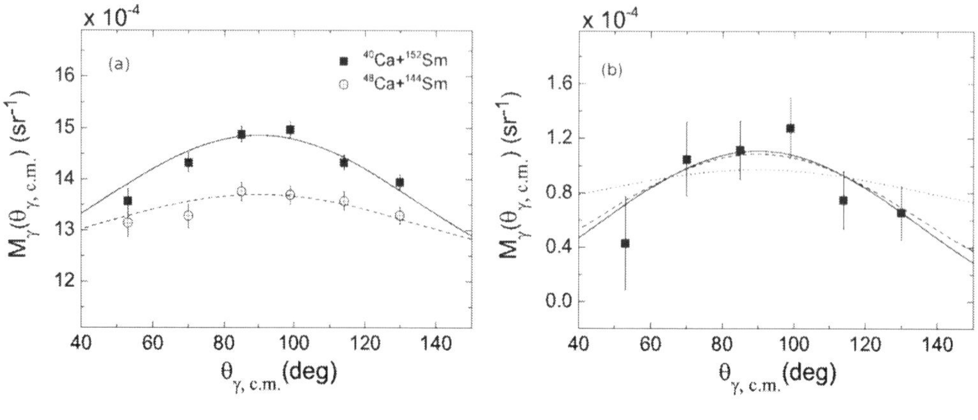

Figure 3. (a) 9 MeV$\leq E_\gamma \leq$15 MeV γ-ray angular distribution of the ^{40}Ca+^{152}Sm (full symbols) and ^{48}Ca+^{144}Sm (open symbols) reactions, corrected for the detection efficiency. The solid and dashed lines represent a fit of the data with a Legendre polynomial expansion (a_2 coefficients given in the text); (b) angular distribution of the difference between the data of the two reactions appearing in panel (a), i.e., DD γ-ray angular distribution energy-integrated in the interval 9 MeV$\leq E_\gamma \leq$15 MeV (symbols). The solid line is the result of a fit of the DD angular distribution with a Legendre polynomial expansion by using a_2=-1, the dashed line a fit with a_2=-0.84 obtained within the BNV framework with the Asystiff choice and the dotted line shows the fit with a_2=-0.25. All the quantities are in the center-of-mass reference frame.

corrected for the detection efficiency. The lines describe the angular distribution of the emitted γ-rays given by the Legendre polynomial expansion $M_\gamma(\theta_\gamma) = M_0[1 + Q_2 a_2 P_2 cos(\theta_\gamma)]$, where a_2 is the anisotropy coefficient and Q_2 is an attenuation factor for the finite γ-ray counter [31] (0.98 in our case). From a best fit to the data, shown with a solid (dashed) line for the ^{40}Ca+^{152}Sm (^{48}Ca+^{144}Sm) reaction, we found a_2 = -0.13 ± 0.03 for the ^{40}Ca+^{152}Sm reaction and a_2 = -0.06 ± 0.02 for the ^{48}Ca+^{144}Sm one. By using the same argument as previously for the spectra, the observed difference in the γ-ray angular distribution of the two systems can only be ascribed to entrance channel effects, namely the DD excitation. Consequently, the experimental angular distribution of the difference between the data of panel (a) (panel (b)) is very anisotropic around 90° and can be reproduced well with a_2 = -1 (solid line) that describes an emission from a dipole oscillation along the beam axis. The dashed line corresponds to a value of a_2 = -0.84 obtained within BNV calculations [4] for evaporation events, while the dotted one shows a more isotropic angular distribution (a_2 = -0.25). The above a_2 values indicate a preferential oscillation axis of the DD along an axis that has not rotated much on the reaction plane during the DD lifetime, confirming the pre-equibrium nature of this emission. This is in agreement with our previous results for evaporation events [13, 14] and with theoretical expectations [6].

By taking into account the DD γ-ray angular distribution (a_2 = -1) for evaporation events and the response function of the experimental setup, the DD yield, integrated over energy and over angle, is (1.2 ± 0.2)*10^{-3} [17]. The experimental results on the DD in ^{40}Ca+^{152}Sm reaction were compared with BNV calculations. These calculations give centroid energy, width and angular distribution of the DD in good agreement with those of the experiment. However, the theoretical DD yield for evaporation events overpredicted the data, calling for further investigation to clarify this aspect. The results of the present work suggest that BNV calculations do not take into account some aspects of the reaction dynamics which could inhibit the pre-equilibrium γ-ray emission. An ingredient neglected in the present calculations is the deformation of the ^{152}Sm target ground state that could influence the DD excitation mechanism. Calculations are under way to evaluate this point.

3.2 Fission: preliminary results

As mentioned before, fission fragments were detected by two position-sensitive PPACs placed at θ = 52.5°. From TOF and position information of the fragments, velocity vectors, masses and Total Kinetic Energy (TKE) in the laboratory and in the center-of-mass reference frame were obtained by using kinematical considerations. Moreover, an iterative procedure was done to compensate for the energy losses in the target, assuming that the reaction takes place at the middle of the target.

By applying appropriate conditions in the bidimensional plot (ΔE, TOF) of the PPACs and selecting the same mass and TKE distributions for two coincident fission fragments, the observables of the two reactions can be compared properly.

As done in fusion-evaporation, the evaluation of the pre-equilibrium particle emission in coincidence with fission fragments demonstrated that the compound nucleus was formed in both reactions with the same average excitation energy and mass in fission events. From this analysis, the pre-equilibrium particle emission was found to be lower than in evaporation channel, showing a dependence from the impact parameter [32] .

γ-ray - fission events were selected with a triple coincidence between γ-rays detected by MEDEA detector and two fission fragments (selected with the already mentioned conditions) detected by the PPACs. The comparison of the fission γ-ray spectra of the two reactions showed that there is a γ-ray excess coming from the more charge asymmetric reaction, in the energy range E_γ=8-15 MeV, but with a yield lower than the fusion-evaporation channel one. This preliminary outcome confirms the DD excitation also in fission events due to the entrance channel charge asymmetry effect but the reduced yield reflects the dependence on the impact parameter and, therefore, on the related reaction mechanism [8, 9].

A more accurate analysis is under way to compare these results and to investigate the impact parameter dependence of the pre-equilibrium emission in both light particles and γ-rays.

3.3 Conclusion

The present work allows to take a step forward in the study of superheavy element formation, demonstrating that the DD γ radiation, a possible cooling mechanism of the composite system along the fusion path, survives in heavy composite systems in fusion-evaporation channel. However, in order to predict evaporation residue cross sections by taking into account the isospin degree of freedom, an appropriate theoretical model is needed together with a large body of data of the DD energy- and angle-integrated multiplicity in composite systems formed under various reaction conditions.

Besides the superheavy element quest, we found that also in fission events the DD is excited with a lower yield. This preliminary observation, never observed before, provides inedited information on the DD excitation at higher partial waves.

The use of radioactive ion beams will offer the opportunity of very large entrance channel charge asymmetries, thus, maximizing the prompt dipole γ-ray and, therefore will allow us to probe the density dependence of the symmetry energy in the Equation of State at nuclear densities lower than the saturation one, where the dynamical dipole mode is active [6].

References

[1] D. Brink, *Nucl. Phys.* A **519**, 3c (1990)

[2] P. Chomaz, M. Di Toro and A. Smerzi, *Nuc. Phys.* A **563** (1993) 509

[3] C. Simenel, P. Chomaz and G. de France, *Phys. Rev. Let.* **86** (2001) 2971

[4] V. Baran *et al.*, *Phys. Rev. Let.* **86** (2001) 182501 and *Nuc. Phys.* A **679** (2001) 373

[5] M. Papa *et al.*, *Phys. Rev.* C **72**, (2005) 064608

[6] V. Baran *et al.*, *Phys. Rev.* C **79** (2009) 021603 (R)

[7] D. Pierroutsakou *et al.*, *Nucl. Phys.* A **687**, 245c (2001)

[8] D. Pierroutsakou *et al.*, *Eur. Phys. Jour.* A **16** (2003) 423

[9] F. Amorini *et al.*,*Phys. Rev.* C **69**, 014608 (2004)

[10] S. Flibotte *et al.*, *Phys. Rev. Let.* **77** (1996) 1448

[11] D. Pierroutsakou *et al.*, *Eur. Phys. Jour.* A **17** (2003) 71

[12] D. Pierroutsakou *et al.*, *Phys. Rev.* C **71** (2005) 054605

[13] B. Martin *et al.*, *Phys. Let.* B **664** (2008) 47

[14] D. Pierroutsakou, *et al. Phys. Rev.* C **80** (2009) 024612

[15] A. Corsi *et al.*, *Phys. Let.* B **679** (2009) 197

[16] A. Giaz *et al.*, *Phys. Rev.* C **90** (2014) 014609

[17] C. Parascandolo, D. Pierroutsakou *et al.*, *Phys. Rev.* C **93** (2016) 044619

[18] F. Puhlhofer, *Nucl. Phys.* A **280**, 27 (1977) and M.N. Harakeh extended version (private communication)

[19] C. Simenel, Ph. Chomaz, and G. de France, *Phys. Rev.* C **76** (2007) 024609

[20] A. Gavron, *Phys. Rev.* C **21** (1980) 230

[21] A. Del Zoppo *et al.*, *Nucl. Instr. Meth. Phys. Res.* A **327**, 363 (1993)

[22] H. Morgenstern *et al.*, *Phys. Rev. Lett.* **52**, 1104 (1984)

[23] M.P. Kelly *et al.*, *Phys. Rev.* C **56**, 3201 (1997)

[24] E. Holub *et al.*, *Phys. Rev.* C **28**, 252 (1983)

[25] E. Migneco *et al.*, *Phys. Lett.* B **298**, 46 (1993)

[26] D.G. d'Enterria *et al.*, *Phys. Lett.* B **538**, 27 (2002)

[27] E. Migneco *et al.*, *Nuc. Inst. Meth.* A **314** (1992) 31

[28] G. Bellia *et al.*, *Nuc. Inst. Meth.* A **329** (1993) 173

[29] A. S. Umar and V. E. Oberacker, *Phys. Rev.* C **76**, 014614 (2007)

[30] H. L. Wu *et al. et al.*, *Phys. Rev.* C **81**, 047602 (2010); Y. G. Ma *et al.*, *Phys. Rev.* C **85**, 024618 (2012); S. Q. Ye *et al.*, *Phys. Rev.* C **88**, 047602 (2013)

[31] M. E. Rose, *Phys. Rev.* **91** (1953) 610

[32] R. Alba, *Phys. Let.* B **322** (1994) 38

Exploring incomplete fusion fraction in ^{6,7}Li induced nuclear reactions

V. V. Parkar[1,*], V. Jha[1], and S. Kailas[1,2]

[1]Nuclear Physics Division, Bhabha Atomic Research Centre, Mumbai - 400085, India
[2]UM-DAE Centre for Excellence in Basic Sciences, Mumbai - 400098, India

Abstract. We have included breakup effects explicitly to simultaneously calculate the measured cross-sections of the complete fusion, incomplete fusion, and total fusion for ^{6,7}Li projectiles on various targets using the Continuum Discretized Coupled Channels method. The breakup absorption cross-sections obtained with different choices of short range imaginary potentials are utilized to evaluate the individual α-capture and d/t-capture cross-sections and compare with the measured data. It is interesting to note, while in case of ^{7}Li projectile the cross-sections for triton-ICF/triton-capture is far more dominant than α-ICF/α-capture at all energies, similar behavior is not observed in case of ^{6}Li projectile for the deuteron-ICF/deuteron-capture and α-ICF/α-capture. Both these observations are also corroborated by the experimental data for all the systems studied.

1 Introduction

In recent years, there is a lot of data available for complete fusion (CF) cross-sections with ^{6,7}Li and ^{9}Be projectiles on various targets [1–13]. The measured complete fusion cross-sections are found to be suppressed compared to the Coupled Channel calculations or One Dimensional barrier penetration model calculations at above barrier energies. It is also observed that the suppression factor is independent of target atomic number and it increases with decreasing the breakup threshold of the projectile; *viz.*, for ^{6}Li (Sα-d=1.47 MeV), the suppression factor is ~ 32% and for ^{7}Li (Sα-t=2.47 MeV), it is ~ 26% [3]. The origin of this suppression is still not known properly. It is possible that the fraction lost in CF can be directly attributed to incomplete fusion (ICF) fraction, where only part of the projectile fuses with the target. However, ICF can also accommodate the processes where there is a transfer of one nucleon to/from the projectile, which then breaks and subsequent absorption in the target. There are recent studies by Canberra, Australia group [14–16] on measurement of life times and the breakup probabilities of ^{6,7}Li and ^{9}Be nuclei on various targets. From their studies they have concluded that the prompt breakup/near target breakup is partially responsible for the suppression in CF.

In the experiments involving light targets, it is difficult to distinguish residues from CF and ICF channels as both can give rise to the same residual nucleus. Hence in that mass region, the measured cross-sections are denoted as total fusion (TF). In the medium and heavy mass targets, it is possible to distinguish between residues from CF and ICF, hence simultaneous measurement of CF, ICF and TF can be possible. The two major methods of online and offline gamma ray measurements are generally adopted for extraction of these residue cross-sections and hence CF and ICF cross-sections. In few studies where the residues are alpha active, *viz.*, in the case of ^{6,7}Li+^{209}Bi and ^{9}Be+^{208}Pb [2], the CF and ICF cross-sections were extracted from offline alpha counting.

In this proceedings, we report some of the intersting observations from the ICF cross-sections with ^{6,7}Li projectiles, which is described in Section 2. We have also performed coupled channels calculations using short range imaginary potential for simultaneous explanation of CF, ICF and TF cross-sections, which is described in Section 3. The summary and future outlook of the present work is given in Section 4.

2 Interesting observations from the literature on the ICF data with ^{6,7}Li projectiles

Although, there is a lot of data available for CF cross-sections with ^{6,7}Li projectiles on various targets, only limited data is available for ICF channel with these projectiles. We have used the available ICF cross-section data for ^{6,7}Li+^{209}Bi [2], ^{7}Li+^{198}Pt [17], ^{6,7}Li+^{197}Au [18] and ^{7}Li+^{159}Tb [19] reactions and discussed some of the striking features. In Fig. 1, experimental t-ICF and α-ICF cross-section data as a function of $E_{c.m.}/V_B$ for (a)^{7}Li+^{209}Bi [2], (b)^{7}Li+^{198}Pt [17], (c)^{7}Li+^{197}Au [18], and (d)^{7}Li+^{159}Tb [19] reactions are compared and shown with filled circle and filled diamond symbols, respectively. It is evident from this figure that t-ICF cross-sections are much more dominant compared to ones for α-ICF cross-sections at all the measured energies with ^{7}Li projectile. Intuitively, we expect this behavior as triton while approaching the target sees lower Coulomb barrier compared to alpha particle approaching the target, hence the cross-

*e-mail: parkarvivek@gmail.com

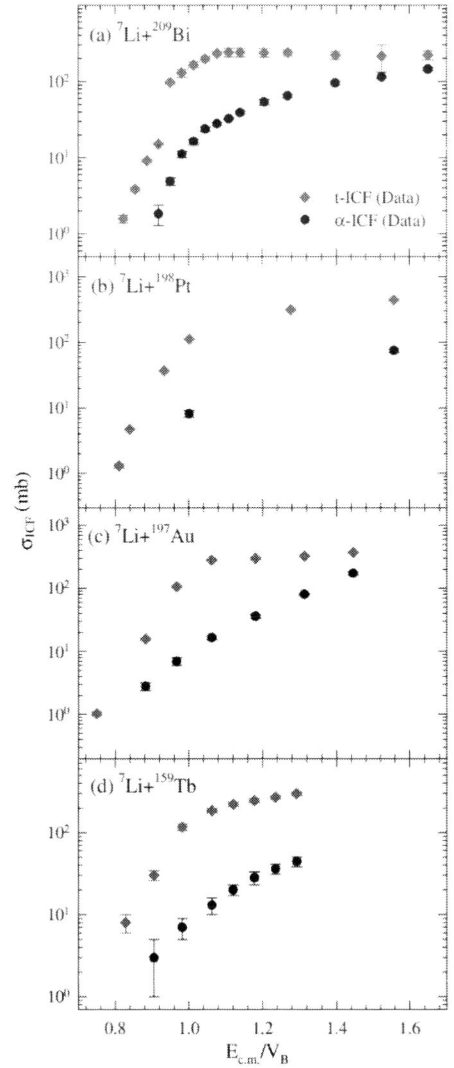

Figure 1. Experimental t-ICF and α-ICF cross-section data as a function of $E_{c.m.}/V_B$ for (a)^{7}Li+^{209}Bi [2], (b)^{7}Li+^{198}Pt [17], (c)^{7}Li+^{197}Au [18], and (d)^{7}Li+^{159}Tb [19] reactions are compared with filled circle and filled diamond symbols, respectively. It is evident from the figure that t-ICF is dominant compared to α-ICF at all the measured energies.

section for t-ICF is more compared to ones for α-ICF. It is to be noted that deuteron and proton pickups in the target would give the same ERs as those following t-ICF process and subsequent few neutron evaporation. Hence, from experiments it is difficult to separate these three processes.

In Fig. 2, experimental d-ICF and α-ICF cross-section data as a function of $E_{c.m.}/V_B$ for (a)^{6}Li+^{209}Bi [2], and (b)^{6}Li+^{197}Au [18] reactions are compared and shown with

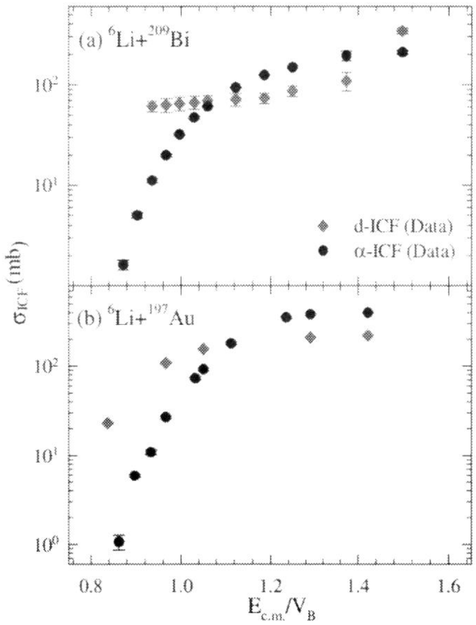

Figure 2. Experimental d-ICF and α-ICF cross-section data as a function of $E_{c.m.}/V_B$ for (a)^{6}Li+^{209}Bi [2], and (b)^{6}Li+^{197}Au [18] reactions are compared with filled circle and filled diamond symbols, respectively. It is evident that d-ICF dominates over α-ICF, but at above barrier energies α-ICF and d-ICF are of similar order.

filled circle and filled diamond symbols, respectively. From this figure, it is seen that around barrier energies, d-ICF dominates over α-ICF, but at above barrier energies α-ICF and d-ICF are of similar order. At above barrier energies, deuteron breaks into proton and neutron and subseqent escape of neutron can be one of the possible reasons for lower cross-section in d-ICF. More data are required with d-ICF and α-ICF with several targets to comment on this behavior.

3 Coupled Channels calculations for simultaneous description of CF, ICF and TF data

Recently, we have reported the method for simultaneous description of CF, ICF and TF data for the ^{6,7}Li+^{209}Bi and ^{6,7}Li+^{198}Pt reactions [20] using Continuum Discretized Coupled Channel (CDCC) calculations involving short range imaginary potential. The details of the calculation method are given in that paper and only short summary is given here. In this paper, we have shown the similar calculations for estimation of CF, ICF and TF cross-sections in ^{7}Li+^{197}Au reaction. The calculations were performed using the code FRESCO version FRESCO 2.9 [21]. The coupling scheme used in CDCC is similar to that described in our earlier works [22]. In ^{7}Li, couplings were included to the 1/2$^-$ first excited state, the 5/2$^-$ and 7/2$^-$ resonances

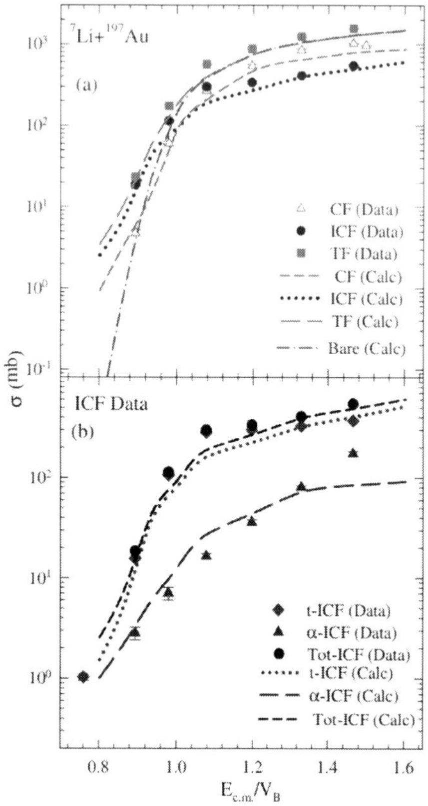

Figure 3. (a) The data of Complete Fusion (CF), Incomplete Fusion (ICF) and Total fusion (TF)=CF+ICF for ^{7}Li+^{197}Au reaction from Ref. [18] are compared with the calculations. (b) Comparison of individual ICF contributions from α-ICF, d-ICF along with Tot-ICF with the calculations. (see text for details).

and the L = 0, 1, 2, 3 α-t continuum. The binding potentials for α-t in ^{7}Li are taken from Ref.[23].

In the CDCC calculations, the fusion cross sections can be obtained as the total absorption cross section, which is equal to the difference of the total reaction cross section σ_R and the cross section σ_D of all explicitly coupled direct reaction channels. The required fragment-target potentials were generated in the cluster folding (CF) model using real potentials, *viz.*, $V_{\alpha-T}$ taken as Sao-Paulo potential [24], while V_{d-T} and V_{t-T} were taken from Refs. [25] and [26], respectively. In the calculations presented here, TF cross-sections were first calculated by including the short-range imaginary (W_{SR}) volume type potentials in the coordinates of both projectile fragments relative to the target, as in Ref. [27], whereas ICF cross-sections were calculated with short range imaginary potentials for only one fragment relative to the target. The fusion cross-section is calculated in terms of that amount of flux which leaves the coupled

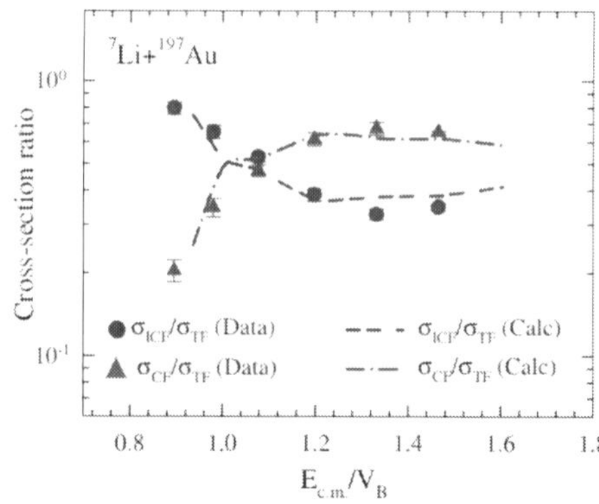

Figure 4. The ratio of cross-section, $\sigma_{\rm ICF}/\sigma_{\rm TF}$ and $\sigma_{\rm CF}/\sigma_{\rm TF}$ derived from the calculations as a function of $E_{c.m.}/V_B$ for ^{7}Li+^{197}Au reaction are shown by dashed and dashed-dot lines, respectively. The symbols denote the experimental data (see text for details).

channels set (total absorption cross-section) because of the short-range imaginary part of the optical potential used for the fragment-target potentials. The use of this short-range imaginary potential is equivalent to the use of an incoming boundary condition inside the Coulomb barrier.

In Fig. 3(a) results of the calculations for the TF, CF and ICF cross-sections are shown with long dashed, short dashed and dotted lines, respectively along with the experimental data for ^{7}Li+^{197}Au reaction. The bare calculations (without breakup couplings) were also performed and the calculated fusion cross-sections are denoted by dashed-dot-dot lines in the above mentioned figure. It is seen that at energies above the Coulomb barrier, the calculations which include the couplings and calculations that omit them have negligible difference, but at energies below the barrier, the coupled TF cross-sections are enhanced in comparison to bare TF cross-sections. The breakup of the projectile nucleus can take place (i) spontaneously in the nuclear or Coulomb field of the target or (ii) it may be triggered by the transfer processes. The latter process is found to be important in some recent works [15, 16] which could explain partially the CF suppression. We have considered the breakup absorption taking place only via the first route in our calculations. To describe the experimental ICF data, the breakup absorption has been calculated with a modified value of imaginary radius parameter and thus it effectively models the full breakup absorption.

The corresponding individual ICF cross-sections, $\sigma_{\alpha\text{-ICF}}$ and $\sigma_{\text{t-ICF}}$, are extracted and shown in Fig. 3(b). The long dashed, dotted and short dashed lines denote the α-ICF, t-ICF and Tot-ICF calculations, respectively. From the figure, it is evident that the t-ICF is dominant than α-ICF and almost equal to the Total ICF cross-section.

The ratio of cross-sections, $\sigma_{\rm ICF}/\sigma_{\rm TF}$ and $\sigma_{\rm CF}/\sigma_{\rm TF}$ derived from the calculations as a function of $E_{c.m.}/V_B$ for ^{7}Li+^{197}Au reaction are shown by dashed and dashed-dot lines in Fig. 4, respectively. The corresponding experimental data of σ_{ICF}/σ_{TF} and σ_{CF}/σ_{TF} are shown with filled circle and filled triangle in Fig. 4, respectively. From the figure it is evident that (i) Ratios of σ_{ICF}/σ_{TF} and σ_{CF}/σ_{TF} remain approximately constant over the energy range above the Coulomb barrier. (ii) For energies below the barrier, the σ_{ICF}/σ_{TF} ratio increase while σ_{CF}/σ_{TF} ratio decrease. This shows the dominance of ICF at below barrier energies in TF over CF cross-sections. The σ_{ICF}/σ_{TF} ratio at above barrier energies gives the value of suppression factor in CF, which is found to be in agreement (~ 30 %) with the literature data with ^{7}Li projectiles from various measurements [3]. Similar observation was also indicated in ^{6,7}Li+^{209}Bi and ^{6,7}Li+^{198}Pt reactions [20].

4 Summary and Futute Outlook

In summary, from the available experimental data, we have highlighted the differences in individual ICF cross-sections for ^{6}Li and ^{7}Li projectiles on various targets. The individual ICF cross-sections imply that the d-ICF cross-section is of similar order that of α-ICF cross-section in case of ^{6}Li, while t-ICF cross-section is much more than α-ICF cross-section in case of ^{7}Li. More data of d-ICF, α-ICF and t-ICF simultaneously for various systems are required to further emphasize this observation.

We have performed Continuum Discretized Coupled Channel calculations using cluster folding potentials in the real part along with short-range imaginary part for calculation of CF, ICF and TF cross-sections for ^{7}Li+^{197}Au reaction. The simultaneous explanation of the measured experimental data for the CF, ICF and TF cross-sections over the entire energy range is obtained using calculations in the full quantum mechanical approach. The calculated ICF fraction which is the ratio of ICF and TF as a function of energy is found to be constant at energies above the barrier and and it increases at energies below the barrier. This ratio which signifies the suppression of CF in TF is constant at above barrier energies and it is in agreement with the available data for several systems. At below barrier, as the ratio increases, it shows the enhanced importance of ICF contribution in TF at below barrier energies. Similar calculations are in progress to study the systematic behavior of individual ICF fractions with ^{6,7}Li projectiles for different target systems.

References

[1] L. F. Canto, P. R. S. Gomes, R. Donangelo, J. Lubian, and M. S. Hussein, Phys. Rep. **596**, 1 (2015) and references therein.
[2] M. Dasgupta *et al.*, Phys. Rev. **C 70**, 024606 (2004).
[3] V. V. Parkar *et al.*, Phys. Rev. **C 82**, 054601 (2010).
[4] C. S. Palshetkar *et al.*, Phys. Rev. **C 82**, 044608 (2010).
[5] Y. D. Fang *et al.*, Phys. Rev. **C 91**, 014608 (2015).
[6] P. R. S. Gomes *et al.*, Phys. Rev. **C 73**, 064606 (2006).
[7] H. Kumawat *et al.*, Phys. Rev. **C 86**, 024607 (2012).
[8] P. K. Rath *et al.*, Nucl. Phys. **A 874**, 14 (2012).
[9] P. K. Rath *et al.*, Phys. Rev. **C 79**, 051601R (2009).
[10] M. K. Pradhan *et al.*, Phys. Rev. **C 83**, 064606 (2011).
[11] A. Mukherjee *et al.*, Phys. Letts. **B 636**, 91 (2006).
[12] S. P. Hu *et al.*, Phys. Rev. **C 91**, 044619 (2015).
[13] Y. W. Wu *et al.*, Phys. Rev. **C 68**, 044605 (2003).
[14] D. H. Luong *et al.*, Phys. Rev. **C 88**, 034609 (2013).
[15] Sunil Kalkal *et al.*, Phys. Rev. **C 93**, 044605 (2016).
[16] K. J. Cook *et al.*, Phys. Rev. **C 93**, 064604 (2016).
[17] A. Shrivastava *et al.*, Phys. Lett. **B 718**, 931 (2013).
[18] C. S. Palshetkar *et al.*, Phys. Rev. **C 89**, 024607 (2014).
[19] R. Broda *et al.*, Nucl. Phys. **A 248**, 356 (1975).
[20] V. V. Parkar, V. Jha and S. Kailas, Phys. Rev. **C 94**, 024609 (2016).
[21] I. J. Thompson, Comput. Phys. Rep. **7**, 167 (1988).
[22] V. Jha and S. Kailas, Phys. Rev. **C 80**, 034607 (2009).
[23] B. Buck and A. C. Merchant, J. Phys. G: Nucl. Part. Phys. **14**, L211 (1988).
[24] L. C. Chamon *et al.*, Phys. Rev. **C** 66, 014610 (2002).
[25] W. W. Daehnick, J. D. Childs, and Z. Vrcelj, Phys. Rev. **C 21**, 2253 (1980).
[26] F. D. Becchetti and G. W. Greenlees, Phys. Rev. **182**, 1190 (1969).
[27] A. Diaz-Torres, I. J. Thompson, and C.Beck, Phys. Rev. **C 68**, 044607 (2003).

 DOI: 10.1051/epjconf/201716300045

Scattering of halo nuclei on heavy targets at energies around the Coulomb barrier

The case of ^{11}Be on ^{197}Au

V. Pesudo[1,2,⋆], M.J.G. Borge[3,4], A.M. Moro[5], J.A. Lay[5], E. Nácher[3], J. Gómez-Camacho[5,6], O. Tengblad[3], L. Acosta[7], M. Alcorta[8], M.A.G. Alvarez[9], C. Andreoiu[10], P.C. Bender[11], R. Braid[12], M. Cubero[13], A. Di Pietro[14], J.P. Fernández-García[5], P. Figuera[14], M. Fisichella[14], B.R. Fulton[15], A.B. Garnsworthy[11], G. Hackman[11], U. Hager[16], O.S. Kirsebom[17], K. Kuhn[12], M. Lattuada[14,18], G. Marquínez-Durán[19], I. Martel[19], D. Miller[11], M. Moukaddam[11], P.D. O'Malley[5], A. Perea[5], M.M. Rajabali[11], A.M. Sánchez-Benítez[19], F. Sarazin[12], V. Scuderi[14], C.E. Svensson[20], C. Unsworth[11], and Z.M. Wang[10,11]

[1]Department of Physics, University of the Western Cape, P/B X17, Bellville ZA-7535, South Africa
[2]iThemba LABS, Somerset West 7129, South Africa
[3]Instituto de Estructura de la Materia, CSIC, 28006 Madrid, Spain
[4]ISOLDE-EP, CERN, CH-1211 Geneva 23, Switzerland
[5]Dept de Física Atómica, Molecular y Nucl., Univ. de Sevilla, 41080 Sevilla, Spain
[6]CN de Aceleradores (U. Sevilla, J. Andalucía, CSIC), 41092, Sevilla, Spain
[7]Instituto de Física. Universidad Nacional Autónoma de México, A.P. 20-364, Cd.Mx. 01000 Mexico
[8]Physics Division, Argonne National Laboratory, Argonne, Illinois 60439, USA
[9]Instituto de Física, Universidade de São Paulo, 05508-090, São Paulo, Brazil
[10]Dept. of Chemistry, Simon Fraser University, Burnaby, BC, Canada V5A 1S6
[11]TRIUMF, 4004 Wesbrook Mall, Vancouver, British Columbia V6T 2A3, Canada
[12]Physics Department, Colorado School of Mines, Golden, Colorado 80401, USA
[13]CICANUM, Universidad de Costa Rica, Apdo. 2060, San José, Costa Rica
[14]Laboratori Nazionali del Sud, INFN, via Santa Sofia 62, 95123, Catania, Italy
[15]Department of Physics, University of York, YO 10 5DD Heslington, York, UK
[16]National Superconducting Cyclotron Laboratory, Michigan State University, East Lansing, MI 48824, USA
[17]Department of Physics and Astronomy, Aarhus University, 8000 Aarhus, Denmark
[18]Dipart. di Fisica e Astronomia, via Santa Sofia 64, 95123, Catania, Italy
[19]Departamento de Ciencias integradas, Universidad de Huelva, 21071 Huelva, Spain
[20]Dept. of Physics, University of Guelph, Guelph, Ontario N1G 2W1, Canada

Abstract. This work reports on the scattering of ^{11}Be on ^{197}Au at energies around and below the Coulomb barrier. By experimentally identifying the elastic scattering, inelastic scattering and breakup channels, and comparing them with different calculations, valuable information on the ^{11}Be structure and its $B(E1)$ distribution to the continuum are obtained. On top of that, a deeper understanding of the scattering process at low energies is achieved for reactions of this kind, making these studies extendable to other loosely-bound systems like ^{17,19}C.

1 Introduction

Low energy nuclear reactions have been a major source of knowledge in the field of nuclear physics since its very inception. With the constant development of new targets and ion sources, more and more nuclei can be produced and accelerated via the ISOL method, so low-energy nuclear reactions can be used to study more exotic nuclei. Hence, experimental techniques and theoretical approaches must be revisited for new data and calculations to continue providing significant contributions to the field.

It is well known that weakly bound nuclei, such as halo nuclei, have a strong $B(E1)$ to the states in the continuum, which means that undergo breakup relatively easily when exposed to a Coulomb field. This has been thoroughly studied at intermediate and high energies using heavy targets and restricting to very small scattering angles, and the $B(E1)$ to the continuum of halo nuclei has been extracted [1–4]. These measurements, however, are not straightforward and depending on the structure model and the treatment of the nuclear component, the results can differ significantly in absolute value, like in the case of [2] and [4].

⋆e-mail: vicentepesudo@tlabs.ac.za

DOI: 10.1051/epjconf/201716300045

This work is part of a series of experiments studying reactions of light halo nuclei on heavy targets at energies close the Coulomb barrier [5–14]. These experiments aim at understanding the dynamics of reactions involving a weakly bound light halo nucleus in a slowly varying intense Coulomb field. The field is felt by the projectile for a long time and adiabatic rearrangements can happen, leaving it in a more favorable configuration to break up and modifying the differential cross sections for the main reaction channels. This work, first, shows the dominance of the Coulomb interaction in this scenario and then validates the method for extracting valuable information on the reaction dynamics and structure properties (such as $B(E1)$ distributions) of weakly bound nuclei. This property had so far mainly been obtained using high energy beams and this is the first time that the full analysis is consistent with former experiments performed at higher energies [1, 4].

The case of ^{11}Be ($J^{\pi} = 1/2^{+}$ in its ground state) is particularly interesting because it has a $1/2^{-}$ bound excited state at $E_x = 320$ keV and because the ^{10}Be core is well deformed ($\beta_2 = 0.67$). The main reaction channels are, hence, elastic scattering, inelastic scattering and breakup ($S_n = 0.5$ MeV [15]). The experimental angular distribution of all of them, compared with different calculations, provides very valuable information on the reaction process and the structure of the ^{11}Be projectile.

2 Experimental setup

The experiment was performed at TRIUMF, Vancouver, Canada. The 500 MeV proton beam from the main cyclotron hit a 20 g/cm^2 Ta target in ISAC-II and the ^{11}Be radioactive beam is extracted using the laser ion source TRILIS. Two beam energies were chosen, one around the Coulomb barrier ($E_{c.m.} = 37.1$ MeV, $V_b = 40$ MeV) and another clearly below the barrier ($E_{c.m.} = 29.6$ MeV). The average current on target was 10^5 pps.

Four silicon ΔE - E telescopes for the identification of the charged ejectiles were placed inside the scattering chamber (see Fig. 1). Three of them consisted on a 40 μm double-sided silicon strip detectors (DSSSD) backed with a 500 μm PAD and the fourth, at backward angles, consisted on a 20 μm Single-sided silicon strip detector backed with a 300 μm DSSSD. Having a DSSSD in every telescope allowed us to have high granularity and resolution, separating the ^{11}Be from the ^{10}Be fragments after breakup in ΔE - E two-dimensional plots (see Fig.2). The detectors were directly mounted on a printed circuit board (PCB) and the PCB had been previously cabled to the feedthroughs, reducing the risk of damage of the detectors during the setup.

Around the scattering chamber there were 12 high-purity germanium clovers of TIGRESS, each of them consisting of four eight-fold segmented crystals. The coincidence of the 320 keV gamma ray with ^{11}Be fragment was used to identify the inelastic scattering (Fig. 3), and the

Figure 1. CAD image of the setup, with the four silicon telescopes mounted on a PCB. The pipe on the right hand side of the caption correspond to the upstream side of the chamber.

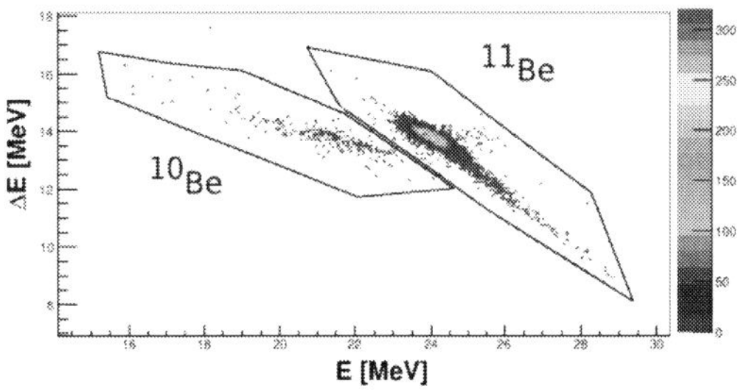

Figure 2. Energy deposited in the back detector in the x axis versus energy deposited in the ΔE detector in the y axis. The ^{10}Be and ^{11}Be fragments, corresponding to breakup and quasielastic scattering can be separated.

coincidence with the 279 gamma ray from ^{197}Au for estimating the excitation of the target. The segmentation of the germanium and silicon detectors allowed for a Doppler correction of the 320 keV gamma ray emitted in-flight (Fig. 4).

3 Theoretical models

The results are compared with optical model (OM) calculations, first-order semiclassical calculations (the equivalent photon method, EPM, from Ref. [16]), continuum-discretized coupled-channels calculations (CDCC) in which the ^{11}Be was described using single-particle states of the neutron with respect to the ^{10}Be core, and CDCC including the ^{10}Be core excitation degree of freedom (XCDCC).

The optical model calculation was performed using central Woods-Saxon potentials for both the real and the imaginary parts. In the semiclassical model the projectile is considered to follow a classical Coulomb trajectory and the $E1$ excitation is treated perturbatively, both for the

 DOI: 10.1051/epjconf/201716300045

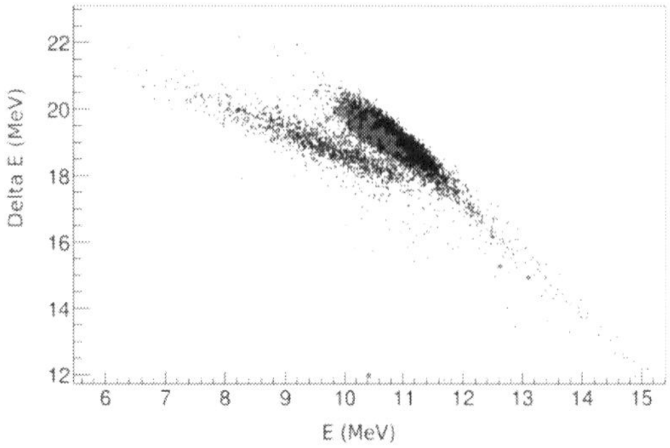

Figure 3. Two dimensional plot with the coincidence with the 320 keV gamma ray depicted in red.

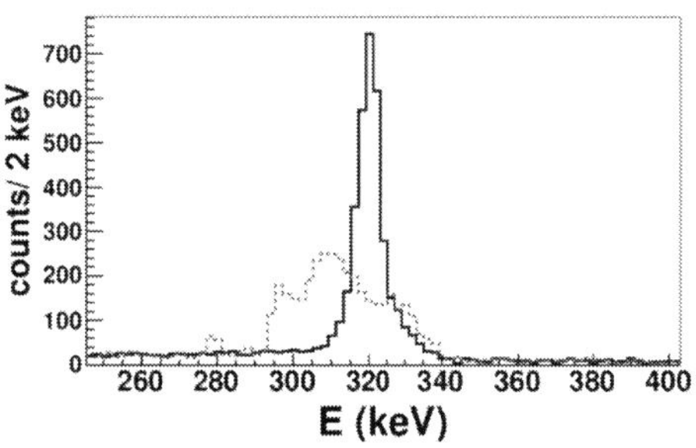

Figure 4. Raw and Doppler-corrected gamma spectrum in coincidence with T1, the telescope at foremost angles.

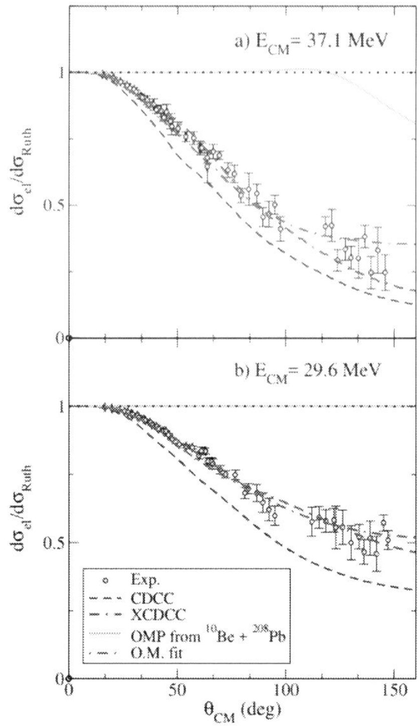

Figure 5. Measured differential elastic scattering cross section at (a) E_{cm} = 37.10 MeV and (b) 29.64 MeV, compared with the CDCC and XCDCC calculations described in the text and within an optical model using central Woods-Saxon potentials.

inelastic scattering and for the breakup. Since the calculation is performed at first order, it does not account for any rearrangement or adiabatic evolution of the incoming nuclei.

Two CDCC calculations were also performed. The first one uses a single particle (SP) structure model for the ^{11}Be, considering only the states of the neutron in a spherical potential of a frozen ^{10}Be core ($j^{\pi} = 0^+$). The second one has an added level of sophistication and takes into account that the neutron can be coupled to the ^{10}Be in its ground state, but also in a 2^+ excited state. In the model used (from Ref. [17]), the ^{10}Be ground state has a permanent deformation of $\beta_2 = 0.67$ [18]. The coupling with the 2^+ state is included within the rotational model. The ground state, for instance, will be an admixture of $|^{10}\mathrm{Be}(0^+) \otimes \nu(s_{1/2})\rangle_{1/2^+}$, $|^{10}\mathrm{Be}(2^+) \otimes \nu(d_{3/2})\rangle_{1/2^+}$ and $|^{10}\mathrm{Be}(2^+) \otimes \nu(d_{5/2})\rangle_{1/2^+}$. It is worth pointing out that the first one is the actual *halo configuration*, since the centrifugal barrier of the *d* orbit hinders the development of the halo. For further details on the calculations, see Ref. [19].

4 Discussion

The first remarkable feature is the deviation of the elastic scattering cross section from Rutherford cross section at very low scattering angles, even at energies well below the barrier. This is an indication of strong dipole polarizability [20] and a pattern often observed in weakly bound nuclei, specially pronounced in halo nuclei [11, 12]. In order to reproduce this behaviour within the OM, unusually large values for the diffuseness were needed (see table 1). The diffuseness gives an idea of the steepness of the potential and such values are a signature of the presence of long-range couplings.

Also within the OM, a study of the radius of sensitivity was performed. It consisted on fixing the value of the imaginary diffuseness to a certain value, and perform the minimization releasing the rest of parameters. The χ^2 of the different minimizations is presented in Fig. 6, and in Fig. 7 the imaginary part of the potentials obtained with the different minimizations are shown. The fact that all these potentials, minimized independently, cross each other at a particular radius evidences that the reaction is specially sensitive to the value of the potential in that point. The sensitivity radius we obtain is $R_s \sim 35$ fm,

Table 1. Values of the parameters for the real and imaginary part of the nuclear potential and χ^2 obtained in the minimization. For comparison, the optical model parameters for the scattering of the core nucleus, ^{10}Be, on ^{208}Pb are also presented.

	Core ^{10}Be+^{208}Pb [21]	$E_{c.m.}$ = 37.10 MeV	$E_{c.m.}$ = 29.64 MeV
V (MeV)	113.	14.0	9.2
r_V (fm)	1.1	1.2	1.2
a_V (fm)	**0.6**	**3.1**	**3.8**
W (MeV)	169	0.21	0.179
r_W (fm)	1.196	1.2	1.2
a_W (fm)	**0.30**	**8.68**	**8.73**
χ^2		1.3	1.0

much larger than the sum of the radii of the colliding nuclei: $R(^{11}\mathrm{Be}) + R(^{197}\mathrm{Au})$ = 7.3 fm + 7.0 fm = 14.3 fm. This result is a clear indication of the importance of long range couplings, i.e., of the dominance of the Coulomb interaction.

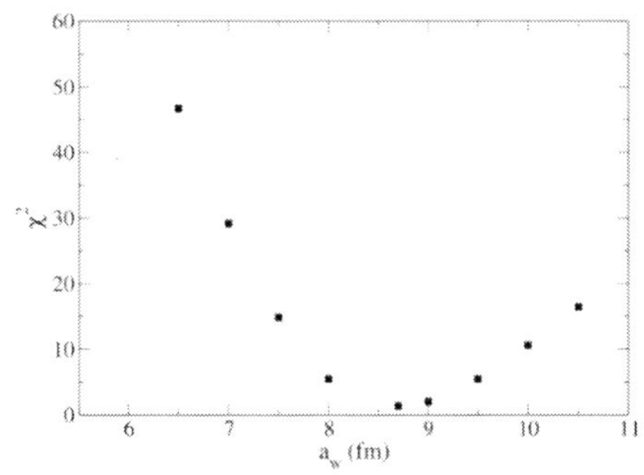

Figure 6. Value of the χ^2 obtained for different values of the imaginary diffuseness around the minimum (a_w = 8.7 fm).

For a proper comprehension of the information that the CDCC and conveyed by the XCDCC, it is convenient to discuss the other two reaction channels first. The inelastic scattering and the breakup cross sections are shown in Fig. 8 and Fig. 9, respectively. One can see that the EPM reproduces reasonably well the inelastic scattering cross section at both energies, but fails in reproducing the breakup cross section. Thus, the excitation to the 320 keV bound excited state can be understood as a pure first-order E1 process, while the excitation to the states in the continuum needs of higher-order couplings with higher-order multipolarities.

The case of the CDCC is the opposite, the breakup cross section is reproduced satisfactorily while the inelastic one is not. For the breakup channel to be reproduced, the model space that had to be considered

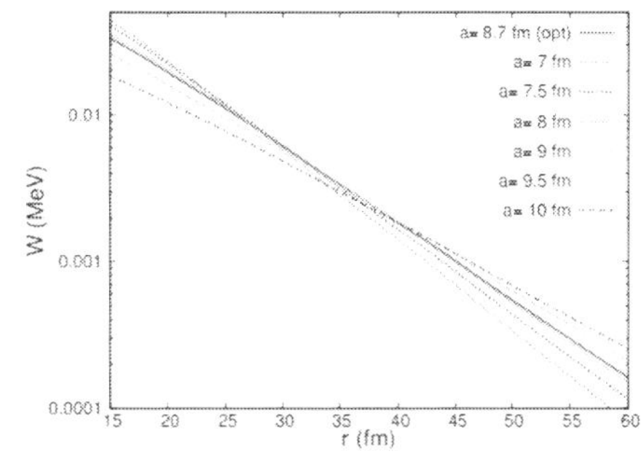

Figure 7. Imaginary part of the potential obtained with the different χ^2 minimization shown in Fig. 6. The plot has been zoomed in order to distinguish the crossing point of the different potentials.

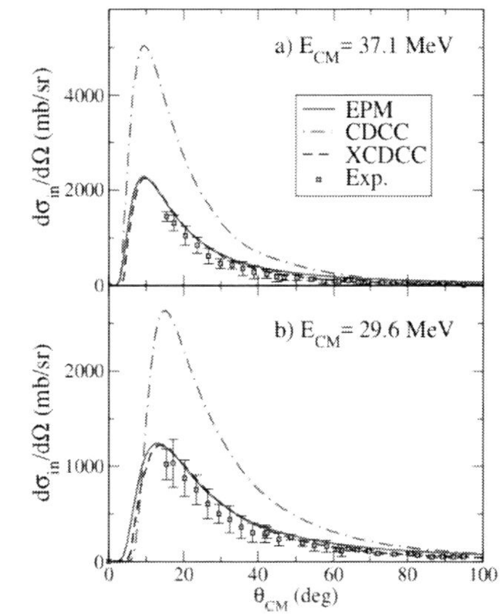

Figure 8. Angular distribution of the inelastic differential cross section of ^{11}Be + ^{197}Au, populating the $1/2^-$ bound excited state in ^{11}Be for (a) $E_{c.m.}$ = 37.10 MeV and (b) 29.64 MeV. Experimental data compared with EPM, CDCC and XCDCC calculations described in the text.

in the calculation was large: states in ^{11}Be with $j^\pi \leq 15/2^\pm$ and excitation energies $E_x \leq 12$ MeV were included. The importance of both, high-order (including continuum-continuum couplings) and high-multipolarity couplings, was apparent. The total angular momentum of the reaction, which is classically related with the impact parameter, that needed to be included in the calculations for obtaining convergence is also large, $J < 1200$, which reinforces the conclusions obtained via the study of the radius of sensitivity. Calculations varying the depth of the nuclear potential by a 20%, were also performed, clarifying the dominance of the Coulomb potential.

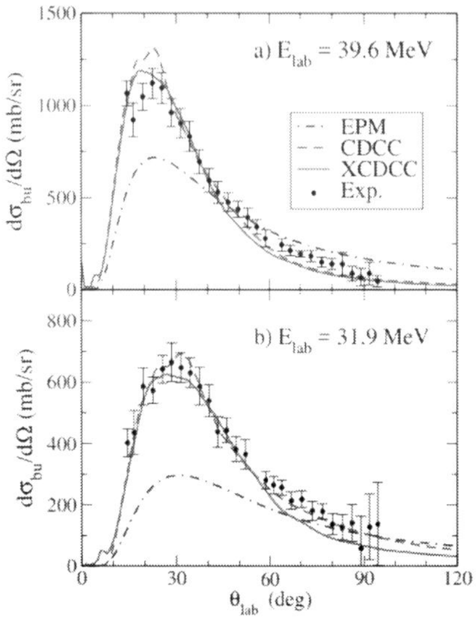

Figure 9. Differential breakup cross section in the laboratory frame at (a) E_{lab} = 39.6 MeV and (b) 31.9 MeV. Experimental data compared with EPM, CDCC and XCDCC calculations described in the text.

On the other hand, the CDCC calculation clearly overestimates the inelastic scattering. This mismatch was expected, since the experimental $B(E1)$ to the bound excited state ($B(E1: 1/2^+ \rightarrow 1/2^-)$= 0.116 e^2 fm^2 [22]) is overestimated by a factor of two by the adopted single-particle model, taken from Ref. [23]. The flux added to this channel is mainly subtracted from the elastic channel, which explains the failure of CDCC in explaining the elastic data. This is not an intrinsic limitation of this parametrization specifically, but of the SP model itself. It is not possible to reproduce the three main outgoing channels: elastic, inelastic and breakup, considering the states of a neutron with respect to a frozen ^{10}Be core. A model that reduces the $B(E1)$ to the bound excited state, also reduces the $B(E1)$ to the continuum. In this context the development of the XCDCC formalism was necessary [24, 25].

Introducing a more sophisticated model for the ^{11}Be structure, both observables can be reproduced within the same coupled-channels calculation. In particular, we used for the ^{11}Be structure a particle-rotor (PR) model with the parametrization of Ref. [17]. The improvement observed in reproducing the experimental data with respect to the CDCC is because the PR model gives a more realistic $B(E1)$than the SP model. The improvement with respect to the EPM is because high-order continuum-continuum couplings are included in the coupled-channels calculations.

The slight overestimation in the inelastic cross section by both the XCDCC and the EPM reinforces the value for the $B(E1: 1/2^+ \rightarrow 1/2^-) = 0.102$ e^2 fm^2 obtained by Kwan et al. in Ref. [26] over the previous value of $B(E1) = 0.116$ e^2 fm^2 obtained by Millener et al. in Ref. [22] and used in these calculations.

Fig. 10 shows the $B(E1)$ extracted from previous works at higher energies (Refs. [2, 4]) and the one extracted from the structure models used in this work. The model from Ref. [17] that explains all our observables, reproduces the $B(E1)$ obtained by Fukuda et al. in Ref. [4]. The data in Ref. [2], despite following the same pattern, differs in absolute magnitude, so this work has the added value of disentangling this incompatibility. Another interesting feature is that the 3.37 MeV gamma ray from the decay of the 2^+ to the 0^+ ground state in ^{10}Be was not observed. Although there is a non-negligible contribution of states above the energy of the 2^+, the average excitation energy is around 1 MeV (well below the 3.87 MeV that would be needed for this decay channel to be open). This observation, however, is consistent with the nucleus being predominantly broken up from its *halo* configuration, with the neutron coupled to the ^{10}Be ground state.

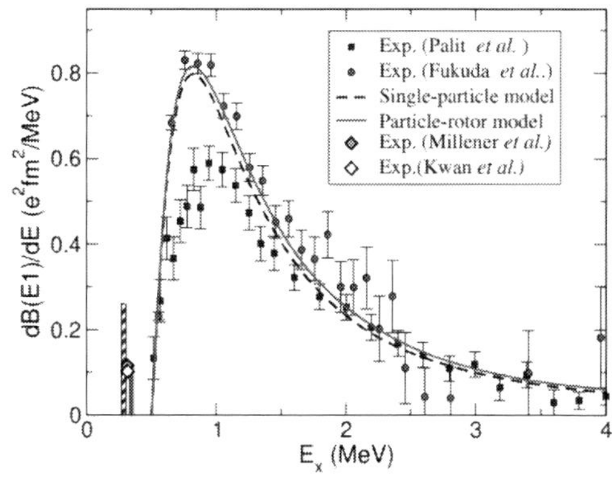

Figure 10. $B(E1)$ strength distributions for ^{11}Be. The experimental data in the continuum are from the Coulomb dissociation experiments of Refs. [2, 4]. The filled diamond represent the experimental $B(E1; 1/2^+ \rightarrow 1/2^-)$ value (in e^{2}fm^2) between the bound states of ^{11}Be [22, 26] and the vertical bars the output of the calculation. The theoretical distributions correspond to the SP and PR models.

5 Conclusions

The three main reaction channels, namely, elastic scattering, inelastic scattering and breakup have been measured and separated at energies around and below the Coulomb barrier for the ^{11}Be + ^{197}Au reaction. The dominance of the Coulomb interaction in this regime over most of the angular range has been inferred independently via optical model and CDCC calculations. The XCDCC calculation reproduces all the observables simultaneously and, together with the experimental procedure presented here, is a powerful tool to understand the dynamics of the reaction and are complementary to high-energy Coulomb breakup measurements for determining structure properties of the weakly bound nuclei involved, such as the $B(E1)$. The $B(E1)$ obtained with Tarutina's model in Ref. [17] reproduces our data and the data measured in RIKEN at intermediate energies, validating these results.

The reaction dynamics of the scattering process are strongly determined by the slowness of the collision, during which the weakly bound nucleus has time to evolve adiabatically into configurations from which breakup is specially favorable, enhancing this cross section with respect to the pure first-order direct breakup.

Acknowledgements

This work was supported by the Spanish Government under the projects FPA2015-64969-P, FIS2014-53448-C2-1-P, FPA2013-47327-c02-01-R, FPA2012-32443, FIS2013-41994-P and Consolider CPAN CSD2007-00042; the EUROCORES project EUI-2009-04162;the European Unions Horizon 2020 research and innovation program under grant agreement No 654002; the Helmholtz Association (HGT) through the Nuclear Astrophysics Virtual Institute (VH-VI-417); ATI Sistemas; The Natural Sciences and Engineering Research Council of Canada; The Canada Foundation for Innovation; and the British Columbia Knowledge Development Fund. TRIUMF receives federal funding via a contribution agreement through the National Research Council of Canada. OSK acknowledges support from the Villum Foundation.

References

[1] T. Nakamura *et al.*, Phys. Lett. B **331B**, 296 (1994)
[2] R. Palit *et al.*, Phys. Rev. C **68**, 034318 (2003)
[3] T. Nakamura *et al.*, Phys. Rev. Lett. **96**, 252502 (2006)
[4] N. Fukuda *et al.*, Phys. Rev. C **70**, 054606 (2004)
[5] M. Mazzocco *et al.*, Eur. Phys. J. A **28**, 295 (2006)
[6] D. Escrig *et al.*, Nucl. Phys. A **792**, 2 (2007)
[7] A. M. Sánchez-Benítez *et al.*, Nucl. Phys. A **803**, 30 (2008)
[8] L. Acosta *et al.*, Eur. Phys. J. A **42**, 461 (2009)
[9] L. Acosta *et al.*, Phys. Rev. C **84**, 044604 (2011)
[10] A. Di Pietro, *et al.*, Phys. Rev. Lett. **105**, 22701 (2010)
[11] M. Cubero, *et al.*, Phys. Rev. Lett. **109**, 262701 (2012)
[12] A. Di Pietro, *et al.*, Phys. Rev. C **85**, 054607 (2012)
[13] J. P. Fernández-García *et al.*, Phys. Rev. Lett. **110**, 142701 (2013)
[14] J. P. Fernández-García *et al.*, Phys. Rev. C **92**, 044608 (2015)
[15] J.H. Kelley *et al.*, Nucl. Phys. A **88**, 880 (2012)
[16] K. Alder, A. Winther, *Electromagnetic excitation* (North Holland, Amsterdam, 1975)
[17] T. Tarutina, L.C. Chamon, M.S. Hussein, Phys. Rev. C **67**, 044605 (2003)
[18] F.M. Nunes *et al.*, Nucl. Phys. A **609**, 43 (1996)
[19] V. Pesudo *et al.*, Phys. Rev. Lett. **118**, 152502 (2017)
[20] M. V. Andres, J. Gomez-Camacho, Phys. Rev. Lett. **82**, 001387 (1999)
[21] J.J. Kolata *et al.*, Phys. Rev. C **69**, 047601 (2004)
[22] D.J. Millener, J.W. Olness, E.K. Warburton, S.S. Hanna, Phys. Rev. C **28**, 497 (1983)
[23] P. Capel, G. Goldstein, D. Baye, Phys. Rev. C **70**, 064605 (2004)
[24] N.C. Summers, F.M. Nunes, I.J. Thompson, Phys. Rev. C **73**, 031603 (2006)
[25] R. de Diego, J.M. Arias, J.A. Lay, A.M. Moro, Phys. Rev. C **89**, 064609 (2014)
[26] E. Kwan *et al.*, Phys. Lett. B **732**, 210 (2014)

Fusion reactions induced by radioactive beams: the $^{18}F(p,\alpha)^{15}O$ case

R.G Pizzone[1,*], B.T. Roeder[2], L. Trache[3], R.E. Tribble[24], C. Spitaleri[15], S. Cherubini[15], M. Gulino[16], I. Indelicato[1], M. La Cognata[1], L. Lamia[1], G.G. Rapisarda[1], and R. Spartá[1]

[1] INFN - Laboratori Nazionali del Sud, Catania, Italy
[2] Cyclotron Institute, Texas A& M University, College Station, TX, USA
[3] IFIN-HH, Bucharest-Magurele, Romania
[4] Brookhaven National Laboratory, USA
[5] Dipartimento di Fisica e Astronomia, Università degli Studi di Catania, Catania, Italy
[6] KORE University, Enna, Italy

Abstract. Gamma ray astronomy has made big strides in the last decades paving the way to a better understanding of explosive nucleosynthesis. In particular, crucial information on novae nucleosynthesis is linked to the abundance of the ^{18}F isotope, which might be detected in explosive environments. Therefore, the reaction network producing and destroying this radioactive isotope has been extensively studied in the last years. Among those reactions, the $^{18}F(p,\alpha)^{15}O$ cross section has been measured by means of several dedicated experiments, both using direct and indirect methods. The presence of resonances in the energy region of astrophysical interest has been reported by many authors. In the present work a report on a recent experiment performed via the Trojan Horse Method (THM) at the Texas A&M Cyclotron Institute is presented and the results are given and compared with the ones known in the literature, both direct and indirect. Data arising from THM measurements are then averaged and the reaction rate calculated in the novae energy range. Hints on future astrophysical applications will also be given.

1 Introduction

Gamma ray emission from novae detected in dedicated satellite-borne experiments has become a probe for understanding novae explosions as well as the structure of such exotic stellar objects. In particular, it was noted [1] that electron-positron annihilation should occur in nova envelopes, since short-lived β^+ unstable radioactive nuclei (i.e., positron emitters) are synthesized during the explosion, according to the present models. The 511 keV line might be one of the main observable features. Specifically, positrons emitted by ^{18}F may have the special feature to be emitted (and then quickly annihilated) at the moment (around 110 minutes, half-life of ^{18}F) when the novae envelope starts to be transparent to the γ-radiation [1–3].

^{18}F appears to be produced in the novae inner shells via the Hot-CNO cycle according to several authors [2]. In particular the production path goes through:

$$^{16}O(p,\gamma)^{17}F(p,\gamma)^{18}Ne(\beta^+)^{18}F$$

or

$$^{16}O(p,\gamma)^{17}F(\beta^+)^{17}O(p,\gamma)^{18}F$$

while its destruction is mainly connected with the following processes: $^{18}F(p,\alpha)^{15}O$ or $^{18}F(p,\gamma)^{19}Ne$, where the first is the main channel. Thus the cross sections and the related reaction rates for the $^{18}F(p,\alpha)^{15}O$ reaction should be measured in the astrophysically relevant Gamow window [4], of the order of few hundreds keV (corresponding to $0.05 \leq T_9 \leq 0.5$).

In the last decade this reaction has been widely studied and, in particular, great efforts have been devoted to its study by means of direct measurements at the relevant astrophysical energies. Such a measurement appears to be very challenging not only for the involved energy range which leads to tiny cross sections but also because the ^{18}F is a radioactive isotope, so it requires dedicated techniques to be produced.

Several data sets are up-to-now available [5–13]. Nevertheless many uncertainties are still present on the low-energy resonance and its width, thus affecting the determination of the reaction rate at the temperatures relevant for astrophysics and, consequently, the novae nucleosynthesis. Therefore new experimental investigation, especially focused in the novae nucleosynthesis Gamow window, are mandatory.

2 Method

Alternative and challenging ways to obtain the bare nucleus cross section, σ_b, for charged-particles at sub-Coulomb energies have been provided by indirect methods. Among them, the Trojan Horse Method (THM) [14, 15] is particularly suited to investigate binary reactions induced at astrophysical energies by neutrons or

* e-mail: rgpizzone@lns.infn.it

charged particles by using appropriate three-body reactions. The THM allows one to avoid both Coulomb barrier suppression and electron screening effects, thus preventing the use of extrapolations. The method has proven very helpful in the last two decades for application to several aspects of nuclear astrophysics research like primordial nucleosynthesis [16–18], AGB nucleosynthesis [19, 20], the lithium problem [21–23], light elements depletion in stars [24, 25]. It has also been used with neutron induced reactions [26].

The basic assumptions of the Trojan Horse Method (THM) have already been reviewed recently in [27]. Most used Trojan Horse nuclides are deuteron and ^{6}Li but also ^{3}He has been successfully used several times showing the polar invariance of the quasi-free process [28, 29] which is at the very basis of the THM. The breakup process can then be thought as occurring within the nuclear region, so that Coulomb repulsion effects are greatly reduced. As a consequence, the method also becomes insensitive to problems connected with the electron screening effect, giving the possibility to measure the bare nucleus cross section. The first measurement with radioactive ion beams by means of THM was discussed in [30] where the ^{18}F(p,α)^{15}O reaction was studied for the first time by means of the THM. In this paper we will apply the method to a new experimental run for the ^{18}F(d,αn)^{15}O measurement in order to obtain relevant information on the ^{18}F(p,α)^{15}O cross section at energies relevant for astrophysics (see Fig. 1). This will help to confirm the indirect data already obtained in [30] and improve the statistics.

In the ^{18}F(d,αn)^{15}O process, the QF break-up is identified and selected, with deuteron splitting into its constituents p and n, whereby n is regarded as the spectator to the ^{18}F(p,α)^{15}O virtual reaction. Moreover, appropriate kinematics conditions can be selected so that the ^{18}F(p,α)^{15}O binary reaction can then take place at low interaction energies, in principle even negligible, according to the post collision prescription:

$$E_{cm} = E_{\alpha-^{15}O} - Q_{2b} \qquad (1)$$

where $E_{\alpha-^{15}O}$ is the relative energy between the detected α and ^{15}O while Q_{2b}(= 2.88 MeV) is the Q-value for the ^{18}F(p,α)^{15}O process.

According to the Plane Wave Impulse Approximation (PWIA), the three-body differential cross section measured in an α-^{15}O coincidence experiment can be expressed in terms of the two-body one as:

$$\frac{d^3\sigma}{dE_{\alpha_1} d\Omega_\alpha d\Omega_O} \propto (KF)\,|\Phi(p_s)|^2 \left(\frac{d\sigma}{d\Omega}\right)^{HOES} \qquad (2)$$

where KF is a kinematical factor. The experimental spectator momentum distribution $|\Phi(\tilde{p}_s)|^2$ is related to the p-n relative motion in the ^{2}H nucleus [31] with $(d\sigma/d\Omega)^{HOES}$ the half off-energy-shell binary cross section of astrophysical interest.

Experimental evidence for a QF contribution in the ^{18}F(d,αn)^{15}O process process has been obtained in a different experimental run in a wide energy range [30].

For the target break-up, one expects a maximum in the QF contribution at the kinematical conditions where the spectator energy is zero, thus reflecting the neutron momentum distribution in ^{2}H, which shows a maximum at p_s=0, since the relative p-n motion is mainly l=0 [31]. This gives rise to the choice of the detection angles for the outgoing α and ^{15}O particles. They are calculated using three-body kinematics under the condition that the spectator energy, E_n, is null and are referred to as the *quasi-free angles* in the standard prescriptions of the THM.

Since the Coulomb barrier is assumed to be overcome in the entrance channel, the obtained half-off-energy-shell cross section, $(d\sigma/d\Omega)^{HOES}$, should be the nuclear part only of the cross section for bare-nuclei, without the Coulomb barrier and also without electron screening effects. Moreover, the nuclear cross section is obtained within an arbitrary normalization constant to be matched to direct measurements, so that direct data have to be available at energies suitable for the normalization procedure. The agreement between the two cross sections at higher energies and the subsequent normalization represents indeed a necessary requirement for the application of the THM to a reaction of astrophysical interest and constitutes a natural step also for reactions induced by radioactive ion beams.

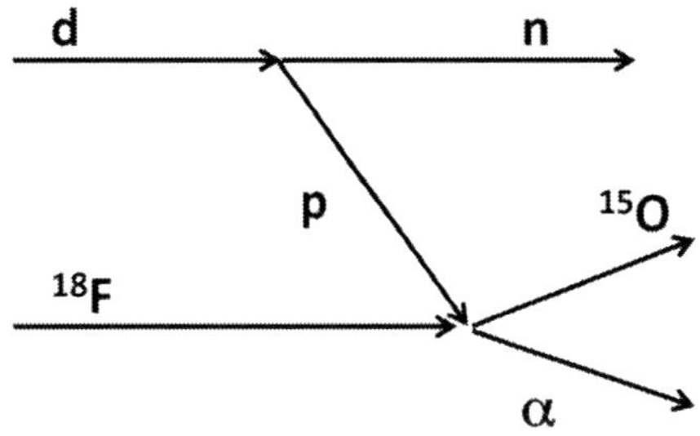

Figure 1. Schematic representation of the quasi-free mechanism of interest in the three-body reaction used in the Trojan horse method. The upper vertex marks the deuteron break-up while the lower vertex marks the half-off energy-shell process ^{18}F(p,α)^{15}O.

3 Results

The experiment, was performed at Cyclotron Institute of the Texas A&M University and it is described in [32]. The THM prescriptions and analysis were applied and the three body reaction of interest was selected. Taking into account the properties of quasi-free processes data were gated as reported ex tensively in [32]. Applying the formalism described above, the two body cross section is derived by dividing the experimental three-body one by the product of the kinematic factor modulated by the momentum distribution of the spectator inside the Trojan Horse nucleus [15].

The extracted $[d\sigma/d\Omega]^{HOES}$ as a function of E_{cm}, was then compared to a previous THM run performed in CNS

Table 1. Energy levels of ^{19}Ne in the energy range explored by the present experiment. Progressive numbers in first column correspond to energy levels. The ^{19}Ne* energy values are taken from [30]

Number	E_{cm} (MeV)	Energy ^{19}Ne* (keV)	$J^{(\pi)}$	Reference
1	-0.34	6070	$3/2^+, 5/2^-$	[9]
3	-0.16	6255	$11/2^-$	[9]
4	0.05	6460	$3/2^+, 5/2^-$	[9, 35]
5	0.13	6537	$5/2^+, 9/2^+$	[35]
6	0.33	6755	$3/2^-$	[9, 11, 35]
7	0.56	6967	$5/2^+$	[35]

at the CRIB facility whose details are reported extensively in [30]. A good agreement shows up, within the experimental uncertainties. The observed levels, correspond to levels in ^{19}Ne which are reported in Table 1, taken from [35] or [9]. Although the energy resolution is poorer than in the previous run (mainly due to the poorer angular resolution of the present experimental apparatus), the agreement between the two data sets confirms once again the applicability of the THM to the present reaction.

The first validity check that standard THM prescriptions do recommend is to reproduce the direct excitation function. This is done by comparing the distributions measured with direct methods to the one measured by means of THM. The latter should be normalized to the direct data. The THM cross section extracted above is corrected for the penetrability factor (below the Coulomb barrier) which also makes the comparison of half-off-energy-shell and on-energy-shell data [15] possible. The penetrability factor is, as usual, described in terms of the regular and irregular Coulomb functions.

THM data are also not affected by suppression effects coming from the centrifugal barrier. Assuming the J^{π} values of the populated ^{19}Ne excited states as in Table 1, the data of each resonance have been integrated over the full angular range by means the corresponding Legendre polynomial. Finally, the data have been corrected also for the penetrability of the centrifugal barriers.

It is then possible to normalize to the direct data (after comparison with the data from [30]) at the higher possible energies (0.5-0.65 MeV) in the present case. The comparison is reported in [32]. Energy states reported in Table 1 were investigated in the present work; in particular the explored energy range makes relevant the contribution from the 6255 (sub threshold state), 6460, 6537, 6755, 6967 keV states of ^{19}Ne . The respective J^{π} were assigned accordingly as reported in the Table 1. A specific discussion should be done for the resonance at E_{cm}=0.05 MeV (corresponding to the 6460 keV state in ^{19}Ne) where two possible values of J^{π} were taken into account. This is clear in Fig. 2. In the upper panel the black circles correspond to the choice of $J^{\pi}= 3/2^+$ for the 6460 keV state in ^{19}Ne following [35]. Direct data from [5, 6] are reported for comparison and normalization purposes. In the lower panel the full dots represents the results for a $J^{\pi}=5/2^-$ assumption for the same level as discussed in [9]. This uncertainty leads therefore to an uncertainty between the S(E) lower limit (corresponding to $J^{\pi}=5/2^-$ for the 6460 keV state in

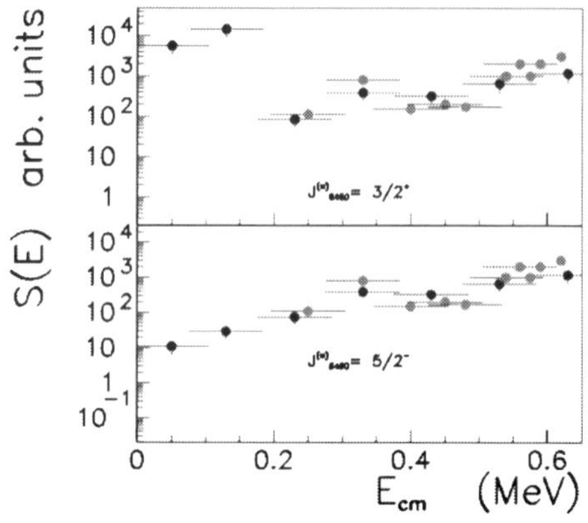

Figure 2. Average of the present data (black dots) and those from [5, 6] in red. Lower panel: S(E)-factor extracted with the choice of $J^{(\pi)}=5/2^-$ for the 6460 keV state in ^{19}Ne. Upper panel: S(E)-factor for $J^{(\pi)}= 3/2^+$.

^{19}Ne) and an upper limit assuming $J^{\pi}= 3/2^+$. Further studies (both with direct and indirect methods) and in particular the angular distribution will be necessary to improve the data quality in the low energy range.

It was possible, with the present experimental run, to confirm the possibility of application of the THM to the ^{18}F(d,αn)^{15}O reaction for studying the ^{18}F(p,α)^{15}O (as reported in [30]) within the experimental errors. This was also possible in the present case where the experimental set-up and the beam production line is much simpler than the ones used in [30] (e.g. simpler detection system and no beam-tracking available in the present case). This also strengthens the role of the THM for nuclear astrophysics in the further years since it may play a leading task in the field of radioactive beams, even in cases where the experimental setup is quite simple, like the present one. In fact, it was possible to extract the astrophysical S(E)-factor by means of the THM for a reaction induced by an unstable beam, thus confirming results from [30] in all the energy range relevant for astrophysics. Future astrophysical applications of the present results will be investigated via the means of R-Matrix theory in a forthcoming paper, with special attention to the role of the involved resonances. Moreover, on the experimental side efforts are necessary to improve the energy and angular resolution of the detection system and therefore reduce the statistical error on the S(E)-factor. The extraction of the angular distribution will also be crucial to assign the J^{π} of the involved levels and will be the aim of a future, dedicated experiment, to be performed in the future with the optimized version of the detection system adopted in [30] which turned out to be the best performing for the THM application.

References

[1] J. Jose & M. Hernanz, J. Phys. G: Nucl. Part. Phys.,

 DOI: 10.1051/epjconf/201716300046

34, R431 (2007)

[2] A. Coc M. Hernanz, J. Jose & J.P. Thibaud, Astron. & Astrophys. 357, 561 (2000)

[3] R. Diehl, AIP Conference Proceedings, 1269, 144-153 (2010)

[4] C. Iliadis, "Nuclear Physics of Stars", Wiley (2007)

[5] D.W.Bardayan, J. C. Batchelder, J. C. Blackmon, A. E. Champagne et al. 2002, Phys. Rev. Lett. 89, 262501

[6] C.E. Beer A. M. Laird, A. St. J. Murphy, M. A. Bentley, et al. 2011, Phys. Rev. C, 83, 042801(R)

[7] N. DeSereville, C. Angulo, A. Coc, N.L. Achouri et al., Phys. Rev. C 79, 015801 (2009)

[8] Graulich JS, Cherubini S, Coszach R, El Hajjami, et al., Phys. Rev. C, 63, (2001)

[9] A. Laird, A. Parikh, A. St. J. Murphy, K. Wimmer et al. , Phys. Rev. Lett., 110 032502, (2013)

[10] Murphy et al., Phys. Rev. C79, 058801 (2009)

[11] Adekola et al., Phys. Rev. C83, 052801(4) (2011)

[12] Adekola et al., Phys. Rev. C84, 054611 (2011)

[13] Mountford et al., Phys. Rev. C85, 022801(R) (2012)

[14] C. Spitaleri Problems of Fundamental Physics, II, 21-36 (1991)

[15] R.E. Tribble et al., Rep. on Progr. Phys., 77,106901 (2014)

[16] R.G. Pizzone, R. Sparta, C.A. Bertulani, C. Spitaleri, et al., Ap. J., 786, 112 (2014)

[17] A. Rinollo et al., Nucl. Phys. A758 146c (2005)

[18] A. Tumino et al. 2014, Ap. J., **785**, 96

[19] S. Palmerini, M.L Sergi, M. La Cognata, L. Lamia, et al., Ap. J., 764, 128 (2013)

[20] R.G. Pizzone, G.D'Agata, C. Spitaleri et al., Ap. J., 836, 57 (2017)

[21] L. Lamia, C. Spitaleri, et al., A. & A. 541, 158 (2012)

[22] Aliotta, M., C. Spitaleri, et al., Eur. Phys. J. 9, 435 (2000)

[23] C. Spitaleri, L. Lamia, S.M.R. Puglia, S. Romano, et al., Phys. Rev. C 90, 035801 (2014)

[24] L. Lamia, C. Spitaleri, et al. 2015, Ap. J., **811**, 99

[25] L. Lamia, C. Spitaleri, R.G. Pizzone, et al., 2013, Ap. J. 768, 65

[26] M. Gulino et al. 2013, Phys. Rev. C87 012801

[27] C. Spitaleri et al. 2016, Eur. Phys. J. A**52**, 77

[28] R.G. Pizzone, C. Spitaleri, L. Lamia, C. Bertulani, et al., Phys. Rev. C, 83, 045801 (2011)

[29] R.G. Pizzone et al., Phys. Rev. C 87, 025805 (2013)

[30] S. Cherubini, M. Gulino, C. Spitaleri, M. La Cognata et al. Phys. Rev. C 92, 015805 (2015)

[31] L. Lamia, M. La Cognata, C. Spitaleri, B. Irgaziev, R.G. Pizzone, Phys. Rev. C 85, 025805, (2012)

[32] R.G. Pizzone, B.T. Roeder et al. 2016, Eur. Phys. J. A52, 24

[33] R.G. Pizzone, C. Spitaleri, S. Cherubini, M. La Cognata, et al., Phys. Rev. C 71, 058801 (2005)

[34] R.G. Pizzone, C. Spitaleri, A.M. Mukhamedzhanov, L.D. Blokhintsev, et al., Phys. Rev. C 80, 025807 (2009)

[35] C.D. Nesaraja, N. Shu, D.W. Bardayan, J.C. Blackmon, et al., Phys. Rev., C 75, 055809, (2007)

Fusion-fission studies on $^{19}F+^{206,208}Pb$ and $^{11}B,^{19}F+^{238}U$ reactions around barrier energies at IUAC facility

Sugathan Pullanhiotan[1,*], *Rakesh* Dubey[1], *Chandrabhan* Yadav[1], *Akhil* Jhingan[1], *Golda* Komalan Satheedas[1], *Saneesh* Nedumbally[1], and *Mohit* Kumar[1]

[1] *Inter University Accelerator Centre, Aruna Asaf Ali Marg, New Delhi, India 110067*

Abstract. Fission process is strongly influenced by entrance channel dynamical variables. Among these, the nuclear charge product, mass asymmetry and deformation play important role in fission dynamics. Reaction characteristics are distinguished by investigating the properties of fission mass and angular distributions. Experiments using actinide targets are challenging due to many conflicting results making unambiguous identification of quasi-fission difficult. At IUAC accelerator facility many experiments have been performed to make a systematic study of fission mechanism and role of entrance channel parameters and deformation. Fragment mass distribution, angular distribution and neutron multiplicity measurements are performed to study reactions using spherical and deformed targets.

1 Introduction

One of the key problems in heavy ion collision is to understand what are the important variables governing the dynamics of colliding heavy nuclei at various energies. In collisions where heavy partners are involved, it is observed that quasi-fission(QF) is a competing process that hinders complete fusion and subsequent formation of residues of heavy nuclei. At high excitation energies process like fast fission dominates and their contribution found to be negligible at moderate excitation energies. The dynamics of the fusion process is strongly influenced by many entrance channel properties such as charge product of nuclei (Z_pZ_t), mass asymmetry (α), deformation of nuclei (β), shell effects and collision energy[1–3]. Classically, QF has been observed in high fissile systems with $Z_pZ_t > 1600$. However there are reaction systems reported where QF were observed in lighter system with $Z_pZ_t \sim 800$[4]. Mass asymmetry and deformation found to be the factors influencing the onset of QF in such light systems. For example, the variance of the mass distribution of fission fragments found to be larger in $^{19}F+^{197}Au$ and $^{30}Si+^{186}W$ reactions as compared to $^{12}C+^{204}Pb$, though all of them formed the same compound nucleus(CN) ^{216}Ra at similar excitation energies[4]. This was attributed to presence of quasi-fission in former two reactions whose entrance channel mass asymmetry (α) is lower than the critical mass asymmetry α_{BG} (Businaro Gallone mass asymmetry for maximum potential energy). It was reported that, more symmetric systems with mass asymmetry lower than the critical mass asymmetry evolves through QF process. On the other hand, asymmetric reactions with mass asymmetry larger than critical mass asymmetry leads to compact compound nucleus equilibrated in all degrees of freedom. Apart from mass asymmetry, deformation of the nuclei also played role in fission dynamics. Reactions involving deformed target showed anomalously large angular anisotropy and enhancement of mass variance at low energy[5–8]. The anomalously large mass variance and angular anisotropy below barrier energies are reported to be due to orientation dependent QF.

Experimental investigation of mass and angular distribution of fission products is the most commonly employed method for identifying compound and non-compound nuclear processes. According to standard statistical model of fission, the mass distribution of fission products follow Liquid Drop Model (LDM) predictions showing symmetric mass distribution and the width of mass distribution vary smoothly as a function of excitation energy. For angular distribution, the basic assumption is that K (the projection of angular momentum on symmetry axis) is fully equilibrated at the saddle point and a symmetric angular distribution indicates the decay from an equilibrated system. However for any non-compound process such as QF, mass asymmetry and K degree of freedom may not be fully equilibrated before the fission, resulting in large mass width and angular anisotropy deviating considerably from model predictions. Large angular anisotropies observed in fission involving actinide targets have been attributed to presence of either fission before K equilibration, i.e pre-equilibrium fission(PEF) or the quasi-fission[9–14]. Presence of QF signify less chance for evaporation residue(ER)formation and one expect reduced ER cross section in such reactions. Another distinct feature of QF is the strong mass angle correlation that has been clearly observed in systems involving heavy projectiles[1]. Though reason for large mass width and

* e-mail: sugathan@iuac.res.in

DOI: 10.1051/epjconf/201716300047

anisotropy have been attributed to QF, the identification of the reaction process is not always unambiguous. For example in reaction $^{16}O + ^{238}U$ around barrier energies, large mass width and anisotropy have been observed, but no suppression of ER observed[3, 7, 15]. So unambiguous identification of QF events becomes a challenging task, especially when lighter projectiles are involved in low Z_pZ_t systems.

Another interesting property of fission is the presence of asymmetric fission in low energy fission of actinides and pre-actinides due to the shell effects. There is strong evidence of multi-mode nature of fission, where a transition from asymmetric to symmetric mass division occurs as excitation energy is increased[16–18]. Mass distribution from such system will reflect contributions from both symmetric and asymmetric mass components. Hence presence of multi-mode fission makes the mass distribution different than single Gaussian at low energies.

In a series of experiments carried out at IUAC heavy ion accelerator facility [19], we have further extended the study of fission properties by measuring fragment mass and angular distributions from fission induced by medium light projectiles on spherical and deformed targets[20, 21]. The reactions were chosen to identify the role of deformation as well as mass asymmetry influencing the fission properties in these reactions.

2 Experiments

Experiments were performed using ^{11}B,^{12}C and ^{19}F projectiles bombarding on $^{206,208}Pb$ and ^{238}U targets at energies around the Coulomb barrier. We have measured the mass distribution of FF in reactions ^{11}B,$^{12}C + ^{238}U$, $^{19}F + ^{206,208}Pb$, and both mass and angular distribution in $^{19}F + ^{238}U$. Pulsed beam of $\sim$ 1.5 ns time width and 4 MHz repetition rate were used in all time of flight measurements. Targets with typical thickness of $\sim 110\,\mu g/cm^2$ were prepared on carbon backing $\sim 20\,\mu g/cm^2$. For mass distribution measurement, a pair of multi-wire proportional counters (MWPCs) were used inside the 1.5m diameter scattering chamber to collect FF in coincidence. Experimental setup and data analysis details are given in Ref.[20]. The data were analyzed to extract mass-angle correlation, mass ratio distribution and variance of mass distribution. From position and time signal, velocity components were reconstructed and mass ratio distribution was extracted from velocity components. Full momentum transfer condition was applied to separate transfer induced fission events. The extracted mass ratio(M_R) distributions were fitted by Gaussian distribution at all energies and corresponding mass variance obtained.

2.1 Mass distribution from reaction $^{19}F + ^{206,208}Pb$

The reactions $^{19}F + ^{206,208}Pb$ forms the light actinide compound nuclei $^{225,227}Pa$. Mass distribution measurements were performed over the excitation energies E_{CN} ranging from 30 to 50 MeV. The results are reported in[20, 21]. Since both $^{206,208}Pb$ targets are spherical, no deformation dependent effects are expected in these reactions. Our measurements showed mass distribution can be described by single symmetric Gaussian fit at all measured energies. A sample of measured mass distribution is shown in Fig. 1 for reaction $^{19}F + ^{206}Pb$ at two beam energies, one at below the barrier $\frac{E_{Cm}}{V_B} = 0.93$ and other at above barrier energy $\frac{E_{Cm}}{V_B} = 1.1$. Though mass distribution could be fitted by single Gaussian, the mass variance obtained from the fit increased at lower energies. The data from $^{19}F + ^{208}Pb$ also showed the same property. The plot of mass variance as a function of excitation energy for $^{19}F + ^{208}Pb$ is given in figure 2a. As the excitation energy decreases, the mass variance falls in relation to energy until it increases again at low excitation energy. Similar increase of mass variance at below barrier energy has been reported in system involving deformed target which has been attributed to onset of QF due to orientation effect[7, 8]. However in the present reaction, QF is minimal as there is no evidence of other signatures of QF here. No mass-angle correlation was observed in this reaction. The angular distribution data for $^{19}F + ^{208}Pb$ as displayed in Fig. 2b already showed normal anisotropy as expected in complete fusion-fission reaction[22]. A systematic evaluation of data from mass distribution of nearby light actinide nuclei showed that, the presence of asymmetric mass division at low excitation energy could be the reason for broadening of mass width [20]. The nuclei in this mass region showed multi-mode nature of fission, where there is a transition from symmetric mass division to asymmetric division as the excitation energy is decreased[16, 23, 24]. The origin of multi-mode fission has been attributed to the shell structure of fragments[25–28].

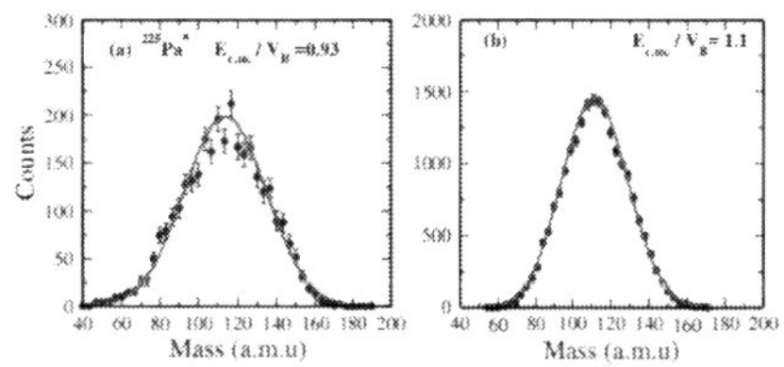

Figure 1. (Color online) The mass distribution of fission fragments from reaction $^{19}F + ^{206}Pb$ (a) at below barrier energy $\frac{E_{Cm}}{V_B} = 0.93$ and (b) above barrier energy $\frac{E_{Cm}}{V_B} = 1.1$. The solid line indicates best fit to data using single Gaussian function.

2.2 Shell effects in mass distribution of $^{12}C + ^{238}U$

The importance of shell effects in fission mass distribution was highlighted in a recent paper where mass distribution from compound nuclei $^{242-247}Cf$ was compared against different entrance channels[29]. Mass distribution of reaction $^{12}C+^{235}U$ around barrier energies showed flat top distribution signifying contribution of more asymmetric fission components in this channel as compared to other reaction such as S+Pb. The mass-asymmetric component here was attributed to low excitation energy fission due

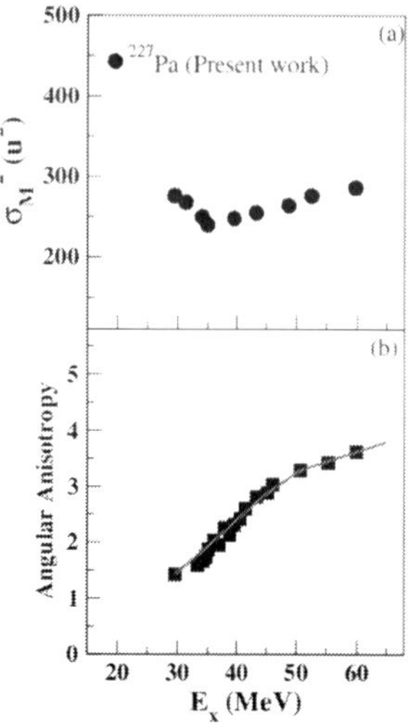

Figure 2. (Color online) (a) Mass variance (σ^2_M) at various excitation energies in reaction ^{19}F+^{208}Pb, from Ref.[20] and (b) the fission fragment angular anisotropy data from Ref.[22].The solid line in (b) corresponds to SSPM predictions.

to multi-chance fission. This observation has been validated with detailed theoretical model calculation using GEF code which predict the fraction of asymmetric fission considering shell effects at low excitation energies quite well[30]. From one of our earlier measurement at IUAC, we have already reported the mass distribution of reaction ^{12}C+^{238}U forming CN ^{250}Cf and compared with another reaction ^{18}O+^{232}Th at around barrier energies[31]. A detailed re-analysis of low energy data from ^{12}C+^{238}U showed flat-topped non single Gaussian mass distribution in the fission of CN ^{250}Cf clearly demonstrating the presence of asymmetric fission arising from shell effects of fragments. The experimental mass distribution from reaction ^{12}C+^{238}U at lowest energy along with GEF prediction is depicted in figure 3. The prediction of GEF calculation when multi-chance fission is included matches closely with experimental data that can be described well by double Gaussian fitting.

2.3 Fission study on ^{11}B, ^{19}F + ^{238}U reactions

The reaction ^{11}B+^{238}U $\longrightarrow$ ^{249}Bk* was performed using pulsed ^{11}B beam with energy varying from 52 MeV to 66 MeV. Fission fragments were detected in coincidence by placing MWPC pairs at folding angles. The forward detector was placed 35 cm away from target and kept 40° with respect to beam whereas the backside detector was placed 27 cm at 130°. For reaction ^{19}F+^{238}U $\longrightarrow$ ^{257}Md*, ^{19}F beam energy was varied from 92 MeV to 120 MeV and MWPC detectors were placed at corresponding folding angles. Event by event data analysis were performed using velocity reconstruction method as described before.

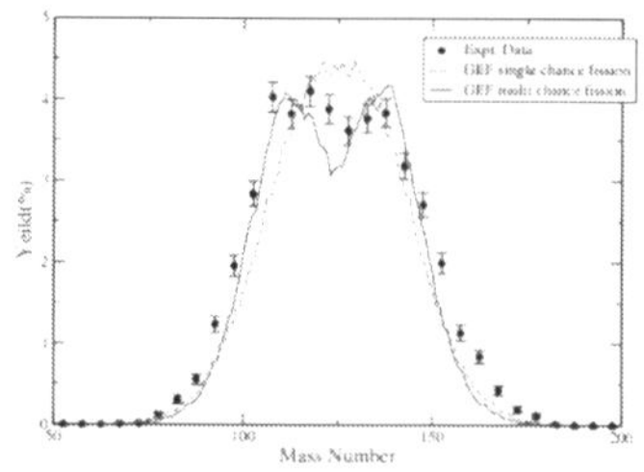

Figure 3. (Color online) Mass distribution of fission fragments from reaction ^{12}C + ^{238}U at excitation energy E_{CN}=34.2 MeV along with GEF predicted distribution for single chance and multi chance fission.

2.3.1 Mass distribution

For both reactions, mass-angle-distribution(MAD) have been extracted at different energies and Fig. 4 shows a typical MAD plot for reaction ^{19}F + ^{238}U at below barrier energy corresponding to $\frac{E_{Cm}}{V_B}$ = 0.97. The emission angle ($\theta_{c.m.}$) span from ≈ 100° to 160°. It is seen that that mass ratio is independent of $\theta_{c.m.}$ suggesting there is no significant correlation between mass and angle of fragments in this reaction at any of the measured energies. No correlation is found in reaction ^{11}B + ^{238}U also.

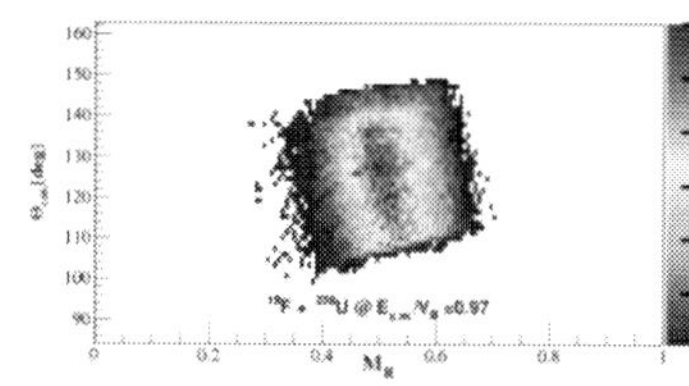

Figure 4. Mass-angle distribution of fragments from reaction ^{19}F + ^{238}U at below barrier energy.

Fragment mass distribution measured in reactions ^{11}B+^{238}U and ^{19}F+^{238}U at two bombarding energies are shown in Fig. 5a-d. These energies correspond to below and above the barrier in respective cases. At all energies, the measured mass spectra from both reactions could be described by a single symmetric Gaussian centered around mass ratio 0.5. As the excitation energy in fissioning nuclei increases, the mass width increase as a function of temperature as expected. There is no significant deviation from symmetric Gaussian shape at any of the measured energy point which signify fully mass equilibrated fusion-fission dominating in both reactions.

In earlier works, the nearby reaction ^{16}O+^{238}U showed skewness in mass distribution and enhancement in mass width at low excitation energies due to deformation dependent QF[7, 22]. Present measurement did not show any such behavior in reaction ^{19}F+^{238}U. The mass vari-

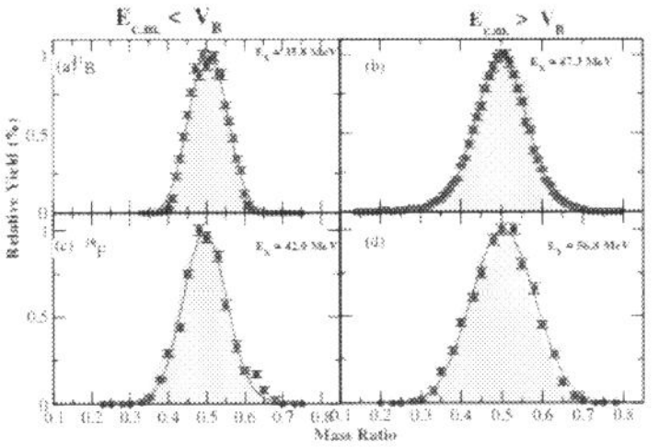

Figure 5. (a)-(b)Mass distribution from reaction $^{11}B+^{238}U$ at below barrier energy(a) and above barrier energy (b). (c)-(d) mass distribution from $^{19}F+^{238}U$ at below barrier (c) and above barrier energy (d). The solid line corresponds to best fit using single Gaussian and shaded region represent the results of GEMINI^{++} predictions.

Table 1. Entrance channel parameters for reactions used in this study. Columns shows the projectile, target, compound nucleus (CN), $\alpha(=\frac{A_T-A_P}{A_T+A_P})$, α_{bg}, fissility (χ), Z_PZ_T.

Projectile	CN	α	α_{bg}	χ	Z_PZ_T
^{11}B	^{249}Bk	0.911	0.892	0.813	460
^{12}C	^{250}Cf	0.904	0.896	0.823	552
^{16}O	^{254}Fm	0.874	0.902	0.841	736
^{19}F	^{257}Md	0.852	0.905	0.849	828

ance did not show anomalous enhancement at low excitation energy region. For these systems, the theoretical mass distribution has been simulated using statistical code *GEMINI*$^{++}$[32] which showed the predicted mass distribution closely matching with experimental data in both cases. We assume that the measured mass distribution in these reactions follow normal fusion-fission at all measured energies. The absence of any mass angle-correlation also suggest that the reactions are mainly dominated by fission events followed from mass equilibrated system. The entrance channel parameters for these reactions are shown in Table 1.

2.3.2 Angular Distribution

Angular distribution measurement of fragments in reaction $^{19}F + ^{238}U$ around barrier energy was performed using the hybrid detector telescope array (HyTAR) facility setup inside the scattering chamber. The array consist nine ΔE(gas ionization) and E (Silicon) detectors for particle identification based on differential energy loss method. The DC beam of ^{19}F ions with energies ranging from 96 MeV to 121 MeV was used to bombard the ^{238}U target. The angular range covered from 70° to 170° with respect to beam direction. The data analysis was performed to determine the anisotropy in angular distribution.

2.3.3 Angular anisotropy

After applying the Jacobian for conversion of data from lab frame to centre-of mass frame, the angular anisotropy has been extracted from the ratio $\frac{W(180°)}{W(90°)}$ where $W(180°)$ and $W(90°)$ are the fission yields at 180° and 90° respectively. The experimental anisotropy has been then compared with anisotropy calculated according to the statistical saddle point model (SSPM). Figure 6 displays the ratio of measured anisotropy to theoretical anisotropy as a function of energy for present reaction $^{19}F + ^{238}U$ along with data from reaction $^{16}O + ^{238}U$[5]. Similar to ^{16}O induced fission, anomalously large anisotropy is observed in present reaction also.

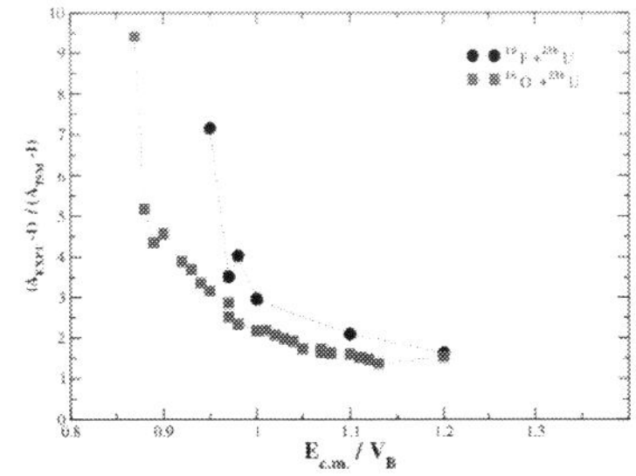

Figure 6. Ratio of measured anisotropy to theoretical anisotropy plotted as a function of energy. Solid circle represent data for reaction $^{19}F+^{238}U$ and solid square corresponds data from $^{16}O+^{238}U$ [5]

Anomalously large angular anisotropy has been observed in many other reactions using medium light projectiles (mass $A_p < 24$) on various actinide targets ^{232}Th, ^{238}U and ^{237}Np around barrier energies[9–11]. Though large angular anisotropy has been attributed to non-compound process such as PEF and QF, there is no clear evidence of which process influence the dynamics at various stages of fission process. Large angular anisotropy in $^{19}F+^{238}U$ imply narrow K distribution suggesting incomplete shape equilibration, however there is no strong evidence of non-equilibration in mass degree of freedom. Symmetric mass distribution, normal mass width and absence of strong mass-angle correlation suggest that presence of QF may not be significant in reaction $^{19}F + ^{238}U$ at measured energies.

2.4 Neutron multiplicity measurement

Neutrons play a major role in the de-excitation process of heavy compound system. Neutron multiplicity is a sensitive observable for time scales and can be used as neutron clock. Since QF is fast process it is expected that average pre-scission neutron multiplicity will be lower in case of QF compared to fusion-fission[33]. Using mass gated neutron multiplicity measurements, extensive work has been carried out to discriminate fusion-fission and quasi- fission events based on the time scale involved in the process[34]. We have recently commissioned a large neutron detector array for studying the multiplicity of neutron emission in coincidence with fission fragments[35]. Using this facility experiment has been performed to extract mass gated neutron multiplicity from reaction $^{19}F+^{208}Pb$ at 110 MeV

Table 2. Pre-scission (Mpre) and post-scission(Mpost) neutron multiplicity gated by different mass split in reaction $^{19}F+^{208}Pb$ at 110 MeV beam energy.

Mass Gate	Mpre	Mpost
Symmetric	2.29 ± 0.20	1.59 ± 0.08
Asymmetric	3.44 ± 0.16	1.05 ± 0.06

beam energy. 50 neutron detectors and a pair of MWPCs were used to detect neutrons in coincidence with fission-fragments. Neutron-gamma discrimination was accomplished by pulse shape discrimination(PSD) technique. Figure 7 shows typical neutron-gamma discrimination from two dimensional plot between PSD and time of flight(TOF). The pre-scission and post-scission neutron multiplicities have been extracted using moving source fit applied to full set of experimental spectra. Global fitting for all neutron spectra were used to extract the parameters.

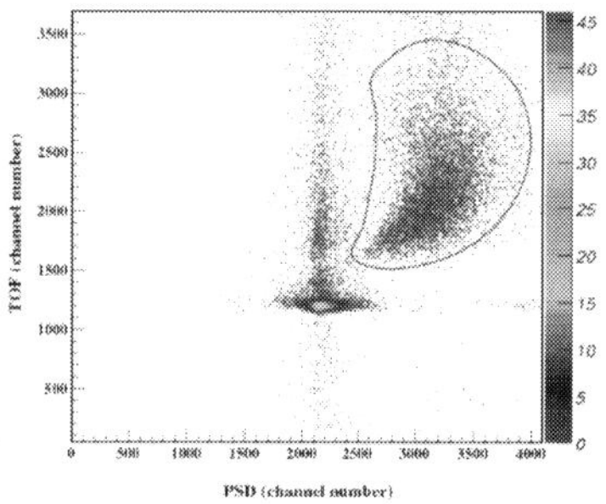

Figure 7. Neutron-gamma separation using PSD vs TOF. The gated region shows neutrons well separated.

Preliminary analysis shows that average pre-scission multiplicity is higher for asymmetric mass splits as compared to symmetric mass splits. Similar observation has been reported in reaction $^{19}F+^{209}Bi$ at 108 MeV beam energy[36]. The higher multiplicity for asymmetric mass region suggest the presence of multi-chance fission leading to asymmetric masses and more neutrons. Preliminary results are shown in Table 2. The facility has been recently used for measuring mass gated neutron multiplicity in heavy systems using projectile ^{48}Ti on ^{208}Pb target.

3 Conclusion

Fission fragment mass distributions have been measured for fission induced by projectiles such as ^{11}B,^{12}C, ^{19}F on Pb and U targets in series of experiments performed at near barrier energies. Fission fragments were detected in coincidence using pair of MWPCs in time of flight setup and fragment's velocity components were extracted to generate mass distribution. For all reactions the mass distributions could be described by symmetric Gaussian. For $^{225,227}Pa$ nuclei formed in fusion of ^{19}F + $^{206,208}Pb$, the mass variance found to increase at below barrier energies. Such enhancement in mass variance of at low energies have been observed in other light actinide nuclei $^{224,226}Th$ and ^{228}U where asymmetric and symmetric mass division were observed.

The role of deformation and mass asymmetry has been explored by measuring mass distribution in ^{11}B, ^{19}F + ^{238}U reaction and angular distribution in ^{19}F + ^{238}U reactions. In both systems mass distribution found to be symmetric and matches with $GEMINI^{++}$ model calculations. No strong evidence of mass asymmetry or deformation influencing the mass equilibration found in these reactions. However measured angular anisotropies of ^{19}F + ^{238}U reaction showed anomalous large anisotropies deviating from SSPM predictions. This may suggest that in ^{19}F + ^{238}U there is evidence of fission events without equilibration in K degree of freedom but fully equilibrated in mass degree of freedom.

The authors wish to thank all collaborators, students, accelerator staff and Director, IUAC for their continuous support.

References

[1] J. Töke *et al.*, Nucl.Phys. A **440**,327(1985).
[2] B. Back *et al.*, Phy.Rev.C, **32**, 195(1985)
[3] D. Hinde *et al.*, Phys.Rev.Lett., **101**,092701(2008)
[4] A. C. Berriman *et al.*, Nature **413**, 144(2001)
[5] D.J. Hinde *et al.*, Phy. Rev. C **53**,1290(1996)
[6] C.J. Lin *et al.*, Phy. Rev. C **85**,014611(1996)
[7] K. Banerjee *et al.*, Phys. Rev. C **83**, 024605 (2011).
[8] T.K.Ghosh *et al.*, Physics Letters B **627**, 26 (2005).
[9] V.S. Ramamurthy and S.S. Kapoor, Phys. Rev. Lett. **54** ,178(1985)
[10] Z. Liu *et al*, Phy. Lett. B **353**,173(1995).
[11] J. P. Lestone *et al.*, Phy. Rev. C **56**, 2907(R) (1997)
[12] D. Hinde *et al.*, Phys.Rev.Lett., **74**, 1295(1995)
[13] B. Back *et al.*, Phy.Rev. Lett, **46**, 1068(1981)
[14] H.Q. Zhang, Phy. Rev. C **81**,034611(2010).
[15] K. Nishio *et al.*, Phys. Rev. Lett **93**, 162701(2004)
[16] U. Brosa, S. Grossmann, A. Müller, Phys. Rep. **197**, 167u (1990).
[17] Y. Nagame *et al.*, Physics Letters B**387**, 26 (1996).
[18] K. H. Schmidt *et al.*, Nucl. Phys. A **665**, 221 (2000).
[19] G.K. Mehta and A.P. Patro, Nucl. Instr. and Meth. A **268**, 334 (1988).
[20] R. Dubey *et al.*, Physics Letters B **752**, 338(2016).
[21] P. Sugathan, Journal of Physics: Conference Series, **755**, 012007(2016).
[22] D. J. Hinde, *et al.*, Nuclear Physics A**685**, 72 (2001).
[23] A. Turkevich and J. B. Niday, Phys. Rev. **84**, 52 (1951).
[24] V.V. Pashkevich, Nucl. Phys. A **169**, 275 (1971).
[25] P. Möller and J. R. Nix, Nucl. Phys. A**229**, 269 (1974).
[26] B.D. Wilkins, E.P. Steinberg, R.R. Chasman, Phys. Rev. C **14**,1832 (1976).
[27] M.G. Mustafa, U. Mosel, H.W. Schmitt, Phys. Rev. C **7**,1519 (1973).

[28] V. Pokrovsky, M. G. Itkis, J. M. Itkis, *et al.*, Phys.Rev. C**62**, 014615 (2000).

[29] J. Khuyagbaatar *et al.*, Phys. Rev. C **91**, 054608 (2015).

[30] B.Jurado and K.H. Schmidt, http://www.khs-erzhausen.de/GEF-2015-2.2.html,GEF code Version 2.2,(2015).

[31] C. Yadav *et al.*, Phy.Rec. C **86**, 034606(2012).

[32] R.J. Charity, Phy. Rev. C **82**,014610(2010).

[33] L. Donadille *et al.*, Nucl. Phys. A **656**, 259 (1999)

[34] D. J. Hinde *et al.*, Phys.Rev. C **30**, 143 (1984).

[35] P. Sugathan *et al*, Pramana Journal of Physics **83**, 807 (2014)

[36] L. M. Pant *et al.*, Eur. Phys. J. A **16**,43(2003)

 DOI: 10.1051/epjconf/201716300048

The Influence of Nuclear Reactions and Nuclear Structure in Astrophysics

K. E. Rehm[1,*]

[1] Physics Division, Argonne National Laboratory, Argonne, IL 60439, USA

Abstract. Nuclear reactions play an important role for the energy production and the nucleosynthesis in stars. New facilities, able to accelerate radioactive nuclei or high-intensity stable beams have allowed us to measure in the laboratory reactions involving short-lived nuclei or processes with very small cross sections, which are crucial for stellar nucleosynthesis. I will discuss some of the recent experiments studying fusion and transfer reactions with radioactive beams which play a critical role in various quiescent and explosive stellar environments.

1 Introduction

Nuclear structure and nuclear reactions involving the whole range from light to heavy nuclei are crucial for determining the format of our universe from its beginning to the present day. A well known example are the mass-5 and mass-8 stability gaps which restrict nucleosynthesis during the first few minutes of the Universe to the production of nuclei with A≤7. Heavier nuclei are then 'cooked' in the high-density environment of stars or during stellar explosions. For this the nuclear shell structure causes the abundance peaks of certain elements around the so-called waiting-point nuclei in the vicinity of closed neutron shells.

The production of carbon, an element crucial for life on Earth, would not have been possible if Nature had not put an excited 0^+ level, the so-called Hoyle state, into the structure of ^{12}C, with an energy located right in the Gamow window for the triple-alpha reaction.

Oxygen, another crucial element, which makes up about 2/3 of the mass of the human body, and is, after hydrogen and helium the third-most abundant element in the universe, is produced in stars via the $^{12}C(\alpha,\gamma)^{16}O$ reaction with a yield that is strongly affected by a subthreshold resonance in ^{16}O. For the subsequent alpha capture reaction, $^{16}O(\alpha,\gamma)^{20}Ne$, we find that the nuclear structure in ^{20}Ne provides no level with the proper quantum numbers, which inhibits this channel and restricts the destruction of ^{16}O. These are just a few examples among the 6000 nuclei and 60,000 nuclear reactions participating in stellar nucleosynthesis showing the critical influence of nuclear structure and reactions in stellar environments.

In this contribution I will discuss three examples of this influence for three long-lived radio isotopes: ^{60}Fe ($T_{1/2}$=2.62 My),^{44}Ti ($T_{1/2}$=59 y) and ^{26}Al ($T_{1/2}$=0.717 My). The fact that all of these isotopes have been detected in our galaxy proves that nucleosynthesis is ongoing in stellar environments.

* e-mail: rehm@anl.gov

2 ^{60}Fe and the ^{12}C + ^{12}C fusion reaction

Among the three examples chosen for this presentation, ^{60}Fe is the isotope with the longest half-life: $T_{1/2}$=2.62± 0.04 My [1, 2], which, however, is still short compared to the age of the solar system (~4.6 By). ^{60}Fe material has been found in the interstellar medium through γ-ray measurements with the INTEGRAL satellite, in cosmic rays with the CRIS detector aboard the Advanced Composition Explorer (ACE) Spacecraft [3] and on Earth in deep-sea sediments, which are thought to originate from recent (~2.2 My) and close (60-130 parsecs) supernovae explosions [4, 5]. Since ^{60}Fe is located on the neutron-rich side of the mass valley its production requires a complex reaction path which is discussed in more detail in Ref. [6]. Although the nuclei produced in the fusion of C+C are far away in mass and Z from ^{60}Fe, this reaction still plays a critical role for the production of ^{60}Fe.

Fusion of ^{12}C + ^{12}C has been studied extensively over the years down to the nb region. The measurements at the lowest energies are quite challenging due to the small yields and the strong influence from various background reactions. For that reason, extrapolation techniques have been developed in order to obtain cross sections at energies where no experimental data are available. This extrapolation became more complicated when fusion hindrance at sub-barrier energies was discovered in 2002 [7]. As discussed in Ref. [8], fusion reactions involving medium-mass nuclei with negative Q-values require a maximum in the astrophysical S-factor ($S(E) = \sigma E exp(2\pi\eta)$ where $\eta = Z_1 Z_2 e^2/(\hbar v)$ is the Sommerfeld parameter), since the fusion cross section has to reach zero at a c.m. energy of E=-Q. For fusion reactions with positive Q-values, this occurs at unphysical (negative) c.m. energies and, thus, no maximumof the S-factor is required. However, several re-

 DOI: 10.1051/epjconf/201716300048

cent experiments involving lighter nuclei with positive Q-values have shown that an S-factor maximum is observed for these systems as well [9].

In order to study the fusion hindrance in the ^{12}C + ^{12}C system, we have recently measured C+C fusion cross sections using the particle-γ-coincidence technique [10]. Details of the detection technique can be found at other places in these proceedings [11]. The resulting C+C S-factors from this and two other recent experiments are shown in Fig. 1.

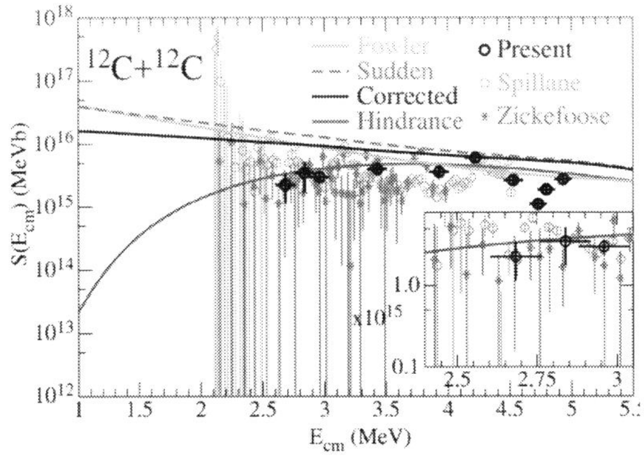

Figure 1. S-factors for ^{12}C + ^{12}C fusion from the particle-γ measurement (black points) in comparison to the results from recent γ (green) or particle (magenta) experiments, respectively ([12, 13]). The lines are S-factor extrapolations from Refs. [14–17].

The predictions of various extrapolation results are given by the solid and dashed lines. While most of these extrapolations increase towards lower energies, the inclusion of fusion hindrance (see red solid line) predicts a maximum of the S-factor at about $E_{c.m.}$=3.5 MeV, in agreement with the solid black points from the particle-γ experiment.

This S-factor maximum and the resulting reduction of the fusion cross sections has important consequences for the production of ^{60}Fe. As discussed in Ref. [6], hindrance of carbon fusion in stellar environments will delay the carbon ignitition by shifting it to higher densities and temperatures. This leads to a higher neutron production and, thus, to an increased ^{60}Fe (as well as ^{26}Al) production yield. In Ref. [18] a comparison of the ^{60}Fe yield extracted from Accelerator Mass Spectrometry measurements of deep-sea sediments and predictions from a variety of supernovae models has been provided. An increase in the ^{60}Fe production cross section will strongly influence the conclusions regarding the age and distance of the supernovae explosions. For that reason more high-quality fusion cross sections for ^{12}C + ^{12}C at even lower energies are clearly needed.

3 Reactions producing and destroying ^{44}Ti

The relatively short half-life of 59.0±0.6 y restricts the use of this radioisotope to more recent stellar events and for that reason so far only two sites are known where ^{44}Ti has been detected. One is the supernovae remnant Cassiopeia A, which was seen on Earth about 350 years ago and is located at a distance of ~3.4 kparsecs. The other is the well-studied recent supernova 1987A in the neighboring Large Magellanic Cloud at a distance of 51.4 kparsecs. First seen by the COMPTEL and INTEGRAL satellites [19, 20], recent measurements by the NUSTAR satellite [21] have provided data with unprecedented position resolution to resolve the globular distribution of ^{44}Ti from the SN explosion.

The isotope ^{44}Ti is produced through the $^{40}Ca(\alpha,\gamma)^{44}Ti$ reaction in the alpha-rich freeze-out phase of the expanding supernova shock front, but it can be destroyed again via the $^{44}Ti(\alpha,p)^{47}V$ process. Sensitivity studies have provided a list of reactions that are critical for interpreting the abundance measurements from the satellite measurements [22, 23].

For the production reaction, $^{40}Ca(\alpha,\gamma)^{44}Ti$, several experiments have been performed at various laboratories during the last 10 years using a variety of detection techniques. The reaction rates from these measurements were found to agree with each other within a factor of about 1.5 [24], but were by a factor of 3 larger than the rates from an earlier measurement [25]. This increased production rate is in agreement with the deduced amounts of ^{44}Ti in the two supernovae sites mentioned above. This comparison, however, also depends on processes that can destroy ^{44}Ti, e.g. via the $^{44}Ti(\alpha,p)^{47}V$ reaction. Since a measurement of this reaction requires the use of a radioactive ^{44}Ti beam, only two experiments have been performed so far. In the first measurement [26], the heavy ^{47}V recoil nuclei from the inverse $^{4}He(^{44}Ti,^{47}V)p$ reaction were detected and identified according to mass, Z and energy in the focal plane of a recoil separator. While this approach allows us to separate contributions from stable beam contaminants (e.g. ^{44}Ca), the detection efficiency (determined by the charge-state selection in the recoil separator) allows measurements down to about 1 mb using present-day radioactive beam intensities. In a more recent experiment [27], the outgoing protons were detected in an array of position-sensitive Si detectors. In this experiment, an upper limit of the energy-integrated cross section of about 60 μb was determined. This technique, however, suffers from the same difficulties encountered in the ^{12}C + ^{12}C fusion experiments, i.e. a sensitivity to background protons produced e.g. in the Al windows of the helium gas target.

In order to extend these measurements to lower energies we have developed a new high-efficiency detector system which is based on the active target technique [28]. A cross section of the gridded multi-sampling ionization chamber (MUSIC) is shown in Fig. 2a. The anode of the ionization chamber consists of 16 strips, which are partitioned into right and left sections. The two additional strips at the entrance and exit of the detector are used for particle identification. The ionization chamber is filled with helium gas, which acts as both target and detection gas.

The incident beam passing through the detector produces a certain amount of electrons in each individual an-

 DOI: 10.1051/epjconf/201716300048

ode strip. If at a given location along the particle trajectory inside the detector an (α,p) reaction occurs, the recoil products, which have a higher nuclear charge, will produce a larger energy loss signal. Since the (α,p) reaction is studied in inverse kinematics, an angle- and energy-integrated cross section is determined in the experiment. Furthermore, this technique is 'self-normalizing' since reaction products as well as the incident beam particles are measured by the same detector. Simulations and first measurements of (α,p) (as well as (α,n)) reactions with ^{17}O and ^{23}Na beams have already been performed [29]. Based on these results we have made Monte Carlo simulations about the expected response of the detector for the $^{44}Ti(\alpha,p)^{47}V$ reaction, which are shown in Fig. 3. First experiments with ^{44}Ti are planned as soon as beams from the new AIRIS separator [30] at ATLAS will become available.

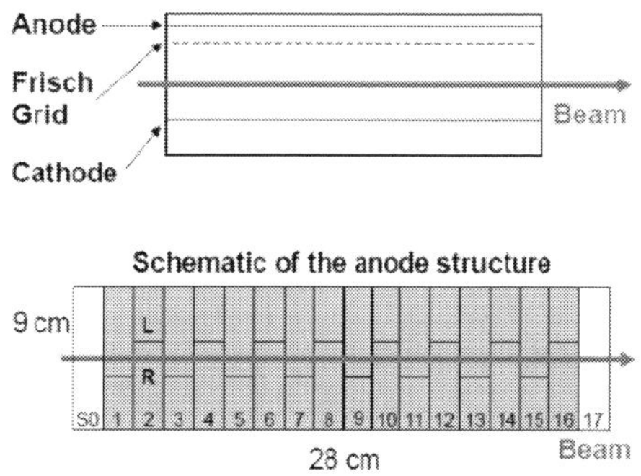

Figure 2. Schematic of the Multi-Sampling Ionization Chamber MUSIC.

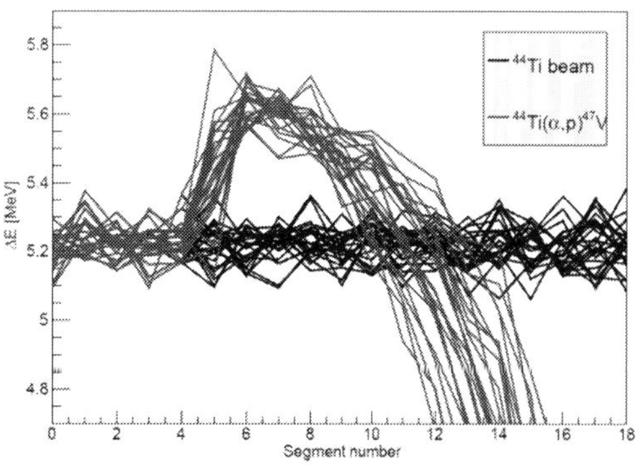

Figure 3. Monte Carlo simulation of the signals from the ^{44}Ti beam and the ^{47}V recoils resulting from the $^{4}He(^{44}Ti,^{47}V)p$ reaction.

4 Production of an isomeric $^{26}Al(0^+)$ beam with the in-flight technique

Among the three radioisotopes discussed in this presentation, ^{26}Al is perhaps the most well studied example. The detection of the characteristic 1.809-MeV γ-ray line from the decay of the long-lived radioisotope ^{26}Al ($T_{1/2}$=7.17x10^5 y) in the interstellar medium by the COMPTEL and INTEGRAL γ-ray satellites [31, 32] has demonstrated that nucleosynthesis is still ongoing in the Galaxy. From the detailed maps of the ^{26}Al distribution, an equilibrium mass of ~2-3 solar masses of ^{26}Al in the Galaxy [32] has been deduced. ^{26}Al can be produced in a number of different sources: in massive Wolf-Rayet (WR) stars either during the hydrogen-burning or the core collapse supernova (CCSN) phase [32], with additional contributions from asymptotic giant branch (AGB) stars and classical novae (CN) [33, 34]. A detailed understanding of the production mechanism, however, is complicated by the existence of a short-lived ($T_{1/2}$ =6.4 s) isomeric 0^+ state in ^{26}Al, located 228-keV above the $J^{\pi}=5^+$ ground state. This state decays by β^+ emission and is therefore 'invisible' to γ-ray satellites. While the communication between the ground state and the isomeric state is inhibited by the large spin difference (ΔJ=5), the two states may communicate with each other via thermal excitations involving higher-lying levels in ^{26}Al which puts constraints on the production and destruction processes of ^{26}Al in astrophysical environments.

Two experiments of the $^{26}Al(5^+)(d,p)^{27}Al$ reaction were conducted recently with the goal to extract spectroscopic information of key resonances in ^{27}Si via its mirror symmetry. From these results the $^{26}Al(5^+)(p,\gamma)^{27}Si$ reaction rate in WR stars and AGB stars is well constrained [27, 35]. For the isomeric state, however, estimates of the $^{26}Al(0^+)(p,\gamma)^{27}Si$ reaction rate are presently based on Hauser-Feshbach calculations by scaling the ground state (p,γ) rate [37].

Several attempts have been made at different laboratories worldwide to produce an isomeric $^{26}Al(0^+)$ beam with sufficient intensity, purity, energy resolution and a high isomer-to-ground state ratio in order to study nuclear reactions induced by the 0^+ isomer in ^{26}Al [38].

At the ATLAS accelerator at Argonne National Laboratory we have recently produced an ^{26}Al (0^+) beam with 70% isomeric content and performed a first $^{26}Al(0^+)(d,p)^{27}Al$ experiment [38]. The $^{26}Al(0^+)$ isomer was produced via the inverse (p,n) reaction $p(^{26}Mg,^{26}Al)n$. For that a 6.1 MeV/u primary beam of ^{26}Mg bombarded a gas cell filled with 1 atm of hydrogen gas. The energy of the primary beam and the target pressure in the gas cell were chosen to maximize the production of the isomeric state over the ground state. The isomeric $^{26}Al^{13+}$ beam was separated from the 12^+ charge state of the ^{26}Mg production beam through its different magnetic rigidity using a bending magnet in the beam line system [39].

In a first step, a test experiment was performed using a rotating stopper setup to measure the purity and isomer content of the beam. The radioactive ^{26}Al beam was stopped in a 100 mg/cm^2 thick Au-catcher foil mounted on a rotating wheel. After an implantation time of 15 sec the foil was rotated by 180° and placed in between two face-to-face NaI detectors where the two coincident 511-keV photons from the β^+ decay of ^{26}Al(0^+) were detected. The measured half-life of 6±1 sec confirmed the isomeric content of the beam. During the γ-counting time the full ^{26}Al beam was transmitted to the Split-Pole Spectrograph where the beam was identified with respect to mass, Z and its total energy. GEANT4 simulations were performed to calculate the detection efficiencies of the NaI detectors.

The purity of the ^{26}Al beam was better than 98% as shown in Fig. 4a in a spectrum measured with a silicon surface barrier detector at 0° . Fig. 4b shows the coincident 511-keV annihilation radiation measured with two NaI detectors. The inset confirms the presence of 0^+ isomer through its known 6.35 s half-life. From this beam development run it was established that, for this specific production energy and target thickness, 70±10% of the radioactive ^{26}Al beam was in the isomeric 0^+ state. The main contaminant in the ^{26}Al^{13+} beam came from the primary ^{26}Mg^{11+} production beam with an energy of 86 MeV while the ^{26}Mg^{12+} charge state was completely eliminated by a RF sweeper. Details about the experiment studying the ^{26}Al(0^+)(d,p)^{27}Al reaction can be found in Ref. [38].

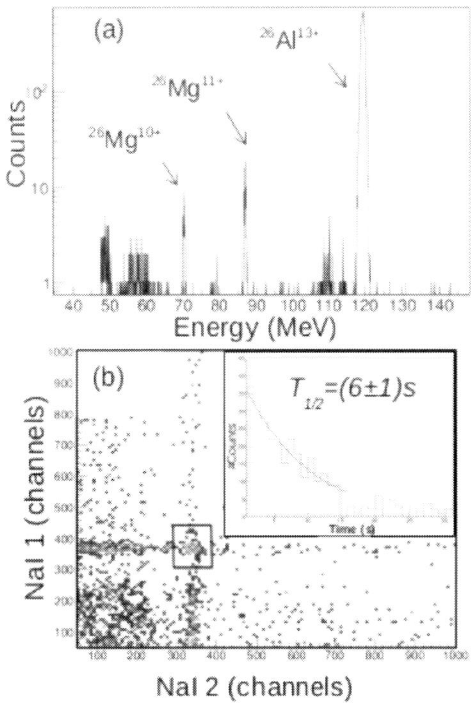

Figure 4. a) Energy spectrum of the secondary ^{26}Al^{13+} beam measured in a Si detector. The main beam contaminant originates from the energy-degraded ^{26}Mg^{11+} beam. b) $\gamma - \gamma$ coincidence spectrum measured with two NaI detectors. The insert shows the time dependence confirming the production of a 6.35 s ^{26}Al(0^+) beam.

5 Summary

The examples presented in this contribution are a clear indication of the close relationship that exists between observational astronomy, nuclear physics and computational astrophysics. New developments in these fields will improve our knowledge of stellar events over the next few years. New Earth and space-based observatories, covering all wavelengths, cutting-edge facilities to accelerate high-intensity stable and radioactive beams and new and faster computers will provide the framework to keep astrophysics in the forefront of science.

6 Acknowledgements

The author wants to thank C. L. Jiang and the collaborators from the ^{12}C + ^{12}C fusion experiment as well as S. Almaraz-Calderon and the collaborators from the ^{26}Al(0^+) experiment for permission to present results before publication. I also want to thank M. Avila and R. Talwar from the ANL MUSIC collaboration for some of the data from the active target system. This work was supported by the US Department of Energy, Office of Nuclear Physics, under Contract No. DEAC02-06CH11357. This research used resources of ANL's ATLAS facility, which is a DoE Office of Science User Facility.

References

[1] G. Rugel *et al.*, Phys. Rev. Lett. **103**, 072502 (2009).
[2] A. Wallner *et al.*, Phys. Rev. Lett. **114**, 041101 (2015).
[3] W. R. Binns *et al.*, Science **352**, 677 (2016).
[4] K. Knie *et al.*, Phys. Rev. Lett. **93**, 171103 (2004).
[5] A. Wallner *et al.*, Nature **532**, 69 (2016).
[6] L. R. Gasques *et al.*, Phys. Rev. C **76**, 035802 (2007).
[7] C. L. Jiang *et al.*, Phys. Rev. Lett. **89**, 052701 (2002).
[8] C. L. Jiang *et al.*, Phys. Rev. C **73**, 014613 (2006).
[9] C. L. Jiang *et al.*, to be published.
[10] C. L. Jiang *et al.*, Nucl. Instrum. Meth. in Phys. Res. A **682**, 12 (2012).
[11] S. Courtin, Proceedings of this conference.
[12] T. Spillane *et al.*, Phys. Rev. Lett. **98**, 122501 (2007).
[13] J. Zickefoose, Thesis, Univ. of Connecticut (2011), unpublished.
[14] W. Fowler, G. Caughlan and B. Zimmerman, Ann. Rev. Astrophys. **13**, 69 (1975).
[15] H. Esbensen, X. D. Tang and C. L. Jiang, Phys. Rev. C **84**, 064613 (2011).
[16] C. L. Jiang *et al.*, Phys. Rev. Lett. **110**, 072701 (2013).
[17] C. L. Jiang *et al.*, Phys. Rev. C **75**, 015803 (2007).
[18] D. Breitschwerdt *et al.*, Nature **532**, 73 (2016).
[19] A. F. Iyudin *et al.*, Astron. Astrophys. **284**, L1 (1994).
[20] S. A. Grebenev *et al.*, Nature **490**, 373 (2012).

[21] B. W. Grefenstette *et al.*, Nature **506**, 339 (2014).
[22] L. S. The, D. D. Clayton, L. Jin and B. S. Mayer, Astrophys. J **504**, 500 (1998).
[23] G. Magkotsios *et al.*, Astrophys. J. Suppl. Ser. **191**, 66 (2010).
[24] D. Robertson *et al.*, Phys. Rev. C **85**, 045810 (2012) and references therein.
[25] E. L. Cooperman, M. H. Shapiro and H. Winkler, Nucl. Phys. A **284**, 163 (1977).
[26] A. A. Sonzogni *et al.*, Phys. Rev. Lett. **84**, 1651 (2000).
[27] V. Margerin *et al.*, Phys. Lett. B **731**, 358 (2014).
[28] P. F. F. Carnelli *et al.*, Nucl. Instrum. Meth. in Phys. Res. A **799**, 197 (2015).
[29] M. L. Avila *et al.*, Nucl. Instrum. Meth. in Phys. Res. A **859**, 63 (2017).
[30] B. B. Back, Proceedings of this conference.
[31] R. Diehl *et al.*, Astron. Astrophys. **298**, 445 (1995).
[32] R. Diehl *et al.*, Nature **439**, 45 (2006).
[33] L. Siess and M. Arnould, Astron. Astrophys. **489**, 395 (2008).
[34] J. Jose, M. Hernanz and A.Coc, Astrophys. J. **479**, L55 (1997).
[35] S. Pain *et al.*, Phys. Rev. Lett. **114**, 212501 (2015).
[36] V. Margerin *et al.*, Phys. Rev. Lett. **115**, 062701 (2015).
[37] C. Iliadis, A. Champagne, A. Chieffi and M. Limongi, Astrophys. J. Suppl. Ser. **193**, 16 (2011).
[38] S. Almaraz-Calderon *et al.*, to be published.
[39] B. Harss *et al.*, Rev. Sci. Instrum. **71**, 380 (2000).

Superfluid effects in collision between systems with small particle number

Guillaume Scamps[1,*] and *Yukio* Hashimoto[1,**]

[1]Center for Computational Sciences, University of Tsukuba, Tsukuba 305-8571, Japan

Abstract. The interpretation of the new effects of the superfluidity in reactions with small number of particles is discussed in a simple model where the exact solution is accessible. It is find that the fluctuations of observable with the gauge angle reproduce well the exact fluctuations. Then a method of projection is proposed and tested to determine the transfer probabilities between two superfluid systems.

1 Introduction

In a precedent contribution [1], we studied the reaction between two superfluid nuclei $^{20}O+^{20}O$ with the Time-dependent Hartree-Fock-Bogoliubov theory. The dependence of the observables with respect to the initial relative gauge angle between the two initial fragments has been studied. A Josephson effect has been found, as well as a dependence of the nucleus-nucleus potential with this gauge angle. Those effects have also been studied in Ref. [2]. As point out by Ref. [3], the interpretation of this result is ambiguous for systems which should have a given number of particles in each fragment initially and so respect the gauge angle symmetry. This motivate us to consider a more simple model that can be solved exactly.

2 Model

Our model is inspired by the Dietrich model of nuclear Josephson effect [4]. Two systems are initially considered S_1 and S_2 both are composed of Ω doubly degenerated levels with N particles. Each level i is paired with a conjugated level $\bar{i}$. The initial Hamiltonian of the system is composed of a pairing interaction,

$$\hat{H}_0 = G \sum_{i\neq j\in S_1} \hat{a}_i^\dagger \hat{a}_{\bar{i}}^\dagger \hat{a}_{\bar{j}} \hat{a}_j + G \sum_{i\neq j\in S_2} \hat{a}_i^\dagger \hat{a}_{\bar{i}}^\dagger \hat{a}_{\bar{j}} \hat{a}_j. \quad (1)$$

For simplicity, we consider all the particles paired. The exact solution of this Hamiltonian is obtained by diagonalizing the Hamiltonian in the space of all the configurations that have N particles in both sides. We consider the case of Ω=4, N=4 and $G = -0.2$ MeV. In that case, the total energy of the system with the exact solution is 1.6 MeV. For the HFB solution, we adjust the interaction $G^{\text{eff}}=\frac{4}{3}G$ in order to obtain the same initial energy.

*e-mail: scamps@nucl.ph.tsukuba.ac.jp
**e-mail: hashimoto@nucl.ph.tsukuba.ac.jp

The initial state is then propagated in time, with the time-dependent Hamiltonian that connect the two systems,

$$\hat{H}(t) = \hat{H}_0 + V(t) \sum_{i\in S_1; j\in S_2} \left(\hat{a}_i^\dagger \hat{a}_{\bar{i}}^\dagger \hat{a}_{\bar{j}} \hat{a}_j + \hat{a}_j^\dagger \hat{a}_{\bar{j}}^\dagger \hat{a}_{\bar{i}} \hat{a}_i\right). \quad (2)$$

For simplicity reason, the time dependent interaction is chosen as,

$$V(t) = V_0 e^{-at^2}, \quad (3)$$

with a=0.3×10^{44}s^{-2}. The system is evolved between time t=-15×10^{-22}s and t=15×10^{-22}s. An effective interaction $V_0^{\text{eff}}=\frac{4}{3}V_0$ is also taken in order to be consistent with the initial calculation. Two values of V_0 are tested, a weak and a strong interaction are chosen respectively with $V_0 = -0.003$ and -0.4 MeV. The two calculations, exact and TDHFB differs at the initialization of the calculation. For the exact case, there is no gauge angle, because the exact state has a good number of particles in both fragments so only one calculation is done. But for the TDHFB dynamic, the initial total state is defined as :

$$|\Psi_\varphi\rangle = \prod_{i\in S_1}(u_i + v_i a_i^\dagger a_{\bar{i}}^\dagger) \prod_{j\in S_2}(u_j + e^{2i\varphi} v_j a_j^\dagger a_{\bar{j}}^\dagger)|-\rangle, \quad (4)$$

with φ the relative gauge angle. 24 initial values of φ are chosen in the interval $[0, \pi]$.

3 Results

Two observables are computed as a function of time, the number of particles in the S_1 system $\langle \hat{N}_{S_1}\rangle$ and the total energy $\langle \hat{H}(t)\rangle$. In this model, we expect to reproduce with TDHFB the Josephson effect and the fluctuations of the internal energy described in the introduction. Note that in this model, the energy is not conserved because the Hamiltonian changes as a function of time. This is what we find in the fig. 1 in the TDHFB case and in fig. 4 for the exact case. The TDHFB average number of particles in one of the fragments changes with respect to φ following a sin(2φ) dependence and the energy changes linearly

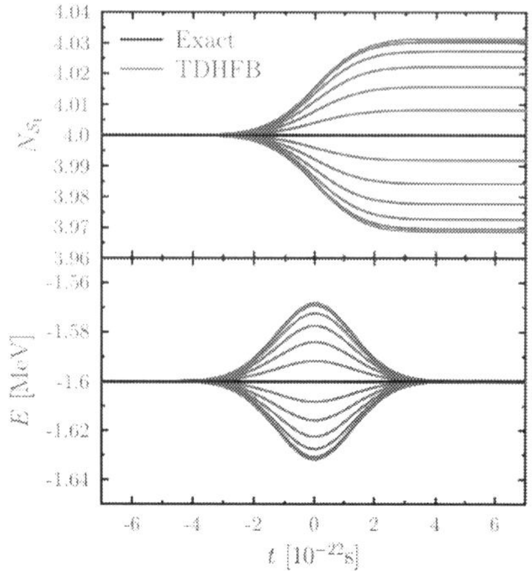

Figure 1. Comparison of the exact and TDHFB average number of particles in the S_1 system (up) and the average energy (down) as a function of time in the weak interaction case. Each red line corresponds to a calculation with a given value of the initial relative gauge angle φ.

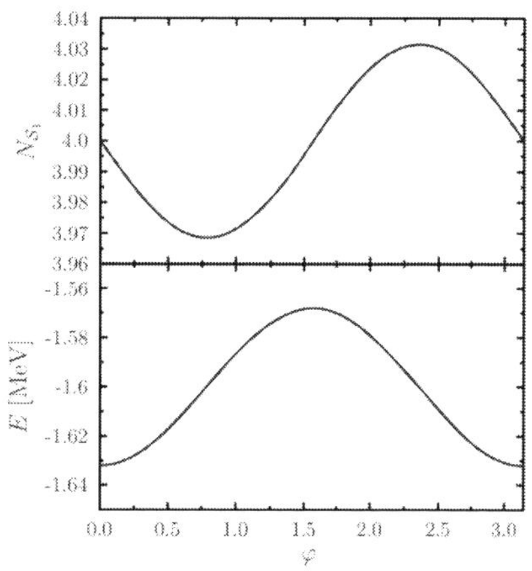

Figure 2. Dependence of the TDHFB results with the relative gauge angle φ. Top : Number of particles in the S_1 system at the time 6×10^{22} s. Down : Energy at time t=0.

with $\cos^2(\varphi)$. The dependences with respect to the relative gauge angle are shown in fig. 2.

If we compare the average value of $\langle\Psi_\varphi|\hat{O}|\Psi_\varphi\rangle$ we find a good agreement with the exact case. But it is more difficult to interpret the fluctuations of those observables. In that aim, we compute the quantal fluctuations of the observable in the exact case,

$$\sigma_{\hat{O}}^{\text{Exact}} = \sqrt{\langle\Psi|\hat{O}^2|\Psi\rangle - \langle\Psi|\hat{O}|\Psi\rangle^2}, \tag{5}$$

to the standard deviation of the TDHFB observable with respect to φ,

$$\sigma_{\hat{O}}^{\text{TDHFB}} = \sqrt{\frac{1}{\pi}\int_0^\pi \langle\Psi_\varphi|\hat{O}|\Psi_\varphi\rangle^2 d\varphi - \left(\frac{1}{\pi}\int_0^\pi \langle\Psi_\varphi|\hat{O}|\Psi_\varphi\rangle d\varphi\right)^2}. \tag{6}$$

These fluctuations are included in fig. 3 and 4 by error bars. A very good agreement is found, showing that this interpretation of the fluctuations of the observable is a good one. Note that we should not take into account the direct calculation of the fluctuations of the observable in each trajectory,

$$\sigma_{\hat{O}}^{\varphi} = \sqrt{\langle\Psi_\varphi|\hat{O}^2|\Psi_\varphi\rangle - \langle\Psi_\varphi|\hat{O}|\Psi_\varphi\rangle^2}. \tag{7}$$

Indeed, at the initial time, these fluctuations are non zero because of the number of particles is not a good quantum number and because the HFB ground state is not an eigenstate of the Hamiltonian. Then, to convolute the distribution of $\sigma_{\hat{O}}^{\varphi}$ with $\sigma_{\hat{O}}^{\text{TDHFB}}$ will only bring spurious results. Then in realistic cases, the present interpretation should be used with caution and a more rigorous method of restauration of the gauge angle symmetry should be consider.

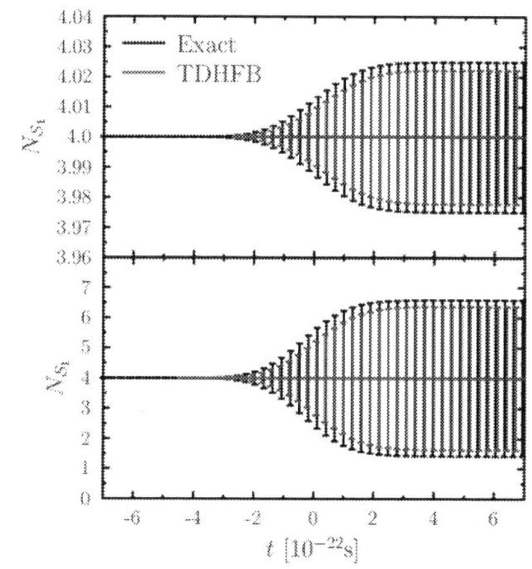

Figure 3. Comparison of the exact and TDHFB results for the number of particles in the S_1 system as a function of time in the weak (up) and strong (down) interaction case. The error bars represent the fluctuations of this number.

4 Projection method

Although fluctuations of observables brings interesting information, we would like to develop a method of projection in order to determine the complete distribution of particles in each fragment. To determine the transfer probabilities, we use the projection method [5] that determines the probability $P_S(N)$ to have a given number of particles

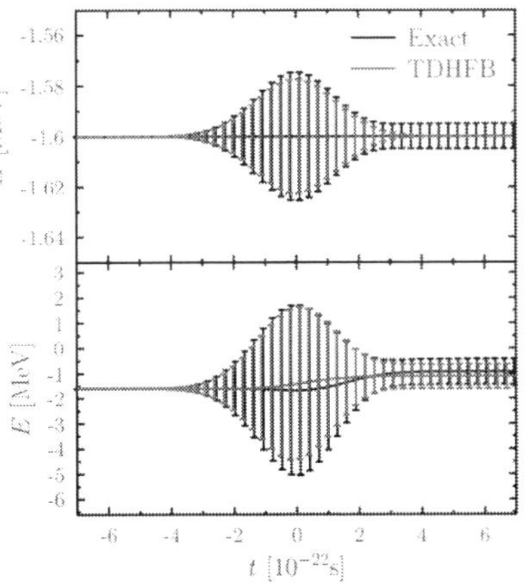

Figure 4. Comparison of the exact and TDHFB results for the energy as a function of time in the weak (up) and strong (down) interaction cases. The error bars represent the fluctuations of the energy.

N in a subspace $\mathcal{S}$,

$$P_{\mathcal{S}}(N,t) = \langle\Psi(t)|\hat{P}_{\mathcal{S}}(N)|\Psi(t)\rangle, \tag{8}$$

with

$$\hat{P}_{\mathcal{S}}(N) = \frac{1}{2\pi}\int_0^{2\pi} e^{i\varphi(\hat{N}_S - N)}d\varphi. \tag{9}$$

$\hat{N}_S$ is the operator that count the number of particles in the subspace $\mathcal{S}$.

As discussed in Ref. [6], the projection method has to be modified when used with theories that mix different number of particles. It is the case here, where we use quasi-particle states that contain initially components with the good number of particles N_{tot} but also components with ..., N_{tot}-4, N_{tot}-2, N_{tot}+2, N_{tot}+4,... Those components will induce spurious results. For exemple, at the initial time, even if the two fragments are separated by an infinite distance, the pair transfer probability will not be zero.

To suppress the spurious component, a double projection technique is applied,

$$P_{\mathcal{S}}(N) = \frac{\langle\Psi(t)|\hat{P}_{\mathcal{S}}(N)\hat{P}(N_{\text{tot}})|\Psi(t)\rangle}{\langle\Psi(t)|\hat{P}(N_{\text{tot}})|\Psi(t)\rangle}, \tag{10}$$

with $\hat{P}(N_{\text{tot}})$ the projector onto the total space. This method only works for collisions where only one of the fragments is in the superfluid phase, the other one has to be in the normal phase. Let's see what happen, if we use this method to collisions where both fragments break the gauge angle symmetry with average number N_1 and N_2 respectively for system 1 and 2. The total wave function after projection onto the good total number $N_{\text{tot}} = N_1 + N_2$ will still contain spurious components with an initial number of particles $N_1 - 2n$ and $N_2 + 2n$ with n an integer. Those components will correspond to pair transfer states, and are spurious because they can happen before the collision.

Then for the collisions between two superfluid systems, we have to compute the probability

$$P_{\mathcal{S}}(N,t) = \frac{\langle\Psi_{N_1,N_2}(t)|\hat{P}_{\mathcal{S}}(N)|\Psi_{N_1,N_2}(t)\rangle}{\langle\Psi_{N_1,N_2}(t)|\Psi_{N_1,N_2}(t)\rangle}, \tag{11}$$

with the state,

$$|\Psi_{N_1,N_2}(t)\rangle = \hat{U}^{TDPHFB}(t_0,t)\hat{P}_{\mathcal{S}}(N_1)\hat{P}_{\bar{\mathcal{S}}}(N_2)|\Psi(t_0)\rangle, \tag{12}$$

with $\bar{\mathcal{S}}$ the complement of the subspace $\mathcal{S}$. The subspace $\mathcal{S}$ and $\bar{\mathcal{S}}$ contain respectively the initial systems 1 and 2 at the initial time. $\hat{U}^{TDPHFB}(t_0,t)$ is the propagator between time t_0 and time t of a projected state. The self-consistent propagation of a quasi-particle state would be a theory interesting to develop. Nevertheless in the present calculation, we choose to propagate independently each state with different gauge angles, with the TDHFB propagator $\hat{U}^{\text{TDHFB}}$,

$$\begin{aligned}\hat{U}(t_0,t_f)\hat{P}_{\mathcal{S}}(N_i)|\Psi(t=0)\rangle \\ = \frac{1}{2\pi}\int_0^{2\pi} d\varphi e^{-i\varphi N_i}\hat{U}^{\text{TDHFB}}(t_0,t_f)e^{i\varphi\hat{N}_S}|\Psi(t=0)\rangle,\end{aligned} \tag{13}$$

$$= \frac{1}{2\pi}\int_0^{2\pi} d\varphi e^{-i\varphi N_i}|\Psi(\varphi,t_f)\rangle, \tag{14}$$

with $|\Psi(\varphi,t)\rangle$ the evolved state using the TDHFB equation of motion of the initially state rotated by an angle φ. Using the properties,

$$\hat{P}_{\mathcal{S}}(N_1)\hat{P}_{\bar{\mathcal{S}}}(N_2) = \hat{P}_{\mathcal{S}}(N_1)\hat{P}(N_1+N_2), \tag{15}$$

$$e^{i\varphi\hat{N}_{\text{tot}}}\hat{U}^{\text{TDHFB}}(t_0,t) = \hat{U}^{\text{TDHFB}}(t_0,t)e^{i\varphi\hat{N}_{\text{tot}}}, \tag{16}$$

we find the expression of the probability,

$$\begin{aligned}P_{\mathcal{S}}(N,t) = \frac{1}{\mathcal{N}}\frac{1}{(2\pi)^4}\iiiint_0^{2\pi} e^{i(\varphi_1-\varphi_4)N_1 - i\varphi_2 N - i\varphi_3 N_{\text{tot}}} \\ \times\langle\Psi(\varphi_1,t)|e^{i\varphi_2\hat{N}_S}e^{i\varphi_3\hat{N}_{\text{tot}}}|\Psi(\varphi_4,t)\rangle d\varphi_{1,2,3,4},\end{aligned} \tag{17}$$

with the norm,

$$\begin{aligned}\mathcal{N} = \frac{1}{(2\pi)^3}\iiint_0^{2\pi} e^{i(\varphi_1-\varphi_4)N_i - i\varphi_3 N_{\text{tot}}} \\ \times\langle\Psi(\varphi_1,t)|e^{i\varphi_3\hat{N}_{\text{tot}}}|\Psi(\varphi_4,t)\rangle d\varphi_{1,3,4}.\end{aligned} \tag{18}$$

The overlap is computed with the Pfaffian method [7],

$$\langle\Psi(\varphi_1,t)|e^{i\varphi_2\hat{N}_S}e^{i\varphi_3\hat{N}_{\text{tot}}}|\Psi(\varphi_4,t)\rangle = (-1)^n\frac{\det C^*\det C'}{\prod_\alpha^n v_\alpha v'_\alpha}\text{pf}\mathcal{M}, \tag{19}$$

$$\mathcal{M} = \begin{bmatrix} V^T U & V^T e^{i\varphi_3}(1+\Theta(i)e^{i\varphi_2})V'^* \\ -V'^\dagger e^{i\varphi_3}(1+\Theta(i)e^{i\varphi_2})V & U'^\dagger V'^* \end{bmatrix}. \tag{20}$$

With C the matrix obtained from the Bloch-Messiah decomposition [8] and v_α the occupation numbers in the canonical basis. The C, v_α, V and U correspond to the bra $\langle\Psi(\varphi_1,t)|$ while the C', v'_α, V' and U' refer to the ket

 DOI: 10.1051/epjconf/201716300049

$|\Psi(\varphi_4, t)\rangle$. The function $\Theta(i)$ is equal to 1 if the state i belongs to the system S and 0 if not.

To test this method, we used a slightly different model than previously, we introduce the single particle energy. The four states of each fragment have an energy e_i=i MeV with i=1,2,3,4. Because we use the projection method, we don't take an effective interaction, we take the case of $G = -1$ MeV, V_0=-0.03 MeV and a=0.3×10^{44}s^{-2}. Using the standard TDHFB equation,

$$i\hbar\frac{\partial}{\partial t}\begin{pmatrix} U(t) \\ V(t) \end{pmatrix} = \mathcal{H}\begin{pmatrix} U(t) \\ V(t) \end{pmatrix}, \quad (21)$$

with

$$\mathcal{H} = \begin{pmatrix} h & \Delta \\ \Delta^* & -h^* \end{pmatrix}, \quad (22)$$

we found a spurious behavior. As figure 5 shows with red solid line, after the reaction happens, when $V(t)$ become small, the pair transfer probability $P_2 = P_{S_1}(6, t)$ is not constant. This is a non-physical behavior that is not present in the exact calculation. We believe that this behavior is due to the approximation that the TDHFB are independently propagate in time. This induce different rotation velocity in the gauge angle plane due to the different chemical potential after the Josephson transfer takes place. To reduce this spurious behavior, our prescription is to remove the chemical potential in both fragments,

$$\mathcal{H} = \begin{pmatrix} h - \lambda_S(t) & \Delta \\ \Delta^* & -h^* + \lambda_S(t) \end{pmatrix}, \quad (23)$$

with λ_S the chemical potential computed in the subspace S as,

$$\lambda_S = \frac{1}{\Omega}\sum_{k>0\in S}^{\Omega} \mathrm{Real}\left(\frac{\Delta_k(2n_k - 1)}{2\kappa_k} + e_k\right). \quad (24)$$

With $n_k = |v_k|^2$ and $\kappa_k = u_k v_k^*$. Note that in this model the quasi-particle basis $\{k\}$ is the same as the particle basis $\{i\}$. This prescription improves the result on fig. 5, but a better agreement with the exact solution is found if one removes also the rotation due to the different quasi-particles energy,

$$\mathcal{H} = \begin{pmatrix} h - \lambda_S(t) - \epsilon_k(t) & \Delta \\ \Delta^* & -h^* + \lambda_S(t) - \epsilon_k(t) \end{pmatrix}, \quad (25)$$

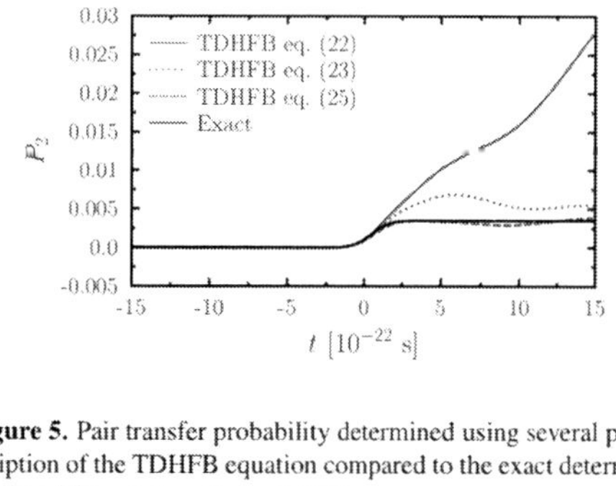

Figure 5. Pair transfer probability determined using several prescription of the TDHFB equation compared to the exact determination of the pair transfer probability.

with the quasi-particle energy,

$$\epsilon_k = (e_k - \lambda_S)(1 - 2n_k) + \Delta_k\kappa_k^* + \Delta_k^*\kappa_k. \quad (26)$$

Using this TDHFB equation of motion, on fig. 5, the transfer probability is almost stable after the two systems are isolated and the average result reproduces well the exact calculation.

5 Conclusion

In conclusion, we used a simple model of reaction between two superfluid systems. We interpret the fluctuations of the TDHFB observables with respect to the initial relative gauge angle as statistical fluctuations. The average value and the standard deviation of the energy and the number of transferred particles from TDHFB are closed to the exact results. We then developed a projection method to determine the transfer probabilities. A spurious behavior is found after the separation of the two systems, a prescription to modify the TDHFB equation in order to cure the problem is proposed and reproduce correctly the exact results. This prescription will be used in the case of a realistic calculation in a future contribution.

References

[1] Y. Hashimoto and G. Scamps, Phys. Rev. C **94**, 014610 (2016).
[2] P. Magierski, K. Sekizawa, G. Wlazłowski, arXiv:1611.10261 [nucl-th] (2016).
[3] A. Bulgac, arXiv:1701.06683 [nucl-th] (2017).
[4] K. Dietrich, Phys. Lett. B **32**, 428 (1970).
[5] C. Simenel, Phys. Rev. Lett. **105**, 192701 (2010).
[6] G. Scamps and D. Lacroix, Phys. Rev. C **87**, 014605 (2013).
[7] G.F. Bertsch, L.M. Robledo, Phys. Rev. Lett. **108**, 042505 (2012).
[8] P. Ring and P. Schuck, *The Nuclear Many-Body Problem*

 DOI: 10.1051/epjconf/201716300050

The nucleon localization function in static and time-dependent DFT

Bastian Schuetrumpf[1,*] and *Chunli* Zhang[1,**]

[1]*NSCL/FRIB Laboratory, Michigan State University, East Lansing, Michigan 48824, USA*

Abstract. Static and time-dependent density functional theory (DFT) calculations are often used to predict fission fragment distributions or fusion cross sections with great success. However, nuclear shell structure and clusterization effects are usually studied using the density distribution of the nucleons which is a poor indicator of these phenomena. In this work, we employ a measure called the localization function, which was first introduced in chemistry to visualize electronic bonds and recently applied to nuclear physics for light nuclei to reveal α-clustering. We show, that the localization function reveals the shell structure of the fragments in fission long before the scission point and also illustrates the shell structure of intermediate states in time-dependent DFT calculations.

1 Introduction

The emergence of cluster states in atomic nuclei is a universal phenomenon with many implications for both nuclear physics and astrophysics [1–3]. Several factors are contributing to clustering, a complete microscopic understanding has still not been achieved. Cluster configurations are often energetically favorable due to the large binding energy per nucleon in constituent clusters, such as α particles and this argument has often been used to explain properties of α-conjugate nuclei [4], cluster emission [5, 6] and fission [7]. Clusters are also coupled to decay channels, which explains [8, 9] the occurrence of cluster states at low excitation energies around cluster-decay thresholds [10]. Clustering is of special relevance in the dynamics of the tripple-α reaction [11].

A microscopic description of cluster states requires the use of an advanced many-body framework. A good starting point is density functional theory (DFT) based on a realistic nuclear energy density functional [12]. Within DFT, cluster states have a simple interpretation in terms of quasimolecular states.

The degree of clustering in nuclei is difficult to assess quantitatively in DFT as the single particle wave functions are spread throughout the nuclear volume. Furthermore the single particle states in the Slater determinant are not fixed since the Slater determinant is invariant under unitary transformations. Therefore, in this study, we utilize a measure - dependent on s.p. orbitals, yet invariant under unitary transformation - called localization, originally introduced for the identification of electronic bonds in atomic and molecular systems (see e.g. [13–15]) and has recently been applied to characterize clusters in light nuclei [16].

In this contribution we introduce the localization measure in Sec. 2. In Sec. 3 we show results for heavy nuclei, in Sec. 4 we review findings for fissioning nuclei and in Sec. 5 we show first results for time-dependent calculations.

*e-mail: schutrum@nscl.msu.edu
**e-mail: zhangch@nscl.msu.edu

2 Method

2.1 The nucleon localization function

A localization measure in the DFT framework was first introduced for electronic systems to visualize electronic bonds in [13], referred to as the electronic localization function (ELF). For nuclear systems the localization was first utilized in [16] for light nuclei. For nuclear matter the nucleon localization function (NLF) is based on the inverse conditional probability of finding a nucleon with same spin and isospin in the vicinity of another knowing with certainty that this one is located at $\boldsymbol{r}$. The measure can be expressed through densities and a natural choice for the normalization is the Thomas-Fermi kinetic energy. The resulting dimensionless NLF reads

$$C_{q\sigma}(\boldsymbol{r}) = \left[1 + \left(\frac{\tau_{q\sigma}\rho_{q\sigma} - \frac{1}{4}|\boldsymbol{\nabla}\rho_{q\sigma}|^2 - \boldsymbol{j}^2_{q\sigma}}{\rho_{q\sigma}\tau^{\mathrm{TF}}_{q\sigma}}\right)^2\right]^{-1} \quad (1)$$

and contains the densities

$$\rho_{q\sigma}(\boldsymbol{r}) = \sum_{\alpha\in q} v_\alpha^2 |\psi_\alpha(\boldsymbol{r}\sigma)|^2, \quad (2a)$$

$$\tau_{q\sigma}(\boldsymbol{r}) = \sum_{\alpha\in q} v_\alpha^2 |\boldsymbol{\nabla}\psi_\alpha(\boldsymbol{r}\sigma)|^2, \quad (2b)$$

$$\boldsymbol{j}_{q\sigma}(\boldsymbol{r}) = \sum_{\alpha\in q} v_\alpha^2 \mathrm{Im}[\psi_\alpha^*(\boldsymbol{r}\sigma)\boldsymbol{\nabla}\psi_\alpha(\boldsymbol{r}\sigma)], \quad (2c)$$

$$\boldsymbol{\nabla}\rho_{q\sigma}(\boldsymbol{r}) = 2\sum_{\alpha\in q} v_\alpha^2 \mathrm{Re}[\psi_\alpha^*(\boldsymbol{r}\sigma)\boldsymbol{\nabla}\psi_\alpha(\boldsymbol{r}\sigma)] \quad (2d)$$

$$\tau^{\mathrm{TF}}_{q\sigma} = \frac{3}{5}(6\pi^2)^{2/3}\rho_{q\sigma}^{5/3} \quad (2e)$$

DOI: 10.1051/epjconf/201716300050

where v_α^2 is the canonical occupation probability, $q \in \{p, n\}$ labels the isospin and $\sigma \in \{\uparrow, \downarrow\}$ labels the spin or signature symmetry. For the time-independent calculations presented here, time-reversal symmetry is conserved; hence, $\boldsymbol{j}_{q\sigma}$ vanishes. For time-dependent calculations, the spin is in general not a conserved quantity and the signature symmetry has to be taken into account. However, since we are only interested in the total localization of neutrons and protons, we show in this case the localizations C_q for average densities $\rho_{q\mathrm{av}} = (\rho_{q\uparrow}(\boldsymbol{r}) + \rho_{q\downarrow}(\boldsymbol{r}))/2$ and for the other densities respectively.

Two important values for the localization should be mentioned at this point:

$C_{q\sigma} = \mathbf{0.5}$: Due to the normalization with the Thomas-Fermi kinetic energy a localization of 0.5 indicates a localization as in a Fermi gas of nuclei.

$C_{q\sigma} = \mathbf{1.0}$: A localization of one indicates perfect localization, i.e. that the probability of finding another nucleon with same spin and isospin is zero. Perfect localization can be found in an α-particle, where only one particle of each spin and isospin combination is present.

In practice especially for light nuclei we found, that if the numerical value the wave function is very small far outside a nucleus, the value of the localization is poorly determined and depends strongly on numerical effects. Since we are only interested in the localization inside a nucleus, we utilize a masking function to suppress the localization outside a nucleus as

$$C(\boldsymbol{r}) \to C(\boldsymbol{r})\rho_{q\sigma}(\boldsymbol{r})/\max[\rho_{q\sigma}(\boldsymbol{r})] \quad . \tag{3}$$

2.2 DFT engines

The model of choice for this study is the Skyrme EDF method. In particular we are using the UNEDF1 interaction [17], which was optimized for fission calculations and largely deformed nuclei.

We use two well established EDF solvers for the calculations. For the ground state calculations and constrained fission calculations in Sec. 3 and Sec. 4 we employ the solver HFBTHO [18], based on a deformed harmonic oscillator basis expansion for the wave functions solving the Hartree-Fock-Bogolyubov (HFB) equations including the pairing effect self-consistently.

For the time-dependent calculations in Sec. 5, we employ the solver Sky3D [19] which operates on a cartesian 3d grid employing the fast Fourier transform for derivatives. It solves the time-dependent Hartree-Fock equations and includes the pairing effect in the BCS limit with frozen occupations.

3 Localization in heavy nuclei

In [16] it has been shown, that the NLF reveals α-clustering in light nuclei very clearly. Here, the NLF for heavier nuclei shall be studied.

In Fig. 1 the neutron and proton densities (left panels) as well as the NLF (right panels) of ^{132}Sn are displayed. While the number densities show no structure, except for a small dip in the center of the proton density, the NLF show concentric rings of enhanced localization. The case of ^{132}Sn is particularly interesting, because it is a doubly magic nuclei with one more closed shell for the neutrons. This also manifests in the NLF with an additional enhancement at the center of the neutron NLF. In contrast to electronic systems, the number of closed shells cannot be determined by the number of radial NLF enhancements. The difference is, that in electronic systems the length scale of the shells is proportional to $\sim n^2$ where n is the principal quantum number and in nuclear systems it is proportional to $\sim \sqrt{n}$ and also dependent on the angular momentum. Therefore the shells overlap and cannot be clearly distinguished. However, the imprint on the NLF of one more closed shell is clearly visible.

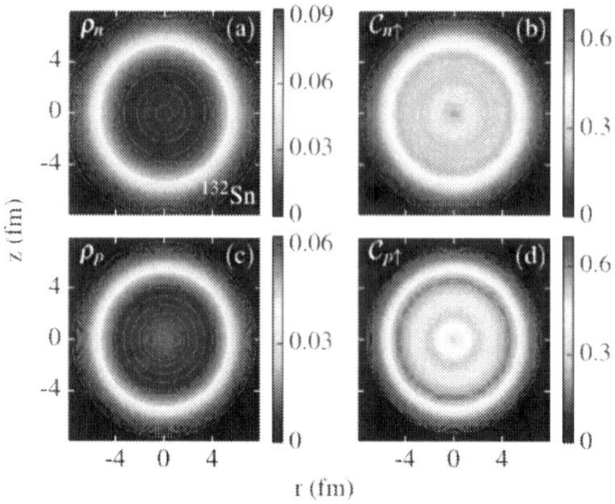

Figure 1: Nucleonic number densities (in fm^{-3}) and spatial localizations for the ground state of ^{132}Sn. This Figure is taken from [20].

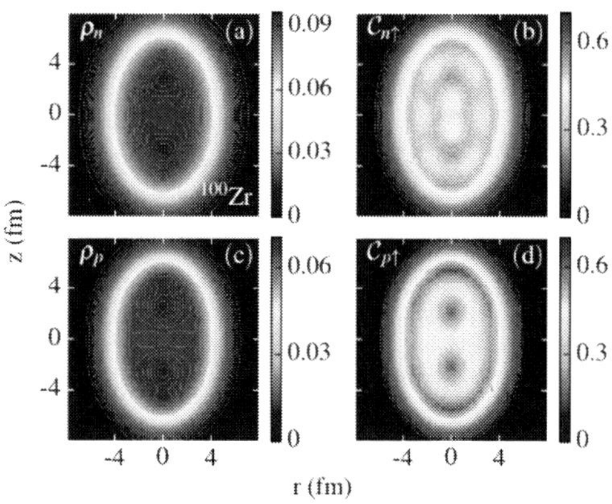

Figure 2: Nucleonic number densities (in fm^{-3}) and spatial localizations for the ground state of ^{100}Zr. This Figure is taken from [20].

For a second example ^{100}Zr is shown in Fig. 2. This nucleus does not have a magic number of either nucleons and is prolately deformed. The proton NLF reveals two

 DOI: 10.1051/epjconf/201716300050

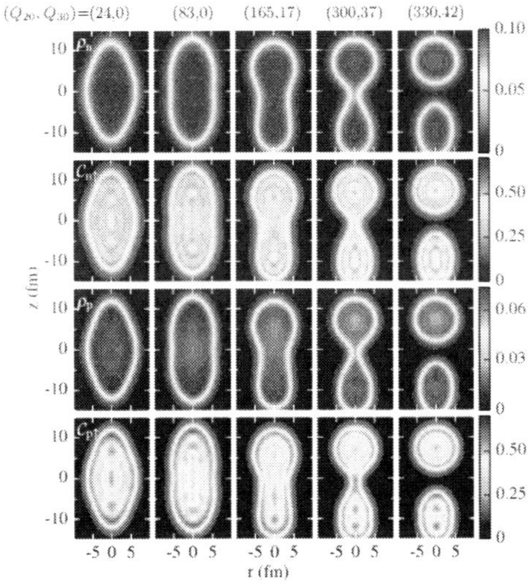

Figure 3: Nucleonic number densities (in fm^{-3}) and spatial localizations for ^{232}Th for five configurations along the fission pathway. This Figure is taken from [20]. The deformations are marked at the top of each column with in units of b and $b^{3/2}$, respectively.

centers and one ring of enhanced localization, whereas the neutron NLF exhibits a two ring structure. Also here the 20 additional neutrons leave a strong imprint on the localization.

4 Localization for fission calculations

As a next step we study the localization during fission in heavy nuclei. As an example we take the asymmetric fission of ^{232}Th. We consider the fission pathway which has been calculated in [21] and plot the densities and NLF for five points on this pathway in Fig 3. This pathway leads to the two fragments ^{132}Sn and ^{100}Zr which were discussed in Sec. 3. This will make it possible to compare the NLFs of the compound nucleus to those of the fragments. For that purpose we performed static calculations with constraints on the quadrupole moment Q_{20} which quantifies the elongation and the octupole moment Q_{30} which quantifies the mass asymmetry as marked at the top of the figure. Rotational symmetry along the z-axis is assumed. The state which is displayed in the left column corresponds to the ground state. The next state is the metastable fission isomer. The third state is a state across the barrier which has already a non-zero mass asymmetry deformed. The fourth state is around the scission point and the fifth state shows already separated fragments.

While the first two compact configurations have a very distinct structure in the NLF, they do not resemble the fragments. In contrast, the first configuration on the asymmetric part of the path shows already the structure of the fragments. It is also worth to notice, that in the neck region,

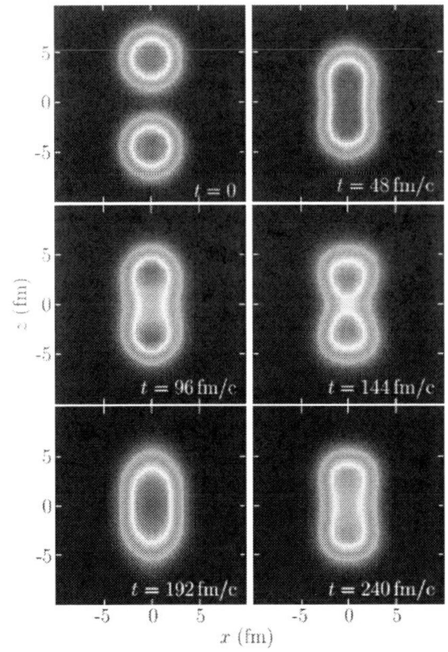

Figure 4: Snapshots of the time evolution of the nucleonic number densities (in fm^{-3}) of the central collision of two ^{16}O nuclei with E_{cm} = 40 MeV.

where both fragments overlap, the NLF has values around 0.5. This indicates, that in this region there is not a distinct shell structure dominant and the structure is similar to a Fermi gas. The density however does not show any resemblance to a specific fragment and the neck contains still a large number of nucleons which makes it impossible to predict the fragments.

The fourth configuration shows even more clearly the NLF of the fragments. The neck is already very small and especially for the neutrons an enhanced localization is visible. This enhancement is the result of very few particles which are present in the neck so the possibility of finding a second with same spin and isospin is very small. The fifth configuration finally exhibits the exact structure of the fragments.

5 Localization in time-dependent DFT

In this section we focus on first time-dependent results for the NLF. As a test case we picked the symmetric collision between two ^{16}O nuclei. First we consider a central collision with a center-of-mass energy of 40 MeV.

The density distribution is shown in Fig. 4. In practice the calculation is performed in a 32 fm × 32 fm × 32 fm box with a larger initial separation as shown in the Figure. We show here only a sector in space and time. In this collision the two oxygen form a compound nucleus. At t = 144 fm/c the compound has the largest deformation

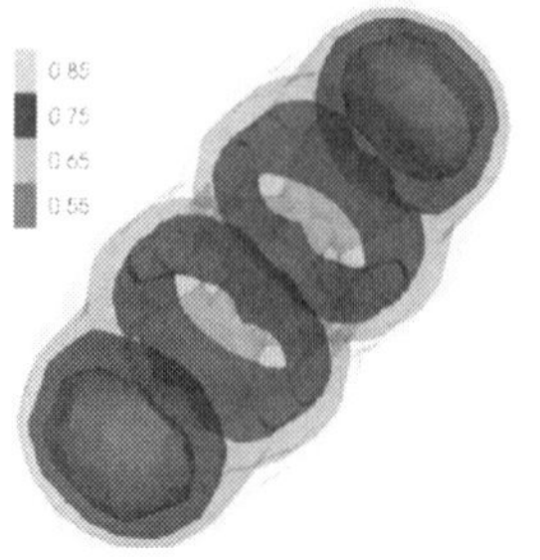

(a) t = 192 fm/c

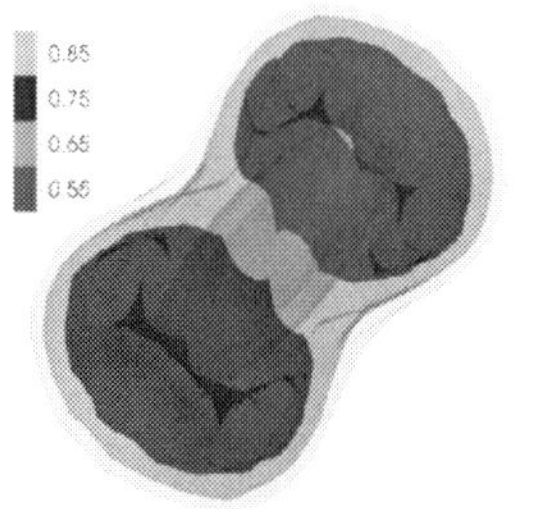

(b) t = 240 fm/c

Figure 5: Snapshots of the time-dependent evolution of the neutron NLF of the central collision of two ^{16}O nuclei with E_{cm} = 40 MeV.

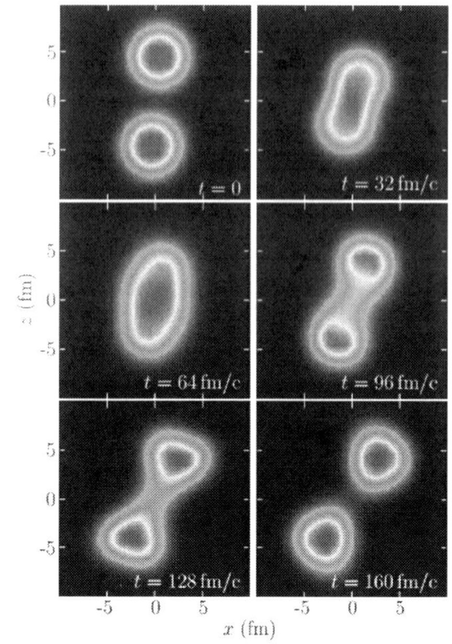

Figure 6: Same as Fig. 4 but for E_{cm} = 100 MeV and b = 2 fm

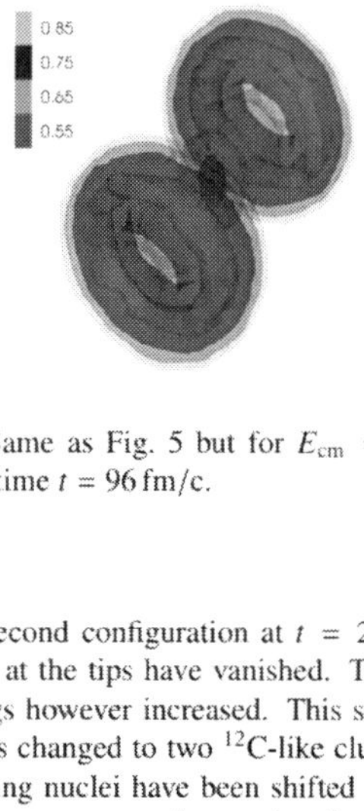

Figure 7: Same as Fig. 5 but for E_{cm} = 100 MeV and b = 2 fm at time t = 96 fm/c.

and the two oxygen are again visible in the density. Afterwards the compound oscillates between two configurations which are shown in the lower panels at t = 192 fm/c and t = 240 fm/c.

In the evolution of the NLF much more structure is visible. In Fig 5 two configurations of the neutron NLF are shown, corresponding to the configurations in the lower panel of Fig. 4. Here, we only show the neutron NLF, since for light isospin symmetric nuclei the difference between the neutron and proton NLF stemming from the small Coulomb contribution is neglectable for our purpose.

In the configuration at t = 192 fm/c large NLF values at the tips of the compound are visible. These are strong indicators of α-clusters. The rings of enhanced localization in between are often indicators for a ^{12}C-like shell structure. It can be regarded as an ^{16}O nucleus without one of the p-shells which create a hole in the spherical NLF in one direction. The particle content of the regions where the rings are present confirms this fact. The NLF ring-like configuration is not the ground state of a ^{12}C nucleus but an excited state. We interpret this state of the ^{32}S compound nucleus as a α-^{12}C-^{12}C-α. Such a ring can also be found in the ground states of nuclei as e.g. in ^{20}Ne. In [16] ^{20}Ne was interpreted as a quasi-molecular α-^{12}C-α configuration.

In the second configuration at t = 240 fm/c the enhancements at the tips have vanished. The separation of the two rings however increased. This suggests, that the structure has changed to two ^{12}C-like clusters at the tips. The remaining nuclei have been shifted into the interior. They do not appear as α-clusters, but the nucleons are delocalized between the rings and therefore do not show an enhancement in the localization.

Figure 6 displays the time evolution of the density distribution of a noncentral collision with an impact parameter of b = 2 fm and a center-of-mass energy of E_{cm} = 100 MeV. After the two fragments form a compound nucleus the two fragments rotate connected with a neck and then separate around t = 160 fm/c.

The most interesting configuration for the NLF is here the configuration at t = 96 fm/c shown in Fig. 7. The two ^{12}C rings are present, similar to the central case, but an enhancement inside the neck is also present. This enhancement results again from the small possibility of finding a second particle with same spin and isospin inside the small neck region. Afterwards, the spherical enhancement of the ^{16}O fragments appear again when the neck vanishes.

6 Conclusion

In our work, we expand the use of the localization from light nuclei, where mainly α-clusters can be exhibited, to heavy nuclei, where imprints of the shell structure can be observed. We show at the example of ^{132}Sn, that the imprint of closed shells are clearly visible in the NLF, although the length scale of the shells in a nucleus is proportional to $\sim \sqrt{n}$ and also dependent on the angular momentum and thus can overlap. The example of ^{100}Zr demonstrates the impact of open shells.

Furthermore we expose the usefulness of the NLF for fission calculations. The NLF in the asymmetric decay of ^{232}Th illustrates, that the structure of the fragments is visible in the calculations long before the scission point. We suggest, that this method can therefore be used in such fission calculations for an earlier identification.

The NLF in time-dependent calculations reveals, that a strong imprint of the shell structure is still visible in highly excited systems. The study of the central collision of two ^{16}O nuclei demonstrates, that the states of the ^{32}S compound nucleus which are occupied are of the nature of α-^{12}C-^{12}C-α and another state with ^{12}C at the tips and the remaining nucleons in the center. It shows clearly, that quantum calculations as performed in this work exhibit much more information than a liquid drop picture and that the shell structure is immensely important. The non-central fusion fission reaction suggests, that intermediate states in reactions can be very important and the NLF provides a powerful access.

With the case of the collision between two ^{16}O nuclei the study of the NLF is at the very beginning. We intend to expand the study to a variety of medium mass nuclei. The effect of isospin asymmetry has still to be investigated. We also plan to perform time-dependent calculations with pairing in the BCS limit.

Useful discussions with P.-G. Reinhard, N. Schunck, and A.S. Umar are gratefully acknowledged. This work was supported by the U.S. Department of Energy, Office of Science under Award Numbers DOE-DE-NA0002847 (the Stewardship Science Academic Alliances program), DE-SC0013365 (Michigan State University), and DE-SC0008511 (NUCLEI SciDAC-3 collaboration). An award of computer time was provided by the Institute for Cyber-Enabled Research at Michigan State University. We also used computational resource provided by the National Energy Research Scientific Computing Center (NERSC).

References

[1] C. Beck, ed., *Clusters in Nuclei, vol 1*, Vol. 818 of *Lecture Notes in Physics* (Springer, Berlin, Heidelberg, 2010)
[2] C. Beck, ed., *Clusters in Nuclei, vol 2*, Vol. 848 of *Lecture Notes in Physics* (Springer, Berlin, Heidelberg, 2012)
[3] C. Beck, ed., *Clusters in Nuclei, vol 3*, Vol. 875 of *Lecture Notes in Physics* (Springer, Berlin, Heidelberg, 2014)
[4] L.R. Hafstad, E. Teller, Phys. Rev. **54**, 681 (1938)
[5] H.J. Rose, G.A. Jones, Nature **307**, 245 (1984)
[6] D. Aleksandrov, A. Belyatskiĭ, Y. Glukhov, E. Nikol'Skiĭ, B. Novatskiĭ, A. Ogloblin, D. Stepanov, ZhETF Pisma Redaktsiiu **40**, 152 (1984)
[7] S. Bjørnholm, J.E. Lynn, Rev. Mod. Phys. **52**, 725 (1980)
[8] J. Okołowicz, M. Płoszajczak, W. Nazarewicz, Prog. Theor. Phys. Suppl. **196**, 230 (2012)
[9] J. Okołowicz, W. Nazarewicz, M. Płoszajczak, Fortschr. Phys. **61**, 66 (2013)
[10] K. Ikeda, N. Takigawa, H. Horiuchi, Prog. Theor. Phys. Suppl. **E68**, 464 (1968)
[11] A.S. Umar, J.A. Maruhn, N. Itagaki, V.E. Oberacker, Phys. Rev. Lett. **104**, 212503 (2010)
[12] M. Bender, P.-H. Heenen, P.-G. Reinhard, Rev. Mod. Phys. **75**, 121 (2003)
[13] A.D. Becke, K.E. Edgecombe, J Chem. Phys. **92**, 5397 (1990)
[14] M. Kohout, Int. J. Quantum Chem. **97**, 651 (2004)
[15] J. Poater, D. M., M. Solà, B. Silvi, Chem. Rev. **105**, 3911 (2005)
[16] P.-G. Reinhard, J.A. Maruhn, A.S. Umar, V.E. Oberacker, Phys. Rev. C **83**, 34312 (2011)
[17] M. Kortelainen, J. McDonnell, W. Nazarewicz, P.-G. Reinhard, J. Sarich, N. Schunck, M.V. Stoitsov, S.M. Wild, Physical Review C **85**, 024304 (2012)
[18] M.V. Stoitsov, N. Schunck, M. Kortelainen, N. Michel, H. Nam, E. Olsen, J. Sarich, S. Wild, Computer Physics Communications **184**, 1592 (2013)
[19] J.A. Maruhn, P.-G. Reinhard, P.D. Stevenson, A.S. Umar, Comput. Phys. Commun. **185**, 2195 (2014)
[20] C.L. Zhang, B. Schuetrumpf, W. Nazarewicz, Physical Review C **94**, 064323 (2016)
[21] J.D. McDonnell, W. Nazarewicz, J.A. Sheikh, Physical Review C **87**, 054327 (2013)

 DOI: 10.1051/epjconf/201716300051

Solitonic excitations in collisions of superfluid nuclei

a qualitatively new phenomenon distinct from the Josephson effect

Kazuyuki Sekizawa[1,*], *Gabriel* Wlazłowski[1,2,**], and *Piotr* Magierski[1,2,***]

[1]*Faculty of Physics, Warsaw University of Technology, ulica Koszykowa 75, 00-662 Warsaw, Poland*
[2]*Department of Physics, University of Washington, Seattle, Washington 98195-1560, USA*

Abstract. Recently, we have reported a novel role of pairing in low-energy heavy ion reactions at energies above the Coulomb barrier, which may have a detectable impact on reaction outcomes, such as the kinetic energy of fragments and the fusion cross section [arXiv:1611.10261, arXiv:1702.00069]. The phenomenon mimics the one studied experimentally with ultracold atomic gases, where two clouds of fermionic superfluids with different phases of the pairing fields are forced to merge, inducing various excitation modes of the pairing field. Although it originates from the phase difference of the pairing fields, the physics behind it is markedly different from the so-called Josephson effect. In this short contribution, we will briefly outline the results discussed in our recent papers and explain relations with the field of ultracold atomic gases.

1 Introduction

Although, almost 60 years have passed after the seminal work by A. Bohr, B.R. Mottelson, and D. Pines [1] that pointed out the existence of the pairing correlations in atomic nuclei, the pairing dynamics in low-energy heavy ion reactions still remains as one of the open problems in nuclear physics. On the analogy of the so-called Josephson effect occuring between superconducting metals [2], researchers envisaged analogous effects, referred to as nuclear Josephson effect, leading to subbarrier nucleon-transfer processes in collisions of two superfluid nuclei [3–6]. Nowadays simulations of heavy ion reactions based on microscopic time-dependent mean-field theories including pairing correlations have become possible, providing valuable insights into the problem [7–10], owing to the continuous increase of computation power. We should note, however, that numerical simulations of nuclear reactions based on full time-dependent Hartree-Fock-Bogoliubov (TDHFB) theory have been achieved, only very recently [11]. It can be expected that now the time is ripe enough for elucidation of qualitatively new phenomena associated with dynamical effects of pairing in low-energy heavy ion reactions.

Due to collaborations of nuclear theory groups at the University of Washington and Warsaw University of Technology, a theoretical framework, which utilizes a local treatment of time-dependent density functional theory (TDDFT) including superfluidity, has been extensively developed. It has been shown that the approach, named time-dependent superfluid local density approximation (TDSLDA), is a powerful tool for accurately describing complex dynamics of strongly-correlated fermionic systems, like ultracold atomic gases [12–18] and nuclear systems [19–23]. In our recent papers [24, 25], we have extended the application of TDSLDA to low-energy heavy ion reactions, employing FaNDF^0 nuclear functional [26, 27] without spin-orbit coupling, and found a qualitatively new phenomenon associated with the pairing field dynamics in collisions of two superfluid nuclei.

In Refs. [24, 25], we have reported the results of TDSLDA simulations for ^{44}Ca+^{44}Ca, ^{90}Zr+^{90}Zr, ^{86}Zr+^{126}Sn, and ^{240}Pu+^{240}Pu reactions at various incident energies (note that neutrons in "^{90}Zr" are in superfluid phase without spin-orbit coupling). It has been shown that a soliton-like excitation mode emerges inside the neck region, in which pairing vanishes, when two superfluid nuclei possessing different phases of the pairing field collide. The situation resembles the one encountered in experiments with ultracold atomic gases, where two clouds of superfluid atomic gases are forced to merge, creating a solitonic excitation (domain wall) which decays through quantum vortices [28, 29]. From the results, it has been shown that the solitonic structure is relatively long-lived object that stays until the composite system splits. Because of this fact, the solitonic excitation effectively hinders energy dissipation from translational motion to internal degrees of freedom as well as the neck formation, resulting in dramatic changes of the reaction dynamics (see Refs. [24, 25] for more details).

In Fig. 1, we show one of the illustrative examples of our TDSLDA simulations for the ^{90}Zr+^{90}Zr reaction at $E \simeq 1.05V_{\rm Bass}$, for two extreme cases: $\Delta\varphi = \pi$ and $\Delta\varphi = 0$, which are shown in upper-half and lower-half of each panel of the figure, respectively. Here, $\Delta\varphi$ denotes

*e-mail: sekizawa@if.pw.edu.pl
**e-mail: gabrielw@if.pw.edu.pl
***e-mail: piotrm@if.pw.edu.pl

DOI: 10.1051/epjconf/201716300051

the relative phase difference of the pairing fields of colliding superfluid nuclei. Snapshots of the total density on the reaction plane at various times are shown in the left column, while those of the absolute value of the neutron's pairing field are shown in the right column. Since the collision energy is higher than the Coulomb barrier, two nuclei collide clearly showing a substantial overlap of the density (see the left column). However, in the $\Delta\varphi = \pi$ case (upper-half), two nuclei do not fuse and eventually reseparate generating binary reaction products; whereas the $\Delta\varphi = 0$ case (lower-half) resulted in fusion.

What makes the dynamics so different? The answer lies in the dynamics of excitation modes of the pairing field, which is depicted in the right column of the figure. From the figure, one can clearly see that a narrow region in which the pairing field vanishes is created on the course of collision in the $\Delta\varphi = \pi$ case (upper-half), which we refer to as solitonic excitation of the pairing field. Since the phase changes so steeply between the colliding nuclei, the system chooses to become normal inside the solitonic structure. The solitonic excitation prevents the system to form a compact (mononuclear) shape (see left column) and, as a result, fusion reaction is hindered, even though the system once overlapped substantially (Note that there is a remarkable difference compared to a lighter system, where two nuclei easily get fused once the system overlapped only slightly [25]). Summarizing, from our comprehensive simulations, we have found that [24, 25]: total kinetic energy (TKE) of outgoing fragments can be changed as large as 25 MeV in ^{240}Pu+^{240}Pu; fusion threshold energy can be affected almost 30 MeV in ^{90}Zr+^{90}Zr; similar effects present also for an asymmetric system, ^{86}Zr+^{126}Sn; whereas the pairing effects turned out to be small in a lighter system, ^{44}Ca+^{44}Ca, because of the strong tendency towards fusion, consistent with an earlier study [11].

Although the effect originates from the relative phase difference of the pairing fields of two colliding superfluid nuclei, it is not related to the Josephson effect. As we feel that it would be useful to clarify the distinct character of the effect we studied, as compared to the Josephson effect, we explain this point in detail.

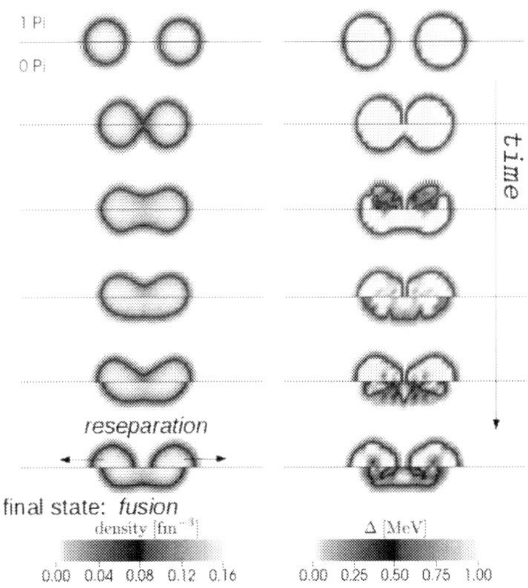

Figure 1. Snapshots of the total density (left column) and the absolute value of the neutron's pairing field (right column) at various times in the TDSLDA simulations for the ^{90}Zr+^{90}Zr reaction at $E \simeq 1.05V_{\rm Bass}$, where $V_{\rm Bass}$ is the phenomenological fusion barrier [30]. Upper-half (lower-half) of each panel shows those for the $\Delta\varphi = \pi$ (0) case. For full movies see Supplemental Material of Ref. [24].

2 Solitonic excitation vs Josephson effect

Let us first remind that the physics of the Josephson effect deals with the phenomenon of the current flowing through a junction in the absence of the chemical potential difference. The first Josephson's paper on the subject entitled "Possible new effects in superconductive tunnelling" [2] states clearly in the introduction that: *"We here present an approach to the calculation of tunneling currents between two metals that is sufficiently general to deal with the case when both metals are superconducting. In that case new effects are predicted, due to the possibility that electron pairs may tunnel through the barrier leaving the quasi-particle distribution unchanged"*.

This current, being only dependent on the mutual phase differences of the pairing fields, is responsible for a variety of effects which are known under a common name of "Josephson effect". However, the crucial point is that the effect appears from the tunneling through the barrier, and therefore, the current arises as a consequence of a *weak coupling* of the two pair condensates. Consequently, its magnitude depends on $\sin(\Delta\varphi)$, where $\Delta\varphi$ is the phase difference between the condensates, as is dictated by the equations responsible for dynamics of the Josephson effect:

$$J(t) = J_c \sin[\Delta\varphi(t)], \tag{1}$$

$$\frac{d(\Delta\varphi)}{dt} = \frac{2eU}{\hbar}, \tag{2}$$

where J_c is the amplitude of the Josephson current, e is the elementally charge, $\hbar$ is the reduced Plank constant, and U is the voltage, which, in nuclear case, will be proportional to the chemical potential difference. Thus, sub-barrier multinucleon transfer processes between superfluid nuclei can be viewed as a nuclear Josephson effect and papers [3–6] focused solely on this aspect.

However, we would like to emphasize here that the Josephson effect described above is markedly different from the effect we studied in Refs. [24, 25]. What we investigated is collisions of superfluid nuclei at energies above the Coulomb barrier, which, on the contrary to the Josephson effect, can be regarded as a *strong-coupling* limit, where two condensates (nuclei) actually merge producing a solitonic excitation (domain wall). Both effects are associated with the pairing field, but are weakly connected in the sense that: no solitonic excitation is necessary for the Josephson current to appear; whereas no

DOI: 10.1051/epjconf/201716300051

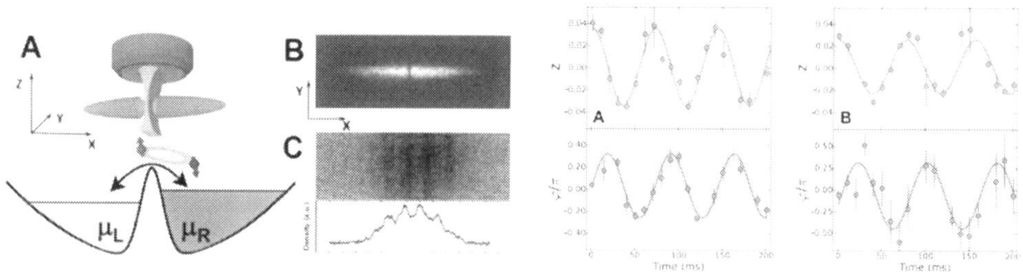

Figure 2. Left: (A) Sketch of the experimental apparatus for studies of Josephson junction realized through introducing potential barrier between two ^{6}Li condensates, and (B) image of the cloud. Right: Oscillation of the population imbalance $z = \Delta N/N$ and phase difference φ between fermionic superfluids due to the coherent flow of the particles through the junction. (A) corresponds to the BEC regime (B) corresponds to the so-called unitary limit. The figure has been reprinted with permission from: G. Valtolina et al., Science **350**, 1505 (2015) [see this reference for more details].

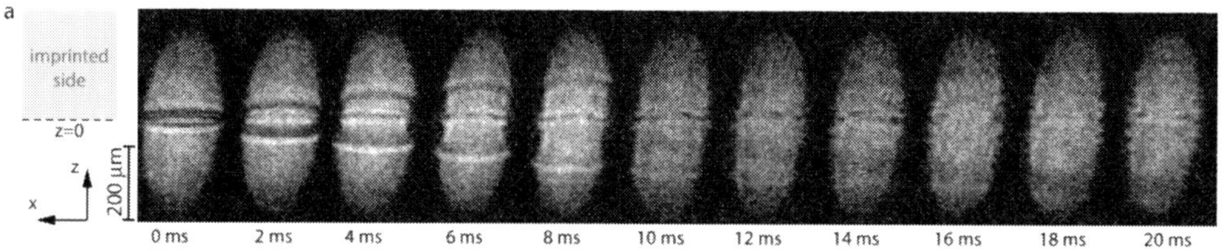

Figure 3. Evolution of the atomic density when two clouds with different phases are put into contact. Just after making contact two sound waves propagating towards the edges are generated. In the center dark soliton (visible as darker line) is excited. There is no external barrier that separates the cloud. The soliton lives about 10 ms and decays into a vortex ring (visible as two dots on the last panel). The vortex ring is only transient, and is replaced by a vortex line in the final state. The figure has been reprinted with permission from: M.J.H. Ku, et al., Phys. Rev. Lett. **116**, 045304 (2016) [see this reference for more details].

Josephson current is necessary for the effect we studied to appear. Indeed, we have shown that the strongest effect is observed actually for the $\Delta\varphi = \pi$ case where no Josephson current is present. [24, 25]. Usually the Josephson effect is studied in the regime of a quasistationary situation, which occur when two condensates couple weakly through the potential barrier, and therefore no topological excitation of the pairing field (soliton or quantum vortex) is expected. In our case, on the other hand, we studied rather violent collisions at energies above the Coulomb barrier, *i.e.*, far from the stationary situation, which trigger the appearance of the localized excitation mode between colliding superfluid nuclei.

To emphasize better the differences between the Josephson effect and the effect that we studied in our papers [24, 25], let us consider the system where both phenomena can be studied with far better accuracy than in nuclear systems. Namely, let us consider the case of ultracold fermionic (atomic) gases. In the latter case, studies of the Josephson effect can be achieved experimentally. The current is realized as a tunneling between two fermionic superfluids separated by an external potential barrier. In Fig. 2 (left), we show a typical experimental setup, which allows to investigate a Josephson alternating current due to the *weak coupling* of two Lithium atomic clouds [31]. As a result an oscillatory current is observed accompanied with an oscillatory evolution of the phase difference between the condensates [see Fig. 2 (right)]. For the flow of the atoms the weak coupling is sufficient and no solitonic excitation is present.

On the other hand, the effect we studied is more closer to the experiments performed by the MIT group [32–34], which consist in *merging* two atomic clouds with different phases and producing topological excitations (see Fig. 3). There is no external potential that separates the cloud at all. Note that this phenomenon, although originating from the pairing field dynamics, is not related to the Josephson effect. The reason is that in this case physics is different, although of course certain flow of particles induced by the pairing field is observed. However, the main feature comes from the nonlinear evolution of the pairing field, which makes the solitonic excitation relatively long lived. With TDSLDA approach, we were able to successfully simulate this experiment and we found that in case of the unitary Fermi gas the excitation decays into a vortex ring [35], which is another type of the topological excitations of the pairing field. This prediction was confirmed by the experiment in Ref. [34].

We understand that in nuclear physics community this distinction has not been emphasized, since all papers, which concern physics of colliding superfluid nuclei, have investigated only the consequences of the supercurrent (or pair transfer) which would manifest in multinucleon transfer processes in heavy ion reactions. Indeed, this effect has been studied, but it is relatively weak and has nothing to do with the phenomenon we studied in Refs. [24, 25]. The latter requires not only the phase difference but also the creation of the solitonic structure which stays relatively long so as to hinder energy dissipation and the neck formation during the collision. In other words, the Josephson effect originates from the weak coupling between two condensates separated by a potential barrier, whereas in our case we consider a strong coupling limit where two condensates actually merge/collide. It would be misleading to refer to the latter as the effect that is already known from the weak coupling limit, in the same way, like it would be misleading to say that properties of a turbulent flow are known because a laminar flow has been studied. Analogously, physics of a weakly-interacting Fermi gas described by the standard Bardeen-Cooper-Schrieffer (BCS) theory is completely different from the one in the strong coupling limit (*i.e.* for the unitary Fermi gas) where a variety of qualitatively new phenomena occur like, *e.g.*, pseudogap or anomalously small ratio of the shear viscosity to the entropy density. One cannot say that the unitary Fermi gas is a well-known system because it is a superfluid just like a typical atomic nucleus.

3 Summary

In our recent papers [24, 25], we have reported a qualitatively new phenomenon in low-energy heavy ion reactions at energies above the Coulomb barrier, which is associated with relative phase difference of the pairing fields of colliding superfluid nuclei. We have found that the complex pairing field dynamics may substantially alter the reaction outcomes, such as the total kinetic energy of the fragments and the fusion cross section, which may be experimentally detectable for carefully selected systems. In our simulations, a "solitonic excitation" (domain wall) is observed in the neck region, inside which pairing vanishes, resulting in hindrance of energy dissipation as well as the neck formation, leading to significant changes of the reaction dynamics. Although this effect originates from the relative phase of the pairing fields of colliding superfluid nuclei, the physics behind it is markedly different form the so-called Josephson effect: the latter is associated with a weak coupling of condensates (separated by a potential barrier) without any solitonic excitation, whereas the effect we studied is associated with a strong-coupling limit where two condensates are actually merge/collide for which no Josephson current is necessary. The ongoing further extensions and applications of the present work will enable us to find qualitatively new phenomena associated with the pairing field dynamics that have never been investigated in nuclear physics community.

Acknowledgments

This work was supported by the Polish National Science Center (NCN) under Contracts No. UMO-2013/08/A/ST3/00708. The code used for generation of initial states was developed under grant of Polish NCN under Contracts No. UMO-2014/13/D/ST3/01940. Calculations have been performed at HA-PACS (PACS-VIII) system—resources provided by Interdisciplinary Computational Science Program in Center for Computational Sciences, University of Tsukuba.

References

[1] A. Bohr, B.R. Mottelson and D. Pines, Phys. Rev. **110**, 936 (1958).

[2] B.D. Josephson, Phys. Lett. **1**, 251 (1962); Rev. Mod. Phys. **36**, 216 (1970).

[3] K. Dietrich, Phys. Lett. **B32**, 428 (1970).

[4] K. Dietrich, Ann. Phys. (N.Y.) **66**, 480, (1970).

[5] K. Dietrich, K. Hara, and F. Weller, Phys. Lett. **B35**, 201 (1971).

[6] J. H., Sorensen and A. Winther, Phys. Rev. C **47**, 1691 (1993).

[7] G. Scamps, D. Lacroix, G.F. Bertsch, and K. Washiyama, Phys. Rev. C **85**, 034328 (2012).

[8] G. Scamps and D. Lacroix, Phys. Rev. C **87**, 014605 (2013).

[9] S. Ebata and T. Nakatsukasa, JPS Conf. Proc. **1**, 013038 (2014).

[10] S. Ebata and T. Nakatsukasa, JPS Conf. Proc. **6**, 020056 (2015).

[11] Y. Hashimoto and G. Scamps, Phys. Rev. C **94**, 014610 (2016).

[12] A. Bulgac, M.M. Forbes, M.M. Kelley, K.J. Roche, and G. Wlazłowski, Phys. Rev. Lett. **112**, 025301 (2014).

[13] G. Wlazłowski, A. Bulgac, M.M. Forbes, and K.J. Roche, Phys. Rev. A **91**, 031602(R) (2015).

[14] A. Bulgac and S. Yoon, Phys. Rev. Lett. **102**, 085302 (2009).

[15] A. Bulgac, Y.-L. Luo, P. Magierski, K.J. Roche, and Y. Yu, Science **332**, 1288 (2011).

[16] A. Bulgac, P. Magierski, and M.M. Forbes, *The Unitary Fermi Gas: From Monte Carlo to Density Functionals* in *BCS-BEC Crossover and the Unitary Fermi Gas*, edited by W. Zwerger, Lecture Notes in Physics, Vol. **836**, pp 305-373 (Springer, Heidelberg, 2012).

[17] A. Bulgac, Y.-L. Luo, and K.J. Roche, Phys. Rev. Lett. **108**, 150401 (2012).

[18] A. Bulgac, Ann. Rev. Nucl. Part. Sci. **63**, 97 (2013).

[19] I. Stetcu, A. Bulgac, P. Magierski, and K.J. Roche, Phys. Rev. C **84**, 051309(R) (2011).

[20] I. Stetcu, C.A. Bertulani, A. Bulgac, P. Magierski, and K.J. Roche, Phys. Rev. Lett. **114**, 012701 (2015).

[21] A. Bulgac, P. Magierski, K.J. Roche, and I. Stetcu, Phys. Rev. Lett. **116**, 122504 (2016).

[22] P. Magierski, *Nuclear Reactions and Superfluid Time Dependent Density Functional Theory*, to be published as a chapter in "Progress of time-dependent nuclear reaction theory" (ed. Yoritaka Iwata) in the ebook series: "Frontiers in nuclear and particle physics" (Bentham Science Publishers); arXiv:1606.02225 [nucl-th].

[23] G. Wlazłowski, K. Sekizawa, P. Magierski, A. Bulgac, and M.M. Forbes, Phys. Rev. Lett. **117**, 232701 (2016).

[24] P. Magierski, K. Sekizawa, and G. Wlazłowski, arXiv:1611.10261 [nucl-th].

[25] K. Sekizawa, P. Magierski, and G. Wlazłowski, Proceedings of Science (INPC2016) 214 (2017); arXiv:1702.00069 [nucl-th].

[26] S.A. Fayans, JETP Letters **68**, 169 (1998).

[27] S.A. Fayans, S.V. Tolokonnikov, E.L. Trykov, and D. Zawischa, Nucl. Phys. **A676**, 49 (2000).

[28] M.J.H. Ku, W. Ji, B. Mukherjee, E. Guardado-Sanchez, L. W. Cheuk, T. Yefsah, and M. W. Zwierlein, Phys. Rev. Lett. **113**, 065301 (2014).

[29] M.J.H. Ku, B. Mukherjee, T. Yefsah, and M.W. Zwierlein, Phys. Rev. Lett. **116**, 045304 (2016).

[30] R. Bass, Nucl. Phys. **A231**, 45 (1974).

[31] G. Valtolina et al., Science **350**, 1505 (2015).

[32] T. Yefsah et al., Nature **499**, 426 (2013).

[33] Mark J.H. Ku et al., Phys. Rev. Lett. **113**, 065301 (2014).

[34] Mark J.H. Ku et al., Phys. Rev. Lett. **116**, 045304 (2016).

[35] A. Bulgac, M. M. Forbes, M. M. Kelley, K. J. Roche, and G. Wlazłowski, Phys. Rev. Lett. **112**, 025301 (2014).

EPJ Web of Conferences **163**, 00052 (2017)
FUSION17 DOI: 10.1051/epjconf/201716300052

First Elastic Scattering Measurement of ^{8}Li on ^{209}Bi at the Australian National University

C. Sengupta[1,*], *I.P. Carter*[1], *K.J. Cook*[1], *E.C. Simpson*[1], *M. Dasgupta*[1], *D. J. Hinde*[1], *D.Y. Jeung*[1], *Sunil Kalkal*[1,2], *K. Vo-Phuoc*[1], *E. Prasad*[1], *D. Rafferty*[1], *C. Simenel*[1], and *E. Williams*[1]

[1]*Department of Nuclear Physics, Research School of Physical Sciences and Engineering, The Australian National University, Canberra, ACT 0200, Australia*
[2]*Thapar University, Patiala, India*

Abstract. The effects of unusual structures of nuclei, such as neutron halos, on nuclear reaction mechanisms are not well understood, particularly at near barrier energies. Using the SOLEROO Radioactive Ion Beam facility at the Australian National University, below-barrier reactions with ^{8}Li incident on ^{209}Bi have been performed. Beam purities of about 95% are achieved by rejecting unwanted beam species using a solenoidal separator along with tracking and tagging the secondary beam with two parallel plate avalanche counters (PPACs) placed immediately after the solenoid. However, the radioactive ion beam exiting the solenoid is not parallel to the primary beam axis. To obtain a precise angular distribution of elastic scattering, the tracking facility is used to deduce the true scattering angle on an event-by-event basis. The elastic cross-section for ^{8}Li on ^{209}Bi is then extracted, verifying the capability of the facility to perform precise cross-section measurements.

1 Introduction

A suppression of complete fusion at above barrier energies has been observed in reactions with light, weakly bound nuclei. A series of fusion cross-section measurements performed for ^{9}Be bombarding on ^{208}Pb and ^{209}Bi [1–4], ^{6,7}Li on ^{209}Bi [2, 3, 5] and ^{9,10,11}Be on ^{209}Bi [6] indicate that the complete fusion cross-sections at energies around and above the barrier are suppressed by 30%, compared with reactions of nuclei having a high energy threshold against breakup [2]. Many measurements have also been performed to understand the role of breakup in suppression of fusion cross-sections [7, 8]. However, the impact of breakup on fusion with weakly bound radioactive nuclides is not yet clear.

In exotic nuclei such as ^{8}Li or ^{6}He, containing weakly bound neutrons around a relatively tightly bound core, the nucleon density distribution has an extensive tail which may help the attractive nuclear forces begin to act at large distances between the projectile and the target. This may lower the barrier and enhance fusion cross-sections, particularly in the sub-barrier energy region [9, 10]. However, these nuclei are weakly bound and it is quite possible that their interaction with the target may cause them to break-up. In turn, this may prevent or hinder complete fusion of the projectile with the target [11, 12].

To achieve a better understanding about these exotic nuclei, the SOLEROO Radioactive Ion Beam (RIB) facility has been developed at the Australian National University [13]. Here, we first briefly describe SOLEROO RIB facility. This has been used to study ^{8}Li elastic scattering from ^{209}Bi as a first step towards measuring reaction cross-section of the above system. To extract a reliable angular distribution, the trajectories of the ^{8}Li ions onto the secondary target have been reconstructed. From the reconstructed quantities, the true scattering angle of the scattered ^{8}Li has been obtained for each incident event. The importance of obtaining the true scattering angle is then discussed.

2 The Radioactive Ion Beam Facility at ANU

The SOLEROO RIB facility at the Australian National University is based on a super-conducting solenoidal separator [14–17] which produces RIBs by in-flight transfer reactions via interactions with a primary target. The primary

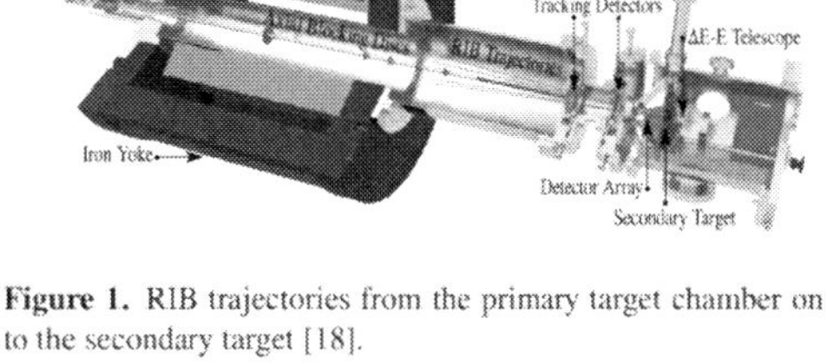

Figure 1. RIB trajectories from the primary target chamber on to the secondary target [18].

*e-mail: Chandrima.Sengupta@anu.edu.au

beam of ^{7}Li was provided by the 14UD tandem accelerator and bombarded a 5.572 mg/cm^2 ^{9}Be primary target, producing a secondary beam of ^{8}Li. This neutron pickup reaction has a positive Q value of 368 keV. Following production, the transfer products enter a magnetic field region (2° - 6° acceptance) and the desired RIBs are separated using a 6.5 T axial magnetic field [14] and focussed on to a secondary target [18]. The direct beam is stopped by a Faraday cup. The unwanted beam species are focussed on to an axial rod or stopped by axial blocking discs depending on their focal lengths.

After the solenoid, there are two position sensitive parallel plate avalanche counters (PPACs) [16, 17]. These detectors are used to track the RIB on its path to the secondary target. Fig. 1 shows a schematic of the device and the beam trajectories originating from the primary target chamber passing through the solenoid, focussed on to the secondary target. The reaction products produced from the secondary target are detected by a large area position sensitive detector array (BALiN) [19].

2.1 Tracking System

Using a single solenoid, a maximum purity of 30% can be achieved in this reaction [13, 18]. The PPACs placed at the exit of the solenoid help to reject the unwanted beam species and a purity greater then 95% can be achieved for ^{6}He and ^{8}Li beam [17] by tagging and tracking the beam species. The PPACs [20] consist of two X-Y position foils and one centre foil. The X-Y position foils are negatively biased aluminised Mylar (50 nm aluminium on 0.9 μm Mylar foil), having x, y position strips spaced by 1 mm, allowing us to achieve position sensitivity of less than 1mm [17]. The signal in the position foils is passed through delay lines and is collected as a difference of time taken by the signal to travel to the two ends of the delay line. The X-Y position information collected from the two PPACs allows tracking of the secondary beam on its path to the secondary target. The centre foil of 0.9 μm Mylar is positively biased and coated with 50 nm Aluminium on both sides. The spacing between each foil is 4 mm and the PPAC has a square active area of 35 x 35 mm^2. The total applied bias was kept constant at 880 V across the detector, where -250 V was applied on the X, Y signal foils and +630 V on the centre foil. The filling gas (Propane) is circulated at 10 Torr between the foils using a gas flow system [18]. The detector window was made up of 0.9 μm Mylar. When charged particles pass through the detector, electron-ion pairs are generated, which are collected by the anodes and cathodes, respectively. Electrons, being lighter than the ions, produce the fastest signal and therefore this signal is used to generate timing information. This allows us to electronically identify and tag the beam species using the energy loss in the centre foil and Time of Flight (ToF) information [18].

2.2 Experimental Setup

The scattered ^{8}Li particles from ^{209}Bi were measured using the Breakup Array for Light Nuclei (BALiN). This array consists of four double sided Silicon strip detectors.

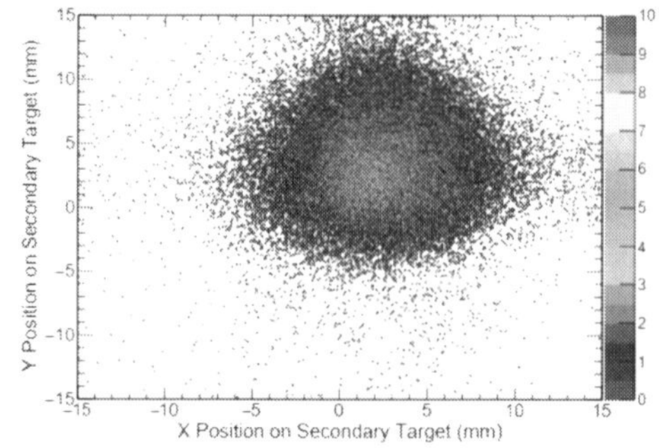

Figure 2. Reconstructed beam profile on the secondary target.

Each detector is 400 μm thick, with 16 arcs and 8 sectors, giving a total of 512 position pixels. It has a total effective area of 80 cm^2, having inner radius of 32.6 mm and outer radius of 135.1 mm. Each sector covers 6.8° in azimuthal angle ϕ. A 0.7 μm Mylar foil was placed in front of the front detectors to stop electrons. One of the detectors in the backward angles was not useful for elastic scattering as the gain of the preamplifier was set too high. The two forward detectors covered 20° to 90° in scattering angle θ and 16° to 164° in ϕ and the backward detector covered 90° to 162° in θ and 273° to 345° in ϕ.

3 Reconstruction of Beam Profile on the Secondary Target

The energy of the RIB produced in the primary target will vary depending on its exit angle from the primary target, the interaction point on the primary target and the excited state in which ^{8}Li is produced. As the acceptance of the solenoid is between 2° and 6° in the laboratory frame, the RIB exits the solenoid at angles between 2° and 6°, and will have a finite beam spot size on the secondary target. Therefore, it is necessary to know the exact interaction point of the beam on the secondary target. Using the two position sensitive tracking detectors, it is possible to define a direction vector of the RIB and extrapolate it to the secondary target plane to reconstruct the beam profile on the secondary target. This calculation is done on an event-by-event basis. Fig. 2 shows the reconstructed beam profile of the elastically scattered events of ^{8}Li on the secondary target ^{209}Bi. The FWHM of the beam profile on the secondary target is around 8 mm. The offset of the beam in y arises due to the misalignment of the solenoid axis with the primary beam axis while the offset in x comes due to the steering elements along the beam line.

4 Reconstruction of focal length of the Secondary Beam

The RIB of ^{8}Li is produced in the primary target chamber in the interaction of ^{7}Li with the ^{9}Be primary target. Along with the desired beam species, some other contaminant ions such as $^1_1H^+$, $^2_1H^+$, $^3_1H^+$, $^4_2He^{2+}$, $^6_3Li^{3+}$, $^7_3Li^{3+}$ have

 DOI: 10.1051/epjconf/201716300052

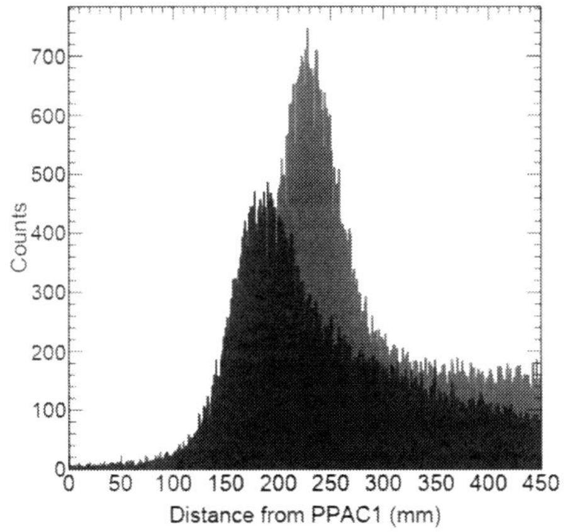

Figure 3. Reconstructed focal point of the elastically scattered ^{8}Li ground state (red) and first excited state (blue) from PPAC1.

also been identified [18]. $^1_1$H$^+$, $^2_1$H$^+$, $^3_1$H$^+$ and $^4_2$He^{2+} produced from ^{7}Li breakup [18] generates a broad energy distribution. As a result, other particles along with ^{8}Li can pass through the solenoid and hit the secondary target.

To separate these unwanted species, a software routine has been developed to calculate the focal point on an event-by-event basis. Applying a cut on the focal length allows selection of desired species of a particular energy. Fig. 3 shows the reconstructed focal point of the elastically scattered ^{8}Li particles (red denotes the ground state and the blue denotes the first excited state of ^{8}Li) on the primary beam axis assuming the origin at PPAC1.

The most problematic species which is transmitted to the secondary target is energy degraded ^{7}Li [18]. If the ^{7}Li primary beam scatters off some surface and then has the same rigidity as ^{8}Li, it will be transported to the secondary target. By carefully adjusting the primary beam tune, the contamination of $^7_3$Li^{3+} can be minimised.

5 Reconstruction of True Scattering Angle

The RIB exiting the solenoid and incident on the secondary target is not parallel to the primary beam axis and makes an angle, θ_{Beam}, with the primary beam axis. The geometric angle of a detector pixel, θ_{Geom} is the angle associated with a vector constructed between the centre of the secondary target and the pixel hit on BALiN, with respect to the primary beam axis. The true scattering angle (θ_{RIB}) will depend on the interaction point of the ^{8}Li projectile on the secondary target and θ_{Beam} for that projectile, and thus can be significantly different from θ_{Geom}. Here, θ_{Geom}, θ_{RIB} and θ_{Beam} are in the laboratory frame.

To derive a relationship between θ_{Geom} and θ_{RIB}, an intermediate angle, θ_{Parallel} is reconstructed using a routine in the software, where θ_{Parallel} is the scattering angle of any event if the incident RIB is parallel to the primary beam axis. Essentially, θ_{Parallel} takes into account the interaction point of the RIB particle on the secondary target.

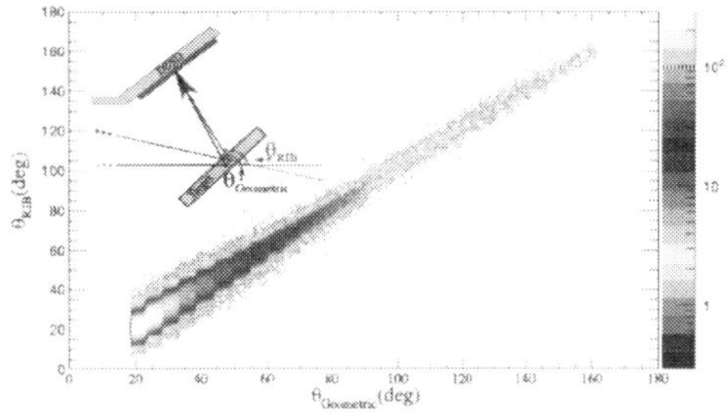

Figure 4. Conversion between θ_{Geom} and θ_{RIB} for the system ^{8}Li on ^{209}Bi at E_{lab}=26.88 MeV [18].

The relationship derived from simple geometry is shown below for conversion between θ_{Parallel} and θ_{RIB}:

$$\cos(\theta_{\text{RIB}}) = \cos(\theta_{\text{Beam}})\cos(\theta_{\text{Parallel}}) - \sin(\theta_{\text{Parallel}})\sin(\theta_{\text{Beam}})\cos(\phi_{\text{Beam}} - \phi_{\text{Parallel}}) \quad (1)$$

Therefore, to reconstruct the true scattering angle we need to know the angle of incidence of the beam (θ_{Beam}) on the secondary target, and the exact interaction point of the beam on the secondary target. These can be found using the tracking system, as discussed previously.

The distribution of events with θ_{RIB} and θ_{Geom} from the experimental data of ^{8}Li on ^{209}Bi at E_{lab}=26.88 MeV is shown in the Fig. 4. It can be seen that for a particular θ_{Geom}, there is a spread in θ_{RIB}. The full width half maximum of the Gaussian distribution of θ_{RIB} corresponding to a given θ_{Geom} is around 7° over the full angular range covered by the detectors. As mentioned earlier in Sec. 2, the acceptance of the solenoid is between 2° and 6°. Therefore, much of the spread in θ_{RIB} comes from this factor. Also there is a finite beam spot size on the secondary target mentioned above and pixelization of the BALiN array, will introduce some spread in θ_{RIB}.

6 Preliminary result of elastic scattering of ^{8}Li by ^{209}Bi

Elastic scattering measurements of ^{8}Li on ^{209}Bi were carried out using the 14UD accelerator at the Australian National University. The primary beam of ^{7}Li was accelerated at an energy of 32 MeV, ^{8}Li was produced in a transfer reaction of ^{7}Li on ^{9}Be. The solenoid field was set to 3.11 T to optimally transport ^{8}Li ions. The ^{209}Bi target, rotated to 45°, had a thickness of 1.22 mg/cm^2. Taking the energy loss in the primary target and in the PPACs into account, the secondary beam energy in the laboratory frame incident on the secondary target is 26.88 MeV. At this below-barrier energy, we would expect the majority of the elastic scattering to be purely Rutherford scattering. One of the forward arcs at 58.5° where there is no edge effect, has been used as the monitor to normalize the elastic cross-section.

To extract elastic cross-sections for the above system, a software gate was applied to select the elastically scattered events of ^{8}Li in its ground state and 0.98 MeV first

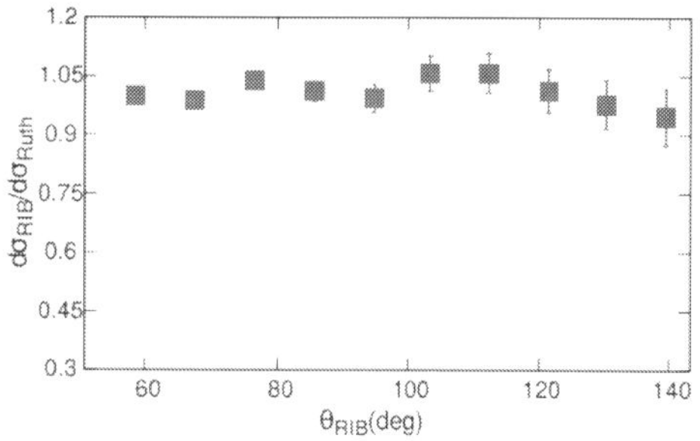

Figure 5. Ratio of elastic scattering cross-section to Rutherford cross-section as a function of θ_{RIB} for the system ^{8}Li on ^{209}Bi at E_{lab}=26.88 MeV.

excited state. At the final stage of data processing, the experimental data is binned in 9° steps in θ_{Geom} and θ_{RIB}.

The next step is to find the yield for each θ_{RIB} bin. In order to calculate the solid angle of each θ_{Geom} bin, we simulate the events over 4π sr, and filter through the known detector geometry. The coordinate system of BALiN pixels is defined with respect to the centre of the secondary target and the geometric solid angle ($d\Omega_{Geom}$) is therefore calculated as a function of θ_{Geom} in the simulation. Within the data binned as a function of θ_{Geom} and θ_{RIB}, a bin by bin efficiency correction is made, and then projected on to the θ_{RIB} axis to extract the true angular distribution. Then to extract the elastic cross-section as a function of θ_{RIB}, we use the following equation:

$$\frac{d\sigma}{d\Omega_{lab}} = \frac{Y(\theta_{det}, E)}{Y(\theta_{Monitor}, E)} \frac{d\sigma}{d\Omega_{Monitor}} \quad (2)$$

Where, $\frac{d\sigma}{d\Omega_{lab}}$ is the elastic cross-section determined as a function of θ_{RIB}, Y(θ_{det},E) is the yield in each θ_{RIB} bin, Y($\theta_{Monitor}$,E) is the yield in the monitor bin and $\frac{d\sigma}{d\Omega_{Monitor}}$ is the elastic cross-section of that monitor bin.

Fig. 5 shows the ratio of elastic cross-section to Rutherford cross-section of ^{8}Li on ^{209}Bi at 26.88 MeV as a function θ_{RIB}. We only show the data that falls within the geometric coverage of the detector having no edge effect. The statistical uncertainty in cross-section is smaller than the points at the forward angles but bigger at the backward angles. This result looks reasonable, but will be refined in future work as described below.

7 Conclusion and Future Work

This paper has discussed elastic scattering of ^{8}Li on ^{209}Bi measured using the SOLEROO RIB facility at the Australian National University. A transformation relation between θ_{RIB} and θ_{Geom} has been established to calculate the true scattering angle. Using this relationship, θ_{RIB} has been reconstructed event-by-event in the experimental data. Then the yield for each θ_{RIB} bin is corrected for efficiency. Then the elastic cross-section is obtained as a function of θ_{RIB}.

The above correction implemented in the simulation is a first order correction. In future, the beam interaction point on the secondary target will be taken into account to calculate the geometric efficiency.

The effect of the misalignment of the solenoid with the primary beam axis is expected to have small effect on reconstruction of the true scattering angle. However, work is under process to test the effect of this misalingnment on the reconstructed quantities.

Implementing this correction in the data analysis, an optical model fit would allow us to get the optical potential parameters using FRESCO [21] and extract the reaction cross-section.

References

[1] M. Dasgupta, D. J. Hinde, R. D. Butt, R. R. M. Anjos, A. Berriman, N. Carlin, P. R. S. Gomes, C. C. R. Morton, J. O. J. Newton, A. Szanto de Toledo, Others, and K. Hagino, Phys. Review Letters, Journal **82**, 1395 (1999).

[2] M. Dasgupta, P. R. S. Gomes, D. J. Hinde, S. Moraes, R. Anjos, A. Berriman, R. Butt, N. Carlin, J. Lubian, C. Morton, and Others, Phys. Review C, Journal **70**, 024606 (2004).

[3] C. Signorini, T. Glodariu, Z. H. Liu, M. Mazzocco, M. Ruan, and F. Soramel, Progress of Theoretical Physics Supplement, Journal **154**, 272 (2004).

[4] M. Dasgupta, D. J. Hinde, S. L. Sheehy, and B. Bouriquet, Phys. Review C, Journal **81**, 024608 (2010).

[5] M. Dasgupta, D. J. Hinde, K. Hagino, S. Moraes, P. R. S. Gomes, R. Anjos, R. Butt, A. Berriman, N. Carlin, C. Morton, J. Newton, and A. Szanto de Toledo, Phys. Review C, Journal **66**, 041602(R) (2002).

[6] D. J. Hinde and M. Dasgupta, Phys. Review C, Journal **81**, 064611 (2010).

[7] K. J. Cook, E. C. Simpson, D. H. Luong, Sunil Kalkal, M. Dasgupta, and D. J. Hinde, Phys. Review C, Journal **93**, no. 6, 1-15 (2016).

[8] Sunil Kalkal, E. C. Simpson, D. H. Luong, K. J. Cook, M. Dasgupta, D. J. Hinde, I. P. Carter, D. Y. Jeung, G. Mohanto, C. S. Palshetkar, E. Prasad, D. C. Rafferty, C. Simenel, K. Vo-Phuoc, and E. Williams, Phys. Review C, Journal **93**, 1-13 (2016).

[9] J. J. Kolata et al., Phys. Review C, Journal **65**, 054616 (2002).

[10] A. M. Moro et al., Phys. Review C, Journal **68**, 034614 (2003).

[11] E. F. Aguilera et al., Revista mexicana de física, Journal **50**, 1–4 (2004).

[12] E. F. Aguilera et al., Phys. Review C, Journal **80**, 044605 (2009).

[13] A. J. Horsley, D. J. Hinde, M. Dasgupta, R. Rafiei, A. Wakhle, M. Evers, D. H. Luong, R. du Rietz, Nucl. Instr. Meth. In Phys. Res. A, Journal **646**, 174 (2011).

[14] R. Rafiei, D. J. Hinde, M. Dasgupta, D. C. Weisser, A. G. Muirhead, A. B. Harding, A. K. Cooper, H. J. Wallace, N. R. Lobanov, A. Wakhle, M. L. Brown, C. J. Lin, A. J. Horsley, R. du Rietz, D. H. Luong, and M. Evers, Nucl. Instruments Methods Phys. Res.

 DOI: 10.1051/epjconf/201716300052

Sect. A Accel. Spectrometers, Journal **631**, 12–21 (2011).

[15] H. Laurent and J. Schapira, Nuclear Instruments and Methods, Journal **162**, 181 – 192 (1979).

[16] I. P. Carter et al., EPJ Web of Conferences,Journal **63**, 02022 (2013).

[17] I. P. Carter et al., EPJ Web of Conferences, Journal **91**, 00001 (2015).

[18] I. P. Carter, *Developing Techniques For High Fidelity Studies of Reactions With Light Weakly Bound Nuclei* (Ian Paul Carter, The Australian National University, 2016).

[19] K. J. Cook, *Zeptosecond dynamics of transfer-triggered breakup: mechanisms, timescales, and consequences for fusion* (Kaitlin Jennifer Cook, The Australian National University, 2016) 82-87.

[20] Glenn F. Knoll, *Radiation Detection and Measurement* (John Wiley & Sons, United States of America, 1999) 160.

[21] I. J. Thompson, Comput. Phys. Rep., Journal **7**, 167 (1988).

DOI: 10.1051/epjconf/201716300053

Pre-compound emission in low-energy heavy-ion interactions

Manoj Kumar Sharma[1,*], Mohd. Shuaib[2], Vijay R. Sharma[3], Abhishek Yadav[3], Pushpendra P. Singh[4], Devendra P. Singh[5], Unnati[6], B. P. Singh[2,**], and R. Prasad[2]

[1]Department of Physics, Sri Varsheny (Post Graduate) College, Aligarh (UP) - 202 001, India
[2]Department of Physics, Aligarh Muslim University, Aligarh 202 002, India
[3]NP-Group, Inter University Accelerator Centre, New Delhi - 110 067, India
[4]Department of Physics, Indian Institute of Technology Ropar, Punjab 140 001, India
[5]Department of Physics, University of Petroleum and Energy Studies, Dehradun - 248 007, India
[6]Department of Physics and Astrophysics, Delhi University, Delhi-110007 India

Abstract. Recent experimental studies have shown the presence of pre-compound emission component in heavy ion reactions at low projectile energy ranging from 4 to 7 MeV/nucleons. In earlier measurements strength of the pre-compound component has been estimated from the difference in forward-backward distributions of emitted particles. Present measurement is a part of an ongoing program on the study of reaction dynamics of heavy ion interactions at low energies aimed at investigating the effect of momentum transfer in compound, pre-compound, complete and incomplete fusion processes in heavy ion reactions. In the present work on the basis of momentum transfer the measurement of the recoil range distributions of heavy residues has been used to decipher the components of compound and pre-compound emission processes in the fusion of ^{16}O projectile with ^{159}Tb and ^{169}Tm targets. The analysis of recoil range distribution measurements show two distinct linear momentum transfer components corresponding to pre-compound and compound nucleus processes are involved. In order to obtain the mean input angular momentum associated with compound and pre-compound emission processes, an online measurement of the spin distributions of the residues has been performed. The analysis of spin distribution indicate that the mean input angular momentum associated with pre-compound products is found to be relatively lower than that associated with compound nucleus process. The pre-compound components obtained from the present analysis are consistent with those obtained from the analysis of excitation functions.

1 Introduction

The pre-compound (PCN) emission process in heavy ion reactions at relatively low energies ranging from 4 to 7 MeV/nucleon has been a topic of recent interest [1], since it is, generally, expected to occur at relatively high energies $\approx$10-15 MeV/nucleon[2]. The PCN process may be understood as the fusion of two heavy nuclei in such a way so that a composite nucleus forms far from the statistical equilibrium, and a large fraction of its energy is considered to be in the form of an orderly translational motion of the nucleons of the projectile and target nucleus[3–8]. This orderly motion transforms slowly into chaotic thermal motion through a series of two-body interactions. The thermalization process completes when the composite nucleus reaches a state of thermal equilibrium referred to as compound nucleus (CN). During the thermalization of an excited composite system, it may be possible that a single nucleon or a cluster of nucleons having considerable energy is ejected into the continuum. As soon as the state of thermal equilibrium is attained, the accumulation of sufficient energy on a single nucleon or a cluster of nucleons may occur in a random sequence of events and hence, may require much longer emission times favoring the emission of low energy particles. The emission of such light particles prior to the establishment of thermodynamic equilibrium of the composite system are termed as the PCN particles and the reaction mechanism is referred to as the PCN process.

The time scale at which PCN emissions occur is very short $\approx 10^{-21}$ sec., while for further evaporations from the equilibrated nucleus take a longer time $\approx 10^{-16}$ sec. The rate of emission of the PCN nucleons depends on the complexity of self-consistent mean-field interaction between the projectile and target nucleus, their structure effects[7, 9] and excitation energy availed by nucleons of composite system. This determines the initial energy distribution among the nucleons in the projectile and the target nuclei, which starts a cascade of nucleon-nucleon interactions as soon as the two nuclei touch each other. Some of the important experimental characteristics of the PCN process are; (i) the presence of a larger number of high-energy particles as compared to the spectrum predicted by the statistical model, (ii) forward-peaked angular distribution of the emitted particles through PCN process, (iii) ob-

*e-mail: manojamu76@gmail.com
**e-mail: bpsinghamu@gmail.com

servation of smaller recoil range/linear momentum of the evaporation residues left over emission of PCN particles as compared to CN particles, (iv) observation of lower value of the spin with PCN process as compared to CN process, (v) slowly decreasing tails of the excitation functions (EFs) etc.,

In the earlier measurements, the relative strengths of the PCN and CN components in the reactions were estimated from the enhancement in the flux of emitted light particles in the forward direction over the backward direction. The other method often employed is to analyze the measured EFs of evaporation residues by comparing with the statistical model predictions. The observed deviation in EFs with statistical model predictions at relatively higher projectile energies side (tail portion of the EFs) is attributed to the PCN emission process. During last few decades the understanding of the PCN and the CN emission in light ion reactions has been well studied but in heavy ion reactions it needs to be further explored particularly for those associated with the emission of light particles in primary stage in a very short reaction time (10^{-21} s) prior to the establishment of equilibrated CN[7, 10–14]. The emission of such PCN particles reduces the momentum of the product residues. As such, the measurements of the momentum transfer during the interaction may provide a promising tool for the characterization of the reaction mechanism involved.

Although, information about the momentum transfer in heavy ion reaction may also be obtained by several methods[2, 15, 16], in the present measurements this information has been obtained from the study of the recoil range distributions (RRDs) and the spin distributions (SDs) of the reaction residues. Since emitted PCN particles takes away a significant part of angular momentum as compared to CN particles, the angular momentum associated with the PCN products is relatively lower than that associated with the CN process. Therefore, in PCN reactions, the residues are populated at relatively lower spin states as compared to those of the residues populated via CN process. This paper is organized as follows. The experimental details and analysis of RRDs, SDs and EFs are given in the subsections of the section 2. The conclusions drawn from the analysis are given in the section 3 of this paper.

2 Experimental details and analysis

In the present paper the following three self-consistent measurements i.e., (i) the recoil range distributions (RRDs), (ii) the spin distributions (SDs) of the product nuclei and (iii) the excitation functions (EFs), have been performed in the interaction of ^{16}O projectile with ^{159}Tb and ^{169}Tm targets. In both RRDs and EFs measurements, the detection of delayed γ-rays of product residues takes place through the recoil catcher off-line spectroscopy based activation technique. However, the measurements of the SDs is based on the detection of prompt γ-rays of product residues in a particle-γ coincidence experiments in forward and backward directions. A brief description on each experiment carried out in the Inter University Accelerator

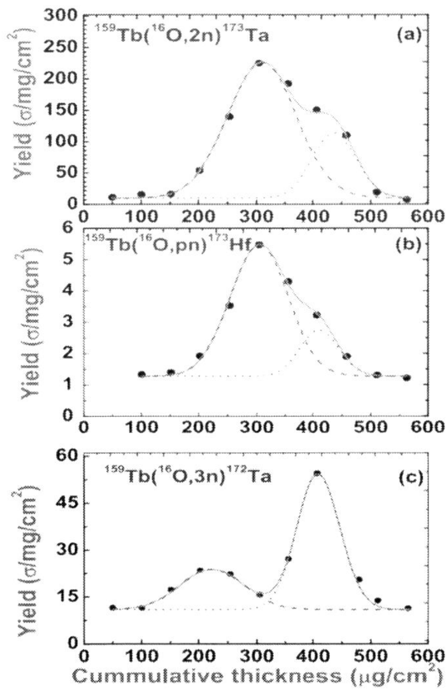

Figure 1. (Color online) The experimentally measured RRDs for the reactions ^{159}Tb(^{16}O,2n)^{173}Ta, ^{159}Tb(^{16}O,pn)^{173}Hf and ^{159}Tb(^{16}O,3n)^{172}Ta at energy ≈90 MeV. In this figure a Gaussian peak (blue dotted curve) obtained from ORIGIN software at higher cumulative thickness represents CN contribution while other Gaussian (red dashed curve) at lower thickness represents PCN contribution.

Centre (IUAC), New Delhi, India along with their results are given in following subsections;

2.1 Measurement and Analysis of Recoil Range Distributions

In order to investigate the role of linear momentum transfer in the PCN emission, the experiments have been carried out at in the General Purpose Scattering Chamber (GPSC) of IUAC, New Delhi, India to measure the RRDs of recoiling residues produced both by the CN and the PCN emission processes in the reactions ^{159}Tb (^{16}O,2n) ^{173}Ta, ^{159}Tb (^{16}O,pn) ^{173}Hf and ^{159}Tb(^{16}O,3n)^{172}Ta for the ^{16}O+^{159}Tb system at 90 MeV and for reactions ^{169}Tm(^{16}O, 2n)^{183}Ir and ^{169}Tm(^{16}O, 3n)^{182}Ir in the ^{16}O+^{169}Tm system at incident energy 88 MeV, respectively. The Coulomb barrier for systems ^{16}O+^{159}Tb and ^{16}O+^{169}Tm are ≈ 64 and 67 MeV respectively. In this experiment, the target (thickness ≈ 500 $\mu g/cm^2$ for ^{169}Tm and ≈ 400 $\mu g/cm^2$ for ^{159}Tb) followed by a stack of nearly fifteen thin Al catcher foils of varying thickness (≈ 16-45 $\mu g/cm^2$ prepared by vacuum evaporation technique) was mounted in the GPSC

 DOI: 10.1051/epjconf/201716300053

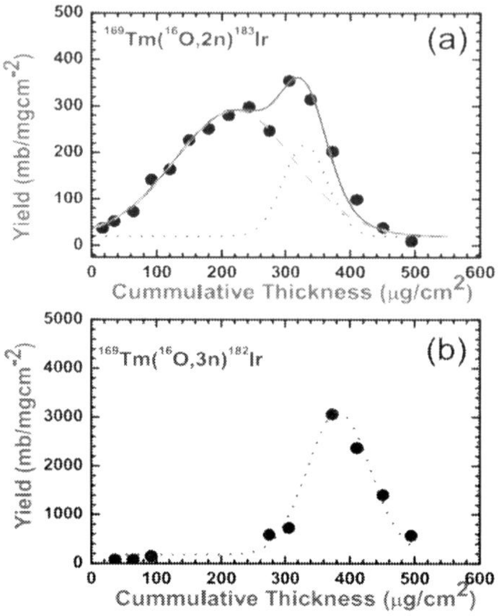

Figure 2. (Color online) (a-c). The experimental RRD for the reactions ^{169}Tm(^{16}O,2n)^{183}Ir and ^{169}Tm(^{16}O,3n)^{182}Ir at energy ≈88 MeV. In this figure a Gaussian peak (blue dotted curve) at higher cumulative thickness represents CN contribution while other Gaussian (red dashed curve) at lower thickness represents PCN contribution.

normal to the beam direction. Furthermore, SRIM[17] calculations of energy loss in the target sample has been performed to calculate the energy of recoiling residues at the mid of the target. Depending on the momentum carried away by the product residues, the recoiling residues were trapped at different ranges in the stack of thin Al-foils. The duration of irradiation was about 15 hrs. The thickness of the catcher foils was measured precisely prior to their use, by measuring the energy loss suffered in each catcher foil by 5.485 MeV α particles from ^{241}Am source. The code SRIM[17] was used for determining the thickness from the energy loss measurements.

The activities induced in each thin catcher were followed off-line for about two weeks using a pre-calibrated high resolution (2 keV for 1.33 MeV γ ray of ^{60}Co) HPGe detector. The experimental cross-section (σ) for the reaction products in different catcher foils were divided by its respective thickness that gives the resulting yield of experimentally measured RRDs. The experimental uncertainties in cross-section is <10% that arise due to various factors viz., uncertainty in the number of target nuclei due to non-uniformity in sample thickness, the fluctuations in the beam current, uncertainty in the determination of detector efficiency, statistical error in counts and the dead time of the counting system. The experimentally measured RRDs for reactions ^{159}Tb(^{16}O,2n)^{173}Ta, ^{159}Tb(^{16}O,pn)^{173}Hf and ^{159}Tb(^{16}O,3n)^{172}Ta at 90 MeV are shown in Fig 1 (a-c). Solid curve in this figure guides the eye to the experimental RRD data. As can be seen from this figure that the experimental RRD data for these reactions has two peaks, one at a relatively lower value (≈ 250-300$\mu g/cm^2$, depending on straggling and number of evaporation residues) of cumulative catcher thickness and the other at ≈ 400-450 $\mu g/cm^2$. The peak at ≈ 400-450 $\mu g/cm^2$ corresponds to the fraction of residue produced through the CN process and is consistent with the full momentum transfer events in complete fusion reactions. The peak at relatively smaller range may be attributed to the fact that the residues ^{183}Ir are produced via the PCN process when emission of neutrons takes place prior to the establishment of thermodynamical equilibrium. The theoretical value (mean range) of the experimental RRD data is calculated with the help of energy loss program of the code SRIM and this value of mean range is fed in the program ORIGIN to fit the Gaussian peak of the experimental RRD. The Gaussian peak (shown by red dashed curve) at lower cumulative catcher thickness of larger width is obtained which shows the contribution of PCN emission. By using similar procedure, the RRDs for reactions ^{169}Tm(^{16}O,2n)^{183}Ir and ^{169}Tm(^{16}O,3n)^{182}Ir have also been measured at ≈88 MeV. The experimental RRDs with their theoretical simulations for these reactions are shown in Fig 2 (a-b), respectively. The peaks in the experimental RRD arise due to the overlap of the heavy residues produced via two different reaction mechanism i.e, the CN and the PCN emission of particles. As can be seen from this figure that the experimental RRD data for the reaction ^{169}Tm(^{16}O,2n)^{183}Ir has two peaks, one at a relatively lower value (≈ 200-250$\mu g/cm^2$) of cumulative catcher thickness and the other at ≈ 330-380 $\mu g/cm^2$. The peak at ≈ 330-380 $\mu g/cm^2$ corresponds to the fraction of residue produced through the CN process. It is pointed out here that due to short half live(≈15 min) of the residue ^{182}Ir, we could measure only the RRD corresponds to CN component.

2.2 Measurement and Analysis of Spin Distributions

The second experiment based on particle-gamma coincidence technique has been performed for measuring the population of spin states during de-excitation of reaction residues. In the present work, the spin-distributions of the reaction ^{159}Tb(^{16}O,2n)^{173}Ta at ≈93 MeV and for ^{169}Tm(^{16}O,2n)^{183}Ir at ≈88 MeV have been measured using Gamma Detector Array (GDA) alongwith Charged Particle Detector Array (CPDA). The Experimental details are given in Ref. [21]. However, we mention them here in brief for ready reference. The GDA is an assembly of 12 Compton suppressed, high resolution HPGe γ-spectrometers arranged at 45^0, 99^0, and 153^0 angles with respect to the beam axis and there are four detectors at each of these angles. The CPDA is a set of 14-phoswich detectors housed in a 14 cm diameter scattering chamber, covering nearly 90% of total solid angle. The reaction residues have been identified from their characteristic prompt γ-transition lines [22]. The values of relative production yields of the residues (observed area under the peak of the experimentally measured prompt gamma

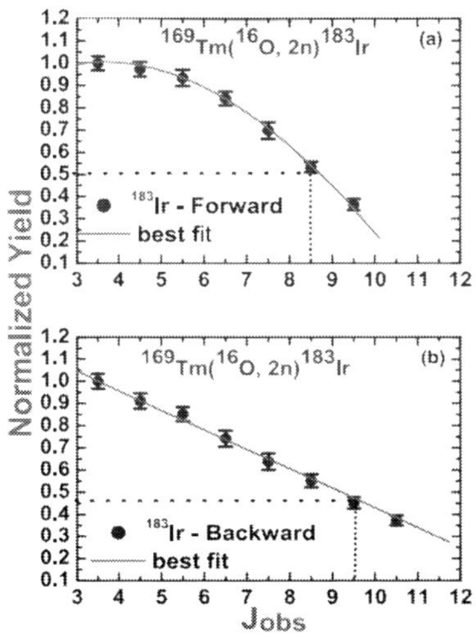

Figure 3. (Color online) The experimentally measured spin distributions for reactions ^{169}Tm(^{16}O,2n)^{183}Ir in forward and backward directions. The curve and line in this figure guide the eye as a best fit to the experimentally normalized yield of spin distributions of PCN and CN processes.

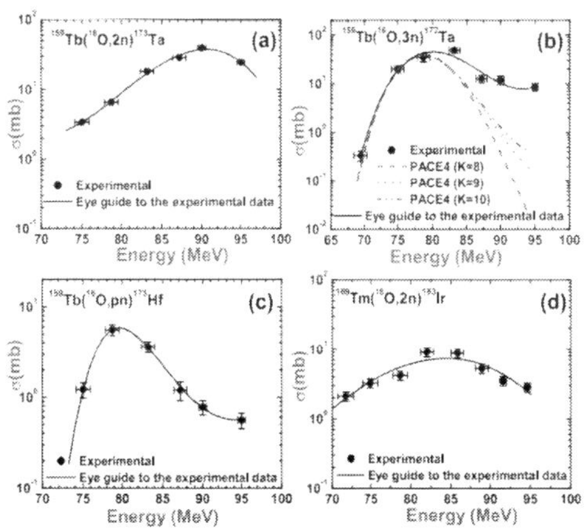

Figure 4. (Color online) (a-d) The experimentally measured EFs for reactions ^{159}Tb(^{16}O,2n)^{173}Ta, ^{159}Tb(^{16}O,pn)^{173}Hf, ^{159}Tb(^{16}O,3n)^{172}Ta, and ^{169}Tm(^{16}O,2n)^{183}Ir, respectively. The theoretically calculated EFs by using code PACE4 gives negligibly small values of cross-sections for all these reactions except for ^{159}Tb(^{16}O,3n)^{172}Ta. The experimentally measured EFs and the PACE4 predictions with different values of parameter K from K=8 to K=10 for reaction ^{159}Tb(^{16}O,3n)^{172}Ta is also shown in Fig 4(b).

lines with proper detector efficiency correction) have been plotted as a function of observed spin J_{obs} corresponding to prompt gamma-transitions[23]. The relative yield has been normalized with minimum observed spin (J^{min}_{obs}) at highest yield (Y^{max}_{obs}). As a representative case the experimentally measured SDs obtained from prompt γ-rays recorded in forward and backward directions for the reaction ^{169}Tm(^{16}O,2n)^{183}Ir are shown in Fig 3 (a) and 3(b) at 88 MeV, respectively. As can be seen from these figures, the measured SD and hence its decay pattern for this reaction obtained in the forward and backward directions are distinctly different from each other indicating widely different reaction mechanisms involved. It may be pointed out the entirely different shapes of SDs in forward and backward directions indicate that the two processes are quite different in nature.

The observed lower value of the mean input angular momentum ($\approx 8.5\hbar$) in forward direction is due to the fact that emission of two PCN neutrons takes away a significant part of the angular momenta. On the other hand a relatively higher observed value of the mean input angular momentum in backward direction ($\approx 11.5\hbar$) is because of the emission of two equilibrated neutrons. As such, it is concluded that distinctly different SDs give a direct evidence of the PCN emission process. Thus, the results of the measurements of the SDs further supplement the conclusions drawn from the RRDs measurements.

2.3 Measurement and Analysis of Excitation Functions

The thirds experiment has been carried out for measuring the EFs by using stacked foil activation technique. In this experiment the stacks consisting of (^{159}Tb and ^{169}Tm) target samples followed by Al foils of suitable thickness have been irradiated for ≈8-10 h in a GPSC of 1.5 m diameter having an in-vacuum transfer facility. The Al foils serve as energy degrader as well as catcher foil where the recoiling residues are trapped. The pertinent decay data required for cross-section measurements of the reaction residue has been taken from Ref.[24]. The activities produced in each target catcher assembly have been measured using a high resolution large volume (100 c.c.) High Purity Germanium Detector (HPGe). The reaction cross sections for the product residues have been obtained from the measured intensities of the characteristic γ-rays using the standard formulation[1].

The analysis of experimental EFs has been performed within the framework of statistical model calculations based on code PACE4[25] that calculates the reaction cross-section for evaporation residues (not for pre-compound residues) using the Bass formula[26]. The level density used in this code is calculated from the expression $a = (A/K)$, where, A is the mass number of the compound nucleus and K is a free parameter known as level density parameter constant. The experimentally measured EFs for the reactions ^{159}Tb(^{16}O,2n)^{173}Ta, ^{159}Tb(^{16}O,3n)^{172}Ta,

$^{159}Tb(^{16}O,pn)^{173}Hf$ and $^{169}Tm(^{16}O,2n)^{183}Ir$ in the energy range ≈70-95 MeV are shown in Figs. 4 (a) to 4(d), respectively. The half-lives of these evaporation residues ^{173}Ta, ^{172}, ^{173}Hf and ^{183}Ir are 3.14 h, 36.8 m, 23.6 h and 57 m, respectively. The solid curves in these figures guide to the eye to the experimentally measured EFs. The theoretically calculated EFs by using code PACE4 gives negligibly small values of cross-sections for all these reactions except $^{159}Tb(^{16}O,3n)^{172}Ta$, hence, the cross-section values obtained from code PACE4 are not shown in figures 4(a), 4(c) and 4(d). In the present calculations, the effect of variation of parameter K from K=8 to K=10 of code PACE4 on measured EFs for reaction $^{159}Tb(^{16}O,3n)^{172}Ta$ is also shown in Fig. 4(b). As can be seen from this figure, by varying the value of K, the experimentally measured EFs could not be reproduced in higher energies (tail portion of EFs) side. The higher values of experimental cross sections in the tail portion of EFs for this reaction as compared to the theoretical calculations may be attributed to the PCN emission process, which is one of the dominant mode reaction channels and is not considered in the PACE4 calculations. It means that analysis of EFs with code PACE4 gives negligible contribution of CN process for the reactions $^{159}Tb(^{16}O,2n)^{173}Ta$, $^{159}Tb(^{16}O,pn)^{173}Hf$ and $^{169}Tm(^{16}O,2n)^{183}Ir$, while the analysis of RRD for the same reactions gives energy dependent contributions of the CN and the PCN processes. This shows that RRD measurements are much sensitive as compared to those of EFs. The analysis of measured EFs for reaction $^{169}Tm(^{16}O,3n)^{182}Ir$ gives a contribution of ≈35% PCN and ≈65% of CN at ≈88 MeV as shown in Fig 4 (b). This data is consistent with the corresponding values obtained from the RRD measurements at ≈88 MeV. Further, the reasonable agreement between the two sets of experiments speaks favorably on the consistency of these measurements.

3 Conclusions

The significant contribution due to PCN process has been observed. The results obtained by RRDs are supported by the SD measurements. The RRD measurements are found to be sensitive tool to decipher the CN and the PCN processes. The auxiliary experiments on the EF measurements is found to be consistent with both the RRD and SD measurements.

4 Acknowledgements

The authors are thankful to the Director, IUAC, New Delhi for extending all the facilities for carrying out the experiments. MKS thanks to the Council of Scientific and Industrial Research (CSIR), New Delhi, India vide the project number 03(1361)16/EMR-11 for financial support.

References

[1] M. K. Sharma, P. P. Singh, D. P. Singh, V. Sharma, A. Yadav, Unnati, Indu Bala, R. Kumar, B. P. Singh, and R. Prasad, Phys. Rev. C **91**, 044601 (2015).

[2] J. Gómez del Campo, D. Shapira, J. McConnell, C. J. Gross, D. W. Stracener, H. Madani, E. Chávez, M. E. Ortíz, Phys. Rev. C **60R**, 021601 (1999).

[3] M. Blann, Ann. Nucl. Sci. **25**, 123 (1975).

[4] M. Blann Nucl. Phys. A **235**, 211 (1974).

[5] E. Holub, D. Hilscher, G. Ingold, U. Jahnke, H. Orf, and H. Rossner, Phys. Rev. C **28**, 252 (1983).

[6] E. Holub, D. Hilscher, G. Ingold, U. Jahnke, H. Orf, H. Rossner, W. P. Zank, W. U. Scroder, H. Germeke, K. Keller, L. Lassen, and W. Lucking, Phys. Rev. C **33**, 143 (1986).

[7] M. Cavinato, E. Fabrici, E. Gadioli, E. Gadioli Erba, P. Vergani, M. Crippa, G. Colombo, L . Redaelli, and M. Ripamonti, Phys. Rev. C **52**, 2577 (1995).

[8] F. Amorini, M. Cabibbo, G. Cardelaa, A. Di Pietro, A. Musumarra, M. Papa, G. Pappalardo, F. Rizzo, and S. Tudisco, Phy. Rev. C **58**, 987 (1988).

[9] H. Zheng, S. Burrello, M. Colonna and V. Baran, Phys. Lett. B **769**, 424 (2017).

[10] H.C. Britt and A.R. Quinton, Phys. Rev. **124**, 877 (1961).

[11] T. Otsuka and K. Haradav, Phys. Lett. B **121**, 106 (1983).

[12] P. Vergani, E. Gadioli, E. Vaciago, E. Fabrici, E. Gadioli Erba, M. Galmarini, G. Ciavola, and C. Marchetta, Phys. Rev. C **48**, 1815 (1993).

[13] C. Birattari et. al, Phys. Rev. C **54**, 3051 (1996).

[14] H. Delagrance, A. Fleury, F. Hubert, and G. N. Simonoff, Phys. Lett. B **37**, 397 (1971).

[15] H. Morgenstern, W. Bohne, K. Grabisch, D. G. Kover and H. Lehr, Phys. Lett. B **113**, 463 (1982).

[16] B. B. Back et. al., Phys. Rev. Lett. **50**, 88 (1983).

[17] The Stopping and Range of Ions in Matter (SRIM) code: [http://www.srim.org/SRIM/SRIMLEGL.htm].

[18] M. K. Sharma et. al., Phys. Rev. C **70**, 044606 (2004).

[19] D. P. Singh et. al, Phys. Rev. C **81**, 054607 (2010).

[20] M. K. Sharma et. al., Phys. Rev. C **91**, 024606 (2015).

[21] P. P. Singh, B. P. Singh, M. K. Sharma et al., Phys. Lett. B 671, 20 (2009).

[22] S. Andrd, J. Genevey-Rivier, et al., Phys. Rev. Lett. 38, 327 (1977).

[23] RADWARE Level Scheme http://radware.phy.ornl.gov/agsdir1.html.

[24] E. Browne and R. B. Firestone, 1986 Table of Radioactive Isotopes (Wiley, New York).

[25] A. Gavron, Phys. Rev. C **21**, 230 (1980).

[26] R . Bass, Nucl. Phys. A **231**, 45 (1974).

Effect of projectile on incomplete fusion reactions at low energies

Vijay R. Sharma[1,*], Mohd. Shuaib[2], Abhishek Yadav[1], Pushpendra P. Singh[3], Manoj K. Sharma[4], R. Kumar[1], Devendra P. Singh[5], B. P. Singh[2,**], S. Muralithar[1], R. P. Singh[1], R. K. Bhowmik[1], and R. Prasad[2]

[1] NP-Group, Inter University Accelerator Centre, New Delhi - 110 067, India
[2] Department of Physics, Aligarh Muslim University, Aligarh 202 002, India
[3] Department of Physics, Indian Institute of Technology Ropar, Punjab 140 001, India
[4] Physics Department, S. V. College, Aligarh - 202 001, India
[5] Department of Physics, University of Petroleum and Energy Studies, Dehradun - 248 007, India

Abstract. Present work deals with the experimental studies of incomplete fusion reaction dynamics at energies as low as ≈ 4 - 7 MeV/A. Excitation functions populated via complete fusion and/or incomplete fusion processes in ^{12}C+^{175}Lu, and ^{13}C+^{169}Tm systems have been measured within the framework of PACE4 code. Data of excitation function measurements on comparison with different projectile-target combinations suggest the existence of ICF even at slightly above barrier energies where complete fusion (CF) is supposed to be the sole contributor, and further demonstrates strong projectile structure dependence of ICF. The incomplete fusion strength functions for ^{12}C+^{175}Lu, and ^{13}C+^{169}Tm systems are analyzed as a function of various physical parameters at a constant v_{rel} ≈ 0.053c. It has been found that one neutron (1n) excess projectile ^{13}C (as compared to ^{12}C) results in less incomplete fusion contribution due to its relatively large negative α-Q-value, hence, α Q-value seems to be a reliable parameter to understand the ICF dynamics at low energies. In order to explore the reaction modes on the basis of their entry state spin population, the spin distribution of residues populated via CF and/or ICF in ^{16}O+^{159}Tb system has been done using particle-γ coincidence technique. CF-α and ICF-α channels have been identified from backward (B) and forward (F) α-gated γ spectra, respectively. Reaction dependent decay patterns have been observed in different α emitting channels. The CF channels are found to be fed over a broad spin range, however, ICF-α channels was observed only for high-spin states. Further, the existence of incomplete fusion at low bombarding energies indicates the possibility to populate high spin states.

1 Introduction

The study of breakup of heavy ions followed by the fusion of one of the fragments with the target has gained a resurgent interest. Several authors reported the competition of complete fusion (CF) and incomplete fusion (ICF) reactions at energies slightly above the Coulomb barrier [1–10], and a substantial ICF fraction has been observed at these energies. In general, the complete fusion (CF) process is considered to be the sole contributor to the total fusion cross section at low energies i.e. starting from threshold to 7 MeV/A.

Though, several studies and models [11–18] are developed to understand ICF phenomenon, but the dynamics of ICF at low incident energies is still not fairly understood and thus continues to be an active area of investigations. High quality data for ICF processes has been obtained and reported so far worldwide [1–8, 19, 20], and conclusively demonstrate the low energy ICF, but are limited only for a few projectile-target combinations. Therefore, for better understanding of ICF dynamics, several inclusive experiments at the Inter-University Accelerator Center (IUAC), New Delhi, India were performed by our group [5, 7, 8, 21, 22]. Based on experimental findings an attempt has been made to correlate data with suitable set of physical parameters viz., (a) projectile energy, (b) mass-asymmetry of interacting partners (μ), (c) α-Q value, and (d) input ℓ values. In the present work, the percentage fraction of ICF were deduced from the analysis of experimental excitation functions of individual reaction residues for different projectile-target combinations. This suggests at the studied energy regime projectile structure plays an important role to understand ICF dynamics. To understand the role of ℓ values, a particle-γ coincidence experiments were performed and spin distribution of individual reaction residues have been measured. The spin distribution of various evaporation residues reveals that the mean input angular momenta (<ℓ>) associated with the ICF observable are found to be higher than those involved with CF. Further, the occurrence of ICF is due to the influence of centrifugal potential in peripheral interactions, where driving angular momentum limits do not allow CF to occur.

The present paper is organized as; the experimental details are presented in section 2. The observation from our recent experiments are given in section 3, where the influence of ICF on CF and its dependencies on various en-

*e-mail: phy.vijayraj@gmail.com
**e-mail: bpsinghamu@gmail.com

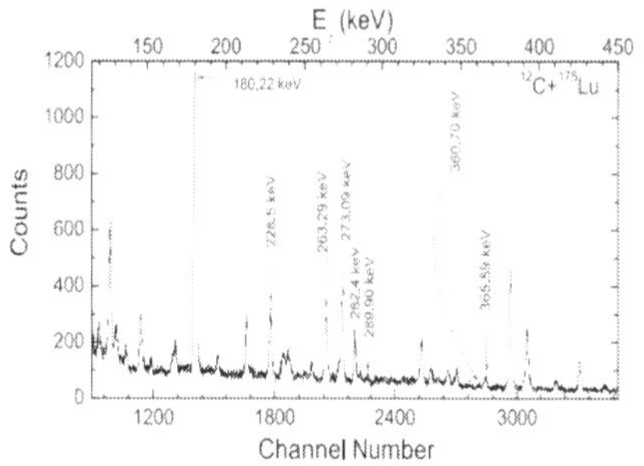

Figure 1. (Color online) Typical γ-ray spectrum of $^{12}C+^{175}Lu$ interaction at ≈ 78.36 ± 1.08 MeV

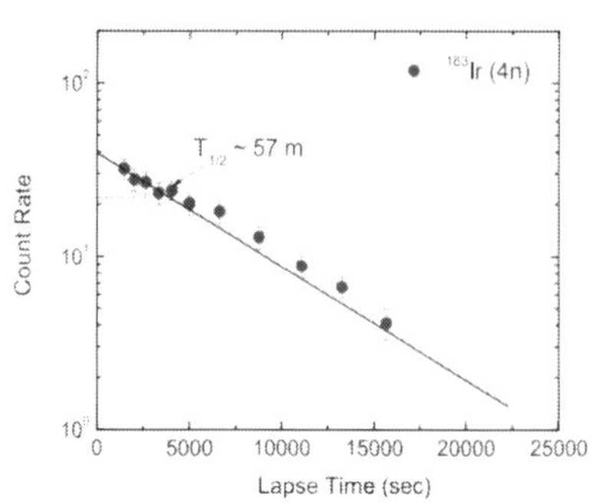

Figure 2. (Color online) Typical decay curve of ^{183}Ir residue observed in $^{12}C+^{175}Lu$ interaction at ≈ 78.36 ± 1.08 MeV beam energy.

trance channel parameters are demonstrated. Summary of the present work is given in Section 4.

2 Experimental details

To explore the dynamics of incomplete fusion reaction processes, experiments were performed using 15 UD pelletron accelerator of Inter University Accelerator Centre (IUAC), New Delhi, India. These experiments were performed using two different setups (a) General Purpose Scattering Chamber (GPSC) and (b) Gamma Detector Array (GDA) coupled with Charged Particle Detector Array (CPDA). The detailed methodology and experimental setups have been discussed in refs.[5, 7, 21, 22]. However, a short account of experimental conditions are given here for the ready reference.

In order to achieve information on the behavior of fraction of ICF observable on various entrance channel parameters, the value of incomplete fusion fraction (F_{ICF}) has been deduced as a function of various physical parameters (discussed in sec 3) in a separate set of experiment. In this experiment, absolute production cross-sections of individual reaction residues populated via CF and/or ICF in $^{12}C+^{175}Lu$, and $^{13}C+^{169}Tm$ systems have been measured at energies starting from threshold to 7 MeV/A [7, 23]. The target foils of isotopically pure (99.9%) ^{175}Lu, and ^{169}Tm along with in ^{27}Al catchers were prepared using rolling technique. The thickness of the samples were determined from the observed change in the energy of the α-particles by using stopping power values and were found to ≈ 1.5 to ≈ 2.5 mg/cm^2 for both targets and ≈ 1.0 to ≈ 3.0 mg/cm^2 for Al-catchers. The thickness of the Al-catchers was chosen keeping in view the fact that even the most energetic residues produced due to the complete momentum transfer may be trapped in the catcher thickness. The irradiation has been carried out with a beam current ≈ 2-3 pnA. The evaporation residues (ERs) produced during the irradiations were counted off-line with two pre-calibrated HPGe detectors. The ERs have been identified by their characteristic γ-lines and are further confirmed by decay-curve analysis. A part of γ-ray spectra obtained at E_{Lab} = 78.36 ± 1.08 MeV in $^{12}C+^{175}Lu$ system is shown in Fig.1. Some of the peaks corresponding to different reaction residues are marked. As a typical example, the observed decay curve for one of the Iridium isotopes ^{183}Ir ($T_{1/2}$ = 57 min) populated in $^{12}C+^{175}Lu$ system is shown in Fig. 2. Further, the production cross-sections (σ_R) of ERs have been calculated using standard formulation [23]. The overall errors in the present measurement are estimated to be ≈ 10%. A detailed discussion on error analysis is presented in Refs. [24].

To probe the role of ℓ-values in the onset of ICF at low projectile energies, the spin-distributions (SDs) of xn/pxn/αxn channels has been measured in $^{16}O+^{159}Tb$ system using particle-γ coincidence technique. The experiment was performed at four set of projectile energies viz. ≈ 83.5± 1.5, 88.5± 1.5, 93.5± 1.5 and 97.6± 1.4 MeVs, using GDA alongwith CPDA facility of IUAC, New Delhi, India. The particle-γ coincidences were recorded using 12 Compton suppressed, high resolution HPGe detectors installed at an angles 45^0, 99^0, 153^0 with respect to the beam axis in Gamma Detector Array (GDA). It may be mentioned that there were four detectors at each of these angles in GDA set up. Further, the GDA set up is coupled with the Charge particle detector array (CPDA) which is a set of 14-phoswich detectors housed in a 14 cm diameter scattering chamber. These particle detectors cover nearly 90% of the total solid angle. All 14 detectors of CPDA have been divided into the angular zones viz.; (i) Forward angle (F) 10^0 to 60^0, (ii) Sideways (S) 60^0 to 120^0, and (iii) Backward angle (B) 120^0 170^0. Depending on the fast and slow components of the CPDA, proton and α-particles in each angular ring can be identified. For the experiment, isotopically pure, self-supporting ^{159}Tb (100%) target of thickness ≈ 1.5 mg/cm^2 (prepared by rolling technique) has been bombarded with $^{16}O^{+7}$ beams delivered from 15UD Pelletron Accelerator. A prompt γ-ray spectra was recorded in event-by-event multi parameter mode, which includes different coincidences i.e. particles (Z=1,2) detected in backward (B), forward (F) and 90^0 (S)-angles. Singles data have also been collected to identify xn-channels (produced via CF). Data analysis was done in two steps. In the first step, spin distributions of xn-channels have been measured by looking into singles spectra. For the identification of pxn-channels, backward (B)-α-gated spectra has been subtracted from backward(B)-particles(Z=1,2)-gated spectra to achieve proton-gated spectra. However, αxn -channels populated via CF have been identified from the backward (B)-α-gated spectra. Further, the fast-α-particles (which are expected to be emitted only in forward cone) produced via ICF have been identified from forward(F)-α-gated spectra obtained by subtracting backward (B)-α-gated spectra from forward(F)-α-gated spectra. The intensity and area under the photo-peak (efficiency corrected) of the characteristic prompt α-transitions were used to determine the relative production yield [21]

3 Projectile Dependence of ICF: Recent Results

In the present work, a systematic investigation have been done to understand the effect of various entrance channel parameters (i.e. projectile energy, entrance channel mass-asymmetry of interacting partners, α-Q-value, and input ℓ-values) on the onset and strength of incomplete fusion. Interpretation of the data and results are presented in this section.

3.1 ICF dependence on incident energy

The excitation functions (EFs), expected to be populated via CF and/or ICF mode have been measured for $^{12}C+^{175}Lu$, and $^{13}C+^{169}Tm$ systems at energies starting from the Coulomb barrier to 7 MeV/A and analyzed in the framework of statistical model code PACE4 to probe the energy dependence of ICF. Details of code PACE4 can be found in ref. [25]. In this code, the nuclear level density (a = A/K) plays an important role which can be tuned to fit the experimental data. In order to choose the suitable value of level density to reproduce fusion EFs, different values of K = 8 - 12 have been tested. The evaporation residues (ERs) populated via xn/pxnchannels have been found to be in good agreement with that estimated from code PACE4. On the other hand, the experimental EFs of α-emitting channels have been found to be significantly enhanced as compared to the PACE4 predictions. This enhanced value of α ERs may be attributed to ICF. To have the enhanced value of ERs (i.e. ICF contribution), experimental EFs of α-emitting channels were subtracted from the PACE4 values (σ_{ICF} = σ_{exp} - σ_{PACE4}) [25]. For better insight into the onset and strength of ICF, the percentage fraction of ICF (F_{ICF}) has been deduced and is plotted as a function of projectile relative velocity (v_{rel}) [5] for $^{12}C+^{175}Lu$, $^{13}C+^{169}Tm$ and $^{12}C+^{169}Tm$ systems in Fig.3. It may be mentioned that F_{ICF} defines empirical probability of ICF.

As shown in this figure, the F_{ICF} increases linearly with normalized projectile energies in terms of relative velocity of projectile (i.e., v_{rel}). This suggests strong projectile energy dependence of ICF reactions. On the other

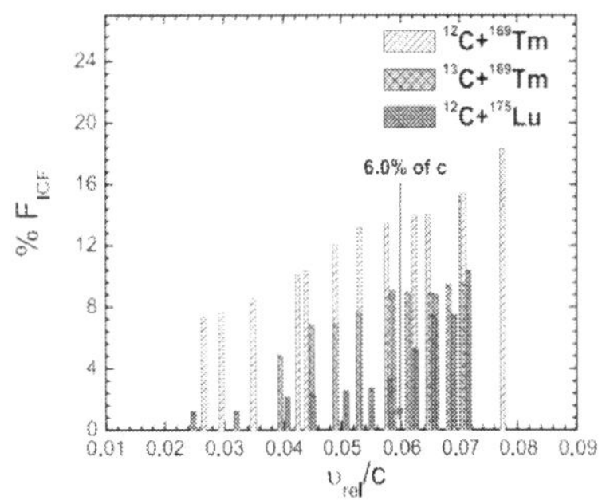

Figure 3. (Color online) The percentage fraction of ICF as a function of projectile relative velocity (v_{rel}) for $^{12}C+^{175}Lu$, $^{12}C+^{169}Tm$ and $^{13}C+^{169}Tm$ systems

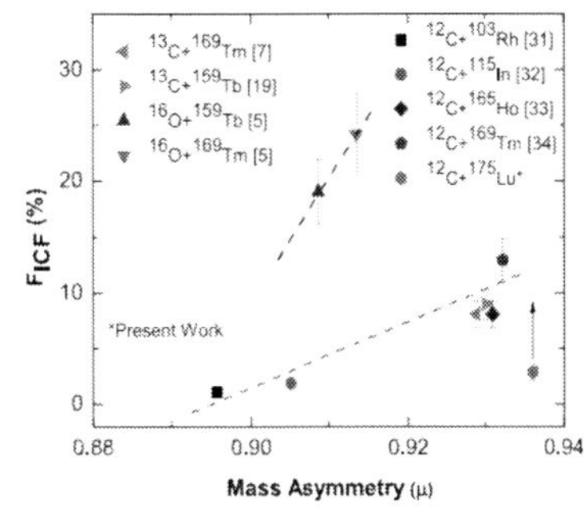

Figure 4. (Color online) The percentage fraction of ICF as a function of μ for 9 projectile-target combinations at a constant value of v_{rel} ≈ 0.053c. Dashed lines are drawn to guide the eyes.

hand, Morgenstern systematics [26, 27] suggests that ICF contributes significantly above $v_{rel} \geq 0.06$ (6 % speed of light). However, in figure 3, the values of v_{rel} are in the range from ≈ 2.7 % of c to ≈ 7.8 % of c for $^{12}C+^{175}Lu$, and $^{12}C+^{169}Tm$ systems and from 2.3 % of c to 7.0 % of c for $^{13}C+^{169}Tm$ system. Hence, the results presented in Fig.3 clearly demonstrate the onset of ICF at relatively lower value of v_{rel} i.e. F_{ICF} is significant at well below the proposed onset value of v_{rel}. Further, for the same target the value of F_{ICF} for ^{12}C-projectile is lower than ^{13}C-projectile for the entire measured energy range. The difference in F_{ICF} for two systems (^{12}C,$^{13}C+^{169}Tm$) clearly points towards projectile structure and/or mass asymmetry effect on ICF.

3.2 Projectile structure dependence of ICF

To understand projectile structure and/or mass asymmetry (μ) dependence of ICF, the deduced percentage fraction of ICF for $^{12}C+^{175}Lu$, and $^{13}C+^{169}Tm$ systems alongwith

 DOI: 10.1051/epjconf/201716300054

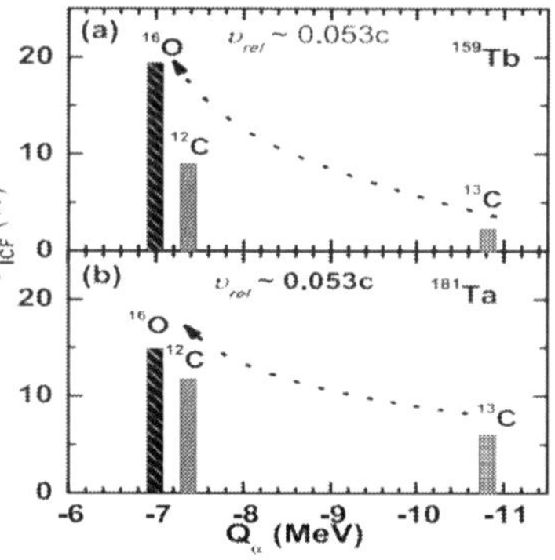

Figure 5. (Color online) Comparison of F_{ICF} on the basis of α-Q value of the projectile at a constant v_{rel} = 0.053c for different projectile-target combinations

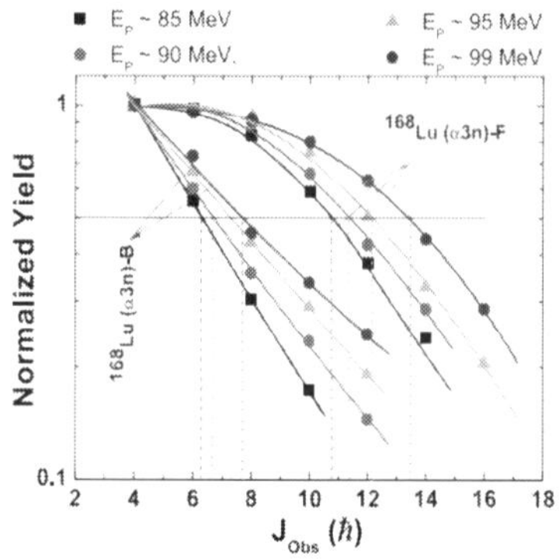

Figure 6. (Color online) Experimentally generated spin-distributions of α3n-(B), and α3n-(F) channels. Lines and curves are the best fit to the experimental data points.

7 projectile-target combinations are plotted as a function of mass asymmetry at a constant value of $v_{rel} \approx 0.053$ in Fig.4. In order to understand projectile dependence on ICF more conclusively, data for the ^{16}O projectile is also added. It may be pointed out that ^{12}C, ^{13}C and ^{16}O are widely different projectiles and hence the probability of breakup depends on the input ℓ-values imparted into system in peripheral interactions. As can be seen from Fig.4, the Morgensterns systematics does not explain the variation of F_{ICF} with μ for given systems. However, the value of F_{ICF} increases with μ for ^{12}C, ^{13}C projectiles and individually for ^{16}O projectile. It is interesting to note that the ^{12}C, ^{13}C+^{169}Tm systems is a more mass asymmetric ($\mu \approx$ 0.9337) system than ^{16}O+^{169}Tm system ($\mu \approx 0.9135$), but the value of F_{ICF} is $\approx$ 18 % higher than that observed for ^{12}C, ^{13}C+^{169}Tm system. The aforementioned observations based on 9 projectile-target combinations strongly contradict Morgensterns mass-asymmetry systematics, and suggest strong projectile dependence of ICF for alpha cluster projectiles. However projectile dependence of mass asymmetry is valid for same atomic number and different mass number projectiles (i.e. ^{12}C, ^{13}C, etc. on the same target) is a question of interest. As can be seen in Fig 3 the values of F_{ICF} for the ^{13}C projectile are less than for the ^{12}C projectile in the studied energy range. The difference in F_{ICF} for two systems (^{12}C, ^{13}C+^{169}Tm) is clearly evident, and indicates the strong projectile dependence of F_{ICF} with some physical parameter other than μ. In order to explore this issue, the ICF strength functions for ^{12}C, ^{13}C+^{169}Tm systems have been studied in terms of projectile α-Q value.

3.3 Projectile α-Q value dependence of ICF

The percentage fraction of ICF is analyzed in terms of projectile α Q-value for 6 projectile-target combinations. The value of F_{ICF} is plotted as a function of Q_α in Fig.5(a-b) at a constant $v_{rel} \approx 0.053$c. As shown in this figure, the probability of ICF for the ^{13}C projectile is smaller than that for the ^{12}C projectile. This clearly point towards the projectile structure effect. It may be pointed out that ^{12}C is a well known α-cluster nucleus with Q_α = -7.37 MeV. However, ^{13}C has a larger Q_α (= -10.64 MeV) than ^{12}C. The higher α Q-value for ^{13}C translates into the smaller breakup probability into constituent α-clusters, resulting smaller ICF fraction than that found in ^{12}C induced reactions. In order to strengthen α Q-value dependence, the value of F_{ICF} for the ^{16}O (Q_α=-7.16 MeV)+^{159}Tb system is found to be 18% which is reduced to only 3% for the ^{13}C (Q_α = -10.64 MeV)+^{159}Tb system. Similar trend was observed for the ^{181}Ta target. Hence, from the data presented in this figure, it can be inferred that the α Q-value is an important entrance channel parameter which essentially dictates the probability of ICF.

3.4 ℓ dependence of ICF

In order to understand the physics of angular momentum involvement and de-excitation pattern of any residual nuclei, the spin distribution of various CF and ICF channels have been measured for ^{16}O+^{159}Tb system at laboratory energies (E_{lab}) $\approx$ 83.5$\pm$ 1.5, 88.5$\pm$ 1.5, 93.5$\pm$ 1.5 and 97.6$\pm$ 1.4 MeV with beam intensities of $\approx$ 4-6 pnA. It is expected that in case of CF residues (high E* and low ℓ) the de-excitation pattern are more likely as statistical transitions, on the other hand, the ICF residues achieve low E* (due to the involvement of partial degrees of excitations) and high ℓ (relatively higher values of impact parameters contribute to the high spin states) at a given projectile energy, therefore, the number of yrast - like transitions are much larger than that of statistical ones.

In view of the above, a comparison of the spin distribution pattern for the α3n channel (^{168}Lu) populated via both CF mode ($\rightarrow$ identified in the backward (B) α-gated spectra) and ICF mode ($\rightarrow$ identified in the forward (F) α-gated spectra) are plotted in Fig.6. As shown in this figure, there is a difference in the spin-distributions which indicate the involvement of entirely different mode of reactions in the production of these residues. The intensity of α3n (B)-

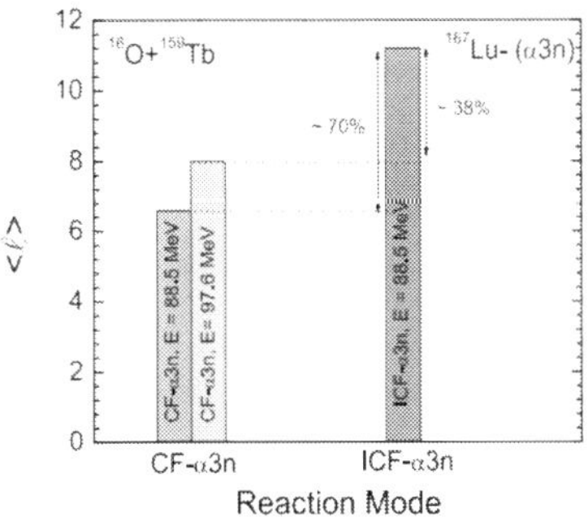

Figure 7. (Color online) The value of the mean angular momenta i.e., the value of $<\ell>$ deduced from the best fitting procedure of spin distributions [21] involved in ^{167}Lu isotope populated via CF and ICF processes as a function of modes of reaction.

channels (CF residues) falls off rather quickly with observed spin (J_{obs}), indicating strong feeding and/or broad spin population during the de-excitation of CN. However, for α3n (F)-channels (ICF residues), the intensity appears to be almost constant up to a certain value of J_{obs}, and then decreases towards band head. This indicates the absence of feeding to the lowest members of the Yrast band and/or the population of low spin states are strongly hindered in ICF-channels.

Concerning the usefulness of incomplete fusion as a tool to populate high-spin states in final reaction products the mean input angular momenta ($< \ell >$) associated with α3n channel (populated via both CF and ICF modes) have been deduced as suggested in Ref [21] and is plotted in Fig.7. As can be seen in Fig. 7, the $< \ell >$ values involved in the production of ^{168}Lu via ICF-α3n channel (i.e. $\approx$ 11.0 $\hbar$) is found to be 70% larger than that via CF-α3n channel (i.e. $\approx$ 6.5 $\hbar$) at the same projectile energy, i.e., $E_{lab} \approx 88.5\pm$ 1.5 MeV. On the other hand, the value of ℓ involved in ICF-α3n channel (i.e. $\approx$ 10.5 $\hbar$) at 83.5$\pm$ 1.5 MeV is found to be $\approx$ 40% more than that involved in CF-α3n channel (i.e. $\approx$ 7.5 $\hbar$) even at highest experimental incident laboratory energy (E_{lab}) $\approx$ 97.6$\pm$ 1.4 MeV. This clearly demonstrates the involvement of higher ℓ-values in the production of incomplete channels which may lead to the population of higher spin-states in the final reaction products which may not be possible otherwise. The present spin distribution measurement for ^{16}O+^{159}Tb system supports earlier findings by Dracoulis *et al* [28], Lane *et al* [29], and Mullins *et al* [30] on the populatation of high spin states in final reaction products at low bombarding energies. Hence, ICF can be used as a spectroscopic tool to study the high spin states even at low energies.

4 Summary

This paper gives the recent experimental results for the excitation function and particle-γ coincidence experiments. The EFs for several residues populated via CF and/or ICF in ^{12}C+^{175}Lu and ^{13}C+^{169}Tm systems have been measured at energies starting from threshold to $\approx$ 7 MeV/nucleon, and analyzed in the framework of statistical model code PACE4. A systematic analysis of ICF dependence on various entrance channel parameters has been performed. It has been found that ICF strongly depends on incident energy, projectile and target type, α Q-values. Results and analysis presented on projectile structure effects suggest more ICF fraction for less negative α Q-value projectiles. Results presented are in good agreement with the existing data.

The spin-distribution(s) associated with ICF are found to be originated from the narrow spin population, localized near and/or above to the critical angular momentum (ℓ_{crit}) for CF, where a given direct-α-fragment is emitted to release excess driving angular momenta. This reveals a competition from successively opened ICF channels for each value of ℓ above the ℓ_{crit} for normal fusion (CF). On the basis of the present analysis of experimental data, the mean angular momenta associated with ICF are found to be $\approx$ 70% higher than that for CF at the same energy. Further, the mean input angular momenta associated with ICF residues at lowest energy is 40 % higher than that for the CF residue at the highest studied energy, this helps to understand the population of high spin states via ICF route.

5 Acknowledgements

The authors thank to the Director, IUAC, New Delhi, for providing all the necessary facilities to carry out this work. One of the author (VRS) thanks to DST for providing travel support.

References

[1] P. R. S. Gomes, R. Linares, J. Lubian, C. C. Lopes, E. N. Cardozo, B. H. F. Pereira, and I. Padron, Phys. Rev. C 84, 014615 (2011).

[2] K. Surendra Babu, R. Tripathi, K. Sudarshan, B. D. Shrivastava, A. Goswami, and B. S. Tomar, J. Phys. G 29, 1011 (2003).

[3] K. Sudarsan Babu et al., Nucl. Phys. A 739, 229 (2004).

[4] D.J. Hinde, M. Dasgupta, Phys. Rev. C 81, 064611 (2010), and the references therein.

[5] P. P. Singh, B. P. Singh, M. K. Sharma, Unnati, D. P. Singh, and R. Prasad, Phys. Rev. C 77, 014607 (2008).

[6] M. Dasgupta, D. J. Hinde, R. D. Butt, R. M. Anjos, A. C. Berriman, N. Carlin, P.R. S.Gomes, C. R. Morton, J. O. Newton, A. Szanto de Toledo, and K. Hagino, Phys. Rev. Lett. 82, 1395 (1999).

[7] Vijay R Sharma et al., Phys. Rev. C 89, 024608 (2014).

[8] Mohd. Shuaib, et al, Phys. Rev. C 94, 014613 (2016).

[9] K. Sudarshan, et al, Phys. Rev. C 69, 027603 (2004).

[10] R. Tripathi, et al, Eur. Phys. J. A 42: 25 (2009).

[11] I. Tserruya et al., Phys. Rev. Lett. 60, 14 (1988).

DOI: 10.1051/epjconf/201716300054

[12] E. Takada, T. Shimoda, N. Takahashi, T. Yamaya, K. Nagatani, T. Udagawa, and T. Tamura, Phys. Rev. C 23, 772 (1981).

[13] J.Wilczynski, K. Siwek-Wilczynska, J. Van Driel, S. Gonggrijp, D. C. J. M. Hageman, R. V. F. Janssens, J. Lukasiak, R. H. Siemssen, and S. Y. Van Der Werf, Nucl. Phys. A 373, 109 (1982).

[14] M. Blann, Phys. Lett. 27, 337 (1971).

[15] J. P. Bondrof et al., Nucl. Phys. A 333, 285 (1980).

[16] B. G. Harvey, Phys. Lett. B 130, 373 (1983).

[17] B. G. Harvey, Nucl. Phys. A 444, 498 (1985).

[18] A. Y. Abul-Magd, Z. Phys. A 298, 143 (1980).

[19] A. Yadav, V. R. Sharma, P. P. Singh, R. Kumar, D. P. Singh, Unnati, M. K. Sharma, B. P. Singh, and R. Prasad, Phys. Rev. C 86, 014603 (2012), and references therein.

[20] Manoj K Sharma Physical Rev. C 91, 024608 (2015).

[21] Vijay R. Sharma, et al., J. Phys. G, Nucl. Part. Phys. 42 055113 (2015).

[22] Pushpendra P. Singh, et al., Phys. Lett. B 671, 20 (2009).

[23] Vijay R. Sharma, Ph.D. Thesis, A.M. University, Aligarh-Unpublished, (2015).

[24] B. P. Singh, M. G. V. Sankaracharyulu, M. A. Ansari, H. D. Bhardwaj, and R. Prasad, Phys. Rev. C 47, 2055 (1993).

[25] O. B. Tarasov and D. Bazin, Nucl. Instrum. Methods Phys. Res., Sect. B 204, 174 (2003).

[26] H. Morgenstern, W. Bohne, W. Galster, and K. Grabisch, Z. Phys. A. Atomic Nuclei 324, 443 (1986).

[27] H. Morgenstern, W. Bohne, W. Galster, K. Grabisch, and A. Kyanowski, Z. Phys. A 313, 39 (1983); H. M. Morgenstern, W. Bohne, K. Grabisch, D. G. Kovar, and H. Lehr, Phys. Lett. B 113, 463 (1982).

[28] G.D. Dracoulis, et al., J. Phys. G, Nucl. Part. Phys. 23, 1191 (1997).

[29] G.J. Lane, G.D. Dracoulis, A.P. Byrne, A.R. Poletti, T.R. McGoram, Phys. Rev. C 60, 067301 (1999).

[30] S.M. Mullins, et al., Phys. Lett. B 393, 279 (1997); S.M. Mullins, et al., Phys. Rev. C 61, 044315 (2000).

[31] Bindu Kumar, et al., Phys. Rev. C 59, 2923 (1999).

[32] S. Mukherjee, et al., Int. J. Modern Phys. E 15, 237 (2006).

[33] S. Gupta, et al., Phys. Rev. C 61, 064613 (2000).

[34] P. P. Singh, et al., Jour. of Phys: Conf. Series 590, 012031 (2015).

 DOI: 10.1051/epjconf/201716300055

Effect of Pauli repulsion and transfer on fusion

C. Simenel[1,*], *K. Godbey*[2], *A. S. Umar*[2], *K. Vo-Phuoc*[1], *M. Dasgupta*[1], *D. J. Hinde*[1], and *E. C. Simpson*[1]

[1] *Department of Nuclear Physics, Research School of Physics and Engineering, The Australian National University, Canberra ACT 2601, Australia*

[2] *Department of Physics and Astronomy, Vanderbilt University, Nashville, TN 37235, USA*

Abstract. The effect of the Pauli exclusion principle on the nucleus-nucleus bare potential is studied using a new density-constrained extension of the Frozen-Hartree-Fock (DCFHF) technique. The resulting potentials exhibit a repulsion at short distance. The charge product dependence of this Pauli repulsion is investigated. Dynamical effects are then included in the potential with the density-constrained time-dependent Hartree-Fock (DCTDHF) method. In particular, isovector contributions to this potential are used to investigate the role of transfer on fusion, resulting in a lowering of the inner part of the potential for systems with positive Q-value transfer channels.

1 Introduction

A dream shared by many theorists working on the quantum many-body problem is to find a way to describe the tunnelling of a many-body wave-function. For instance, this would enable a fully microscopic description of sub-barrier fusion, without other parameters than those of the energy density functional describing the interaction between the nucleons.

Such a tool could then be used to investigate the long-standing deep sub-barrier fusion hindrance puzzle [1–3] (see Ref. [4] for a review). It is also crucial to predict fusion cross-sections in systems such as $^{12}C+^{12}C$ at astrophysical energies in order to get a deeper insight into stellar nucleosynthesis mechanisms (see, e.g., Ref. [5] and the contributions to this conference from S. Courtin, G. Fruet, E. Rehm, and N. T. Zhang for recent works). In addition, it would help guiding experimental programs aiming at studying the impact of exotic structures (e.g., neutron-skins and pigmy dipole resonances) and of the continuum on fusion (see Refs. [6–8] and contributions from D. Bazin, J. Kolata, R. T. de Souza and G. Colucci for new or recent experimental programs).

However, such a theory is not yet available. Indeed, microscopic descriptions of fusion reactions are based on mean-field approximations, such as the time-dependent Hartree-Fock (TDHF) theory, which do not account for tunnelling of the many-body wave function. This is because more than one mean-field is required to describe the outcome of a near-barrier reaction (one for the fused system and one for the outgoing fragments).

Figure 1 shows TDHF density evolutions for $^{16}O+^{16}O$ central collisions at near barrier energies, one leading to fusion (just above the barrier) and one leading to two outgoing fragments (below the barrier). We see that both collisions lead to density distributions occupying different regions of space. Thus, sub-barrier fusion, associated with non-zero fusion and scattering probabilities, cannot be described with a single local mean-field. In TDHF, fusion probabilities are then either 0 or 1 for a given initial condition.

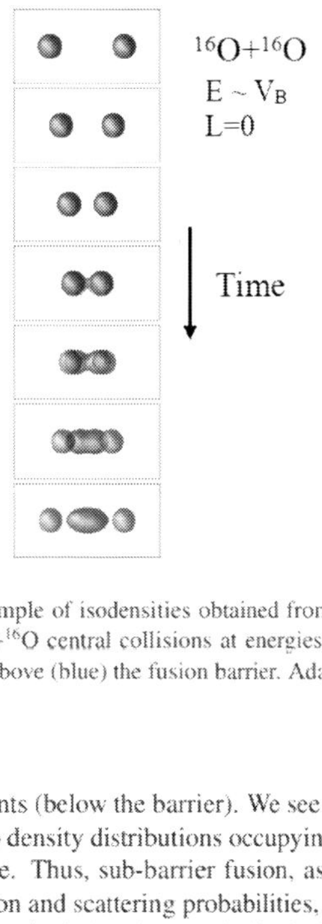

Figure 1. Example of isodensities obtained from TDHF calculations of $^{16}O+^{16}O$ central collisions at energies slightly below (red) and just above (blue) the fusion barrier. Adapted from [9].

A (temporary) solution is to reduce the many-body description to a two-body system (though it often oversimplifies the problem) where the fusion probability is computed from the transmission through a microscopically derived nucleus-nucleus potential (see, e.g., Refs. [10–13]).

* e-mail: cedric.simenel@anu.edu.au

In the present contribution we review recent microscopic methods to determine the nucleus-nucleus potential and use them to predict fusion cross-sections. First, we discuss the standard frozen Hartree-Fock (FHF) approach in section 2. We then extend this approach in section 3 to incorporate the effect of the Pauli exclusion principle between nucleons belonging to different collision partners using a new density-constrained extension of the Frozen-Hartree-Fock (DCFHF) technique. The effect of Pauli repulsion on deep sub-barrier fusion is then discussed in section 4. Finally, dynamical effects are studied in section 5 using the density-constrained time-dependent Hartree-Fock (DCTDHF) method. In particular, the isovector contributions to the dynamical potential are used to get a deeper insight into the effect of transfer on fusion.

2 The frozen Hartree-Fock (FHF) method

A standard approach is to compute the bare nucleus-nucleus potential from frozen static Hartree-Fock (HF) ground-state densities [14, 15] using the energy density functional approach of Brueckner *et al.* [16]. The HF ground-state of a nucleus is obtained by solving the variational principle

$$\delta\langle\Phi|\hat{H}|\Phi\rangle = 0, \tag{1}$$

where $|\Phi\rangle$ is an independent many-fermion state which can be written as a Slater determinant of the occupied single-particle states, ensuring a full account of the Pauli exclusion principle in the ground-state density.

At the mean-field level, the state $|\Phi\rangle$ contains the same information on the system as the one-body density matrix ρ. The energy of the system can then be written as an energy density functional (EDF) $E[\rho]$. In practice, the Skyrme EDF [17] is often used, where the energy density $\mathcal{H}(\mathbf{r})$ only depends on the local part of ρ and of its derivatives,

$$E[\rho] = \int d^3r\ [\rho(\mathbf{r}), \tau(\mathbf{r}), \mathbf{J}(\mathbf{r}) \cdots], \tag{2}$$

where $\rho(\mathbf{r})$, $\tau(\mathbf{r})$, and $\mathbf{J}(\mathbf{r})$ are the local particle, kinetic, and spin-orbit densities, respectively. These densities are symmetric under time-reversal transformations. For systems for which this symmetry does not hold, other densities such as the current density $\mathbf{j}(\mathbf{r})$ need to be taken into account.

The method proposed by Brueckner [16] to get the potential (assumed to be central) between the nuclei at a distance R is to compute the total energy of the system from the sum of the densities and subtracting the individual ground-state energies:

$$V(R) = E[\rho_1 + \rho_2] - E[\rho_1] - E[\rho_2] \tag{3}$$

where ρ_1 and ρ_2 are the HF ground-state one-body density matrices of the nuclei separated by a distance R.

It is important to note that, although the Pauli exclusion principle is properly accounted for between nucleons belonging to the same nucleus, it is neglected between nucleons of different nuclei. The problem comes from the fact that the total system is described by summing densities instead of building a properly antisymmetrised many-body wave-function.

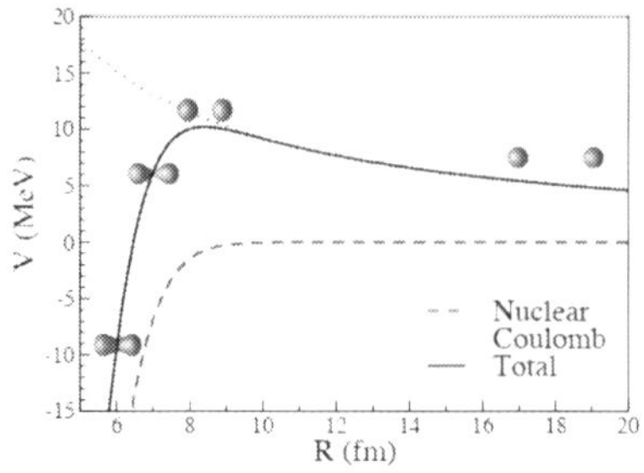

Figure 2. Frozen Hartree-Fock nucleus-nucleus potential of $^{16}O+^{16}O$ with the SLy4*d* Skyrme parameterization.

An example of such frozen Hartree-Fock (FHF) potential is shown in Fig. 2 for $^{16}O+^{16}O$. The nuclear part is computed from the SLy4*d* parametrisation [18] of the Skyrme functional and the Coulomb potential includes both the direct and exchange (using the Slater approximation) contributions.

The surfaces represent isodensities at half the saturation density $\rho_0/2 = 0.08$ fm^{-3}. We see that, at the barrier, the nuclei are still relatively far from each other, indicating that the overlap between their densities is small. The Pauli exclusion principle is thus expected to play a minor role near the barrier in such light systems. Well inside the barrier, however, we see that the spatial overlap between the nuclei is more significant, and is then expected to lead to a "Pauli repulsion" [19].

3 Pauli repulsion

In the past, Pauli repulsion has been accounted for in the nucleus-nucleus bare potential with various methods. For instance, a direct antisymmetrisation of the overlapping wave-functions (e.g., with a Gram-Schmidt algorithm), has been considered [19–21]. The problem with this technique is that it can potentially reduce the neck density inducing too large Pauli repulsion [22]. Another traditional method is to increase the kinetic density $\tau(\mathbf{r})$ (e.g., via the Thomas-Fermi model) [20, 21]. This method, however, neglects the effect of the Pauli exclusion principle on other terms of the functional, such as the spin-orbit term which has been shown to absorb a large part of the Pauli repulsion [22]. Thus, the Pauli exclusion principle has a more complicated effect than just increasing the kinetic energy of the nucleons.

3.1 The density-constrained frozen Hartree-Fock (DCFHF) method

The density-constrained frozen Hartree-Fock (DCFHF) approach was proposed in Ref. [22]. It is a new method to calculate microscopic bare nucleus-nucleus potentials

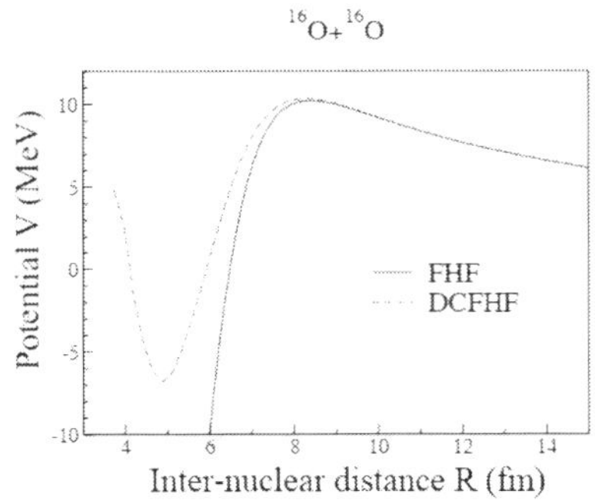

Figure 3. Comparison of the FHF (solid line) and DCFHF (dashed line) potentials in $^{16}O+^{16}O$ with the SLy4d parameterization.

taking into account the Pauli exclusion principle exactly. It can be seen as the static counterpart of the well established density-constrained time-dependent Hartree-Fock (DCTDHF) method to compute instantaneous dynamical potentials in heavy-ion collisions [23] (see section 5).

The DCFHF method is based on a similar variational principle as in Eq. (1) with additional constraints on the proton and neutron local densities,

$$\delta\langle\Phi|\left[\hat{H} - \int d^3r\ \lambda(\mathbf{r})\rho(\mathbf{r})\right]|\Phi\rangle = 0, \quad (4)$$

where the distinction between proton and neutron densities has been omitted for clarity. The Lagrange parameters $\lambda(\mathbf{r})$ constrain the density at each point $\mathbf{r}$ to be the sum of the HF ground-sate densities of the nuclei at a distance R.

The independent particle state $|\Phi\rangle$ thus describes the entire system and is fully antisymmetrised so that the Pauli exclusion principle is accounted for exactly. It is used to compute the DCFHF potential via

$$V_{DCFHF}(R) = \langle\Phi|\hat{H}|\Phi\rangle - E[\rho_1] - E[\rho_2]. \quad (5)$$

Figure 3 shows the FHF and DCFHF potentials in $^{16}O+^{16}O$. As expected, the inclusion of the Pauli exclusion principle has little effect near the barrier due to the small overlap between the nuclei. However, a Pauli repulsion is observed inside the barrier, increasing its width.

3.2 Charge product dependence

It is well known that the Coulomb barrier in light systems is obtained for small overlap between the nuclei (see Fig. 2), while in heavy systems more overlap is required for the strong nuclear interaction to counterbalance the larger Coulomb repulsion. It is then interesting to investigate the charge product dependence of the Coulomb repulsion.

Figure 4 shows a comparison of DCFHF potentials in various systems. We observe an important increase of the Pauli repulsion in the heavier systems. The pocket becomes shallower with increasing charge product Z_1Z_2 and

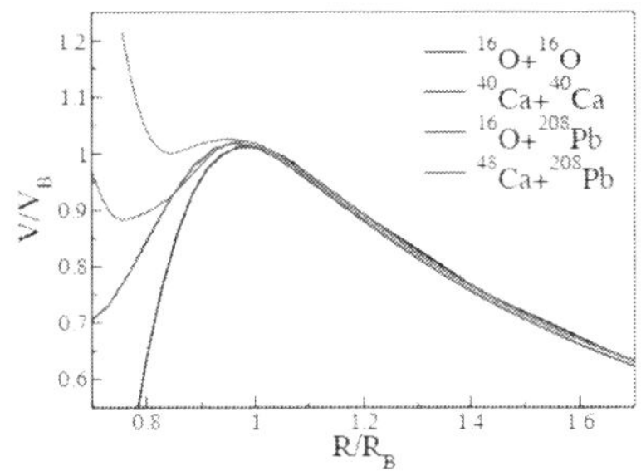

Figure 4. DCFHF potentials in various systems. R_B and V_B denote the FHF barrier radius and height, respectively.

almost disappears in $^{48}Ca+^{208}Pb$. Note that the two-body picture for such heavy systems is questionable. Indeed, the $^{48}Ca+^{208}Pb$ case is extreme as the DCFHF calculation predicts that fusion is impossible at 3% below the barrier. In fact, a smooth transition toward an adiabatic potential for the compound system is expected [24] which would allow fusion to occur at lower energies.

4 Deep sub-barrier fusion hindrance

Wether the pocket inside the barrier is physical or not, one expects an increase of the barrier width due to Pauli repulsion and a resulting reduction of the tunnelling probability. Pauli repulsion could then provide a possible contribution to the deep sub-barrier fusion hindrance observed experimentally [1–3].

4.1 Couplings to low-lying collective states

In order to predict deep sub-barrier fusion cross-sections, it is important to take into account couplings to low-lying collective states. This is done traditionally within the coupled-channel (CC) framework (see Ref. [25] for a recent review). The CC approach starts with an ion-ion potential whose origin does not include any excitations of the nuclei, that is, the "bare" potential. In that sense, FHF and DCFHF potentials, being computed from ground-state densities, can in principle be used in CC calculations.

However, in addition to the repulsion at short distance, the Pauli exclusion principle is expected to change the internal structure of the reactants and could then affect the coupling to low-lying collective states (see discussion in the supplemental material of Ref. [22]). In particular, it could induce a damping mechanism of collective vibration during the fusion process which requires extension of the standard coupled-channel method [26].

Moreover, potentials with Pauli repulsion in medium-mass and heavy systems are often shallow (see Fig. 4), inducing numerical instabilities in CC calculations. This is illustrated in Fig. 5 where CC calculations of $^{16}O+^{208}Pb$ fusion cross-sections down to deep sub-barrier energies have been performed with the CCFULL code [28] using the DCFHF potential with different choices of absorbing

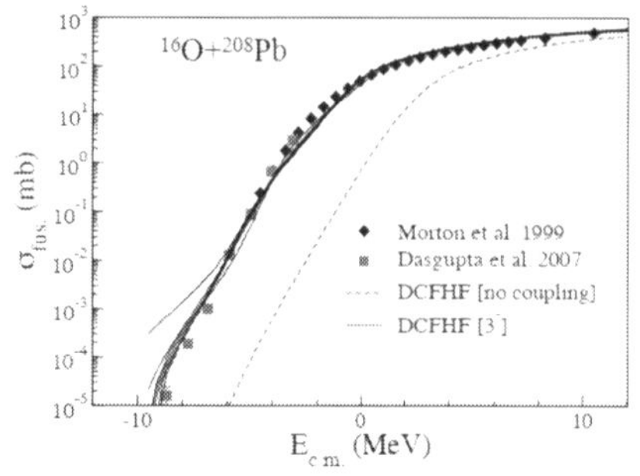

Figure 5. Fusion cross-sections from CCFULL with the DCFHF potential without (dashed line) and with couplings (solid lines) to the first collective 3^- state in target and projectile. The solid lines have been obtained with various choices of absorbing boundary conditions. Experimental data are from Refs. [2, 27].

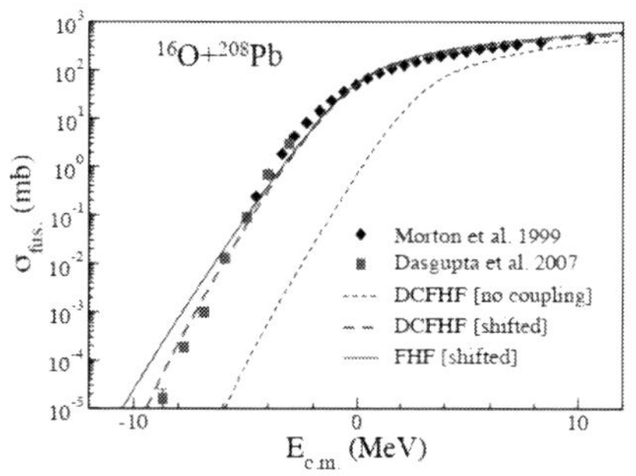

Figure 6. Same as Fig. 5 with shifted cross sections from DCFHF (long dashed line) and FHF (solid lines) potentials to account for couplings.

boundary condition parameters. Although the coupling to the first octupole phonon (3^- state) in projectile and target produces a relatively good overall agreement with experimental data [2, 27], we see that the behaviour at deep sub-barrier energies strongly depends on the choice of absorbing boundary conditions.

Thus, to calculate fusion cross-sections at deep sub-barrier energies, in particular their logarithmic slopes, we use a simpler one-barrier penetration model and an overall shift of the fusion cross-sections in order to account for the overall effect of the couplings as suggested in Ref. [2].

The results are shown in Figs. 6 and 7. We observe a better agreement with experimental data at deep sub-barrier energies for calculations with the DCFHF potential. In particular the logarithmic slope is improved. Nevertheless, discrepancies with experimental data remain, showing that the inclusion of the Pauli repulsion provides only part of the solution to the deep sub-barrier fusion hindrance problem.

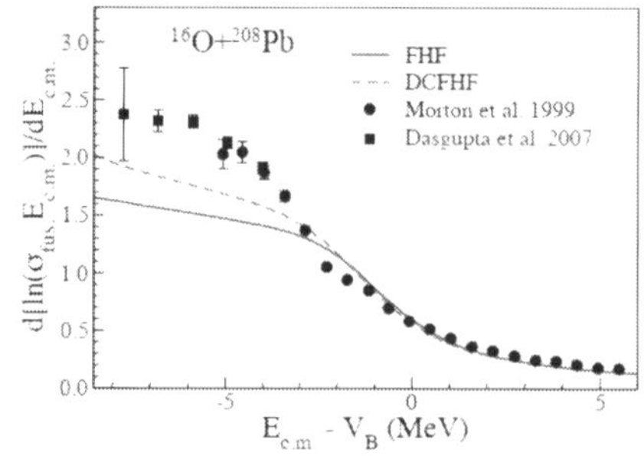

Figure 7. Logarithmic slopes of the fusion cross-sections shown in Fig. 6.

4.2 Other hindrance mechanisms

Let us pause to review briefly the other mechanisms invoked in the literature to explain deep sub-barrier fusion hindrance.

- Mişicu and Esbensen have argued that a pocket was formed inside the barrier due to the incompressibility of nuclear matter and the large density overlap between the nuclei [29].
- Ichikawa, Hagino and Iwamoto suggested that the hindrance could be due to a smooth transition from sudden to adiabatic potentials [30].
- Dasgupta and collaborators invoked a possible decoherence mechanism reducing the effect of the couplings to low-lying collective states [2].
- The ANU group has investigated dissipation mechanisms induced by cluster transfer which could reduce the fusion probability [31, 32].
- Ichikawa and Matsuyanagi have suggested the possibility of a reduction of the collectivity of vibrational modes on the way to fusion [26].

The Pauli exclusion principle induces two additional possible fusion hindrance mechanisms. The first one is due to the widening of the barrier from Pauli repulsion. The second one is the alteration of the couplings and collectivity of the low-lying states (an effect analogous to Ichikawa and Matsuyanagi's idea) due to the change of the internal structure induced by Pauli blocking.

It is likely that more than one of the above effects contribute to the observed fusion hindrance at deep sub-barrier energies. However, the incompressibility is unlikely to play a strong role. Indeed, standard FHF calculations properly account for this effect as they are based on a Skyrme EDF fitted to reproduce the incompressibility of infinite nuclear matter. Yet, these calculations do not show any hindrance (see Figs. 6 and 7). It is more likely that the repulsion potential introduced phenomenologically by Mişicu and Esbensen in [29] is in fact simulating the effect of the Pauli repulsion [22].

5 Dynamical nucleus-nucleus potential

Apart from incompressibility, all the fusion hindrance mechanisms proposed above are in fact dynamical effects, involving time evolution in one way or another. We then naturally turn to a time-dependent generalisation of the DCFHF method to incorporate dynamical effects in the calculation of the potential.

5.1 The density-constrained time-dependent Hartree-Fock method

The density-constrained time-dependent Hartree-Fock (DCTDHF) method has been widely discussed and used in the literature [10, 13, 33–37]. To get a nucleus-nucleus DCTDHF potential, the starting point is a TDHF calculation of a collision above the fusion barrier.

The TDHF equation is the time dependent generalisation of the variational principle (1)

$$\delta \langle \Phi | (\hat{H} - i\partial_t) | \Phi \rangle = 0. \qquad (6)$$

Modern TDHF codes [18, 33, 38–42] have been used in a large number of fusion studies. See Refs. [43–48] for recent applications to fusion. See also recent reviews [49, 50] and contributions to this conference from A. S. Umar, L. Guo, K. Vo-Phuoc, K. Washiyama, G. Scamps, P. D. Stevenson, B. Schuetrumpf, K. Sekizawa and A. Bulgac for applications of TDHF and its extensions.

Figure 1 shows an example of density evolution in $^{16}O+^{16}O$ central collisions. At above barrier energies, a set of densities leading to the compact fused system is obtained. The idea behind DCTDHF is then to use these densities as constraints in Eq. (4) and to compute the potential from Eq. (5). The Pauli exclusion principle is included exactly as the TDHF density is obtained from a fully antisymmetrised state $|\Phi\rangle$ describing the entire system. Note that, as the DCTDHF potential already incorporates dynamical effects, it is not a bare potential and should not be used in coupled-channels calculations. An effect of the couplings between relative motion and internal excitations is to induce an energy dependence of the potential [13].

5.2 (Isovector) transfer couplings

In a recent work [51], we used the following decomposition of the energy density

$$\mathcal{H}(\mathbf{r}) = \frac{\hbar^2}{2m}\tau(\mathbf{r}) + \mathcal{H}_0(\mathbf{r}) + \mathcal{H}_1(\mathbf{r}) + \mathcal{H}_C(\mathbf{r}), \qquad (7)$$

where $\mathcal{H}_0$, $\mathcal{H}_1$ and $\mathcal{H}_C$ are the isoscalar, isovector, and Coulomb contributions, respectively, in order to separate the isoscalar (v_0) and isovector (v_1) contributions to the DCTDHF potential,

$$V(R) = v_0(R) + v_1(R) + V_C(R), \qquad (8)$$

where V_C is the Coulomb potential.

In the static case (e.g., FHF and DCFHF), the isovector potential v_1 vanishes. It is thus entirely induced by

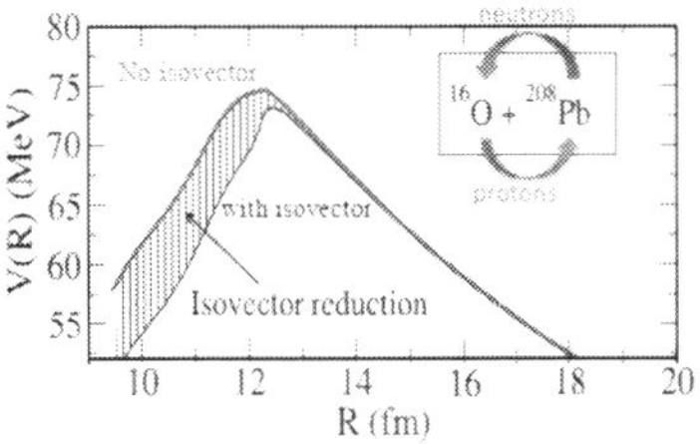

Figure 8. DCTDHF potential in $^{16}O+^{208}Pb$ with (black) and without (red) isovector contribution v_1 from a near-barrier TDHF evolution at $E_{c.m.} = 75$ MeV.

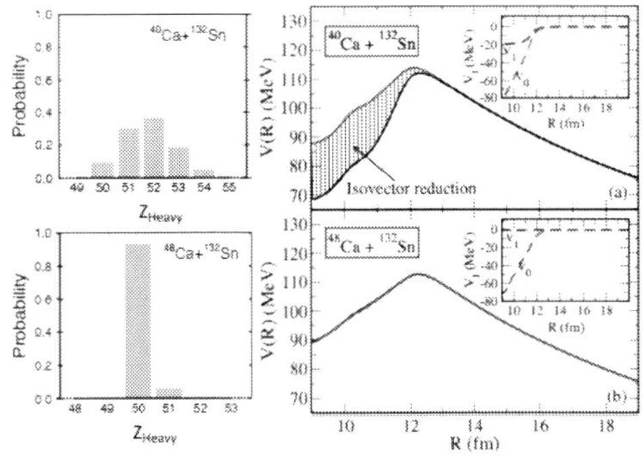

Figure 9. (Right) Same as Fig. 8 for $^{40,48}Ca+^{132}Sn$ reactions at $E_{c.m.} = 120$ MeV. (Left) final proton number probability distributions in the heavy fragment just below the barrier.

dynamical effects. For instance, systems with N/Z asymmetries encounter a rapid charge equilibration (transfer of protons and neutrons in opposite directions) [52] which has a strong impact on v_1 [51]. This is illustrated in Fig. 8 for the $^{16}O+^{208}Pb$ system, where a large reduction of the potential inside the barrier is induced by isovector transfer. The resulting effect on the inner part of the potential is then opposite to the Pauli repulsion.

We then used this technique to investigate the impact of transfer on heavier systems where experimental signatures are not so clear [4, 53]. The isovector reduction of the potential due to transfer channels depends naturally on the presence of positive Q−value transfer channels. As shown in Fig. 9, isovector reduction is observed in $^{40}Ca+^{132}Sn$ which has several positive Q−value transfer channels, but not in $^{48}Ca+^{132}Sn$ which has no positive Q−value transfer channel. This is also confirmed by the proton number distributions in the heavy fragments obtained just below the barrier from a particle number projection technique [54] which show almost no proton transfer in $^{48}Ca+^{132}Sn$ (left panels in Fig. 9; see also Ref. [46] and K. Vo-Phuoc's contribution). A similar effect (not shown in Fig. 9) is observed for neutron transfer.

The above analysis provides an explanation for the fact that fusion in the $^{48}Ca+^{132}Sn$ system is relatively well

 DOI: 10.1051/epjconf/201716300055

explained by standard CC calculations neglecting transfer, whereas they underpredict the fusion cross sections in ^{40}Ca+^{132}Sn [55]. Indeed, transfer is expected to increase sub-barrier fusion in the latter system thanks to a narrowing of the barrier (see Fig. 9-a). As DCTDHF potentials account both for transfer effects and Pauli repulsion, the calculated sub-barrier fusion cross-sections from such potentials are in relatively good agreement with experiment [35], despite the fact that these calculations have no adjustable parameters.

This work has been supported by the Australian Research Council Grants Nos. FT120100760, and by the U.S. Department of Energy under grant Nos. DE-SC0013847.

References

[1] C.L. Jiang *et al.*, Phys. Rev. Lett. **89**, 052701 (2002)
[2] M. Dasgupta *et al.*, Phys. Rev. Lett. **99**, 192701 (2007)
[3] A.M. Stefanini *et al.*, Phys. Rev. C **82**, 014614 (2010)
[4] B.B. Back, H. Esbensen, C.L. Jiang, K.E. Rehm, Rev. Mod. Phys. **86**, 317 (2014)
[5] D. Santiago-Gonzalez *et al.*, EPJ Web of Conferences **117**, 09011 (2016)
[6] R.T. deSouza, S. Hudan, V.E. Oberacker, A.S. Umar, Phys. Rev. C **88**, 014602 (2013)
[7] J. Kolata *et al.*, Nucl. Instr. Meth. A **830**, 82 (2016)
[8] V. Singh *et al.*, Phys. Lett. B **765**, 99 (2017)
[9] C. Simenel, J. Phys. G **41**, 094007 (2014)
[10] A.S. Umar, V.E. Oberacker, Phys. Rev. C **74**, 061601 (2006)
[11] C. Simenel, R. Keser, A.S. Umar, V.E. Oberacker, Phys. Rev. C **88**, 024617 (2013)
[12] C. Simenel, M. Dasgupta, D.J. Hinde, E. Williams, Phys. Rev. C **88**, 064604 (2013)
[13] A.S. Umar, C. Simenel, V.E. Oberacker, Phys. Rev. C **89**, 034611 (2014)
[14] C. Simenel, B. Avez, Intl. J. Mod. Phys. E **17**, 31 (2008)
[15] K. Washiyama, D. Lacroix, Phys. Rev. C **78**, 024610 (2008)
[16] K.A. Brueckner, J.R. Buchler, M.M. Kelly, Phys. Rev. **173**, 944 (1968)
[17] T.H.R. Skyrme, Phil. Mag. **1**, 1043 (1956)
[18] K.–H. Kim, T. Otsuka, P. Bonche, J. Phys. G **23**, 1267 (1997)
[19] T. Fliessbach, Z. Phys. **247**, 117 (1971)
[20] D.M. Brink, F. Stancu, Nucl. Phys. A **243**, 175 (1975)
[21] P.G. Zint, U. Mosel, Phys. Lett. B **56**, 424 (1975)
[22] C. Simenel, A.S. Umar, K. Godbey, M. Dasgupta, D.J. Hinde, Phys. Rev. C **95**, 031601(R) (2017)
[23] A.S. Umar, V.E. Oberacker, Phys. Rev. C **74**, 021601(R) (2006)
[24] T. Ichikawa, K. Hagino, A. Iwamoto, Phys. Rev. Lett. **103**, 202701 (2009)
[25] K. Hagino, N. Takigawa, Prog. Theo. Phys. **128**, 1001 (2012)
[26] T. Ichikawa, K. Matsuyanagi, Phys. Rev. C **92**, 021602 (2015)
[27] C.R. Morton *et al.*, Phys. Rev. C **60**, 044608 (1999)
[28] K. Hagino, N. Rowley, A. Kruppa, Comput. Phys. Commun. **123**, 143 (1999)
[29] Ş. Mişicu, H. Esbensen, Phys. Rev. Lett. **96**, 112701 (2006)
[30] T. Ichikawa, A. Iwamoto, P. Möller, Phys. Rev. C **79**, 014305 (2009)
[31] M. Evers *et al.*, Phys. Rev. C **84**, 054614 (2011)
[32] D.C. Rafferty *et al.*, Phys. Rev. C **94**, 024607 (2016)
[33] A.S. Umar, V.E. Oberacker, Phys. Rev. C **73**, 054607 (2006)
[34] A. S. Umar and V. E. Oberacker and C. J. Horowitz, Phys. Rev. C **85**, 055801 (2012)
[35] V.E. Oberacker, A.S. Umar, Phys. Rev. C **87**, 034611 (2013)
[36] X. Jiang, J.A. Maruhn, S. Yan, Phys. Rev. C **90**, 064618 (2014)
[37] A.S. Umar, V.E. Oberacker, C. Simenel, Phys. Rev. C **92**, 024621 (2015)
[38] C. Simenel, P. Chomaz, G. de France, Phys. Rev. Lett. **86**, 2971 (2001)
[39] L. Guo, T. Nakatsukasa, EPJ Web. Conf. **38**, 09003 (2012)
[40] J.A. Maruhn, P.G. Reinhard, P.D. Stevenson, A.S. Umar, Comp. Phys. Comm. **185**, 2195 (2014)
[41] K. Sekizawa, K. Yabana, Phys. Rev. C **88**, 014614 (2013)
[42] G. Scamps, D. Lacroix, Phys. Rev. C **87**, 014605 (2013)
[43] K. Washiyama, Phys. Rev. C **91**, 064607 (2015)
[44] M. Tohyama, A.S. Umar, Phys. Rev. C **93**, 034607 (2016)
[45] D. Bourgin, C. Simenel, S. Courtin, F. Haas, Phys. Rev. C **93**, 034604 (2016)
[46] K. Vo-Phuoc, C. Simenel, E.C. Simpson, Phys. Rev. C **94**, 024612 (2016)
[47] A.S. Umar, V.E. Oberacker, C. Simenel, Phys. Rev. C **94**, 024605 (2016)
[48] S. Long, L. Guo, Nucl. Phys. Rev. **34**, 41 (4) (2017)
[49] C. Simenel, Eur. Phys. J. A **48**, 152 (2012)
[50] T. Nakatsukasa, K. Matsuyanagi, M. Matsuo, K. Yabana, Rev. Mod. Phys. **88**, 045004 (2016)
[51] K. Godbey, A.S. Umar, C. Simenel, Phys. Rev. C **95**, 011601(R) (2017)
[52] C. Simenel *et al.*, Phys. Lett. B **710**, 607 (2012)
[53] J.F. Liang *et al.*, Phys. Rev. C **94**, 024616 (2016)
[54] C. Simenel, Phys. Rev. Lett. **105**, 192701 (2010)
[55] J.J. Kolata *et al.*, Phys. Rev. C **85**, 054603 (2012)

 DOI: 10.1051/epjconf/201716300056

Classical dynamical modelling of near-barrier breakup

E. C. Simpson[1,*], *K. J.* Cook[1], *Sunil* Kalkal[1], *D. H.* Luong[1], *I. P.* Carter[1], *M.* Dasgupta[1], and *D. J.* Hinde[1]

[1] *Department of Nuclear Physics, Research School of Physics and Engineering, Australian National University, Canberra, ACT 2601 Australia*

Abstract. The complete fusion of light, weakly-bound nuclides is known to be significantly suppressed with respect to comparable well-bound projectiles or with respect to single barrier penetration model calculations. Strong α-clustering in these light systems mean that they very easily disintegrate into clusters, either via direct excitation of their intrinsic cluster continuum, or via transfer reactions which connect to unbound states in neighbouring nuclides. This breakup is thought to reduce the probability for complete fusion. Here we discuss which processes cause breakup, whether or not breakup happens fast enough, and the interpretation of measurements made at the Australian National University of breakup using classical dynamical models. Understanding the intimate details of breakup, and the resonances through which it proceeds, will be crucial in determining its likely influence on fusion.

1 Introduction

Heavy-ion fusion is necessarily a many-body process. It involves dissipation of the collision kinetic energy into a multitude of internal degrees of freedom, in order to form a single compound system. It is an ideal testing ground for quantum dynamics and many-body tunnelling. Despite this, couplings to low energy collective states, or transfer of individual nucleons between the projectile and target, can have profound effects on fusion cross sections [1]. To understand fusion, it is crucial to understand the interplay between these low excitation, few-body degrees of freedom, and the complex dissipative effects ultimately required for fusion.

Few-body effects are particularly prevalent in near-barrier reactions of light weakly-bound nuclides. The strong binding of the α-particle makes nuclides with $A <$ 10 intrinsically fragile. The lowest cluster separation thresholds for the stable nuclides ^{6}Li, ^{7}Li and ^{9}Be are all less than 2.5 MeV [2]. This makes them liable to breakup in nuclear collisions, which in turn is believed to have importance consequences for fusion. Indeed, above barrier complete fusion of light, weakly-bound projectiles with heavy targets is known to be suppressed by 25-35%, when compared to either coupled channels calculations or comparable examples with well-bound projectiles [3, 4]. It is thought that breakup processes reduce the probability for complete capture of the projectile's charge. A reduction of the complete fusion cross sections is associated with increased incomplete fusion — usually interpreted as capture of one of the breakup fragments following breakup.

Exactly how and why fusion is suppressed in these nuclides is a topic of very active research (see Ref. [5] for a recent review). FUSION17 encompassed a large number of talks addressing different aspects of the reactions of these light weakly-bound nuclides, including measurements of fusion and reaction cross sections (Aguilera [6]), quasielastic scattering (Mukherjee), charged-particle coincidences (Kalkal [7], Cook [8], G. L. Zhang [9]), and attempts to model incomplete fusion (Parkar [10]).

As we progress towards more exotic nuclides, understanding the consequences of weak-binding on fusion becomes increasingly important. At this conference seen talks experimental measurements of reactions of ^{7}Be (Mazzocco), ^{8}B (Kolata [11]), and the halo nucleus ^{11}Be (Pesudo [12]), in addition to time-dependent Hartree-Fock calculations for exotic calcium isotopes (Vo-Phouc [13]). Innovative technical developments were also presented, such as new methods for extraction of fusion excitation functions with stacked targets, which will soon be utilized in experiments with exotic beams (Di Pietro [14, 15]).

What has emerged from recent works with stable beams is that the breakup mechanisms of these nuclides are complex and diverse [16–18]. Here we discuss these reactions, not aiming to quantify to what degree fusion is suppressed, but to understand the manner in which breakup could suppress fusion. We consider sub-barrier breakup reactions, where absorption of the breakup fragments through incomplete fusion is (nearly) eliminated, allowing us to focus exclusively on breakup. In Section 2 we discuss the mechanisms which cause breakup, before considering whether or not breakup happens fast enough to influence fusion in Section 3. Then in Section 4 we discuss classical dynamical models of breakup, which can provide further insights into the detail and timescales of breakup. In Section 5 we discuss these results and the out-

*e-mail: edward.simpson@anu.edu.au

DOI: 10.1051/epjconf/201716300056

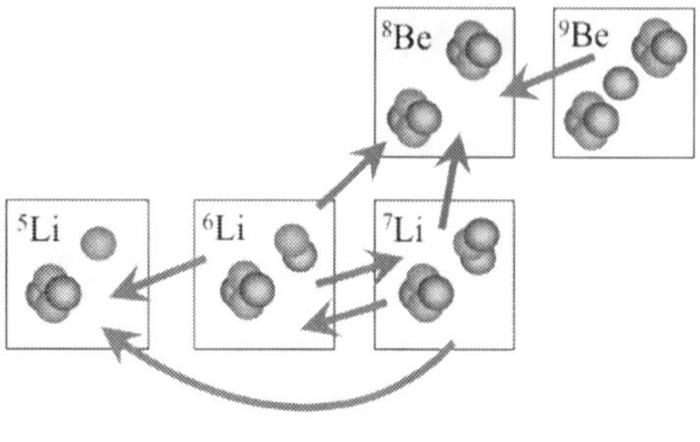

Figure 1. (Colour online) Diagram illustrating the possible transfers that lead to breakup in the reactions of ^{6}Li, ^{7}Li and ^{9}Be. Blue lines indicate neutron transfers, red lines indicate proton transfers, and the purple line indicates np transfer. ^{5}Li and ^{8}Be are both unbound, and though ^{6,7}Li and ^{9}Be are bound, their cluster breakup thresholds are low.

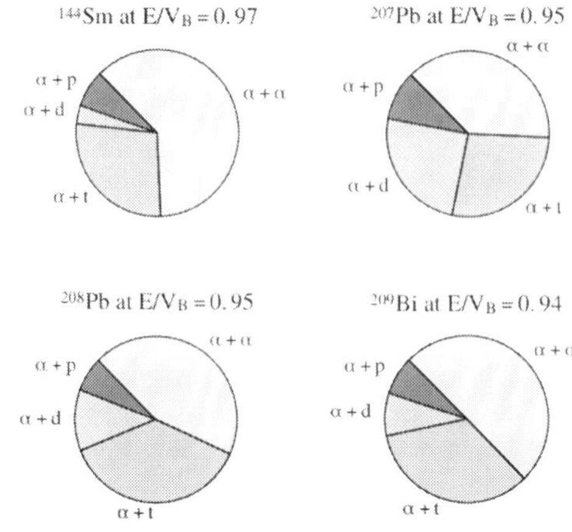

Figure 2. Fractional contribution to breakup for interactions of ^{7}Li with ^{144}Sm, ^{207}Pb, ^{208}Pb and ^{209}Bi [18]. The modes shown are direct breakup (teal), proton pickup (green), neutron stripping (pink) and two-neutron stripping (red). Note that the effect of narrow resonaces has been excluded — see the text for further details.

look for classical dynamical models and investigations of fusion suppression.

2 What processes cause breakup?

The first question we must ask is what mechanism causes light weakly-bound nuclides break up? The strong binding of the α-particle means that light nuclides with $A < 10$ have low cluster breakup thresholds: ^{6}Li$\rightarrow \alpha + d$ at 1.474 MeV, ^{7}Li$\rightarrow \alpha + t$ at 2.468 MeV, and ^{9}Be$\rightarrow \alpha + \alpha + n$ at 1.665 MeV. Direct excitation of the intrinsic cluster continuum of the projectile is one possible trigger of breakup. Indeed, many works have noted a correlation between the direct breakup threshold and the degree of fusion suppression [19, 20]. This correlation is apparent not only for weakly-bound stable nuclides such as ^{6,7}Li, ^{9}Be and ^{6}He, but also for more well-bound nuclides such as ^{10,11}B, ^{12,13}C and ^{16}O. So on the face of it we might conclude that direct breakup causes fusion suppression.

The reality is not quite so simple. Whilst direct excitation of the intrinsic cluster continua does contribute to breakup, strong α-clustering in this region means that many neighbouring nuclides are also very weakly bound or unbound, and transfer reactions can connect these different cluster states. This is illustrated in Figure 1 for ^{6}Li, ^{7}Li and ^{9}Be. For ^{7}Li projectiles, neutron stripping can give unbound states in ^{6}Li resulting in $\alpha + d$, two-neutron stripping gives unbound ^{5}Li and results in $\alpha + p$, and proton pickup gives unbound ^{8}Be and $\alpha + \alpha$. Similar transfer-induced breakup modes exist for ^{6}Li and ^{9}Be.

Recent measurements of charged particle coincidences have revealed that these transfer-induced breakup modes are indeed important for ^{6,7}Li and ^{9}Be [16–18]. The fractions for the different breakup modes are illustrated in Figure 2. These results include an efficiency correction which depends on the breakup mode, and though this efficiency correction was improved upon in later work (see Ref. [8], we do not anticipate that this change will significantly alter the deduced breakup fractions. It should also be noted that the fractions shown exclude breakup from long-lived resonances, as only prompt breakup is expected to result in fusion suppression. In all cases direct breakup into $\alpha + t$ is significant, but the strongest prompt breakup mode is proton pickup to produce $\alpha + \alpha$. As the energy increases towards the barrier energy, the fractional contribution of these transfer-induced breakup modes tends to increase [18].

Whilst the fraction of direct breakup is reasonably constant across the targets presented in Figure 2, the relative strengths of the transfer-induced breakup mechanisms vary markedly. Significant differences occur even when the target is changed by just a single nucleon, driven by changes in transfer Q-values. For lighter targets, where the Coulomb field is much smaller, direct breakup is found to be considerably reduced for ^{6}Li, and negligible for ^{7}Li [7].

3 Does breakup happen fast enough?

Knowing the relevant breakup mechanism is, however, insufficient. In addition to know how breakup occurs, we must also consider whether it occurs fast enough to be able to influence fusion. In a classical sense, if breakup is to have a chance of suppressing fusion, the distinct clusters must be produced before the projectile-like object reaches the fusion barrier at R_B. If breakup occurs only once the projectile (or projectile-like reaction product) has passed inside the barrier, it is unlikely that either breakup fragment will escape fusion with the target, effectively resulting in complete fusion.

Here our focus will be on sub-barrier breakup measurements. At sub-barrier energies, the classical trajectory of the projectile and target never reaches the fusion barrier at R_B. As a result, the probability of fusion of the fragments after breakup (i.e. incomplete fusion) is neg-

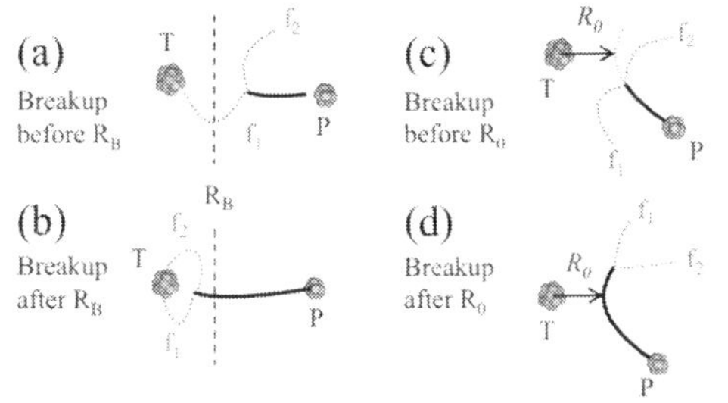

Figure 3. Illustration of breakup at above-barrier energies [(a) and (b)] and sub-barrier energies [(c) and (d)]. (a) At above-barrier energies breakup prior to the projectile reaching the fusion barrier has the capacity to suppress fusion, whilst (b) breakup once the projectile has already passed inside R_B cannot. (c) At sub-barrier energies breakup prior to R_B [i.e. (a)] corresponds to breakup prior to the distance of closest approach, whereas (d) breakup after R_B corresponds to breakup after the distance of closest approach has been passed.

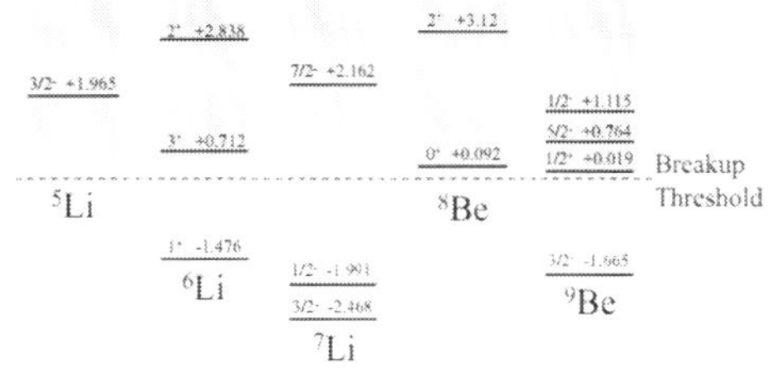

Figure 4. Sketch of the structure of light nuclei relative to their lowest breakup threshold. Horizontal black lines indicate known bound states and resonances. For resonances where the width is substantial, the approximate width is indicated by the red shaded region around the state.

ligible, allowing the clearest possible picture of breakup to be obtained. To affect fusion at above-barrier energies, breakup must occur prior to R_B. This loosely translates, at sub-barrier energies, into breakup occurring when the projectile is still approaching the distance of closest approach R_0. This is illustrated in Figure 3.

Given the variety of different mechanisms that can trigger breakup, it is necessary to consider not only the structure of the light nuclides in question (^{6}Li, ^{7}Li and ^{9}Be), but also their close neighbours (^{5}Li and ^{8}Be). This is illustrated in Figure 4. In each case, known bound states and resonances are plotted relative to the lowest breakup threshold. Unbound states, found in the pink shaded section are marked by a black line and, if their width is substantial, a shaded red region indicates their approximate width. The density of resonant states is low, but it is possible also that non-resonant partial waves may contribute to breakup. The relative importance of different resonant states for a given breakup mode, may also depend on the mechanism — $\alpha + d$ produced via direct breakup of ^{6}Li may not be the same as $\alpha + d$ produced via neutron stripping from ^{7}Li.

What has emerged from recent work is that the structure of these nuclides is, in fact, crucial. For example, suppose the breakup occurs via a (moderately) narrow resonant state, such as the 3^+ resonance in ^{6}Li. This state, with a width of 24 keV, has a mean lifetime of 27×10^{-20} s, more than an order of magnitude longer than the interaction timescale. In an above-barrier collision, it is unlikely to break into fragments (in this case $\alpha + d$) before reaching the fusion barrier. This breakup is referred to as asymptotic since, at sub-barrier energies, narrow resonances will only decay into fragments once they have travelled far from the target. This cannot cause fusion suppression. Indeed, if the state is quasi-bound in this sense, it should act as any other coherent inelastic coupling. We are therefore interested in identifying breakup that occurs promptly, on the timescale of the close interaction of the projectile and target. Prompt breakup may proceed through these wide resonant states, or via non-resonant parts of the continuum — for example, in the case of ^{6}Li, where there are no broad resonances.

One way to experimentally distinguish prompt and asymptotic breakup is via the relative energy distribution E_{rel} of the two breakup fragments. As asymptotic breakup occurs far from the target, the interaction of the fragments and target is weak, and their relative energy remains characteristic of the state in question and can be cleanly gated out. This was the method used in Ref. [18] to exclude asymptotic breakup from the fractions given in Figure 2. We emphasize that identifying and excluding asymptotic breakup fundamentally requires coincidence measurements — singles measurements cannot help here.

Separating prompt and asymptotic breakup is, however, insufficient. As discussed earlier, at sub-barrier energies we wish to distinguish prompt breakup occurring before and after the distance of closest approach. Prompt breakup that occurs after the projectile is receding from the target corresponds to breakup after R_B and will thus not contribute to fusion suppression.

4 Classical dynamical models of breakup

The complexity of the reactions of ^{6}Li, ^{7}Li and ^{9}Be make them difficult to treat in a complete fashion (i.e. including all channels) with quantum models. Not only must we treat nucleon transfer, but also the coupling between different continuum states of the resulting breakup product. Furthermore, we cannot describe incomplete fusion with such models — fusion, treated using an absorptive potential, makes no distinction between complete and incomplete fusion.

One alternative approach is to use stochastic classical dynamical models [21]. In the model, the classical trajectories of the projectile and target are propagated in time, treating the breakup stochastically. The trajectories of the fragments and target-like reaction product are tracked, allowing the energy and angle correlations of the breakup fragments to be determined, and the cross sections for incomplete fusion to be estimated. Due to some important

 DOI: 10.1051/epjconf/201716300056

physical and technical limitations of existing implementations of the model [22], a new code has been written. Here we give a brief overview of the model and recent developments of it, and discuss recent results.

4.1 Recent developments

The initial condition for each simulation is set by stochastically sampling an impact parameter b below some user defined maximum b_{max}. This differs from the previous implementation [22], where a quantized angular momentum L was sampled. The use of an impact parameter b is more consistent with the classical nature of the model. The projectile and target trajectories are then tracked, assuming their interaction is described with a Coulomb plus Woods-Saxon nuclear potential. The parameters of these potentials are usually constrained to fit the Sao Paulo potential [23], or, in the case of the breakup fragments, light-ion scattering (e.g., $\alpha + \alpha$).

The reaction causing breakup is treated as a stochastic process, with the probability of the reaction assumed to depend exponentially on the distance between the projectile and target. The details of this exponential dependence (the slope and normalisation) are determined to match the experimentally measured breakup function, which is just the breakup probability as a function of distance of closet approach [18, 24]. At the point of reaction, a distribution of excitations E_x of the projectile-like nucleus, and corresponding lifetimes τ for the intermediate nucleus (e.g., ^{8}Be), are randomly sampled. In the case of proton pickup by ^{7}Li, the lifetime represents how long it takes the ^{8}Be* resonance state to decay into two α particles. In the case of the ^{8}Be ground state this time is long — of the order of 10^{-16} s — but it is much shorter in the case of the ^{8}Be 2^+. Here, these distributions are described using a one-level one-channel limit of R-matrix theory [25, 26]. This is an important improvement over Ref. [22], which assumed either a flat or exponential distribution with E_x, and did not explicitly account for the lifetimes of the intermediate nucleus.

The positions of the projectile-like and target-like nuclei are then propagated until the randomly sampled lifetime τ has elapsed. There is therefore a distinction between the point of reaction and point of decay (or fragment production). At the latter point, the projectile-like nucleus spontaneously decays into two fragments. The initial condition of the two fragments is such that their initial velocity corresponds to an asymptotic relative energy given by E_x, and their initial position is at their classical turning point for that energy, or their mutual barrier radius, whichever is larger. The fragments and target-like nucleus are then propagated until all nuclei are sufficiently well separated.

The simulations build up a picture of the expected fragment-fragment correlations by Monte Carlo sampling the collision impact parameter, the reaction point, the excitation and lifetime of the intermediate system, and the initial orientation of the breakup fragments. Applications to transfer-triggered breakup through ^{8}Be [8] and direct breakup of ^{6}Li and ^{7}Li [7] have been made, with quite good agreement with experimental observables.

4.2 Results for ^{7}Li+^{58}Ni

We now focus on the case of ^{7}Li+^{58}Ni, disintegrating via the proton-pickup breakup mode ($\alpha + \alpha$) in near-barrier collisions. Experimental data from Ref. [27] and the results of initial calculations are shown in Figure 5. The calculations are filtered by the detector coverage from the experiment such that they can be directly compared with the experimental data. The data is plotted as a function of two angles: θ_{12}, the opening angle of the velocity vectors of the two fragments, and β, the orientation of their relative motion with respect to the motion of their centre of mass. These angles are illustrated in the bottom left of Figure 5, and give an accessible way to view the fragment correlations, and their sensitivity to target proximity in particular [27]. Distant from the target nucleus, and for a well-defined ^{8}Be excitation energy, there is a distinctive correlation between β and θ_{12}. For prompt breakup, this correlation is distorted by the interaction of the α-particles and the target. The experimental data [Figure 5(a)] show two distinctive bands — one at $\theta_{12} < 15°$, and one extending from ($\theta_{12} = 0°, \beta = 0°$) to ($\theta_{12} = 70°, \beta = 90°$). The first of these is associated with the ^{8}Be ground state, and follows exactly what one would expect of asymptotic breakup. The second band is associated with prompt breakup of the ^{8}Be 2^+ state. For further details, see Ref. [27, 28].

The previous work [27] illustrated the results of the model by focusing on breakup occuring at fixed distances prior to and after the distance of closest approach, without explicitly considering the lifetime of the ^{8}Be state involved. Here we show complete Monte Carlo calculations where the impact parameters, reaction distance, the ^{8}Be excitation energies and lifetimes, and the initial orientation of the α fragments are all stocastically sampled. In Figure 5(d) we show the coordinates at which the reaction was determined to occur ($T_{Reaction}, R_{Reaction}$) and in Figure 5(e) the point at which ^{8}Be decayed into two α particles (T_{Decay}, R_{Decay}). The calculations are consistent with earlier calculations from Ref. [27] which concluded that much of the breakup observed comes from events where the ^{8}Be is receding from the target. Here, however, this is explicitly because the ^{8}Be is allowed to propagate for some time — the duration of which is consistent with the structural properties of the 3 MeV 2^+ state — before decaying into two α particles. Crucially, it seems essential to delay the breakup in order to reproduce the band of events seen.

The agreement with the experimental data is not perfect. One can clearly see an excess of events near $\beta = 90°$ in Figure 5(c) when compared to the experimental data shown in Figure 5(a). If one compares Figure 5(c) to Figure 5(b) (instantaneous breakup), it suggests that the excess of events is caused from too much breakup occurring near the distance of closest approach, despite the delay of breakup incorporated into the simulations. We discuss some of the possible causes and consequences of this below.

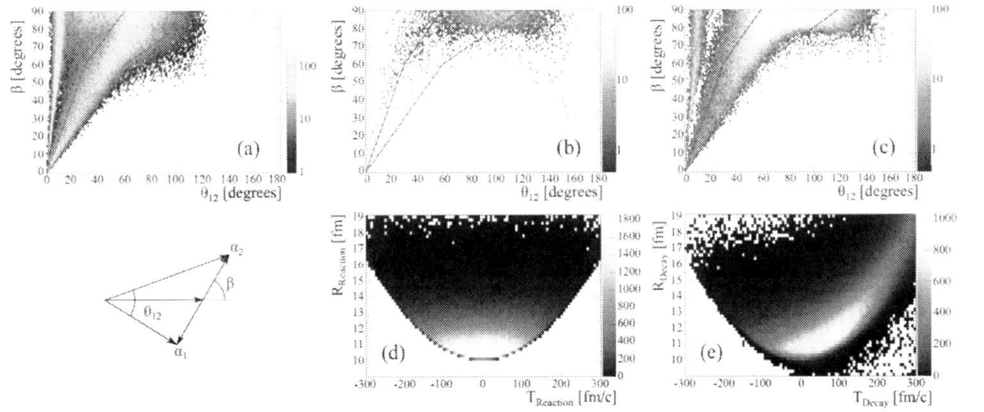

Figure 5. Experimental results and simulations for the ^{7}Li+^{58}Ni reaction. (a) Experimental data (see Ref. [27] for further details). (b) Simulations assuming that breakup is instantaneous at the point of reaction. (c) Simulations incorporating the lifetime of ^{8}Be. (d) Simulated time (with respect to the time of distance of closest approach) and projectile-target distance of the transfer reaction forming ^{8}Be. In the simulations (b) where breakup is instantaneous, this also corresponds to the breakup point. (e) Times and positions of breakup for the case where the ^{8}Be lifetime is included. In plots (a-c) red lines are drawn to guide the eye. See text for further discussions.

5 Discussion and outlook

As demonstrated by the abundance of talks at this conference, breakup of the weakly-bound stable nuclides ^{6}Li, ^{7}Li and ^{9}Be, and the relationship between breakup and fusion suppression, is a very active topic of research. Here we have focused on the nature of breakup, and exactly how and why it can influence fusion. The broad conclusion is that each projectile is different, with a variety of mechanisms triggering breakup. Inferring the likely influence on fusion requires a very detailed understanding of the mechanisms triggering the breakup, and on the structure of any intermediate nuclides in particular. We discussed classical dynamical simulations of breakup, focusing on the case of breakup with ^{7}Li projectiles via proton pickup. The simulations indicate that even short-lived resonant states, such as the ^{8}Be 2^{+}, may survive for a sufficiently long time so as to not be able to influence fusion significantly at above barrier energies.

The results also hint that there is some stabilisation of ^{8}Be that prevents it breaking up when near the distance of closest approach. The nature of this, and the cause is unclear. It may be that the present structure model for ^{8}Be is not good enough, or that the Coulomb field of the target changes the structure of ^{8}Be, possibly causing it to be stabilised. A similar phenomenon has been hinted at in ^{8}Be emission following heavy-ion collisions [29].

Another possible cause of disagreement could be polarization of the ^{7}Li cluster projectile by its interaction with the target. Classically, the different charge to mass ratios of the two intrinsic clusters (α and triton) could cause them to orient themselves with respect to the Coulomb field. This may have some impact on the observed distribution of events with β and θ_{12} A related question would be that of tidal effects once the reaction triggering breakup has occurred, but prior to production of the breakup fragments. In the present purely classical approach the intermediate resonance state is treated as immutable until it decays into two fragments. For narrow states, this is perhaps a reasonable approximation, but it is less clear that broad and non-resonant states — i.e. those contributing to the prompt breakup — will be similarly unaffected.

It is not possible to sensibly answer all these questions within the scope of classical dynamical models, which are inherently limited by their reliance on classical mechanics, and cannot predict the strength of these various effects. Classical dynamical models can, however, give some idea of the likely impacts of these effects on the breakup fragment correlations and on above-barrier complete and incomplete fusion. What is clear is that we must go beyond identifying those cases where fusion suppression occurs, and develop a more sophisticated understanding of the near-barrier breakup of weakly-bound nuclides. In near-future reaction studies with dripline nuclides, this will be absolutely essential.

Acknowledgements

This work was supported by Australian Research Council Grant Nos. FT120100760 and DP170102423.

References

[1] J.R. Leigh, M. Dasgupta, D.J. Hinde, J.C. Mein, C.R. Morton, R.C. Lemmon, J.P. Lestone, J.O. Newton, H. Timmers, J.X. Wei et al., Phys. Rev. C **52**, 3151 (1995)

[2] G. Audi, M. Wang, A. Wapstra, F. Kondev, M. MacCormick, X. Xu, B. Pfeiffer, Chinese Physics C **36**, 1287 (2012)

DOI: 10.1051/epjconf/201716300056

[3] M. Dasgupta, D.J. Hinde, K. Hagino, S.B. Moraes, P.R.S. Gomes, R.M. Anjos, R.D. Butt, A.C. Berriman, N. Carlin, C.R. Morton et al., Physical Review C **66**, 041602(R) (2002)
[4] M. Dasgupta, P.R.S. Gomes, D.J. Hinde, S.B. Moraes, R. M.Anjos, A.C. Berriman, R.D. Butt, N. Carlin, J. Lubian, C.R. Morton et al., Physical Review C **70**, 024606 (2004)
[5] L.F. Canto, P.R.S. Gomes, R. Donangelo, J. Lubian, M.S. Hussein, Physics Reports **596**, 105 (2015)
[6] E.F. Aguilera, E. Martinez-Quiroz, D. Lizcano, A. Gómez-Camacho, J.J. Kolata, L.O. Lamm, V. Guimarães, R. Lichtenthäler, O. Camargo, F.D. Becchetti et al., Phys. Rev. C **79**, 021601 (2009)
[7] S. Kalkal, E.C. Simpson, Physical Review C **93**, 044605 (2016)
[8] K.J. Cook, E.C. Simpson, D.H. Luong, S. Kalkal, M. Dasgupta, D.J. Hinde, Phys. Rev. C **93**, 064604 (2016)
[9] S.P. Hu, G.L. Zhang, J.C. Yang, H.Q. Zhang, P.R.S. Gomes, J. Lubian, J.L. Ferreira, X.G. Wu, J. Zhong, C.Y. He et al., Phys. Rev. C **93**, 014621 (2016)
[10] V.V. Parkar, V. Jha, S. Kailas, Phys. Rev. C **94**, 024609 (2016)
[11] E.F. Aguilera, P. Amador-Valenzuela, E. Martinez-Quiroz, J. Fernández-Arnáiz, J.J. Kolata, V. Guimarães, Phys. Rev. C **93**, 034613 (2016)
[12] V. Pesudo, M.J.G. Borge, A.M. Moro, J.A. Lay, E. Nácher, J. Gómez-Camacho, O. Tengblad, L. Acosta, M. Alcorta, M.A.G. Alvarez et al., Phys. Rev. Lett. **118**, 152502 (2017)
[13] K. Vo-Phuoc, C. Simenel, E.C. Simpson, Phys. Rev. C **94**, 024612 (2016)
[14] M. Fisichella, A.C. Shotter, A. Di Pietro, P. Figuera, M. Lattuada, C. Marchetta, V. Privitera, L. Romano, C. Ruiz, M. Zadro, Phys. Rev. C **92**, 064611 (2015)
[15] M. Fisichella, A.C. Shotter, P. Figuera, J. Lubian, A. Di Pietro, J.P. Fernandez-Garcia, J.L. Ferreira, M. Lattuada, P. Lotti, A. Musumarra et al., Phys. Rev. C **95**, 034617 (2017)
[16] R. Rafiei, R. du Rietz, D.H. Luong, D.J. Hinde, M. Dasgupta, M. Evers, A. Diaz-Torres, Phys. Rev. C **81**, 024601 (2010)
[17] D.H. Luong, M. Dasgupta, D.J. Hinde, R. du Rietz, R. Rafiei, C.J. Lin, M. Evers, A. Diaz-Torres, Physics Letters B **695**, 105 (2011)
[18] D.H. Luong, M. Dasgupta, D.J. Hinde, R. du Rietz, R. Rafiei, C.J. Lin, M. Evers, A. Diaz-Torres, Physical Review C **88**, 034609 (2013)
[19] L.R. Gasques, D.J. Hinde, M. Dasgupta, A. Mukherjee, R.G. Thomas, Physical Review C **79**, 034605 (2009)
[20] B. Wang, W.J. Zhao, P.R.S. Gomes, E.G. Zhao, S.G. Zhou, Physical Review C **90**, 034612 (2014)
[21] A. Diaz-Torres, D.J. Hinde, J.A. Tostevin, M. Dasgupta, L.R. Gasques, Physical Review Letters **98**, 152701 (2007)
[22] A. Diaz-Torres, Computer Physics Communications **182**, 1100 (2011)
[23] L.C. Chamon, D. Pereira, M.S. Hussein, M.A. Cândido Ribeiro, D. Galetti, Phys. Rev. Lett. **79**, 5218 (1997)
[24] D.J. Hinde, M. Dasgupta, B.R. Fulton, C.R. Morton, R.J. Wooliscroft, A.C. Berriman, K. Hagino, Phys. Rev. Lett. **89**, 272701 (2002)
[25] A.M. Lane, R.G. Thomas, Reviews of Modern Physics **30**, 257 (1958)
[26] F.C. Barker, Australian Journal of Physics **41**, 743 (1988)
[27] E.C. Simpson, K.J. Cook, D.H. Luong, S. Kalkal, I.P. Carter, M. Dasgupta, D.J. Hinde, E. Williams, Physical Review C **93**, 024605 (2016)
[28] E.C. Simpson, K.J. Cook, M. Dasgupta, S. Kalkal, D.H. Luong, I.P. Carter, D.J. Hinde, E. Williams, EPJ Web Conf. **123**, 03002 (2016)
[29] A.B. McIntosh, S. Hudan, C.J. Metelko, R.T. de Souza, R.J. Charity, L.G. Sobotka, W.G. Lynch, M.B. Tsang, Phys. Rev. Lett. **99**, 132701 (2007)

EPJ Web of Conferences **163**, 00057 (2017)
FUSION17 DOI: 10.1051/epjconf/201716300057

Superheavy Nuclei in the Quark-Meson-Coupling Model

Jirina Stone[1,*], *Pierre* Guichon[2,**], and *Anthony* Thomas[3,***]

[1] *Department of Physics, University of Oxford, OX1 3PU United Kingdom, Department of Physics and Astronomy, University of Tennessee, TN 37996 USA*
[2] *CEA/IRFU/SPhN Saclay, F91191 France*
[3] *CSSM and CoEPP, Department of Physics, University of Adelaide, SA 5005 Australia*

Abstract. We present a selection of the first results obtained in a comprehensive calculation of ground state properties of even-even superheavy nuclei in the region of 96 < Z < 136 and 118 < N < 320 from the Quark-Meson-Coupling model (QMC). Ground state binding energies, the neutron and proton number dependence of quadrupole deformations and Q_α values are reported for even-even nuclei with 100 < Z < 136 and compared with available experimental data and predictions of macro-microscopic models. Predictions of properties of nuclei, including Q_α values, relevant for planning future experiments are presented.

1 Introduction

One of the main difficulties in solving the nuclear many-body problem is to incorporate nuclear medium effects into the calculation. In contrast to, for example, elementary quantum electrodynamics, where the force between two electric charges is quantified by the Coulomb law and forces among many electrons can be calculated precisely using the principle of superposition, no equivalent approach exists in nuclear physics. Scattering experiments with free nucleons provide information about bare N-N potentials, but an involved numerical treatment is necessary to transform these potentials to a form which desribes the forces between hadrons in the hadronic environment. There have been many theoretical attempts to overcome this problem but they all suffer from the uncertainty associated with many empirical parameters which are not uniquely constrained, thus preventing reliable predictions of nuclear properties in regions not accessible to experiment. One particular area of interest concerns the transuranic superheavy nuclei, where one may hope to discover new chemical elements, not existing in nature but synthesized in the laboratory. Theoretical predictions guiding such experiments are of major interest, especially information about spherical and deformed shell closures and related islands of stability. Given the ambiguity in models based on nucleonic degrees of freedom, alternative approaches have been sought.

The idea of modeling the effective nuclear interaction in-medium using quark degrees of freedom originated in the late 1980s with work by Guichon and collaborators [1–3]. They suggested that the origin of nuclear many body forces and of the saturation of nuclear forces can be found in the modification of the structure of a nucleon when it is imbedded in a medium consisting of other nucleons. The model, referred to as the Quark-Meson-Coupling (QMC) model, has been applied to finite nuclei [4, 5], nuclear matter and neutron stars (see e.g. [6, 7] as well as hypernuclei [8] and a variety of other problems [9]. The model is based on the idea that, instead of the usual treatment of modeling nuclear forces through exchange of mesons coupled to nucleons, taken as point-like particles, this exchange takes place directly between quarks in different nucleons, taken as a cluster of confined valence quarks having a structure in the form, for example, of the MIT bag. When the quarks in one nucleon interact *self-consistently* with the quarks in the surrounding nucleons by exchanging a σ meson, the effective mass M^*_N of the nucleon is no longer linear in the scalar mean field (σ). It is expressed as

$$M^*_N = M_N - g_{\sigma N}\sigma + (d/2)(g_{\sigma N}\sigma)^2, \qquad (1)$$

where $g_{\sigma N}$, the σ nucleon coupling constant in free space, is a parameter of the model. By analogy with electromagnetic polarizabilities, the coefficient d, calculated in terms of the nucleon internal structure, is known as the "scalar polarizability" [1]. The appearance of this term in the nucleon effective mass is sufficient to lead to nuclear saturation. This demonstrates a clear link between the internal structure of the nucleon and fundamental properties of atomic nuclei.

Models of nuclei in the region above Z = 100 can be divided into two classes [10], mean field models with effective density dependent interactions in the Hartree-Fock approximation [11–17], and macro-microscopic models based on the liquid-drop model combined with Strutinsky-type shell corrections (see e.g. [18–21]. Although these models yield similar results for some bulk properties of superheavy nuclei, they differ in important details. Su-

*e-mail: jirina.stone@physics.ox.ac.uk
**e-mail: pierre.guichon@cea.fr
***e-mail: anthony.thomas@adelaide.edu.au

Table 1. Neutron numbers corresponding to proton and neutron drip lines, derived from the Fermi energy for isotopes of elements 96 < Z < 136

Z	N(p)	N(n)	Z	N(p)	N(n)
96	132	224	118	174	278
98	134	226	120	180	286
100	138	230	122	184	290
102	138	236	124	188	296
104	146	240	126	192	298
106	146	242	128	196	302
108	154	246	130	202	306
110	158	250	132	208	310
112	164	256	134	214	314
114	168	260	136	218	314
116	170	268			

perheavy nuclei exist mainly because of their shell structure, in turn, dependent on level densities close to the Fermi level. It is therefore very important to understand their single-particle spectrum, in particular the location and magnitude of proton and neutron shell gaps. In addition, it may be expected that the competition between the Coulomb repulsion and the surface tension in many nucleon systems would play an important role and the sharp shell closures, familiar in lighter nuclei, may be diluted to form regions of a distinct spherical or deformed nature around a particular (N,Z) shell closure.

The various models predict different proton and neutron shell closures above ^{208}Pb. The macroscopic–microscopic models yield a 'canonical' island of stability at Z = 114, N = 184 [10, 11], while the self-consistent mean-field models (SC) non-relativistic models predict Z = 126, N = 172 - 184 [11, 12] and the relativistic mean-field models (RF) favour Z = 120, N = 172 [11, 16]. The most recent comparison of traditional models applied to superheavy nuclei may be found in [14] but no way to reconsile these differences is offered.

It is therefore interesting to examine the predictions of the QMC model which belongs to the class of mean field models but differs in key ways from both non-relativistic models with the Skyrme or Gogny interaction and classical relativistic field models in important details [5]. In this paper we present a selection of the first results obtained in a comprehensive study of the ground state properties of spherical and axially symmetrical even-even superheavy nuclei in the region 96 < Z < 136 and 118 < N < 320. A new version of the QMC model, QMCπ, is used in this work, where the published model [5] has been extended to include pion-exchange. The calculational procedure is described in section 2 and the resulting ground state binding energies of selected isotopes and the evolution of the deformation parameter β_2 for bound isotopes of elements with 100 < Z < 128 and isotones with 148 < N < 204, along with the Q_α values are presented in section 3. Section 4 contains predictions of the QMCπ model for properties of superheavy nuclei considered in planning future experiments, followed by section 5 summarizing the current results and outlining future work.

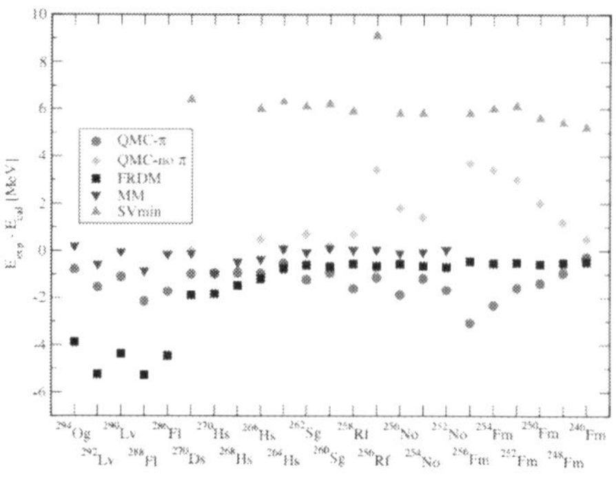

Figure 1. (Color online). Ground state binding energies of selected 'benchmark' even-even superheavy nuclei. The experimental data were taken from [27, 28].

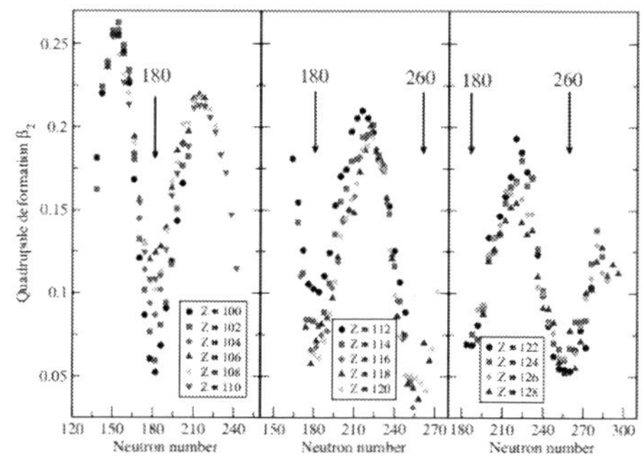

Figure 2. (Color online). Quadrupole deformation calculated in QMCπ for isotopes with proton number 100 < Z < 128.

2 Method of calculation

The QMCπ energy density functional (EDF) has been constructed in the same way as described in [5], except that the explicit pion exchange has been included in the Hamiltonian. We used HF+ BCS code SKYAX allowing for axially symmetric and reflection-asymmetric shapes, adapted by P.-G. Reinhard [22] for use with the QMC EDF. The best parameter set was sought using the experimental data set by Klüpfel et al. [23] and the fitting package POUNDERS [24, 25]. The volume pairing in the BCS approximation as been adopted with proton and neutron pairing strength fitted to data in [23]. We note that the addition of the explicit pion exchange in the model did not increase the number of parameters beyond the four used in the previous work, namely G_σ, G_ω, G_ρ and M_σ, but its addition was reflected in slight changes (less than 5%) from the values reported in [5]. The new parameter set is compatible with nuclear matter properties E_0 =-15.8 MeV, ρ_0 = 0.153 fm $^{-3}$, K_0 = 319 MeV, S_0 = 30 MeV and L = 27 MeV.

The selection of isotopes studied was limited by the requirement that only bound single particle states are occupied; that is, the Fermi level does not increase to the

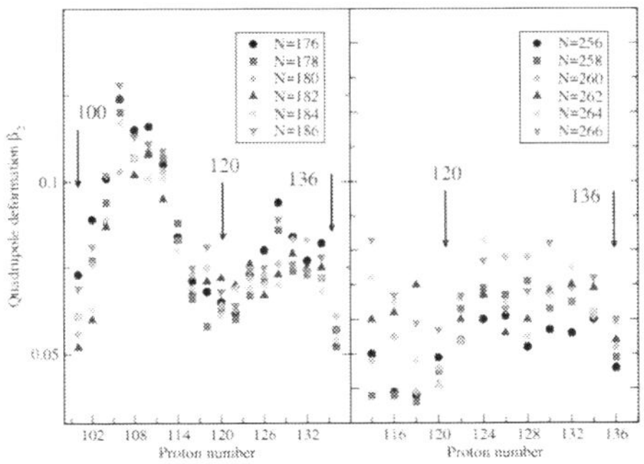

Figure 3. (Color online). The same as figure 2 but for isotones with neutron number 176< N < 186 (left panel) and 256< N < 266 (right panel).

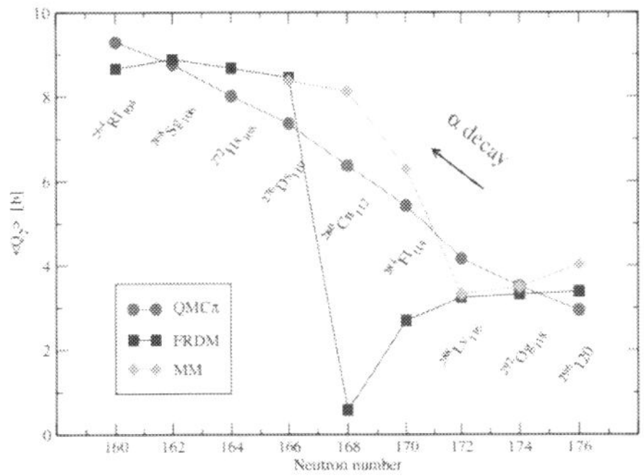

Figure 4. (Color online). Intrinsic quadrupole moments of even-even nuclei in the α-decay chain of ^{296}120, as calculated in the QMCπ, FRDM and MM models. See Fig. 1 in [29] and Ref. [30] for a comparison with other models and the text for more explanation.

continuum. This condition limited the range of neutron number for each element as given in Table 1.

3 Ground state properties and decay of selected superheavy nuclei

3.1 Binding energies and shapes

Figure 1 illustrates the difference between experimental and calculated ground state binding energies of selected superheavy nuclei for which the experimental data is known well enough that they may be regarded as 'benchmark' data [26] to calibrate theoretical models for subsequent use in uncharted areas of the nuclear landscape, notably in the transuranic region. These results illustrate a rather dramatic change from a significant underbinding, yielded by the mean field model with the Skyrme SVmin parameter set [23] and, to a lesser extent, predicted by QMC-without-pion model in the Z=100 - 104 region [5], to a slight overbinding with the present QMCπ, microcroscopic-macroscopic models Finite-Range-Droplet-Model (FRDM) [18] and of Muntian et al. (MM) [20, 21]. It is interesting to observe that the inclusion of pion-exchange in the QMC model has a significant effect on the ground state binding energies. As mentioned already in the introduction, one of the main motivations for modeling the superheavy region is to locate regions of relative stability. These regions would be the best targets for synthesis of either new elements or unknown isotopes for elements already known. The deformation parameters, β_2, determined in the QMCπ model by minimizing the ground state binding energies, are plotted with respect to neutron (proton) number in Figs. 2 and 3, respectively. The results in Fig. 2 clearly identify two regions of low deformation. Values of β_2 less than about 0.1 are predicted for isotopes of all elements with 100 < Z < 128 with neutron number in the region around N~ 180 and, for Z > 114 also in the vicinity of N = 260. The calculation of isotones with 176 < N < 186 reveals a strong indication for low deformation close to Z = 100, 120 and 136 (or possibly higher), as illustrated in Fig. 3 (left panel). The isotones with 256 < N < 266 generally exhibit low deformation in the region of 116 < Z < 136 (right panel). The effect is less prominent at Z = 120, but still clearly present at Z = 136. The analysis of the quadrupole deformation in QMCπ shows no evidence for the Z = 114 shell closure predicted in the FRDM and MM models nor at Z = 126, as suggested in SC models. However, it does agree with the results of RF models finding a shell closure at Z = 120. The neutron shell closures at N = 172, found in the SC and RF models is also not confirmed in this work, while the N=180 region is somewhat close to the value N = 184 found in the FRDM, MM and SC models. The region of sphericity we predict at N = 260 and Z = 136 is outside the reach of current experimental techniques but may potentially lead to a new island of stability. Further calculation, in particular of spectra of single-particle states, is needed to take this problem further.

Next we examine in more detail the development of the quadrupole deformation of isotopes along α decay chains of special interest, as they are a prime tool for identification of new elements (keeping in mind that spontaneous fission is in competition with α decay). In Fig. 4 we show the predicted mean intrinsic quadrupole moment calculated in this work. This is related to β_2 as $< Q_2 > = (3/(4\pi)AR_0^2\beta_2$, with R_0=1.2 $A^{1/3}$. We note that this definition [13] is consistent with the procedure used to calculate the deformation parameters in this work. There are alternative definitions in the literature, used for example in [29]. Therefore, only the qualitative neutron number dependence of $< Q_2 >$ can be compared between different calculations. We see a remarkably smooth decrease in deformation with increasing neutron number along the decay chain in the QMCπ model prediction, similar to that found by Shi et al. [29] in calculations using UNEDF1 and UNDEF1^{SO} Skyrme forces. In contrast, the FDRM microscopic-macroscoscopic model predicts a distinctive spherical shape at N = 168. The MM model more or less shows a decrease in deformation with increasing neutron number with a slight tendency to sphericity at N = 172.

Table 2. Q_α in MeV calculated in the QMCπ, FRDM and MM models. Experimental data without errors have been taken from nuclear mass tables [42], while entries with errors are from original papers [33, 35–37].

Z	N	Q_α^{exp}	$Q_\alpha^{QMC\pi}$	Q_α^{FRDM}	Q_α^{MM}
102	150	8.548	8.20	8.82	8.53
102	152	8.226	8.10	7.97	8.06
102	154	8.581	8.00	8.57	8.36
104	152	8.926	9.40	8.75	8.93
104	154	9.190	9.10	9.32	9.29
106	154	9.901	10.10	9.93	9.95
106	156	9.600	10.00	9.61	9.49
108	156	10.591	11.00	10.57	10.59
108	158	10.346	10.60	9.69	10.04
108	160	9.62± 0.16	10.30	9.00	9.49
108	162	9.02± 0.03	10.20	8.69	8.78
110	160	11.117	11.10	10.30	11.36
112	172	10.230	10.40	8.69	9.76
114	172	10.33±0.08	11.30	9.39	9.76
114	174	10.08±0.06	11.00	9.16	10.32
116	174	11.00±0.08	11.60	11.12	11.08
116	176	10.80±0.07	11.40	10.82	11.06
118	176	11.81±0.06	12.20	12.28	12.11

It is important to realize that deformation is treated differently in mean-field and micro-macro models (for details see [14]). Thus experimental data on quadrupole moments of superheavy elements may provide important information, leading to some distinction between the two classes of model to be used in the superheavy region. The recent progress in laser resonance ionization spectroscopy [31, 32] offers a prospect that electromagnetic moments, charge radii and isotope shifts of superheavy isotopes will be accessible experimentally in the near future.

3.2 Q_α values

Knowledge of α decay life-times is crucial to predicting properties of the α decay chains of superheavy elements, which, as pointed out above, are crucial in the search for and detection of new elements and their isotopes. The α-decay life-times are exponential functions of the energy release, Q_α, in the decay, which, in turn, depends on the mass difference between the parent and daughter states. This means that while the absolute values of the nuclear masses are not crucial in this context, the differences are essential. We adopt here the expression for the α-decay half-life given in Ref. [19] for an even-even parent with the atomic number Z

$$\log(T_{1/2}/s) = (aZ + b)(Q_\alpha/MeV)(cZ + d) \quad (2)$$

with a=1.66175, b=-8.5166, c=-0.20228 and d=-33.9069. It follows, for example, that a difference in Q_α of the order of 1 MeV in a nucleus with Z = 118 would make a difference in $T_{1/2}$ of three orders of magnitude. Thus the calculation of Q_α as close to reality as possible is vital for planning experiments. The neutron number dependence of Q_α obtained in this work is illustrated in Fig. 5. It is

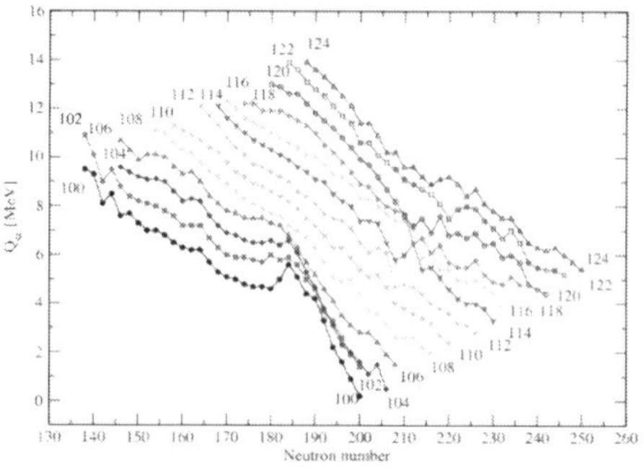

Figure 5. (Color online). Values of the α particle separation energy, Q_α, calculated in QMCπ for isotopes with 100 < Z < 120 in the region of neutron numbers 138 < N < 252.

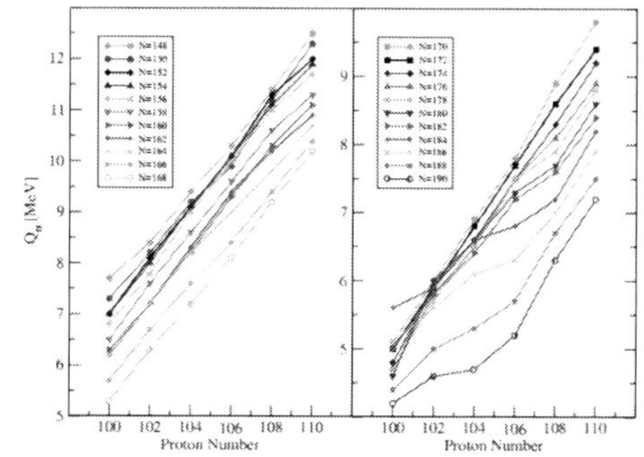

Figure 6. (Color online). The same as Fig. 5 but for isotones with 148 < N < 190 in the region of proton numbers 100 < Z < 110. The N=152, 162, 172 and 184 systems, corresponding to the suggested shell closures in different models, are highlighted by thicker lines in both panels.

beyond the scope of this work to compare the QMCπ predictions for Q_α with all the models in the literature. We refer the reader to, for example, [14, 39–41] and references therein. The important result, which has not been observed in any of the other models, is that the (weak) effects of the N = 152 and 162 shell closures disappear in nuclei with Z > 108, while the effects are enhanced for N∼180. We predict a smooth neutron number dependence of Q_α for N<200 for all elements with Z up to 124, not showing any effects of shell structure. Some variations may be indicated for higher N but at this stage we can draw no systematic conclusion. To our knowledge such behaviour of Q_α in this region has not been predicted before and a follow up would be very interesting.

Values of Q_α in the region Z≤110 are given in Fig. 6, showing the proton number dependence of Q_α for 148 < N < 190. We see a relatively smooth increase in Q_α with increasing Z for N ≤ 168, with a minor sensitivity to N = 152 and N = 162 (highlighted by thicker lines in the left panel of Fig. 6. This pattern prevails to N = 170 -174 (top part of the right panel) before a notable development of

Table 3. Predicted properties of selected isotopes of Rf, Hs, Fl, Og, El-120 and El-124, as calculated in the QMCπ, FRDM and MM models. Experimental binding energies are taken from [27, 28]. All energies are in MeV.

Isotope	Z	N	E_{bin}^{exp}	E_{bin}	β_2	Q_α	E_{bin}	β_2	Q_α	E_{bin}	β_2	Q_α
				QMCπ			FRDM			MM		
^{252}Rf	104	148		1860.20	0.245	9.4	1860.29	0.251	9.54	1859.29	0.245	9.85
^{268}Sg	106	162	1963.37	1965.30	0.229	9.4	1965.39	0.232	7.59	1964.48	0.233	7.89
^{262}Hs	108	154		1911.80	0.228	11.1	1912.32	0.243	10.88	1911.21	0.244	11.00
^{272}Hs	108	164	1981.79	1983.80	0.236	9.8	1984.48	0.220	9.20	1982.98	0.225	9.80
^{272}Ds	110	162	1973.36	1974.2	0.238	10.90	1975.56	0.221	10.04	1973.90	0.226	10.74
^{284}Fl	114	170	2034.01	2035.40	0.129	11.6	2037.98	0.064	9.44	2033.85	0.149	11.53
^{296}Og	118	178		2095.80	0.095	11.9	2099.12	-0.063	12.29	2094.67	0.039	12.06
^{296}120	120	176		2082.90	0.065	13.2	2085.70	0.075	13.69	2081.83	0.085	13.23
^{298}120	120	178		2097.30	0.063	13.0	2100.16	0.040	13.35	2095.48	0.054	13.44
^{306}124	124	182		2124.80	0.076	15.0	2126.39	0.000	13.43			

lower Q_α values, varying systematically with Z (bottom part of the right panel).

The outcome of the QMCπ model indicates that there is a subtle interplay between proton and neutron degrees of freedom in developing regions of nuclei with increased α-decay half life. As already discussed in the introduction and in the literature (e.g. [11]), it is likely that the sharp shell closures and shape changes observed in lighter nuclei, will instead be manifest as smoother patterns around the expected "shell closures". These patterns have their origin in the competition between the Coulomb repulsion and surface tension of the large nuclear systems in which the single-particle structure is only one of the critical ingredients.

4 Future experiments

Following discussions during the Fusion2017 meeting [26, 38] we provide predictions for binding energies, quadrupole deformations and Q_α values for selected nuclei summarized in Table 3. Corresponding results from the FRDM and MM models have been added for comparison. The reason for interest in these nuclei comes from their calculated shell structure in some traditional models, for example ^{260}Hs (N = 152, Z = 108 both supposed to be deformed shell closures). The presence of the N = 162 shell closure should manifest itself in ^{268}Sg and hence influence the Q_α value of ^{272}Hs. The Z = 108 shell closure could be studied by examining the properties of ^{272}Ds. The existence of an α-branch from the decay of ^{272}Ds is expected but has not yet been seen, the likelihood depending on the Q_α value. The prospect of observing ^{296}Og, ^{296,298}120 and, possibly, ^{306}124, is considered to be realistic using available beams and targets.

Although all three models in Table 3 agree reasonably well in their predictions of binding energies and deformations, the differences in Q_α are significant (see section 3.2). The most striking result is that QMCπ gives a much higher value of Q_α than those yielded by the FRDM and MM models. For example, the calculated α-decay half-live of ^{268}Sg in the QMCπ, FRDM and MM models are 0.21 s, 3.6 d and 5.9 h, respectively. One of the reasons for these variations is the presence of the N=162 shell closure in the models. The higher Q_α value is consistent with existence of the shell closure. This result also supports the suggestion of a higher Q_α for ^{272}Hs. The calculated values of Q_α for ^{284}Fl, ^{296}Og, ^{296,298}120 and ^{306}124 lead to very short $T_{1/2}$, 1.4x10^{-5} s, 4.0x10^{-4} s, 2.3x10^{-6}s, 5.8x10^{-6} s and 1.0x10^{-8} s, respectively. More model predictions in this regions can be found in [30].

5 Summary and conclusions

We have presented the first results of the application of the QMCπ model to superheavy nuclei. Having in mind that the model is dependent on only four, well constrained variable parameters with a clear physical meaning, it is encouraging to observe that predictions for ground state binding energies, axially symmetrical shapes, regions of shell closure and Q_α values are in good agreement with experimental data where available. The results presented in this work are of a very similar quality as the outcomes of other models, mean-field and macroscopic-microscopic which depend on many more parameters. Of course, the most interesting observation is that some of the predictions of the QMCπ model differ significantly from those in the traditional models and only experimental data can decide which model is closest to reality. Naturally there are some limitations in this work. As the energy surfaces as a function of quadrupole deformation may be complicated in the superheavy region (see e.g. [14]), it would be desirable to confirm the present results for nuclear shapes and the corresponding binding energies by repeating the minimization of the total binding energy with a constraint on the quadrupole moment. This procedure would help to avoid local minima and make a decision in cases of coexisting shapes with very similar binding energies. In addition, the predicted regions of shell closures, deduced in the present work from the neutron and proton number dependence of quadrupole deformation may be made more precise by examination of the spectrum of proton and neutron energy levels. These extensions of the current calculation are in progress.

6 Acknowledgement

It is a pleasure to thank Chris Düllmann for suggestions and comments, concerning experimental data, their significance and prospects. The contribution of Kay Marie Martinez, who performed the fitting of the QMCπ parameters using the POUNDERS package, is gratefully acknowledged. This work was supported by the University of Adelaide and the Australian Research Council through the ARC Centre of Excellence for Particle Physics at the Terascale and grant DP150103101.

References

[1] P.A.M. Guichon, Phys. Lett. B **200**, 235 (1988).
[2] P.A.M. Guichon, K. Saito, E.N. Rodionov, A.W. Thomas, Nucl.Phys. A **601**, 349 (1996).
[3] P.A.M. Guichon, A.W. Thomas, Phys. Rev. Lett. **93**, 132502 (2004).
[4] P.A.M. Guichon, H.H. Matevosyan, N. Sandulescu, A.W. Thomas, Nucl. Phys. A **772**, 1 (2006).
[5] J. R. Stone, P. A. M. Guichon, P. G. Reinhard, A. W. Thomas,Phys. Rev. Lett. **116**, 092501 (2016).
[6] J.R. Stone, P.A.M. Guichon, H.H. Matevosyan, A.W. Thomas, Nucl. Phys. A **792**, 587 (2007).
[7] D.L. Whittenbury, J.D. Carroll, A.W. Thomas, K. Tsushima, J.R. Stone, Phys. Rev. C **89**, 065801 (2014).
[8] P.A.M. Guichon, A.W. Thomas, K. Tsushima, Nucl.Phys. A **814**, 66 (2008).
[9] K. Saito, K. Tsushima and A. W. Thomas, Prog. Part. Nucl. Phys. **58**, 1 (2007).
[10] A. Sobiczewski, K. Pomorski, Prog. Part. Nucl. Phys. **58** 292 (2007).
[11] M. Bender, W. Nazarewicz, P.-G. Reinhard, Phys. Lett. B **515**, 42 (2001).
[12] M. Bender, P.-H. Heenen, P.-G. Reinhard, Rev. Mod. Phys. **75**, 121 (2003).
[13] M. Bender, P.-H. Heenen, J. Phys. Conf. Series **420**, 012002 (2013).
[14] P.-H. Heenen, J. Skalski, A. Staszczak, D. Vretenar, Nuclear Physics A **944** 415 (2015).
[15] J. Dobaczewski, A. Afanasjev, M. Bender, L. Robledo, Y. Shi, Nucl. Phys. A **944**, 388 (2015).
[16] A. Afanasjev, T. Khoo, S. Frauendorf, G. Lalazissis, I. Ahmad, Phys. Rev. C **67**, 024309 (2003).
[17] V. Prassa, T. Niksic, D. Vretenar, Phys. Rev. C **88**, 044324 (2013).
[18] P. Møller, A.J. Sierk, T. Ichikawa, H. Sagawa, Atomic Data and Nuclear Data Tables **109–110**, 1 (2016).
[19] P. Møller, J.R. Nix and K.-L. Kratz, Atomic Data and Nuclear Data Tables, **66**, 131 (1997).
[20] I. Muntian, Z. Patyk, A. Sobiczewski, Physics of Atomic Nuclei, **66**, 1015 (2003).
[21] I. Muntian, S. Hofmann, Z. Patyk, A. Sobiczewski, Acta Phys. Pol. B **34**, 2073 (2003).
[22] Private communication and Ref.[5]
[23] P. Klüpfel, P.-G. Reinhard, T. J. Bürvenich, and J. A. Maruhn, Phys. Rev. C **79**, 034310 (2009).
[24] M. Kortelainen et al., Physical Review C **82**, 024313 (2010).
[25] S. M. Wild, J. Sarich, N. Schunck, J. Phys. G: Nucl. Part. Phys. **42**, 034031 (2015).
[26] C.E. Düllmann, Private communication (2017)
[27] W.J. Huang, G. Audi, M. Wang, F.G. Kondev, S. Naimi, X. Xu, Chinese Physics C, **41**, 030002 (2017).
[28] M. Wang, G. Audi, F.G. Kondev, W.J. Huang, S. Naimi and X. Xu, Chinese Physics C, **41**, 030003 (2017).
[29] Y. Shi et al., Phys.Rev. C **90**, 014308 (2014)
[30] C. E. Düllmann, in print in Proceedings of ICFN6, (2017) (http://proton.phy.vanderbilt.edu)
[31] M. Laatiaoui et al., Nature 538, 495 (2016)
[32] M. Block, in this Proceedings
[33] Yu.Ts. Oganessian et al., J. Phys. G: Nucl. Part. Phys. **34**, R165 (2007).
[34] Yu.Ts. Oganessian et al., Phys. Rev. C **87**, 034605 (2013).
[35] Yu.Ts. Oganessian, V.K. Utyonkov, Nuclear Physics A **944** 62 (2015).
[36] J. Dvorak et al., Phys. Rev. Lett. **97**, 242501 (2006).
[37] K. Nishio et al., Phys. Rev. C **82**, 024611 (2010).
[38] C. E. Düllmann, EPJ Web of Conferences **131**, 08004 (2016).
[39] B. G. Carlsson , D. E. Ward, S. Aberg, EPJ Web of Conferences **131**, 08002 (2016).
[40] N. Wang, M. Liu, X. Wu, J. Meng, Phys. Rev. C **93**, 014302 (2016).
[41] P. Jachimowicz, M. Kowal, J. Skalski, Phys. Rev. C **89**, 024304 (2014).
[42] M. Wang, G. Audi, A. H. Wapstra, F. G. Kondev, M. Mac-Cormick, X. Xu, and B. Pfeiffer, Chin. Phys. C **36**, 1603 (2012).

Nucleon-nucleon correlations probed at far sub-Coulomb energies

S. Szilner[1,⋆], L. Corradi[2], D. Montanari[3,4], T. Mijatović[1], G. Pollarolo[5], E. Fioretto[2], A. Goasduff[3], D. Jelavić Malenica[1], D. Mengoni[3], G. Montagnoli[3], F. Scarlassara[3], N. Soić[1], A. M. Stefanini[2], and J. J. Valiente-Dobón[2]

[1] Ruđer Bošković Institute, Zagreb, Croatia,
[2] Istituto Nazionale di Fisica Nucleare, Laboratori Nazionali di Legnaro, Italy,
[3] Universitá di Padova, and Istituto Nazionale di Fisica Nucleare, Padova, Italy,
[4] IPHC, CNRS-IN2P3, Université de Strasbourg, Strasbourg, France,
[5] Universitá di Torino, and Istituto Nazionale di Fisica Nucleare, Torino, Italy

Abstract. Excitation functions of one- and two-neutron transfer channels have been measured for the closed shell ^{40}Ca+^{96}Zr and superfluid ^{60}Ni+^{116}Sn systems from the Coulomb barrier energy to energies far below. By using the unique PRISMA performance in terms of resolution and efficiency, measurements at very low bombarding energies have been performed. In the case of the ^{60}Ni+^{116}Sn system, a complementary particle-γ coincident measurements provide the ground and excited state populations.
The experimental transfer probabilities have been compared with semiclasical microscopic calculations that incorporate nucleon-nucleon correlations, together with known structure information of entrance and exit channels nuclei and reaction dynamics. These calculations well reproduce, in the whole energy range, one- and two-neutron transfer channels in shape and magnitude.

1 Introduction

Pairing interaction induces particle-particle correlations that are essential in defining the properties of finite quantum many body systems in their ground and neighbouring states. These structure properties may influence in a significant way the evolution of the collision of two nuclei, i.e. the effects of pair correlations affect a variety of nuclear processes. In this sense, low energy transfer reaction is one of the most important tool used to probe nucleon nucleon correlations in nuclear systems [1].

The pair correlations have been studied in measurements of two-particle transfer reactions by extraction of enhancement coefficients, defined as the ratio of the actual cross section to the prediction of models using uncorrelated states. Such enhancement factors, in principle, should provide a direct measurement of the correlation of the populated states. Unfortunately, the experimental evidence obtained by these factors is often obscured by the fact that almost all existing studies involve inclusive cross sections at energies higher than the Coulomb barrier, where the reaction mechanism is complicated by the interplay between nuclear and Coulomb trajectories.

On the contrary, at energies below the Coulomb barrier nuclei interact at very large distances, thus the interacting nuclei are only slightly influenced by the nuclear potential and Q values are restricted to few MeV for the open transfer channels. These conditions diminish the complexity of calculations where only the overlap of the tails of the intrinsic wave functions that are involved in the transfer process have to be taken into account. This energy regime is, on the other hand, characterized by low transfer cross sections. Only with the advent of large solid angle magnetic spectrometers the detection efficiency and resolution to identify the fragments in the heavy-ion collisions reached a sufficient level to make these experiments feasible [2–5].

In this work we present the results of a sub-barrier transfer measurements performed with the PRISMA spectrometer. Transfer probabilities for the closed shell ^{40}Ca+^{96}Zr [6] and superfluid ^{60}Ni+^{116}Sn [7] systems have been recently measured from the Coulomb barrier energy to energies far below it. Transfer yield has been detected at the level of 10^{-4} with respect to the elastic channel. We also performed a particle-γ coincidence experiment for the ^{60}Ni+^{116}Sn system in order to investigate the ground and excited state population [8]. By taking advantage of the high energy resolution of γ detectors, such measurements allowed to define how the transfer strength is distributed over the single states.

2 Sub-barrier measurements

In direct kinematics the transfer probability at large distances of closest approach correspond to fragments that are scattered at forward angles. In this angular range the overwhelming yield of elastically scattered particles often prevents a good identification for isotopes close to the projectile. At energies below the barrier, measurements of

⋆e-mail: szilner@irb.hr

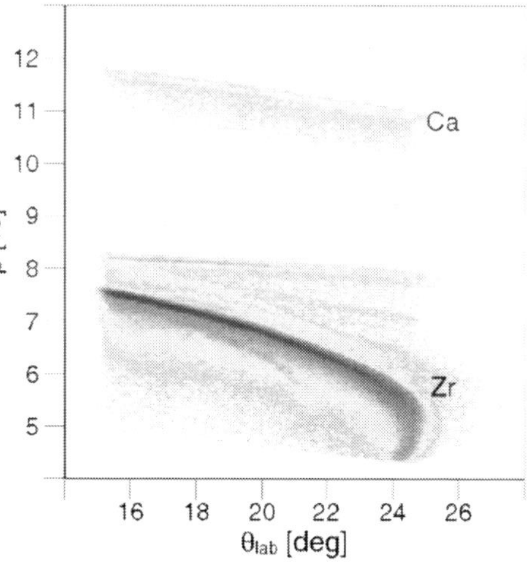

Figure 1. Velocity ($\beta = v/c$) vs in-plane scattering angle (θ_{lab}) for the reaction ^{96}Zr+^{40}Ca at $E_{lab}(^{96}Zr)$ = 330 MeV. The magnetic fields of the spectrometer were set to bring the Ca ions with maximum yield near the center of the focal plane area. The events at large β correspond to Ca-like recoils, and those at lower β to Zr-like ions scattered from Ca, as well as from other elements in the target (from Ref. [6]).

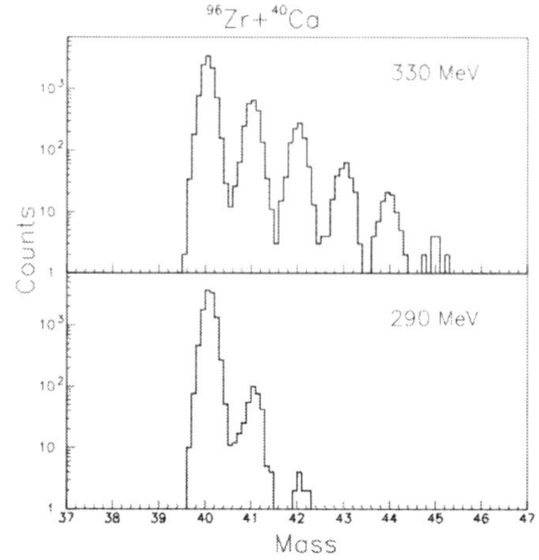

Figure 2. Mass distributions for pure neutron transfer channels obtained in the reaction ^{96}Zr+^{40}Ca at the indicated bombarding energies of the ^{96}Zr beam, i.e. $E_{lab}(^{96}Zr)$ = 330 MeV (top) and 290 MeV (bottom), and with the Ca-like recoils detected in PRISMA.

heavy-ion-transfer reaction products present other significant technical difficulties. Backscattered ions have low kinetic energies which severely limit their identification. This can be partly overcome by the use of inverse kinematics, where the lighter target-like fragments are detected with a magnetic spectrometer at forward angles. In this case, the energy of the recoiled light fragment is usually large enough to obtain a good resolution in mass and nuclear charge.

In the measured reactions, ^{96}Zr+^{40}Ca and ^{116}Sn+^{60}Ni, (lighter) target recoils have been detected with PRISMA, and the excitation functions at several bombarding energies have been obtained from the Coulomb barrier to 20-25% below, reaching ~ 15.5 fm of distance of closest approach. The identification of fragments has been done on an event-by-event basis by using, for the atomic number, the range of the ions as a function of the total energy released in the ionization chamber and, for the mass, by reconstructing the trajectories of the ions inside the magnetic elements of PRISMA [3, 9–12].

A fingerprint of the inverse kinematic is the behaviour of the velocity as a function of the focal plane position. Such matrix, the velocity vs in-plane scattering angle for the ^{96}Zr+^{40}Ca reaction, is displayed in Fig. 1. One sees the events corresponding to Ca-like recoils as well as those corresponding to Zr-like ions. The ridges of these Zr-like ions are due to scattering on Ca (main ridge presenting a bending around the kinematic limiting angle), on lighter elements (with smaller limiting angles), and on other heavier elements. The almost horizontal ridge, with high yield and with a $\beta \sim 0.08$, is due to Zr scattering in direct kinematics.

To illustrate the obtained mass resolution and the clear separation of different isotopes we show, as an example, the mass spectra at selected energies for the ^{96}Zr+^{40}Ca reaction in Fig. 2. At the higher energies, close to and above the Coulomb barrier, more than four neutrons have been transferred, while at the sub-barrier energies only one or at most two neutron transfers survive. Very similar mass distributions have been also observed for the ^{116}Sn+^{60}Ni reaction.

In order to have an overall description of the experimental data it is convenient to extract the transfer probability P_{tr}, defined as the ratio of the transfer cross section to the corresponding Rutherford cross section, and to plot these probabilities as a function of the distance of closest approach D for a Coulomb trajectory. The extraction of the transfer probability P_{tr} is rather alluring because the angular distributions obtained at different bombarding energies will be all superimposed, if the Q-value window does not change significantly. Further, the transfer probability at large distances D may be obtained by measuring an excitation function at fixed center of mass angle or by measuring angular distributions at fixed energy. The transfer probabilities for the ^{116}Sn+^{60}Ni measurement are presented (full symbols), together with the calculations, in Fig. 3 for the $(1n)$ and $(2n)$ neutron transfer channels. While the transfer probabilities for the ^{96}Zr+^{40}Ca measurement are presented in Fig. 4.

To inspect in more details the distribution of the transfer strength we show in Figs. 5 and 6 the total kinetic energy loss (TKEL) distributions for the ^{116}Sn+^{60}Ni and ^{96}Zr+^{40}Ca reactions, respectively. The TKEL were con-

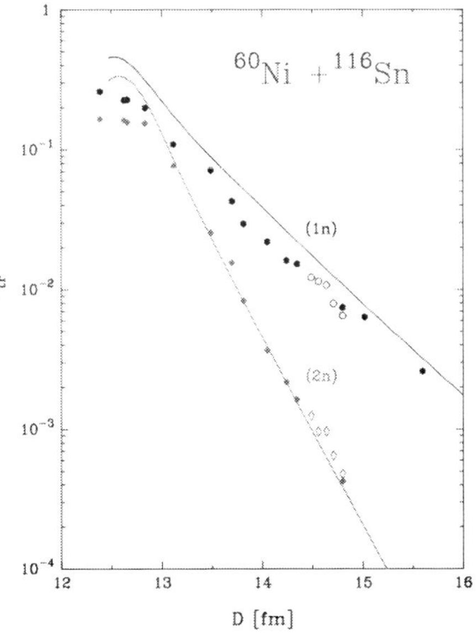

Figure 3. Experimental (points) and microscopically calculated (lines) transfer probabilities P_{tr} as a function of the distance of closest approach D for one- (^{61}Ni) and two-neutron (^{62}Ni) transfers in the ^{60}Ni+^{116}Sn system (from Ref. [8]).

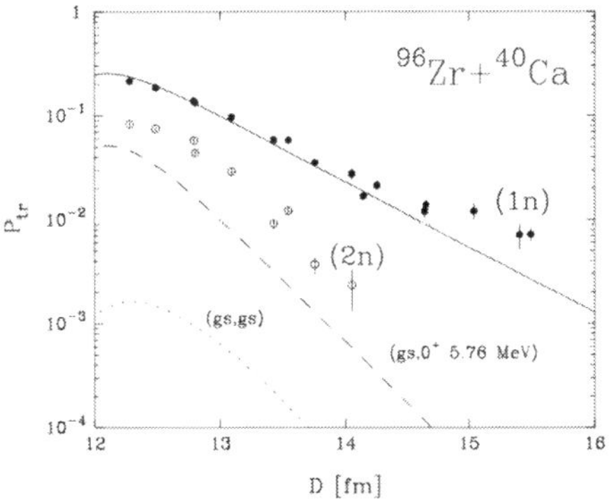

Figure 4. Experimental transfer probabilities (P_{tr}) as a function of the distance of closest approach for the ^{96}Zr+^{40}Ca system. The P_{tr} for one- and two-neutron transfer channels are represented with the full and empty symbols, respectively. The full line represents the inclusive transfer probability for one-neutron transfer, and the dotted line represents the calculation where only the ground-to-ground state transition for the two-neutron transfer has been included. For the dash line, the transition to the 0^+ excited state at ~6 MeV in ^{42}Ca has been included in the calculations (from Ref. [6]).

structed assuming a binary reaction and imposing the conservation of momentum. The ground-state Q values are indicated in figures. The width of the distributions is compatible with the intrinsic detector resolution, and the trajectory reconstruction, beam position and angle indetermination on target and target straggling effects.

In the case of the ^{116}Sn+^{60}Ni reaction, the experimental energy distribution for the elastic(+inelastic), one- and two-neutron transfer channels are concentrated around the ground state (see top panels of Fig. 5). In this very well Q-value matched system the reaction is dominated by the ground-to-ground state transition in (0n) and (2n) channels, and is close to the ground state in the (1n) channel. The situation is different in the ^{96}Zr+^{40}Ca reaction, in particular for the (2n) channel. One should notice that in this system for the (2n) channel the ground state Q-value is at +5.5 MeV. The TKEL spectra (see Fig. 6) display maxima at ~6 MeV at all measured energies, leaving the ground state unpopulated. These maxima, together with the known structure of the low energy states, suggest that in this channel the excited 0^+ states that were identified with the pair mode [13, 14] have been populated. Within the presently reached energy resolution, it is clear that a definite assignment of the populated single states was not feasible.

In general, with heavy ions, due to the illustrated limitation in the energy resolution with particle detectors, it is mandatory to combine such detectors with the gamma detection in order to distinguish the single states. Thus, we performed a particle-gamma coincidence experiment for the ^{60}Ni+^{116}Sn system by using the PRISMA spectrometer coupled to the Advanced Gamma Tracking Array (AGATA) demonstrator [15–17]. The demonstrator configuration included four triple cluster modules providing a simulated full-absorption efficiency of 2.64% for 1.3 MeV γ rays. This detector combination provides unique features for the selection of transfer products and the efficient detection of associated γ rays.

2.1 The particle-γ coincident measurement

We used a ^{60}Ni beam at 245 MeV on the ^{116}Sn target, and detected the Ni-like fragments in PRISMA placed at 70°. This laboratory angle corresponds to large distances of closest approach, where the effect of absorption is negligible, with a still sufficient transfer yield. The transfer probabilities extracted from the angular distribution are shown (open symbols) in Fig. 3, demonstrating a consistent matching of the two experiments. In Fig. 5 we also report the TKEL (bottom) at distances comparable with the inverse kinematic measurement, where one can appreciate the similarity of the two distributions, which are clearly peaked around the ground state Q-value.

The particle-γ coincidence measurement allowed to estimate the fraction of the total cross section going into excited states. In the inelastic channel the dominant transitions were from the 2^+ states of the binary partners (see Fig. 7). In the one-neutron transfer channel, a strong population of the low lying states of the single particle character has been observed, in particular the population of the $1/2^-$ state in ^{61}Ni and of the $5/2^+$ state in ^{115}Sn. For the two-neutron transfer channel we observed only a few events compatible with the transition from the 2^+ state to

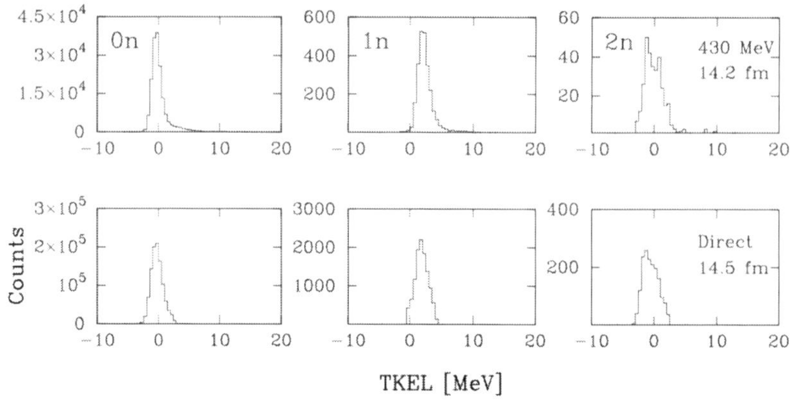

Figure 5. TKEL spectra obtained for the elastic(+inelastic) (0n), one- (1n) and two-neutron (2n) transfer channels at comparable distances of closest approach in the inverse kinematic (top) [7] and direct kinematic (bottom) [8] measurements for the $^{60}Ni+^{116}Sn$ system. The position of the ground state Q values: Q^{1n}_{gs}= - 1.7 MeV and Q^{2n}_{gs}= + 1.3 MeV.

the 0^+ ground state of ^{62}Ni. For this two-neutron transfer channel, from the experimental intensities of the γ lines, we concluded that the strength going to excited states is less than 24% of the total transfer strength of this channel. This finding demonstrates a dominance of the ground-to-ground state transition in this two-neutron transfer channel.

3 Nucleon-nucleon correlations

The transfer probabilities have been analyzed by employing a microscopic theory, which, for the first time in a heavy-ion collision, provided a consistent description of one- and two-neutron transfer channels, in shape and magnitude.

In more details, to compute the inelastic and one-neutron transfer channels the calculations have been performed by employing the same optical potential. Taking into account the collectivity of the low-lying states of the target and the projectile and the strong Coulomb field, a coupled-channels calculation have been performed for the inelastic channel by including the states which have been strongly populated in the particle-γ measurement in the vibrational approximation and with the tabulated values for the deformation parameters. For the one-neutron transfer channel the calculations have been performed in the DWBA approximation. These calculations provide, in good approximation, the direct population of states which can be compared with the experimental intensities obtained in the particle-γ measurement. Such comparisons provide a direct check on the one-particle form factors that have been used in the successive approximation for the two-neutron transfer. The transfer probability for the inclusive ($1n$) channel (full black lines in Figs. 3 and 4 for the $^{60}Ni+^{116}Sn$ and $^{40}Ca+^{96}Zr$ systems, respectively) has been calculated by summing over all possible transitions that can be constructed from the single particle states in projectile and target (see Refs. [6, 7]), and by weighting each transition with the corresponding spectroscopic factor. For both systems, the calculations agree well with the experimental transfer probabilities.

To calculate the ($2n$) channel one has to solve the well-known system of semiclassical coupled equations up to the second-order Born approximation. The amplitude consists of three terms, the simultaneous transfer of the pair of nucleons, the nonorthogonality term, and the term which represents the successive process via an intermediate channel. For the calculation of the ($2n$) channel the ground-to-ground state transition has been included following the formalism of Ref. [18]. In this formalism one can employ the same single particle form factors used for the ($1n$) channel. The ground states have been described in the BCS approximation with a standard state-independent pairing force.

For the ($2n$) channel of the $^{60}Ni+^{116}Sn$ system (see Fig. 3) the experimental points are very well described by the calculation. This fact indicates that the two-neutron transfer channel in this system is populating essentially only the ground state, in agreement with the experimental total kinetic energy loss distributions where the transfer strength was clearly concentrated around the ground state. Even more, the particle-γ measurement provided the upper limit for the possible excited state population in the ($2n$) channel.

For the $^{96}Zr+^{40}Ca$ system [6] (see Fig. 4) the calculation describes well the one-neutron transfer channel, while in the case of the two-neutron channel the calculation which includes (dotted line) only the ground-to-ground state transitions underestimates the observed probabilities. Clearly, this transition does not contribute to the total transfer strength in agreement with what was experimentally observed in the TKEL spectra (see Fig. 6). As already discussed, the maxima of the TKEL for the ($2n$) channel have been connected with the population of the excited 0^+ states. Thus, in Fig. 4 with a dash line, we show the predicted transfer probability for the transition to

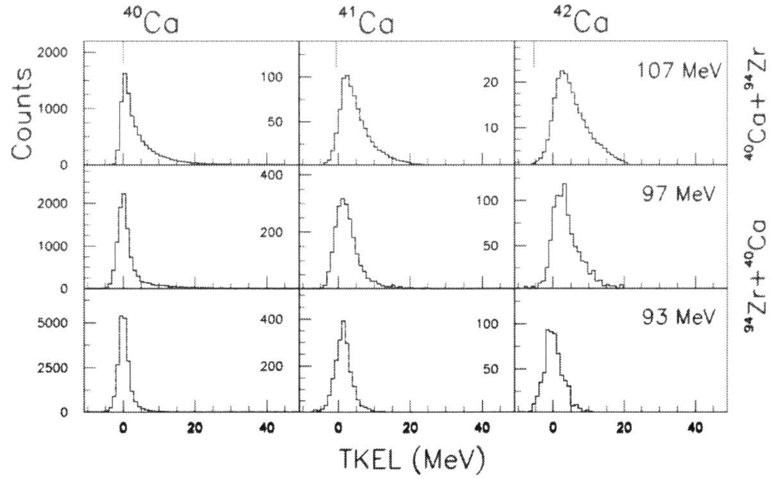

Figure 6. Experimental TKEL distributions for the elastic(+inelastic), one- and two-neutron transfer channels obtained in the "direct" ^{40}Ca+^{96}Zr [3] and inverse kinematic ^{96}Zr+^{40}Ca [6] measurements at the indicated center of mass energies. These energies correspond to about 10% above, at, and 10% below the Coulomb barrier. The vertical lines represent the ground state Q values: Q^{1n}_{gs}= + 0.5 MeV and Q^{2n}_{gs}= + 5.5 MeV.

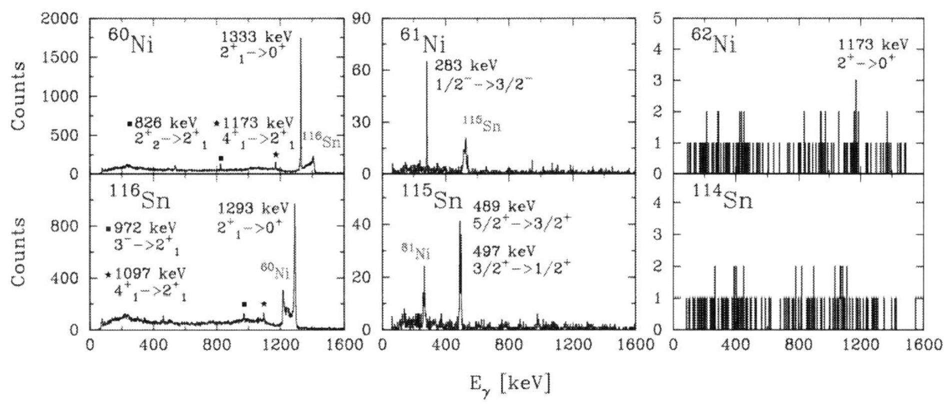

Figure 7. Top panels: Doppler-corrected γ spectra for ^{60}Ni, ^{61}Ni, and ^{62}Ni detected in PRISMA for the ^{60}Ni+^{116}Sn measured in the direct kinematic. Bottom panels: Doppler-corrected spectra for the heavy binary partners ^{116}Sn, ^{115}Sn, and ^{114}Sn. The strongest transitions are labeled with the γ-ray energy and spin and parity of the initial and final states. The broader peaks, corresponding to the wrongly Doppler-corrected reaction partner, are also labelled (red color) in each frame.

the 0^+ state at ~6 MeV in ^{42}Ca. It is apparent that the contribution of this transition is much larger than the ground state one. Besides the good match with the Q-value, this is mostly due to the $2p_{3/2}$ component that dominates its wave function and that has a much larger form factor. We stressed that the still missing strength was attributed to states with large angular momentum populated in the reaction, while only 0^+ to 0^+ transitions have been included in the theoretical calculation. At variance with the ^{96}Zr+^{40}Ca system, in the ^{116}Sn+^{60}Ni system the ground-to-ground state Q values for one- and two-neutron transfer channels are very close to optimum (Q_{opt} ~0), representing a beautiful Q-value matched case. One sees that in this case the employed microscopic theory very well reproduces the experimental data in all energy range, in particular the transfer probability for two neutrons is very well reproduced by considering solely the ground-to-ground state transition. It is very important to stress that the shown results are not

DOI: 10.1051/epjconf/201716300058

coming from best fit procedure, but that they employed a microscopic theory that incorporates nucleon-nucleon correlations.

4 Summary and outlook

The advent of the large solid angle magnetic spectrometers, coupled to large γ arrays, ensured significant advances in the field of multinucleon transfer reactions at energies close to the Coulomb barrier. We presented transfer probabilities for one- and two-neutron transfer channels in the ^{40}Ca+^{96}Zr and ^{60}Ni+^{116}Sn systems from the Coulomb barrier energy to energies corresponding to very large distances of closest approach where the nuclear absorption is negligible. The employed microscopic theory, that incorporates nucleon-nucleon correlations, very well reproduces the experimental data in the whole energy range, in particular, the transfer probability for two neutrons in the Q-value matched ^{60}Ni+^{116}Sn system is very well reproduced, in magnitude and slope, by considering solely the ground-to-ground state transition. It is very important to continue these studies, and to compare the experimental observables with microscopic calculations where nucleon-nucleon correlations are treated together with the relevant structure and dynamics information.

As a follow up of these results, it will be important to study whether and to what extent the effect of neutron-neutron correlations, in the evolution of the reaction, is modified in the presence of high Coulomb fields. In the very heavy-ion collisions, the population of final states with high excitation energies and large angular momenta may significantly change the transfer strength of the ground-to-ground state transitions.

In this context, it is also important to investigate the role played by proton-proton and neutron-proton correlations. As known, multinucleon transfer reactions allow the transfer of large number of nucleons, and in order to study proton-neutron correlation one has to use systems where the population of the ($\pm np$) and ($\pm 2p$) channels are allowed by the Q-value considerations.

The issue connected with the pair correlations is presently at the focus of a renewal of interest [19–23], in particular in ongoing research with radioactive beams, where the pairing interaction may be significantly modified in nuclei with extended neutron distributions [24].

This work was partly supported by the EC FP7 ENSAR (262010), by Croatian Science Foundation (project 7194), and by the Scientific center of excellence for advance materials and sensors.

References

[1] D. M. Brink and R. A. Broglia, Nuclear Superfluidity: *Pairing in Finite Systems* (Cambridge University Press, Cambridge, England, 2005).
[2] L. Corradi, G. Pollarolo, and S. Szilner, J. of Phys. G **36**, 113101 (2009).
[3] S. Szilner *et al.*, Phys. Rev. C **74**, 024604 (2007).
[4] H. Savajols and the VAMOS collaboration, Nucl. Phys. A **654**, 1027c (1999).
[5] A. Cunsolo *et al.*, Nucl. Instr. Methods Phys. Res. A **481**, 48 (2002).
[6] L. Corradi *et al.*, Phys. Rev. C **84**, 034603 (2011).
[7] D. Montanari *et al.*, Phys. Rev. Lett. **113**, 052501 (2014).
[8] D. Montanari *et al.*, Phys. Rev. C **93**, 054623 (2016).
[9] T. Mijatović *et al.*, Phys. Rev. C **94**, 064616 (2016).
[10] D. Montanari *et al.*, Eur. Phys. J. A **47**, 4 (2011).
[11] L. Corradi *et al.*, Nucl. Instrum. Methods Phys. Res. Sect. B **317**, 743 (2013).
[12] T. Mijatović *et al.*, Eur. Phys. J. A **52**, 113 (2016).
[13] S. Szilner *et al.*, Eur. Phys. J. A **21**, 87 (2004).
[14] R.A. Broglia, O. Hansen and C. Riedel, Advances in Nuclear Physics, edited by M. Baranger and E. Vogt, Plenum, New York, 1973, Vol. 6.
[15] A. Gadea *et al.*, Nucl. Instrum. Methods Phys. Res., Sect. A **654**, 88 (2011).
[16] E. Farnea *et al.*, Nucl. Instrum. Methods Phys. Res., Sect. A **621**, 331 (2010).
[17] S. Akkoyun *et al.*, Nucl. Instrum. Methods Phys. Res., Sect. A **668**, 26 (2012).
[18] J. H. Sorensen and A. Winther, Nucl. Phys. A **550** (1992) 306.
[19] C. Simenel, Phys. Rev. Lett. **105**, 192701 (2010).
[20] G. Potel *et al.*, Phys. Rev. Lett. **105**, 172502 (2010).
[21] G. Potel *et al.*, Phys. Rev. Lett. **107**, 092501 (2011).
[22] A. Lemasson *et al.*, Phys. Lett. B **697**, 454 (2011).
[23] G. Scamps and D. Lacroix, Phys. Rev. C **87**, 014605 (2013).
[24] J. Dobaczewski *et al.*, Phys. Rev. Lett. **72**, 981 (1994).

EPJ Web of Conferences **163**, 00059 (2017)
FUSION17 DOI: 10.1051/epjconf/201716300059

Perspectives of Super-Heavy Nuclei research with the upcoming separator-spectrometer setup S^3 at GANIL/SPIRAL2 - The VAMOS Gas-Filled separator and AGATA

Christophe THEISEN[1,*] on behalf of the S^3 and VAMOS-GFS collaborations

[1] *Irfu, CEA, Université Paris-Saclay, 91191 Gif-sur-Yvette, France*

Abstract. Several facilities or apparatus for the synthesis and spectroscopy of the Super-Heavy Nuclei (SHN) are presently under construction in the world, which reflect the large interest for this region of extreme mass and charge, but also for the need of even more advanced research infrastructures. Among this new generation, the GANIL/SPIRAL2 facility in Caen, France, will soon deliver very high intense ion beams of several tens of particle μA. The Super Separator Spectrometer S^3 has been designed to exploit these new beams for the study of SHN after separation. It will provide the needed beam rejection, mass selection and full arsenal of state-of-the art detection setups. Still at GANIL, the AGATA new generation gamma-ray tracking array is being operated. The VAMOS high acceptance spectrometer is being upgraded as a gas-filled separator. Its coupling with AGATA will lower the spectroscopic limits for the prompt gamma-ray studies of heavy and super-heavy nuclei. In this proceeding, these new devices will be presented along with a selected physics case.

1 Introduction

The synthesis of new elements has reached up to $Z = 118$ with the IUPAC/IUPAP recently assigning the naming rights for the elements 113, 115, 117 and 118 to groups at the FLNR Dubna, Russia and at RIKEN, Japan [1–4]. Despite tremendous worldwide efforts in the production of the heaviest nuclei, the region of SHN is still not well known nor delineated. Indeed, the production cross-sections of the heaviest known nuclei are extremely low, at the pb level in the $Z = 114 - 118$ region, which corresponds with present state-of-the-art facilities to the synthesis of a few atoms per month. This low production is itself related to the extreme fragility of the species: the nuclei beyond rutherfordium ($Z = 104$) are solely stabilised by quantum mechanics. A charged liquid drop would fall apart due to Coulomb repulsion. Therefore those species provide an ideal laboratory for nuclear structure studies. On the theoretical side, there is not yet any consensus concerning the position and properties of an hypothetical island of stability around a doubly magic spherical nucleus. While microscopic-macroscopic models predict $Z = 114$, $N = 184$ as doubly magic numbers, models based on effective forces predict either $Z = 120$ or 126, $N = 172$ or 184: see e.g. [5, 6]. These nuclei are however inaccessible using a stable beam, even when using a radioactive actinide target.

Besides the synthesis of the heaviest species approaching the island of stability, advances in the field can be made performing detailed studies of lighter nuclei for which production cross-sections are higher. Decay spectroscopy after separation of deformed nuclei around $(N, Z) = (152, 100)$ and $(162, 108)$ provides direct links to the next heavier spherical closed shell nuclei, by investigating single particle levels which are, according to model calculations, relevant in both regions [7, 8]. A particularly interesting feature, accessible with this experimental approach, are meta-stable states created by nuclear deformation, so called K isomers. Following the trend of vanishing deformation they can be used to trace the spherical SHN and to locate the island of stability. A second route consists in prompt spectroscopy studies of deformed nuclei, still with the idea to trace orbitals active both in deformed nuclei and in heavier spherical nuclei [9]. Such studies can be made down to the 10 nb level, ^{256}Rf being the heaviest nuclei for which collective properties have been deduced so far [10].

Progress and success in the field will be obtained by lowering the cross-section limits to reach heavier nuclei in the direction of the island of stability, and by performing detailed studies. This will be done improving experimental means like new high intensity ion accelerators, efficient in-flight separators and spectrometers, and highly efficient detector systems with fast electronics. The new SPIRAL2 facility and, in particular, the separator-spectrometer setup S^3, presently under construction at the accelerator laboratory GANIL in Caen, France, will offer great perspectives for the field. Still at GANIL, the upgrade of the VAMOS vacuum spectrometer as a Gas-Filled Separator (GFS) will provide, when coupled to e.g. AGATA, a very efficient tool for prompt spectroscopy.

* e-mail: christophe.theisen@cea.fr

2 The GANIL/SPIRAL2 facility

The "historical" part of the GANIL facility is based on three cyclotrons, 2 of which serve for stable beams acceleration, plus the CIME cyclotron for reacceleration of radioactive species using the ISOL method (SPIRAL1 facility). Among the different instruments, the AGATA gamma-ray spectrometer is hosted at GANIL until 2019 and will be coupled to the VAMOS-GFS high acceptance separator, as detailed in section 4.

The new forthcoming SPIRAL2 facility is based on a superconducting linear accelerator that will deliver ions up to $\approx$ 15 MeV/A. The S^3 Super Separator Spectrometer is one of the major experimental device that will benefit from the high intense beams. In the first phase of the project "phase1", the RFQ injector $A/Q = 3$ will allow production of mostly $A \lesssim 50$ beams. This RFQ has been optimised for the deuterium beams used for neutron production at the Neutrons For Science facility, and for the "phase2" of the project. In the "phase1++" stage, a second RFQ $A/Q = 7$, which construction is foreseen from 2019, will deliver ultra-high intense beams up to uranium. This phase will provide a world-class facility for SHN research, competitive with other ongoing projects such as the SHE factory in Dubna [1]. As an example, ^{48}Ca beam of $\approx$ 2.5 pμA intensity will be available with the RFQ $A/Q = 3$, raised up to $\approx$ 15 pμA with "phase1++".

The 30 meter long S^3 Super Separator Spectrometer, shown in figure 1, has been optimised for selecting from the high background of parasitic reactions the rare heavy nuclei produced after fusion-evaporation reactions. In order to sustain the beam intensity, targets will be arranged on a rotating wheel installed in the target cave, with the possibility of using actinide material. The first separator section, a momentum achromat, has a QQQDQQQ-QQQDQQQ optics configuration with the purpose of getting rid of most of the direct beam (more than 99.9 %). The spectrometer section with a QQQEQQQ-QQQDQQQ configuration, combining an electric and a magnetic dipole, acts as a mass separator. The electric dipole will also separate the fusion-evaporation residues from the direct beam according to their electric rigidity. According to simulations, the direct beam should be fully suppressed. A rejection factor of about 10^{12} is expected for the beam-like particles. Ions will be dispersed at the focal plane according to their mass over charge state ratio and provide a resolution $\Delta M/M \approx 1/300$. A transmission of $\approx 40\%$ is expected for asymmetric reaction such as ^{208}Pb(^{48}Ca,2n)^{254}No while a lower transmission of $\approx 15\%$ is expected for more asymmetric reactions such as ^{238}U(^{22}Ne,5n)^{255}No.

As an option, devices can be installed at the achromatic point of S^3, such as the Fast Ion-Slow Ion Collision (FISIC) station for atomic ion-ion collisions studies, or a target surrounded by a photon detection array for prompt studies of exotic nuclei produced after two-step reactions. Decay spectroscopy after separation will be performed at the focal plane using an array known as SIRIUS (Spectroscopy and Identification of Rare Ions Using S^3). The first detection stage consists of one or two trackers that will provide the time-of-flight measurement for kinematic selection, and the position needed for the M/Q identification. Ions will be implanted in a segmented 10×10 cm^2 double-sided silicon strip detector (128 strips on both vertical and horizontal directions) 300 μm thick, where subsequent decays will also be measured (alpha, fission, electrons). A tunnel made of four 1 mm thick, 10×10 cm^2 Si pad detectors will be placed upstream for conversion-electron spectroscopy, escaped alpha and fission fragments detection. The Si detectors will be surrounded by 5 HPGe detectors for gamma-ray spectroscopy. All detectors will be equipped with digital, fully time stamped electronics to minimise the dead-time and allow measurement of fast decays. The S^3 and SIRUS coupling is expected to be completed in 2018 and then will be ready for commissioning. More details can be found in [11]. As an alternative to SIRIUS, it will be possible to send ions in a gas cell. There, in-gas jet laser spectroscopy can be performed with a very high resolution (200 MHz). Isotopic shifts, electric or magnetic moments can be measured. Provided the atomic scheme is known or measured, laser ionisation can selectively send the nuclei of interest (or their isomers) to a low energy line. Their mass can be measured with high precision with a multi-reflection time-of-flight spectrometer [12]. In a latter phase expected in 2023, the option to send the ions in the DESIR hall [13] will be available for ground state and atomic properties studies: mass measurement, decay studies, collinear laser spectroscopy, etc. S^3 is not only designed for SHN, but for other regions where mass selection and/or tagging is needed for the spectroscopy and synthesis of rare nuclei, which corresponds to a large panel of isotopes from $N \sim Z$ nuclei up to the heaviest nuclei.

3 Selected physics case with S^3

3.1 "First day" experiments

As "first day" experiments following the commissioning phase, physics cases of technically moderate difficulty but with potentially high scientific outcome have been selected by the collaboration. Among the possible candidates, there is a strong interest in the study of high-K isomers in ^{254}No and ^{256}Rf. High-K isomerism is a rather general phenomenon in the transfermium region due to the presence of high-K single particle states around the Fermi level. In even-even deformed nuclei, it corresponds to the breaking of a pair promoted towards two different single-particle states coupled with a high-K projection along the symmetry axis. Therefore their study allows to track the properties of the corresponding orbitals, to provide information on pairing correlations, and eventually allows the spectroscopy of states fed along the isomer decay path. Because of the K-forbiddness, the decay of these states is largely hindered and provides a clear experimental signal that allows to isolate the corresponding exotic states using the so-called calorimetric technique [14]. Isomeric states in the transfermium region were first postulated in ^{254}No and ^{250}Fm by A. Ghiorso et al. in the early 1970's at the Lawrence Berkeley National Laboratory (LBNL) [15].

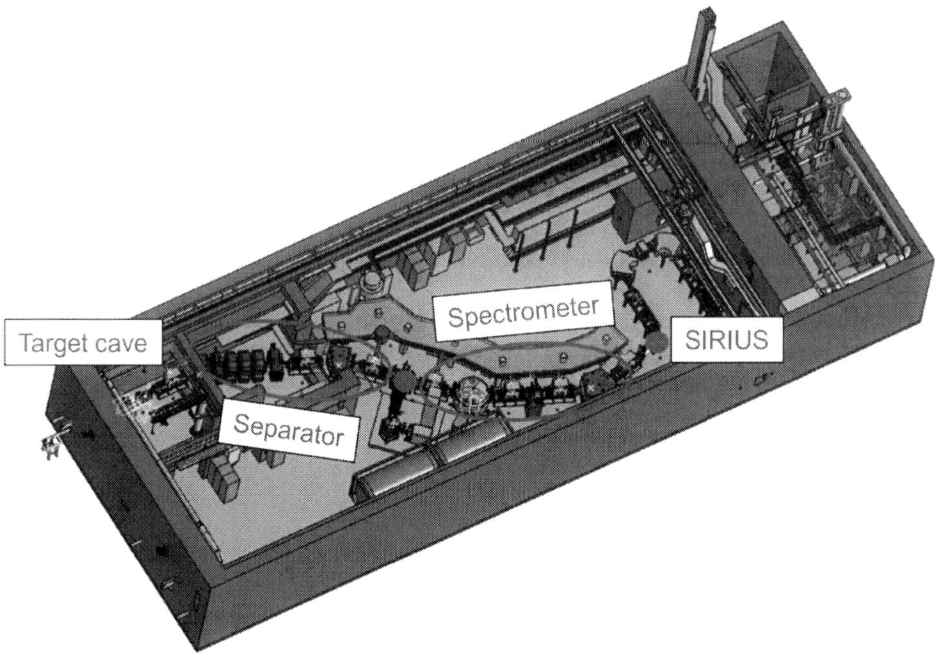

Figure 1. Design view of S^3. The position of the achromatic point between the separator and spectrometer sections, and of the SIRIUS station, are represented by red dots.

While the correct half-life was deduced, neither the excitation energy nor the decay path could be measured. It is only in 2006 that a detailed level scheme could be obtained in two experiments performed at the University of Jyväskylä [16] and at the Argonne National Laboratory (ANL) [17]. The experiment was later repeated at GSI and LBNL [18, 19]. Despite the large cross section of about 2 μb, there are significant differences in the level schemes and interpretations: the isomer is either understood as the breaking of a pair of protons $\pi[514]7/2^- \otimes \pi[624]9/2^+$ or neutrons $\nu[613]7/2^+ \otimes \nu[734]9/2^-$, which has consequences on the underlying single-particle scheme and deformed shell gap. Also a higher-lying isomeric state presumably of 4qp character has been measured in the four above-mentioned experiments, but without precise excitation energy, only tentative decay scheme, spin, parity and single-particle configuration. In the same mass region, three isomeric states have been observed recently at LBNL in ^{256}Rf [20]. Due to the lower cross-section of $\approx$ 17 nb, only tentative placement and single-particle configuration could be proposed. Moreover an experiment performed at ANL provides a different interpretation [21].

Consequently, the two $N = 152$ isotopes ^{254}No and ^{256}Rf deserve comprehensive new investigations. These cases are feasible already in the first operation period of S^3, using the reactions ^{208}Pb(^{48}Ca,2n)^{254}No and ^{208}Pb(^{50}Ti,2n)^{256}Rf, respectively. Assuming a beam intensity of $\approx$ 2 pμA with the $A/Q = 3$ RFQ, targets of 500 μg/cm^2 and a transmission of 50%, a statistics of 10^7 and 90×10^3, respectively, is expected in a one week experiments, which exceeds the previous obtained figures by more than one order of magnitude.

3.2 Further studies with S^3

With the availability of the RFQ injector $A/Q = 7$, a large panel of nuclides will be accessible with decent statistic for spectroscopic studies. To take an example, cross-sections at the 10 pb level reached for some nuclides in the $Z = 114 - 116$ region will correspond to a statistics of several hundred tagged nuclei per week at S^3, assuming a transmission of $\approx$ 15% for very asymmetric reactions using actinide targets.

Besides single-particle spectroscopy, direct experimental proof of the atomic number could also be the measurement of characteristic X-rays where attempts have been recently made with limited success so far for e.g. $Z = 115$ [22, 23]. In the case of ^{288}Mc ($Z = 115$) synthesised using the reaction ^{48}Ca+^{243}Am, about 1800 L X-rays are expected in a one week experiment using S^3 [8].

Returning to the K-isomer cases, their role in an enhanced stability of SHN has been pointed-out [24]. In several cases, nuclides are characterised by a higher stability and longer lifetime of the metastable state compared to the ground-state. Such an inversion was observed in ^{270}Ds already during its first synthesis experiment: isomer half-life $t_{1/2} = 6.0^{+8.2}_{-2.2}$ ms compared to $t_{1/2} = 100^{+140}_{-40}$ μs for the ground-state [25]. Because of the low cross-section of 15(5) pb, only 3 events were observed for each decaying state, which does not allow a robust interpretation. With S^3, a total of $\approx$ 400 implanted nuclei are expected per week. Other examples of recently observed isomer – ground state lifetime inversion are ^{254}Rf and ^{250}No.

The nuclide ^{254}Rf has been recently studied at ANL and LNBL [26]. Two isomeric states with half-lives of 4.7(1.1) μs and 247(73) μs have been observed, and interpreted as 2qp neutron $K^\pi = 8^-$, and 4qp $K^\pi = 16^+$ configuration, respectively. Surprisingly, the half-live of the 4qp isomer exceeds that of the ground-state by one order of magnitude. Because of the short half-live of the 2qp isomer, utilisation of digital signal processing was a key of the success in order to disentangle the signals corresponding to the implantation, electromagnetic and alpha decays. Because of the modest cross-section of $\approx$ 2.4 nb for the fusion-evaporation reaction ^{50}Ti(^{206}Pb,2n)^{254}Rf, neither the precise excitation energy of the isomer nor their de-excitation could be measured. With S^3 in the 1++ phase, an implantation rate of $\approx$ 8/min is expected corresponding to $\approx 80 \times 10^3$ events per week, which is comfortable for decay studies after separation.

The lifetime of ^{250}No has been a long-standing puzzle since its first synthesis by Oganessian et al. with the Dubna gas-filled recoil separator [27], studies by Belozerov et al. using VASSILISSA [28], and by Peterson et al. using the fragment mass analyser at ANL [29]. The first experiment by Oganessian et al. gave the surprising result that going from ^{252}No to ^{250}No, the spontaneous fission half-life drops by seven orders of magnitude. In the subsequent experiment using VASSILISSA, two spontaneous fission half-lives $t_{1/2} = 5.9$ μs and 54 μs were measured but it was not clear whether the long component was due to the neighbouring ^{249}No or to an isomeric state. Finally, Peterson et al. assigned the shorter-lived decay ($t_{1/2} = 3.7^{+1.1}_{-0.8}$ μs) to the ground state and the longest to a $K^\pi = 6^-$ isomer ($t_{1/2} = 43^{+22}_{-15}$ μs). It is only recently in an experiment performed at the University of Jyväskylä that the electromagnetic branch from the isomeric state could be detected [30]. Again, using digital signal processing was a key ingredient for disentangling the implantation, de-excitation and decay. In the phase1++ of SPIRAL2, about 40 implanted nuclei are expected per minute, leading to a statistics of $\approx 420 \times 10^3$ implanted nuclei in a one week experiment.

There are obviously a wealth of other interesting studies to be performed with S^3 using either the SIRIUS decay station, the low energy branch or the forthcoming DESIR hall, not only in the region of the heaviest elements, but along the nuclear chart where separation selection and/or tagging techniques are required to study rare nuclei immersed in a dominant background of parasitic reactions. More details can be found in e.g. [8, 31].

4 The VAMOS Gas-Filled Separator and AGATA

The decay spectroscopy after separation discussed above is an ideal tool for the study of isomeric and low-lying single-particle states, since the decay process selectively populates states in the daughter nucleus at low excitation energy and low angular momentum. On the other hand, fusion-evaporation reactions populate medium to high spin states. Therefore prompt spectroscopy of the transitions emitted straight after the collision allows to study collective properties. However, using a detection array at the target position induces severe restrictions on the beam intensity and therefore on the cross-section limits for the nuclei to be studied. So far ^{256}Rf is the heaviest nucleus for which collective properties have been studied using prompt gamma-ray spectroscopy [10]. Lowering the spectroscopic limits can be done in three directions: improving the transmission for the recoiling nuclei, increasing the prompt array efficiency, and improving the prompt array counting rate capabilities.

The VAMOS-GFS project is an upgrade of the existing VAMOS spectrometer [32] that adds a Gas-Filled Separator (GFS) mode. This upgrade was guided by mainly two motivations. First VAMOS is a large acceptance device therefore a huge transmission is expected in particular for the heaviest elements leaving the target with a large angular distribution. The second motivation is related to the opportunity for its coupling with state-of-the-art gamma-ray detectors (the new generation gamma-tracking array AGATA [33, 34], and other arrays such as EXOGAM2, PARIS...). Hence VAMOS-GFS and AGATA which is operated at GANIL until 2019, should provide one of the most efficient devices (in terms of transmission and selectivity) for the spectroscopy of rare nuclei produced by fusion-evaporation reactions. Thanks to the concept of tracking and fully digital signal processing, AGATA will provide a large efficiency and a high counting rate capability. The feasibility and the potential of VAMOS-GFS have been established in a test experiment performed in 2009 [35]. Excellent performances in terms of fusion-evaporation residue transmission and background rejection have been obtained for the ^{40}Ca + ^{150}Sm collisions around the Coulomb barrier: a large transmission of $\approx$ 95% was measured for the neutron evaporation channels. However, the modifications, mandatory to transform the vacuum VAMOS spectrometer into a gas-filled separator, were of provisional character. The aim of the VAMOS-GFS project is to make available to a widespread community a fully operational device, with the best performances for a large range of reactions and a smooth operation.

A sketch of the VAMOS-GFS upgrade is shown in Fig. 2. VAMOS has natively a $Q_v Q_h (E \times B) D$ optics configuration, restricted in most experiments to $Q_v Q_h D$. Since the dipole is the last optics element, the magnetic field, focal plane detection and beam dump geometries can be optimised for each reaction such as to increase the physical separation between the beam and the fusion-evaporation residues. In practice, the beam dump will be at a fix position on the low magnetic rigidity side, namely $\approx 30°$ with

 DOI: 10.1051/epjconf/201716300059

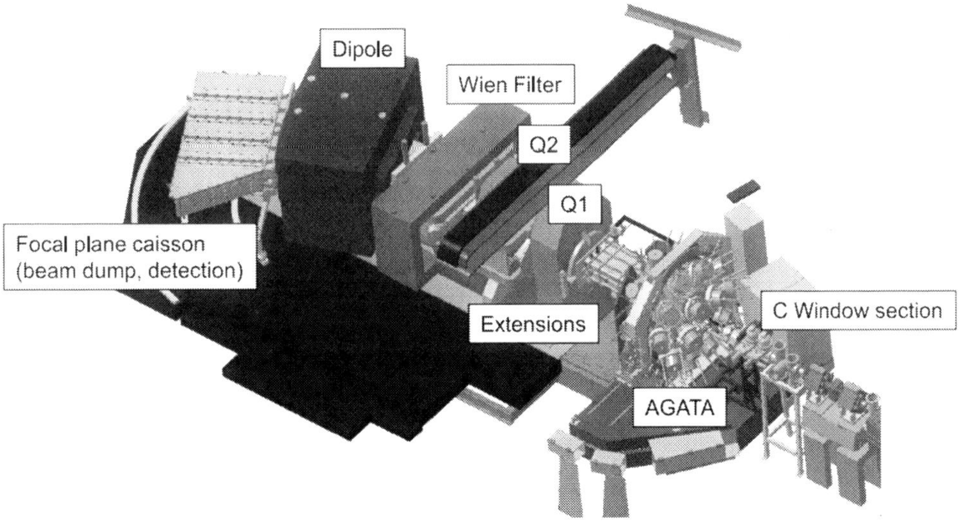

Figure 2. Design view of the VAMOS Gas-Filled separator coupled to the AGATA gamma-ray tracking array.

respect to the dipole exit central angle (therefore constraining the magnetic field to be used), while the recoil detection will be placed according to the magnetic rigidity of the fusion-evaporation residues. This contrasts with existing or foreseen separators for which the detection is at a fixed position on the optics axis. To give an example, the optimum detection position for VAMOS-GFS is at $\approx -5°$ and $\approx 15°$ with respect to the dipole exit central angle for the fusion-evaporation reactions ^{208}Pb(^{48}Ca,2n)^{254}No and ^{58}Ni(^{54}Fe,2n)^{110}Xe, respectively. The beam rejection and transmission will be optimised in such manner. A transmission of $\approx 70\%$ is expected for the two reactions mentioned above.

The physics case covered by VAMOS-GFS includes experiment where it is mandatory to physically filter the nuclei of interest lost in a huge background of parasitic reactions, plus eventually to tag the nuclei using their characteristic decay (Recoil Tagging and Recoil Decay Tagging techniques). Besides the region of the actinides and SHN, the foreseen experiments will also cover the ^{100}Sn region, nuclei at the proton drip-line, and the neutron-deficient lead region. The physics case gathering a large community has been discussed in several collaboration meetings and reviewed by the GANIL program advisory committee. In a longer term, fusion-evaporation reactions used in conjunction with exotic beams will permit to reach exotic isotopes, and/or still un-explored regions of the (angular momentum L, excitation energy E^*) phase space. The interest in fusion-evaporation for reaction dynamics studies is obvious, giving access to the evolution of the system as driven by rotation and/or temperature.

The modifications to upgrade VAMOS imply the following items. The separation between the incoming vacuum beam line and the gas-filled region is performed using a thin Carbon window. At its nominal position, the angular acceptance of VAMOS is ≈ 70 msr. Pushing VAMOS-GFS back with respect to the target position using extensions will help to achieve a better beam control and to increase the maximum $B\rho$ capability, with only a slight decrease of the transmission. The transmission of $\approx 70\%$ quoted above is estimated pushing VAMOS-GFS back by 60 cm. The new focal plane chamber needed to accommodate the variable detection angle has been built. The detection will be based on a modified version of MUSETT, an array of segmented double-sided silicon strip detectors [36], and on a multi-wire proportional chamber for time-of-flight and energy loss measurements. A set of masks will be used to minimise the parasitic ions reaching the detection. Moreover, a shield will be placed between the tantalum beam dump and the focal plane detectors to avoid contamination from the radiations coming from the beam dump. The shield will stop all charged particles, and the fraction of 100 keV photons transmitted will be less than 10^{-6}. No contamination of the prompt detection array by the beam dump is expected since the latter is placed at ≈ 8 m from target point, and since it is shielded by the VAMOS two quadrupoles and by the Wien filter. The VAMOS-GFS project is in its last completion phase.

It should be noted that the VAMOS-GFS project has large similarities with the AGFA project (see the contribution of B.B. Back in these proceedings), with also the magnetic dipole being the last optics element, a similar transmission, and the opportunity to be coupled with the GRETA gamma-ray tracking array.

The laboratories primarily involved in this work are: Grand Accélerateur National d'Ions Lourds (GANIL), Institut de recherche sur les lois fondamentales de l'Univers (CEA, Irfu), Institut de Physique Nucléaire d'Orsay

(IPNO), Institut Pluridisciplinaire Hubert Curien (IPHC), Centre de Sciences Nucléaires et de Sciences de la Matière (CSNSM), Argonne National Laboratory (ANL), KU Leuven, Laboratoire de Physique Corpusculaire de Caen (LPC), and the Institut des NanoSciences de Paris (INSP).

S^3 has been funded by the French Research Ministry, National Research Agency (ANR), through the S^3 EQUIPEX (EQUIPment of EXcellence) reference ANR-10EQPX-46, the FEDER (Fonds Européen de Développement Economique et Régional), the CPER (Contrat Plan Etat-Région), and partially supported by the U.S. Department of Energy, Office of Nuclear Physics, under contract No. DE-AC02-06CH11357, the KU Leuven by the European Commission FP7-INFRASTRUCTURES 2007, SPIRAL2 Preparatory Phase Grant agreement No. 212692. SIRIUS has been funded by the CPIER (Contrat de Plan Interrégional Etat-Région). MUSETT has been funded by the ANR reference ANR-05-BLAN-0150. We benefited from a support of the P2IO labex for performing some optics simulations for VAMOS-GFS and for signal processing studies for SIRIUS.

References

[1] Yu. Ts. Oganessian et al., Nucl. Phys. A **944**, 62 (2015)
[2] K. Morita et al., J. Phys. Soc. Jpn. **73**, 2593 (2004)
[3] P.J. Carol et al., Pure and Applied Chemistry **88**, 139153 (2016)
[4] P.J. Carol et al., Pure and Applied Chemistry **88**, 155160 (2016)
[5] S. Ćwiok, J. Dobaczewski, P.-H. Heenen, P. Magierski and W. Nazarewicz, Nucl. Phys. A **611**, 211 (1996)
[6] J. Dobaczewski, A.V. Afanasjev, M. Bender, L.M. Robledo and Yue Shi, Nucl. Phys. A **944**, 388 (2015)
[7] M. Asai, F.P. Heßberger and A. Lopez-Martens, Nucl. Phys. A **944**, 308 (2015)
[8] D. Ackermann, Nucl. Phys. A **944**, 376 (2015)
[9] Ch. Theisen, P.T. Greenlees, T.-L. Khoo, P. Chowdhury and T. Ishii, Nucl. Phys. A **944**, 333 (2015)
[10] P.T. Greenlees, J. Rubert, J. Piot, B.J.P. Gall et al., Phys. Rev. Lett. **109**, 012501 (2012)
[11] F. Déchery et al., Eur. Phys. J. A **51**, 66 (2015)
[12] R. Ferrer et al., Nucl. Instr. Meth. B **317**, 570 (2013)
[13] DESIR: Désintégration, Excitation et Stockage des Ions Radioactifs = Decay, Excitation and Storage of Radioactive Ions. http://www.cenbg.in2p3.fr/desir/
[14] G.D. Jones, Nucl. Instr. Meth. A **488**, 471 (2002)
[15] Albert Ghiorso, Kari Eskola, Pirkko Eskola, and Matti Nurmia, Phys. Rev. C **7**, 2032 (1973)
[16] R.-D. Herzberg et al., Nature **442**, 896 (2006)
[17] S. K. Tandel et al., Phys. Rev. Lett. **97**, 082502 (2006)
[18] F. P. Heßberger et al., Eur. Phys. J. A **43**, 55 (2010)
[19] R.M. Clark et al., Phys. Lett. B **690**, 19 (2010)
[20] H. B. Jeppesen et al., Phys. Rev. C **79**, 031303(R) (2009)
[21] A.P. Robinson et al., Phys. Rev. C **83**, 064311 (2001)
[22] D. Rudolph et al., Phys. Rev. Lett. **111**, 112502 (2013)
[23] J.M. Gates et al., Phys. Rev. C **92**, 021301 (2015)
[24] F. Xu, E. Zhao, R. Wyss and P. Walker, Phys. Rev. Lett. **92**, 252501 (2004)
[25] S. Hofmann et al. Eur. Phys. J. A **10**, 5 (2001)
[26] David et al., Phys. Rev. Lett. **115**, 132502 (2015)
[27] Yu. Ts. Oganessian et al., Phys. Rev. C **64**, 054606 (2001)
[28] A.V. Belozerov et al., Eur. Phys. J. A **16**, 447 (2003)
[29] D. Peterson et al., Phys. Rev. C **74**, 014316 (2006)
[30] B. Sulignano et al., to be published
[31] D. Ackermann and Ch. Theisen, Accepted for publication in Physica Scripta
[32] S. Pullanhiotan, M. Rejmund, A. Navin, W. Mittig, S. Bhattacharyya, Nucl. Instr. Meth. A **593**, 343 (2008)
[33] S. Akkoyun et al., Nucl. Instr. Meth. A **668**, 26 (2012)
[34] E. Clement et al., Nucl. Instr. Meth. A **855**, 1 (2017)
[35] Ch. Schmitt et al., Nucl. Instr. Meth. A **621**, 558 (2010)
[36] Ch. Theisen et al., Nucl. Instr. Meth. A **747**, 69 (2014)

 DOI: 10.1051/epjconf/201716300060

Fusion and the Discovery of Isotopes

M. Thoennessen[1,*]

[1] National Superconducting Cyclotron Laboratory, Facility for Rare Isotope Beams, and Department of Physics & Astronomy, Michigan State University, East Lansing, MI 48824, USA

Abstract.
Heavy-ion fusion-evaporation has been the most productive method to form and identify new isotopes. Of the presently known over 3200 isotopes, almost 25% were discovered with heavy-ion induced reactions. Since the development of the first heavy-ion beam at Berkeley in 1950 most of the nuclides on the neutron-deficient side of the line of stability and all isotopes of the superheavy elements were discovered utilizing fusion reactions. In addition, some isotopes were first produced in heavy-ion transfer, charge-exchange, incomplete fusion or deep inelastic reactions. The discovery of isotopes relies on new advances in accelerator and detector technology. The continuous development of pioneering and innovative separation and detection techniques have pushed the limit towards – and in many cases beyond – the proton-dripline. A review of the discovery of neutron-deficient and super-heavy nuclides in heavy-ion induced reactions as well as an outlook for the discovery potential in the future is presented.

1 Introduction

The discovery of new isotopes is the first and necessary step towards the study and exploration of more and more exotic nuclei. Exploring the nuclear landscape and pushing towards the limits of nuclear existence is important for the understanding of the strong force and the nucleosynthesis processes in the universe. More than 3200 different isotopes have been discovered until now. A recent compilation summarized details of the discovery of all isotopes [1–5] including the year, laboratory and country of the discovery as well as the production mechanism used to produce the isotopes. The compilation revealed that more than 25% of these isotopes were formed in heavy-ion induced fusion reactions making this method the most productive way to discover new isotopes. Figure 1 shows the accumulated number of isotopes discovered as a function of year for the different production mechanisms. Fusion became the dominant production method in the mid 1980s when it overtook light particle reactions. At the present time the only two mechanisms able to reach new isotopes are fusion evaporation and projectile fragmentation reactions.

The current paper presents an overview of the discovery of neutron-deficient and super-heavy nuclides in heavy-ion induced reactions as well as an outlook for the discovery potential in the future. Previously, summaries focusing on the role of fission [6] and the proton dripline [7] have been presented.

* e-mail: thoennessen@nscl.msu.edu

2 Heavy-Ion Accelerators

The first time the acceleration of heavy-ions was presented at a conference was at the Seattle meeting of the American Physical Society (APS), June 18–21, 1940. Luis Alvarez from the University of California at Berkeley described the acceleration of carbon ions: "The 37-inch cyclotron chamber was filled with CH_4 and a beam of 50 Mev $_6C^{12++++++}$ ions was detected with a linear amplifier. To resolve these ions from alpha-particles, it was necessary to reduce the dee voltage and to adjust the magnetic field to the low side of the alpha-particle peak. Under these conditions, about 500 carbon nuclei entered the ionization chamber per minute" [8].

Subsequent progress at Berkeley was reported by Richard Condit two years later: "It has been found possible to produce $^{12}C^{6}+$ of energy 85 Mev and $^{16}O^{8}+$ of energy 113 Mev with the 60" Berkeley cyclotron. Very small beams of these high speed particles were led from the cyclotron through a thin window into a Wilson cloud chamber of conventional design" [9].

Another four years later, York *et al.* presented further improvements at the APS meeting in Berkeley, July 12–13 1946: "Experiments have been done to produce accelerated stripped light nuclei with the 60" Berkeley cyclotron for use in nuclear experiments. Two types of sources have been investigated: an arc source such as that normally used in cyclotrons and a spark source similar to that used in spectroscopic investigations of highly ionized atoms. Typical yields from an arc source are 1000 $C^{12,+6}$ 146-Mev ions per second and 100,000 $C^{12,+6}$ 135-Mev ions per second..." [10].

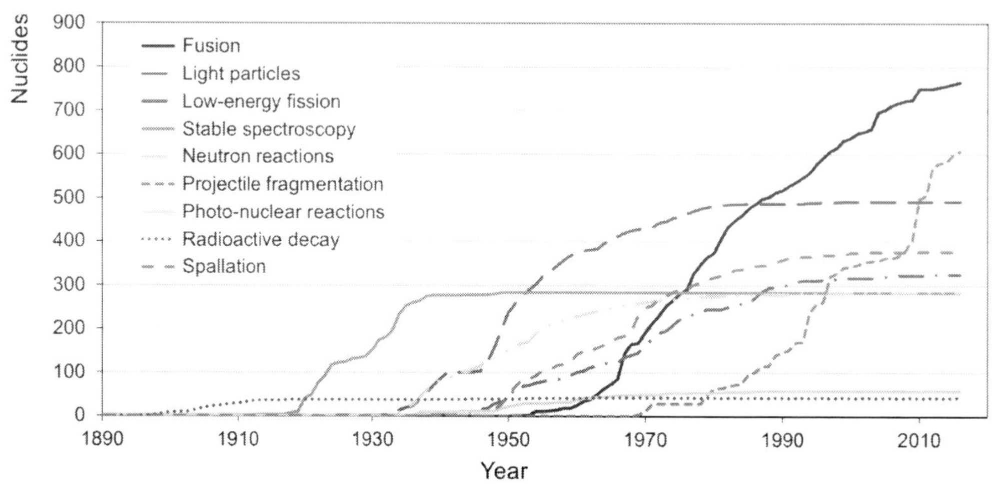

Figure 1. Discovery of nuclides as a function of year. The cumulative sum of the discovered nuclides is shown for different production mechanisms. The line-types and colors are explained in the legend. (Adapted from [4]).

3 Fusion Evaporation Reactions

The first successful fusion reactions of two heavy-ions were reported by Miller *et al.* in the 1950 article published in Physical Review. They bombarded internal aluminum and gold targets and identified ^{34}Cl and ^{205}At, respectively: "The nuclear reactions for the production of these two radio-isotopes are presumably Al^{27}(C^{12},αn)Cl^{34}, Au^{197}(C^{12},4n)At^{205}" [11].

Only a couple of months later Ghiorso discovered the first new isotope with this new technique. They identified ^{246}Cf by irradiating ^{238}U with a beam of 120 MeV ^{12}C. The new californium isotope was then formed in the 4n evaporation channel [12]. It turned out to be the first correct identification of a californium isotope because the earlier published discovery of the element 98 (californium) [13] had misidentified the observed isotope [14].

Soon thereafter it was realized that heavy-ion fusion-evaporation reactions were not only an excellent tool to produce new elements but that they also could be used to populate light neutron-deficient nuclei. As an example for a region of current research interest, Figure 2 shows the neutron deficient section of the chart of nuclides in the medium mass region between antimony and bismuth. Proton and β^+-emitters are indicated by the blue and red boxes, respectively. The calculated proton dripline is shown by the thick black line and unknown isotopes which should be in reach to be discovered in the near future are shown as white boxes.

The figure shows that there are several proton-bound even-Z elements that have not been discovered yet. They range from ^{119,120}Ce and ^{123,124}Nd, to ^{147,148}Yb and ^{151,152}Hf. In addition, for the odd-Z elements there are still unknown isotopes between the last β-decaying isotope and first proton-emitting isotope. They are ^{118,119}La, ^{122,123}Pr, ^{132,133}Eu, and ^{136,137}Tb. They are difficult to detect because the sensitivity for identifying the β^+ decay is not sufficient and the proton decay branch is too weak. In this mass region, promethium is the only odd-Z element for which no proton-emitting isotope has been identified.

The figure also indicates a few isotopes which so far have only been reported in conference proceedings or internal reports (yellow boxes). Souliotis had presented the identification of ^{126}Nd, ^{136}Gd, ^{138}Tb, and ^{143}Ho, ^{150}Yb and ^{153}Hf at the International Conference on Achievements and Perspectives in Nuclear Structure held in Aghia Palaghia, Crete, Greece, July 11-17, 1999 [15], however, the results were subsequently not published in the refereed literature. The observation of ^{143}Ho was also reported in an annual report by Seweryniak *et al.* [16]. In addition, although the discovery of ^{144}Tm was presented at several conferences [17–19] it was never published in a refereed journal.

Over the years fusion-evaporation reactions have become one of the most productive reaction mechanisms to discover new nuclides. In addition to the almost 200 transuranium and superheavy nuclei well over 500 neutron-deficient nuclides were first identified in heavy-ion fusion evaporation reactions. Since 1951, the 750 isotopes discovered by fusion were observed at 34 different institutions in 16 countries.

Figure 3 shows the chart of nuclides highlighting the isotopes discovered by fusion reactions (red). In addition to a large fraction of neutron-deficient nuclides, all isotopes of the superheavy elements (defined as Z≥102) were first produced in reactions induced by heavy ions. The figure also shows the isotopes first observed by projectile fragmentation (green). This method dominates the production of exotic neutron-rich nuclides and is the only competing method to produce new nuclides along the proton dripline, especially in the medium mass region.

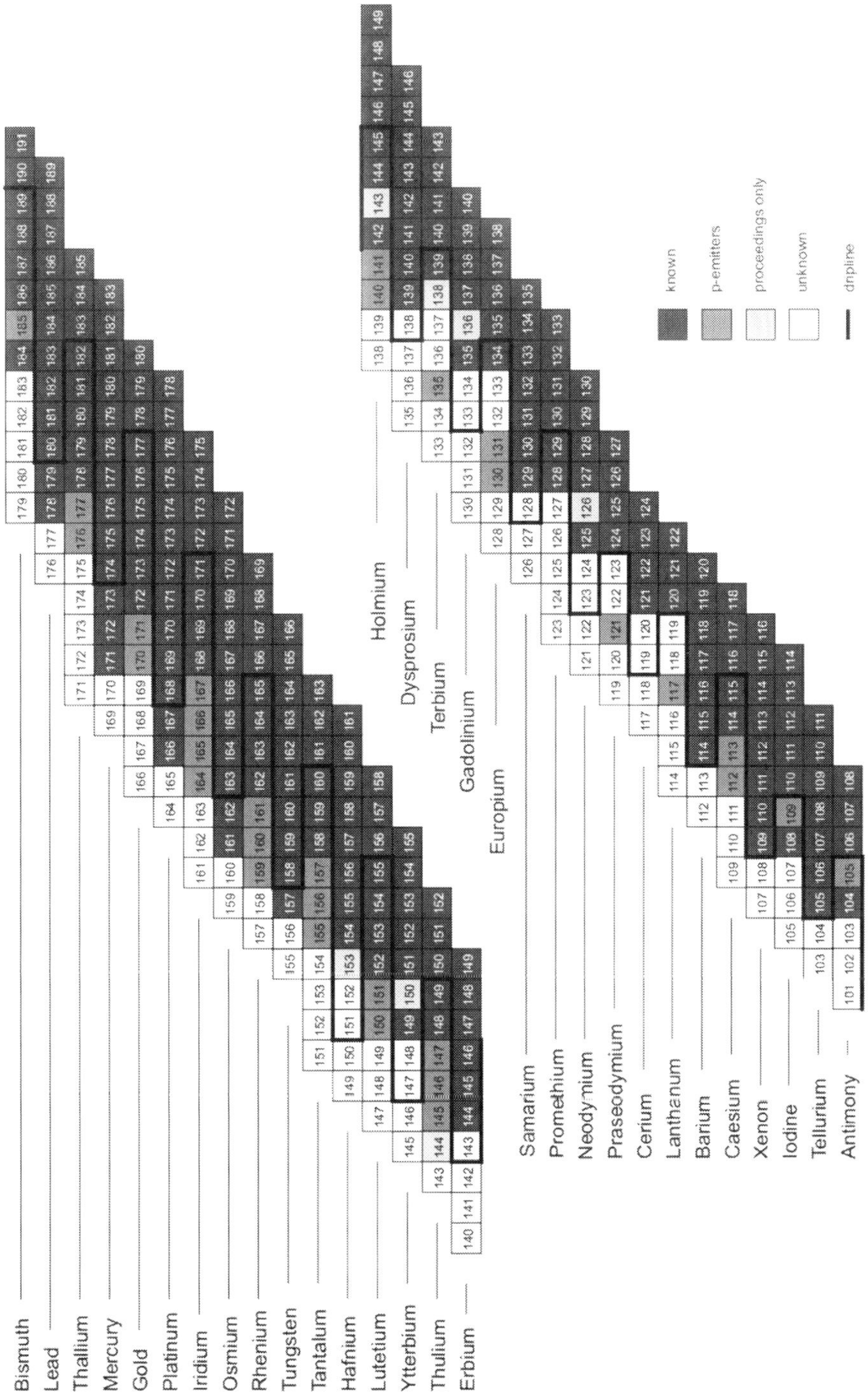

Figure 2. Neutron-deficient chart of the nuclides between antimony and bismuth. The thick black line indicates the proton dripline and the color coding of the isotopes is shown in the legend (Adapted from [3]).

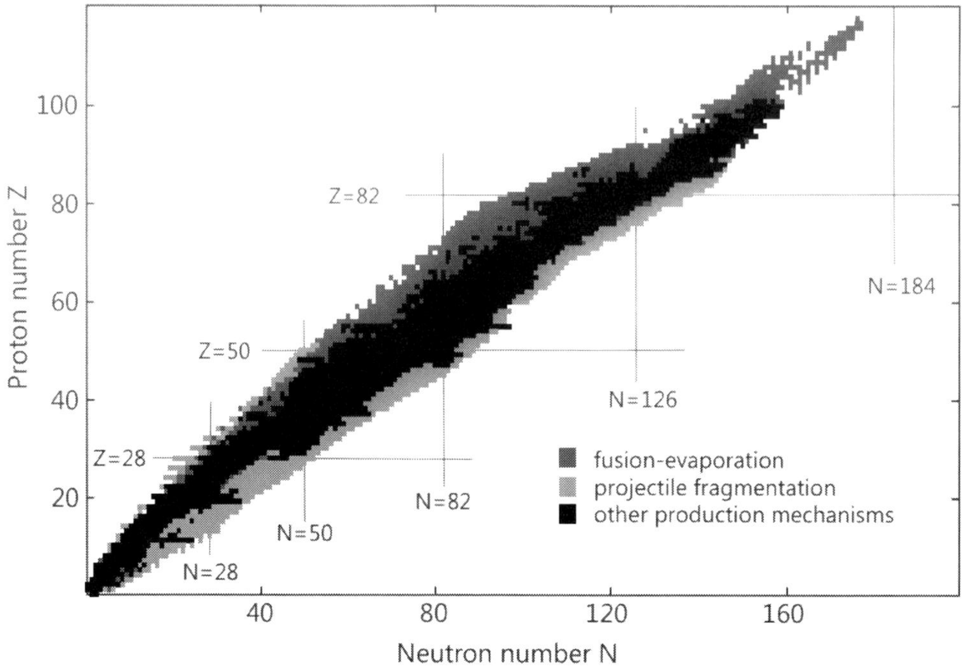

Figure 3. Chart of nuclides. Isotopes discovered by fusion and projectile fragmentation reactions are shown in red and green, respectively. All other known isotopes are shown in black.

Examples of recent discoveries in the heavy mass region are ^{236}Bk and ^{240}Es at Jyväslylä in Finland [20] and ^{230}Am, ^{234}Cm, and ^{234}Bk at RIKEN in Japan [21]. In addition, ^{178}Pb was also observed last year at Jyväslylä [22]. Most recently the proton unbound nuclide ^{17}Na was discovered at Michigan State University [23].

4 Discovery potential

Although most of the yet to be discovered isotopes are on the neutron-rich side of the valley of stability, there are probably still several hundred unknown isotopes at and beyond the proton dripline that should be able to be observed and identified in the future. Most of them should be reachable with fusion-evaporation reactions. Figure 4 shows the chart of nuclides indicating all compound nuclides that can be populated in heavy-ion fusion reactions with all possible combinations of stable projectiles and stable targets [24]. These compound nuclides decay then by multiple neutron evaporation populating even more neutron-deficient isotopes.

It was recently shown that in addition to complete fusion evaporation reactions, low-energy heavy-ions can also be used to induce multi-nucleon transfer reactions in order to reach regions of new isotopes. Devaraja *et al.* reported the observation of five new isotopes by bombarding curium targets with 270 MeV ^{48}Ca ions at GSI in Germany [25]. The population of ^{233}Bk required the addition of one proton and the removal of 16 neutrons from ^{248}Cm. ^{229}Am was then observed in the subsequent α-decay chain. The third isotope, ^{233}Am was populated by removing one proton and 24 neutrons. Although the subsequent α-decay populated the previously unobserved ^{293}Np no properties of this isotopes were measured. In addition Devaraja *et al.* claimed the first observation of ^{216}U, however, the discovery of this isotope had been submitted for publication only a month earlier by Ma *et al.* from Lanzhou, China [26].

Multi nucleon transfer reactions had been proposed for the population of neutron-rich isotopes of superheavy elements by Zagrebaev and Greiner [27]. Thus low-energy heavy-ions might also play a role in the extension of the nuclear chart in this region which is essentially not reachable by any other method. Although fusion of radioactive beams has been suggested as an alternative method to populate isotopes in this region, the current proposed intensities for near-term radioactive beam facilities are not sufficient [28].

5 Summary

In conclusion, fusion-evaporation reactions have been the most productive method to produce new isotopes. Although since about 2010 most of the newly discovered isotopes per year have been observed by projectile fragmentation, fusion reactions continue to make significant contributions. Further developments and improvements of ac-

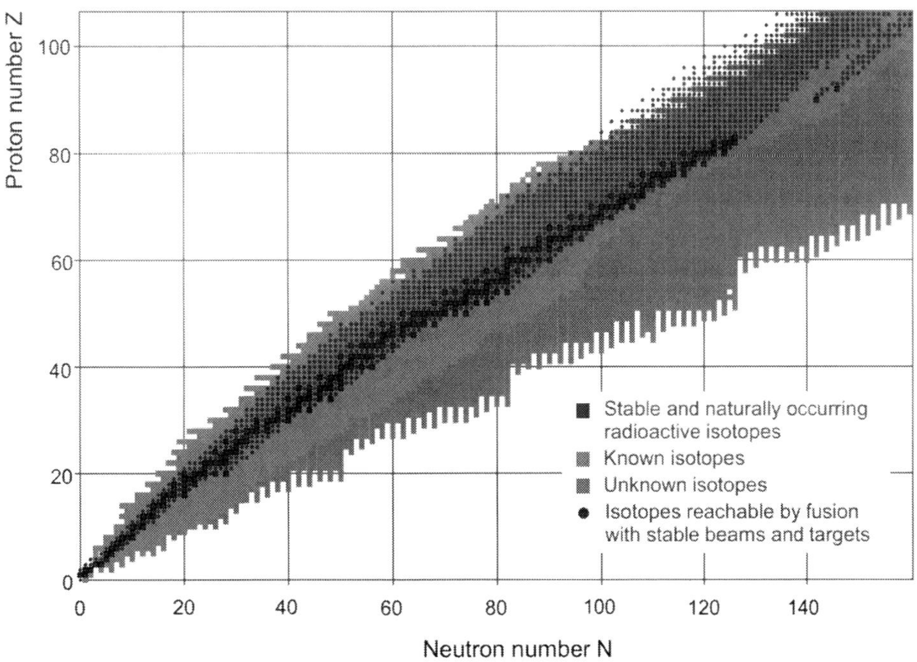

Figure 4. Chart of nuclides: stable (black), presently observed (green) and still unknown (red) nuclides are shown. The small black dots represent compound nuclides that can be populated by fusion reactions with all possible combinations of stable projectiles and stable targets. Dripline nuclides can then be reached by multiple neutron evaporation (Figure adapted from [24]).

celerators and detection and identification techniques are critical for future discoveries to expand the known regions of the nuclear chart. In addition, multi nucleon transfer reactions could also contribute to the observation of new isotopes.

Acknowledgements

Support of the National Science Foundation under grant No. PHY15-65546 is gratefully acknowledged.

References

[1] M. Thoennessen, Nucl. Phys. News **22(3)**, 19 (2012)
[2] M. Thoennessen, Rep. Prog. Phys. **76**, 056301 (2013)
[3] M. Thoennessen, Int. J. Mod. Phys. E **25**, 1630004 (2016)
[4] M. Thoennessen, *The Discovery of Isotopes, A Complete Compilation*, Springer International Publishing 2016, doi:10.1007/978-3-319-31763-2
[5] http://www.nscl.msu.edu/~thoennes/isotopes/
[6] M. Thoennessen, Pramana **85**, 457 (2015)
[7] M. Thoennessen, Nucl. Phys. Rev. **33**, 117 (2016)
[8] L. W. Alvarez, Phys. Rev. **58**, 192 (1940)
[9] R. Condit, Phys. Rev. **62**, 301 (1942)
[10] H. York *et al.*, Phys. Rev. **70**, 446 (1946)
[11] J. F. Miller *et al.*, Phys. Rev. **80**, 486 (1950)
[12] A. Ghiorso *et al.*, Phys. Rev. **81**, 154 (1951)
[13] S. G. Thompson *et al.*, Phys. Rev. **78**, 298 (1950)
[14] A. Chetham-Strode, G. R. Choppin, and B. G. Harvey, Phys. Rev. **102**, 747 (1956)
[15] G. A. Souliotis, Physica Scripta **T88**, 153 (2000)
[16] D. Seweryniak *et al.*, ANL-03/23 (Physics Division Ann. Rept., 2002), p.31 (2003)
[17] R. Grzywacz *et al.*, Eur. Phys. J. A **25**, s145 (2005)
[18] K. P. Rykaczewski *et al.*, AIP Conf. Proc. **764**, 223 (2005)
[19] C. R. Bingham *et al.*, Nucl. Instrum. Meth. B **241**, 185 (2005)
[20] J. Konki *et al.*, Phys. Lett. B **764**, 265 (2017)
[21] D. Kaji *et al.*, J. Phys. Soc. Japan **85**, 015002 (2016)
[22] H. Badran *et al.*, Phys. Rev. C **94**, 054301 (2016)
[23] K. W. Brown *et al.*, Phys. Rev. C **94**, 044326 (2017)
[24] M. Thoennessen, Rep. Prog. Phys. **67**, 1187 (2004)
[25] H. M. Devaraja *et al.*, Phys. Lett. B **748**, 199 (2015)
[26] L. Ma *et al.*, Phys. Rev. C **91**, 051302 (2015)
[27] V. Zagrebaev and W. Greiner, Phys. Rev. C **78**, 034610 (2008)
[28] W. Loveland, Phys. Rev. C **76**, 014612 (2007)

DOI: 10.1051/epjconf/201716300061

Time-dependent mean-field investigations of the quasifission process

A.S. Umar[1,*], C. Simenel[2,**], and S. Ayik[3,***]

[1]Department of Physics and Astronomy, Vanderbilt University, Nashville, TN 37235, USA
[2]Department of Nuclear Physics, Research School of Physics and Engineering, The Australian National University, Canberra ACT 2601, Australia
[3]Physics Department, Tennessee Technological University, Cookeville, TN 38505, USA

Abstract. We demonstrate that the microscopic Time-dependent Hartree-Fock (TDHF) theory provides an important approach to shed light on the nuclear dynamics leading to the formation of superheavy elements. In particular, we discuss studying quasifission dynamics and calculating ingredients for compound nucleus formation probability calculations. We also discuss possible extensions to TDHF to address the distribution of observables.

1 Introduction

Synthesis of superheavy elements (SHE) [1] with fusion-evaporation reactions is strongly hindered by the quasifission (QF) mechanism which prevents the formation of an equilibrated compound nucleus and depends on the structure of the reactants. Within the last few years the time-dependent Hartree-Fock (TDHF) approach [2–4] has been utilized for studying the dynamics of quasifission [3, 5–14] and scission dynamics [15–20]. Such calculations are now numerically feasible to perform on a 3D Cartesian grid without any symmetry restrictions and with much more accurate numerical methods [21–24].

Much progress has been obtained by the community working on developments and applications of the time-dependent mean-field approaches to nuclear dynamics since the last edition of the FUSION conference [25–30]. Particularly, the study of quasifission is showing a great promise to provide insight based on very favorable comparisons with experimental data [8, 10]. These include comparison with experimental kinetic energy (TKE) and mass-angle distributions (MAD). In particular, the dependence of quasifission on the orientation of deformed nuclei [31–34], time-scales for the quasifission process [35–37], the connection between the sticking times and the rotation of the compound nucleus [35, 38], influence of magic shells [8, 39–41] are some of the experimental insights obtained from the theory. Similarly, an extension of TDHF called the density-constrained TDHF [42, 43] (DC-TDHF) has been used to obtain microscopic potential barriers, capture cross-sections, and excitation energies for superheavy [44, 45] and lighter systems [46, 47].

*e-mail: umar@compsci.cas.vanderbilt.edu
**e-mail: cedric.simenel@anu.edu.au
***e-mail: Ayik@tntech.edu

2 Insights from TDHF

2.1 Capture cross-sections

The experiments to discover new elements are notoriously difficult, with fusion evaporation residue (ER) cross-section in pico-barns. This cross-section is commonly expressed in the product form

$$\sigma_{\mathrm{ER}} = \sum_{L=0}^{J_{\max}} \sigma_{\mathrm{cap}}(E_{\mathrm{c.m.}}, L) P_{\mathrm{CN}}(E^*, L) W_{\mathrm{sur}}(E^*, L) , \quad (1)$$

where $\sigma_{\mathrm{cap}}(E_{\mathrm{c.m.}}, L)$ is the capture cross section at center of mass energy $E_{\mathrm{c.m.}}$ and angular momentum L. P_{CN} is the probability that the composite system fuses into a compound nucleus (CN) rather than breaking up via quasifission, and W_{sur} is the survival probability of the fused system against fission.

The TDHF time-evolution coupled with the DC-TDHF approach have been used to calculate capture cross-sections (cross-section for the system to be trapped in a potential pocket) for a variety of systems [44, 45]. Calculation of capture cross-sections for reactions involving deformed nuclei are particularly difficult since an average over all orientations of the deformed nucleus with respect to the beam direction has to be taken [46]. In Fig. 1 we plot the calculated capture cross-sections with the corresponding experimental ones. We see that there is a very reasonable agreement with the experiment for these systems. The dashed line shows a linear dependence for systems with ^{208}Pb as the target. The systems with ^{238}U target fall on another line with different slope [35, 48, 49].

2.2 Mass-angle distributions

Study of quasifission together with capture are intimately related to understanding the process for forming a compound nucleus, the quantity named P_{CN} in Eq. 1 [50]. Figure 2 shows the time-evolution of the ^{50}Ti + ^{249}Bk reaction

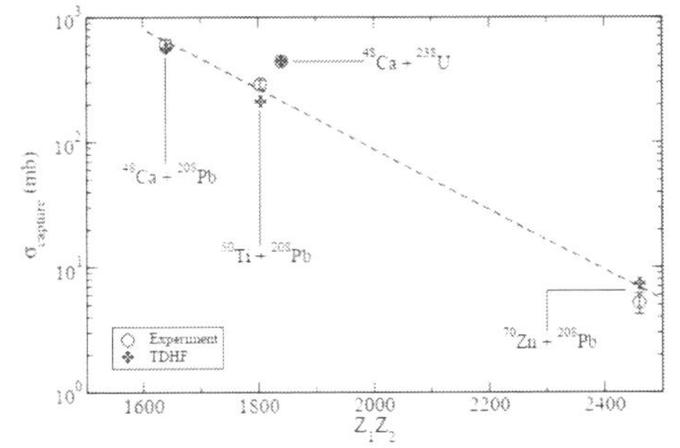

Figure 1. Capture cross-sections calculated using TDHF and the corresponding experimental values plotted as a function of Z_1Z_2. The dashed line shows a linear dependence for systems with ^{208}Pb as the target. The systems with ^{238}U target fall on another line with different slope [35, 48, 49].

at $E_{\mathrm{c.m.}} = 233$ MeV and impact parameter $b = 1.0$ fm [14]. For this impact parameter and energy TDHF theory predicts quasifission. As the nuclei approach each other, a

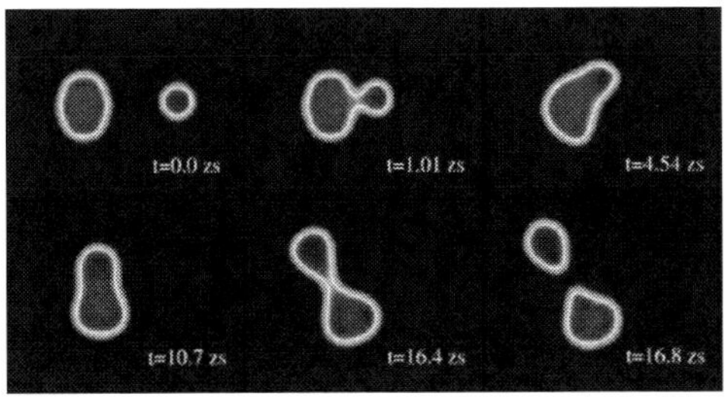

Figure 2. Quasifission in the reaction ^{50}Ti +^{249} Bk at $E_{\mathrm{c.m.}}$ = 233 MeV with impact parameter $b = 1.0$ fm. Shown is a contour plot of the time evolution of the mass density. Time increases from left to right and top to bottom.

neck forms between the two fragments which grows in size as the system begins to rotate. Due to the Coulomb repulsion and centrifugal forces, the dinuclear system elongates and forms a very long neck which eventually ruptures leading to two separated fragments. In this case, the contact time is found to be 16 zs.

The utilization of TDHF to study quasifission can serve to gain insights about this reaction process that is not easily available from the experiments. Experimentally, the produced fragments in the collisions of relatively heavy systems originate from deep-inelastic, quasifission, and fusion-fission events. Fragments produced from deep-inelastic collisions correspond to smaller mass/charge exchange, while the fragments from the fusion-fission peak around the symmetric breakup of the equilibrated excited compound nucleus. Fragments coming from the quasifission process fill the region in between but can also contribute to the symmetric breakup region. Sometimes these are referred to as fast-quasifission (t≈few-35 zs) versus slow-quasifission (t>35 zs). We observe that the quasifission is a dynamical process and does not correspond to an equilibrated system. One of the best ways to visualize the distribution of produced fragments is through the so-called mass-angle distribution (MAD) [32, 35]. In Fig. 3 we show MADs corresponding to two reactions (a) ^{40}Ca+^{238}U at energy $E_{\mathrm{c.m.}}/V_B = 1.142$ [8], and (b) ^{54}Cr+^{186}W at energy $E_{\mathrm{c.m.}}/V_B = 1.130$ [10], where V_B corresponds to the Bass barrier height. The plot shows the detected fragments with respect to their c.m. detection angle and mass-ratio $M_R = M_1/(M_1 + M_2)$. The filled symbols show the fragments obtained from TDHF calculations marked by the value of the initial orbital angular momentum in units of $\hbar$ (a) or the impact parameter in units of fm (b). The multiplicity of the experimental masses are shown in the adjoint legends. These reactions illustrate very clearly the deformation dependence of the quasifission products. For the highly asymmetric reactions involving an actinide nucleus such as the ^{40}Ca+^{238}U system, the large deformation of Uranium has a major impact on quasifission products. As we see in Fig. 3(a) the TDHF results originating from collisions with the tip of the Uranium (purple symbols) contribute mainly to the mass asymmetric regions, whereas collisions with the side of the Uranium (cyan symbols) provide significant contribution to the more mass symmetric region. For the less mass asymmetric and neutron-rich system ^{54}Cr+^{186}W shown in Fig. 3(b) TDHF contributes less to the mass symmetric region and the effects of deformation has largely disappeared. The primary reason for the tip orientation resulting in a more asymmetric breakup is due to the fact that for this orientation nuclei come into contact earlier and rapid excitation builds up while the system is still in a less compact configuration [45]. We also note that the calculations correctly produce the transition from deep-inelastic to the quasifission regime. In addition, TDHF calculations also explain the dependence of MADs on the beam energy [8], as well as the dependence on the neutron-richness of the target and/or projectile [10].

Other quantities that can be studied are the relationship between the mass/charge transfer and rotation angle with the contact-time (time interval from initial contact to scission). In Fig. 4 we show these quantities for ^{54}Cr+^{186}W at energy $E_{\mathrm{c.m.}}/V_B = 1.130$. In Fig. 4 we plot the rotation angle (a) and mass/charge transfer (b) as a function of contact time (zs) for ^{54}Cr+^{186}W at energy $E_{\mathrm{c.m.}}/V_B = 1.130$. The numbers adjoint to points indicate the impact parameter in units of fm. We note from Fig. 4(a) that the rotation angle generally increases for smaller impact parameters with the caveat that for small impact parameters the relationship may not be linear i.e. the rotation angle for $b = 2.4$ fm is longer than that for $b = 2.0$ fm. This was also observed for other systems [14]. A similar behavior is observed in the mass and charge transfer for the tip orientation as seen in Fig. 4(b).

2.3 Fragment TKEs

The total kinetic energy of the fission fragments has been investigated experimentally for several systems [39–41] The quasi-fission contact times are long enough to enable the conversion of the initial relative kinetic energy

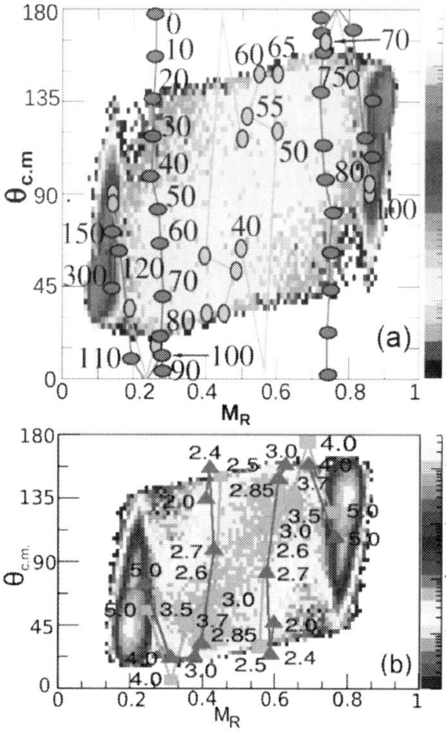

Figure 3. MADs corresponding to two reactions (a) $^{40}Ca+^{238}U$ at energy $E_{c.m.}/V_B = 1.142$ [8], and (b) $^{54}Cr+^{186}W$ at energy $E_{c.m.}/V_B = 1.130$ [10], where V_B is the corresponding Bass barrier height. The filled symbols show the fragments obtained from TDHF calculations marked by the value of the initial orbital angular momentum in units of $\hbar$ (a) or the impact parameter in units of fm (b). The density of the experimental masses are shown in the adjoint legends.

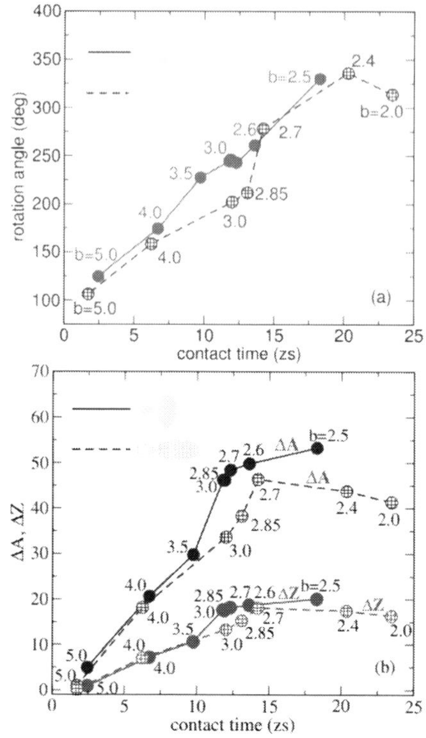

Figure 4. Shown are the (a) rotation angle and (b) mass/charge transfer as a function of contact time (zs) for $^{54}Cr+^{186}W$ at energy $E_{c.m.}/V_B = 1.130$. The numbers adjoint to points indicate the impact parameter in units of fm.

in to internal excitations. Experimentally, the measured total kinetic energy (TKE) of the quasi-fission fragments in $^{40,48}Ca+^{238}U$ reactions is in relatively good agreement with the Viola systematics [34, 35]. The TDHF approach contains one-body dissipation mechanisms which are dominant at near-barrier energy. It can then be used to predict the final TKE of the fragments. The TKE of the fragments formed in $^{40}Ca+^{238}U$ have been computed for a range of central collisions up to 10% above the barrier. Figure 5 shows that the TDHF predictions of TKE are in excellent agreement with the Viola systematics [51, 52]. This indicates that the relative TKE of the quasi-fission fragments are primarily due to their Coulomb repulsion and do not carry a fraction of the initial TKE as is the case for deep-inelastic collisions.

However, a detailed study of the $^{48}Ca,^{50}Ti+^{249}Bk$ systems also shows an influence of the orientation on the TKE [14]. This is illustrated in Fig. 6 which shows that collisions with the tip are less dissipated than collisions with the side of the actinide. Recent calculations for the

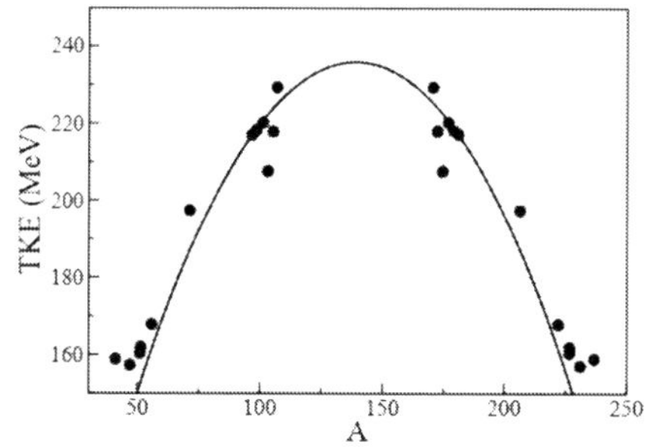

Figure 5. TKE of both the light and heavy fragments formed in $^{40}Ca+^{238}U$ central collisions at $E_{c.m.}/V_B = 1.0 - 1.1$. The solid line represents TKE values based on the Viola formula [51].

$^{64}Ni+^{238}U$ system [53] system compare favorably to the experiment [48].

2.4 Shape evolutions

The proper characterization of fusion-fission and quasifission is one of the most important tasks in analyzing reac-

 DOI: 10.1051/epjconf/201716300061

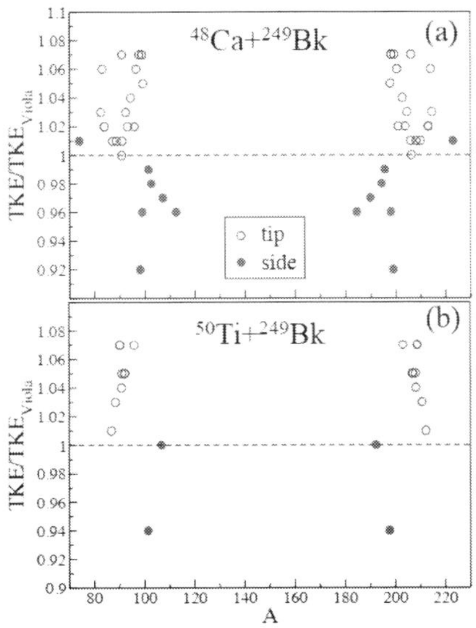

Figure 6. TKE-mass correlations in (a) ^{48}Ca+^{249}Bk and (b) ^{50}Ti+^{249}Bk. Tip (side) orientations are plotted with open (filled) symbols. The TKE is normalized to the TKE from Viola systematics using the masses and charges of the fragments in the exit channel. Adapted from [14].

tions leading to superheavy elements. Experimental analysis of fusion-fission and quasifission fragment angular distributions $W(\theta)$ is commonly expressed in terms of a two-component expression [54–58],

$$W(\theta) = \sum_{J=0}^{J_{CN}} \mathcal{F}_J^{(FF)}(\theta, K_0^2(\alpha)) + \sum_{J=J_{CN}}^{J_{max}} \mathcal{F}_J^{(QF)}(\theta, K_0^2(\alpha)), \quad (2)$$

where $\alpha \equiv FF$ (fusion-fission) or QF (quasifission). Here, J_{CN} defines the boundary between fusion-fission and quasifission, assuming a sharp cutoff between the angular momentum distributions of each mechanism.

The quantum number K is known to play an important role in fission [59]. The latter is defined as the projection of the total angular momentum along the deformation axis. In the Transition State Model (TSM) [59], the characteristics of the fission fragments are determined by the K distribution at scission. The argument K_0 entering Eq. 2 is the width of this distribution which is assumed to be Gaussian. It obeys $K_0^2 = T\Im_{eff}/\hbar^2$, where the effective moment of inertia, $\Im_{eff}$, is computed from the moments of inertia for rotations around the axis parallel and perpendicular to the principal deformation axis $\frac{1}{\Im_{eff}} = \frac{1}{\Im_\parallel} - \frac{1}{\Im_\perp}$, and T is the nuclear temperature at the saddle point. The physical parameters of the fusion-fission part are relatively well known from the liquid-drop model [60, 61]. In contrast, the quasifission process never reaches statistical equilibrium. In principle, it has to be treated dynamically, while Eq. (2) is based on a statistical approximation. In addition, the usual choice for the nuclear moment of inertia for the quasifission component, $\Im_0/\Im_{eff} = 1.5$ [50, 55, 62], is somewhat arbitrary. Here, $\Im_0$ is the moment of inertia of an equivalent spherical nucleus.

We have developed methods to extract these ingredients directly from TDHF time-evolution of collisions resulting in quasifission [12]. The main collective observable of interest for fission and quasifission (both dynamical and statistical) studies is the moment of inertia of the system. The proper way to calculate the moment-of-inertia for such time-dependent densities (particularly for non-zero impact parameters) is to directly diagonalize the moment-of-inertia tensor represented by a 3×3 matrix with elements

$$\Im_{ij}(t)/m = \int d^3r\, \rho(\mathbf{r},t)(r^2\delta_{ij} - x_i x_j) , \quad (3)$$

where ρ is the local number-density calculated from TDHF evolution, m is the nucleon mass, and $x_{i=1,2,3}$ denote the Cartesian coordinates. Numerical diagonalization the matrix $\Im$ gives three eigenvalues. One eigenvalue corresponds to the moment-of-inertia $\Im_\parallel$ for the nuclear system rotating about the principal axis. The other two eigenvalues define the moments of inertia for rotations about axes perpendicular to the principal axis.

Using the time-dependent moment-of-inertia obtained from the TDHF collision one can calculate the so-called effective moment-of-inertia defined above. It is standard to compute the effective moment of inertia relative to a spherical system using the mass independent quantity $\Im_0/\Im_{eff}$, where $\Im_0$ is the moment-of-inertia of a spherical nucleus with the same number of nucleons [50, 58].

In our publication [12] we have calculated the moment-of-inertia ratio for the ^{48}Ca + ^{249}Bk non-central collisions at $E_{c.m.} = 218$ MeV. At the point of final touching configuration the moment-of-inertia ratios are in the range 1.4 – 1.8, suggesting a relatively strong impact parameter dependence. We have also studied the energy, impact parameter, and orientation dependence of this ratio and shown that a simple impact parameter modeling of the ratio may not be appropriate. Some of these methods have been utilized in experimental papers [63].

3 Beyond TDHF Calculations

The mean-field description of reactions using TDHF provides the mean values of the proton and neutron drift. It is also possible to compute the probability to form a fragment with a given number of nucleons [22, 64–66], but the resulting fragment mass and charge distributions are often underestimated in dissipative collisions [67, 69]. Much effort has been done to improve the standard mean-field approximation by incorporating the fluctuation mechanism into the description. At low energies, the mean-field fluctuations make the dominant contribution to the fluctuation mechanism of the collective motion. Various extensions have been developed to study the fluctuations of one-body observables. These include the TDRPA approach of Balian and Vénéroni [68], the time-dependent generator coordinate method [70], or the stochastic mean-field (SMF) method [71]. The effects of two-body dissipation on reactions of heavy systems using the TDDM [72, 73], approach have also been recently reported [74, 75]. Here we discuss some recent results using the SMF method [76].

In the stochastic mean-field (SMF) approach, the fluctuations in the initial state are incorporated in a stochastic manner by introducing a proper distribution the initial single-particle density matrices [71]. This results in an ensemble of single-particle density matrices generated by evolving each density in its own mean-field Hamiltonian. By calculating the expectation values of an observable in each event, it is possible to determine probability distributions of observables. In a number of studies, it has been shown that the SMF approach provides a good approximation for fluctuations of the collective motion. In particular, in small amplitude limit, the approach gives rise to the same expression for the dispersion of one-body observables familiar from the variational approach of Balian and Vénéroni [68]. In the di-nuclear regime, the SMF approach gives rise to Langevin description for nucleon exchange between projectile-like and target-like nuclei characterized by diffusion and drift coefficients [77]. It is well known that the Langevin description is equivalent to the evolution of the distribution function of the collective variable according to the Fokker-Plank equation. As a result of this equivalence, instead of carrying out stochastic simulations, it is more convenient to calculate the transport coefficients and employ the Fokker-Plank approach. When the drift coefficients are linear functions of mass and charge asymmetry variables, the primary fragment charge and mass distribution $P(N, Z, t)$, is given by a correlated Gaussian in the neutron-proton plane. The distribution function is specified by the mean neutron, $\overline{N}$, and mean proton, $\overline{Z}$, numbers of target-like or project-like fragments, and co-variances σ_{NN}, σ_{ZZ}, and σ_{NZ}. The co-variances are determined by a set of coupled differential equations in which the inputs are provided by the neutron and proton diffusion and the drift coefficients. In the SMF approach these transport coefficients are calculated in terms of the solutions of the TDHF equations. Calculations take into account the full collision geometry and do not involve any fitting parameter other than the standard parameters of the Skyrme interaction. In the earlier investigations, transport coefficients were calculated in the semi-classical approximation in Markovian limit [76, 78]. Recently, we were able to calculate these transport coefficients in the quantal framework by including the shell structure and the Pauli blocking in an exact manner [79]. As an example, in Fig. 7 we show the result of nucleon diffusion calculations in the central collisions of $^{40}Ca+^{238}U$ system at bombarding energies $E_{cm} = 202$ MeV. Figure 7 illustrates a few equal probability lines for primary population of the target-like fragments at the exit channel in the (N, Z) plane. Probability of populating a fragment with neutron and proton numbers (N, Z) relative to the fragment with mean neutron numbers $(\overline{N}, \overline{Z})$ is determined by e^{-C}, where C indicates numbers on the equal probability lines shown in Fig. 7. In this figure dot at the centers of ellipses indicates the elements with the mean neutron and proton number at the exit channel. The mean values of neutron and proton numbers at the exit channel are $(\overline{N} = 125, \overline{Z} = 82)$ and $(\overline{N} = 137, \overline{Z} = 86)$ in Ca induced collisions. For example, the probability of populating a heavy trans-uranium primary fragment with $(\overline{N} = 155, \overline{Z} = 98)$ relative to the probability of populating the element with mean neutron and proton numbers is about $exp(-0.5) = 0.6$. We note that the correlated Gaussian function specified by the first two moments provides an approximate description of the primary fragment population. The approximation is reasonable within the ellipse with $C = 1.0$ around the center point, but becomes gradually unreasonable as we move out from the center points near to the tail of the distribution function. For example, as seen beyond the upper ends of the ellipse with $C = 1.5$, we observe finite but small probabilities for populating fragments even exceeding the total mass of the system. Therefore, in particular near the tail region, more accurate description of the fragment population probability is required. Also we, note that the primary fragments de-excite by fission and by light particle emission. This de-excitation process is not incorporated into the calculations.

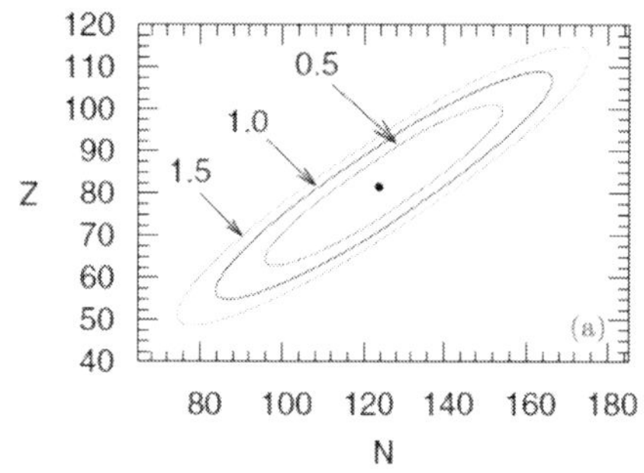

Figure 7. Primary distribution of the target-like fragments produced in central collision of $^{40}Ca+^{238}U$ at $E_{cm} = 202$ MeV. Lines indicate equal production probability $exp(-C)$ relative to the central fragment with $C = 0.5$, 1.0, and 1.5. Calculated in Stochastic mean-field approach in semi-classical limit.

This work has been supported by the Australian Research Council Grants Nos. FT120100760 and DP160101254, and by the U.S. Department of Energy under grant Nos. DE-SC0013847 and DE-SC0015513.

References

[1] Ch. E. Düllmann *et al.*, Nucl. Phys. A **944**, 1 (2015), Special Issue on Superheavy Elements
[2] J.W. Negele, Rev. Mod. Phys. **54**, 913 (1982)
[3] C. Simenel, Eur. Phys. J. A **48**, 152 (2012)
[4] T. Nakatsukasa *et al.*, Rev. Mod. Phys. **88**, 045004 (2016)
[5] C. Golabek, C. Simenel, Phys. Rev. Lett. **103**, 042701 (2009)

DOI: 10.1051/epjconf/201716300061

[6] D.J. Kedziora, C. Simenel, Phys. Rev. C **81**, 044613 (2010)
[7] C. Simenel *et al.*, Phys. Lett. B **710**, 607 (2012)
[8] A. Wakhle *et al.*, Phys. Rev. Lett. **113**, 182502 (2014)
[9] V.E. Oberacker *et al.*, Phys. Rev. C **90**, 054605 (2014)
[10] K. Hammerton *et al.*, Phys. Rev. C **91**, 041602(R) (2015)
[11] A.S. Umar, V.E. Oberacker, Nucl. Phys. A **944**, 238 (2015)
[12] A.S. Umar *et al.*, Phys. Rev. C **92**, 024621 (2015)
[13] K. Sekizawa, K. Yabana, Phys. Rev. C **93**, 029902 (2016)
[14] A.S. Umar *et al.*, Phys. Rev. C **94**, 024605 (2016)
[15] C. Simenel, A.S. Umar, Phys. Rev. C **89**, 031601(R) (2014)
[16] G. Scamps *et al.*, Phys. Rev. C **92**, 011602 (2015)
[17] C. Simenel *et al.*, EPJ Web. Conf. **107**, 07001 (2016)
[18] P.M. Goddard *et al.*, Phys. Rev. C **92**, 054610 (2015)
[19] P.M. Goddard *et al.*, Phys. Rev. C **93**, 014620 (2016)
[20] A. Bulgac *et al.*, Phys. Rev. Lett. **116**, 122504 (2016)
[21] A.S. Umar, V.E. Oberacker, Phys. Rev. C **73**, 054607 (2006)
[22] K. Sekizawa, K. Yabana, Phys. Rev. C **88**, 014614 (2013)
[23] J.A. Maruhn *et al.*, Comp. Phys. Comm. **185**, 2195 (2014)
[24] B. Schuetrumpf *et al.*, Phys. Rev. C **93**, 054304 (2016)
[25] G. Scamps, D. Lacroix, EPJ Web. Conf. **86**, 00042 (2015)
[26] K. Sekizawa, K. Yabana, EPJ Web of Conferences **86**, 00043 (2015)
[27] C. Simenel *et al.*, EPJ Web. Conf. **86**, 00047 (2015)
[28] P.D. Stevenson *et al.*, EPJ Web. Conf. **86**, 00058 (2015)
[29] A. Wakhle *et al.*, EPJ Web of Conferences **86**, 00061 (2015)
[30] K. Washiyama, EPJ Web of Conferences **86**, 00062 (2015)
[31] D.J. Hinde *et al.*, Phys. Rev. C **53**, 1290 (1996)
[32] D.J. Hinde *et al.*, Phys. Rev. Lett. **100**, 202701 (2008)
[33] K. Nishio *et al.*, Phys. Rev. C **77**, 064607 (2008)
[34] K. Nishio *et al.*, Phys. Rev. C **86**, 034608 (2012)
[35] J. Töke *et al.*, Nucl. Phys. A **440**, 327 (1985)
[36] R. du Rietz *et al.*, Phys. Rev. Lett. **106**, 052701 (2011)
[37] R. du Rietz *et al.*, Phys. Rev. C **88**, 054618 (2013)
[38] R.G. Thomas *et al.*, Phys. Rev. C **77**, 034610 (2008)
[39] M.G. Itkis *et al.*, Nucl. Phys. A **734**, 136 (2004)
[40] G.N. Knyazheva *et al.*, Phys. Rev. C **75**, 064602 (2007)
[41] E.M. Kozulin *et al.*, Phys. Rev. C **89**, 014614 (2014)
[42] A.S. Umar, V.E. Oberacker, Phys. Rev. C **74**, 021601(R) (2006)
[43] A.S. Umar *et al.*, Phys. Rev. C **80**, 041601(R) (2009)
[44] V.E. Oberacker *et al.*, Phys. Rev. C **82**, 034603 (2010)
[45] A.S. Umar *et al.*, Phys. Rev. C **81**, 064607 (2010)
[46] A.S. Umar, V.E. Oberacker, Phys. Rev. C **74**, 061601 (2006)
[47] A.S. Umar *et al.*, Phys. Rev. C **85**, 055801 (2012)
[48] E.M. Kozulin *et al.*, Phys. Rev. C **94**, 054613 (2016)
[49] R. Bock *et al.*, Nucl. Phys. A **388**, 334 (1982)
[50] R. Yanez *et al.*, Phys. Rev. C **88**, 014606 (2013)
[51] V.E. Viola *et al.*, Phys. Rev. C **31**, 1550 (1985)
[52] D.J. Hinde *et al.*, Phys. Rev. C **45**, 1229 (1992)
[53] K. Sekizawa, K. Yabana, Phys. Rev. C **93**, 054616 (2016)
[54] J.R. Huizenga *et al.*, Phys. Rev. **177**, 1826 (1969)
[55] B.B. Back, Phys. Rev. C **31**, 2104 (1985)
[56] B.B. Back *et al.*, Phys. Rev. C **32**, 195 (1985)
[57] J.G. Keller *et al.*, Phys. Rev. C **36**, 1364 (1987)
[58] M.B. Tsang *et al.*, Phys. Lett. B **129**, 18 (1983)
[59] R. Vandenbosch, J.R. Huizenga, *Nuclear Fission* (Academic Press, New York, 1973)
[60] A.J. Sierk, Phys. Rev. C **33**, 2039 (1986)
[61] S. Cohen *et al.*, Ann. Phys. **82**, 557 (1974)
[62] W. Loveland, Eur. Phys. J. A **51**, 120 (2015)
[63] E. Prasad *et al.*, Phys. Rev. C **93**, 024607 (2016)
[64] S.E. Koonin *et al.*, Phys. Rev. C **15**, 1359 (1977)
[65] C. Simenel, Phys. Rev. Lett. **105**, 192701 (2010)
[66] G. Scamps, D. Lacroix, Phys. Rev. C **87**, 014605 (2013)
[67] C.H. Dasso *et al.*,Z. Phys. A **289**, 395 (1979)
[68] R. Balian, M. Vénéroni, Ann. Phys. **216**, 351 (1992)
[69] C. Simenel, Phys. Rev. Lett. **106**, 112502 (2011)
[70] H. Goutte *et al.*, Phys. Rev. C **71**, 024316 (2005)
[71] S. Ayik, Phys. Lett. B **658**, 174 (2008)
[72] M. Tohyama, Phys. Lett. B **160**, 235 (1985)
[73] M. Tohyama, A.S. Umar, Phys. Lett. B **549**, 72 (2002)
[74] Marlène Assié *et al.*,Phys. Rev. Lett. **102**, 202501 (2009)
[75] M. Tohyama, A.S. Umar, Phys. Rev. C **93**, 034607 (2016)
[76] S. Ayik *et al.*, Phys. Rev. C **92**, 064615 (2015)
[77] S. Ayik *et al.*, Phys. Rev. C **79**, 054606 (2009)
[78] B. Yilmaz *et al.*, Phys. Rev. C **90**, 024613 (2014)
[79] S. Ayik *et al.*, Phys. Rev. C **94**, 044624 (2016)

Fusion with exotic nuclei using a microscopic approach

K. Vo-Phuoc[1,*], C. Simenel[1], and E. C. Simpson[1]

[1] Department of Nuclear Physics, Australian National University, ACT 2601, Australia

Abstract. Fusion reactions near the barrier energy are studied in this proceedings with the Hartree–Fock (HF) method, on systems involving $^{40-54}$Ca+^{116,132}Sn. Both static and time-dependent (TDHF) versions of the method are used to probe what structure effects play a role in the fusion barrier energy. In static HF calculations for fusion, ground state properties of the nuclei influence the barrier energy. When dynamics are added in with TDHF, some effects from static properties (for example, the neutron skin) disappear. To understand the role of vibrations in dynamic reactions, TDHF is used in conjunction with the coupled-channels approach. Discussion on the effects of transfer is also presented.

1 Introduction

Properties of nuclides near the neutron dripline are often not well known. Moreover, how neutron-rich nuclides interact in reactions remains an open question as it is often unfeasible to study this experimentally. How the nuclear structure of these nuclei affects heavy-ion fusion reactions is of strong interest, with upcoming exotic beam facilities and SHE formation studies.

Effects of internal nuclear structure on heavy-ion fusion can be seen by studying features of experimental fusion barrier distributions [1, 2]. Experiments have shown that dynamic effects such as low-lying vibrational couplings [3–5] and transfer reactions [6–10] strongly affect fusion reactions. Other dynamic effects, such as breakup [11–14] and rotational couplings [4], have also been seen to play a role.

The conventional choice of method for studying heavy-ion fusion was the coupled-channels (CC) approach, which needs input parameters of the structure of the nuclei in the reaction. When these parameters are not known or experimental data is unavailable, especially for very neutron rich nuclei, a microscopic approach is more appealing.

The Hartree–Fock (HF) method is a self-consistent theory which is a useful tool to study both nuclear structure and reactions. By contrast to CC methods, the only input is the nucleon-nucleon interaction. The time-dependent version (TDHF) allows for the study of dynamical effects on fusion reactions. TDHF has been used for several purposes for nuclear structure and reactions (see Refs. [15–17] for reviews). Modern codes solve the TDHF equation in three dimensions and include spin-orbit interaction [18–21]. Recent applications include vibration [22–26], fusion [27–34], transfer [35–39], quasifission [40–46] and fission [47–52] studies.

The TDHF method for studying reactions includes many dynamical effects simultaneously. This makes it difficult to isolate the effect of a single dynamical coupling on a fusion reaction. The method of including microscopic inputs to coupled-channels (CC) analysis [28, 53] is a way to isolate certain dynamic effects, and is used to unravel the TDHF dynamics and the influence of these on the fusion reactions in this proceedings.

2 Numerical details

The Hartree–Fock (HF) method assumes nucleons in a nucleus are independent particles that generate a self consistent mean-field. The only input into the HF method is the nucleon-nucleon interaction, often taken to be the Skyrme interaction. This is a major advantage of the method as no prior knowledge of the structure or dynamical modes of nuclei are needed for calculations. In this proceedings, two Skyrme parametrisations were used, the SLy4d [18] and UNEDF1 [54] parametrisations. These two in particular were chosen as they have no correction for centre of mass motion and are therefore suitable for dynamic calculations involving two different nuclei [16, 55].

The TDHF equation reads

$$[h[\rho], \rho] = i\hbar \frac{d\rho}{dt}, \quad (1)$$

where ρ is the one-body density matrix of the independent particle system and $h[\rho]$ is the HF hamiltonian. Before any dynamical calculations using equation (1) can be made for reactions or vibrations, individual nuclei in their HF ground state must first be calculated using the static version of equation (1) where the right hand side is equal to zero. The EV8 code [56] is used for calculating HF ground states. The box size we used for a HF ground state nucleus is $(2 \times 11.2)^3$ fm^3 assuming three planes of symmetry.

* e-mail: kirsten.vo-phuoc@anu.edu.au

Pairing in the HF ground state is accounted for at the BCS level with a density-dependent delta pairing force [57, 58] in the form of

$$\hat{v}_{\text{pair}}(\bar{\mathbf{r}}_1, \bar{\mathbf{r}}_2) = \tilde{t}_0 \delta(\bar{\mathbf{r}}_1 - \bar{\mathbf{r}}_2)\left(1 - \frac{\rho(\bar{\mathbf{R}})}{\tilde{\rho}_0}\right), \quad (2)$$

with parameter values $t_0 = 1000$ MeV.fm^3 and $\tilde{\rho}_0 = 0.16$ fm^{-3}.

As for reactions calculations (both static and dynamic), the box size is increased to 67.2×22.4×(2×11.2) fm^3 in the $x-y-z$ orientation with the x axis being the collision axis. Only central collisions are studied in this work. There is now only one plane of symmetry, $z = 0$. The maximum starting distance between two nuclei was 44.8 fm.

In dynamic reaction calculations, the TDHF equation (1) is solved by the TDHF3D code [18]. A mean field is now defined for the entire system, generated by all the nucleons. The single-particle occupation numbers induced by pairing correlations are kept constant in the dynamics. A Galilean boost, $e^{i\mathbf{k}_j\cdot\hat{\mathbf{r}}}$, giving each nucleon in nucleus j momentum, is applied to the system at the initial time which moves the nuclei towards each other.

In dynamic vibration calculations, the box size is for a single nucleus, with dimensions 22.4×22.4×(2×11.2) fm^3. Only one plane of symmetry is present again, at $z = 0$. The TDHF3D code is also used to evolve the nucleus with response to a multipole boost at initial time, $e^{-i\varepsilon\hat{Q}_\lambda}$ where

$$\hat{Q}_\lambda = \sum_{i=1}^{A} r^\lambda Y_{\lambda 0}(\hat{r}), \quad (3)$$

and with boost intensity ε small enough to be in the linear regime. More details on the vibrational boost is given in Section 4.2.

3 Static effects

For static HF reactions, the density distributions of the two nuclei, ρ_1 and ρ_2, are frozen in their HF ground state at all points of the reaction. The frozen HF potential [59, 60] between them, $V(\mathbf{r})$, is calculated. For the nuclear part of the potential, the sum of the energies is used,

$$V_N(\mathbf{r}) = E(\mathbf{r}) - E_{\text{HF}}[\rho_1] - E_{\text{HF}}[\rho_2], \quad (4)$$

where the energy of the total system is

$$E(\mathbf{r}) = \int \mathcal{E}[\rho_1(\mathbf{r}') + \rho_2(\mathbf{r} - \mathbf{r}')]d\mathbf{r}'. \quad (5)$$

The HF energy for each nucleus is

$$E_{\text{HF}}[\rho_j] = \int \mathcal{E}[\rho_j(\mathbf{r})]d\mathbf{r}, \quad (6)$$

where $\mathcal{E}$ is the Skyrme energy density. This approach does not account for Pauli repulsion which can be important inside the barrier [53]. Here, we neglect this effect on the barrier energy, V_B.

In Fig. 1, the static barriers, V_B, for $^{40-54}$Ca+^{116}Sn are shown by the dashed lines for the SLy4d (red, closed circles) and UNEDF1 (orange, open circles) parametrisation alongside the dotted line of Akyüz–Winther (AW) potential [61] barriers (purple crosses) for comparison. The most noticeable difference between the static HF barriers and the AW barriers is that the latter shows a near constant decrease of the fusion barrier energy as neutron number is increased and the static HF calculations show a change of gradient in the barrier energies before and after the ^{48}Ca projectile. Also note that the UNEDF1 parametrisation yields a higher barrier than the SLy4d parametrisation, by 2 MeV for a ^{40}Ca projectile and by 1 MeV for a ^{54}Ca projectile.

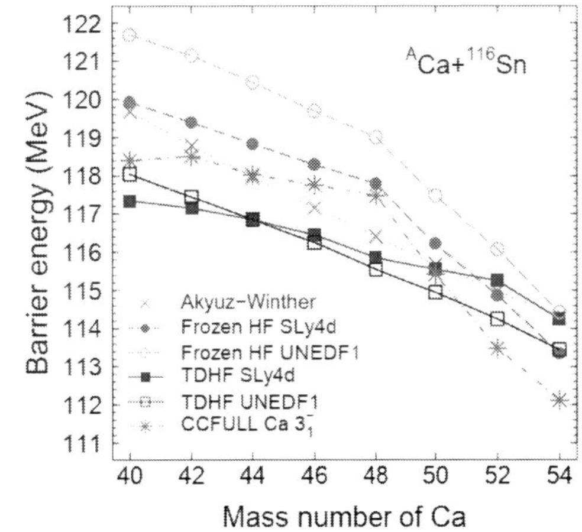

Figure 1. Barrier energies for reactions $^{40-54}$Ca+^{116}Sn comparing static HF, TDHF, AW and CCFULL (with the 3^-_1 state of calcium) calculations. For the HF calculations, both SLy4d and UNEDF1 parametrisations are shown.

To explain what causes this changing gradient in both the static HF lines for V_B, we consider the root-mean-square (rms) radius in the calcium projectiles. This is a static property arising from the HF ground state. Both neutron and proton rms radii for the calcium chain are shown in Fig. 2 (solid and dashed lines respectively) for both parametrisations. It can be seen that in the neutron rms radius (solid line), there is a change of gradient before and after ^{48}Ca. An increase in nuclear radius induces a decrease of the barrier energies. This is because the point at which the fusion barrier occurs between the nuclei will be at a larger distance thereby decreasing the Coulomb potential and hence the overall fusion barrier. The single particle shells of the neutrons helps to explain why the neutron radius increases more rapidly after ^{48}Ca. As ^{48}Ca is a doubly magic nucleus the energy gap between these shells is substantial. The neutrons in the $2p$ shells filled after $N = 28$ are less bound, meaning their radial wavefuncion extends to a larger radius than the neutrons in the $1f_{\frac{7}{2}}$ shell ($N \leq 28$) which gives rise to a more rapid increase of neutron rms radius.

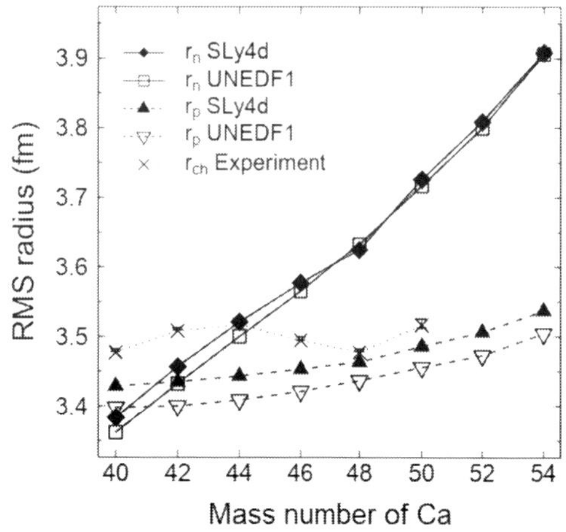

Figure 2. Calcium isotope proton (dashed line), neutron (solid line) rms radii in the SLy4d and UNEDF1 parametrisations. Also shown are the experimental charge rms radii.

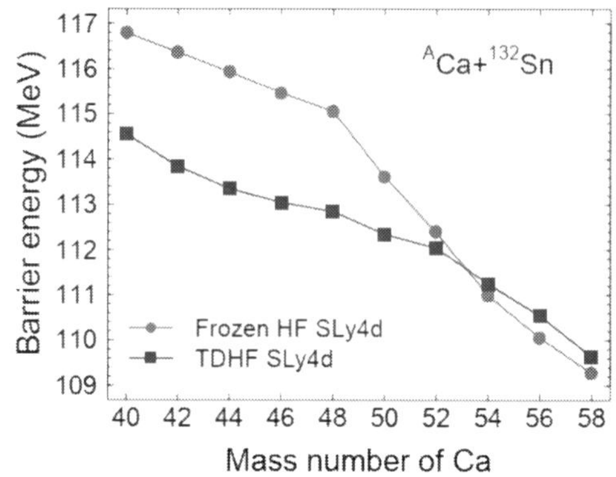

Figure 3. Static and dynamic HF barrier energies with ^{132}Sn as the heavy fragment, using the SLy4d parametrisation.

Also included in Fig. 2 are the experimental charge rms radii (dotted line, crosses) [62, 63]. These values are relatively close to the calculated proton radii though the deviations are larger for mid-shell nuclei due to correlations not accounted for at the mean-field level [64]. If HF reproduced these values for the proton radii of these calcium isotopes then the expected fusion barrier for these systems would be at slightly lower energies.

4 Dynamic effects

4.1 Barriers

Freezing nuclei in their ground state is not a realistic way to simulate fusion reactions. Dynamical couplings must be included to allow nuclei to be excited from their ground state, so for calculating reactions the TDHF equation (1) is used, outlined in Section 2.

In TDHF, as there is only a single, average trajectory for a reaction, the fusion probability is either 0 or 1. For reactions involving calcium on tin, if a compound nucleus has formed and remained compact for 5 zs, it is classified as having successfully fused. The TDHF fusion barrier or "threshold" is found by incrementally increasing the energy for a system until fusion is observed [60].

The TDHF fusion thresholds are shown by the solid lines in Fig. 1, for SLy4d (blue, closed squares) and UNEDF1 (black, open squares). We see that including dynamics, in most cases, reduces the static fusion barrier. The exception is for ^{52,54}Ca+^{116}Sn in the SLy4d parametrisation, where the dynamic barrier is actually higher than the static barrier. Whilst the dynamic barrier for UNEDF1 is not higher than its static counterpart for the two heaviest systems, the difference between the two HF barriers as neutron number increases reduces consistently. Overall, the barrier is reduced more for UNEDF1 parametrisation. This could indicate that the dynamical modes in this parametrisation are stronger than for the SLy4d parametrisation. We also observe that including dynamics washes out the static structure effects as the TDHF barriers no longer have a change of gradient after the ^{48}Ca nucleus.

The expected dynamical effects present in these TDHF calculations are vibrations and transfer. As all the nuclei ground states were spherical or near-spherical, we do not expect coupling to rotational states to play a major role in these reactions. The goal now is to unravel the dynamics in the TDHF calculations to see which dynamical effects have an impact on fusion thresholds.

Although the main focus of the work is for $^{40-54}$Ca+^{116}Sn, we performed similar calculations for the systems $^{40-58}$Ca+^{132}Sn shown in Figure 3. There is also a crossover present between static and dynamic barriers from ^{54}Ca+^{132}Sn onwards.

4.2 Vibration effects

It is well known that low-lying octupole phonons play an important role in fusion, in particular in reactions with ^{40}Ca [65]. Bearing this in mind, coupling to the 3^-_1 state of the calcium isotopes was considered for these reactions within the CC approach. Quadrupole vibrations could also impact fusion, however, the low-lying 2^+ states in ^{40}Ca are not collective. The study of the couplings to 2^+_1 states in the other calcium isotopes will be the subject of future work.

An octupole boost taking $\lambda = 3$ for the operator in equation (3) is applied to the HF ground state of the calcium isotopes. The expectation value of $\hat{Q}_3$, also known as the octupole moment, is seen in Fig. 4 (top panel) for ^{40}Ca with boost intensity $\varepsilon = 1 \times 10^{-8}$ fm^{-3}. When a sine Fourier transformation of this is taken, the octupole strength function (bottom panel in Fig. 4) is produced, resolving the energy spectrum of the 3^- states (see Ref. [16] for details).

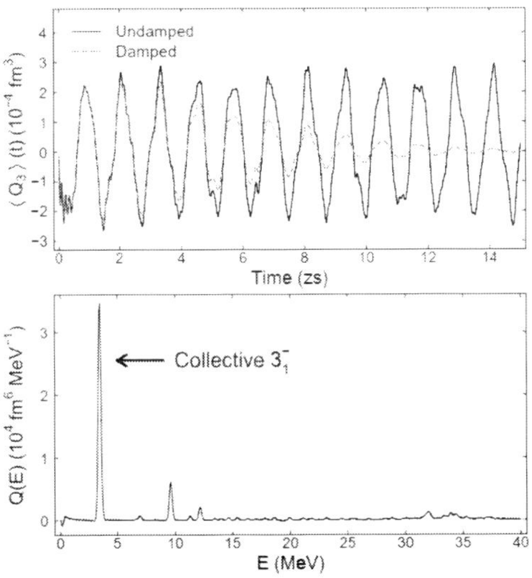

Figure 4. Octupole moment and strength function for ^{40}Ca.

The first intense peak of the strength function corresponds to the first collective 3_1^- state. The position of the peak is the excitation energy and the area under the peak corresponds to the transition probability, $|q_3|^2$, where $q_3 = \langle 3^-|\hat{Q}_3|0\rangle$ is the transition amplitude between the ground state $|0\rangle$ and an excited state $|3^-\rangle$. This can be used to determine the deformation parameter according to [66]

$$\beta_3 = \frac{4\pi|q_3|}{3R_A^3 A}, \tag{7}$$

where R_A was taken to be the HF ground state half density radius of the nucleus. In practice, $\langle\hat{Q}_3\rangle(t)$ is multiplied by a damping function to avoid spurious oscillations in the Fourier spectrum due to the finite time of the TDHF calculation.

In Table 1 are the 3_1^- state excitation energy and deformation parameter for ^{40}Ca, comparing the experimental and TDHF results. Shown in addition to SLy4d and UNEDF1 parametrisations is the SLy4 parametrisation [67] which has the same fitting procedure as for SLy4d but includes the centre of mass correction. This illustrates the large variation in excitation energy amongst different parametrisations. The deformation parameter is slightly underestimated by TDHF in all cases compared to the experimental values. Only the results from the SLy4d parametrisation will be presented from here onwards.

To see what effect just the 3_1^- state of the calcium isotopes has on the fusion barriers of the systems, the CC approach [68], using the CCFULL code [69] is used since dynamical effects can be added in one at a time unlike in TDHF. Couplings to low-lying states could induce a complex structure in the distribution of fusion barriers [2]. A main feature, however, is the lowering of the main peak in the barrier distribution which often results in lowering the average barrier. These average barriers, calculated with

	$E_x(3_1^-)$ (MeV)	β_3
Exp.	3.737	0.3-0.4
SLy4	3.916	0.184
SLy4d	3.442	0.224
UNEDF1	3.363	0.171

Table 1. Comparison of excitation energy and deformation parameter between experimental values [66] and TDHF values for three parametrisations for ^{40}Ca.

coupling to the 3_1^- state in the calcium isotopes, are shown in by the brown dot-dashed line with stars in Fig. 1. As expected, this coupling lowers the static barrier in all cases. Couplings to low-lying 2^+ states (that are not included here) are expected to produce a similar effect.

The inclusion of couplings to low-lying vibrations is helpful for explaining the difference between the static and dynamic cases for the systems with $^{40-50}$Ca as projectiles but not for the two most neutron-rich systems. Indeed, we see in Fig. 1 that for the ^{52,54}Ca+^{116}Sn systems, including dynamics actually increased the fusion barrier as the TDHF fusion threshold is higher than the frozen HF bare potential barrier. This is also the case for $^{54-58}$Ca+^{132}Sn with very little difference (less than 0.5 MeV) between the two barriers for $^{52-58}$Ca+^{132}Sn. Therefore, another dynamical effect aside from vibrations must be considered to help explain this observation.

4.3 Transfer effects

Our consideration of transfer is motivated by the fact that the HF dynamic barrier is not consistently lowering the static barrier in the case for the most neutron rich systems. The effect of nucleon transfer acting to balance out the N/Z ratio between two fragments in a reaction is accounted for in TDHF calculations [37, 70].

The N/Z ratio for $^{40-54}$Ca ranges from 1 to 1.70 and those of ^{116,132}Sn lie within this range, at 1.32 and 1.64 respectively. For this proceeding, the focus is on proton transfer for these systems. It is expected from the argument of balancing the N/Z ratios that proton transfer would happen in the direction from the light to heavy fragment for the lighter systems, and from heavy to light fragments in neutron rich systems.

Our observation of nucleon transfer in TDHF calculations for the systems $^{40-54}$Ca+^{116,132}Sn occuring at centre of mass energies just below the TDHF fusion threshold ($0.999\% V_B$) is from the final fragments. It is quantified by both average numbers of transferred neutrons and protons, and transfer probabilities for protons using the particle number projection method [35]. The average transfer at the end of a reaction, shown by the final Z and N number of the heavy fragment for the systems $^{40-54}$Ca+^{116,132}Sn can be seen in Fig. 5. For a reaction showing near zero values for average transfer (e.g. ^{46}Ca+^{116}Sn and ^{52}Ca+^{132}Sn) there are still possibilities of having non-zero probabilities for other transfer channels.

In the case of ^{116}Sn, we can see that the average Z is close to 51 in a reaction with ^{40}Ca, corresponding to

DOI: 10.1051/epjconf/201716300062

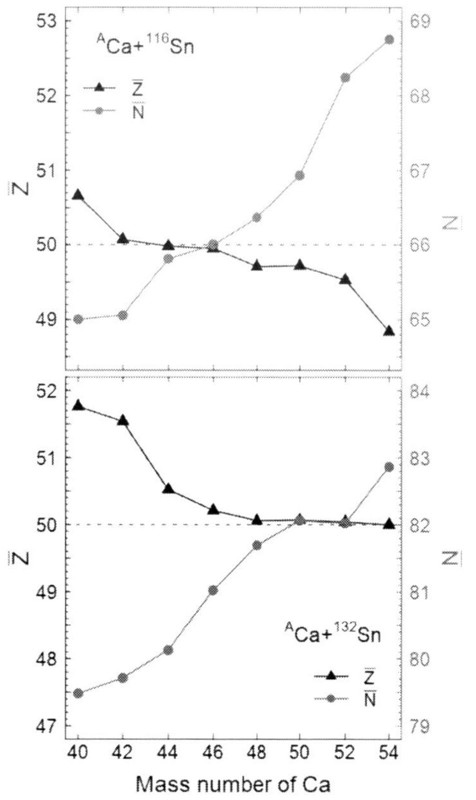

Figure 5. Average Z and N numbers of the heavy fragment at the end of reactions ^{A}Ca+^{116,132}Sn. The original N and Z number of the heavy fragment is indicated by the horizontal dashed line.

proton transfer from calcium to tin. The average N in the same reaction is 65, indicating one neutron transfer from tin to calcium. The most neutron rich system has transfer of protons and neutrons in the opposite direction and at larger magnitudes.

With ^{132}Sn, we see the same direction of proton and neutron transfer as for ^{116}Sn but by different amounts. Very little average proton transfer occurs after ^{48}Ca+^{132}Sn in agreement with Ref. [34], perhaps because the N/Z ratio of ^{132}Sn is much higher and is a closer match to the heavier calcium nuclei. It was found that subbarier fusion is enhanced for ^{40}Ca+^{132}Sn more than what can be accounted for by vibrational states [71]. The Q-values of nucleon transfer plays an important role in these reactions at sub-barrier energies [71] [ref Back, Liang]. In Ref. [34], the isovector contribution to the DC-TDHF fusion potentials for ^{40,48,54}Ca+^{132} is also related to neutron transfer, compatible with the directions in Fig. 5 and the resulting TDHF fusion barriers in Fig. 3.

In addition to the isovector contribution to the potential, proton transfer in either direction will change the Coulomb repulsion. In particular, one proton transfer from ^{54}Ca to ^{116}Sn increases the Coulomb potential, $V_C = e^2 Z_1 Z_2 / r$ by 2.7% which in turn increases the fusion barrier. This is consistent with what is seen in the TDHF fusion barrier thresholds and is a possible explanation for why the dynamical TDHF fusion threshold is higher than the static frozen HF barrier for this system.

5 Conclusion

We have used a microscopic approach to probe fusion reactions involving neutron-rich nuclei. Effects of some static and dynamic structures of the nuclei on fusion were studied using HF methods. In static fusion reactions, properties such as rms neutron radius of the calcium isotopes arising from the HF ground state have an effect on the fusion barrier. Nuclides with less bound neutrons and an increasing neutron skin (calcium isotopes heavier than ^{48}Ca) lead to a more rapid decrease of the fusion barrier. Including dynamics into reactions, it was seen that for most of the cases this decreased the fusion barrier. However, for the most neutron rich systems, it increased or was very close to the static fusion barrier. The static structure effects disappeared completely with the inclusion of dynamics. Coupled-channels analysis showed that coupling to the 3_1^- state of the calcium nuclei lowered the fusion barrier in all cases. Vibrational states are not enough to describe the fusion barrier differences for the most neutron rich systems, so the significance of other dynamical effects such as proton transfer was discussed. Future works will consider the effect of couplings to low-lying quadrupole vibrations, further understanding the role of neutron transfer on fusion and also incorporating dynamical pairing in the reactions.

Acknowledgement

This work has been supported by the Australian Research Council under Grant Nos. FT120100760, DP170102423 and FL110100098. The calculations in this proceedings were done with the assistance of resources from the National Computational Infrastructure (NCI) which is supported by the Australian Government.

References

[1] N. Rowley et al., Phys. Lett. B **254**, 25 (1991)
[2] M. Dasgupta et al., Annu. Rev. Nucl. Part. Sci. **48**, 401 (1998)
[3] C.R. Morton et al., Phys. Rev. Lett. **72**, 4074 (1994)
[4] J.R. Leigh et al., Phys. Rev. C **52**, 3151 (1995)
[5] A.M. Stefanini et al., Phys. Rev. Lett. **74**, 864 (1995)
[6] M. Beckerman et al., Phys. Rev. Lett. **45**, 1472 (1980)
[7] A. Stefanini et al., Nucl. Phys. A **456**, 509 (1986)
[8] H. Timmers et al., Phys. Lett. B **399**, 35 (1997)
[9] C.L. Jiang et al., Phys. Rev. C **89**, 051603 (2014)
[10] C.L. Jiang et al., Phys. Rev. C **91**, 044602 (2015)
[11] M. Dasgupta et al., Phys. Rev. Lett. **82**, 1395 (1999)
[12] A. Chatterjee et al., Phys. Rev. Lett. **101**, 032701 (2008)

[13] A. Lemasson et al., Phys. Rev. C **82**, 044617 (2010)
[14] A. Lemasson et al., Phys. Lett. B **697**, 454 (2011)
[15] J.W. Negele, Rev. Mod. Phys. **54**, 913 (1982)
[16] C. Simenel, Eur. Phys. J. A **48**, 152 (2012)
[17] T. Nakatsukasa et al., Rev. Mod. Phys. **88**, 045004 (2016)
[18] K.H. Kim et al., J. Phys. G: Nucl. Part. Phys. **23**, 1267 (1997)
[19] A.S. Umar, V.E. Oberacker, Phys. Rev. C **71**, 034314 (2005)
[20] T. Nakatsukasa, K. Yabana, Phys. Rev. C **71**, 024301 (2005)
[21] J.A. Maruhn et al., Phys. Rev. C **71**, 064328 (2005)
[22] C. Simenel, P. Chomaz, Phys. Rev. C **80**, 064309 (2009)
[23] I. Stetcu, A. Bulgac et al., Phys. Rev. C **84**, 051309 (2011)
[24] S. Fracasso et al., Phys. Rev. C **86**, 044303 (2012)
[25] B. Avez, C. Simenel, Eur. Phys. J. A **49**, 76 (2013)
[26] G. Scamps, D. Lacroix, Phys. Rev. C **88**, 044310 (2013)
[27] C. Simenel et al., Phys. Rev. C **88**, 024617 (2013)
[28] C. Simenel et al., Phys. Rev. C **88**, 064604 (2013)
[29] A.S. Umar et al., Phys. Rev. C **89**, 034611 (2014)
[30] G.F. Dai et al., Phys. Rev. C **90**, 044609 (2014)
[31] K. Washiyama, Phys. Rev. C **91**, 064607 (2015)
[32] D. Bourgin et al., Phys. Rev. C **93**, 034604 (2016)
[33] K. Vo-Phuoc et al., Phys. Rev. C **94**, 024612 (2016)
[34] K. Godbey et al., Phys. Rev. C **95**, 011601 (2017)
[35] C. Simenel, Phys. Rev. Lett. **105**, 192701 (2010)
[36] C. Simenel, Phys. Rev. Lett. **106**, 112502 (2011)
[37] K. Sekizawa, K. Yabana, Phys. Rev. C **88**, 014614 (2013)
[38] K. Sekizawa, K. Yabana, Phys. Rev. C **90**, 064614 (2014)
[39] G. Scamps, D. Lacroix, Phys. Rev. C **87**, 014605 (2013)
[40] A. Wakhle et al., Phys. Rev. Lett. **113**, 182502 (2014)
[41] V.E. Oberacker et al., Phys. Rev. C **90**, 054605 (2014)
[42] K. Hammerton et al., Phys. Rev. C **91**, 041602 (2015)
[43] A.S. Umar et al., Phys. Rev. C **92**, 024621 (2015)
[44] E. Prasad et al., Phys. Rev. C **93**, 024607 (2016)
[45] K. Sekizawa, K. Yabana, Phys. Rev. C **93**, 054616 (2016)
[46] A.S. Umar et al., Phys. Rev. C **94**, 024605 (2016)
[47] C. Simenel, A.S. Umar, Phys. Rev. C **89**, 031601 (2014)
[48] G. Scamps et al., Phys. Rev. C **92**, 011602 (2015)
[49] P. Goddard et al., Phys. Rev. C **92**, 054610 (2015)
[50] P. Goddard et al., Phys. Rev. C **93**, 014620 (2016)
[51] A. Bulgac et al., Phys. Rev. Lett. **116**, 122504 (2016)
[52] Y. Tanimura et al., Phys. Rev. Lett. **118**, 152501 (2017)
[53] C. Simenel et al., Phys. Rev. C **95**, 031601 (2017)
[54] M. Kortelainen et al., Phys. Rev. C **89**, 054314 (2014)
[55] M. Kortelainen et al., Phys. Rev. C **85**, 024304 (2012)
[56] P. Bonche et al., Comp. Phys. Com. **171**, 49 (2005)
[57] S.J. Krieger et al., Nucl. Phys. A **517**, 275 (1990)
[58] M. Bender et al., Phys. Rev. C **60**, 034304 (1999)
[59] K. Washiyama, D. Lacroix, Phys. Rev. C **78**, 024610 (2008)
[60] C. Simenel, B. Avez, International Journal of Modern Physics E **17**, 31 (2008)
[61] R.A. Broglia, A. Winther, *Heavy Ion Reaction Lecture Notes, Vol. I: Elastic and Inelastic Reactions* (Benjamin/Cummings, Reading, MA, 1981)
[62] H.D. Wohlfahrt et al., Phys. Rev. C **23**, 533 (1981)
[63] L. Vermeeren et al., Phys. Rev. Lett. **68**, 1679 (1992)
[64] E. Caurier et al., Phys. Lett. B **522**, 240 (2001)
[65] K. Hagino, N. Takigawa, Phys. Rev. Lett. **79**, 2014 (1997)
[66] T. Kibedi, R.H. Spear, Atomic Nuclear and Data Tables **80**, 35 (2002)
[67] E. Chabanat et al., Nucl. Phys. A **635**, 231 (1998)
[68] K. Hagino, N. Takigawa, Prog. Th. Phys. **128**, 1061 (2012)
[69] K. Hagino et al., Comp. Phys. Com. **123**, 143 (1999)
[70] C. Simenel et al., Phys. Lett. B **710**, 607 (2012)
[71] J.J. Kolata et al., Phys. Rev. C **85**, 054603 (2012)

 DOI: 10.1051/epjconf/201716300063

Near-barrier Fusion Evaporation and Fission of ^{28}Si+^{174}Yb and ^{32}S+^{170}Er

Dongxi Wang[1], *Chengjian* Lin[1,2,*], *Huiming* Jia[1], *Nanru* Ma[1], *Lijie* Sun[1], *Xinxing* Xu[1,3], *Lei* Yang[1], *Feng* Yang[1], *Huanqiao* Zhang[1], and *Pengfei* Bao[1]

[1]China Institute of Atomic Energy, P.O. Box 275(10), Beijing 102413, China
[2]College of Physics and Technology, Guangxi Normal University, Guilin 541004, China
[3]Department of Physics, The University of Hong Kong, Hong Kong, China.

Abstract. Fusion evaporation residues and fission fragments have been measured, respectively, at energies around the Coulomb barrier for the ^{28}Si+^{174}Yb and ^{32}S+^{170}Er systems forming the same compound nucleus ^{202}Po. The excitation function of fusion evaporation, fission as well as capture reactions were deduced. Coupled-channels analyses reveal that couplings to the deformations of targets and the two-phonon states of projectiles contribute much to the enhancement of capture cross sections at sub-barrier energies. The mass and total kinetic energy of fission fragments were deduced by the time-difference method assuming full momentum transfer in a two-body kinematics. The mass-energy and mass-angle distributions were obtained and no obvious quasi-fission components were observed in this bombarding energy range. Further, mass distributions of fission fragments were fitted to extract their widths. Results show that the mass widths decrease monotonically with decreasing energy, but might start to increase when $E_{\rm c.m.}/V_{\rm B} < 0.95$ for both systems.

1 Introduction

Heavy-ion induced fusion-fission reactions at energies around the Coulomb barrier have attracted much attention [1, 2], due to many degrees of freedom involved in the interaction processes. It has been experimentally proved that the coupled interactions of some intrinsic motions and relative motion play a significant role in the process that two nuclei fuse into one, and it is commonly interpreted theoretically as coupled-channels (CC) effects. It is expected that the deformation [3], vibrational state [4] and transfer channel [5–7] may be involved in the capture reaction. After two nuclei come into contact with each other, the system evolves along the potential energy surface, the path of which may be complicated. The system may achieve equilibrium of all the degrees of freedom and whereafter the compound nucleus (CN) can survive through evaporating light particles. The system can also go to fission via some equilibrium or non-equilibrium processes, such as complete fusion-fission, pre-equilibrium fission [8], quasi-fission [9] and fast fission [10–12]. These processes may strongly depend on some factors of the entrance channel, for example, deformations of projectile and target, the mass asymmetry, and so on. Comprehensive research on fusion-fission reactions can cast useful light on the understanding of heavy ion induced reactions.

Previously, we have measured the angular distributions for the fission fragments of the ^{32}S+^{184}W system [13] and, the mass and energy distributions for the fission fragments of the ^{34}S+^{186}W system [14]. Anisotropy deduced from the measured angular distributions reveals the appearance of non-equilibrium fission. It can be interpreted as quasi-fission by the dinuclear system model or pre-equilibrium fission [14]. In order to further explore on the fission mechanism, study on extensive systems is desired.

In the present work, the ^{28}Si+^{174}Yb and ^{32}S+^{170}Er systems were chosen because both systems form the same compound nucleus ^{202}Po. It is noted that mass-angle correlations and mass ratio distributions of fission fragments have been measured systematically for the systems of ^{16}O+^{186}Os, ^{24}Mg+^{178}Hf, ^{34}S+^{168}Er, and ^{48}Ti+^{154}Sm [15], also forming the same compound nucleus ^{202}Po. Therefore, the entrance-channel dependence can be investigated in a systematic manner. Moreover, both fusion evaporation residues (ERs) and fission fragments have been measured for the ^{34}S+^{168}Er system [16], giving a good reference to the present work. It is worth to point out that both fusion evaporation and fission have observable cross sections at near-barrier energy region due to the higher fission barrier of ^{202}Po ($\approx$ 15 MeV) [17]. It provides a good case to explore the evolving paths of both fusion and fission along the potential energy surface, to strictly test as well as help to develop relevant theoretical models, and to comprehensively understand the complete fusion-fission dynamics.

In the following Sec. 2, details of the experimental procedure for fusion evaporation and fission measurements are present. Experimental results and preliminary discussions are reported in Sec. 3 and conclusions are given in Sec. 4 finally.

*e-mail: cjlin@ciae.ac.cn

DOI: 10.1051/epjconf/201716300063

2 Experimental procedure

The beams of ^{32}S and ^{28}Si, at several energies in the range of 133.0 - 181.0 MeV and 119.2 - 162.1 MeV, respectively, were supplied by the HI-13 tandem accelerator of the China Institute of Atomic Energy, Beijing, China. Highly isotopically-enriched ^{170}ErO$_3$ (95 μg/cm^2) and ^{174}Yb (130 μg/cm^2) targets sputtered onto carbon backings (20-23 μg/cm^2) were used. Four silicon (Si) detectors placed symmetrically at θ_{lab} = $\pm 20°$ (right-left and up-down) with respect to the beam direction were used to monitor the Rutherford scattering and to provide a normalization for the fusion evaporation and fission cross sections. Fusion ERs and fission fragments were measured independently with the same beam.

2.1 Measurement of fusion ERs

Fusion ERs mainly concentrated within a few degrees of the incident beam direction were separated from the beam-like particles (BLPs) by using an electrostatic deflector setup [18]. It consists of two pairs of electrodes followed by two micro-channel plate (MCP) detectors and a 48 × 48 mm^2 quadrant Si detector. Distances of MCP1-MCP2 and MCP2-Si are 41.0 and 22.7 cm, respectively. Time-of-flight (ToF) versus E spectra, ToF1-E and ToF2-E, are useful to reduce the spurious backgrounds from the random coincidences between the Si and one of the MCPs, especially for the measurements of low fusion evaporation cross sections at sub-barrier energies and for asymmetric systems studied in this work. The reaction products were limited by a collimator of 2.5 mm in diameter before entering the electric fields, corresponding to an opening angle θ_{lab} = $\pm 0.38°$. Fig. 1 shows a typical ToF1 versus E spectrum for the ^{32}S+^{170}Er system at an energy near the Coulomb barrier, where BLPs and ERs can be separated clearly.

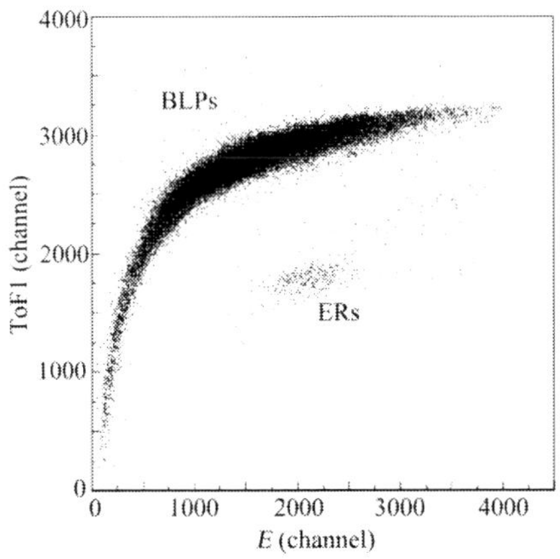

Figure 1. Two-dimensional plot ToF1-E taken at E_{lab} = 169 MeV and θ_{lab} = 2° for ^{32}S+^{170}Er. A group of BLPs and ERs are indicated.

The ERs angular distributions were measured in the range θ_{lab} = $-4°$ to $+8°$ with a step of 0.5° or 1° for both systems at E_{lab} = 143 MeV and 163 MeV, respectively. Their shapes do not change appreciably with the beam energy, resulting from the dominant neutron and/or proton evaporation from the CN [19], which is consistent with the calculation of the code PACE2 [20]. Typical angular distributions of ERs for the ^{32}S+^{170}Er system as well as their fittings by a single Gaussian function are shown in Fig. 2.

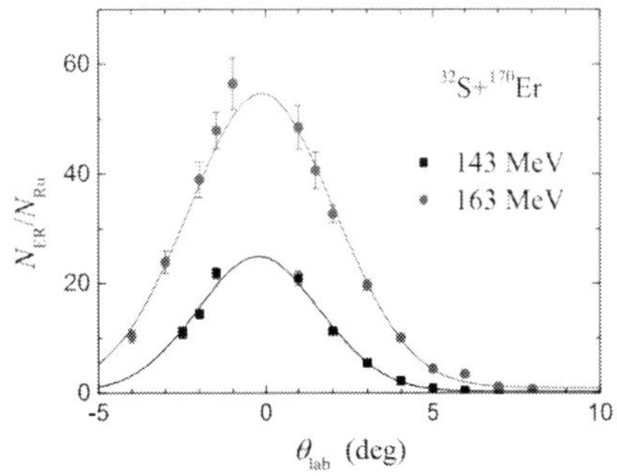

Figure 2. Angular distributions of ERs for the ^{32}S+^{170}Er system at E_{lab} = 143 MeV and 163 MeV. Curves are single Gaussian fits used to obtain the total fusion evaporation cross sections.

For other energies, ERs were measured only at θ_{lab} = 1.5°. Total fusion evaporation cross sections were obtained by integration of the angular distribution with the same shape. Meanwhile, corrections were made for the transmission efficiency of the electrostatic deflector, which is about 40%.

2.2 Measurement of fission fragments

Two fission fragments were detected in coincidence by using two Parallel Plate Avalanche Counters (PPACs). The PPAC1 and PPAC2 have an active area ($X \times Y$) of 180× 100 mm^2 and 250 × 100 mm^2 with X and Y position resolution of about 1 and 2 mm, respectively. The emission angles θ_1 and θ_2 projected on the reaction $X - Z$ plane and the out-of-plane angles φ_1 and φ_2 for the fission fragment 1 and 2 were transformed from the XYZ coordinate, assuming the beam direaction as the Z axis. The detectors were located around the target, centering at θ_1 = 45° and θ_2 = $-105°$ with distances of d_1 = 155 mm and d_2 = 125 mm from the target to the centers of entrance windows and covered angles of $\Delta\theta_1$ = $\pm 27°$ and $\Delta\theta_2$ = $\pm 30°$ as well as 14.9° $\leq \varphi_1 \leq$ 75.7° and 60.0° $\leq \varphi_2 \leq$ 150.0° for PPAC1 and PPAC2, respectively. The detectors were operated with isobutane gas at a pressure of about 3 Torr.

The time difference ΔT between the two fission fragments was measured with start and stop signals obtained from PPAC2 and PPAC1, along with the energy losses ΔE_1 and ΔE_2 in both PPACs by passing nuclei. Fig. 3 (a) gives an example of measured coincident events in the

 DOI: 10.1051/epjconf/201716300063

$\Delta E_1+\Delta E_2$ versus ΔT coordinate plane, for ^{28}Si+^{174}Yb reactions at the $E_{\rm c.m.}/V_{\rm B}$ = 1.25. A clear separation between fission fragments and scattered particles can be noted.

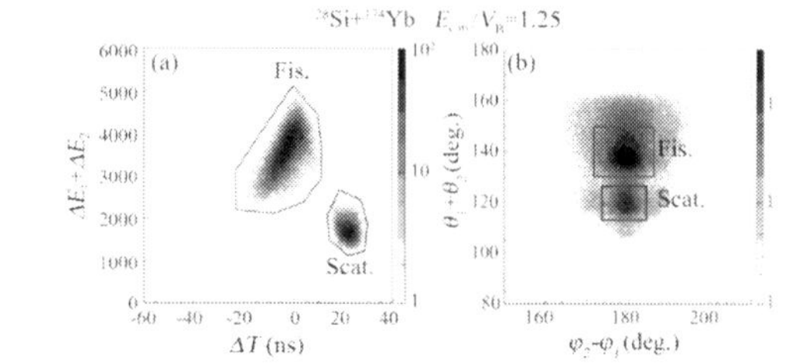

Figure 3. (a) Events from the ^{28}Si+^{174}Yb reaction, mapped on the $\Delta E_1+\Delta E_2$ versus ΔT axes. Fission events within the polygon and scattering events are clearly separated. (b) Same as in (a) but in the coordinates of $\theta_1+\theta_2$ versus $\varphi_2-\varphi_1$.

An additional way to distinguish fission fragments from the scattered projectile and recoiled nuclei is to exploit the difference in the reaction kinematics. This is demonstrated in Fig. 3 (b), where events are plotted in the $\theta_1+\theta_2$ versus $\varphi_2-\varphi_1$ coordinate plane. The calibration for the time difference ΔT was made from the elastic-recoil peak positions appearing in the ΔT spectrum. The timing resolution of a single PPAC pixel was determined to be 0.3 ns. The experimental mass resolution obtained from the elastic scattering peak is about $\sigma_{\rm exp}$ = 2 − 4 u, depending on the measured angle.

In the present work, the ΔT method was employed to deduce the velocity and mass of fission fragment by applying the momentum conservation as well as the mass conservation in a two-body kinematics. In the laboratory system, we have

$$m_1 v_1 sin\theta_1 = m_2 v_2 sin\theta_2, \tag{1}$$

$$m_1 v_1 cos\theta_1 + m_2 v_2 cos\theta_2 = m_{\rm P} v_{\rm P} = m_{\rm CN} v_{\rm CN}, \tag{2}$$

$$m_1 + m_2 = m_{\rm P} + m_{\rm T} = m_{\rm CN}, \tag{3}$$

$$\Delta T = \frac{r_1}{v_1} - \frac{r_2}{v_2}, \tag{4}$$

where m is the mass, v is the velocity and r is the flight distance of outgoing particles before being detected, subscript P, T, CN and 1, 2 represent the projectile, target, CN and fission fragments, respectively. Finally, the velocity and mass of outgoing particles can be expressed as

$$v_1 = \frac{r_1 sin\theta_1 + r_2 sin\theta_2}{r_2 sin(\theta_1+\theta_2) + \Delta T v_{\rm CN} sin\theta_1} \cdot v_{\rm CN}, \tag{5}$$

$$v_2 = \frac{r_1 sin\theta_1 + r_2 sin\theta_2}{r_1 sin(\theta_1+\theta_2) - \Delta T v_{\rm CN} sin\theta_2} \cdot v_{\rm CN}, \tag{6}$$

$$m_1 = \frac{r_2 sin(\theta_1+\theta_2) + \Delta T v_{\rm CN} sin\theta_1}{r_1 sin\theta_1 + r_2 sin\theta_2} \cdot \frac{sin\theta_2}{sin(\theta_1+\theta_2)} \cdot m_{\rm CN}, \tag{7}$$

$$m_2 = \frac{r_1 sin(\theta_1+\theta_2) - \Delta T v_{\rm CN} sin\theta_2}{r_1 sin\theta_1 + r_2 sin\theta_2} \cdot \frac{sin\theta_2}{sin(\theta_1+\theta_2)} \cdot m_{\rm CN}, \tag{8}$$

and then the total kinetic energy (TKE) can be calculated,

$$\text{TKE} = \frac{1}{2} m_1 v_1^2 + \frac{1}{2} m_2 v_2^2. \tag{9}$$

Table 1. The relevant parameters used in the CCFULL calculations.

Nuclei	E_{2^+} (MeV)	B($E2$) (W.u.)	β_2
^{28}Si	1.779	13.2	−0.411
^{32}S	2.230	10.1	0.315
^{170}Er	0.079	208	0.336
^{174}Yb	0.076	201	0.321

3 Results and discussions

3.1 Excitation function

The excitation functions of evaporation were determined by doing the integrals with respect to angular distribution and performing the correction of the transmission efficiency. As for fission, position and time difference of fission fragments detected by PPACs were used to deduce their mass, emission angle and TKE. Integral of the angular distribution leads to the fission cross section, and the fission excitation function was then determined. Adding the excitation functions of ERs and fission together, the excitation function of capture reaction was obtained.

The measured fusion ERs, fission and capture excitation functions are shown in Fig. 4 for ^{28}Si+^{174}Yb and ^{32}S+^{170}Er, respectively. The CC code CCFULL [21] was used to reproduce the capture cross sections and the relevant parameters used in the CC calculations are listed in Table 1. It can be seen from the figure that the single-channel (SC) results (dashed lines) under-predict the sub-barrier data. However, experimental capture excitation functions could generally reproduced by the CC calculations with couplings to the vibrational two-phonon 2^+ states of projectiles and the rotational $2^+, 4^+, 6^+, 8^+$ states of targets, except a small deviation at energies around the barrier.

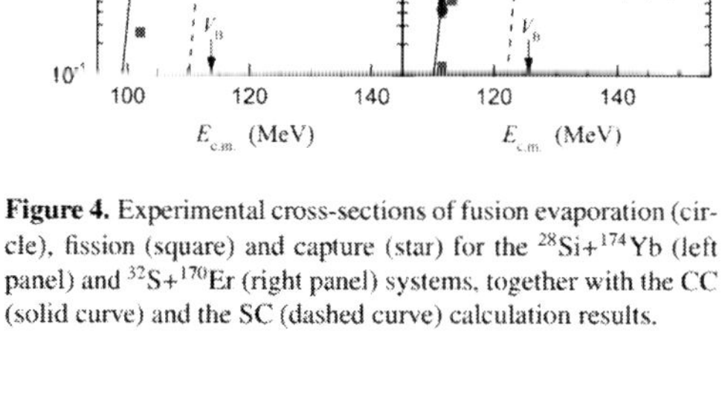

Figure 4. Experimental cross-sections of fusion evaporation (circle), fission (square) and capture (star) for the ^{28}Si+^{174}Yb (left panel) and ^{32}S+^{170}Er (right panel) systems, together with the CC (solid curve) and the SC (dashed curve) calculation results.

3.2 Mass-TKE correlations

The contour diagrams representing the angle integrated fragment mass-TKE correlations for a selected subset of the $^{28}Si+^{174}Yb$ and $^{32}S+^{170}Er$ systems are shown in Fig. 5. In both figures, one can distinguish two regions of a scattering region close to the target and projectile masses and a region of fission in the direction of symmetry. The almost independence of the fission TKE on the bombarding energy indicates that the initial excess kinetic energy is completely damped out and transformed into internal excitation in all capture processes.

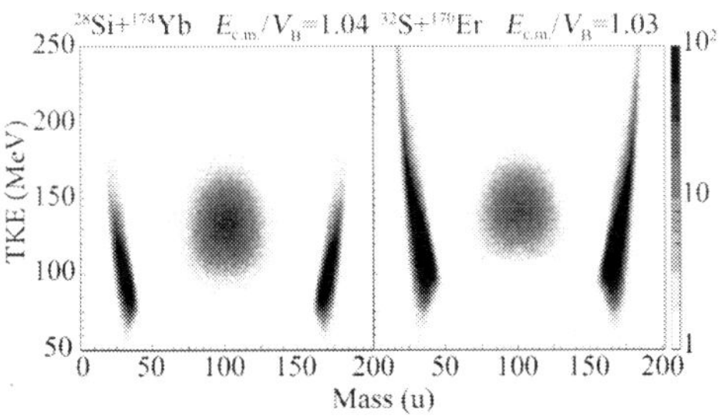

Figure 5. The correlations between fragment mass and TKE for the reactions of $^{28}Si+^{174}Yb$ (left) and $^{32}S+^{170}Er$ (right) at $E_{c.m.}/V_B$ = 1.04 and 1.03, respectively.

3.3 Mass-angle correlations

The correlations between fragment mass and center-of-mass angle for the $^{28}Si+^{174}Yb$ and $^{32}S+^{170}Er$ systems are shown as contour diagrams in Fig. 6. Also in these diagrams, the separation between scattering and fission reactions is clearly apparent. Both the two systems show symmetric fission fragment mass distributions at all angles, which is a characteristic feature for the CN fission.

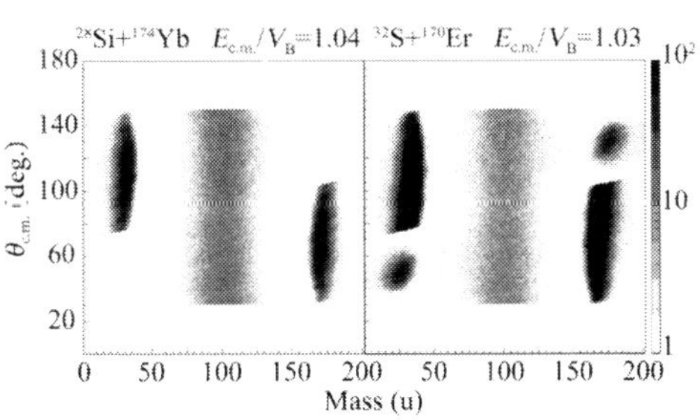

Figure 6. The correlations between fragment mass and $\theta_{c.m.}$ for the reactions of $^{28}Si+^{174}Yb$ (left) and $^{32}S+^{170}Er$ (right) at $E_{c.m.}/V_B$ = 1.04 and 1.03, respectively.

3.4 Mass distributions

As mentioned above, no obvious correlations were observed in mass-angle and mass-TKE distributions for the two systems at energies around the barrier. The existing systematic study shows that the width of mass distribution for the quasi-fission is larger than that for the fusion-fission [22]. Therefore an increased mass width can be used as an indicator for the presence of quasi-fission. Fig. 7 shows the mass distributions of fission fragments for the $^{28}Si+^{174}Yb$ and $^{32}S+^{170}Er$ systems at three reaction energies, respectively, where the fission yields are normalized to 200%.

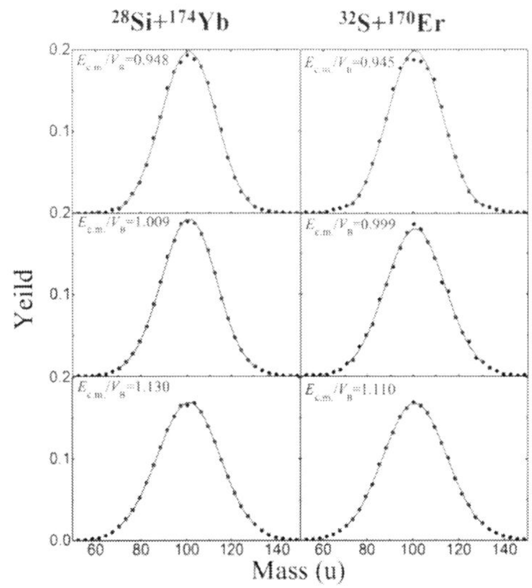

Figure 7. Mass distributions of fission fragments for the reactions of $^{28}Si+^{174}Yb$ (left) and $^{32}S+^{170}Er$ (right) at different reaction energies.

The mass spectra can be described by a single Gaussian distribution with the center at $A_{CN}/2$. The standard deviation σ_M of the mass distributions was obtained at each reaction energy, as shown in Fig. 8. It is evident from the figure that σ_M decreases monotonically with decreasing energy, but seems to increase when $E_{c.m.}/V_B < 0.95$ for both systems.

4 Conclusions

The fusion evaporation and fission reactions of $^{32}S+^{170}Er$ and $^{28}Si+^{174}Yb$ were measured at the energy ranges of $0.91 \leq E_{c.m.}/V_B \leq 1.23$ and $0.93 \leq E_{c.m.}/V_B \leq 1.25$, respectively. The excitation functions of fusion evaporation, fission as well as capture reactions were obtained. The coupled-channels calculations, where both the 2^+ two-phonon vibrational-states of projectiles and the 2^+, 4^+, 6^+, 8^+ rotational-states of targets are considered, could roughly reproduce the capture excitation functions for the two systems. For fission, the mass-TKE and mass-angle distributions were deduced and no obvious quasi-fission components were observed in this bombarding energy range. The mass widths decrease monotonically with decreasing energy, but might start to increase when

$E_{c.m.}/V_B < 0.95$ for both systems. Further study is still in progress.

5 Acknowledgments

This work is supported by the National Key Basic Research Program of China (Grant No. 2013CB834404) and by the National Natural Science Foundation of China (Grants No. 11635015, No. 11375268, No. 11475263, No. U1432246, No. U1432127, and No. 11505293).

References

[1] K. Banerjee *et al*, Phys. Rev. C **83**, 024605 (2011).
[2] D.J. Hinde *et al*, Phys. Rev. Lett. **89**, 282701 (2002).
[3] C. Y. Wong, Phys. Rev. Lett. **31**, 766 (1973).
[4] H. Esbensen, Nucl. Phys. A **352**, 147 (1981).
[5] R. A. Broglia *et al*, Phys. Lett. B **133**, 34 (1983).
[6] C. H. Dasso *et al*, Nucl. Phys. A **407**, 221 (1983).
[7] C. H. Dasso *et al*, Nucl. Phys. A **405**, 381 (1983).
[8] Z. H. Liu *et al*, Phys. Lett. B **353**, 173 (1995).
[9] W. J. Swiatecki *et al*, Phys. Scr. **24**, 113 (1981).
[10] B. Heusch *et al*, Z. Phys. A **288**, 391 (1978).
[11] C. Lebrun *et al*, Nucl. Phys. A **321**, 207 (1979).
[12] C. Gregoire *et al*, Phys. Lett. B **99**, 17 (1981).
[13] H. Q. Zhang *et al*, Phys. Rev. C **81**, 034611 (2010).
[14] C. J. Lin *et al*, J. Phys. Conf. Ser. **420**, 012126 (2013).
[15] R. Rafiei *et al*, Phys. Rev. C **77**, 024606 (2008).
[16] C. R. Morton *et al*, Phys. Rev. C **62**, 024607 (2000).
[17] P. Möller *et al*, Phys. Rev. C **79**, 064304 (2009).
[18] H. Q. Zhang *et al*, Chin. Phys. C **34**, 1628 (2010).
[19] D. J. Hinde *et al*, Nucl. Phys. A **385**, 109 (1982); *ibid.* **398**, 308 (1983).
[20] A. Gavron, Phys. Rev. C **21**, 230 (1980).
[21] K. Hagino *et al*, Comput. Phys. Commun. **123**, 143 (1999).
[22] C. J. Lin *et al*, Phys. Rev. C **85**, 014611 (2012).

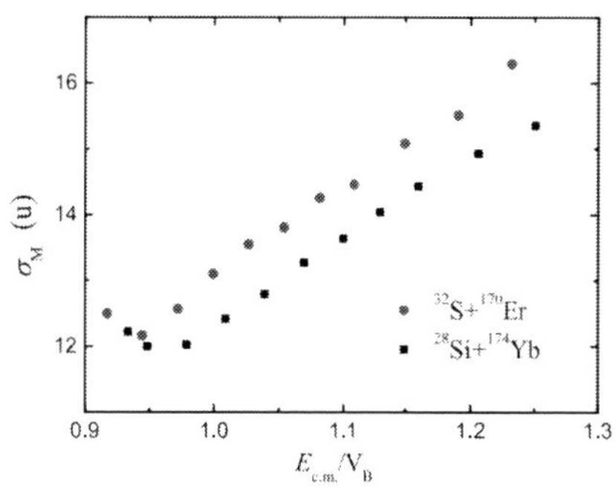

Figure 8. Mass widths of fission fragments as a function of $E_{c.m.}/V_B$ for the $^{32}S+^{170}Er$ (circle) and $^{28}Si+^{174}Yb$ (square) systems.

Fusion hindrance in heavy systems with time-dependent Hartree-Fock

Kouhei Washiyama[1,*]

[1] *Center for Computational Sciences, University of Tsukuba, Tsukuba 305-8577, Japan*

Abstract. We analyze fusion hindrance in heavy systems, where the fusion probability around the Coulomb barrier is hindered compared with that in light and medium-mass systems. We perform simulations of central collisions around the Coulomb barrier in heavy systems with time-dependent Hartree–Fock (TDHF) and find that the fusion hindrance is realized in TDHF simulations. We extract nucleus–nucleus potential and energy dissipation in heavy systems by a method combining a microscopic TDHF evolution to a macroscopic collective equation of motion. We find that the extracted potentials exhibit a dynamical increase at small relative distances, while the extracted friction coefficients show rather a behavior similar to that in light and medium-mass systems. We find from our analysis that the dynamical increase in potential is a main contribution to this fusion hindrance.

1 Introduction

The synthesis of superheavy elements by using heavy-ion fusion reactions has been a major subject in nuclear physics. The discovery of superheavy elements gives an impact in our society. Especially in Japan, a great impact has been given to a wide area of Japanese society through major newspapers and TV news since Nihonium was discovered and named.

The discovery of such superheavy elements has been realized by a recent progress on experimental technique in heavy-ion fusion reactions. From the theory side, we need to predict suitable combinations of projectile and target, and suitable bombarding energies to help the experimental side to design future experiments. However, lack of understanding the whole process of superheavy element synthesis that is from the entrance channel to survival of evaporation residue against fission, is still significant.

In this proceedings, we focus on fusion process from the capture (touching of projectile and target) and compound nucleus formation. It is known for long as a fusion hindrance[1] that fusion probability of heavy systems (whose charge products are larger than approximately 1,600) at energies around the Coulomb barrier is significantly hindered compared with that of light and medium-mass systems [1–3]. This hindrance was predicted and extensively discussed by Swiatecki and his collaborators with a model based on the macroscopic liquid-drop model [4–6]. In these seminal works, potential landscape and energy dissipation were key ingredients to understand the mechanism of the fusion hindrance. Later on, Langevin dynamical models have been developed and applied to the analysis of the synthesis of superheavy elements such as competition of the compound nucleus formation and quasi-fission process [7, 8].

We would like to address the origin of the fusion hindrance from the microscopic point of view. The aim of the present work is therefore to analyze the fusion hindrance phenomena by a microscopic reaction model capable of treating fusion reactions around the Coulomb barrier energy. For this aim, we employ the time-dependent Hartree–Fock method as a microscopic reaction model. Furthermore, to understand the fusion hindrance from the potential landscape and energy dissipation mentioned above, we employ a method that maps the time evolution of a microscopic system that has many internal degrees of freedom to a macroscopic equation of motion that has only a few collective degrees of freedom, which we proposed [9, 10]. As a result, this method extracts nucleus–nucleus potential and energy dissipation from collective degrees of freedom to internal degrees of freedom. In the following, we will explain our method of extracting potential and energy dissipation and show these results in heavy systems. Then, we will show that the obtained dynamical increase in potential is a main origin of the fusion hindrance.

2 Method

The time-dependent Hartree-Fock (TDHF) method is a powerful method to describe nuclear collective motion far from equilibrium. The method can treat both static and dynamical properties of nuclei in a self-consistent framework. Since the first application of TDHF [11] in 1970, many applications of TDHF to nuclear reactions have been performed. Recent progress due to powerful computational facilities allows us to perform TDHF simulations with full Skyrme energy density functionals and without restriction on spacial symmetries. Along this line, we

*e-mail: washiyama@nucl.ph.tsukuba.ac.jp

[1] Note that the fusion hindrance explained in this proceedings is different phenomenon from the fusion hindrance appearing at deep subbarrier energies in light and medium-mass systems, which were discussed by several authors during the conference.

briefly explain our method of extracting nucleus–nucleus potential and friction coefficient from TDHF simulations, which consists of the following steps: i) First, we perform TDHF simulations for central collisions to obtain the time evolution of densities of a colliding system. ii) From the obtained densities, we divide the colliding system into projectile-like and target-like fragments by properly defining a neck position. We define a neck position as the position at which one-dimensional density $\rho(x, y = 2\,\text{fm}, z = 0\,\text{fm})$ along the collision axis (x axis) becomes minimum [12]. iii) We compute the center-of-mass positions of each fragment, denoted by R_1 and R_2, and associated momenta of each fragment, denoted by P_1 and P_2. We also compute the masses of each fragment by $m_i = P_i/\frac{dR_i}{dt}$ ($i = 1, 2$), and the reduced mass μ of the system. iv) We construct time evolution of relative distance R and relative momentum P from R_1, R_2, P_1 and P_2. v) We assume that the obtained time evolution of the collective variables R and P obeys the following classical macroscopic equation with a friction term,

$$\frac{dR}{dt} = \frac{P}{\mu}, \tag{1}$$

$$\frac{dP}{dt} = -\frac{dV}{dR} - \frac{d}{dR}\left(\frac{P^2}{2\mu}\right) - \gamma\frac{P}{\mu}, \tag{2}$$

where V and γ denote nucleus–nucleus potential and friction coefficient, respectively, which are considered as unknown quantities. These quantities may depend on the relative distance R. vi) To obtain two unknown quantities $V(R)$ and $\gamma(R)$, we prepare two trajectories for two slightly different center-of-mass energies under the assumption that $V(R)$ and $\gamma(R)$ do not change by this slight change of energy. Then, to solve the system of two equations at R, $\gamma(R)$ is obtained as

$$\gamma(R) = -\frac{\dot{P}_I - \dot{P}_{II}}{\dot{R}_I - \dot{R}_{II}} + \frac{\dot{R}_I + \dot{R}_{II}}{2}\frac{d\mu}{dR}, \tag{3}$$

where subscripts I and II indicate quantities at two slightly different energies E_I and E_{II}, respectively. Accordingly, $V(R)$ can be obtained. We take $E_{II} - E_I = 0.4\,\text{MeV}$ when E_I is the fusion threshold energy above which fusion occurs and below which fusion does not occur. In the following, the nucleus–nucleus potentials, friction coefficients, and reduced masses are extracted at the fusion threshold energy, except for $E_{\rm cm} = 250\,\text{MeV}$ in Fig. 1 and ^{70}Zn + ^{208}Pb system in Fig. 3 where we take $E_{II} - E_I = 2.0\,\text{MeV}$.

We would like to note that the extraction described above is not possible when the following situations are encountered: (i) the denominator in Eq. (3) becomes close to zero, (leading to divergence of the friction coefficient), (ii) the velocity of a fragment becomes close to zero, (leading to divergence of its mass), (iii) the neck position can not be properly determined (leading to difficulty of constructing two-body kinematics). We have encountered these situations in many systems considered when the overlap of projectile and target nuclei becomes significantly large at the later stage of the fusion process. Once we face one of the above situations at a certain relative distance, we stop the extraction. We denote this relative distance as $R_{\rm min}$.

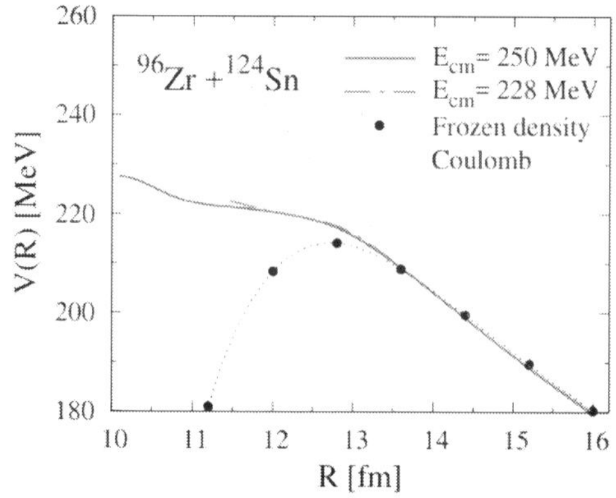

Figure 1. Nucleus–nucleus potentials as a function of R. The solid and dot-dashed lines are from our method at two different energies, while the solid circles are the result of the frozen density potential. The dotted line shows the pure Coulomb potential for reference.

We perform the TDHF simulation with Skyrme SLy4d parametrization [13]. Pairing correlations are neglected in both static and dynamical calculations. We define fusion in TDHF as a reaction where a colliding nuclei keeps a compact shape for a sufficiently long time ($\sim 2000\,\text{fm}/c$). More details can be found in Ref. [12].

3 Result

3.1 Nucleus–nucleus potential

First, we show in Fig. 1 nucleus-nucleus potentials obtained from our method for ^{96}Zr + ^{124}Sn, whose charge product is 2,000. The solid and dot-dashed lines are the potentials extracted at $E_{\rm cm} = 250, 228\,\text{MeV}$, respectively. The filled circles are so-called frozen density potential where the interaction energy is calculated from the energy density functionals with keeping the target and projectile densities frozen to their respective ground-state densities, while all the dynamical effect during the collision and the Pauli effect from the overlap of the projectile and target densities are neglected. One can clearly see that the potentials extracted from our method do not show a barrier structure at smaller distances, and rather show monotonic increase as two nuclei approach to each other. This observation is rather general [12]. We do not observe such a dynamical increase in the potentials extracted in light and medium-mass systems [9] and in the frozen density potentials. We think that this property is related to the consideration that in heavy systems the conditional saddle point appears inside the touching configuration of two nuclei because of large Coulomb energy in the extra-push model [4–6]. Furthermore, we do not see energy dependence of the extracted potential. This is also different from the potential extracted in light and medium-mass systems that show strong energy dependence [9]. Other important effect that increases the nucleus–nucleus potential comes

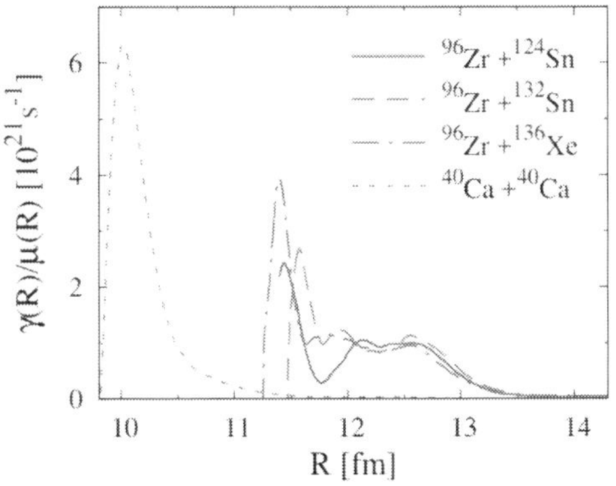

Figure 2. Reduced friction coefficient γ/μ as a function of R for different systems.

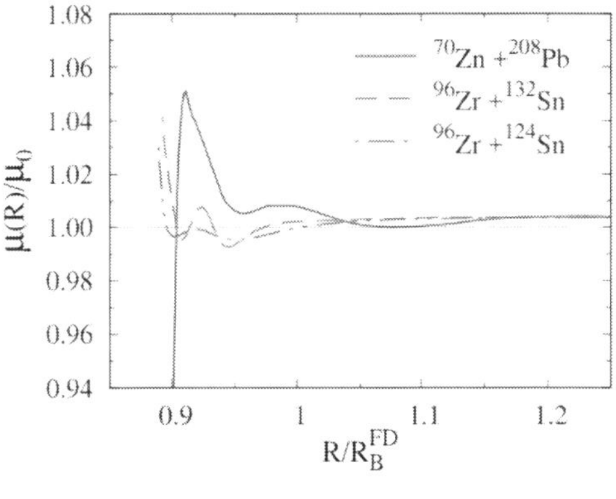

Figure 3. Reduced mass divided by its initial reduced mass as a function of R scaled by its barrier radius for three heavy systems.

from the Pauli effect at overlap of colliding nuclei. As stated above, the Pauli effect is neglected in the frozen density potential, while this is included in TDHF dynamics.

3.2 Friction coefficient

As is shown before, our method simultaneously extracts potentials and friction coefficients. Here, we discuss the property of friction coefficients extracted for heavy systems. Figure 2 shows the extracted friction coefficients scaled by the corresponding reduced mass (called reduced friction coefficient) as a function of R for three heavy systems as well as that for medium-mass systems ^{40}Ca + ^{40}Ca as a comparison. Though the extracted potentials in heavy systems are very different from that in light and medium-mass systems, the extracted frictions in heavy systems show a similar behavior to that in ^{40}Ca + ^{40}Ca, that is, the strength of the friction coefficients in heavy systems is of the same order of magnitude as that in ^{40}Ca + ^{40}Ca, and the friction coefficients become larger as R decreases and has a peak around $R_{\rm min}$.

3.3 Collective mass

Our method gives the reduced mass of two-body kinematics during fusion reactions. Figure 3 shows the reduced mass divided by its initial mass μ_0 as a function of relative distance scaled by the barrier radius of the frozen density potential denoted by R_B^{FD} for three heavy systems. At larger R, the reduced mass is close to its initial mass, and as R decreases, the reduced mass starts to oscillate and then increase at the distance close to $R_{\rm min}$. For the ^{70}Zn + ^{208}Pb, sudden decrease is seen after the increase. Thus the reduced mass obtained from our method shows a dynamical property and a relative distance dependence. Recently, the property of the inertial mass for the collective path of nucleus–nucleus collisions has been extensively discussed by the adiabatic self-consistent coordinate method [14, 15]. In that study, the inertial mass is close to its reduced mass at large R, then increases as R decreases, which is similar to the behavior of our reduced mass, though the method of deriving the inertial mass is different from our method.

3.4 Extra-push energy by TDHF

TDHF simulations of fusion reactions in heavy systems show fusion hindrance [12, 16, 17]. As a consequence, the extra-push energy $E_{\rm extra}$ by TDHF can be defined as $E_{\rm extra} = E_{\rm thres} - V_{\rm FD}$, where $E_{\rm thres}$ is the fusion threshold energy, while $V_{\rm FD}$ is the barrier height of the frozen density potential. The top panel of Fig. 4 shows the comparison between $E_{\rm extra}$ by TDHF and the experimental one [18] for the ^{90,92,94,96}Zr + ^{124}Sn systems. Our TDHF simulations underestimate the experimental extra-push energies. The discrepancy becomes larger as the neutron number in Zr becomes larger. The discrepancy can be attributed by lack of dissipation beyond the mean-field level. (One-body dissipation is treated self-consistently in our TDHF simulations.) Other reason of this discrepancy might be due to neglecting pairing correlations, which K. Sekizawa pointed out in his talk during the conference.

3.5 Origin of fusion hindrance

We have calculated nucleus–nucleus potentials and friction coefficients as well as extra-push energies in heavy systems. Next, we will consider the origin of fusion hindrance. First, we focus on the fusion trajectory at the fusion threshold energy. We will use in the following the nucleus–nucleus potential and friction coefficient extracted at the fusion threshold energy. Then, we define the increase in potential energy ΔV as the energy between the extracted potential at $R = R_{\rm min}$ and the barrier $V_{\rm FD}$ of the frozen density potential, which is $\Delta V = V(R_{\rm min}) - V_{\rm FD}$, and accumulated dissipated energy $E_{\rm diss}$ at $R = R_{\rm min}$ as

$$E_{\rm diss}[R_{\rm min}(t)] = \int_{t_0}^{t} dt' \gamma[R(t')]\dot{R}(t')^2 , \qquad (4)$$

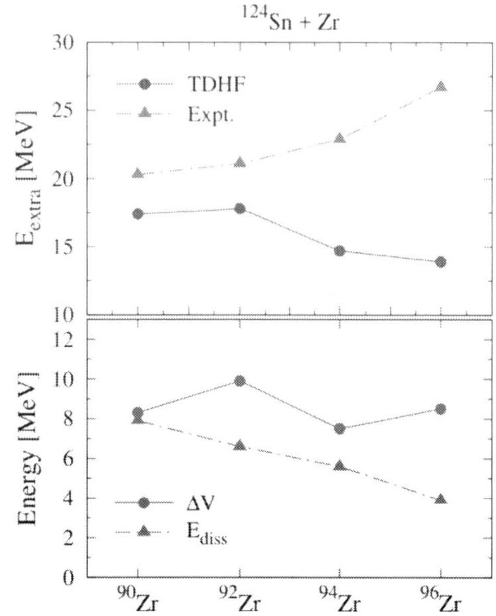

Figure 4. (Top) Comparison between TDHF and experimental [18] extra-push energies. (Bottom) Comparison between the increase in potential energy ΔV and the accumulated dissipated energy for the ^{90,92,94,96}Zr + ^{124}Sn systems.

where the integration is performed from an initial time t_0 to the time t at which our extraction procedure stops. From the energy conservation and the fact that the kinetic energy at $R_{\min}$ is very small from our observation, the extra-push energy is simply represented by the sum of the increase in potential energy ΔV and accumulated dissipated energy $E_{\rm diss}$. The bottom panel of Fig. 4 shows the comparison between the increase in potential energy ΔV and accumulated dissipated energy $E_{\rm diss}$ for the ^{90,92,94,96}Zr+^{124}Sn systems. ΔV is larger than $E_{\rm diss}$. We found that $\Delta V > E_{\rm diss}$ holds for other heavy systems [12]. We conclude that the dynamical increase in potential energy induced from the large Coulomb energy and the Pauli effect at short relative distances is the main contribution to the presence of the extra-push energy and the origin of the fusion hindrance from the present analysis.

4 Summary

We have discussed the rather historical fusion hindrance in heavy systems. We performed the analysis of the fusion hindrance by using TDHF dynamics. We extracted nucleus–nucleus potential and energy dissipation from our method combining TDHF dynamics and a macroscopic equation of motion. We showed that potentials in heavy systems show monotonic increase as the relative distance of two nuclei approaches. This is a unique feature seen in heavy systems in our analysis. Then, we analyzed the extra-push energy by using the properties of potentials and energy dissipation extracted from our method. Dynamical increase in potential has more significant contribution to the fusion hindrance.

This work was funded by ImPACT Program of Council for Science, Technology and Innovation (Cabinet Office, Government of Japan).

References

[1] H. Gäggeler, T. Sikkeland, G. Wirth, W. Brüchle, W. Bögl, G. Franz, G. Herrmann, J. V. Kratz, M. Schädel, K. Sümmerer et al., Z. Phys. A **316**, 291 (1984)

[2] C.-C. Sahm, H.-G. Clerc, K.-H. Schmidt, W. Reisdorf, P. Armbruster, F. P. Heberger, J. G. Keller, G. Münzenberg, D. Vermeulen, Z. Phys. A **319**, 113 (1984)

[3] C.-C. Sahm, H.-G. Clerc, K.-H. Schmidt, W. Reisdorf, P. Armbruster, F. P. Heberger, J. G. Keller, G. Münzenberg, D. Vermeulen, Nucl. Phys. A **441**, 316 (1985)

[4] W. J. Swiatecki, Prog. Part. Nucl. Phys. **4**, 383 (1980)

[5] W. J. Swiatecki, Phys. Scr. **24**, 113 (1981)

[6] W. J. Swiatecki, Nucl. Phys. A **376**, 275 (1982)

[7] Y. Aritomo, M. Ohta, Nucl. Phys. A **744**, 3 (2004)

[8] V. Zagrebaev, W. Greiner, J. Phys. G **31**, 825 (2005)

[9] K. Washiyama, D. Lacroix, Phys. Rev. C **78**, 024610 (2008)

[10] K. Washiyama, D. Lacroix, S. Ayik, Phys. Rev. C **79**, 024609 (2009)

[11] P. Bonche, S. Koonin, J.W. Negele, Phys. Rev. C **13**, 1226 (1976)

[12] K. Washiyama, Phys. Rev. C **91**, 064607 (2015)

[13] K.-H. Kim, T. Otsuka, P. Bonche, J. Phys. G **23**, 1267 (1997)

[14] K. Wen, T. Nakatsukasa, Phys. Rev. C **94**, 054618 (2016)

[15] K. Wen, T. Nakatsukasa, arXiv:1703.04319 (2017)

[16] L. Guo, T. Nakatsukasa, EPJ Web Conf. **38**, 09003 (2012)

[17] C. Simenel, Eur. Phys. J. A **48**, 152 (2012)

[18] K. H. Schmidt, W. Morawek, Rep. Prog. Phys. **54**, 949 (1991)

DOI: 10.1051/epjconf/201716300065

Production and investigation of heavy neutron rich nuclei

Sergey Zemlyanoy[1,*], *Konstantin* Avvakumov[1], *Eduard* Kozulin[1], *Valentin* Fedosseev[2], *Robert* Bark[3], and *Zenon* Janas[4]

[1]*Joint Institute for Nuclear Research, 141980 Dubna, Russia*
[2]*CERN, 1211, Geneva 23, Switzerland*
[3]*iThemba LABS, Nat. Research Foundation, Somerset West 7129, South Africa*
[4]*Faculty of Physics, University of Warsaw, Hoza 69, PL-00681 Warsaw,*

Abstract. A project devoted to the production and study of neutron rich heavy nuclei (GALS - project) is being realized at Flerov Laboratory for Nuclear Reactions (FLNR) - JINR. GALS is planned to exploit available beams from the U-400M cyclotron in low energy multi-nucleon transfer reactions to study exotic neutron-rich nuclei located in the "north-east" region of nuclear map. Products from 4.5 to 9 MeV/nucleon heavy-ion collisions, such as ^{136}Xe on ^{208}Pb, are to be captured in a gas cell and selectively laser-ionized in a sextupole (quadrupole) ion guide extraction system.

1 Introduction

The closed neutron shell N=126 – the so called last "waiting point" for the r-process – is located in the region of heavy neutron rich nuclei, "north-east" part of the nuclide chart, which is very important for nuclear physics investigations, especially for the understanding of astrophysical nucleosynthesis and r-process. (Fig. 1, 2). Neutron shell N=126 is a classical shell closures for nuclei along the line of stability. However, as one moves away from the line of stability, abrupt changes in nuclear structure can arise. Especially, appearance of new magic numbers and/or disappearance of traditional ones as well as the onset of deformation are expected, as e.g. already observed in the region of lighter nuclei.

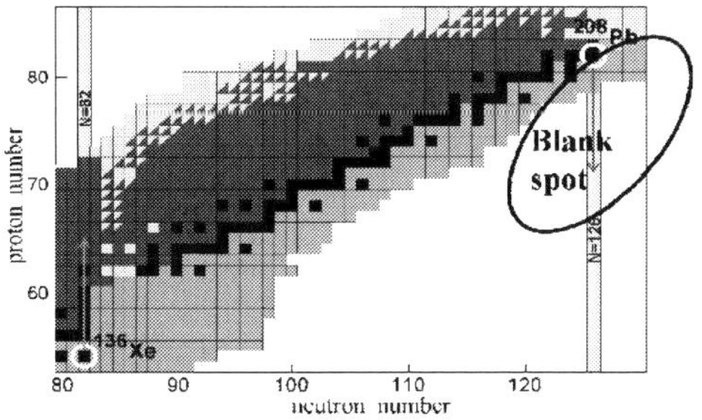

Figure 1. Upper "north-east" part of nuclear map.

Study of the structural properties of nuclei along the neutron shell N=126 could contribute to the current discussion about the quenching of shell gaps in nuclei with large neutron excess. The creation and launch of this facility will open a new field of research in low-energy heavy-ion physics, and new horizons in the study of unexplored "north-east" area of the nuclear map. It could be helpful also for finding a new way for the production of heavy and superheavy nuclei.

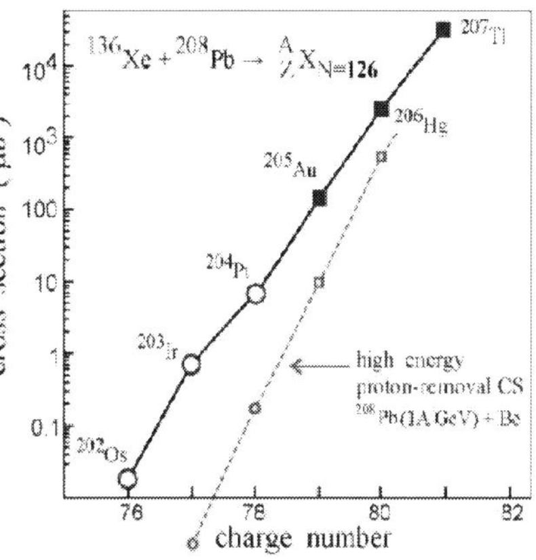

Figure 2. Cross sections for the formation of heavy nuclei in collisions of ^{136}Xe on ^{208}Pb at center of mass energy of 700 MeV. Open circles indicate unknown isotopes [1]. The dashed curve illustrates the yield of the same isotopes in the proton removal reaction from closed shell nucleus ^{208}Pb.

Experiments at GALS are intended for production of heavy neutron rich nuclei in the above mentioned region by using multi-nucleon transfer reactions. As suggested by Zagrebaev and Greiner [1], several tens of new nuclides in

* Corresponding author: e-mail: zemlya@jinr.ru

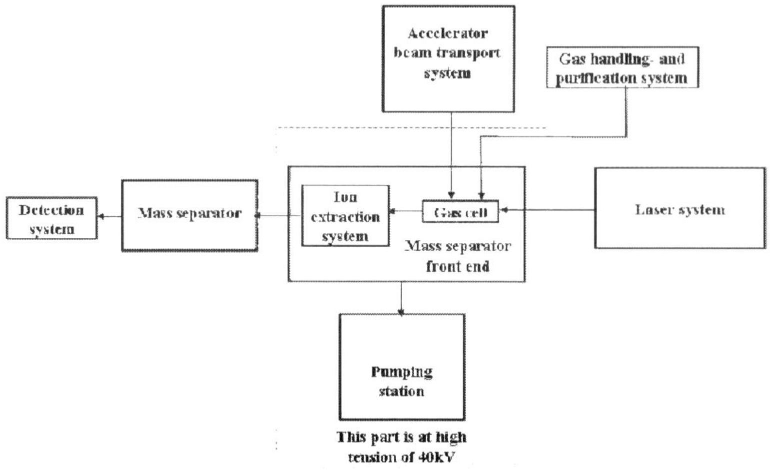

Figure 3. Block scheme of GALS-facilities.

the region of N=126 and $Z\sim75$ can be produced, for example, in the near-barrier collision of ^{136}Xe with ^{208}Pb. Even higher cross sections have been predicted for the production of new neutron rich nuclei in collisions of ^{198}Pt beam with ^{238}U target. At present this seems to be the only realistic method [1-3] that could fill the "blank spot" of the nuclear map (Fig. 1).

2 Experimental method

The experimental approach, fulfilling the purpose of this project, is a combined method of Z and A/Q separation. This method has been proposed in 1992 [4] and subsequently intensively studied (first at LISOL) and developed [5-13]. Known as In-Gas Laser Ionization and Spectroscopy (IGLIS), it is based on stopping of nuclear reaction products in a gas cell and subsequent selective resonance laser ionization [5-13]. This technique allows extracting nuclei with a given atomic number Z, while a separation of the single-ionized isotopes by their mass number can be done rather easily by a magnetic field.

Half-lives of heavy neutron rich nuclei of interest (as a rule, β–decaying), are much longer than the extraction time of ions from a gas cell. Detailed layout of the proposed facility can be found in [2, 3]. Here we present only a brief description of the GALS facility (Fig. 3). Neutron rich isotopes of heavy elements are produced in multinucleon transfer reactions with heavy ions accelerated up to 5-10 MeV/nucleon (depending on projectile-target combination). The target is a foil of about 300 μg/cm^2 thickness (or larger). It is placed at the window of the gas cell (or inside it).

Nuclear reaction products recoiling out from the target as multi-charged ions are thermalized and neutralized by collisions with highly-purified argon or helium buffer gas. Then the atoms of interest (with a given Z) are ionized by means of two or three-step resonance laser irradiation and are extracted by the gas flow through the exit into the vacuum chamber as singly charged ions (Q = +1) with low energies of about 0.2 eV. Subsequently, the ions are confined in a radiofrequency ion guide system, which allows pumping out the residual buffer gas while transporting the ions towards the mass separator. Then the ions are accelerated up to 30-60 keV and selected by the mass-separator. In this way a low-energy beam of singly-ionized ions with a good optical quality (small emittance, energy spread less than 1 eV) is produced. This allows obtaining typical mass resolution of 1500 after the dipole magnet. After mass separation a beam of radioactive ions with a definite atomic number and a previously chosen mass value is obtained. The background due to unwanted isobar and isotope admixture is significantly suppressed that leads to an enhanced sensitivity. This gives the possibility to perform subsequently high sensitive analysis of spectroscopic and decay properties of these nuclei, as well as measurements of their spins, magnetic dipole and electric quadrupole moments and charge radii by means of laser spectroscopy. The operation principle of the different parts of the set up can be found in a number of papers (see e.g. [14] and the references therein).

3 Current status

The ion beams available at FLNR fully satisfy the requirements for this project: the ions that can be used are quite different depending on the problem to be solved - from ^{16,18}O to ^{238}U, beam energies 4.5 - 9 MeV/nucleon (slightly above the Coulomb barrier) and beam intensity up to 10^{13} pps are accessible with a beam size at the target of 3-10 mm and beam emittance of 20·π·mm·mrad. Different heavy targets, including those of actinides, are expected to be used. Therefore the new facility can be developed and coupled directly to the available U-400M accelerator. The mass separator could be a standard magnet

 DOI: 10.1051/epjconf/201716300065

separator similar to GPS at ISOLDE II (see [2, 15]) but nevertheless substantial efforts are needed for its realization.

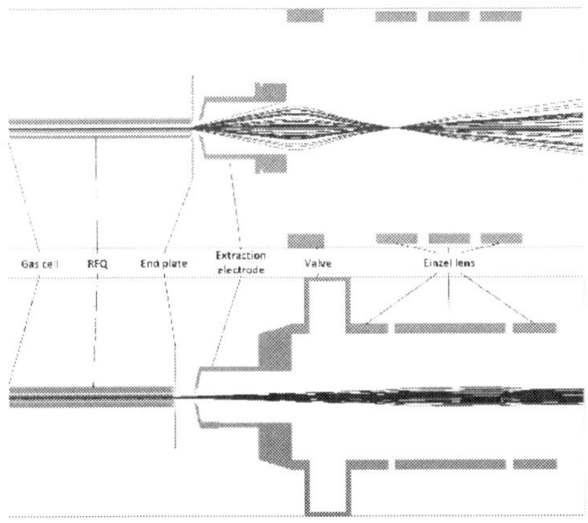

Figure 4. The LISOL-like setup (above) and the FLNR setup (below) in enlarged transverse scale with trajectories of the ions passing through.

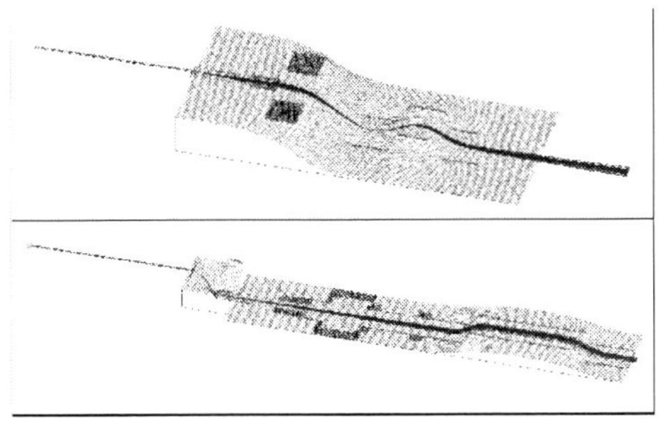

Figure 5. 4 The distribution of potential energy: above – LISOL-like; below – JINR.

3.1 The ion extraction system simulation

The ion extraction system is used for carrying the ions of interest (which are ^{202}Os, ^{203}Ir, ^{204}Pt with an electric charge of +1) coming out from the gas cell nozzle, accelerating them in an electric field and focusing in the detector system plane. Two system options were proposed:

1. The system of the existing LISOL setup in Leuven, Belgium.
2. The system designed in Flerov Laboratory of Nuclear Reactions, JINR, Dubna.

The systems are very similar and contain the same main elements, Fig. 3. The differences are the geometries after the extraction electrode and the voltages applied to the extraction electrode and the end plate (U_{EE}=0 V, U_{EP}=39.54 kV in the JINR system, and U_{EE}=38.34 kV, U_{EP}=39.75 kV in a LISOL-like system). An SPIG (sextupole ion guide), or quadrupole ion guide as an option, is the same in both systems and has on its rods a voltage up to 500 V and a frequency of 4.7 MHz. Both of the systems were simulated using SIMION 8.0 package (http://simion.com/). The initial speed of the singly charged ions of ^{204}Pt at the nozzle exit was set to 560 m/s, which corresponds to kinetic energy of 0.332 eV. The simulation results are shown in Fig. 4, 5 and Table 1.

Table 1. Output ions kinetic energy parameters and overall electrostatic system efficiencies (without considering gas influence) at the different types of ion guide system. Statistics of 1000 ions used in each case.

LISOL-based system	SPIG, U_{rf}=150 V		Quadrupole ion guide, U_{rf}=159 V	
Gas cell exit hole diameter	0.5 mm	0.5 mm	0.5 mm	1 mm
Mean value, eV	39707.55	39710.78	39710.78	39709.99
Standard deviation, eV	7.96	8.22	8.22	6.14
Efficiency, %	98	99	99	98
FLNR system	SPIG, U_{rf}=150 V		Quadrupole ion guide, U_{rf}=159 V	
Gas cell exit hole diameter	0.5 mm	0.5 mm	0.5 mm	1 mm
Mean value, eV	38481.88	38484.78	38484.78	38485.67
Standard deviation, eV	5.06	5.67	5.67	4.81
Efficiency, %	100	100	100	100

In the LISOL-based setup beam divergence after the end plate is significantly higher than in FLNR setup. This leads to beam losses in collisions with the inner surface of the extraction electrode (about 1% of particles were lost there in LISOL-based setup model compared to no losses in the FLNR setup model).

Moreover, the voltage of 20 kV on the middle Einzel lens electrode in the LISOL-based setup is not enough to focus the whole output beam (the voltage of about 35 kV could be needed). The main acceleration gap in the FLNR setup is between the end plate and the extraction electrode which results to better collimated beam through the extraction electrode and the Einzel lens. On the other hand, such a high electric field (10 kV/cm) could increase the possibility of an electric breakdown. The gas pressure in that region of the vacuum chamber is about 10^{-6} Torr, but in case of gas cell foil breakthrough it can increase significantly. It was calculated that in case of adiabatic argon expansion the pressure in this region would be still less than $1.5{\cdot}10^{-3}$ Torr which, according to the Paschen curves (the distance between the electrodes is 4 cm), remains in safe from electric breakdown area. Yet, some additional safety measures should be taken.

The efficiencies (see Table 1) were calculated without taking into account residual gas influence. Basing on the LISOL SPIG efficiency of 66% [8], we can expect the efficiency of the FLNR system with SPIG to be 67% or higher. Thus, the system developed in FLNR looks more preferable due to the less beam losses and less ion energy dispersion.

Table 2. Laser to be used in multistep ionization. The lasers already delivered and installed are presented in bold.

Type	Output power main&harmonic, W (2nd),3rd, 4th	Puls frequency, Hz	Puls length, ns	Wave length, nm
Dye laser	**3, (0.3)**	**10^4**	**10-30**	**213-850**
Ti:Sapphire	2, (0.2), 0.04	10^4	30-50	680-960
Nd-YAG	**(80-100),20-40**	**10^4**	**10-50**	**532**
Matisse system				
Ring dye	0.8-6	cw	cw	540-900
Ti:Sapphire	0.8-6.5	cw	cw	700-1000

3.2 Laser system and excitation schemes

The choice of specific laser ionization scheme, the type and number of lasers is determined by the ionization potentials and and the atomic level schemes of the elements under study. In our case a three-step scheme of ionization looks more favorable. Such a scheme allows choosing more effective optical transitions to increase the yield of resonance-ionized ions, although the use of two-step ionization scheme is as not excluded. Dye laser systems pumped by second and third harmonics of Nd:YAG can provide tuning in a broad spectral range: from near UV to near IR [16]. The generation of a Ti:Sapphire lasers, shifted to the red and infrared edge of the spectrum (680-960 nm), can be also used as complementary to that of the dye lasers. Thus, the installation of two laser-ionization schemes-with dye lasers and Ti:Sapphire lasers - would be of a great benefit allowing to meet diverse experimental demands. Such a double RILIS scheme is already realized in CERN [17]. The parameters of the delivered (and already installed) laser as well as of the ordered ones are summarized in Table 2.

At the first stage of our experiments, excitation schemes using 3-step ionization with a non-resonant transition to continuum will be tested with the available lasers: two dye lasers Credo (Sirah) and Nd:YAG laser. The Credo laser has maximal average power of 20 W at fundamental wavelength and 2 W at the second harmonics; its line width is 1.8 GHz at pulse duration about 7 ns. It is an option that allows remote control of wavelength with stabilization to an external laser wavelength meter. The Nd:YAG laser (product of Edge Wave GmbH) generates maximal average power of 90 W and 36 W for the second and third harmonic respectively with repetition rate 10 -15 kHz and pulse duration of 8 - 10 ns. The divergence parameter of the green beam is m^2=1.4. It needs electrical power supply of 3.6 kW including 1.6 kW for the water chiller.

A Matisse series of a ring cw-dye laser and cw-Ti:Sapphire laser pumped by cw-Nd:YAG laser is planned as a seeding system. Such a system has an advantage in the cases when tuning of the laser wavelength is necessary, e.g. for laser spectroscopy.

As osmium is one of most prominent candidates to achieve the neutron magic line N=126 we have checked ionization schemes of this element (Table 3). Although there are many observed spectral lines, as the atomic spectra information is not yet complete: most of the transitions are not assigned [16-19] and the atomic strengths are not known. Nevertheless several of the possible RIS scheme for Os have been previously studied [20, 21]. The most favorable ionization scheme is expected to be the resonance three steps one via an auto-ionizing state. At present, no literature information is available about such states except the one at 71032 cm^{-1} corresponding to the third step of case C [20]. For this reason at the second stage of the GALS activities a search for auto-ionizing states is planned. At the next phase of experiments, an off-line study of the optimal ionization scheme will be conducted. It supposes a detailed experimental study the efficiency of the different ionization schemes.

4 Outlook

A new GALS facility at FLNR cyclotron U-400M is in a stage of build-up. GALS will apply the highly selective and efficient technique of stepwise resonant ionization in a gas cell (of the element of interest) with subsequently mass separation. In fact, it is a mass separator from a new type as yet not used in JINR. GALS is addressed to the production and study of new neutron rich nuclei located along the neutron shell N=126. This is the so called "north-east" part of the nuclear map. Nuclei from this region are as yet "terra incognita" but of high fundamental interest mainly due to their large impact of on the astrophysical r-process. The most efficient method to produce such nuclei, as motivated in [1], are the multi-nucleon transfer reactions. Our estimations show that at target thickness 0.3 mg/cm, ion beam of 0.1 pμA and setup efficiency of 10% we would be able to measure decay properties of 1 new isotope per day.

References

[1] V. Zagrebaev and W. Greiner, Phys. Rev. Lett. **101**, 122701 (2008)

[2] V. I. Zagrebaev, S. G. Zemlyanoy, E. M. Kozulin et al., Hyperfine Interacs. **227**, 181 (2014)

[3] V. I. Zagrebaev, S. G. Zemlyanoy, E. M. Kozulin et al., J. Phys.Conf. Series **533**, 012025 (2014)

[4] P. Van Duppen, P. Dendooven, M. Huyse et al., Hypfine Interacs. **74**, 193 (1992)

[5] L . Vermeeren, N. Bijnens, M. Huyse et al., Phys. Rev. Lett. **73**, 1935 (1994)

[6] Y. Kudryavtsev et al., Nucl. Instrum. Methods Phys. Res. **B 114**, 350 (1996)

[7] Y. Kudryavtsev et al., Nucl. Instrum. Methods Phys. Res. **B 204**, 336 (2003)

[8] M. Facina et al., Nucl. Instrum. Methods **B 226**, 401 (2004)

[9] I. D. Moore et al., J. Phys. G: Nucl. Part. Phys. **31**, S1499 (2005)

[10] Y. Kudryavtsev et al., Nucl. Instrum. Methods Phys. Res. **B 267**, 2908 (2009)

[11] R. Ferrer et al., Nucl. Instrum. Methods Phys. Res. **B 291**, 29 (2012)

Table 3. Suggested two ($\lambda_3=\lambda_1$ – case A or $\lambda_3=\lambda_2$ – case B) and three step (case C) ionization schemes of OsI.

λ_1 (air), Å	E_1, cm^{-1}	State I	J_1	λ_2 (air), Å	E_2, cm^{-1}	State II	J_1	case
2909.06	34365.33	6s6p $^5F^0$	5	4752.16	55402.47	6s7s ?	?	C
3018.04	33124.48	6s6p $^7P^0$	3	5580.66	51038.49	6s7s e^{5}D	0	A or B
3267.94	30591.45	6s6p $^7P^0$	4	5509.33	48737.34	unknown	4	A
3301.56	30279.95	6s6p $^7F^0$	5	4815.96	51038.49	6s7s e^{5}D	4	A or B
4260.85	23462.90	6s6p $^7D^0$	5	3157.24	55127.00	unknown	?	A or B
4420.47	22615.69	6s6p $^7D^0$	4	4066.69	48737.44	6s7s e^{7}D	4	A or B
4420.47	22615.69	6s6p $^7D^0$	4	3827.14	47198.7	6s7s e^{7}D	5	A or B

[12] R. Ferrer et al., Nucl. Instrum. Methods Phys. Res. **B 317**, 570 (2013)

[13] Y. Kudryavtsev et al., Nucl. Instrum. Methods Phys. Res. **B 297**, 7 (2013)

[14] P. Campbell, I. D. Moore, M. R. Pearson, Prog. Particle Nucl. Phys. **86**, 127 (2016)

[15] H. L. Ravn and B. W. Allardyce In: Treatise on Heavy-Ion Science, **8**, 363 (1989)

[16] V. N. Fedosseev, L. E. Berg, D. V. Fedorov et al., Rev. Sci. Instrum. **83**, 02A903 (2012)

[17] S. Rothe, B. A. Marsh, C. Mattolat et al., J. Phys.: Conf. Series **312**, 052020 (2011)

[18] C. H. Corliss, W. R. Bozman, *Experimentsl transition probabilities for spectral lines of seventy elements*, (NBS Monograph 53 1962)

[19] W. F. Meggers, C. H. Corliss, B. F. Scribner: *Tables of spectral lines intensities (second edition)*, (NBS Monograph 145 1975)

[20] https://www.nist.gov/pml/atomic-spectra-database (version 5, updated October 2015)

[21] J. E. Sansonetti, W.C. Martin, Handbook of Basic Atomic Spectroscopic Data. J. Phys. Chem. Reference Data 34, 1559 (2005)

[22] J. D. Blum, M. J., Pellin, W. F. Calaway, et al., Anal. Chem. **62**, 209 (1990)

[23] W. F. Calaway, R. C. Wien, D. S. Burnett, et al., J. Vac. Sci. Technol. **A 13**, 1310 (1995)

 DOI: 10.1051/epjconf/201716300066

Study of breakup and transfer of weakly bound nucleus ^{6}Li to explore the low energy reaction dynamics

G. L. Zhang[1,*], G. X. Zhang[1], S. P. Hu[2], H. Q. Zhang[3], P. R. S. Gomes[4], J. Lubian[4], C. L. Guo[1], X. G. Wu[3], J. C. Yang[1], Y. Zheng[3], C. B. Li[3], C. Y. He[3], J. Zhong[2,3], G. S. Li[3], Y. J. Yao[1], M. F. Guo[1], H. B. Sun[2], J. J. Valiente-Dobòn[5], A. Goasduff[6], M. Siciliano[6], F. Galtarosa[6], R. Francesco[6], D. Testov[6], D. Mengoni[6], D. Bazzacco[6], P. R. John[7], W. W. Qu[1], F. Wang[1], L. Zheng[1], L. Yu[1], Q. M. Chen[3], P. W. Luo[2,3], H. W. Li[3,8], Y. H. Wu[3,8], W. K. Zhou[3], B. J. Zhu[3], E. T. Li[2], and X. Hao[2]

[1]School of Physics and Nuclear Energy Engineering, Beihang University, Beijing 100191, China
[2]School of Physics and Energy, Shenzhen University, Shenzhen 518060, China
[3]China Institute of Atomic Energy, Beijing 102413, China
[4]Instituto de Física, Universidade Federal Fluminense, Avenida Litoranea s/n, Gragoatá, Niterói, Rio de Janeiro 24210-340, Brazil
[5]INFN, Laboratori Nazionali di Legnaro, Legnaro (Padova), Italy
[6]Dipartimento di Fisica and INFN, Sezione di Padova, Padova, Italy
[7]Institut für Kernphysik der Technischen Universität Darmstadt, Germany
[8]College of Physics, Jilin University, Changchun 130012, China

Abstract. Investigation of the breakup and transfer effect of weakly bound nuclei on the fusion process has been an interesting research topic in the past several years. However, owing to the low intensities of the presently available radioactive ion beam (RIB), it is difficult to clearly explore the reaction mechanisms of nuclear systems with unstable nuclei. In comparison with RIB, the beam intensities of stable weakly bound nuclei such as ^{6,7}Li and ^{9}Be, which have significant breakup probability, are orders of magnitude higher. Precise fusion measurements have already been performed with those stable weakly bound nuclei, and the effect of breakup of those nuclei on the fusion process has been extensively studied. Those nuclei indicated large production cross sections for particles other than the $\alpha + x$ breakup. The particles are originated from non-capture breakup (NCBU), incomplete fusion (ICF) and transfer processes. However, the conclusion of reaction dynamics was not clear and has the contradiction.

In our previous experiments we have performed ^{6}Li+^{96}Zr and ^{154}Sm at HI-13 Tandem accelerator of China Institute of Atomic Energy (CIAE) by using HPGe array. It is shown that there is a small complete fusion (CF) suppression on medium-mass target nucleus ^{96}Zr different from about 35% suppression on heavier target nucleus ^{154}Sm at near-barrier energies. It seems that the CF suppression factor depends on the charge of target nuclei. We also observed one neutron transfer process. However, the experimental data are scarce for medium-mass target nuclei.

In order to have a proper understanding of the influence of breakup and transfer of weakly bound projectiles on the fusion process, we performed the ^{6}Li+^{89}Y experiment with incident energies of 22 MeV and 34 MeV on Galileo array in cooperation with Si-ball EUCLIDES at Legnaro National Laboratory (LNL) in Italy. Using particle-particle and particle-γ coincidences, the different reaction mechanisms can be clearly explored.

1 Introduction

Investigation of the breakup effect of weakly bound nuclei on the fusion process has been an interesting research topic in the past several years [1–3]. Several experiments and theoretical calculations have focused on this subject. The situation is more complicated when weakly bound nuclei are involved with respect to tightly bound nuclei. Different processes can occur after the breakup of weakly bound nuclei (usually the projectiles). When all the fragments fuse with target nucleus one by one, this process is called sequential complete fusion (SCF). When only part of the fragments fuses with the target nucleus, this process is called incomplete fusion (ICF). When no any fragments are captured, this process is called non-capture breakup (NCBU). During the fusion process, the whole projectile can also fuse with the target nucleus without breakup, and this process is named as direct complete fusion (DCF). Experimentally, the residues formed in SCF cannot be distinguished from ones from DCF, since the compound nucleus is the same. Therefore, in the experiment the complete fusion (CF) cross sections include the contributions of both SCF and DCF. The total fusion (TF) cross section is equal to the sum of CF and ICF, $\sigma_T = \sigma_{CF} + \sigma_{ICF}$. Besides the above processes, transfer process can also occur such as

* e-mail: zgl@buaa.edu.cn

one neutron stripping and pickup. In ICF and transfer processes, the same residual nuclei can be formed. However, in these two processes the light charged particles with different energies can be emitted. In coincidence with the light charged particles, these two processes can be separated.

With the availability of radioactive ion beam (RIB) facilities in a few laboratories, one can produce several kinds of unstable nuclei including neutron- and proton-rich nuclei. Experimental fusion cross sections of unstable nuclei on different target nuclei have been reported. However, owing to the low intensities of the presently available RIB, it is difficult to clearly explore the reaction mechanisms of nuclear systems with unstable nuclei. In comparison with RIB, the beam intensities of stable weakly bound nuclei such as ^{6,7}Li and ^{9}Be, which have significant breakup probability, are orders of magnitude higher. Precise fusion measurements have already been performed with those stable weakly bound nuclei, and the effect of breakup of those nuclei on the fusion process has been extensively studied. It was found that the CF suppression factors at energies slightly above the Coulomb barrier for reactions induced by the same projectile are independent of the charge of target nuclei and mainly depend only on the breakup threshold energy values of the projectile [4, 5]. Wang *et al.* [6], for ^{6}Li with a breakup threshold energy of 1.474 MeV, found an average suppression around 40%. In a recent and very interesting paper [7], an estimate of CF was obtained for the ^{6}Li+^{64}Ni by measuring TF. The derived CF suppression was 13±7%, much smaller than the 40% suppression found for heavier targets.

In ^{8}He+^{197}Au experiment [8], by measuring γ rays which are emitted from the evaporated residual nuclei, fusion and transfer cross sections were obtained. In comparison with the cross sections of these two processes, it is indicated that the sub-barrier total reaction cross section is dominated by one and two neutron stripping. In the case of reactions with neutron-rich radioactive ion beams, the coupling to transfer is found to be important. Recently for the first time in ^{7}Li+^{198}Pt reaction [9] at energies near the Coulomb barrier the α-capture cross sections were measured by in-beam exclusive measurements of prompt γ - rays from the heavy-residues with various light charged particles (α, t, d and p). The cross sections of the residues resulting from the processes of fusion, t-capture and neutron transfer channels were measured by off-beam γ methods. The integrated cross sections for the above several channels were obtained. The different processes were distinguished including breakup fusion and the nucleon transfer. The massive transfer of ^{5,6}He for ^{7}Li was also observed and the breakups of ^{6}He+p and ^{5}He+d were not observed due to the high breakup threshold. The result of a classical dynamic model [10–12] gave a good agreement with the experimental data and explained the critical role of different cluster structures of ^{7}Li in the dynamics of reaction mechanism for the first time. These calculations clearly illustrated a two step process, breakup followed by fusion, in case of the capture of t and α clusters. The present results are useful for the theoretical developments. It is important to carry out a systematic investigation on various targets especially for unstable weakly bound nuclei. Very recently Diaz-Torres *et al* start to develop an unified quantum dynamic model to describe the relevant processes (CF, ICF, NCBU and transfer). So this kind of measurements will further push the theoretical development.

2 Experiments and results

Recently at HI-13 Tandem of CIAE, Beijing, we performed the experiments ^{6}Li+^{96}Zr [13] and ^{6}Li+^{154}Sm [15]. The results are shown in Figure 1 (Left) and Figure 2 (Left), respectively. It is found that there is CF suppression of the order of 25%, at energies above the barrier, when compared with the coupled channel calculations for ^{6}Li+^{96}Zr. In Ref. [2] it is remarked that these studies were based on data for relatively heavy targets and for one of the lightest system investigated, ^{6}Li+^{96}Zr, a smaller suppression was found. Thus, the behavior of the suppression factor in collisions of weakly bound projectiles with light and medium mass targets is not fully established. For a reliable conclusion about this subject, new measurements of CF for lighter targets are required. There is an average suppression of the order of 35% for ^{6}Li+^{154}Sm. The present results show that the systematics of the suppression factor observed for ^{6}Li-induced fusion in the heavy mass region may not be consistent in lighter target mass region. The small breakup probability for lighter targets may produce a smaller effect on the fusion cross section than for heavier targets. Because some γ-ray lines were too weak to be used for the cross section determination of ICF, we only can determine the lower limit for the cross sections of ICF. So the lower limit for cross section of TF is given. The lower limit of the TF cross section data is roughly in agreement with the calculations at the highest energies, which indicates that the ICF seems to be responsible for the loss of flux going to CF, and consequently suppressing the CF at this energy region. Since the experimental data in light target mass region are very scarce, then the systematic behavior needs to be further explored.

Figure 1 (Right) shows the cross sections of 1n-stripping in comparison with CF cross sections [14]. At energies around the barrier, the cross sections of 1n-stripping are similar to CF cross sections. As its excitation function does not drop as fast as the CF excitation function, it is expected that transfer cross sections should be larger than the CF cross section at sub-barrier energies. Here we measured γ -rays feeding the excited states of ^{97}Zr, we only can determine the lower limit of the one-neutron stripping cross section, the other processes were not distinguished yet. Our CRC calculations give results that are slightly lower than the data, suggesting that the reaction mechanism is complicated. Experimentally, one may perform γ-particle coincidence, as we plan to do in the near future. Theoretically, it is still a challenge for a computer code to take into account all possible reaction mechanisms. Figure 2 (Right) shows the comparison of ^{6}Li+^{154}Sm with ^{6}Li+^{144,152}Sm. One can observe

DOI: 10.1051/epjconf/201716300066

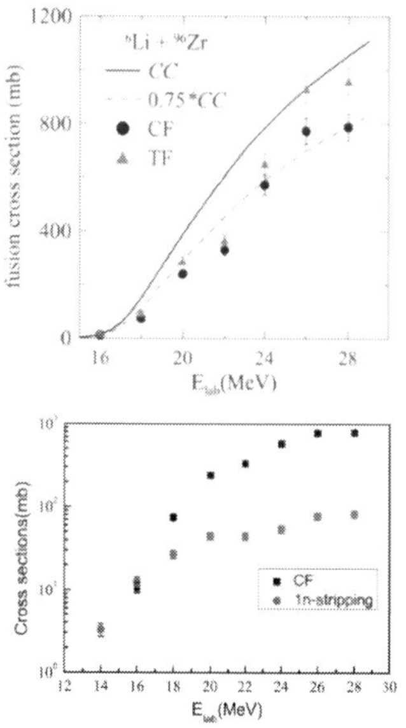

Figure 1. (Up) Complete and total fusion cross sections at energies near Coulomb barrier for ^{6}Li+^{96}Zr [13]; (Down) Measured complete fusion and direct plus two-step 1n-stripping to excited states cross sections corresponding to 1103.1 keV and 1400.1 keV energies for ^{6}Li+^{96}Zr system at near the Coulomb barrier energies [14].

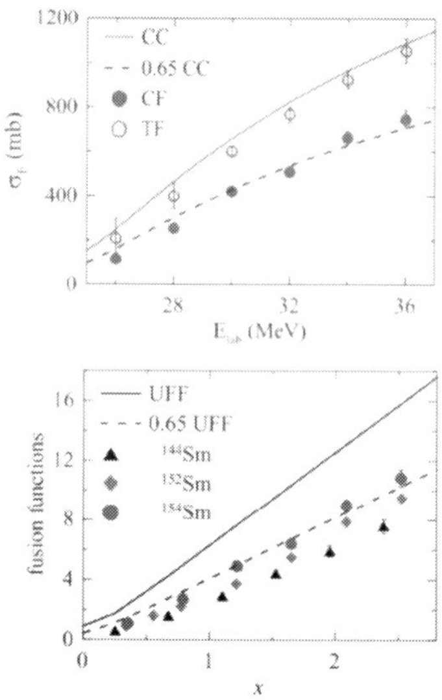

Figure 2. (Up) Complete and lower limit of total fusion cross sections for ^{6}Li+^{154}Sm [15]; (Down) Renormalization experimental fusion functions for the CF of ^{6}Li with three Sm isotopes [15].

that the CF suppressions for the three systems are similar at energies very close to the Coulomb barrier, but when the energy increases, the vibrational ^{144}Sm isotope presents a larger suppression than the deformed ^{152}Sm, and the smallest suppression is found for the most deformed ^{154}Sm. Here we have already taken into account the fact that as the mass A is larger, the size of the system is larger. So, the effect is due to dynamic effects. It is indicated that the breakup effect is larger for the more spherical isotope. The reason for this behavior has to be further investigated both theoretically and experimentally. So the measurement of fusion for more systems is needed.

An experiment using a ^{6}Li beam delivered by the Tandem-XTU accelerator at the LNL (Italy) at E_{lab}=34 MeV and 22 MeV impinging on a ^{89}Y target was performed to study the competition between complete fusion (CF), incomplete fusion (ICF) and transfer reactions. The ^{89}Y target of 550 μg/cm^2 backed on a 340 μg/cm^2 ^{12}C foil was placed at the center of the Galileo γ-ray array [16] coupled to 4π Si-ball EUCLIDES [17–19]. The GALILEO array is made of 25 Compton-Suppressed high-purity germanium detectors distributed on 4 rings: 10 detectors at 90 degrees, 5 detectors at each of the following angles 119, 129 and 152 degrees.

In order to protect the silicon detectors from the extremely high counting rates of elastic scattering, a aluminium cylinder of 200 μm was placed in the center of the

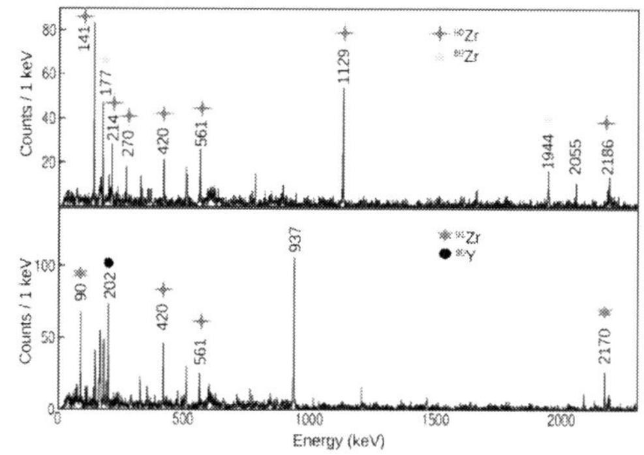

Figure 3. The γ spectra in coincidence with the α particles of the backward (up) and forward (down) angles..

Si-ball in order to completely stop the scattered beam at forward angles. The backward detector of the EUCLIDES Si-ball was left without any absorber. In this setup, the different reaction mechanisms can be easily identified by particle-gamma coincidence. Silicon detectors can be safely run with the beam intensity of ~1.2 enA. The identification of the light-charged particle allows a clear selection of the reaction channel to perform the γ spectrascopy. At the forward angles, the α particles are from transfer when selecting the α particles, the different γ-ray spectrum can be obtained with respect to that of backward angles. In this test experiment we were able to identify the products from transfer, which is shown in Figure 3. The γ rays of ^{90,91}Zr and ^{90}Y can be observed. Then we can derive the cross sections of 1n stripping (^{90}Y), 1p stripping (^{90}Zr) and 1d stripping (^{91}Zr) from the α - γ coincidence. Meanwhile, in the backward alpha-gated γ spectrum, ^{90}Zr, ^{89}Zr and ^{90}Y can be clearly identified. They are related to the components of fusion evaporation residual calculated by PACE4. In two spectra, the γ rays of ^{90}Zr always exists. However, in the backward alpha-gated γ spectrum, the yellow peaks are related to the high lying states of ^{90}Zr, which disappear in the forward alpha-gated γ spectrum. It indicates that the populated states of ^{90}Zr are from the different reaction mechanisms (high lying states from fusion evaporation and low lying states from 1p - stripping reaction). A strong argument is shown that the different reaction mechanisms have been clearly identified by particle - gamma coincidence. According to the above analysis, the present facilities and technology can be applied for this type of experiments.

References

[1] L. F. Canto et al., Phys. Rep., **424**, 1 (2006)
[2] L. F. Canto et al., Phys. Rep., **596**, 1 (2015)
[3] N. Keeley et al., Prog. Part. Nucl. Phys., **59**, 579 (2007)
[4] H. Kumawat et al., Phys. Rev. C., **86**, 024607 (2012)
[5] M. K. Pradhan et al., Phys. Rev. C., **83**, 064606 (2011)
[6] B. Wang et al., Phys. Rev. C., **90**, 034612 (2014)
[7] Md. Moin Shaikh et al., Phys. Rev. C., **90**, 024615 (2014)
[8] A. Lemasson et al., Phys. Rev. Lett., **103**, 232701 (2009)
[9] A. Shrivastava et al., Phys. Lett. B., **718**, 931 (2013)
[10] A. Diaz-Torres et al., Phys. Rev. Lett., **98**, 152701 (2007)
[11] A. Diaz-Torres, J. Phys. G: Nucl. Part. Phys., **37**, 075109 (2010)
[12] A. Diaz-Torres, Comp. Phys. Comm., **182** 1100, (2011)
[13] S. P. Hu et al., Phys. Rev. C., **91**, 044619 (2015)
[14] S. P. Hu et al., Phys. Rev. C., **93**, 014621 (2016)
[15] C. L. Guo et al., Phys. Rev. C., **92**, 014615 (2015)
[16] J. J. Valiente-Dobòn et al., INFN-LNL Annual Report 2014.
[17] D. Testov et al., INFN-LNL Annual Report 2015.
[18] A. Gadea et al., INFN-LNL Annual Report 1996, 225.
[19] A. Gadea et al., INFN-LNL Annual Report 1997, 118.

Fusion17 summary talk: windsurfing on a cross section profile

Dieter Ackermann[1,*,**] and *Lorenzo* Corradi[2,***]

[1] *GANIL - Bd Henri Becquerel BP 55027 - 14076 Caen Cedex 05 - France*
[2] *INFN - Laboratori Nazionali di Legnaro - Viale dell'Universita' 2, 35020, Legnaro (Padova) - Italy*

Abstract. The spectrum of physics, discussed in the 7th edition of the "Fusion" conference series, included a number of topics, ranging from reaction dynamics to structure studies, from the lightest to the heaviest nuclei, from inelastic reactions to fusion and fission, with applications of astrophysical interest and the newest developments and findings in theory and experiment. With this paper we try to summarize the most relevant issues discussed during the five days of Fusion17.

1 Introduction

Before starting our considerations, we wish to express our compliments to the organizers for their excellent job. We enjoyed very much the professional organization of the conference, the stimulating and friendly atmosphere and the high quality level of speakers and talks. Fusion17 was an excellent continuation of the good tradition of the "Fusion" conference series, where a wealth of important new experimental and theoretical work has always been at the focus of intense debate.

The history of conferences dealing with heavy ion collision at the Coulomb barrier had started with the first meeting in 1984 at MIT in Cambridge U.S.A. The first conference carrying "Fusion" in its name was organized in 1997 in South Darras, New South Wales - Australia. After stations in Matsushima Miyagi - Japan (2003), Venice - Italy (2006), Chicago - U.S.A. (2008), Saint Malo - France (2011) and New Delhi - India (2014), it came back after exactly 20 years to Australia and Hobart on the Tasmanian island for the 2017 edition. In the following we report on the second experience "down under" and the rich and comprehensive program, trying to trace the relevant issues raised during the conference. Of course the choice of subjects reflects our personal view and taste. We felt it unnecessary to list specific figures or references, most of which will be found in the individual contributions to the proceedings.

2 An overview

In an attempt to establish the general frame and the main skeleton of fusion conferences, we can roughly distinguish three main energy domains, as shown in Fig. 1.

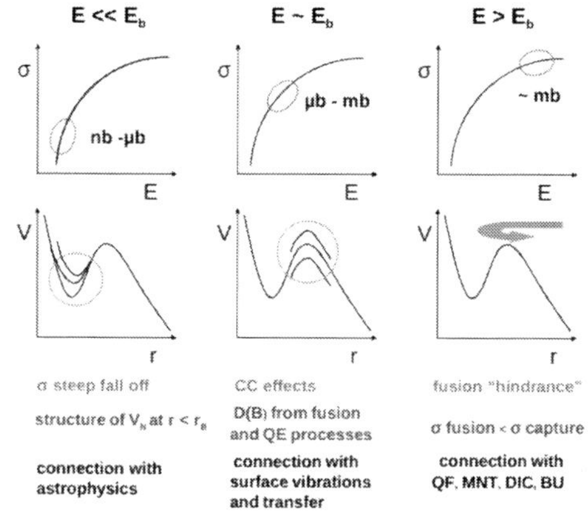

Figure 1. Schematic view of fusion reactions in terms of cross section and associated parts of the nuclear potential, based on bombarding energy ranges.

a) the energy range well below the Coulomb barrier, where cross sections undergo a steep fall off. Here one can explore the shape of the nuclear potential at distances significantly shorter than the fusion barrier radius. Moreover, this is, where a strong connection with astrophysics is established.

b) the energy range near the Coulomb barrier, where coupled channel effects due to surface vibrations and nucleon transfer show up. This is, where the concept of barrier distributions that one can extract from the excitation function for fusion or for quasi-elastic scattering, best applies.

c) the energy range above the Coulomb barrier, where fusion cross sections can be hindered, especially for heavy

* e-mail: ackermann@ganil.fr
** the corresponding author was supported by the European Commission in the framework of the CEA-EUROTALENT program
*** e-mail: corradi@lnl.infn.it

DOI: 10.1051/epjconf/201716300067

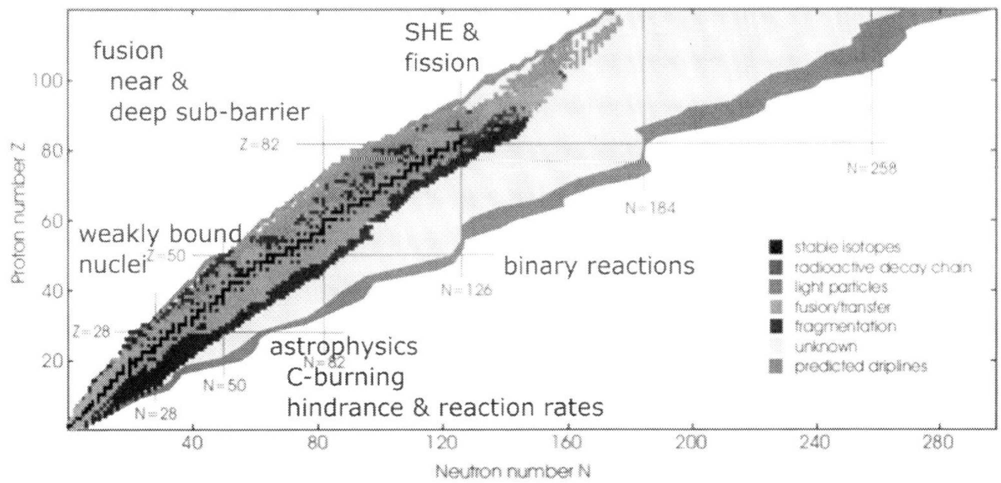

Figure 2. Chart of nuclides illustrating which nuclei have been discovered so far by the various production mechanisms listed in the agenda (courtesy of M. Thoennessen; Rep. Prog. Phys. 76 (2013) 056301). In addition the topics discussed during FUSION17 are indicated (in blue). They are partly localized in various areas of the nuclear chart like the superheavy element (SHE) research at the upper right part or the fusion reactions with weakly bound nuclei at the proton and neutron driplines, mainly in the low mass region. Others, such as binary reactions or the fusion barrier distribution studies, span almost the whole nuclear landscape.

systems. Here the fusion cross section is lower than the one for capture, as a consequence of the competing reaction channels like quasi-fission (QF), multinucleon transfer (MNT) and deep inelastic (DIC) or (especially for weakly bound nuclei) break-up (BU) processes.

Within this frame, many questions were debated and continue to be important issues in the field. Among these questions are the origin of the steep fall off of fusion cross sections, and the connection between fusion, binary processes and other reaction channels (e.g. break-up, especially for weakly bound systems). Challenges are the understanding of the dynamics of fusion reactions for very heavy nuclei, and how this can be quantitatively treated within microscopic theories. These questions were at the core of Fusion17, emerging through the various presentations in the different sub-fields explored during the conference, from astrophysics to fission dynamics and heavy element formation.

In Fig. 2 a compilation of the chart of nuclides is shown, illustrating which nuclei have been discovered through the different (reaction) mechanisms employed so far. Fusion and binary reactions are mainly aiming at nuclei located left from the valley of stability. However, more neutron rich species entered the focus of transfer studies recently with the employment of neutron rich reaction partners, like the region below ^{208}Pb close to the N=126 neutron shell closure. The extremes of the chart of nuclides in the lighter mass area close to the proton and neutron drip lines play a decisive role in reaction studies using weakly bound reaction partners. The topics discussed at Fusion17 summarized in this report are also indicated in Fig. 2, some of them referring to clearly localized regions like the SHE research and fission studies in the upper right part or the just mentioned weakly bound nuclei in the lower left part. Others are involved in investigations for which the whole chart of nuclides is relevant, like fusion barrier studies or binary reactions. The Fusion17 contributions dealing with these topics will be summarized in the following sections.

3 Heavy and superheavy elements

Y. Aritomo, M. Block, Ch.E. Düllmann, J.M. Gates, J. Khuyagbaatar, E. Litvinova, W. Loveland, W. Mittig, P. Schwerdtfeger, J.R. Stone, T. Tanaka, Ch. Theisen, M. Thoennessen

The frame of the field of research concerning heavy and superheavy nuclei, but not only, was set by M. Thoennessen by presenting the trends of isotope production and element production as function of time. He was conveying a rather positive message showing a still linear growth of these numbers throughout the last 80 years. At the forefront of the superheavy element synthesis, one faces, however, extreme difficulties according to Ch.E. Düllmann, when trying to analyse the model predictions for synthesis cross sections for the next heavier elements to be synthesized with Z = 119 and 120. Some way out of this was shown by J. Khuyagbaatar who presented the correlation between the minimum emission angle of fission fragments and the compound nucleus production for heavy systems, and by T. Tanaka who finds that the barrier distributions he extracts from quasi-elastic back scattering show the barrier positions at energies, where the maximum production cross sections are experimentally observed (see also section 5).

Y. Aritomo analysed, how his model predictions, based on the Langevin equation, depend on parameters like the fission barrier or friction for different systems, illustrating its range of applicability. While W. Loveland reminded of the importance of the compound nucleus formation probability for the successful evaporation residue creation for heavy systems, W. Mittig illustrated among other features the possibility of the production of superheavy nuclei in deep inelastic binary reactions of two very heavy colliding nuclei in the high Z/high A tail of the populated isotopic distribution, as proposed by V. Zagrebaev and W. Greiner.

In the last decade of the last century W. Greiner and others had suggested exciting features of atomic systems of very high atomic number, including the particle production from the vacuum, when negative energies for the inner shell electrons of the "super-atom" should be reached. They suggested, in particular, a limit for the highest possible Z related to the fine structure constant $\alpha = 1/137$. P. Schwerdtfeger presented a critical analysis of this in the view of a possible end of the periodic table of Mendeleev, elaborating on features which would motivate an extension of this limit like e.g. screening effects. He also illustrated the state-of-the art approaches to this question by quantum chemical theory approaches, including P. Pykkö's predictions which are advertising 172 as the critical Z.

The relativistic nuclear field theory approach extended by quasi particle vibrational coupling was shown by E. Litvinova to be capable of predicting single particle level properties, possible shell gaps and excitation spectra of superheavy nuclei in the region of the predicted island of shell stabilized superheavy nuclei (SHN), being bench marked for lighter species like doubly magic tins. J.R. Stone could show that her results, obtained by calculations employing quark-meson coupling for various nuclear properties like e.g. masses and binding energies, agreed well with the ones obtained by the finite range liquid droplet model by P. Möller.

The study of decay properties of the heaviest species will be pursued at the LBNL in the near future with the help of a new mass selection device, FIONA, as presented by J.M. Gates. Decay spectroscopy of the heaviest species in the region of deformed shell stabilized nuclei, from fermium up to darmstadtium isotopes, was also discussed by Ch. Theisen who highlighted special nuclear structure features like K-isomers, being prevalent for these nuclei. M. Block illustrated with the example of the heaviest nuclei for which such investigations had been performed up to now, in the nobelium-lawrencium region, that atomic physics methods like laser spectroscopy represent a powerful alternative approach to nuclear structure properties like nuclear spin, magnetic and hence quadrupole moments, and isotopic shifts and hence changes in mean square charge radii.

4 Fusion at near barrier energies

R.T DeSouza, L. Guo, K. Hagino, H.M. Jia, Khushboo, C.J. Lin, K. Vo-Phuoc, P.D. Stevenson, K. Washiyama

K. Hagino gave an overview of the recent studies on near and sub barrier fusion reactions, pointing out the present theoretical challenges and classifying the various employed theoretical models on the basis of the macroscopic or microscopic treatment. On one side we have mixed microscopic-macroscopic concepts, based for instance on coupled channels, which were quite successfull in the explanation of a large body of experimental data since one can incorporate inelastic and transfer degrees of freedom. On the other hand one aims at describing sub-barrier fusion within a fully microscopic frame. Microscopic theories like Time Dependend Hartree Fock (TDHF) are being used more and more extensively, especially in connection with the integration of other mechanisms, like deep inelastic or fission. However many problems need to be overcome, like the inclusion of nucleon-nucleon correlations and, for fusion reactions, the missing treatment of tunnelling. Important, in general, is the need to strongly interconnect structure and dynamics, while different models often overemphasize only one of the two aspects.

R.T. DeSouza presented a comparison between fusion cross sections of $^{18,19}O+^{12}C$, and showed that for the more neutron rich (radioactive) ^{19}O case, the cross sections above the barrier are larger by 20%, while below the barrier they are enhanced by a factor three compared with the ^{18}O case. This underlines the importance to properly take into account nuclear structure properties. Structure properties were also studied by Khushboo who presented new data on sub-barrier fusion of $^{40}Ca+^{70}Zn$ and $^{28}Si+^{94}Zr$ measured with the HIRA spectrometer of IUAC in New Delhi. She discussed how the change of deformation after neutron transfer can affect fusion.

H.M. Jia measured fusion cross sections for the $^{32}S+^{112-124}Sn$ systems, where Sn isotopes have similar inelastic collective properties, and concluded that although one can find a correlation between transfer channels and fusion enhancements, the role of positive Q-value transfer channels is not yet understood. C.J. Lin showed new results on simultaneous measurements of fusion evaporation and fusion-fission cross sections for the $^{32}S+^{170}Er$ and $^{28}Si+^{174}Yb$ systems, where capture cross sections have been reproduced by coupled channel and dinuclear system models.

Different theoretical talks focused on calculations based mainly on TDHF theory. L. Guo discussed the role of the tensor force in heavy-ion fusion, incorporating it into the TDHF frame and showing that it can increase the barrier height. This finding was partially corroborated by K. Washiyama who analyzed the effect of fusion hindrance near the barrier, proposing that the main contribution to the extra-push energy in heavy systems can possibly originate from a dynamical increase of the potential barrier. K. Vo-Phuoc discussed the effect of the excitation of vibrational states and of transfer on fusion within both static HF and TDHF, while P.D. Stevenson analysed the implications of effective interactions (Skyrme) on fusion dynamics near the barrier.

5 Fusion at deep sub-barrier energies

G. Colucci, F. Galtarossa, T. Ichikawa, C. Simenel, T. Tanaka

Fusion studies at sub-Coulomb energies continue to be an interesting issue for two main reasons. The first is the not yet understood microscopic origin of the hindrance phenomenon and the second is their implications for nuclear reaction rates in astrophysically relevant sites. Which theory better suites for the explanation of experimental cross sections is still under debate. We should keep in mind that, when the position of the inner turning point becomes comparable to the touching point, the frozen density approximation may break down. As a matter of fact the predictions of the models based on a two-body frozen density approximation (coupled channels) and a one-body adiabatic approximation cannot be easily disentangled.

To deal with the hindrance phenomenon, T. Ichikawa discussed the importance of introducing in standard coupled channels a phenomenological damping factor, correlated to the reduced strength of the octupole vibration mode. C. Simenel showed the possibility to overcome the limitations of the TDHF theory, with no tunnelling effects, by incorporating the Pauli repulsion which may generate a proper potential well. G. Colucci and F. Galtarossa presented new cross section measurements at far sub-barrier energies in the ^{28,30}Si+^{30}Si and ^{12}C+^{30}Si systems, respectively, in the context of searching for hindrance effects related with positive Q-values for compound nucleus formation. The signature for hindrance, a maximum in the S-factor, was observed for ^{12}C induced reaction but not for the Si+Si systems.

T. Tanaka presented barrier distributions D(B) extracted from quasi-elastic backscattering, which were measured at the gas-filled separator GARIS of RIKEN for the ^{48}Ca,^{50}Ti+^{208}Pb and ^{48}Ca+^{248}Cm systems, which were used to synthesize heavy and superheavy isotopes with atomic numbers 102, 104 and 116, respectively. To accomplish this job it is important to properly separate the quasi-elastic from the deep-inelastic components, as already shown in previous work. Interestingly, the maxima of the D(B) correspond roughly to the energies of the maxima of the fusion cross sections for the various isotopes, a fact which requires some explanation (see also section 3).

6 Reactions of astrophysical interest

A. Caciolli, S. Courtin, M. La Cognata, T. Kawabata, R.G. Pizzone, K.E. Rehm, A. Wallner

The colourful world of nuclear astrophysics was introduced by K.E. Rehm who gave an overview over the possible locations and sites of the various stellar nucleosynthesis processes, pointing to the importance and the difficulties of deep sub-barrier fusion studies. Deep sub-barrier fusion studies were then also the subject of many contributions to the conference ranging from the ^{7}Li problem discussed by T. Kawabata to the fusion reactions in medium-light systems presented by G. Colucci and F. Galtarossa (see also section 5), this last contribution providing an additional data point to the systematics established by C.L. Jiang.

Important for the stellar C-burning, new results on ^{12}C+^{12}C with consequences for stellar reaction rates and the observed ^{26}Al abundance in our Galaxy were presented by S. Courtin. For ^{4}He(^{8}Li,n)^{11}B, M. La Cognata observed a step like structure of the cross section, which he proposes to possibly relate to the partial wave contributions to the fusion excitation function. A. Caciolli reported on low energy fusion studies for the reactions ^{17}O(p,α)^{14}N, ^{22}Ne(p,γ)^{23}Na and ^{2}H(p,γ)^{3}He at the LUNA underground laboratory in the present LUNA400 configuration and with the upgraded accelerator LUNA MV (see also sections 10).

Based on the alternative approach using the Trojan Horse Method and a rare isotope beam (RIB) projectile, G. Pizzone discussed the reaction ^{18}F(p,a)^{15}O studied via the surrogate reaction ^{18}F(d,α^{15}O)n.

As a last day highlight of the conference, the search for traces of radionuclides, stemming from stellar events like nearby super novae, in terrestrial deep sea archives were illustrated by A. Wallner.

7 Reactions with weakly bound nuclei and cluster properties

E.F. Aguilera, K.J. Cook, A. Di Pietro, D. Fabris, F. Gramegna, M. Hussein, S. Kalkal, J. Kolata, M. La Commara, M. Mazzocco, A. Mukherjee, M. Nakao, V.V. Parkar, V. Pseudo, B. Schuetrumpf, E.C. Simpson, G.L. Zhang,

Reactions involving weakly bound nuclei, mainly with Li, B and Be projectiles on heavy targets, have been presented by many authors. A main issue was the role of break-up channels on the fusion process. Due to the many ways in which a break-up process may occur (elastic break-up, break-up followed by partial fusion etc.), disentangling the various components from other effects mimicking the same final partition is notoriously very difficult. Ambiguities in the data interpretation are often amplified with radioactive beams, due to low beam intensity. In spite of these difficulties many groups are actively working in the field and new data have been accumulated, allowing to make more quantitative comparisons with employed theoretical models.

J. Kolata critically reviewed reactions with ^{8}B, showing how break-up cross sections are reproduced by continuum-discretized coupled channels (CDCC) calculations, with the total fusion cross sections being enhanced near and below the barrier. S. Kalkal presented new data using beams of ^{6}Li, discussing how the different break-up mechanisms, including resonant and non-resonant states, and their sensitivity to different targets, can be experimentally investigated. E.C. Simpson and K.J. Cook discussed new experiments, mainly in Li induced reactions, aiming at disentangling through angle-velocity correlations whether the break-up process which may involve short or long lived resonant states mainly occurs before or after the minimum distance between the interacting nuclei in Coulomb trajectories. An interesting outcome of these new studies is that the break-up alone cannot account for the observed suppression of complete fusion, as it depends critically on the lifetime.

V.V. Parkar, M. Hussein and G.L. Zhang presented work connecting the break-up process with incomplete and complete fusion reactions. M. Mazzocco and V. Pseudo showed results of cross section measurements performed with the radioactive ^{7}Be and ^{11}Be beams, respectively. A. Mukherjee discussed the barrier distributions derived from studies of ^{7}Li induced reactions, while E.F. Aguilera showed fusion and elastic scattering data for reactions in the ^{6}Li+^{58}Ni system, and M. La Commara presented results of elastic scattering in ^{8}B+^{208}Pb. A. Di Pietro underlined the importance to carefully take into account the effect of the ion energy loss in the target on correlating energy and fusion cross sections, especially with radioactive beams, where the low beam intensity may require to employ a stack target arrangement.

Light ion systems were also discussed in the context of cluster structures. M. Nakao discussed calculations of cluster structures in mirror cluster systems, while B. Schuetrumpf presented results, where localization and cluster structures are generated by calculations based on density functional theory. Cluster structures have been also discussed by D. Fabris and F. Gramegna through experiments on pre-equilibrium mechanisms performed with the GARFIELD set-up at LNL.

8 Binary reactions

E. Fioretto, A. Karpov, C. Parascandolo, D.C. Rafferty, G. Scamps, K. Sekizawa, S. Szilner, N. Wang, E. Williams

Binary reactions continue to be an important item in our field, due to the tight connection with sub-barrier fusion, to the possibility to learn about the degrees of freedom (single particle, pair or even cluster) acting in the transfer process and, more recently, to the peculiar mechanism leading to the production of neutron rich nuclei.

S. Szilner reported on sub-Coulomb studies with closed shell and superfluid systems performed with the PRISMA spectrometer, where the experimental probabilities of one and two nucleon transfer could be well reproduced by microscopic calculations. She stressed the importance to treat, on the same footing, structure and dynamics to compare experiment and theory in absolute scale and to learn about correlations. The problematic issue connected with the pair correlations is of current interest in ongoing research with radioactive beams, where for example the pairing interaction is expected to be significantly modified in nuclei with extended neutron distributions. The newly accumulated data are also attracting a renewed interest by theorists. The identification of the effects of the pairing interaction partially motivated the work presented in the talk by G. Scamps, who showed a new kind of calculations performed in the framework of models based on TDHF + BCS theory. The effect of pairing correlations was discussed also by K. Sekizawa who presented the results of calculations performed within time dependent density functional theory incorporating superfluidity, suggesting how the particle flow induced by the phase difference between the colliding nuclei (analogue of the Josephson current) is originating from soliton excitations.

How the process evolves for the quasi-elastic to the deep inelastic regime is an issue, as it involves the amount of energy loss in the reaction. One should stress that dissipative processes play a critical role in channels like quasi-fission or deep inelastic (see also Fig. 1), which contribute significantly to the total reaction cross section and which must be taken into account when computing capture. New measurements of transfer cross sections have been shown by D.C. Rafferty, pointing to the gradual onset of energy dissipation, and by E. Williams who discussed new data on ^{58}Ni+^{60}Ni, where deep inelastic and quasi-fission components were compared with TDHF. C. Parascandolo showed new results on the onset of the dynamical dipole mode, comparing two cases with different entrance channel mass asymmetry.

Energy dissipation has also important consequences for the onset of secondary processes. E. Fioretto discussed how the "heavy" partners produced in binary reactions are presently receiving peculiar attention. In fact, certain regions of the nuclear chart, like that below ^{208}Pb (important also for astrophyisics) or in the actinides, can hardly be accessed by fragmentation or fission reactions. Here multinucleon transfer may be a suitable mechanism to approach those neutron rich areas. New data have been shown on the ^{197}Au+^{130}Te system, where from mass-mass correlations, obtained via a kinematic coincidence technique and employing PRISMA, one could get absolute cross sections of heavy transfer products, which were compared with the GRAZING model. Emphasis on the production yield of heavy transfer products has been put also in the talk of A. Karpov who discussed the predictions of models based on the Langevin equation. He showed the kind of agreement presently reached with the (scarce) available data. N. Wang presented an extension of the Quantum Molecular Dynamics model, originally developed for reactions in the Fermi regime, to the energy region close to the Coulomb barrier, where the main features of deep inelastic processes have been reproduced.

9 Fission reactions

A. Bulgac, M. Caamaño, D. Hinde, J. Khuyagbaatar, W. Mittig, K. Nishio, E. Prasad, S. Pullanhiotan, B. Schuetrumpf, K. Sekizawa, A.S. Umar, S.G. Zhou

With *'to fuse or not to fuse'* S. Pullanhiotan introduced the fusion-fission competition in the entrance channel in heavy ion collisions between fusion and quasi-fission, and he reported on fission measurements at IUAC.

A quite substantial part of the conference was dedicated to fission and related issues both from the experimental as well as from the theoretical point of view.

K. Nishio had employed nucleon transfer to induce fission for the projectile target combinations $^{18}O+^{238}U,^{237}Np$, resulting in a comprehensive body of data for a large number of isotopes from uranium to californium. In these investigations he succeeded in systematically studying features like fission barriers, multi chance fission and excitation energy dependent mass distributions. Mass angular distributions had been also systematically studied by D. Hinde. He was pointing out the importance of the "magicity grade" in the entrance channel, i.e. the total number of magic nucleon numbers of target and projectile, and the N/Z ratio matching, as well as he was discussing nuclear deformation and implications for SHN. Conclusions from fission fragment angular distribution studies relevant for compound nuclear formation and SHE synthesis were also presented by J. Khuyagbaatar (see also section 3).

Apart from the possibility to produce SHE in the high Z/high A tail of the isotopical distribution (see section 3), W. Mittig discussed features like ternary fission and the role of the magic Z=208 in $^{238}U+^{238}U$ collisions, a classic reaction in the context of strong atomic Coulomb fields. Fission in inverse kinematics studied at instruments like the FRS and ALADIN at GSI induced by fragmentation or as photo-fission, or VAMOS at GANIL induced by multinucleon transfer, was presented by M. Caamaño. E. Prasad was looking for the role of fission modes, possibly governing fission in the A=200 mass region and below.

Fission dynamics was studied using time dependent density functional theory by A.S. Umar, while S.G. Zhou was elaborating on fission barriers in the framework of multidimensionally-constrained covariant density functional theory, elucidating e.g. a third barrier in light actinides. Using an extension of energy density functional theory employing superfluidity, A. Bulgac demonstrated the dynamical development following proton and neutron densities towards fission of a heavy nuclear system like ^{240}Pu, performing real time calculations. A similar idea was developed by K. Sekizawa in a time-dependent superfluid local density approximation, considering the superfluidity of nuclear matter including features like e.g. the nuclear Josephson effect (see also section 8). Using density functional theory B. Schuetrumpf localized nucleonic clusters to describe the density distribution inside the nucleus modelling fission also in a real time approach (see also section 7).

10 Instrumentation

B.B. Back, D. Bazin, A. Caciolli, I.P. Carter, G. Fruet, J.M. Gates, Ch. Theisen

New instrumentation and the respective research and development are mandatory for the progress of a field and some of them have been already mentioned like the new mass spectrometer FIONA at the BGS of LBNL presented by J.M. Gates (see section 3) or the upgrade of the underground laboratory LUNA MV illustrated by A. Caciolli (see section 6).

With STELLA a new tool for astrophysical studies being installed at the 4 MV pelletron facility, Andromède, of the IPN Orsay in France was introduced by G. Fruet. The set-up is dedicated to study stellar nucleo-synthesis processes like C-burning, employing particle and photon detection with an annular Si detector and the scintillators from the FATIMA array. The rotating target used in first tests had been developed in collaboration with GANIL and will play a major role also in the new separator-spectrometer set-up S^3 of GANIL's SPIRAL2 facility, discussed by Ch. Theisen. The new set-up will open new opportunities for various research lines including SHN (see section 3), the region of N=Z nuclei and nuclear astrophysics served by the high intensity beams from the new LINAC.

The new gas-filled separator AGFA designed among others also for SHN research is presently being put into operation at ANL. B.B. Back showed a compact ion-optical configuration which will be used for decay spectroscopy after separation, employing a comprehensive particle and photon detection set-up, as well as for in-beam nuclear structure studies in conjunction with the efficient γ detection array GAMMASPHERE. A second device, presently under preparation at the physics division of ANL, is the Argonne In-flight Radioactive Isotope Separator AIRIS which is being developed to enhance the radioactive beam capabilities of the ATLAS facility.

The solenoidal exotic rare isotope separator SOLEROO developed for RIB investigations at ANU was presented by I.P. Carter, and the active target time projection chamber designed for a wide field of applications, using the rare isotope beams being available at the new FRIB facility at NSCL/MSU presently being started with ReA3, was discussed by D. Bazin.

11 Concluding remarks

An exciting and comprehensive program perfectly organized in a stimulating environment made this edition of the Fusion conference series a great success. Theoretical approaches were elucidated in all depth, experimental findings confronted with possible explanations in lively discussions and new instrumentation presented, based on most advanced technologies, all together creating a sound picture of the present status of a vivid field.

Apart from the intense scientific debate, the human aspect had not been forgotten. Two colleagues and friends, Paolo Gomes and Valery Zagrebaev, known for their engaged and pioneering contributions to the progress of the field, had left the community during the last years, way too early as many of us sustain. The organizers had honoured their memory by assigning two prizes, carrying their names, for the best presentations given by students or early career researchers.

To conclude, we wish to thank once more the organizers for their warm hospitality and great efforts which were the key of success for the conference. We would also like to thank all of our community for their excellent presentations and lively discussions.

See all of you at the next Fusion20 conference in Shizuoka!

AUTHOR INDEX

AUTHOR INDEX

AUTHOR INDEX

AUTHOR INDEX

AUTHOR INDEX